RENEWALS

B. P. Tissot   D. H. Welte

Petroleum Formation and Occurrence

Second Revised and Enlarged Edition

With 327 Figures

Springer-Verlag
Berlin Heidelberg New York Tokyo 1984

Professor BERNARD P. TISSOT
Institut Francais du Pétrole
and Ecole Nationale Supérieure du Pétrole
4, avenue de Bois-Préau, 92506 Rueil, France

Professor DIETRICH H. WELTE
Kernforschungsanlage Jülich GmbH
Institut für Erdöl und Organische Geochemie
Postfach 1913, 5170 Jülich, FRG
and
IES Gesellschaft für Integrierte Explorationssysteme mbH
Kartäuserstr. 2, 5170 Jülich, FRG

ISBN 3-540-13281-3 2. Auflage Springer-Verlag Berlin Heidelberg New York Tokyo
ISBN 0-387-13281-3 2nd edition Springer-Verlag Berlin Heidelberg New York Tokyo

ISBN 3-540-08698-6 1. Auflage Springer-Verlag Berlin Heidelberg New York Tokyo
ISBN 0-387-08698-6 1st edition Springer-Verlag Berlin Heidelberg New York Tokyo

Library of Congress Cataloging in Publication Data.
Tissot, B. P. (Bernard P.), 1931-. Petroleum formation and occurrence. Includes bibliographies and index. 1. Petroleum – Geology. 2. Gas, Natural – Geology. 3. Geochemical prospecting. I. Welte, D. H. (Dietrich H.) 1935-. II. Title. TN870.5.T53. 1984. 553.2'8. 84-10484.

This work is subject to copyright. All rights are reserved, whether the whole or part of the material is concerned, specifically those of translation, reprinting, re-use of illustrations, broadcasting, reproduction by photocopying machine or similar means, and storage in data banks. Under § 54 of the German Copyright Law where copies are made for other than private use, a fee is payable to "Verwertungsgesellschaft Wort", Munich.

© by Springer-Verlag Berlin Heidelberg 1978 and 1984.
Printed in Germany.

The use of registered names, trademarks, etc. in this publication does not imply, even in the absence of a specific statement, that such names are exempt from the relevant protective laws and regulations and therefore free for general use.

Typesetting and printing: Beltz Offsetdruck, Hemsbach/Bergstr.
Bookbinding: Konrad Triltsch, Graphischer Betrieb, Würzburg
2132/3130-543210

Foreword

The publication of this book *Petroleum Formation and Occurrence* by Bernard Tissot and Dietrich Welte will indeed be welcomed by petroleum geologists, petroleum geochemists, teachers and students in these fields, and all others who are interested in the origin and accumulation of hydrocarbons in nature. It is indeed a privilege for us to have the opportunity of sharing with these two eminent scientists the wealth of information they have acquired and developed during long careers devoted to concentrated scholarly study and practical investigation of the nature, origin, and occurrence of petroleum.

Professor Bernard Tissot graduated from the Ecole Nationale Supérieure des Mines in 1954 and from the Ecole Nationale Supérieure du Pétrole in 1955. In 1955 he received a D.E.S. in geology from the University of Grenoble and then began research work on petroleum geology at the Institut Français du Pétrole. He was made head of the Department of Geochemistry in 1965, and since 1970 has also been teaching organic geochemistry at the Ecole Nationale Supérieure du Pétrole where he became Professor in 1973. Professor Tissot has had a broad and varied background of practical experience. He has been a member of exploration teams in France, New Caledonia, and Sahara. In 1960–1963 he headed a mission of the IFP to the Department of National Development of Australia. Outstanding among his achievements has been the use of the Paris basin as a laboratory in the development of an understanding of the relation between temperature and petroleum genesis. In recent years he has devoted considerable attention to North America and has published studies on the origin of the Athabasca tar sands and the Uinta Basin oil. Together with J. Espitalié he is responsible for the development of new pyrolysis techniques and instrumentation for the identification of petroleum source rocks and their stage of maturation and a mathematical model for the thermal evolution of organic matter in sediments. He is author or co-author of numerous outstanding scientific papers on origin and migration of petroleum published in both French and English in various periodicals and books.

Professor Dietrich Welte received his Ph. D. Degree in 1959 from the University of Würzburg in geology and chemistry. He

then worked for three years as a research geochemist in the Hague for the Shell International Oil Company on the origin of oil and gas. In 1963 he returned to Würzburg where he established a research laboratory in organic geochemistry. In 1966 he received the President's award from the American Association of Petroleum Geologists for his outstanding paper on *Relation between Petroleum and Source Rock*. During 1966 he visited various academic and industrial laboratories in the USA and in 1967 took a position as Senior Research Geochemist and toward the end of his stay was given the function of Research Coordinator for Exploration with Chevron Oil Field Research Company. While being with Chevron he was working on geological-geochemical research projects in California, the Gulf Coast, and other areas in America. In 1970–1972 he taught at the University of Göttingen and since 1972 has been Professor of Geology, Geochemistry, and Oil and Coal Deposits at the Institute of Technology at Aachen. During his scientific career Professor Welte has made numerous outstanding contributions to the knowledge of the geochemistry of petroleum, published in German and in English in various periodicals and books, in addition to numerous private reports resulting from his oil company connections. Most recently Professor Welte accepted the post of Director of the newly founded Institute for Petroleum and Organic Geochemistry at the Nuclear Research Center Jülich, although he will continue some teaching in Aachen.

Although resident in Europe and presently working in European institutions, Professors Tissot and Welte have both been frequent sojourners in America and have long participated actively in meetings of American geochemists and geologists. Both have worked extensively on problems of petroleum genesis in America and have published a number of papers in the Bulletin of the American Association of Petroleum Geologists. Many readers will recall the outstanding conferences on the Geology of Fluids and Organic Matter in Sediments in Banff, Alberta in May 1973 at which for several days the authors presented specific topics in this subject area and fielded questions and comments from a picked audience of petroleum geologists. Both Tissot and Welte have participated and aided greatly in work on the hydrocarbons of cores from the Joides Deep Sea Drilling Project, and are members of the Joides Panel on Organic Geochemistry.

The origin of petroleum has challenged scientists as far back as the 17th century when it was supposed by many to be associated with the mysterious "phlogiston". From being simply a matter of scientific curiosity it became, with the development of the petroleum industry, a subject of vital practical importance, knowledge of which has often meant the difference between success and failure in exploration for petroleum. Much progress has been made by the combined efforts of geologists, geochemists, and geophysicists, but the subject is still a dynamic one with many baffling unknowns and uncertainties. Perhaps even more filled with uncertainties and unknowns at the present day, and clearly just as critically important to practical oil exploration, is the manner of migration and accumulation of petroleum.

Tissot and Welte approach their subject not only as geochemists but also well-armed with geological knowledge. The book starts (Part I) with the production and accumulation of organic matter — the basic feed stock for petroleum. Then follows the very critical Part II on the transformation of this organic matter to kerogen and then to oil, or gas. Important aspects of the formation of coal and oil shales are also treated. Part III deals with the formation of oil and gas pools and the knotty problem of migration of petroleum from source to reservoir. Part IV is concerned with the composition and classification of petroleum, "geochemical fossils", and the relation between character of petroleum and geological environment and subsequent alteration of oil in the reservoir. Finally, Part V covers the practical identification and evaluation of source rocks, the correlation between oils and between oils and source rocks, and an evaluation of the practical uses of geochemistry in petroleum exploration.

The Tissot-Welte volume is a clear, practical, readable account, by two of the world's leading researchers in the field, of the formation, migration and accumulation of petroleum, primarily from the geochemical viewpoint, but adequately seasoned with geology. The text is further clarified by a wealth of tables and figures and is a mine of reference information. The topics are developed in a logical and reasonable way with full credit for other and sometimes divergent views. One of the many merits of the Tissot-Welte book is its open-minded approach to problems and its avoidance of a dogmatic attitude — the true measure of a really deep knowledge of a subject. A particularly strong point is the attention which the book gives to practical procedures and to the complex instrumentation required in modern petroleum geochemical investigations, as well as to the more theoretical aspects of the subject

The book will fill a very much needed place both as an advanced classroom text, and as a guide and reference work on the subject, which should be in the hands of all practicing petroleum geologists and geochemists. These will undoubtedly emerge as better oil finders for the reading of this volume.

HOLLIS D. HEDBERG
Professor of Geology (Emeritus)
Princeton University, 1978

Preface to the Second Edition

The first edition of *Petroleum Formation and Occurrence* was published about 6 years ago. In the Preface, we expressed the hope that this book may help to promote the integration of organic geochemistry into the geosciences and improve communication between geologists and more chemically oriented researchers involved in practical petroleum exploration. We believe we have indeed contributed toward attaining those goals. Furthermore, we are very pleased to see both how rapidly this field of petroleum sciences is growing and to note its influence on geosciences in general. It has become necessary to expand the scope of the book in order to include important new topics, such as biological markers, gas generation, heavy oils and tar sands, migration and modeling, particularly geological and geochemical modeling with high speed computers. It is evident that computer modeling is here to stay, and may very well revolutionize the field. The computer can be used as an experimental tool to test geological ideas and hypotheses whenever it is possible to provide adequate software for normally very complicated geological processes. The enormous advantages offered by computer simulation of geological processes are that no physical or physico-chemical principles are violated and that for the first time the geological time factor, always measured in millions of years rather than in decades, can be handled with high speed computers with large memories. Thus, the age of true quantification in the geosciences has arrived. We believe that this computer-aided, quantitative approach will have an economic and intellectual impact on the petroleum industry, mainly on exploration. Our concepts of petroleum generation, migration, and accumulation are becoming increasingly quantifiable and hence can be used as predictive tools in the search for oil and gas. This approach helps to cut costs not only in mature exploration areas when looking for "overlooked" petroleum, but also in high-risk frontier areas. The intellectual result could be a better chance for a much-needed synthesis of the principal fields in geosciences such as geology, geophysics, and geochemistry. With this second edition of our book, we hope to contribute both to scientifically more advanced and hence risk-diminishing exploration strategies, and also to the growing demand for a synergetic view of the geosciences.

We have not only updated the previous text, but have added completely new chapters on gas, heavy oils and tar sands, distribution of world petroleum reserves, case histories on habitat of petroleum, and geological and geochemical modeling. Other chapters or sections have been enlarged and rearranged, such as those on geochemical fossils (biological markers), primary migration, asphaltenes and resins, coal as a possible source rock, new techniques of source rock characterization, and oil-source rock correlations. We have also included more than 270 new references, added 84 new illustrations, and modified others. The index has been considerably improved and enlarged.

Both of us are indebted to co-workers and colleagues in our organizations, the Institut Français du Pétrole (IFP), Rueil Malmaison, and the Kernforschungsanlage-Jülich (KFA). Special thanks for scientific advice and critical comments go to P. Albrecht, B. Durand, D. Leythaeuser, R. Pelet, M. Radke and J. Rullkötter. One of us (D.H.W) wants to acknowledge the very fruitful cooperation with scientists from the Shengli Oil Field Research Institute in the People's Republic of China. We are both thankful that permission was granted by the Shengli Oil Field Research Institute to publish information on the common work about the Linyi Basin in NE China.

The great help we received from Mrs. R. Didelez (IES), Mrs. B. Hartung (KFA) and Mrs. G. Tramblay (IFP) for organization, coordination and editing of the manuscript of the second edition of our book is also very thankfully acknowledged. Finally, we want to thank our wives again for their patience with us during our work on the second edition of our book.

July, 1984

B. P. Tissot
D. H. Welte

Preface to the First Edition

The subject treated in this book *Petroleum Formation and Occurrence* is truly interdisciplinary. It has its roots in such diverse fields of science as biology, oceanography, and, most important, in various branches of chemistry and geology. Those concerned either in academic or industrial work, with research, or practical problems associated with petroleum, have long recognized this fact. A steadily increasing number of these people is looking for a comprehensive source of information on the various aspects of petroleum exploration.

For a number of years we had been giving seminars in different countries on the origin, migration and accumulation of petroleum. It was from these seminars that the idea to write a book was born. After discussing this matter, it was clear that such a book should not be written by many authors, who might each be expert in a particular field, but that it should be written just by the two of us as generalists, even if it meant an enormous amount of work. In this way we had hoped the book would be easier to read, and that the many facets of this difficult subject could be better understood.

It is our wish that people in the academic world, advanced students in geosciences and chemistry or other branches of science, would benefit from this book. The book may also help to integrate organic geochemistry more into the geosciences than has been the case up to now. Most of all, however, we hope that this book is of use for those working in petroleum exploration and fields related to it. For a long time there has been a lack of communication between geologists active in practical exploration, and researchers more involved in chemically oriented laboratory work: we hope to have bridged that gap.

Through our book we want to demonstrate that the search for petroleum can benefit greatly from the integration and application of the principles of petroleum generation and migration. For many years the decision to drill a well was largely taken on the basis of the recognition of suitable structures. Then the selection of a structure was based mainly on intuition and general experience, because very little information was available whether or not a trap would contain hydrocarbons. A systematic utilization of the new comprehensive understanding of petroleum formation and occurrence, as presented in this book, can improve the success rate in predicting petroleum-filled structures and hence decrease the financial risk of drilling. To follow this concept requires that in the future petroleum geologists must acquire some knowledge of petroleum geochemistry. To this end the teaching of organic geochemistry has to be developed. We hope that this book can serve as a basic text for the petroleum geochemistry covered in such a course.

The completion of the book would have been impossible without the help of our co-workers, colleagues and friends, especially in our organizations at the Institut Français du Pétrole (IFP), Rueil-Malmaison, the Kernforschungsanlage-Jülich (KFA) and the Rheinisch-Westfälische Technische Hochschule Aachen (RWTH). We are highly indebted to P. Albrecht, Ch. Cornford, W. Dow, B. Durand, G. Eglinton, A. Hood, R. Pelet, J. Williams and M. A. Yükler, who critically read and reviewed parts of the manuscript. We thank Mrs. R. Didelez for her indefatigable assistance during preparation and organization of the final manuscript. Last but not least we want to thank our wives for their patience with us while working on the book.

Finally special thanks go to H. D. Hedberg, who read all the text, gave very valuable advice, and encouraged us throughout the work.

April, 1978   B. P. Tissot
                                                            D. H. Welte

Contents

Part I   Production and Accumulation of Organic Matter: A Geological Perspective

Chapter 1   Production and Accumulation of Organic Matter: The Organic Carbon Cycle 3

1.1   Photosynthesis – The Basis for Mass Production of Organic Matter 3
1.2   The Organic Carbon Budget During the History of the Earth 7
1.3   The Organic Carbon Budget in the Black Sea 11
       Summary and Conclusion 13

Chapter 2   Evolution of the Biosphere 14

2.1   Phytoplankton and Bacteria 14
2.2   Higher Plants 17
2.3   Geological History of the Biosphere 19
       Summary and Conclusion 20

Chapter 3   Biological Productivity of Modern Aquatic Environments 21

3.1   Primary Producers of Organic Matter 21
3.2   Factors Influencing Primary Productivity 23
3.3   Present Primary Production of the Oceans 28
       Summary and Conclusion 30

Chapter 4   Chemical Composition of the Biomass: Bacteria, Phytoplankton, Zooplankton, Higher Plants 31

4.1   Proteins and Carbohydrates 31
4.2   Lipids 34
4.3   Lignin and Tannin 44
4.4   Qualitative and Quantitative Occurrence of Important Chemical Constituents in Bacteria, Phytoplankton, Zooplankton and Higher Plants ... 45

4.5	Natural Associations and Their Effects on Biomass Composition	50
	Summary and Conclusion	53

Chapter 5	Sedimentary Processes and the Accumulation of Organic Matter	55
5.1	Fossil and Modern Sediments Rich in Organic Matter, and Their Geological Implication	55
5.2	The Role of Dissolved and Particulate Organic Matter	57
5.3	Accumulation Mechanisms for Sedimentary Organic Matter	59
	Summary and Conclusion	61

References to Part I	63

Part II The Fate of Organic Matter in Sedimentary Basins: Generation of Oil and Gas

Chapter 1	Diagenesis, Catagenesis and Metagenesis of Organic Matter	69
1.1	Diagenesis	69
1.2	Catagenesis	71
1.3	Metagenesis and Metamorphism	72
	Summary and Conclusion	73

Chapter 2	Early Transformation of Organic Matter: The Diagenetic Pathway from Organisms to Geochemical Fossils and Kerogen	74
2.1	Significance and Main Steps of Early Transformations	74
2.2	Biochemical Degradation	75
2.3	Polycondensation	81
2.4	Insolubilization	85
2.5	Isotopic Composition of Organic Matter in Young Sediments	89
2.6	Result and Balance of Diagenesis	90
	Summary and Conclusion	92

Chapter 3	Geochemical Fossils and Their Significance in Petroleum Formation	93
3.1	Diagenesis Versus Catagenesis: Two Different Sources of Hydrocarbons in the Subsurface	93

3.2	Hydrocarbons Inherited from Living Organisms, Directly or Through an Early Diagenesis: Geochemical Fossils (Biological Markers)	98
3.3	n-Alkanes and n-Fatty Acids	100
3.4	Iso- and Anteiso-Alkanes	110
3.5	C_{10}-branched Alkanes	110
3.6	Acyclic Isoprenoids	111
3.7	Tricyclic Diterpenoids	116
3.8	Steroids and Pentacyclic Triterpenoids: Occurrence in Recent and Ancient Sediments	117
3.9	Fate of Steroids and Triterpenoids During Diagenesis and Catagenesis	121
3.10	Other Polyterpenes	126
3.11	Aromatics	126
3.12	Oxygen and Nitrogen Compounds	127
3.13	Kerogen, the Polar Fraction of Sediments, and Asphaltenes of Crude Oils as Possible Sources of Fossil Molecules	129
	Summary and Conclusion	130

Chapter 4 Kerogen: Composition and Classification . . . 131

4.1	Definition and Importance of Kerogen	131
4.2	Isolation of Kerogen	132
4.3	Microscopic Constituents of Kerogen	133
4.4	Chemical and Physical Determination of Kerogen Structure	139
4.5	Chemical Analysis	140
4.6	Physical Analysis	142
4.7	General Structure of Kerogen	147
4.8	Depositional Environment and Composition of Kerogen: the Evolution Paths	151
4.9	Conclusion	159
	Summary and Conclusion	159

Chapter 5 From Kerogen to Petroleum 160

5.1	Diagenesis, Catagenesis and Metagenesis of Kerogen	160
5.2	Experimental Simulation of Kerogen Evolution	169
5.3	Structural Evolution of Kerogen	174
5.4	Formation of Hydrocarbons During Catagenesis	176
5.5	Isotope Fractionation and Kerogen Evolution	189
5.6	Experimental Generation of Hydrocarbons from Organic Material	192
	Summary and Conclusion	198

Chapter 6 Formation of Gas 199

6.1 Constituents and Characterization of Petroleum Gas . 199
6.2 Gas Generated During Diagenesis of Organic Matter 201
6.3 Gas Generated During Catagenesis and Metagenesis of Organic Matter 204
6.4 Gas Originating from Inorganic Sources 207
6.5 Occurrence and Composition of Gas in Sedimentary Basins: Example of Western Europe 208
6.6 Distribution of Gases in Sedimentary Basins 213
     Summary and Conclusion 214

Chapter 7 Formation of Petroleum in Relation to Geological Processes. Timing of Oil and Gas Generation 215

7.1 General Scheme of Petroleum Formation 215
7.2 Genetic Potential and Transformation Ratio 218
7.3 Nature of the Organic Matter. Gas Provinces Versus Oil Provinces 219
7.4 Temperature, Time and Pressure 222
7.5 Timing of Oil and Gas Generation 223
7.6 Comparison Between the Time of Source Rock Deposition and the Time of Petroleum Generation . . 225
     Summary and Conclusion 228

Chapter 8 Coal and Its Relation to Oil and Gas 229

8.1 General Aspects of Coal Formation 229
8.2 The Formation of Peat 230
8.3 Coalification Process 234
8.4 Coal Petrography 241
8.5 Petroleum Generation 245
     Summary and Conclusion 253

Chapter 9 Oil Shales: A Kerogen-Rich Sediment with Potential Economic Value 254

9.1 Historical 254
9.2 Definition of Oil Shales. Oil Shale Versus Petroleum Source Rock 254
9.3 Composition of Organic Matter 256
9.4 Conditions of Deposition 258
9.5 Oil Shale Density 259
9.6 Pyrolysis of Oil Shales 259
9.7 Oil Yield; Composition of Shale Oil 260
9.8 Oil Shale Distributions and Reserves 261
     Summary and Conclusion 266

References to Part II 267

Part III The Migration and Accumulation of Oil and Gas

Chapter 1 An Introduction to Migration and
Accumulation of Oil and Gas 293

Summary and Conclusion 295

Chapter 2 Physicochemical Aspects of Primary
Migration 296

2.1 Temperature and Pressure 296
2.2 Compaction 301
2.3 Fluids 307
2.4 Possible Modes of Primary Migration 309
    Summary and Conclusion 323

Chapter 3 Geological and Geochemical Aspects of
Primary Migration 325

3.1 Time and Depth of Primary Migration 325
3.2 Changes in Composition of Source Rock Bitumen
    Versus Crude Oil 330
3.3 Evaluation of Geological and Geochemical Aspects of
    Primary Migration 333
3.4 Conclusions and Suggestions on Primary Migration . 338
    Summary and Conclusion 340

Chapter 4 Secondary Migration and Accumulation . . . 341

4.1 The Buoyant Rise of Oil and Gas Versus Capillary
    Pressures 342
4.2 Hydrodynamics and Secondary Migration 344
4.3 Geological and Geochemical Implications of
    Secondary Migration 347
4.4 Termination of Secondary Migration and
    Accumulation of Oil and Gas 351
4.5 Distances of Secondary Migration 354
    Summary and Conclusion 356

Chapter 5 Reservoir Rocks and Traps, the Sites of Oil and
Gas Pools 357

5.1 Reservoir Rocks 358
5.2 Traps 360
    Summary and Conclusion 365

References to Part III 366

Part IV **The Composition and Classification of Crude Oils and the Influence of Geological Factors**

Chapter 1 Composition of Crude Oils 375

1.1 Petroleum Versus Source Rock Bitumen 375
1.2 Analytical Procedures for Crude Oil Characterization 375
1.3 Main Groups of Compounds in Crude Oils 379
1.4 Principal Types of Hydrocarbons in Crude Oils . . . 382
1.5 Sulfur Compounds 398
1.6 Nitrogen Compounds 401
1.7 Oxygen Compounds 403
1.8 High Molecular Weight N, S, O Compounds: Resins and Asphaltenes 403
1.9 Organometallic Compounds 408
1.10 Covariance Analysis of Main Crude Oil Constituents 411
Summary and Conclusion 414

Chapter 2 Classification of Crude Oils 415

2.1 General 415
2.2 Historical 416
2.3 Basis of Proposed Classification of Crude Oils 416
2.4 Classification of Crude Oils 417
2.5 Characteristics of the Principal Classes of Crude Oils . 419
2.6 Concluding Remarks 422
Summary and Conclusion 423

Chapter 3 Geochemical Fossils in Crude Oils and Sediments as Indicators of Depositional Environment and Geological History 424

3.1 Significance of Fossil Molecules 424
3.2 Geochemical Fossils as Indicators of Geological Environments 426
3.3 Geochemical Fossils as Indicators of Early Diagenesis 432
3.4 Geochemical Fossils as Indicators of Thermal Maturation 433
3.5 Present and Future Development in the Use of Geochemical Fossils 436
Summary and Conclusion 437

Contents   XIX

Chapter 4   Geological Control of Petroleum Type 439

4.1   General and Geochemical Regularities of
      Composition 439
4.2   Geochemical Regularities Related to the
      Environment of Deposition 440
4.3   Geochemical Regularities in Relation to Thermal
      Evolution 450
4.4   Concluding Remarks on Crude Oil Regularities ... 457
      Summary and Conclusion 457

Chapter 5   Petroleum Alteration 459

5.1   Thermal Alteration 460
5.2   Deasphalting 461
5.3   Biodegradation and Water Washing 463
      Summary and Conclusion 469

Chapter 6   Heavy Oils and Tar Sands 470

6.1   Definitions 470
6.2   Composition of Heavy Oils 472
6.3   Specific Gravity and Viscosity 475
6.4   Origin and Occurrence of Heavy Oils ... 477
6.5   World Reserves and Geological Setting . 480
6.6   Valorization of Heavy Oils 482
      Summary and Conclusion 483

References 484

Part V   *Oil and Gas Exploration: Application of the
           Principles of Petroleum Generation and Migration*

Chapter 1   Identification of Source Rocks 495

1.1   Amount of Organic Matter 495
1.2   Type of Organic Matter 497
1.3   Maturation of the Organic Matter 515
1.4   Conclusions on Characterization of Potential Source
      Rocks 540
      Summary and Conclusion 546

Chapter 2   Oil and Source Rock Correlation 548

2.1   Correlation Parameters 549
2.2   Oil–Oil Correlation Examples 551
2.3   Oil–Source Rock Correlation Examples .. 561
      Summary and Conclusion 570

Chapter 3 Locating Petroleum Prospects: Application of Principle of Petroleum Generation and Migration – Geological Modeling 571

3.1 Acquisition of the Geochemical Information 573
3.2 First Conceptual Model of Petroleum Generation in a Basin 575
3.3 Numerical Simulation of the Evolution of a Sedimentary Basin – Geological Modeling 576
Summary and Conclusion 581

Chapter 4 Geochemical Modeling: A Quantitative Approach to the Evaluation of Oil and Gas Prospects 583

4.1 Necessity of a Quantitative Approach to Petroleum Potential of Sedimentary Basins 583
4.2 Mathematical Model of Kerogen Degradation and Hydrocarbon Generation 585
4.3 Genetic Potential of Source Rocks. Transformation Ratio 589
4.4 Validity of the Model 590
4.5 Significance of the Activation Energies in Relation to the Type of Organic Matter 590
4.6 Application of the Mathematical Model to Petroleum Exploration 593
4.7 Reconstruction of the Ancient Geothermal Gradient 596
4.8 Migration Modeling 604
4.9 Conclusion 607
Summary and Conclusion 608

Chapter 5 Habitat of Petroleum 610

5.1 Habitat of Petroleum in the Arabian Carbonate Platform 611
5.2 Habitat of Petroleum in Young Delta Areas 615
5.3 The Linyi Basin in the People's Republic of China . . 624
5.4 Habitat of Gas in the Deep Basin of Western Canada . 628
Summary and Conclusion 639

Chapter 6 The Distribution of World Oil and Gas Reserves and Geological – Geochemical Implications 641

6.1 Introduction 641
6.2 Geological Setting of Oil and Gas Reserves 642
6.3 Age Distribution of Petroleum Reserves 648

6.4	Significance of the Age and Geotectonic Distribution of Petroleum and Coal	650
6.5	Richness of Sedimentary Basins. Role of Giant Fields and Giant Provinces	656
6.6	Ultimate World Oil and Gas Resources	660
6.7	Paleogeography as a Clue to Future Oil and Gas Provinces	662
	Summary and Conclusion	666

References to Part V 667

Subject Index 679

Part I

Production and Accumulation of Organic Matter:
A Geological Perspective

Chapter 1
Production and Accumulation of Organic Matter
The Organic Carbon Cycle

Production, accumulation and preservation of undegraded organic matter are prerequisites for the existence of petroleum source rocks. The term "organic matter" or "organic material", as used in this book, refers solely to material comprised of organic molecules in monomeric or polymeric form derived directly or indirectly from the organic part of organisms. Mineral skeletal parts, such as shells, bones, and teeth are not included. First, organic matter has to be synthesized by living organisms and thereafter it must be deposited and preserved in sediments. Depending on further geological events, part of the sedimentary organic matter may be transformed into petroleum-like compounds. It is important to realize that during the history of the earth the conditions for synthesis, deposition, and preservation of organic matter has changed considerably.

1.1 Photosynthesis — The Basis for Mass Production of Organic Matter

The emergence of photosynthesis as a worldwide phenomenon is a noteworthy historical event with respect to the formation of potential source rocks. The photosynthetic process converts light energy into chemical energy. Photosynthesis is basically a transfer of hydrogen from water to carbon dioxide to produce organic matter in the form of glucose and oxygen. The oxygen is freed from the water molecule and not from carbon dioxide. From glucose, autotrophic organisms can synthetize polysaccharides, such as cellulose and starch, and all other necessary constituents. A simple form of the equation of photosynthesis reaction is given in Figure I.1.1.

$$6\,CO_2 + 12\,H_2O^* \xrightleftharpoons[674\text{ kcal}]{h \cdot v} C_6H_{12}O_6 + 6\,O_2^* + 6\,H_2O$$

(Glucose)

↓

Polysaccharides

Fig. I.1.1. Equation of photosynthesis. Glucose, relatively rich in energy, is formed by green plants with the help of sunlight ($h \cdot v$). Oxygen is by-product of this process

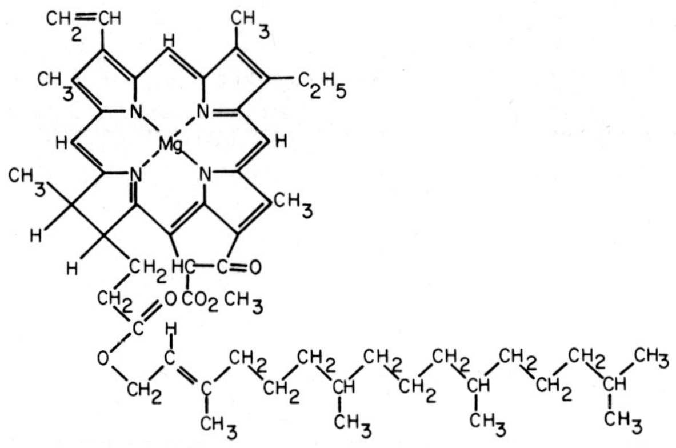

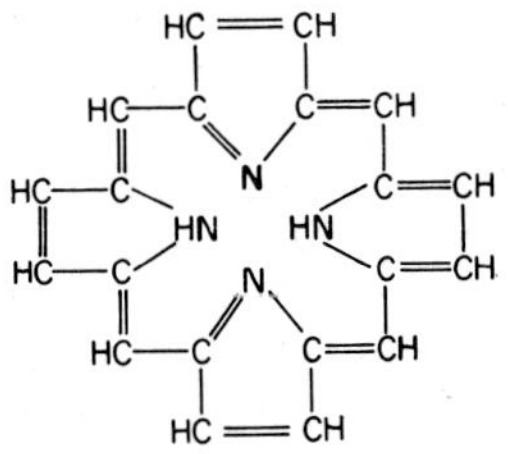

Fig. I.1.2. (a) Molecule of chlorophyll, the light-absorbing green pigment of plants and a prerequisite for photosynthesis. The phytyl side chain of the chlorophyll molecule is an important source for chain-like isoprenoid compounds in sediments and oils. Likewise, the nitrogen-containing porphin nucleus is the precursor for porphyrins. (b) Structure of porphin nucleus

Photosynthesis is the basic process that accomplishes the mass production of organic matter on earth. Primitive autotrophic organisms, such as photosynthetic bacteria and blue-green algae, were the first organisms responsible for this mass production. A basic prerequisite for photosynthesis is the light-absorbing green pigment chlorophyll (Fig. I.1.2). In primitive autotrophs it occurs in a relatively free state in the cell of the organism. In more highly evolved plants it is concentrated in chloroplasts in green leaves. These chloroplasts are complete photosynthetic factories.

The oldest recorded forms of organic life are about 3.1 to 3.3 billion years old and are bacteria and algae-like bodies from the Swaziland Group in South Africa (Schopf et al., 1965). However, it is possible that life on earth is at least as old as the oldest known rocks — 3.7 billion years.

It is assumed that approximately 2 billion years ago, photosynthetic production of organic matter was fairly well established worldwide, and this time serves as a zero reference point. Before it was reached, another billion years probably was required for the isolated occurences of most primitive organisms to spread sufficiently for photosynthesis and, hence, for mass production of organic matter to be universally prevalent.

A summary of geological events considered to be important in the organic carbon cycle is presented in Figure I.1.3. Without water there is no life. Therefore, abundant life, even on a most primitive level, was not possible on earth prior to about 4 billion years ago, when water became a common substance on the earth's surface. During that primordial time, the atmosphere was reducing, i.e., there was practically no free oxygen.

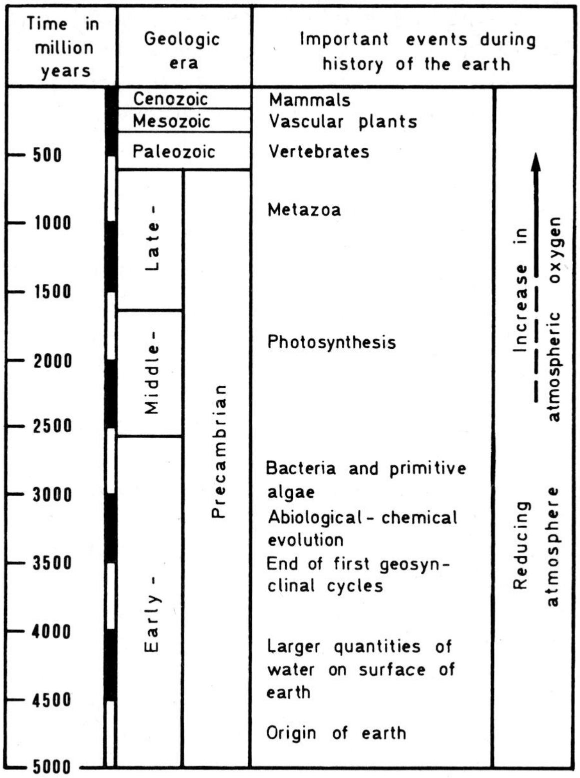

1.1 Photosynthesis — The Basis for Mass Production of Organic Matter

Time in million years	Geologic era		Important events during history of the earth	
−500	Cenozoic		Mammals	
	Mesozoic		Vascular plants	
	Paleozoic		Vertebrates	
−1000	Precambrian	Late	Metazoa	Increase in atmospheric oxygen
−1500				
−2000		Middle	Photosynthesis	
−2500				
−3000		Early	Bacteria and primitive algae	Reducing atmosphere
−3500			Abiological-chemical evolution	
			End of first geosynclinal cycles	
−4000				
−4500			Larger quantities of water on surface of earth	
−5000			Origin of earth	

Fig. I.1.3. Events supposed to be of importance for the evolution of life during the earth's history. Mass production of organic matter on earth did not occur prior to 2 billion years ago when photosynthesis was established as a worldwide phenomenon

It is generally agreed that the early earth's atmosphere was devoid of free oxygen, and that it contained H_2, CH_4, NH_3, N_2 and H_2O. However, this view is not unopposed, the methane-ammonia hypothesis being especially questioned. In connection with this hypothesis, Calvin (1969) refers to an abiological or chemical evolution that started more than 4 billion years ago. When primitive organisms first appeared about 3 billion years ago, they probably utilized the abiologically produced organic molecules as a source of energy to maintain metabolism. Therefore, the first organisms are assumed to have been heterotrophic. However, the growing population of heterotrophs probably could not be supported forever in this way. It is argued that, by the time these organisms had almost depleted the reservoir of abiologically produced organic matter, photosynthesis developed as a second source of energy.

In this way, heterotrophic organisms that were able to use sunlight as an extra source of energy could become independent, and with further evolution could escape the food shortage. Certain purple-colored bacteria living today show these properties. They can act like heterotrophs and utilize organic compounds, and they also contain the green pigment chlorophyll to carry on photosynthesis. The oldest form of photosynthesis, as performed by bacteria, did not produce oxygen. Photosynthetic bacteria are anaerobic. Instead of using H_2O as a hydrogen donor, they can use H_2S and excrete sulfur rather than oxygen.

Certain blue-green algae that evolved from photosynthetic bacteria probably were the first oxygen-producing organisms. Although there are a number of photosynthetic pigments, none can completely replace chlorophyll (Fig. I.1.2) in photosynthetic organisms. Chlorophyll molecules absorb light energy, which elevates electrons to a higher energy level. This gain in energy is transferred to other molecules.

Oxygen is believed to have been toxic to organisms of that time. However, a reducing environment assured that divalent iron was abundant in aqueous solutions. This iron could act as a sink for the oxygen produced as a by-product of photosynthesis. It is very likely that the well-known banded ironstones of the Precambrian are formed by this interplay between photosynthesis and a subsequent oxidation of iron to a trivalent form, with precipitation of the insoluble oxides (Cloud, 1968).

Autotrophic, photosynthetic organisms were superior to heterotrophs and consequently soon dominated the biological realm. As stated before, about 2 billion years ago, photosynthesis emerged as a worldwide phenomenon. Herewith the foundation for the food pyramid and the evolution of higher forms of life was laid. It is argued that after this event the atmosphere of the earth slowly became oxidizing, i.e., free molecular oxygen became available.

Photosynthesis makes use of the energy coming from sunlight, employing only a narrow band of the sun's total radiation. The portion of the spectrum utilized by most photosynthetic organisms is between 4000 to 8000 Å, which nearly equals the portion of light visible to the human eye. Rays with shorter wavelengths and higher energy are even harmful to life. Different parts of the visible light spectrum may be utilized by different photosynthetic organisms. The portion used is determined by the type of pigment an organism employs (Fig. I.1.4). It enables photosynthetic algae and bacteria to live at different depth levels in the same body of water. Life in deeper water is correlative to use of longer wavelengths.

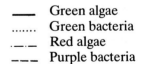

——— Green algae
……… Green bacteria
—·—· Red algae
— — — Purple bacteria

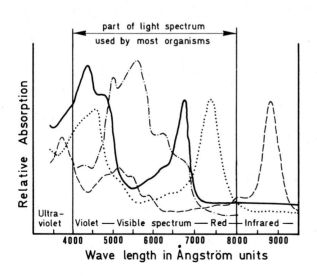

Fig. I.1.4. Different parts of the light spectrum may be utilized by different photosynthetic organisms. This enables photosynthetic algae and bacteria to live at different depth levels in the same body of water. (Adapted from Tappan and Loeblich, 1970; reprinted with permission and from publication of the Geological Society of America)

1.2 The Organic Carbon Budget During the History of the Earth

For a mass balance of carbon used in photosynthesis during the history of the earth, it is necessary to add up all organic carbon present on earth in various repositories, such as ocean waters and sediments. The total estimated amount of organic carbon and graphite, which formerly represented sedimentary organic carbon, is approximately 6.4×10^{15}t (Welte, 1970). A more recent estimate by Hunt (1972) is about twice as high. However, Hunt includes in his balance calculation "organic" carbon in basaltic and other volcanic rocks, as well as in granitic and all metamorphic rocks. The biological origin of much of this "organic" carbon is questionable.

Most of the carbon on earth is concentrated in sedimentary rocks of the earth's crust. Part of it is fixed as organic carbon, and a greater part as carbonate carbon. It is estimated that 18% of total carbon in sedimentary rocks is organic carbon and that 82% of sedimentary carbon is bound in the form of carbonates (Schidlowski et al., 1974).

A relationship of course exists between organic carbon and carbonate carbon. The atmospheric CO_2-reservoir is in a constant exchange with the hydrospheric CO_2-reservoir. From aquatic environments, carbonates may be precipitated or deposited by organisms (shells, skeletons etc.) to form carbonate sediments. Conversely, carbonate rocks may be dissolved to contribute to the equilibrium reaction between CO_3^{2-}, HCO_3^- and CO_2 in waters. Primary organic matter is formed directly from the atmospheric reservoir by terrestrial plants, or by photosynthesis of marine plants from dissolved CO_2 in the hydrosphere. Terrestrial and marine organic matter, in turn, is largely destroyed by oxidation. Thus CO_2 is returned for re-circulation in the system. A simplified sketch showing the main processes and pathways concerning the element carbon in the earth's crust is given in Figure I.1.5. Only an almost negligible portion of the organic carbon in the earth's crust, including the hydrosphere, is found in living organisms and in a dissolved state. The major part of the organic carbon (5.0×10^{15}t) is fixed in sediments. Another sizeable part of the organic carbon (1.4×10^{15}t), mainly, in the form of graphite-like material or metaanthracite, is fixed in metamorphic rocks of sedimentary origin (Table I.1.1).

If it is correct that all this organic carbon has been formed either directly or indirectly by photosynthesis during the earth's history, there should have been a corresponding amount of oxygen liberated simultaneously according to the equation of photosynthesis (Fig. I.1.1). This amount must be accounted for by free oxygen, together with formerly free oxygen presently used by oxidation processes of substance other than biological organic matter. At present, we find free oxygen in the atmosphere (in air 20.95 vol %) and varying amounts dissolved in ocean waters (general range 2–8 ml O_2 per l). Formerly free oxygen is found in both dead and living organic matter. However, most formerly free oxygen has been utilized by the oxidation of various forms of sulfur and iron. This oxygen is bound today in sulfates and in oxides of trivalent iron, and is distributed throughout the earth's crust, including the hydrosphere (Table I.1.2).

As stated before, it is believed that the primordial atmosphere of the earth was reducing and the elements sulfur and iron occurred only in divalent form. Oxygen

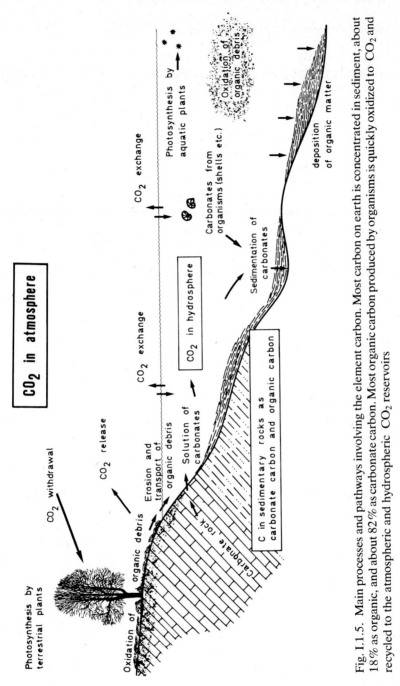

Fig. I.1.5. Main processes and pathways involving the element carbon. Most carbon on earth is concentrated in sediment, about 18% as organic, and about 82% as carbonate carbon. Most organic carbon produced by organisms is quickly oxidized to CO_2 and recycled to the atmospheric and hydrospheric CO_2 reservoirs

produced by photosynthesis, therefore, was used to oxidize sulfides to sulfates and divalent to trivalent iron. The total free and formerly free oxygen found on earth amounts to approximately 16.9×10^{15} t.

1.2 The Organic Carbon Budget During the History of the Earth

Table I.1.1. Organic carbon in the earth's crust expressed in 10^{15} t. (After Welte, 1970)

Organisms and dissolved organic C	0.003
Sediments	5.0
Metamorphic rocks of sedimentary origin (80% of all metamorphic rocks)	1.4
Total organic carbon	6.4

Table I.1.2. Free and formerly free oxygen in the earth's crust, expressed in 10^{15} t exclusive of oxygen in carbonates and silicates. (After Welte, 1970)

Atmosphere	1.18
Oceans	0.02
Biological CO_2	0.16
Dissolved marine SO_4^{2-}	2.6
Evaporitic SO_4^{2-}	10.2
FeO $\longrightarrow$ Fe_2O_3	2.7
Total oxygen	16.9

The ratio of the total quantities calculated for oxygen (16.9×10^{15} t) and organic carbon (6.4×10^{15} t) is similar to the mass ratio of these elements in the CO_2 molecule:

$$\left(\frac{O_2}{C} = \frac{32}{12} = 2{,}66 \right)$$

$$\left(\frac{\text{formerly free oxygen} \times 10^{15} \text{t}}{\text{organic carbon in rocks} \times 10^{15} \text{t}} = \frac{16.9}{6.4} = 2.64 \right)$$

This balance calculation for oxygen and organic carbon on the basis of photosynthesis indicates to us that most of the oxygen not bound in carbonates and silicates has indeed been produced by photosynthesis. Therefore, there should be a relationship between organic carbon in fossil sediments and oxygen levels of paleoatmospheres.

Making use of so-called half-mass ages of sedimentary rocks, as given by Garrels and Mackenzie (1969), an accumulation rate for organic carbon of approximately 3.2×10^6 t y^{-1} was calculated on the basis of the previously mentioned numbers. The present annual marine production of organic carbon is estimated to be 6×10^{10} t (Vallentyne, 1965). With this annual marine production, the total global preservation of organic carbon of 10^{-4}, or 0.01% during the earth's history can be calculated. Although it is difficult to estimate the true preservation, it seems to be safe to assume that it is less than 0.1%. Menzel and Ryther (1970) also estimated that about 0.1% of the annual production of organic matter is buried in surface sediments. Only this tiny fraction of organic matter is preserved in sediments, whereas the remainder is recycled, mainly in the euphotic zone of the top water layer in the oceans. This is why oceanographers speak of a closed system with respect to living phytoplankton and CO_2 in ocean waters. In a study on the origin and fate of organic matter in the Black Sea, Deuser (1971) found a preservation rate of 4%. However, this value of 4% has to be considered as an upper limit, which is reached only under such favorable conditions as are found in the Black Sea. These conditions are an oxygen-free and fairly calm water body without scavenging benthic life at the bottom except for anaerobic bacteria. The sedimentation of certain petroleum source rocks has very likely taken place under similar conditions.

In this connection there is frequently an alternation of environments favorable for production and preservation of organic matter and those in which much less organic matter is preserved in sediments. A good example of this is the series of finely laminated sediments, with alternating layers respectively rich and poor in organic carbon, described by Ross and Degens (1974) in young Black Sea sediments.

The cycle of organic carbon in nature is shown in Figure I.1.6. There is a primary, small cycle (1) with a turnover of about 2.7 to 3.0×10^{12}t of organic carbon, and a half-life of days up to tens of years. There is also a secondary, large cycle (2) comprising an estimated quantity of 6.4×10^{15}t, and with a half-life of several million years. The two cycles are interconnected by the tiny leakage of

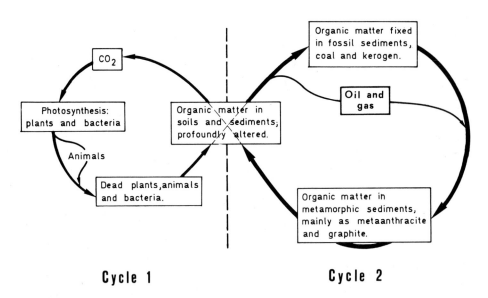

Fig. I.1.6. The tow major cycles of organic carbon on earth. Organic carbon is mainly recycled in cycle 1. The crossover from cycle 1 to cycle 2 is a tiny leak that amounts only to 0.01–0.1% to the primary organic productivity. (After Welte, 1970)

about 0.01% to 0.1% of the total organic carbon, representing oxidation of sedimentary organic matter to CO_2. For our considerations, the larger secondary cycle is of greater importance. Once the organic matter has entered a sediment, its long-term fate is mainly governed by tectonic events. In other words, phases of subsidence and increase in burial, or phases of uplift and erosion, determine whether the organic content of a sediment is preserved and transformed into petroleum, or is eroded and oxidized. If organic matter completes the second cycle during the birth, evolution, and end of a geosyncline, it undergoes increasing burial, and is subjected to diagenesis, catagenesis, and finally metamorphism. The processes of diagenesis and catagenesis are of prime importance in the formation of petroleum.

1.3 The Organic Carbon Budget in the Black Sea

The Black Sea may serve here as a model case for conditions prevailing during the formation of source rock-type sediments. The subsequent considerations follow mainly the findings and thoughts of Deuser (1971).

In the Black Sea, the main source of organic matter is in situ photosynthesis (Fig. I.1.7). A dominant role is played by the smallest unicellular algae. The main groups of marine life depend directly or indirectly upon their rate of production. About 100 g organic carbon $m^{-2}yr^{-1}$ have been produced by photosynthetic organisms over the entire water-covered area during the last 2000 years. In addition, a certain amount of organic carbon (less than 10%) is introduced mainly as detrital material by rivers and the Seas of Azov and Marmara. Shimkus and Trimonis (1974) estimated the contribution of organic carbon to the Black Sea by rivers to be about one third of the total input. Furthermore, chemosynthesis serves as a source of organic carbon. This is the organic carbon resynthesized by autotrophic bacteria, the amount of which is not known exactly. It is estimated to be less than $15 g C_{org.} m^{-2} yr^{-1}$.

Most of the organic carbon produced in and carried into the Black Sea is oxidized to CO_2 in the top 200 m due to respiration, and is returned to the hydrosphere–atmosphere system. The amount immediately returned to the primary organic carbon cycle (Fig. I.1.6) and thus available again for photosynthesis, is probably close to 80%. A small portion is also carried away into the Seas of Marmara and Azov. The remainder is transferred into the anoxic waters below the top 200 m zone. There it is subject mainly to chemical and microbial attack. It seems safe to assume that generally 80% to 95% of the organic carbon is recycled in the top water column, where there is photosynthetic activity.

In the anoxic zone, Deuser (1971) assumed a steady-state situation. The organic matter being oxidized due to sulfate reduction, and due to losses from solubilization and fossilization is balanced by an equal influx of organic matter from above, where oxygen-bearing waters prevail.

About one fourth of the organic carbon introduced into the anoxic zone of the Black Sea is buried in the sediments, and thus fossilized. This amounts to as much as 4% of the total organic carbon input of the Black Sea. This is certainly more

Sources of organic matter in Black Sea

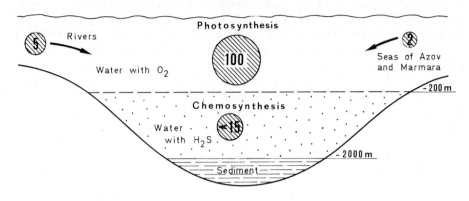

Fate of organic matter in Black Sea

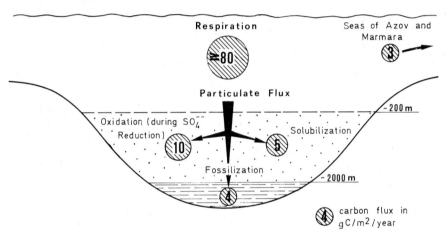

Fig. I.1.7. The organic carbon budget of the Black Sea during the last 2000 years. *Top:* sources of organic matter calculated as g m^{-2} yr^{-1}. *Bottom:* fate of organic matter. About 4% of the total input of organic carbon is fixed in sediments. This value is considerably higher than in the open ocean. (Modified after Deuser, 1971)

than the average amount of organic carbon normally fixed in marine sediments. The main reason for the higher preservation rate in the Black Sea is probably the slower degradation of organic matter in the absence of oxygen. Up to a certain point, a high sedimentation rate can also help to preserve organic material. The importance of those two factors in the preservation of organic matter is hard to quantify, and certainly varies in different environments and localities. It is obvious that in areas such as deltas, the high rate of sedimentation is more important than an anoxic water zone, which should be of greater influence in closed, stagnant bodies of water. As a basis for comparison with the Black Sea, the Amazon River discharges about 10^{10} t organic carbon per year (Williams, 1968). This is about 100 times more than the total annual production of organic carbon in the Black Sea. The Amazon River represents approximately 20% of the total worldwide river runoff.

Organic matter in waters is arbitrarily subdivided into particulate (living and dead) and "dissolved" organic matter (in general particles $<1\mu$). Approximately 10^7 t of the total organic carbon discharge of the Amazon River is in a dissolved state; the rest is particulate organic matter. Whereas the particulate organic matter carried into the sea is more significant locally, the dissolved organic matter finds a wider distribution due to mixing of water masses. No matter what the processes are that lead to preservation and fossilization of organic matter, they are of prime importance for the formation of potential source rocks.

Summary and Conclusion

Photosynthesis is the basis for mass production of organic matter. About 2 billion years ago in the Precambrian, photosynthesis emerged as a worldwide phenomenon. Herewith the foundation for the food pyramid and the evolution of higher forms of life was laid. The enrichment of molecular oxygen in the atmosphere of the earth is a direct consequence of photosynthesis and the mass production of organic matter.

During the earth's history and on a global scale, the average preservation rate of the primary organic production, expressed as organic carbon, is estimated to be less than 0.1%. The upper limit of the preservation rate of organic carbon to be found in certain oxygen-deficient environments favorable for deposition of source rock-type sediments is about 4%.

Chapter 2
Evolution of the Biosphere

We know of the occurrence of petroleum and petroleum precursors (kerogen and bitumen) in Precambrian time (Nonesuch Shale, Michigan, USA). Through the Cambrian and up to the Devonian, mainly marine phytoplankton and bacteria, and to some extent benthonic algae and zooplankton could have served as source material for petroleum. Thereafter, terrestrial organic matter derived from land plants offers an alternative source. The evolutionary level and kind of contributing source organisms may be of decisive influence on the type and amount of petroleum generated in a certain source rock. Therefore, one must consider the evolution of the biosphere in connection with the formation of petroleum.

2.1 Phytoplankton and Bacteria

Some 2 billion years ago in the Precambrian, the main producers of organic carbon were blue-green algae and photosynthetic bacteria. Throughout the Cambrian, Ordovician and Silurian, a variety of marine phytoplanktonic organisms, bacteria, and blue-green algae were the dominant sources of organic carbon until land plants appeared on the continents and spread sufficiently by the Middle Devonian (Zimmermann, 1959). It is estimated that even today marine phytoplankton and bacteria are responsible for 50% to 60% of the world organic carbon production (Vallentyne, 1965).

On the basis of countless analyses, Tappan (1968) and Tappan and Loeblich (1970) have estimated the abundance of fossil phytoplankton through geological time. A diagramatic representation is given in Figure I.2.1. Phytoplankton production started in the Precambrian, increased throughout the Early Paleozoic, then decreased sharply in Late Devonian. During Permo-Carboniferous and Triassic times, production was generally low. Another maximum occurred in Late Jurassic-Cretaceous, dropping abruptly at the end of the Cretaceous. In the Early Paleocene, production was still very low. It increased rapidly in Late Paleocene and Eocene, and decreased again in the Oligocene. Finally, a maximum in the Miocene was followed by a decline to the present level of production.

The first period of high phytoplankton productivity (Precambrian-Early Paleozoic) was governed by characteristic organic-walled plankton such as blue-green algae, various "acritarchs", and green algae. They had no skeletons composed of carbonate, silica, or other mineral substances. The name "acritarch"

2.1 Phytoplankton and Bacteria

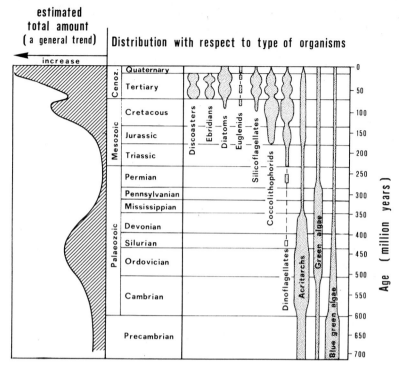

Fig. I.2.1. Variation in abundance of fossil phytoplankton groups and total phytoplankton during geological past. (Adapted from Tappan and Loeblich, 1970); reprinted with permission and from publication of the Geological Society of America)

refers to a purely morphological group of rounded, polygonal or elongate, smooth or ornamented, acid-resistant organic-walled vesicles which are probably algal cysts. Their origin is still being debated. The second maximum in Late Jurassic-Cretaceous was dominated by calcareous nannoplankton, including coccolithophorids (small unicellular plants with a calcareous skeleton) and dinoflagellates. Siliceous phytoplankton, especially silicoflagellates and diatoms, appeared in the Late Cretaceous and became increasingly important in the Cenozoic.

Fossil records are inadequate to quantify the productivity of bacteria throughout geological time. Because of microscopic (or submicroscopic) size, and the lack of hard parts, they are rarely fossilized. However, examples of fossilized bacteria have been recorded from all geological systems including the Precambrian. Fossilized bacteria are often associated with organic matter such as plant tissues, and animal and insect remains (Moore, 1969). Most fossilized bacteria were generically similar to present-day forms occurring in comparable environments (Schopf et al., 1965; Moore, 1967).

Bacteria and blue-green algae, both unicellular, are the only organisms that do not possess membrane-bound organelles such as nuclei within the cell. They are, therefore, called *prokaryotes,* and are distinct from all other organisms with cell nuclei, which are called *eukaryotes.*

Bacteria and algae undoubtedly are and always have been ecological pioneers. Bacteria especially show an enormous versatility in their physiology. This enabled them to live almost everywhere and guaranteed their ubiquity. Bacteria may be heterotrophic, autotrophic (photosynthesis without oxygen production), or both. They are an outstanding example of evolutionary success, relying entirely upon their versatility, and apparently unrestricted throughout geological time. According to Zobell (1946, 1947, 1964), more than 100 species of bacteria and related organisms are presently known to attack organic matter in soils and sediments. There is no reason to assume that this situation has changed drastically since the Precambrian, when larger amounts of organic debris were first available. As will be discussed later, dead bacteria are second only to phytoplankton in contribution of organic matter ultimately buried and preserved in sediments.

The fundamental relationship in the food chain within the pyramid of life causes a direct correlation in occurrence and distribution between autotrophic phytoplankton and heterotrophic zooplankton. The biomass of zooplankton shows a tendency to be high in areas of high phytoplankton productivity (Bogorov and Vinogradov, 1960). This relationship has existed since the emergency of zooplanktonic organisms in the Precambrian, such as unicellular foraminifera and radiolarii. It also applies to other organisms of the animal kingdom, like worms, mollusks, and arthropods. There are a few landmarks with respect to the occurrence of zooplankton and other invertebrates, including the emergence and extinction of graptolites during the Early Paleozoic (Ordovician–Silurian), an occasionally massive occurrence of trilobites during the Cambrian, Ordovician

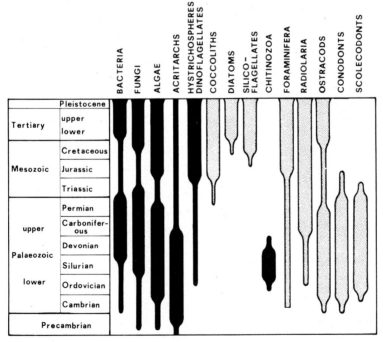

Fig. I.2.2. Main groups of microfossils in the marine environment. (After Moore, 1969)

2.2 Higher Plants

and Silurian, as well as the explosive emergence of foraminifers during the Late Jurassic. Planktonic foraminifera must be considered as major contributors of organic matter to certain marine sediments. Their quantity and occurrence, however, seem to be controlled primarily by phytoplankton productivity, i.e., by the availability of a food-source (Tolderlund and Bé, 1971). More highly organized animals, such as fishes, contribute so little organic matter to sediments that they can practically be ignored. However, the larval stages of most invertebrates have probably contributed varying amounts of organic matter since the Cambrian time.

A quantitative estimate of the contribution of various groups of organisms with high productivity and biomass during geological history is difficult. It exists for only a few categories of organisms, such as phytoplankton (Fig. I.2.1). Main groups of microfossils and their possible organic contribution to sediments in the marine and nonmarine environments are shown in Figure I.2.2 and Figure I.2.3. These figures are not intended to show the continuity of an environment, but rather the kind of organism available when the environment did occur (Moore, 1969). Microfossils rich in organic matter are black, while forms in which inorganic skeletal material predominates are stippled. The appearance of a given group is represented by a narrow area and a more common occurrence by a wider area. No attempt is made to show relative abundance.

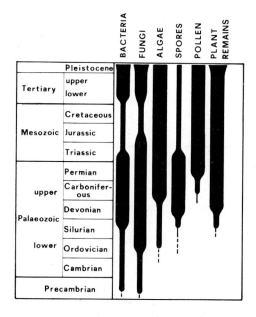

Fig. I.2.3. Main groups of microfossils in the nonmarine environment. (After Moore, 1969)

2.2 Higher Plants

Following phytoplankton and bacteria, higher plants are the third important contributor to organic matter in sediments. As can be seen in Figure I.2.4, remains of higher plants appear in sediments of Silurian age and have been

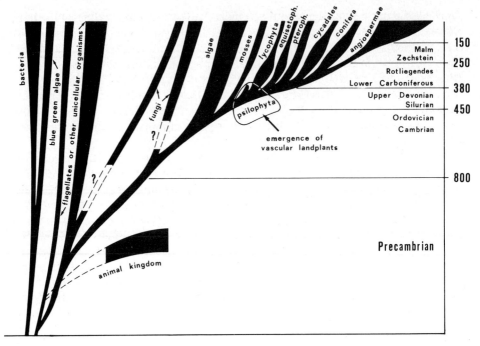

Fig. I.2.4. The evolution of vascular land plants. During Late Silurian the first higher plants, psilophyta, emerged from the marine environment and conquered the continents. (Modified after Zimmermann, 1969)

common relics since Devonian time. The precursors of higher plants evolved throughout the Precambrian, Cambrian and Ordovician. In order of development, these precursors include blue-green algae, green algae, and finally higher algae such as seaweed and kelp. They lived in the marine environment. Evolution of land plants started in the Silurian. The spore record suggests a small number of land-plant types in the Silurian, with continuously increasing diversity through the Devonian.

In this connection, relatively rich spore assemblages in material of Silurian age (Wenlock-Ludlow) from North Africa are of special interest, because of their early age, with respect to the evolutionary history of land plants (Richardson and Ioannides, 1973).

According to the macrofossil record, it was not until latest Silurian that primitive Psilopsida *(Cooksonia)*, belonging to the division of Pteridophyta, conquered the continents. Some Psilopsida also lived in the marine environment. These primitive plants were possibly leafless and rootless, but certainly had a vascular system.

During the Early Devonian, other groups of Pteridophyta evolved (Zimmermann, 1959). In the Middle Devonian, due probably to an explosive evolution, most classes of vascular plants had already appeared as primitive representatives. In the Late Devonian, Psilopsida became scarce, while other Pteridophyta, such as Lycopsida *(Cyclostigma)*, Sphenopsida *(Sphenophyllum)*, and early ferns

(Pteropsida) dominated the land flora. Contrary to the Silurian land plants, Late Devonian plants had small leaves, roots and secondary wood. The land flora of the Late Devonian is in many respects similar to the flora of the Early Carboniferous. During Early Carboniferous, the first Equisetales and seed ferns (pteridosperms) appeared. *Lepidodendron* became a common plant. During the Late Carboniferous, this type of land flora reached a culmination in variety of forms and quantity. Shrubs and larger trees occurred more and more frequently during Late Devonian and Carboniferous, eventually forming dense forests and hence large masses of wood, now partly fixed in extensive coal seams.

Toward the end of the Paleozoic, during the Permian Age, the gymnosperms, a new division of plants, emerged from earlier beginnings. They included primarily such classes as Coniferales, Ginkgoales, Cycadales and Bennettitales, and dominated the flora until the Cretaceous. Because of this dominance of gymnosperms, the time-span between Late Permian and Early Cretaceous is called the "era of gymnosperms". Pteridophyta did not play an important role during this time.

A final important turning point in the evolution of plants was reached during the Early Cretaceous. The character of land vegetation was severely changed by the sudden appearance of angiosperms, which quickly dominated the flora. Although present-day vegetation shows a wider variation in angiosperms than that during the Upper Cretaceous, the same type of plants still covers vast areas of the continents. The sketched evolution of land plants shown in Figure I.2.4 is taken from Zimmermann (1969).

2.3 Geological History of the Biosphere

The evolution of the biosphere and the geological history of the earth are intimately interrelated, and cannot be considered separately. For a short comprehensive understanding of both, partly on a hypothetical basis, we follow largely the ideas and interpretations of Bülow (1959). Throughout geological history, the evolution of flora was always one step ahead of faunal evolution. At all evolutionary levels, plants invaded new ecological systems first, and were later followed by animals. During the Early Paleozoic, when algae dominated the biosphere, only minor quantities of marine invertebrates existed compared to the plant biomass. The tremendous surplus in productivity of lower plants at that time is represented by dark and black marine shales, rich in organic matter, which are the "normal" open marine sediments of the Cambrian, Ordovician and Silurian times.

In later periods, similar shales were less common in marine environments, and their occurrences were limited to special paleogeographic situations. Examples of these special situations are land-locked marine basins such as lagoons, and larger bodies of water, such as the early Atlantic Ocean in the Middle Cretaceous time. In the open marine environment after the Silurian time, a kind of equilibrium had developed between plant production and a plant-consuming fauna. The normal surplus in plant (mainly algae) productivity was thus terminated. The overall evolutionary level of the Late Silurian marine flora and fauna does not seem to be much different from today's marine level of evolution. With the conquest of the continents by land plants during late Silurian and Devonian, the dominance of marine organic productivity slowly began to disappear, until some sort of equilibrium was reached between marine and terrestrial productivity in the Cretaceous time. After the Silurian, the surplus of productivity shifted from open marine environment to coastal areas, and to the paralic basins where most of the Late Paleozoic coal deposits have been found. With the advent of gymnosperms near the end of the Paleozoic, and especially with the appearance of the very adaptable, and hence superior, angiosperms during the Early Cretaceous, plant surplus productivity shifted to continental areas. Large coal deposits of the Cretaceous and Tertiary Age, which originated in inland basins, attest to this shift.

Summary and Conclusion

Starting in the Precambrian till the Devonian, the sole primary producer of organic matter was marine phytoplankton. Since the Devonian an increasing amount of primary production has been contributed by higher terrestrial plants. At present marine phytoplankton and terrestrial higher plants are estimated to produce about equal amounts of organic carbon. Quantitatively, the four most important contributors to organic matter in sediments are phytoplankton, zooplankton, higher plants and bacteria. Higher organized animals, such as fishes, contribute on the average so little to organic matter in sediments that they can practically be neglected.

According to the laws of the food chain, zooplankton shows a tendency to be high in areas of high phytoplankton productivity. Heterotrophic bacteria are abundant where organic matter is available as a source of food.

Chapter 3
Biological Productivity of Modern Aquatic Environments

The biological productivity of aquatic environments, especially marine environments, is of great importance to the formation of potential oil source rocks. Although the primary productivity of organic matter in aquatic environments is presently in the same range as in subaerial environments — due to the wide-spread occurrence of land plants — the chance for preservation of organic matter in sub-aquatic environments is far greater. In sub-aerial environments, free access of air, together with the presence of moisture, allows growth and action of bacteria, and hence a breakdown and destruction of organic matter. However, in sub-aquatic environments, the deposition of fine-grained sediments limits access of dissolved molecular oxygen. Thus, the activity of aerobic bacteria comes to a halt when the limited amount of oxygen trapped in the sediments is exhausted. In this connection, it is also important to realize that air contains 21% (vol) of oxygen, whereas water normally contains only a few ml of oxygen per l. Therefore, in practical sense, organic matter is only preserved and fossilized in sub-aquatic sediments.

3.1 Primary Producers of Organic Matter

Primary production of organic matter from inorganic components in the present-day marine realm may serve as a key to the geological past. In addition, in certain areas, large amounts of organic matter have been contributed by terrestrial higher plants. This, however, was not possible prior to the emergence of land plants during Silurian-Devonian time.

As stated before, the main producers of organic matter in the marine environment are the various groups of unicellular, microscopic phytoplanktonic organisms. The main groups are diatoms, dinoflagellates, Cyanophyceae (blue-green algae) and very tiny, naked phytoflagellates or nonmotile cells, the nannoplankton. They generally range in size from 0.002 mm to 1 mm and may form colonies up to 5 mm in diameter. The nannoplankton and ultraplankton comprise species measuring about 10µ or less (Krey, 1970). All these organisms are equipped with mechanisms such as flagellae or spiny extensions of their skeletons to prevent or slow down their sinking into the darkness of greater water depths. Some organisms, such as certain diatoms, even produce oil droplets to lower their average specific weight, giving them additional buoyancy. Examples of the most important phytoplanktonic and zooplanktonic organisms presently found in the oceans are shown in Figure I.3.1.

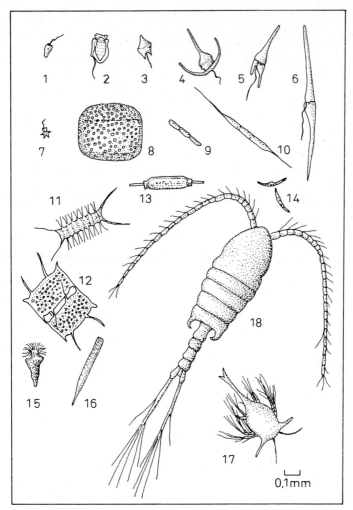

Fig. I.3.1. Microplanktonic organisms of the oceans. (After Krey, 1970). I. Dinoflagellatae: 1 *Prorocentrum micans*, 2 *Dinophysis acuta*, 3 *Peridinium divergens*, 4 *Ceratium tripos* var. *atlantica*, 5 *Ceratium furca*, 6 *Ceratium fusus*. II. Silicoflagellatae: 7 *Distephanus speculum*. III. Diatomeae: 8 *Coscinodiscus concinnus*, 9 *Rhizosolenia faerösensis*, 10 *Rhizosolenia setigera*, 11 *Chaetoceras decipiens*, 12 *Biddulphia mobiliensis*, 13 *Ditylum brightwellii*, 14 *Nitschia closterium*. IV. Protozoae: 15 *Tintinnopsis campanula*, 16 *Tintinnopsis subulata*. V. Cirripediae: 17 *Nauplius* var. *verruca Strömia*. VI. Copepodae: 18 *Temora longicornis*

Benthonic organisms, such as large kelp and rock weed, are by definition attached to the bottom and are restricted to areas of shallow water depths usually not exceeding 50 m. Phytobenthos may be of local importance but on a worldwide scale it can practically be ignored. It is estimated that there is a present annual production of 550×10^9t of phytoplankton and 0.2×10^9t phytobenthos in the ocean (Krey, 1970).

Although still controversial, bacteria are considered to be of importance, along with phytoplankton, with respect to biomass and organic production in the marine environment. However, the overall production of heterotrophic bacteria should be less than the production of phytoplankton, since the production of consumers of organic matter cannot exceed the level of primary production (Datsko, 1959). Exceptions to this are cases where other sources of food are available for bacteria, such as dissolved and particulate organic matter derived from continents. Sorokin (1971), for example, observed that the production of bacterioplankton under 1 m^2 in tropical ocean waters may exceed the primary production by phytoplankton. The role of bacteria in the biological productivity is to decompose dead organic matter and convert its decay products into forms suitable for assimilation by aquatic plants. Furthermore, bacteria themselves serve as food for aquatic animals. Bacteria in the ocean are partly suspended in the water and partly attached to plankton organisms and particles of detritus. Great quantities of bacteria occur also on the sea floor, probably with the exception of very great water depths, and in the top layer of sediment (Bordovskiy, 1965) in shallow to moderate water depths.

Phytoplankton, the primary producer of organic matter, forms the basic member of the food chain and hence of the pyramid of life. Diatoms, dinoflagellates, and coccolithophores are the main producers in this sequence. Next members in the food chain are vegetarians such as small crustaceans (e.g., copepodes, see Fig. I.3.1), that live on phytoplanktonic organisms. Then follow larger animals, such as herring and mackerel. In a food chain of this type the quantitative relationship between phytoplankton and mackerel is about 1000:1. It is interesting to compare this with the relationship between primary organic production and organic matter buried and preserved in marine sediments. This relationship also shows a ratio of about 1000:1.

Tasch (1967) presents information on the geological age range of existing primary producers. Dinoflagellates probably occupied first place, and coccolithophores second place among primary producers in Mesozoic oceans, except toward the end of the Cretaceous, when marine diatoms rose to first place among primary producers. Acritarchs are supposed to be remains of other unicellular algae, which were also primary producers throughout the Palezoic and in Late Precambrian time.

3.2 Factors Influencing Primary Productivity

Biological productivity in the marine environment is mainly controlled by light, temperature, and chemical composition of sea water, especially with regard to mineral nutrients such as phosphate and nitrate. These parameters are interrelated in a very complex manner to physiographic features of the ocean, such as morphology of the ocean basins, ocean currents and mixing of different water bodies.

A vertical profile or hypsographic curve of the earth's surface, showing relative distribution of surface area, is useful for an understanding of the following

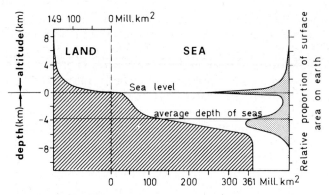

Fig. I.3.2. Hypsographic curve of the earth's surface *(left)* and relative distribution of elevations and depth levels *(right)*. The two maxima of the curve to the right reflect the dualism between continents and oceans, i.e., continental and oceanic crust. (After Kuenen, 1950)

discussion. Figure I.3.2 shows the relative proportions of surface areas at various elevations on land and depths beneath the oceans over the entire earth. Two distinct maxima appear on this curve, one at about 100 m above sea level, and another between 4000 and 5000 m below sea level. The first one includes the old continental masses, and the second the bottom of the deep sea. This reflects the well-known dualism between continents and oceans, and between continental and oceanic crust (Seibold, 1974). The class and distribution of seas by water depth are as follows: areas with a water depth of 0–200 m are conventionally called shelf (7.5% of the sea floor): water depths of 200–4000 m are characteristic of the continental slopes (8%) and include the continental rise (5%); from 4000 m–5500 m is the deep sea abyssal plain (77.5%); and below 5500 m are the so-called trenches (2%). Shelf, continental slope, and continental rise are together termed continental margin. If primary productivity of the oceans is seen in relation to the above distribution of oceanic areas (Table I.3.1), the importance of coastal waters, which include primarily the shelf regions and part of the continental slopes, is evident. A map of the shelf areas is shown in Figure I.3.3.

Table I.3.1. Productivity of coastal waters as compared to the open ocean. (Adapted after Krey, 1970)

Area	Area in $km^2 \times 10^6$	Averaged productivity in $g\ Cm^{-2}yr^{-1}$	Total production in 10^9 t Cyr^{-1}
Open ocean	326	50	16.3
Coastal waters	36	100	3.6
Areas with upwelling	0.4	300	0.1

3.2 Factors Influencing Primary Productivity

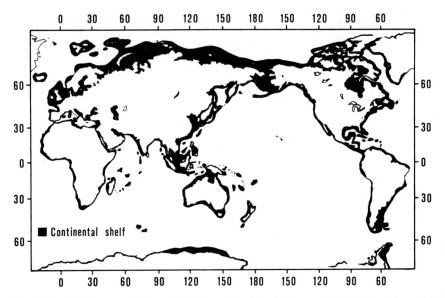

Fig. I.3.3. Shelf areas of the world. Water depth in shelf areas is 0–200 m. About 7.5% of the sea floor belongs to shelf areas

All organic matter in the oceans is ultimately produced by unicellular algae through photosynthesis, except for very small contributions by pelagic macrophytes. The spatial variation in photosynthetic production is mainly controlled by the supply of light and nutrients. In general, growth is limited by supply of light at the ocean surface only in polar regions, at depth in the open ocean, and in turbid coastal waters (Menzel, 1974). Outside these regions, rates of production are controlled by the supply of nutrients, and this in turn is influenced by recycling and circulation that replenish or remove nutrient supply from the euphotic zone of the water column. The interrelationship between occurrence of phosphate, an important nutrient, and growth of phytoplankton in the South Atlantic ocean is shown in Figure I.3.4. In the oceans, 90% or more of the produced plant matter is eaten by grazing organisms (Menzel, 1974). As plants will grow only in the lighted upper layers of the ocean, the concept of the euphotic zone has been elaborated. The euphotic zone is defined as the zone of highest photosynthetic productivity near the surface of the oceans, and takes into account such factors as illumination and nutrients. An exact depth cannot be given, as it varies from one area to another according to local conditions. Normally, the euphotic zone is situated in the top 200 m of surface waters. The majority of organic matter production, however, is concentrated in the upper 60 to 80 m of the water column.

If thermal or salinity stratification stabilizes the water in the euphotic zone, the rate of nutrient replenishment is very slow (e.g., in subtropical regions). Rapid mixing, on the other hand, may continually fertilize the euphotic zone, but plant cells may be transported by the mixing currents out of the light zone before cell division and reproduction can take place. Balanced optimal conditions are frequently encountered in regions with water upwelling (Table I.3.1), such as in

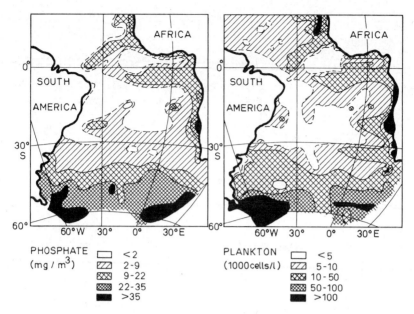

Fig. I.3.4. Distribution of phosphate and plankton in surface waters of the South Atlantic ocean. (After Dietrich, 1963; taken from Debyser and Deroo, 1969)

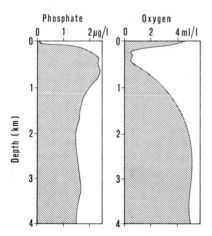

Fig. I.3.5. Depth distribution of phosphate and oxygen in an area of upwelling off southwest Africa. There is a characteristic depletion of nutrients, here exemplified by the depletion of phosphate, due to intense phytoplankton growth in near-surface waters. It is replenished from below by upwelling. (After Emery, 1963)

certain areas of western continental shelves. The resulting depth distribution of sea water properties, as found in an area of upwelling off Southwest Africa (Fig. I.3.5), has been summarized by Emery (1963). It is characterized by a depletion of nutrients due to intense phytoplankton growth in near-surface waters and a replenishment of these by upwelling, locally supplemented by addition of nutrients from rivers. Good situations for production may also occur where there are alternating periods of water stabilization, when phytoplankton blooms in the euphotic zone, and periods of remixing by storm or tidal activity, when nutrients

3.2 Factors Influencing Primary Productivity

are replenished to allow the next bloom. These latter conditions occur frequently in the protected coastal areas of British Columbia (Strickland, 1965). Water mixing, either in short or longer intervals, is thus of great influence on biological productivity.

The relative constancy of chemical composition of the ocean waters, 3.5% salinity by weight (Table I.3.2) indicates good mixing on longer terms, at least of the top 200 m of surface water. In these superficial water layers, wind-driven currents on the average complete the mixing within a few decades. In deeper waters, wind is not the dominating factor, but rather it is sinking masses of cold water from the polar regions. This cold, heavy water creeps underneath water of lesser density and moves toward lower latitudes. This process also causes a long-term mixing, but it takes 1000 or 2000 years to be completed. Temperature in ocean waters is particularly interrelated with illumination and water mixing. Hence, it is difficult to isolate the effect of temperature on primary productivity. However, there are "optimal" temperatures for various phytoplankton species. Dinoflagellates, for instance, prefer warm tropical waters with temperatures of 25°C or more, whereas silicious diatoms and radiolaria dominate the cold waters (5–15°C), especially in the polar regions. The differentiation into various species of organisms is by far greater in warm tropical waters than in cold waters. The same is observed with salinity changes. Areas with lower salinities, such as regions of brackish water with salinities of 5–7$^0/_{00}$, are characterized by minima in number of species. However, a reduction in the number of species does not necessarily mean a reduction in productivity.

When considering chemical composition of sea water, it must be kept in mind that it is normally not salinity as a whole, but availability of nutrients that is most important for productivity. When the concentration of a nutrient becomes too low, it will influence photosynthesis, rate of cell division and even chemical composition of plant cells. Aside from CO_2, nitrogen, phosphorus and trace organics (vitamins) are essential for plant growth. Nitrogen is consumed as ammonia (NH_4^+) or nitrate (NO_3^-), and phosphorus as phosphate (PO_4^{3-}).

Table I.3.2. Main inorganic components of sea water, expressed in molecular form, showing quantity of salt dissolved in 1 l

Formula	g
NaCl	27.2
$MgCl_2$	3.8
$MgSO_4$	1.6
$CaSO_4$	1.3
K_2SO_4	0.9
$CaCO_3$	0.1
$MgBr_2$	0.08
	34.98

Among the trace organics, vitamin B_{12} is important. Generally in the marine environment, nitrogen as nitrate is the most important element, because it is usually the first to be stripped from the water by growing organisms. In a few locations, phosphate may be depleted before nitrate, but this is more common in lakes and rivers than in the marine environment (Strickland, 1965). Marine basins, isolated more or less from the open oceans, show gross water properties that deviate more or less from those of open ocean waters (Fig. I.3.6a, b).

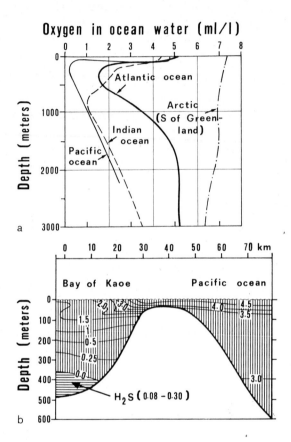

Fig. I.3.6. (a) Vertical distribution of oxygen in various ocean basins. (After Dietrich, 1963). (b) Vertical distribution of oxygen (ml l^{-1}) in waters of a closed bay (Kaoe, Halmahera Islands). (After Kuenen, 1942)

3.3 Present Primary Production of the Oceans

As described in the preceding pages, primary productivity in the oceans is influenced by many interrelated parameters, especially light, nutrients, and water mixing. Knowledge of these parameters, together with direct productivity measurements, provided the information needed for preparation of a map showing intensities of primary biological production in the world oceans (Fig. I.3.7).

3.3 Present Primary Production of the Oceans

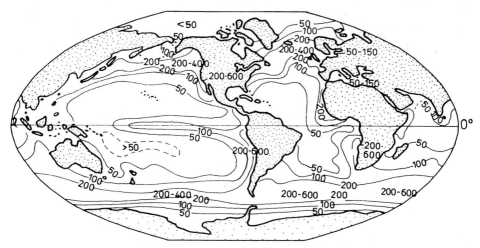

Fig. I.3.7. Primary biological productivity (org. carbon m^{-2} yr^{-1}) in the world oceans. Productivity is relatively high in coastal areas, especially in zones of upwelling along the west coast of the American and African continents. (Adapted from Debyser and Deroo, 1969)

Many researchers have observed a so-called landmass or island effect, where, as a broad generalization, primary productivity is greater near the coast of continents and islands, or even over submerged ridges and banks than in the open ocean (Strickland, 1965). The reason for this phenomenon, which occurs at all latitudes, is not yet fully understood. It could be that grazing, and hence apparent reduction of productivity, are for some reason more continuous and more effective in the open ocean than in coastal areas. Strickland (1965) has summarized the general picture of productivity on a regional basis. The two most important extremes, barren oceans and zones of upwelling, are cited below (see also Fig. I.3.7):

Barren oceans: Most of the tropical oceans between latitudes 10° and 40°, and stagnant tropical gyres like the Sargasso Sea, are very unproductive (about 0.1 g C m^{-2} day^{-1}). However, as this rate persists most of the year, annual productions of 50 g C m^{-2} can be expected.

Continental shelf upwelling: The effect of prevailing winds and of the Coriolis force on ocean currents off the west coast of continents may cause upwelling, and therefore great fertility (2 g C m^{-2} day^{-1} and more).

The primary productivity in freshwater lakes principally obeys the same laws as in the marine environment. Due to the fact that smaller water bodies and shallower depth are involved, climatic influences and influences from the surrounding land are generally much more pronounced.

Summary and Conclusion

The biological productivity in the marine environment is mainly controlled by light, temperature, and mineral nutrients, such as phosphate and nitrate. The larger part of biological production is concentrated in the upper 60 to 80 m of the water column. The biological productivity of coastal waters (100 g $C_{org}m^{-2}a^{-1}$) is on the average about twice as high as that of the open ocean. Most productive are areas with water upwelling (300 g $C_{org}m^{-2}a^{-1}$), such as in certain parts of Western continental shelfs.

Chapter 4
Chemical Composition of the Biomass: Bacteria, Phytoplankton, Zooplankton, Higher Plants

In the preceding chapters, we have learned that bacteria, phytoplankton, zooplankton (especially foraminifers and certain crustaceans), and higher plants are the main contributors of organic matter in sediments. Natural associations of these various groups of organisms in various facies provinces thus determine composition and type of organic matter to be deposited and incorporated in sediments.

From our point of view, we are mainly interested in biomass composition from such natural associations occurring in open waters of the shelf areas, the continental slopes and continental rises. In general, these include the continental margins and large land-locked, isolated basins (marginal seas) such as the Black Sea, the Baltic Sea, the Mediterranean Sea, and smaller lagoonal-type water bodies. As will be shown later, the lipid and lipid-like fractions of organisms are among the chemical constituents playing a dominant role in the formation of petroleum. Hence, in the following chapter, we will concentrate on the composition of the biomass, with special emphasis on lipids and lipid-like constituents of bacteria, phytoplankton, zooplankton and higher plants. However, at present, only limited quantitative information is available on natural associations of organisms and their gross chemical composition. When discussing composition of the biomass, it must be kept in mind that a sediment does not contain a true image of the original population of organisms in the water column above. Moreover, the chemical composition of organisms changes prior to incorporation in a sediment. As far as their soft parts are concerned, basically all organisms are composed of the same groups of chemical constituents, i.e., proteins, lipids, carbohydrates and lignins in higher plants.

4.1 Proteins and Carbohydrates

Proteins are highly ordered polymers, made from individual amino acids. They account for most of the nitrogen compounds in organisms. Proteins constitute such different kinds of materials as muscle fibers, silk and sponge. In the form of enzymes, they catalyze biochemical reactions within and outside of cells. They are of special interest in biological mineralization processes, such as the formation of shells, where they act as organic matrices, and are thus intimately associated with the mineral phase. In the presence of water, insoluble proteins may be broken down, due to the action of enzymes, to their water-soluble monomers, amino acids. In Figure I.4.1, this process of hydrolysis of the peptide bond (a) is shown together with some structural examples of amino acids (b).

Carbohydrates is a collective name for individual sugars and their polymers. They include the monosaccharides, di-, tri- (oligosaccharides) and polysaccharides. The name carbohydrate is derived from the empirical formula $C_n(H_2O)_n$, which suggests hydrated carbon. This empirical formula, however, has no particular chemical significance, since carbohydrates are in fact polyhydroxylated compounds and not just hydrated carbonaceous substances. Carbohydrates are among the most abundant constituents of plants and animals. They are sources of energy and form the supporting tissues of plants and certain animals. Cellulose and chitin are among the most prominent polysaccharides occurring in nature. The polysaccharide cellulose consists of 2000 to 8000 monosaccharide units. Wood is composed of cellulose (40–60%) and lignin. Higher plants synthesize the largest amount of cellulose, whereas certain algae, seaweeds and diatoms

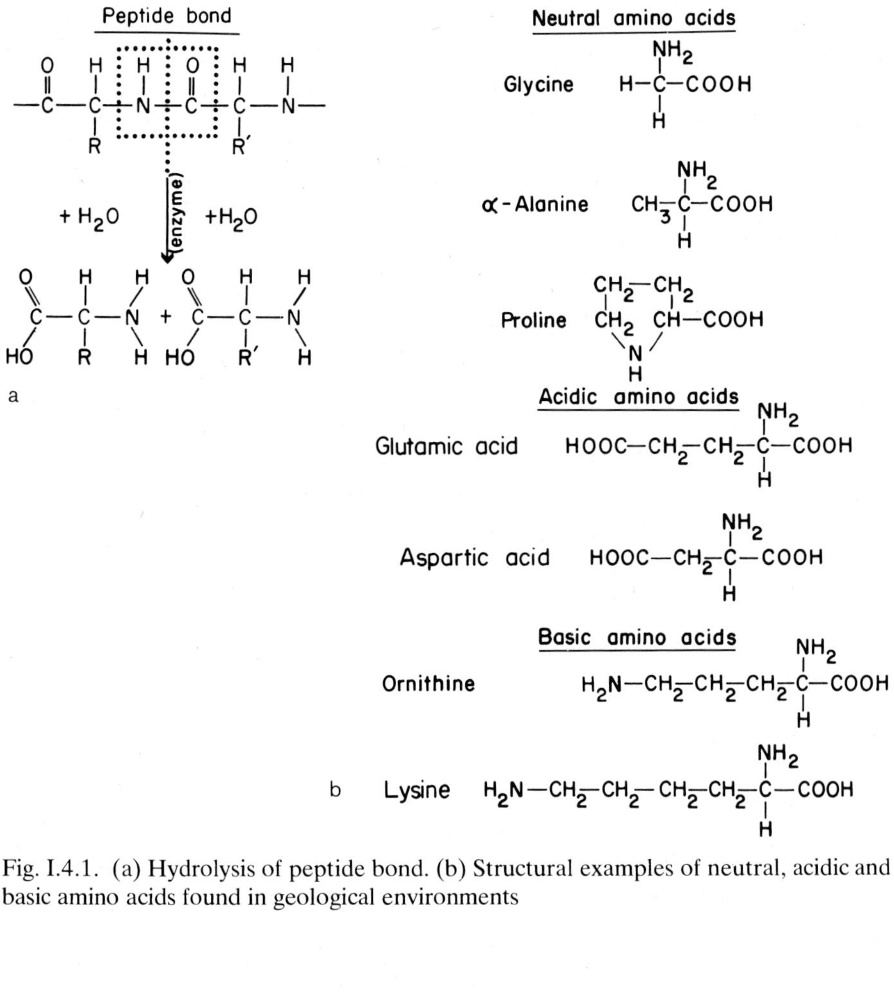

Fig. I.4.1. (a) Hydrolysis of peptide bond. (b) Structural examples of neutral, acidic and basic amino acids found in geological environments

4.1 Proteins and Carbohydrates

appear to be devoid of cellulose (Percival, 1966). Cellulose and chitin have analogous structures consisting basically of glucose and glucosamine units, respectively (Fig. I.4.2). Some algae, especially brown algae (Phaeophyta), contain as much as 40% (dry wt.) of alginic acid, a carbohydrate derivative. Pectin, a similar carbohydrate and like alginic acid, a polyuronid, is a frequent component of higher plants and bacteria (Fig. I.4.2). Polysaccharides are largely water-insoluble substances. They are, however, converted by hydrolysis into water-soluble C_6- and C_5-sugars.

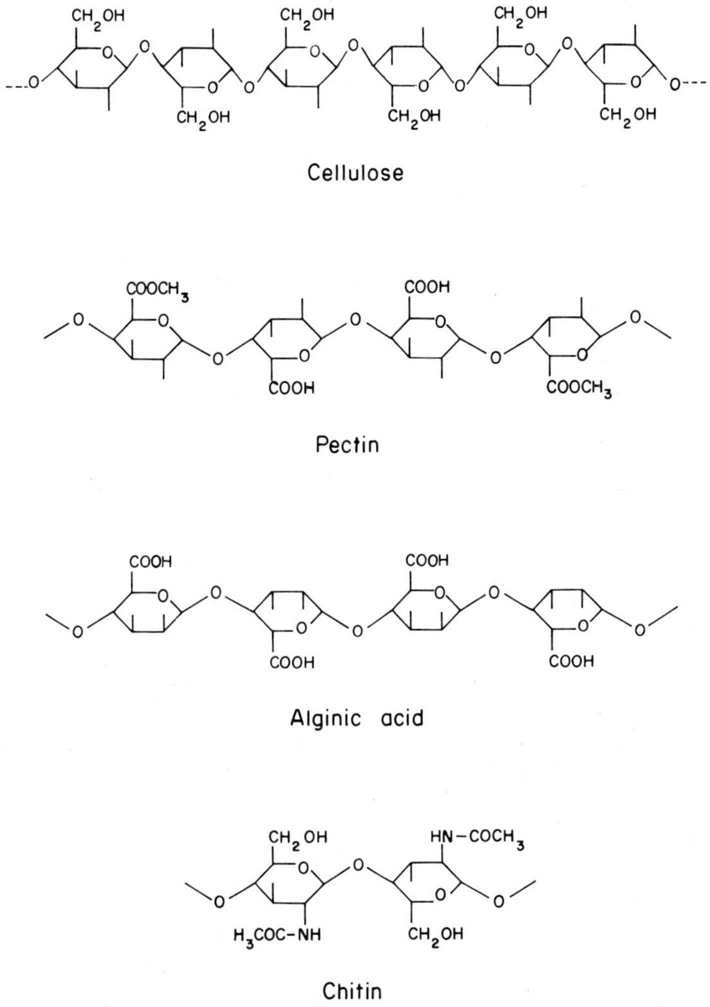

Fig. I.4.2. Macromolecular carbohydrate derivatives (e.g., cellulose, pectin, alginic acid and chitin) occurring in supporting tissues of plants and animals (free vertical bars in formula represent OH-groups)

4.2 Lipids

The term *lipids,* as defined by Bergmann (1963), is applied to all organism-produced substances that are practically insoluble in water, but are extractable by one or another of the so-called fat solvents. These solvents include chloroform, carbon tetrachloride, ethers, aliphatic and aromatic hydrocarbons, and acetone. Lipids encompass fat substances, such as animal fat, vegetable oils, and waxes such as those occurring as surface coatings of leaves. Fats are utilized in the energy budget of organisms. Waxes, however, are primarily designed for protective functions.

Naturally occurring fats are mixtures of various triglycerides, which are classified chemically as esters. Glycerides can be hydrolyzed, usually by aqueous sodium hydroxide, to give glycerol and the component fatty acids, which are obtained as sodium salts (Fig. I.4.3a). Fatty acids of natural fats have an even number of carbon atoms because they are synthesized biochemically from C_2 units (acetate units). Most frequently occurring are fatty acids with 16 and 18 carbon atoms, palmitic and stearic acid respectively (Fig. I.4.3b). Unsaturated fatty acids (Fig. I.4.3c) are common, especially in vegetable oils. From a physiological point of view, fats and oils have a high energy content. Therefore, in animals and plants they are used as energy storage. Seeds, spores and fruits are especially rich in lipids. Among algae, diatoms are known to contain great quantities of lipids, sometimes up to 70% on a dry weight basis. A high lipid content results, especially if the algae grow under nitrogen-deficient conditions and in cold water.

Natural waxes, such as bees' wax and plant waxes, are mixtures of various constituents. Waxes differ from fats in that glycerol is replaced by complex alcohols of the sterol series or by higher even-carbon-number aliphatic alcohols in the C_{16} to C_{36} range. These alcohols are often found in excess of the acids, which are also even-numbered and range from C_{24} to C_{36}. Plant waxes also contain hydrocarbons, especially long-chain n-alkanes having a predominance of molecules with odd carbon numbers.

In addition to these typical lipids, there are a number of lipid-like components such as oil-soluble pigments, terpenoids, steroids, and many complex fats, such as certain phospholipids (phosphatides). A prominent biochemical building block, the basic unit of many of these components, is the isoprene unit, consisting of five C-atoms (Fig. I.4.4).

Isoprene units, due to the conjugated double bonds, may polymerize to form chains and rings. A great variety of these chain-like and cyclic isoprenoid compounds is known to occur in plants and animals. For a long time, the individual monomer isoprene was not found in nature. However, it was recently discovered in small quantities in certain plants. It is a highly volatile, reactive hydrocarbon. Molecules containing 2 isoprene units are called terpenes (or monoterpenes, C_{10}); those with 3 units sesquiterpenes (C_{15}); with 4 units diterpenes (C_{20}); with 6 units triterpenes (C_{30}); and those with 8 units tetraterpenes (C_{40}). Natural gums, like caoutchouc and gutta, which are produced by certain higher plants (Euphorbiaceae and Sapotaceae, respectively), are polyterpenes. There are indications (Brooks and Shaw, 1972) that sporopol-

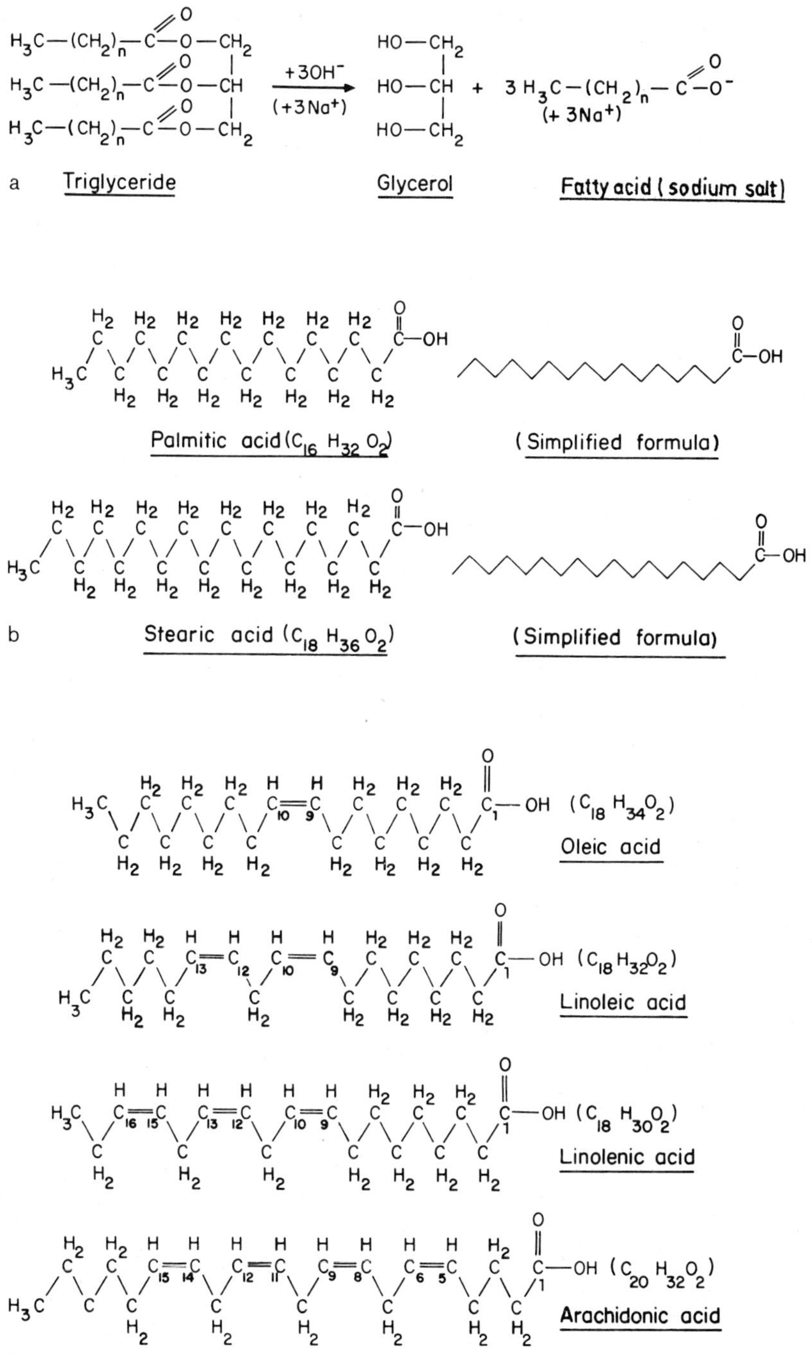

4.2 Lipids

Fig. I.4.3a–c. Main fat and oil components in organisms. (a) Triglyceride, (b) palmitic acid and stearic acid, (c) various unsaturated acids

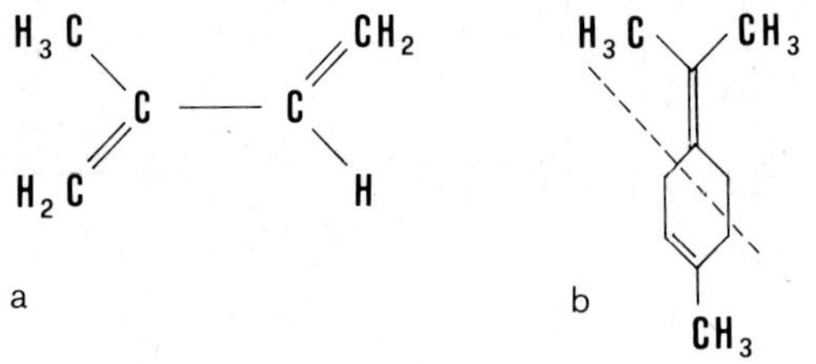

Chemical Composition of the Biomass: Bacteria, Plankton, Higher Plants

![Isoprene and Terpinolen structures]

Fig. I.4.4. (a) Isoprene unit. Building block of many natural biological components, such as steroids, terpenes and carotenoids. (b) Terpinolen, a naturally occurring cyclic monoterpene, consisting of two isoprene units

lenin, an important constituent of spore and pollen membranes, is a polyterpene. The various types of isoprenoids or terpenoids may occur as a linear arrangement of isoprene units or cyclized into from 1- to 5-ring systems. Terpenoids may occur as hydrocarbons, alcohols or their esters, aldehydes, ketones, acids and lactones. They exhibit various degrees of unsaturation.

Monoterpenes (C_{10}) occur abundantly in higher plants and in algae. They are considered as metabolically unstable compounds (Francis, 1971). It is worth mentioning that monoterpenes are especially prominent in plants of arid-type climates, enriched in essential oils. An example is rose oil, a costly flower essence,

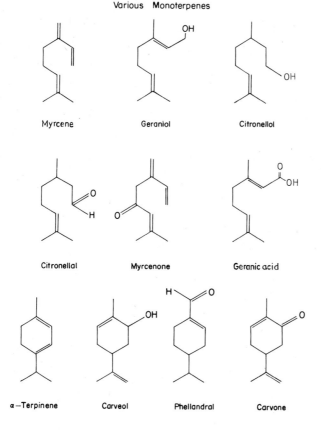

Fig. I.4.5. Examples of monoterpenes, mainly occurring in essential oils of higher plants

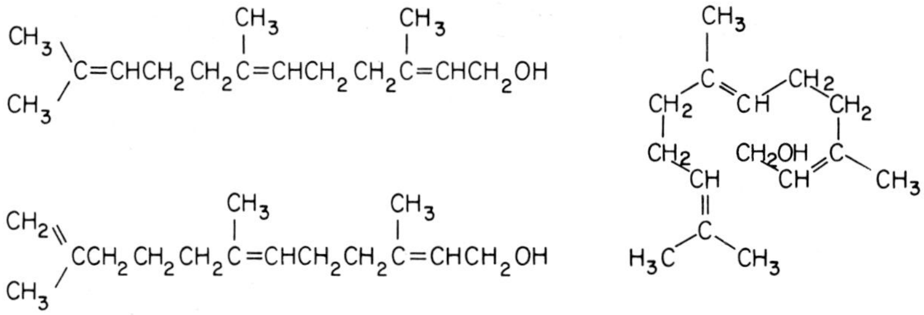

4.2 Lipids

Fig. I.4.6 a and b. Structural example of a sesquiterpene, farnesol, a common isoprenoid alcohol of higher plants and of bacteria (in a bound form). (a) Naturally occurring isomers of farnesol. (b) Farnesol as precursor of dicyclic sesquiterpenes

which is mainly composed of geraniol and citronellol. The chemical structures of these and other monoterpenes in various oxidation stages are given in Figure I.4.5. In addition to being derived from plants, they are also formed by certain arthropods (Francis, 1971).

Sesquiterpenes (C_{15}), consisting of 3 isoprene units, are represented by farnesol, an acyclic isoprenoid alcohol (Fig. I.4.6), and the corresponding olefin farnesin, both of which are found in many plants. Farnesol is a wide-spread compound, but it occurs only in very small concentrations. In certain bacteria, chlorophyll is esterified with farnesol instead of phytol (Figure I.1.2). Monocyclic and dicyclic sesquiterpenes are commonly occurring components of the plant kingdom. Compounds with a farnesol-type skeleton are considered as biochemical precursors of numerous cyclic sesquiterpenoids. Sesquiterpenoid compounds probably fulfil hormonal regulations. They occur mainly in plant material.

Diterpenes (C_{20}) are common constituents of higher plants. They form the main part of resins, especially in conifers. In resins, they occur together with phenylpropane derivatives, such as coniferyl alcohol (Fig. I.4.15a). As will be seen, coniferyl alcohol is a basic constituents of lignins. Most diterpenes exhibit a cyclic structure with two or three rings (Fig. I.4.7). Among the acyclic diterpenes, phytol (Fig. I.4.7a) is the most important component. It is part of the very abundant chlorophyll molecule. Many phases of plant growth and development are affected by the so-called "gibberellins" (Fig. I.4.8), which can be grouped

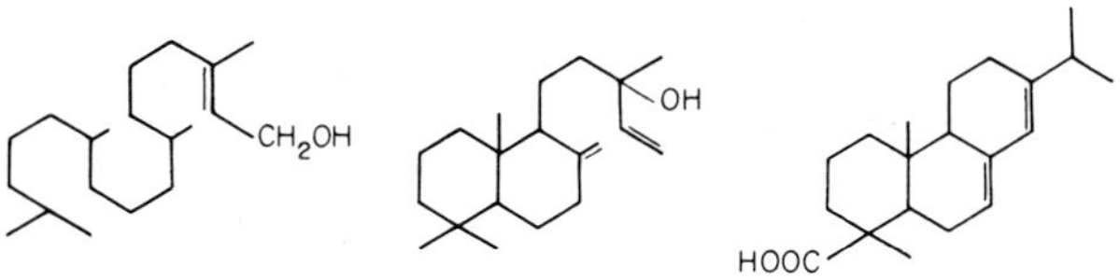

Fig. I.4.7 a–c. Structural examples of acyclic (a), dicyclic (b) and tricyclic (c) diterpenoids. Diterpenes are common constituents of higher plants

a Phytol
b Manool
c Abietic acid

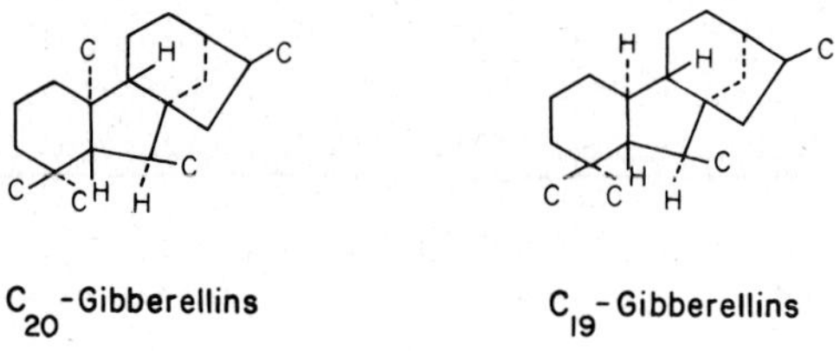

Fig. I.4.8. Structural examples of gibberillins, which occur in small quantities in all higher plants. (After Macmillan, 1971)

among the cyclic diterpenoids. They occur, usually in small quantities, in all higher plants (Macmillan, 1971).

Triterpenes and related steroids are a very important group of cyclic isoprenoid compounds containing 6 isoprene units. Most triterpenoids are pentacyclic, while steroids are all tetracyclic compounds. Triterpenoids, as compared to steroids, generally occur more commonly in higher plants. Steroids are very abundant in the animal and plant kingdom.

A common biochemical precursor for the various triterpenes is squalene ($C_{30}H_{50}$). Squalene (Fig. I.4.9) is a poly-unsaturated hydrocarbon, frequently occurring in plant and animal tissue. It was first isolated by Tsujimoto (1906) from the unsaponifiable fraction of liver oils of various sharks. Thereafter,

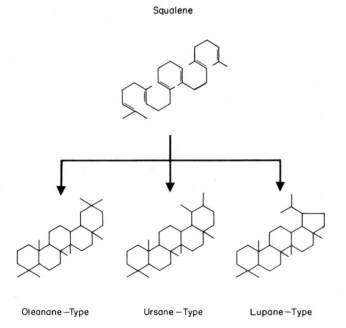

Fig. I.4.9. Squalene and origin of most important types of pentacyclic triterpenes. Squalene occurs in relatively large quantities in most organisms

4.2 Lipids

squalene has been found in varied quantities (sometimes in concentrations of several percent) in the unsaponifiable matter of most organisms, that have been analyzed for its presence. A scheme for the origin of the most important types of pentacyclic triterpenes (ursane, oleanane, lupane) from squalene is given in Figure I.4.9. A number of prominent triterpenes, found in organisms, are considered as direct precursors of hydrocarbons in fossil sediments and petroleum (Fig. I.4.10). Although the various triterpenoids are very abundant in higher

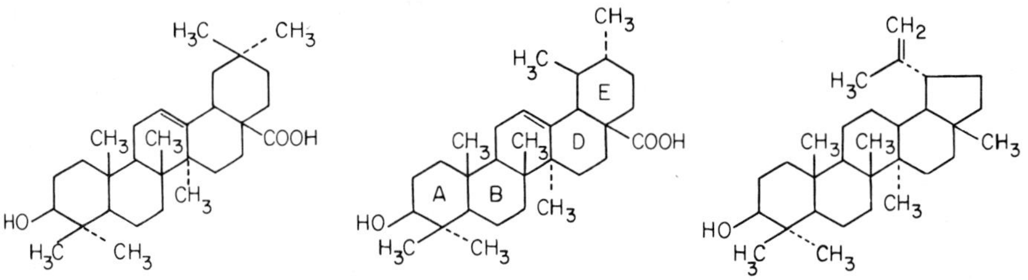

C_{30}, Oleanolic acid (plants)

Oleanane series

C_{30}, Ursolic acid (plants)

Ursane series

C_{30}, Lupeol (plants)

Lupane series

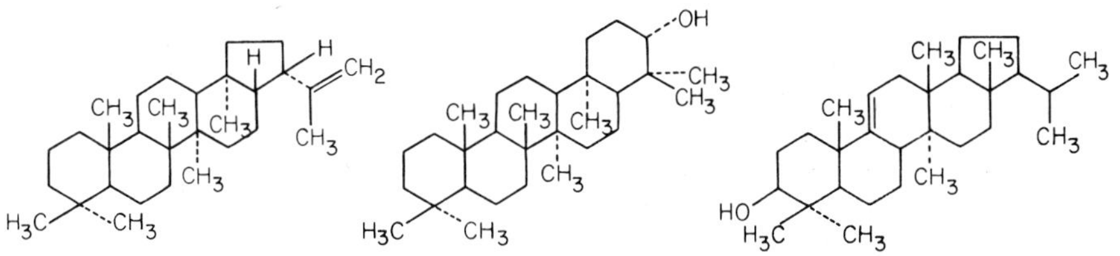

C_{30}, Diploptene

(ferns, blue-green algae, bacteria)

$(17\beta, 21\beta)$H-Hopane series

C_{30}, Tetrahymanol (protozoan)

Gammacerane series

C_{30}, Isoarborinol (plants)

Arborane series

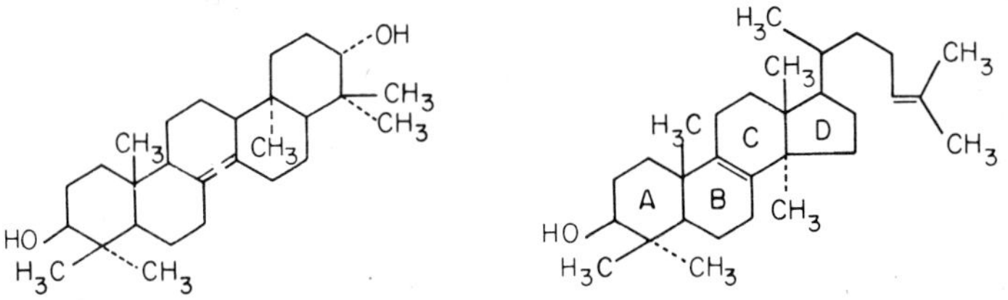

C_{30}, α-Onocerin (plants)

C_{30}, Lanosterol (animal tissues)

Tetracyclic triterpenes

Fig. I.4.10. Prominent triterpenes of living organisms as precursors of fossil triterpenoids

plants, it should be emphasized that one kind, the hopane series (Fig. I.4.10), is found in relatively large concentrations in bacteria and blue-green algae (Dorsselaer, 1975). The other triterpenoids, especially those with five 6-membered rings, seem to be restricted largely to higher plants.

The tetracyclic triterpenoid lanosterol and its derivatives, like other terpenes, are modified biochemically, usually by addition or loss of

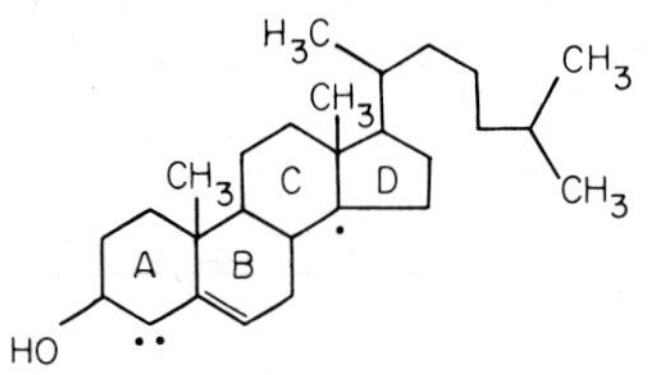

C_{27}, Cholesterol
(animals, plants)

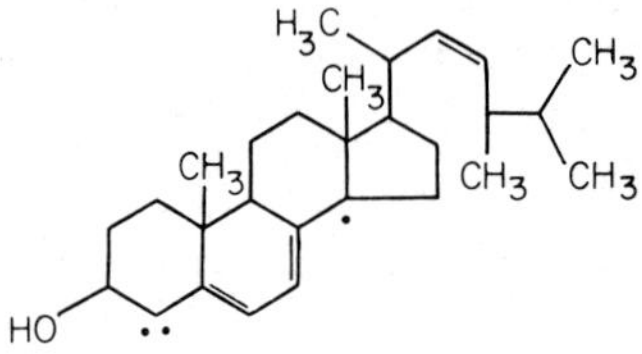

C_{28}, Ergosterol
(yeast)

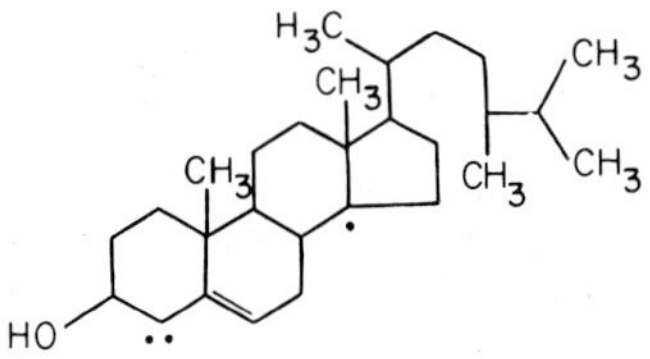

C_{28}, Campesterol
(plants)

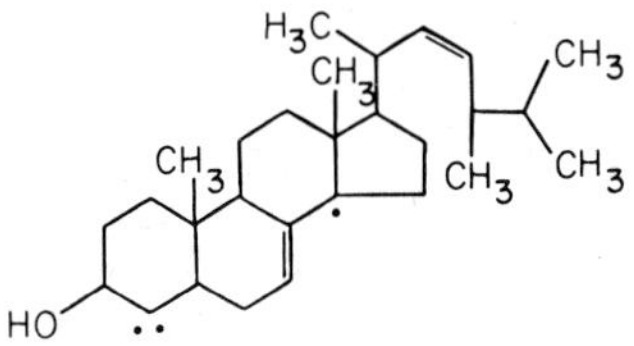

C_{28}, Dihydroergosterol
(algae)

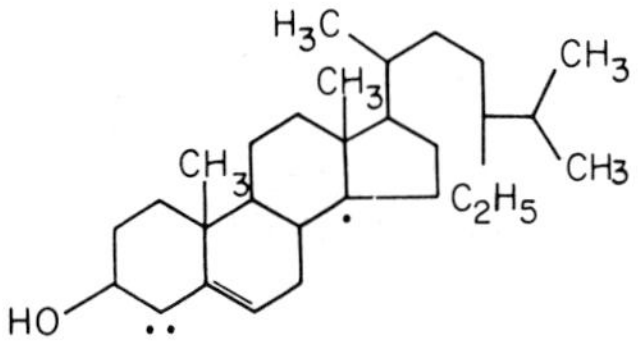

C_{29}, β-Sitosterol
(plants)

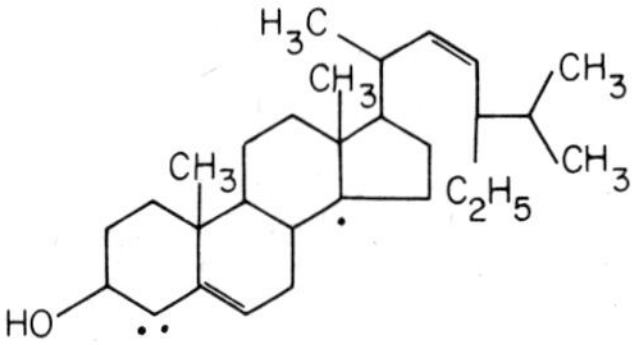

C_{29}, Stigmasterol
(plants)

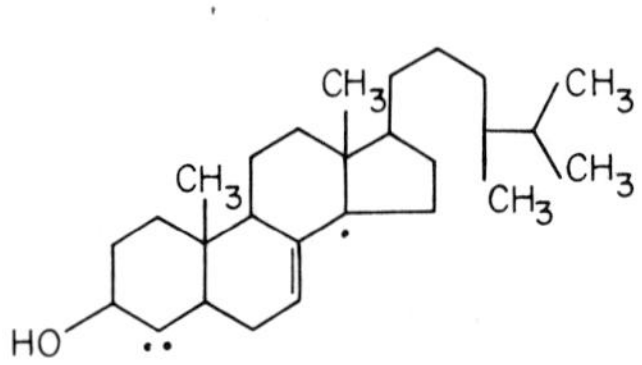

C_{27}, Fungisterol
(fungi)

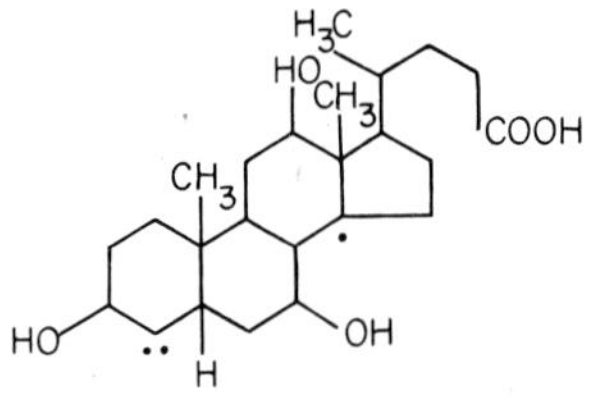

C_{24}, Cholic acid
(bile of higher animals)

Fig. I.4.11. Structural examples of prominent steroids. Positions marked (·) denote sites where a methyl group has been lost methyl groups. The loss of methyl groups leads to the important group called steroids. These are found throughout the plant and animal kingdom. Steroids are rare, but not absent, in bacteria and blue-green algae (Ensminger et al., 1973; Metzner, 1973). However, it is not yet completely clear whether these prokaryotic organisms synthetize steroids themselves, or whether they contain them because they were consumed with their food while they lived heterotrophically. Structural examples of commonly occurring steroids are given in Figure I.4.11. Positions where methyl groups have been lost are marked by dots. Cholesterol (C_{27}) is quantitatively the most important sterol of animals; however, it has also been found in many other organisms even including bacteria. Ergosterol (C_{28}) is most abundant in yeast and other fungi, but has also been discovered in certain algae. Fucosterol (C_{29}) is very common in diatoms and various kinds of algae, such as Phaeophyceae, Chrysophyceae and especially Rhodophyceae (Metzner, 1973). Campesterol (C_{28}) is found in various higher plants (Karrer, 1958). Sitosterol (C_{29}) and stigmasterol (C_{29}) are typical of, and very abundant in, higher plants. Sterols having the same carbon number and structural arrangement, such as fucosterol, sitosterol and stigmasterol (all C_{29} steroids) cannot be differentiated once they have been transformed to saturated steranes.

The most important representatives of the group of tetraterpenes (C_{40}), made from 8 isoprene units, are the carotenoid pigments. There is a large variety of *carotenoid* pigments in nature. Because of their similarity in structure, they are named after the most common member of this group, carotene. Carotene is a polyunsaturated hydrocarbon, composed of isoprene units. Oxygen-containing compounds of this type are called xanthophylls. The carotenoids usually have either a red or yellow color because of the great number of conjugated double bonds in their molecules. Figure I.4.12 shows structural examples of carotenoids such as lycopene and β-carotene, which occur in plants. Carotenoids are abundant in algae, often reaching concentrations of about 0.2% to 0.8% (dry wt.). In some species, concentrations may even rise to 5% or more. Carotenoids are concentrated in certain parts of higher plants, such as in fruits and leaves.

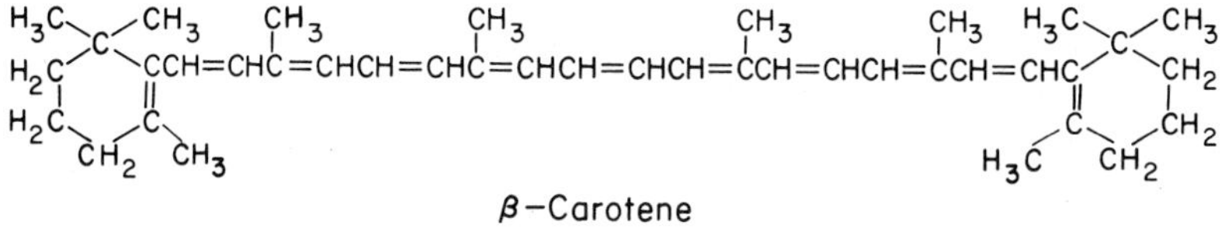

β-Carotene

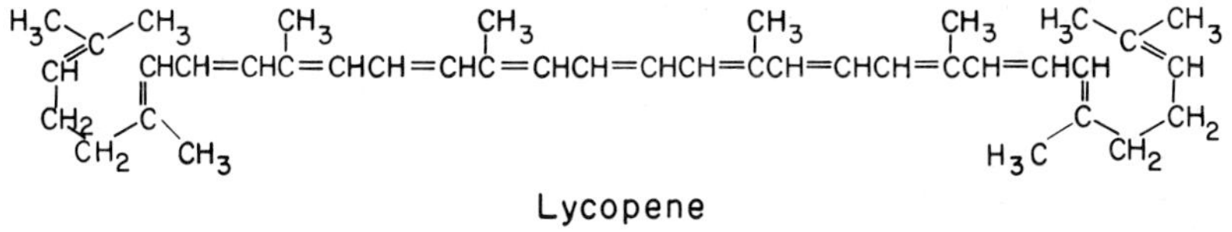

Lycopene

Fig. I.4.12. Structures of important carotenoids, lycopene and β-carotene. Lycopene, $C_{40}H_{56}$, is the red pigment of the tomato and many other fruits. β-carotene, $C_{40}H_{56}$, is one of several carotene isomers occuring in many lower and higher plants

Another important group of pigments, which are also oil-soluble and bear an isoprenoid chain, are the chlorophyll derivatives. *Chlorophylls* are concentrated in the parts of plants where photosynthesis occurs (see Chap. I.1.1). In algae and in leaves of higher plants, average contents of 0.2% to 1% (dry wt.) chlorophyll are reported. The chemical structure of chlorophyll is shown in Figure I.1.2a. Chlorophylls may be considered as derivatives of the porphin nucleus. The structure consists of four pyrrole rings, linked together by methine ($-CH=$) bridges (Fig. I.1.2b). Petroleum porphyrins are also derivatives of the porphin nucleus, i.e., of chlorophylls.

The *phospholipids* or *phosphatides* constitute an integral and quantitatively important part of many membranes at all evolutionary levels. They are also concentrated in the more active tissues (liver, kidney etc.) of animals. Many membranes consist of protein-phospholipid doublelayers. A diagramatic sketch of the architecture of a membrane is given in Figure I.4.13a. Phospholipids differ chemically from simple fats in that they contain a phosphoric acid grouping, and frequently a basic nitrogen in addition to fatty acid components (Fig. I.4.13b).

Suberin and cutin play an important role in protective tissues of plants, such as in cuticles, cork, and spore walls. Cutin and waxes are deposited mainly on the outer side of plant cells, i.e., on the surface, whereas suberin is formed and deposited inside plant tissues and cell walls. Suberin and cutin act as protection against evaporation and promote wound healing. Like waxes, they are not homogeneous substances. They are accompanied by lignins, tannins, long-chain aliphatic compounds (e.g., alcohols) and triterpenes. Suberin and cutin are highly resistant to microbial or enzymatic attack, oxidation, and chemical treatment. They consist mainly of polymerized and cross-linked structures of fatty acids and alcohols. Dicarboxylic and hydroxy acids with 12 to 26 carbon atoms play an important role among suberin constituents. Cutin seems to be a mixture of highly cross-linked hydroxy acids with a dominance of 16, 18 and 26 carbon atoms. Plant cuticulae contain from 50% to 90% cutin (Metzner, 1973). The detailed chemical

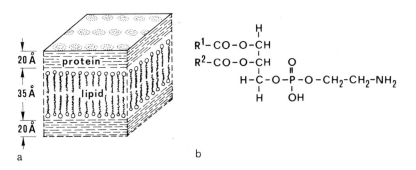

Fig. I.4.13. (a) Example of architecture of membranes, consisting of protein-phospholipid double layers. Membranes are required on all evolutionary levels. Especially primitive organisms, such as bacteria, consist to a large extent of membrane material. (b) Structural example of phospholipid (α-cephalin; R_1 and R_2 are hydrocarbons)

structures of suberin and cutin are not yet known. Matic (1956) proposed a cutin model with fairly random intermolecular esterifications between hydroxy acids and the carboxyl groups, as shown in Figure I.4.14. Hydroxy acids typical of cutin and suberin seem to be found only in higher terrestrial plants ranging from Psilopsida to angiosperms (Eglinton, 1976).

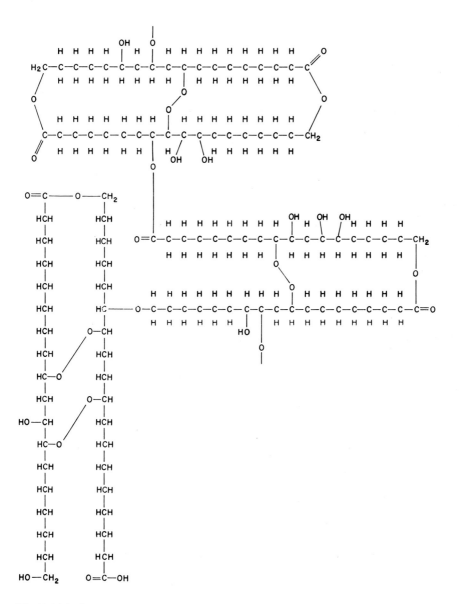

Fig. I.4.14. Structure of cutin (taken from Hitchcock and Nichols, 1971). Cutin occurs in protective tissues of plants. Plant cuticulae consist largely of cutin

4.3 Lignin and Tannin

Two related substances, *lignin* and *tannin,* need emphasis here because of their wide-spread occurrence and, hence, their geochemical significance. Both are characterized by aromatic (phenolic) structures. Aromatic compounds are usually not synthesized by animals, but are very common in plants tissues. Most photosynthetic organisms derive aromatic ring systems from monosaccharides, whereas in heterotrophic bacteria and fungi the synthesis from C_2-units seems to predominate. However, both ways of synthesis are possible in either case (Metzner, 1973). The names lignin and tannin are collective terms, and do not refer to clearly defined substances.

Lignin occurs as a three-dimensional network located between the cellulose micelles of supporting tissues of plants. Basically, lignin is a high molecular weight polyphenol, consisting of units constructed from phenyl-propane derivatives. From an evolutionary point of view, lignin appeared for the first time in mosses. Apparently, typical lignins exist that show a phylogenetic differentiation. In plants, lignins are synthesized by dehydration and condensation of aromatic

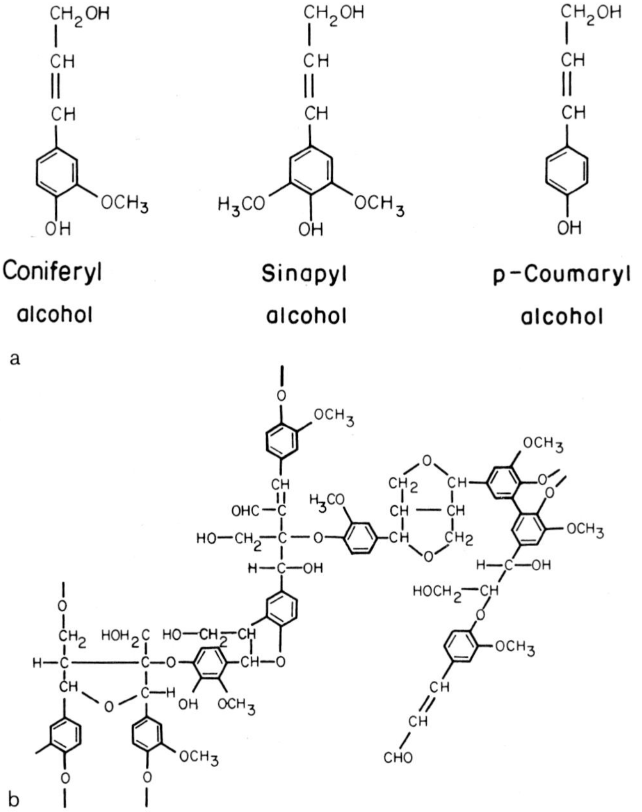

Fig. I.4.15. (a) Aromatic plant alcohols, important building blocks of lignin. (b) Part of a lignin molecule. Lignin occurs as a three-dimensional network between cellulose micelles of supporting tissues of plants. It is very abundant in higher plants. (After Metzner, 1973)

alcohols such as coniferyl, sinapyl and coumaryl alcohol (Fig. I.4.15a). A schematic structure of part of a lignin molecule is shown in Figure I.4.15b.

Tannins are quantitatively much less important than lignins, but are widespread. Their name is derived from their ability to tan hides. In composition and behavior, they are intermediate between lignin and cellulose. They possess carboxyl and phenolic groups on their aromatic structures. The chemical composition of tannins varies considerably, but all are complex derivatives of gallic or ellagic acid (Francis, 1954). Structural formulas of both acids are given in Figure I.4.16. According to Francis (1954), tannins are found principally in higher plants, but also in fungi and algae. In higher plants, they are especially concentrated in the bark (up to 17%) and in leaves (up to 6.5%).

Aside from these quantitatively more important aromatic compounds, there are entire series of phenols and aromatic acids and their derivatives which occur throughout the plant kingdom.

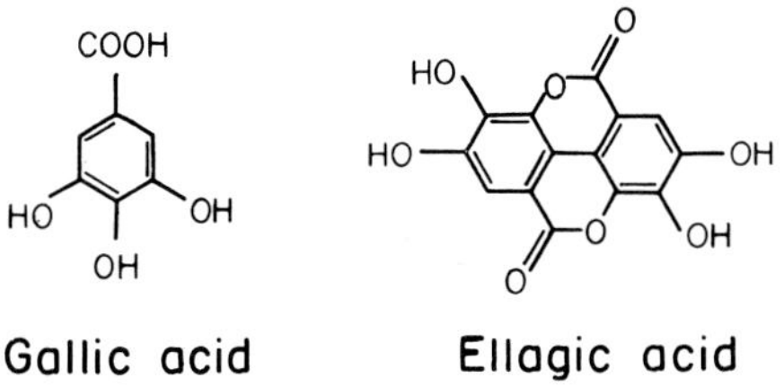

Fig. I.4.16. Structure of gallic and ellagic acids, principal building blocks of tannins. (After Francis, 1954)

4.4 Qualitative and Quantitative Occurrence of Important Chemical Constituents in Bacteria, Phytoplankton, Zooplankton and Higher Plants

Although all organisms are made up of the same group of chemical constituents, there are considerable differences in chemical composition between certain kinds of organism. For instance, there is quite a difference between the most prominent marine plants such as tiny unicellular planktonic algae, and the vast majority of terrestrial higher plants. The organic matter of marine plankton is composed

Table I.4.1. The main chemical constituents of marine plankton in percent of dry weight. (After Krey, 1970)

	Protein	Lipids	Carbohydrates	Ash
Diatoms	24–48%	2–10%	0–31%	30–59%
Dinoflagellates	41–48%	2– 6%	6–36%	12–77%
Copepods	71–77%	5–19%	0– 4%	4– 6%

mainly of proteins (up to 50% or more), a variable amount of lipids (generally between 5 and 25%) and variable amounts of carbohydrates (generally not more than 40%). Table I.4.1 lists average quantities of the main chemical constituents of common marine plankton.

Compared to these quantitatively important marine organisms, higher terrestrial plants, including the majority of trees, are composed largely of cellulose (30 to 50%) and lignin (15 to 25%). Both constituents are incorporated mainly in skeletal structures that fulfil supporting functions for plants, and hence are not needed in aquatic, planktonic organisms. The protein content of higher terrestrial plants may be as high as 10%, but averages 3% or less. In the following discussion, basic information about the chemical composition of relevant classes of organisms will be described in so far as it is available.

Bacteria, the most primitive organisms, are extremely adaptable and, hence, relatively variable in their chemical composition. They contain about 80% or more water; the rest is organic matter. On a dry weight basis, bacteria consist of about 50% carbon, 10–15% nitrogen, 2–6% phosphorus and 1% sulfur. With respect to types of organic compounds, bacteria are composed of about 50% proteins, 20% cell wall material (membranes) and 10% lipids; the rest is ribonucleic and desoxyribonucleic acids (Schlegel, 1969). Membranes are composed of a considerable proportion of lipids and lipid-like constituents (Table I.4.2). Many bacteria are able to store fat substances, polysaccharides, polyphosphates and sulfur. In the lipid fraction of bacteria, a variety of sterols with 27 to 29 carbon atoms have been found (Laskin and Lechevalier, 1973). These sterols belong mainly to the cholesterol (C_{27}), ergosterol (C_{28}) and stigmasterol (C_{29}) types. Recently, triterpenes of the hopane series (Fig. I.4.10) have been found in bacteria (Dorsselaer, 1975). Fatty acids from bacteria are generally in the range of C_{10} to C_{20}. The most abundant ones seem to be branched fatty acids of iso- or anteiso-configuration (methyl branching 1 or 2 atoms, respectively, from terminal C-atom) with carbon numbers from 14 to 18. Hydrocarbons occur in the form of open chains, mainly with 10 to 30 carbon atoms.

Table I.4.2. Membrane constituents of bacteria. (After Schlegel, 1969)

Type of organic compound	Percent of membranes of	
	Micrococcus lysodeikticus	purple bacteria
Lipid	28–37	40–50
(neutral)	(9)	(10–20)
(phospholipids)	(28)	(30)
Protein	50	50
Polysaccharide	15–20	5–30

Phytoplankton, as pointed out before, is the most important producer of organic matter in the aquatic environment. The main organic constituents of two important algal groups, diatoms and dinoflagellates, were presented in Table I.4.1. Environmental conditions, such as water temperature and nutrients, may influence gross composition of algae more than either class or species differences. Composition of the lipid fraction of diatoms is given in Table I.4.3. The amount of free uncombined fatty acids, which comprises more than half of the total lipid fraction, is relatively very high, especially when compared to animal tissue.

Table. I.4.3. Composition of lipid fraction of diatoms. (After Clarke and Mazur, 1941)

Uncombined fatty acids	59–82%
Combined fatty acids	1–17%
Nonsaponifiable fraction	12–29%
Alcohols	3–7%
Hydrocarbons	3–14%

The distribution of major sterols and carotenoids varies considerably among algal classes (Goodwin, 1971). Chlorophytae (green algae) preferentially contain ergosterol, but also contain other sterols, such as cholesterol and fucosterol. In Rhodophytae (red algae) cholesterol is the main sterol, while in Phaeophytae (brown algae), it is fucosterol. The concentration of sterols in algae may be as high as 0.3% on a dry weight basis. The relative amount of carotenoids is highest in Phaeophytae where it may be as much as 6 mg g^{-1} (0.6%) dry weight (Goodwin, 1965).

The major fatty acids present in algae are all saturated or unsaturated monocarboxylic acids with straight even-numbered carbon chains in the C_{12} to C_{20} range. Long-chain fatty acids are prominent components of both stored fats and of polar lipids (phospholipids) of biomembranes in algae. Branched fatty acids of the iso- and anteiso-type, if present, are usually minor components of algae, in contrast to their concentration in bacteria. Palmitic (C_{16}) and stearic acid (C_{18}), together with their unsaturated counterparts with the same carbon numbers, are the most abundant (Erwin, 1973). Polyunsaturated fatty acids are much more common in algae than in higher plants.

The concentration of hydrocarbons in the lipid fraction of algae is generally 3–5%. They consist of saturated and unsaturated hydrocarbons having both straight and branched chains. Marine algae synthetize n-alkanes in the range from *n*-C_{14} to *n*-C_{32} (Clark, Jr., 1966). Although the ratio of even- to odd-numbered homologs is frequently close to one, there are numerous examples where *n*-C_{15} or *n*-C_{17}, or both, are the dominating alkanes (Youngblood et al., 1971). These two often represent more than 90% of the entire homologous series. Stransky et al. (1968) have investigated the hydrocarbon content of the freshwater alga *Scenedesmus quadricanda.* Over 60% of the hydrocarbon fraction consisted of unidentified saturated hydrocarbons. The rest was composed of members of the *n*-alkane homologous series, of 2-methyl (iso) and 3-methyl (anteiso) branched alkanes, and of olefins, mostly 1-alkenes. The n-alkane distribution was dominated by n-C_{17}. Gelpi et al. (1968) observed olefinic straight-chain hydrocarbons with 17 to 33 C-atoms in two algae, *Botryococcus braunii* (golden-brown alga) and *Anacystis montana* (blue-green alga). These hydrocarbons exhibited a predominance of the odd-numbered C_{17}, C_{27}, C_{29} and C_{31} molecules. The total content of these hydrocarbons was in the range 0.1–0.3% dry weight. Maxwell et al. (1968) found that the hydrocarbons extracted from *Botryococcus* consisted solely of two olefinic straight-chain compounds ($C_{34}H_{58}$).

Table I.4.4. Composition of lipid fraction (wt. %) of calanoid copepods. (After Lee et al., 1970)

Fraction	*Calanus helgolandicus* wild type		*Gaussia princeps* wild type
	a	b	
Hydrocarbons	Trace	3	Trace
Wax esters (incl. any sterol esters)	37	30	73
Triglycerides	5	4	9
Polar lipids (free acids, cholesterol, monoglycerides etc.)	14	17	
Phospholipids	44	45	17
Total lipid (percent dry wt.)	12.4	15	28.9

There is relatively little information on the detailed chemical composition of prominent zooplankton, such as foraminifers and copepods (see also Table I.4.1). However, since zooplankton, especially copepods, feed directly on phytoplankton, there is a certain resemblance between the lipid fraction of phytoplankton and zooplankton. Copepods constitute the largest single fraction of zooplankton in most waters. Composition of the lipids of calanoid copepods was investigated by Lee et al. (1970). The lipid fraction of certain wild-type copepods was unusually high, with almost 30% on a dry weight basis. The effect of nutrition on the total amount of fat and the rapid depletion of this fat during starvation led to the conclusion that lipids, especially wax esters, serve in copepods as a reserve energy store. The composition of the copepod lipids, according to Lee et al. (1970), is given in Table I.4.4. The wax esters in the lipid fraction are of special interest, because both their alcohols and their fatty acid constituents are made of long-chain carbon skeletons with 18 or more carbon atoms. Total carbon

4.4 Occurrence of Important Chemical Constituents in Organisms

numbers, alcohol plus acid, in these wax esters were found to range from 30 to 44. According to Lee et al. (1970), the wax esters of marine copepods are very probably the main source of long-chain alcohols observed in marine sediments and in some ocean surface lipids.

Blumer et al. (1963, 1964) reviewed occurrence and distribution of the isoprenoid hydrocarbon pristane (2,6,10,14-tetramethyl-pentadecane) in marine organisms. They found that pristane is a major component of the body fat of copepods, and may be used for buoyancy to conserve energy that would otherwise be used to maintain a suitable position in the water column by swimming. Pristane from calanoid copepods may be the major primary source of this hydrocarbon in the marine environment.

Terrestrial *higher plants* are the fourth major source of organic matter in sediments. As indicated before, the bulk material of higher plants, especially shrubs and trees, is composed mostly of cellulose and lignin (together 50–70%). Quantitatively, lipids and proteins are only of secondary importance. However, certain parts of higher plants, such as barks, leaves, spores, pollen, seeds and fruits, may be highly enriched in lipids and lipid-like substances. This latter fact, together with the resistance of these plant parts to mechanical, chemical and biochemical degradation, is largely responsible for the contribution of terrestrial plant material to aquatic sediments, and especially to marine sediments. The fat content of seeds and fruits varies greatly among various plants (about 1–50%); hence, average numbers are meaningless. Pollen usually contain between 2 and 8% fats. Leaves contain considerable amounts of lipids and lipid-like substances (waxes, cutin, suberin, etc.).

Lipids derived from higher plants may be characterized by a number of features. The *n*-alkanes in the range from C_{10} to C_{40} show a strong predominance of odd-carbon-numbered over even-numbered *n*-alkanes (by a factor of 10 or more). This predominance is especially apparent from n-C_{23} to n-C_{35}, with a strong predominance of the *n*-alkanes n-C_{27}, n-C_{29}, and n-C_{31}. The lowest molecular weight *n*-alkane reported in plants (Koons et al., 1965), is *n*-heptane (C_7H_{16}), and the highest is dohexacontane ($C_{62}H_{126}$). Even-numbered aliphatic alcohols with 24 to 36 carbon atoms are relatively common, especially in plant waxes. Other typical components of higher plants are phenolic compounds, such as coniferyl alcohol (Fig. I.4.15a) and others. Straight-chain saturated fatty acids with an even number of carbon atoms in the range from C_8 to C_{26} are quite common. However, by far the most prominent fatty acids (Fig. I.4.3b) are palmitic (C_{16}) and stearic acid (C_{18}). Straight-chain unsaturated fatty acids most commonly found (Fig. I.4.3c) have 14, 16, 18, and 20 carbon atoms. Again acids with 18, and to a lesser degree with 16, carbon atoms are the most prominent. Various kinds of even-numbered hydroxy acids with 12 to 26 C-atoms, resulting from degradation of cutin and suberin, are also typical contributions from terrestrial higher plants (Eglinton, 1976).

In leaves and fruits, two kinds of pigments (Fig. I.4.12) may occur in concentrations up to 1%. Among the terpenoids built from isoprene units (C_5), the monoterpenes (Fig. I.4.5), e.g., in essential oils, and the diterpenes (Fig. I.4.7), e.g., in resins, are frequently observed. Most types of pentacyclic triterpenoids (Fig. I.4.10) are essentially restricted to higher plants, especially to the more specialized angiosperms. The main sterols in higher plants are sitosterol and stigmasterol, both with 29 C-atoms, but the latter having one double bond in the alipathic side chain (Fig. I.4.11).

4.5 Natural Associations and Their Effects on Biomass Composition

The chemical composition of the biomass in a given area, and at a given time, is mainly dependent on the physical and chemical environment of the biological habitat and the evolutionary level of the organisms. Determining environmental factors include light, temperature, nutrients, and water currents, as far as the aquatic realm is concerned.

Conditions in limnic water bodies will be quite different from either restricted marine basins or the open ocean. The more important environmental factors on land are climatic conditions, morphology of the landscape, and the availability of nutrients. The time factor, in a geological sense, is introduced by the stratigraphic age, or in other words, by the evolutionary level of the organisms in question. More or less typical natural associations of plants and animals result from various combinations of environmental factors through time and space.

Before the Devonian time, the bulk of the biomass was composed of bacteria, blue-green algae, and higher algae living in the marine environment. A Precambrian algal coal, the so-called shungite (Lake Onega, USSR), may be considered as a respresentative fossil example of one of the earliest natural associations of organisms.

Natural associations throughout the Cambrian, Ordovician and even in the Silurian (see also Chap. I.2), bear the imprint of the marine environment, and are practically devoid of higher plants. This means that, in the fossil counterpart in sediments, we do not expect great quantities of lignin and its derivatives. Also we expect to find no cuticules and plant waxes, rarely spores, no pollen, no diatoms, and no foraminifers. From a structural chemical point of view, we expect no pronounced odd predominance among the higher n-alkanes, and only a limited quantity of long-chain aliphatic molecules with more than 25 C-atoms. We expect to find a variety of steroid compounds, but relatively few and only very small quantities of triterpenoids, with the exception of the nearly ubiquitous hopane series (Fig. I.4.10).

Approximately since the beginning of the Devonian, a greater variation in natural associations of organisms can be observed. This is due to a gradually increasing differentiation in the plant and animal kingdom (Fig. I.2.2 to I.2.4) and to the conquest of the continents by plants. A limited correlation exists between the geological facies of a given sedimentary unit and natural associations. However, for a number of reasons, it is merely a similarity and not a match. Aside from the evolutionary factor that influences natural associations, there are the factors of plant and animal geography, and the complicated interrelationships of food chains that do not necessarily leave an imprint on the facies. The law of the food chain, and the quantitative importance of primary producers such as phytoplankton are the two principal reasons why natural associations, as

4.5 Natural Associations and Their Effects on Biomass Composition

discussed here, have to be considered as alterations of, or additions to the basic stock of organisms, which is always formed by phytoplankton and bacteria. In general, three basic natural associations among remnants of biologically produced organic matter can be observed in aquatic sediments:
1. Remnants of algae and zooplankton together with microorganisms (mainly bacteria).
2. Remnants of higher plants, strongly degraded and partly oxidized, and deeply reworked by microorganisms.
3. Remnants of higher plants, little to moderately altered.

Of course, combinations between these natural associations are possible, especially by the addition of terrestrial higher plants to nearshore marine sediments. A schematic summary of natural associations of organic matter during geological history is given in Figure I.4.17.

So far, there is little precise knowledge of difference in chemical composition of the biomass due to various natural associations. Only a few extremes recognized in the fossil record can be sketched. Varying contributions from higher terrestrial plants to an adjacent marine water body with autochthonous bacterial, phytoplankton, and zooplankton communities may serve as an example. The chemical consequences of such a change have been described earlier. Aside from differences in lipid- and hydrocarbon-type composition (e.g., wax alcohols, alkanes), a change in the gross chemical composition of the biomass, i.e., a change in the atomic H/C ratio, may be observed. Predominantly land-derived organic matter with a high content of lignins and carbohydrates has a H/C ratio in the range of 1.3 to 1.5, whereas autochthonous marine plankton with large protein and lipid fractions reaches a H/C ratio of about 1.7 to 1.9 (see also Figs. II.4.12 and II.7.6). The principal reason for this difference is that the land-derived organic matter is more aromatic and richer in oxygen than the marine planktonic biomass.

A consequence of such differences with respect to fossil sediments has been described by Breger and Brown (1968), who interpret the variable hydrogen content of kerogen in shales with wide regional distribution as related to a varying terrestrial (plant) influence. Contributions from terrestrial plants to the marine environment can also be recognized by a shift in the δ^{13}C-range toward lighter ^{13}C/^{12}C ratios (Sackett, 1964; Nissenbaum and Kaplan, 1972) for the organic carbon in marine sediments. Marine organic carbon generally is isotopically heavier than contemporaneous terrestrial organic carbon. The range of this difference varies. On the average, marine organic carbon is heavier by about 5 per mille.

Quantitative data on composition of the biomass in certain bodies of water, with respect to the type of organisms found there, is scarce. Two examples are given in Table I.4.5 and I.4.6, for the Black Sea and the Caspian Sea, respectively. Phytoplankton and/or bacteria have to be considered as the main producers of organic matter, as can be deduced from Tables I.4.5 and I.4.6. The phytoplankton in the Black Sea are composed mainly of diatoms, dinoflagellates and coccolithophorids. The relative distribution of these organisms changes seasonally and regionally (Shimkus and Trimonis, 1974). In shallow water areas of the Black Sea (less than 100 m), large quantities of organic material are also produced

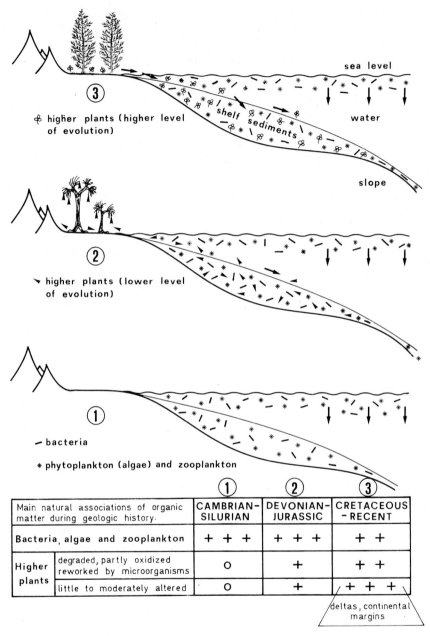

Fig. I.4.17. Main natural associations of organic matter in aquatic sediment during geological history. Cambrian through Silurian sedimentary rocks mainly contain remnants of bacteria, algae and zooplankton. Devonian through Jurassic sedimentary rocks generally contain remnants of algae, zooplankton and bacteria, but in addition some higher plant matter, especially in near-shore environments. Cretaceous to Recent sedimentary rocks generally contain relatively less planktonic organisms and bacteria and frequently a greater part of higher plant remains

Table I.4.5. Composition of biomass and productivity of various types of organism in the Black Sea. (After Wassojewitch, 1955)

Organisms	Biomass, 10^6 t	Production per year 10^6 t	%
Plankton	15.0	2,745	13.22
Phytoplankton (without bacteria)	13.5	2,700	13.0
Zooplankton	1.5	45	0.22
Benthos	40.0	80	0.38
Makrophytae	20.0	40	0.19
Zoobenthos	20.0	40	0.19
Bacteria in water	30.0	12,000–18,000	57.60
Bacteria in sediment	10.0	6,000– 8,000	28.80
Fish	1.0	0.17	—
Total	96.0	> 23,000	100

Table I.4.6. Composition of biomass and annual production in the Caspian Sea. (Taken from Bordovskiy, 1965, after Datsko, 1959)

Group of organisms	Biomass dry wt. (1000 t)	Annual poduction dry wt. (1000 t)
Phytoplankton	650	200,000
Bacteria	320	80,000
Zooplankton	500	15,000
Zoobenthos	4,500	18,000
Phytobenthos	375	375
Fish	1,800	900

by phytobenthos. In the northwest area, up to 9% of the entire biomass is formed by a single kind of algae *(Phyllopora nevrosa)*.

Finally, it is stressed that we have merely scratched the surface of a vast amount of undiscovered information pertaining of biomass composition and how it is related to sediment of various facies.

Summary and Conclusion

The type of organic matter deposited and incorporated in sediments depends largely on the natural association of the various groups of organisms in different facies provinces. From the point of view of petroleum formation, such natural associations which occur on the continental margins and in large land-locked, isolated basins (marginal seas) deserve the greatest attention.

All organisms are basically composed of the same chemical constituents: lipids, proteins, carbohydrates and lignins in higher plants. However, there are very characteristic differences with respect to relative abundance of compound and detailed chemical structure. With respect to the formation of petroleum, the lipids are most important. Lipids encompass fat substances, waxes and lipid-like components, such as oil-soluble pigments, terpenoids, steroids and many complex fats. A prominent building block of many of these components is the isoprene unit (5 C-atoms), consisting of four C-atoms in sequence, and one methyl side chain.

A fundamental difference exists between the chemical composition of marine planktonic algae and terrestrial higher plants. The organic matter of marine plankton is mainly composed of proteins (up to 50% and more), a variable amount of lipids (5 to 25%), and generally not more than 40% carbohydrates. Higher terrestrial plants are largely composed of cellulose (30 to 50%) and lignin (15 to 25%). Both constituents fulfil mainly supporting functions, and are not needed in aquatic, planktonic organisms. Lignin is the major primary contributor of aromatic structures in organic matter of Recent sediments.

Predominantly land-derived organic matter with high contents of lignin and carbohydrates has H/C ratios around 1.0 to 1.5 and is more of an aromatic nature. Organic material mainly derived from autochthonous marine plankton reaches H/C ratios around 1.7 to 1.9 and is more of an aliphatic or alicyclic nature.

Chapter 5
Sedimentary Processes and the Accumulation of Organic Matter

The accumulation of organic matter in sediment is controlled by a number of geological boundary conditions. It is practically restricted to sediment deposited in aquatic environments, which must receive a certain minimum amount of organic matter. This organic matter can be supplied either in the form of dead or living particulate organic matter or as dissolved organic matter. The organic material may be autochthonous to the environment where it is deposited, i.e., it originated in the water column above or within the sediment in which it is buried, or it may be allochthonous, i.e., foreign to its environment of depositon. Both the energy situation in the water body in question and the supply of mineral sedimentary particles must be such as to allow a particular kind of sedimentation. If the energy level in a body of water is too high, either there is erosion of sediment rather than deposition, or the deposited sediment is too coarse to retain low-density organic material. An example is a beach area with strong wave action. Furthermore, in coarse-grained sediment, ample diffusion of oxygen is possible through the wide open pores. On the other hand, if the level of energy is very low, too little sediment is supplied, and there is, like-wise, no appreciable organic sedimentation. Examples of this type occur in certain parts of the deep sea.

Once these boundary conditions are satisfied, the accumulation of organic matter in sediment is dependent on the dualism between processes that conserve and concentrate and those that destroy and dilute organic matter.

5.1 Fossil and Modern Sediments Rich in Organic Matter, and Their Geological Implication

Here, we are primarily interested in analyzing and understanding the formation of those sedimentary rocks that are potential source sediments for petroleum. Although we do not intend at this moment to define potential source rocks of petroleum, only fossil sediment with a minimum amount of organic carbon is considered.

For various reasons, a lower limit is set at about 0.5% organic carbon for detrital rocks and at about 0.3% for carbonate and evaporite-type sediment. A large variety of sediment derived from widely differing geographic and stratigraphic provenances has been analyzed by numerous researchers. From their data, a number of facts can be deduced with respect to the formation of sediment rich in organic matter.

In a comprehensive study of bituminous rock sequences, Bitterli (1963) concluded that paleogeographic turning points, which resulted in either transgressions or regressions, were especially favorable for the deposition of such sediment. According to him, most bituminous rock sequences belong to a paleogeographic setting where transitional (brackish) or alternating marine/freshwater facies prevailed. However, fully marine conditions were not excluded. Furthermore, Bitterli stated that organic matter richness is not inherent in any particular lithofacies, but tends to be associated with fine-grained sediments. This latter fact was emphasized by Hunt (1963), who demonstrated in a study of one type of sediment that the smaller particles, apparently due to their greater adsorption capacity, are associated with a larger amount of organic matter (Table I.5.1).

Table I.5.1. Variation in organic matter with particle size in viking shale of Canada. (After Hunt, 1963)

Particle size	Average wt. % organic matter
Siltstone	1.79
Clay (2–4 μ)	2.08
Clay (less than 2 μ)	6.50

In the case of modern sediment, a number of additional facts can be deduced concerning the deposition of sediment rich in organic matter.

Shimkus and Trimonis (1974) observed that areas of high concentration of organic carbon (more than 3%) in modern Black Sea sediment do not coincide with areas of high primary organic production (more than 0.2 g Cm^{-20} day^{-1}). However, there was a good correlation between large amount of organic matter and high concentration of $CaCO_3$ (more than 30%). From this, it was concluded that the accumulation of organic matter and $CaCO_3$ were both controlled by the changing contributions of clastic-argillaceous materials from terrigenous sources. Thus, the concentration of organic matter and $CaCO_3$ would be low in areas with high rates of deposition of clastic-argillaceous material, due to a dilution effect. This effect is strongest at the periphery of the depositional basin, where the production of organic matter is relatively high. In addition, the data of Shimkus and Trimonis (1974) show that high organic carbon content in sediment is correlatable with high concentrations of subcolloidal particles (< 1 μm), and with the distribution of illite plus montmorillonite. This relationship probably should be investigated in connection with the abundance of suspended material (> 0.7–1.0 μm), which is composed of an appreciable amount of organic carbon (15–75%, average 30%).

The importance of suspended matter with respect to transportation and sedimentation of organic matter was also recognized by Meade et al. (1975) and

Milliman et al. (1975). In coastal waters of the Atlantic ocean, off the coast of the northeastern USA and off the Amazon River delta, suspended matter was found to contain large amounts of organic matter, generally from 40 to 80%. In nearshore areas of the Amazon River delta, the organic part of the suspended matter ranged from 10 to 25% and in offshore areas from 50 to 80%. This suspended matter, being highly enriched in organic material, shows a density gradation. The more organic material, the lower is its density. Low-density material can more easily stay in suspension. Therefore, the settling out of these particles will be inversely related to the energy level of a body of water. Low-density organic matter will be carried away into quiet waters, unless it is either oxidized or consumed by bacteria or other scavengers prior to deposition. The chances for destruction of organic matter are greater in longer water columns, especially if there is enough oxygen (Figs. I.3.5, I.3.6a). However, if there is a stratified water column with oxygen restricted to the surface water layer and H_2S in most of the remaining water column, as occurs in the Black Sea or other landlocked basins (Fig. I.3.6b), there is a greater chance for preservation of organic matter. There are numerous examples of sediment, rich in organic matter, which were deposited at great depths in water columns exceeding 1000 m, as in the Black Sea, Mediterranean Sea, Gulf of Mexico and the Cariaco Trench (McIver, 1975). In some of these cases, the absence of oxygen apparently is not necessary for the preservation of a considerable part of the organic matter.

The importance of allochthonous organic material in modern sediment is exemplified by the occurrence of great quantities of pollen in bottom sediment of the Black Sea near the western shore (Roman, 1974). Although clearly marine, this sediment contains more land-derived than marine particles among the microscopically identifiable fraction of organic matter. Distribution of pollen types is related to the distance of the sample from the shore line.

5.2 The Role of Dissolved and Particulate Organic Matter

Relatively large amounts of dissolved organic carbon (DOC) are found in both sea water and freshwater (see Chap. I.1.2). Dissolved organic carbon, which actually means dissolved organic matter that is determined as carbon, is defined by oceanographers as organic matter that passes through sieves with 0.4 or 0.8 μm pore size. The organic material retained on the sieves is called particulate organic carbon (POC). Menzel and Ryther (1970) state that the vertical distribution of POC and DOC in the oceans is practically homogeneous below a depth of 200 to 300 m; POC is present in concentrations from 3 to 10 μg $C\,l^{-1}$ and DOC in concentrations from 0.35 to 0.7 mg $C\,l^{-1}$ (Fig. I.5.1). They also state that above this depth concentrations are variable, with POC directly proportional and DOC inversely proportional to primary production of organic matter. From these data, together with the observation that an oxygen minimum generally occurs in shallow depths usually in the range of 200 to 300 m, it is concluded that the cycling of organic matter is almost entirely restricted to relatively shallow surface waters. Below this depth, apparently no cycling of organic matter occurs, which may be

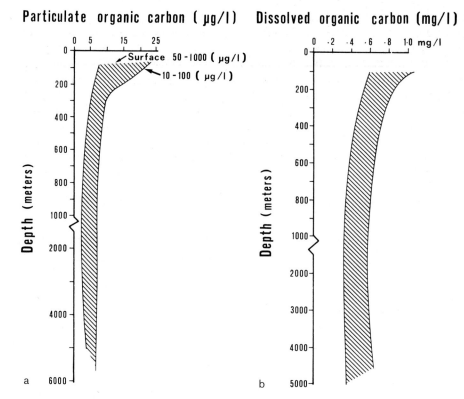

Fig. I.5.1a and b. Average depth distribution of particulate organic carbon (a) and of dissolved organic carbon (b) in various areas of the Atlantic and Pacific Ocean. (Modified after Menzel and Ryther, 1970)

due either to the biochemically inert nature of the compounds, or to a concentration below a threshold level for utilization by bacteria. It may also result from bacteria becoming inactive below a certain depth level, due to increasing pressure (Pelet, 1976).

Mineral particles, especially clay minerals, settling through a water column containing dissolved organic matter may get coated with (polar) organic compounds. The adsorption of amino acids, sugars, phenolic substances and other organic compounds on surfaces of clay minerals and carbonate particles in seawater has been demonstrated by Bader et al. (1960), Chave (1970), Degens (1970), Mitterer (1971) and many others. This adsorption process can also be considered as a conversion of dissolved organic matter to a particulate form. Breger (1970) points out that complex high molecular weight organic materials in seawater may account for a very high percentage of total organic matter. He further indicated that interactions among these substances may lead to phenomena such as coprecipitation, adsorption, flocculation, or protection. The frequently observed inverse relationship between grain size of sedimentary particles and amount of organic matter for both modern and ancient sediment certainly has to be considered in connection with these interactions. Carbon isotope studies by Gormly and Sackett (1977) on DOC, POC, and organic matter deposited in modern sediment of the Gulf of Mexico point in the same direction. According to these authors, the adsorption of dissolved organic matter by suspended mineral particles may play an extremely important role in the extraction of organic matter from water and its subsequent incorporation in bottom sediments.

All these observations are in line with a proposed path of kerogen formation; (1) degraded celluar material ⟶ (2) water-soluble complex, containing amino acids and carbohydrates ⟶ (3) fulvic acids ⟶ (4) humic acids ⟶ (5) kerogen (Nissenbaum and Kaplan, 1972; Welte, 1974). Partly degraded cellular material, or even whole fragments of organisms may be incorporated in this process of kerogen formation. Altered particulate organic matter resulting directly from degradation of organisms may also serve as an adsorbent for organics.

5.3 Accumulation Mechanisms for Sedimentary Organic Matter

From the foregoing discussion, in addition to information presented in previous chapters, a concept for the accumulation mechanisms of sedimentary organic material has been developed.

Generally, marine phytoplankton represents the main source of organic matter in marine sediments, whereas in some shallow water regions with sufficient light for photosynthesis the main source is marine phytobenthos. In both instances, bacteria that rework dead organisms have to be considered as an additional major source of organic carbon. In certain types of sediment, especially those deposited in nearshore areas and deltas, allochthonous land-derived organic material, represented by pollen, spores and other plant debris, may be another dominant contributor.

An apparent but by no means compelling interrelationship exists between the quantity of organic matter supplied to a depositional environment and the concentration of organic matter deposited in the sediment. The supply of organic matter is greatest in areas of high primary productivity, i.e., along continental margin, around "land masses", and especially areas of upwelling (see also Chap. I.3.3, Figs. I.3.4 and I.3.7). However, this supply factor becomes modified by those processes responsible for conservation and concentration of organic matter on one hand, and by those responsible for destruction and dilution on the other.

Under geological conditions prevailing on the earth's surface, all organic matter is unstable. The conservation of organic matter is aided by a number of factors. Great productivity of organic matter has a tendency to create environments with low oxygen concentration (Eh below + 100 mV), or with oxygen deficiency accompanied by the production of H_2S (Eh negative, below 0 mV) by sulfate-reducing bacteria. Decay of organic material and its conversion to CO_2 are responsible for this drop in oxygen concentration. This process frequently is the cause for an oxygen minimum at the bottom of the euphotic zone of seawater.

Low oxygen concentrations, or absence of free oxygen and presence of H_2S, depress the rate of destruction of organic matter. Under such conditions, it is not further mineralized by direct oxidation to CO_2. Nevertheless, the destruction of organic matter is continued at a lower rate by anaerobic, heterotrophic bacteria.

The adsorption of either dissolved or particulate organic matter on surfaces of mineral particles enhances its stabilization in two ways. First, it is better protected against biological consumption, and second, it settles more rapidly through the water column. The residence time an organic particle spends in an oxygen-containing water column is also important. Therefore, in shallow waters up to a few hundred meters depth, conditions are better for accumulation of organic matter than in very deep waters. A similar effect may result from stratification in deeper waters, e.g., due to a pycnocline or thermocline that limits the oxygen exchange and thus can cause oxygen deficiency in lower levels. Lack of oxygen also prevents the consumption of organic matter by aerobic, heterotrophic organisms, which may be active in areas of grazing or bioturbation in bottom sediment. The consumption of organic matter by heterotrophic organisms is a major factor in destruction of organic matter.

The rate of deposition of sedimentary particles plays a key role in all sedimentary processes and, hence, also in the preservation and concentration of organic matter. Assuming a constant supply of organic matter, its concentration in sediment will be inversely related to rate of deposition of mineral particles. The resulting dilution effect is well known. The relationship, however, is not that simple, because it is modified by other factors, such as grain size distribution of mineral particles, length of water column and residence time in it, and consumption of organic matter by heterotrophic organisms. Empirical observations indicate that, with increasing rate of deposition, the concentration of organic matter in sediment passes through a maximum, and only at higher rates of deposition dilution is effective.

Quantitative information on this complicated problem is not available. Good examples of the dilution effect are depocenters of deltas such as the Niger where a relatively large supply of both allochthonous and autochthonous organic material provides comparatively low average organic carbon values (0.3–0.8%) in fine-grained sediments. A depocenter typically is an area where most of the sedimentary load of a river is deposited over a prolonged period of time. In adjacent areas that do not receive so much clastic terrigenous material, average organic carbon values of equivalent sediment are higher (0.7–1.2%). A similar general situation with respect to organic matter sedimentation and dilution by mineral matter was noted in the Gulf of Mexico, where similarly low average organic carbon values were reported (Dow and Pearson, 1975).

The lipid fraction of organic matter, which is of special importance for petroleum source rocks, behaves somewhat differently in sedimentary processes than nonlipid material. The bulk of the lipid fraction is highly water-insoluble, whereas the bulk of the nonlipid fraction, i.e., carbohydrates and proteinaceous compounds and their derivatives, are water-soluble or hydrolyzable. In addition, many highly resistant parts of organisms, such as membranes, cuticles, wax coatings, spores, pollen and cork, are enriched in lipid-like substances (see also Chap. I.4.2). As a result, there is a general tendency for the lipid fraction to survive in particulate organic matter (POC) rather than in the dissolved organic material (DOD). Organisms deposited directly in sediment, of course, include their lipid content as well. Therefore, it may be deduced that the contribution of particulate organic matter to sediment may be more important for a potential source rock than is the contribution from dissolved organic matter. Data by Gormly and Sackett (1977) point in the same direction. They found that the $\delta^{13}C$-values of the lipids of POC are more similar to the average lipid composition of sediment than are the $\delta^{13}C$-values of the dissolved lipids. Likewise, they found the $\delta^{13}C$-value of total DOC to be very similar to the value for average total organic carbon of underlying sediment. Average $\delta^{13}C$-values of lipids in sediment were lighter by about 6 per mille than the average total organic carbon.

Bitterli's (1963) observation that transgressions and regressions are favorable for the deposition of sediment rich in organic matter, is explained in the foregoing discussion about supply and preservation of organic matter. Depositional transgressions and regressions combine such favorable conditions as relatively shallow waters with an ample supply of autochthonous and allochthonous organic matter, and zones of deposition of medium- to fine-grained sediment.

In a series of papers, Emery (1960, 1963, 1965) has comprehensively studied recent environments, which are of interest in determining how a potential source sediment of petroleum is formed. On the continental shelves, the following areas of quiet water are important: lagoons, estuaries, and deep basins of restricted circulation. Examples of such environments are found in southern California, the Gulf of California, parts of the East Indies, the Caribbean, and perhaps parts of northern Canada and the Eurasiatic shelf between Norway and the Taimir Peninsula. Another type of environment favorable for the accumulation of organic matter includes the continental slopes. There, the organic matter may reach very high concentrations, because the rate of deposition of detrital silt and clays is such that an optimal balance between dilution and conservation of organic matter is maintained. Seaward of the continental slope, concentrations of organic matter in sediment commonly decrease. A special situation may exist just beyond the base of the continental slope, where organic-rich sediment can accumulate as a result of sliding down the slope.

Finally, it is concluded that organic-rich sediment may occur wherever there is sufficient supply of organic matter, reasonably quiet waters, and an intermediate rate of sedimentation of fine-grained mineral particles.

Summary and Conclusion

The deposition of sediments rich in organic matter, i.e., such as contain more than about 0,5% by weight of organic carbon, is restricted to certain boundary conditions. Such sediments are deposited in aquatic environments receiving a certain minimum amount of organic matter. In subaerial sediments, organic matter is easily destroyed by chemical or microbial oxidation. The supply of organic matter is high along the continental margins, because of high primary productivity of coastal waters and/or a high input of allochthonous land-derived terrestrial plant material.

Balanced optimum conditions between the energy level in a body of water and rate of sedimentation are needed to preserve and concentrate organic matter in sediments. The clay-size mineral fraction becomes easily coated with organic matter. This fine-grained fraction and suspended particulate organic matter are of a low density, and are carried away from water bodies with a high energy level to more quiet waters. There, deposition of fine-grained sediments limits the access of dissolved molecular oxygen, and therefore increases the chances for preservation of organic matter. If, however, the rate of sedimentation is too high, organic matter is diluted, and a sediment low in organic matter is deposited.

Favorable conditions for the deposition of sediments rich in organic matter are found on the continental shelfs in areas of quiet waters, such as in lagoons, estuaries, and deep basins of restricted circulation. Another environment favorable for the accumulation of organic matter is the continental slopes.

The lipid fraction of organic matter behaves differently in sedimentary processes than nonlipid material. Contrary to the nonlipid material, the bulk of the lipid fraction is highly water-insoluble. Furthermore, many resistant parts of organisms, such as membranes, cuticles, spores, pollen etc. are enriched in lipid-like substances. Therefore, the lipid fraction of organisms tends to survive in the particulate organic matter rather than in the dissolved organic matter. It may thus be deduced that the contribution of particulate organic matter is more important for a potential source rock than that of dissolved organic matter.

References to Part I

Bader, R. G., Hood, D. W., Smith, J. B.: Recovery of dissolved organic matter in seawater and organic sorption by particulate material. Geochim. Cosmochim. Acta **19**, 236–243 (1960)

Bergmann, W.: Geochemistry of lipids. In: International Series of Monographs on Earth Science. Berger, I. A. (ed.). Oxford-London-New York-Paris: Pergamon Press, 1963, Chap. 12, pp. 503–542

Bitterli, P.: Aspects of the genesis of bituminous rock sequences. Geologie en mijnbouw **42**, 183–201 (1963)

Blumer, M., Mullin, M. M., Thomas, D. W.: Pristane in zooplankton. Science **140**, 974 (1963)

Blumer, M. Mullin, M. M. Thomas, D. W.: Helgoländer Wiss. Meeresuntersuchung **10**, 187 (1964)

Bogorov, V. G., Vinogradov, M. Y. E.: Distrubtion of the zooplankton biomass in the central Pacific. Tr. Vses. Gidrobiol. Obshchestva Akad. Nauk S.S.S.R. **10**, (1960)

Bordovskiy, O. K.: Sources of organic matter in marine basins. Marine Geol. **3**, 5–31 (1965)

Breger, I. A.: What you don't know can hurt you: Organic colloids and natural waters. In: Organic Matter in Natural Waters. Hood, D. W. (ed.). Symp. Univ. Alaska Sept. 2–4 1968, Inst. Mar. Sci.: Occasional Publ. No. 1, 563–574 (1970)

Breger, I. A., Brown, A.: Distribution and types of organic matter in a barred marine basin. Trans. N.Y. Acad. Sci. **25**, 741–755 (1963)

Brooks, J., Shaw, G.: Geochemistry of sporopollenin. Chem. Geol. **10**, 69–87 (1972)

Bülow, K. v.: Die Verflechtung von Erd- und Lebensgeschichte. In: Die Entwicklungsgeschichte der Erde. Autorenkollektiv (ed.), Leipzig: Kosmograph, 1959, pp. 438–451

Calvin, M.: Chemical Evolution. Oxford: Clarendon Press, 1969

Chave, K. E.: Carbonate-organic interactions in seawater. In: Organic Matter in Natural Waters. Hood, D. W. (ed.). Symp. Univ. Alaska Sept. 2–4 1968, Inst. Mar. Sci.: Occasional Publ. No. 1, 373–385 (1970)

Clark, R. C., Jr.: Occurrence of normal paraffin hydrocarbons in nature. Woods Hole Oceanogr. Inst. Woods Hole, Mass., Ref.no. 66–34, unpubl. manuscript (1966)

Clarke, H. T., Mazur, A.: The lipids of diatoms. J. Biol. Chem. **126**, 283–289 (1941)

Cloud, P. E., Jr.: Atmospheric and Hydrospheric Evolution on the Primitive Earth. Science, **160**, 729–736 (1968)

Datsko, V. G.: Organic Matter in Soviet Southern Waters. Moscow: Izd. Akad. Nauk. S.S.S.R., (in Russian) 1959

Debyser, J., Deroo, G.: Faits d'observation sur la genèse du pétrole. Rev. Inst. Fr. Pétr., **24**, 1, 21–48 (1969)

Degens, E. T.: Molecular nature of nitrogenous compounds in seawater and recent marine sediments. In: Organic Matter in Natural Waters. Hood, D. W. (ed.). Symp. Univ. Alaska Sept. 2–4 1968, Inst. Mar. Sci.: Occasional Publ. No. 1, 77–105 (1970)

Deuser, W. G.: Organic-carbon budget of the Black Sea. Deep-Sea Res., **18,** 995–1004 (1971)
Dietrich, G.: General Oceanography. New York-London: Wiley Interscience, 1963
Dorsselaer, A. v.: Triterpènes de sédiments. Ph. D.thesis, Univ. Strasbourg 1975
Dow, W. G., Pearson, D. B.: Organic matter in Gulf Coast sediments. Offshore Technol. Conf. Dallas. Paper OTC 2343 (1975)
Eglinton, G.: Personal communication, University of Bristol (1976)
Emery, K. O.: The Sea off Southern California. New York-London: Wiley Interscience, 1960
Emery, K. O.: Oceanographic factors in accumulation of petroleum. Proc. 6th World Petr. Congr. Frankfurt, Sec. I, Paper 42, PD2, 483–491 (1963)
Emery, K. O.: Characteristics of continental shelves and slopes. Am. Assoc. Petr. Geol. Bull. **49,** 1379–1384 (1965)
Ensminger, A., Dorsselaer, A. v., Spyckerelle, Ch., Albrecht, P., Ourisson, G.: Pentacyclic triterpenes of the hopane type as ubiquitous geochemical markers: origin and significance. In: Advances in Organic Geochemistry, Tissot, B., Bienner, F. (eds.). Paris: Technip, 1973, pp. 245–260
Erwin, J. A.: Comparative biochemistry of fatty acids in eukaryotic microorganisms. In: Lipids and Biomembranes of Eukaryotic Microorganisms. Erwin, J. A. (ed.), pt. 2. New York-London: Academic Press, 1973
Francis, M. J. O.: Monoterpene biosynthesis. In: Aspects of Terpenoid Chemistry and Biochemistry. Goodwin, T. W. (ed.), pt. 2. London-New York: Academic Press, 1971, pp. 29–48
Francis, W.: Coal, its Formation and Composition. London: Edward Arnold, 1954
Garrels, R. M., Mackenzie, F. T.: Sedimentary rock types: Relative proportions as a function of geological time. Science **163,** 570–571 (1969)
Gelpi, E., Oro, J., Schneider, H. J., Bennet, E. O.: Olefins of high molecular weight in two microscopic algae. Science **161,** 700–701 (1968)
Goodwin, T. W.: Distribution of carotenoids. In: Chemistry and Biochemistry of Plant Pigments. Goodwin, T. W. (ed.), pt. 4. London-New York: Academic Press, 1965, pp. 127–132
Goodwin, T. W.: Algal carotenoids. In: Aspects of Terpenoid Chemistry and Biochemistry. Goodwin, T. W. (ed.). London-New York: Academic Press, 1971, Chap. 11, pp. 315–356
Gormly, J. R., Sackett, W. M.: Carbon isotope evidence for the maturation of marine lipids. Proc. 7th Int. Meet. Organ. Geochem. Madrid (1975)
Hitchcock, C., Nichols, B. W.: Plant Lipid Biochemistry. London-New York: Academic Press, 1971
Hunt, J. M.: Geochemical data on organic matter in sediments. In: Vorträge der 3. int. wiss. Konferenz für Geochemie, Mikrobiologie und Erdölchemie, Budapest. Bese, V. (ed.), **394,** 1963
Hunt, J. M.: Distribution of carbon in crust of earth. Am. Assoc. Petr. Geol. Bull. **56,** 2273–2277 (1972)
Karlson, P.: Kurzes Lehrbuch der Biochemie. Stuttgart: Thieme, 1970
Karrer, W.: Konstitution und Vorkommen der organischen Pflanzenstoffe. Basel-Stuttgart: Birkhäuser, 1958
Koons, B., Jamieson, G. W., Cierezko, L. S.: Normal alkane distribution in marine organisms: possible significance of petroleum origin. Am. Assoc. Petr. Geol. Bull. **49,** 301–316 (1965)
Krey, J.: Die Urproduktion des Meeres. In: Erforschung des Meeres. Dietrich, G. (ed.), Frankfurt: Umschau, 1970, pp. 183–195

References to Part I

Kuenen, Ph. H. (1942): Cited in: Sallé and Debyser: Formation des gisements de pétrole. Paris: Technip, 1976

Kuenen, Ph. H.: Marine Geology. New York: John Wiley and Sons, 1950

Laskin, A. J., Lechevalier, H. A.: Handbook of Microbiology, Cleveland, Ohio: CRC Press (a division of the Chemical Rubber Co.), 1973

Lee, R. F., Nevenzei, J. C., Pfaffenhöfer, G. A.: Wax esters in marine copepods. Science **167**, 1510–1511 (1970)

Macmillan, J.: Diterpenes — the gibberellins. In: Aspects of Terpenoid Chemistry and Biochemistry. Goodwin, G. (ed.), pt. 6. London-New York: Academic Press, 1971, pp. 153–178

Matic, M. (1956): Cited in: Hitchcock and Nichols: Plant lipid Biochemistry. London-New York: Academic Press, 1971

Maxwell, J. R., Douglas, A. G., Eglinton, E., McCormick, A.: The botryococcenes-hydrocarbons of novel structure from alga *Botryococcus braunii* Kutzing. Phytochemistry **7**, 2157–2171 (1968)

McIver, R. D.: Hydrocarbon occurrences from JOIDES deep sea drilling project. In: Panel Discussion 5, petroleum prospects in the deep ocean regions, preprint of Proc. 9th World Petr. Cong. 1975. Barking, Essex: Applied Science Publishers Ltd., 1975, pp. 1–13

Meade, R. H., Sachs, P. L., Manheim, F. T., Hathaway, J. C., Spencer, D. W.: Sources of suspended matter in waters of the middle Atlantic bight. Sedim. Petrol. **45**, 171–188 (1975)

Menzel, D. W.: Primary productivity, dissolved and particulate organic matter, and the sites of oxidation of organic matter. In: The Sea. Marine Chemistry, Goldberg, D. (ed.). New York-London-Toronto: Wiley, 1974, Vol. 5, pp. 659–678

Menzel, D. W., Ryther, J. H.: Distribution and cycling of organic matter in the oceans. In: Organic Matter in Natural Waters. Hood, D. W. (ed.). Symp. Univ. Alaska Sept. 2–4 1968, Inst. Mar. Sci., Occasional Publ. No. 1, 31–54 (1970)

Metzner, H.: Biochemie der Pflanzen. Stuttgart: Enke, 1973

Milliman, J. D., Summerhayes, C. P., Barretto, H. T.: Oceanography and suspended matter off the Amazon river. Sedim. Petrol. **45**, 189–206 (1975)

Mitterer, R. M.: Calcified proteins in the sedimentary environment. In: Advances in Organic Geochemistry 1971. Oxford-Braunschweig: Pergamon Press, 1972, pp. 441–451

Moore, L. R.: Fossil bacteria. In: The Fossil Record. Swansea Symp. Geol. Soc. London, pt. II, 164–180 (1967)

Moore, L. R.: Geomicrobiology and geomicrobiological attack on sedimented organic matter. In: Organic Geochemistry, pt. 11. Eglinton, G. Murphy, M. T. J. (eds.). Berlin-Heidelberg-New York: Springer, 1969, pp. 265–303

Nissenbaum, A., Kaplan, I. R.: Chemical and isotopic evidence for the in situ origin of marine substances. Limnol. Oceanogr. **17**, 570–582 (1972)

Pelet, R.: Personal communication "on growth of bacteria at great water depth" (1976)

Percival, E.: The Natural Distribution of Plant Polysaccharides. In: Comparative Phytochemistry. Swain, T. (ed.), pt. 8. London-New York: Academic Press, 1966, pp. 139–155

Richardson, J. B., Ioannides, N.: Silurian palynomorphs from Tanezzuft and Acacus formations, Tripolitania, North Africa. Micropaleontology **19**, 257–307 (1973)

Roman, St.: Palynoplanktologic analysis of some Black Sea cores. In: The Black Sea — Geology, Chemistry and Biology. Degens, E. T., Ross, D. A. (eds.). Memoir 20. Tulsa, Oklahoma: Am. Assoc. Petr. Geol., 1974, pp. 396–406

Ross, D. A., Degens, E. T.: Recent sediments of Black Sea. In: The Black Sea — Geology, Chemistry and Biology. Degens, E. T., Ross, D. A. (eds.). Memoir 20. Tulsa, Oklahoma: Am. Assoc. Petr. Geol. 1974, pp. 183–199

Sackett, W. M.: The depositional history and isotopic Organic Carbon composition of marine sediments. Mar. Geol. **2,** 173–185 (1964)

Schidlowski, M., Eichmann, R., Junge, C. E.: Evolution des iridischen Sauerstoff-Budgets und Entwicklung der Erdatmosphäre. Umschau **22,** 703–707 (1974)

Schlegel, H. G.: Allgemeine Mikrobiologie. Stuttgart: Thieme, 1969

Schopf, J. W., Barghorn, E. S., Maser, M. D., Gordon, R. O.: Electron microscopy of fossil bacteria two billion years old. Science **159,** 1385 (1965)

Seibold, E.: Das Meer. In: Lehrbuch der Allgemeinen Geologie. Brinkmann, R., Louis, H., Schwarzbach, M., Seibold, E. (eds.). Band I, Festland–Meer, Kap. 9–16. Stuttgart: Enke, 1974, pp. 291–511

Shimkus, K. M., Trimonis, E. S.: Modern sedimentation in Black Sea. In: The Black Sea — Geology, Chemistry and Biology. Degens, E. T., Ross, D. A. (eds.). Memoir 20. Tulsa, Oklahoma: Am. Assoc. Petr. Geol., 1974, pp. 249–278

Sorokin, J. I.: On the role of bacteria in the productivity of tropical oceanic waters. Int. Rev. Ges. Hydrobiol. **56,** 1–48 (1971)

Stransky, K., Streibl, M., Sorm, F.: Lipid hydrocarbons of the alga. Collection Czechoslov. Chem. Commun. **33,** 416–424 (1968)

Strickland, J. D. H.: Production of organic matter in the primary stages of marine food chain. In: Chemical Oceanography. Riley, J. P., Skirrow, G. (eds.). London-New York: Academic Press, 1965, Vol I, pt. 12, pp. 478–595

Tappan, H.: Primary production isotopes, extinctions and the atmosphere. Paleogeogr. Paleoclimatol. Paleocol. **4,** 187–210 (1968)

Tappan, H., Loeblich, A. R., Jr.: Geobiologic implications of fossil phytoplankton evolution and time-space distribution. In: Symp. Palynology of the Late Cretaceous and Early Tertiary. Kosanke, R. M., Cross, S. T. (eds). East Lansing: Geol. Soc. Am. Michigan State Univ., 1970, pp. 247–340

Tasch, P.: The problem of primary production in the seas through geologic time. Rev. Paleobot. Palynol. **1,** 283–290 (1967)

Tolderlund, D. S., Bé, A. W.: Seasonal distribution of planktonic foraminifera in the western North Atlantic. Micropaleontology **17,** 297–329 (1971)

Tsujimoto (1906): Cited in: Organic Chemistry. Fieser, L. F., Fieser, M. (eds). New York: Reinhold, 1956

Vallentyne, J. R.: Net Primary Productivity in Aquatic Environments. Goldman, C. R. (ed.). Berkeley: California Univ. Press, 1965, p. 309

Wassojewitsch, N. B. (1955): Cited in: Seibold, E.: Das Meer. In: Lehrbuch der Allgemeinen Geologie. Brinkmann, R. et al. (eds.). Stuttgart: Enke, Band I, pt. 9, 1974, pp. 291–511

Welte, D. H.: Organischer Kohlenstoff und die Entwicklung der Photosynthese auf der Erde. Naturwissenschaften **57,** 17–23 (1970)

Welte, D. H.: Recent advances in organic geochemistry of humic substances and kerogen, a review. In: Advances in Organic Geochemistry 1973. Tissot, B., Bienner, F. (eds.). 1974, pt. 1, pp. 3–13

Williams, P. M.: Organic and inorganic constituents of the Amazon River. Nature (London) **218,** 937–938 (1968)

Youngblood, W. W., Blumer, M., Guillard, R. L., Fiore, F.: Saturated and unsaturated hydrocarbons in marine benthic algae. Mar. Biol. **8,** 190–201 (1971)

Zimmermann, W.: Die Phylogenie der Pflanzen. Stuttgart: Fischer, 1959

Zimmermann, W.: Geschichte der Pflanzen, eine Übersicht. Stuttgart: Thieme, 1969

Zobell, C. E.: Action of microorganisms on hydrocarbons. Bacteriol. Rev. **10,** 1 (1946)

Zobell, C. E.: Soil and water microbiology, 4th Int. Congr. Microbiol., Copenhagen, Sect. VII, 453 (1947)

Zobell, C. E.: Geochemical aspects of the microbial modification of carbon compounds, Scripps Inst. Oceanogr. Contrib. 1703, **34,** 1653 (1964)

Part II

The Fate of Organic Matter in Sedimentary Basins: Generation of Oil and Gas

Chapter 1
Diagenesis, Catagenesis and Metagenesis of Organic Matter

The physicochemical transformation of organic matter during the geological history of sedimentary basins cannot be regarded as an isolated process. It is controlled by the same major factors that also determine the variations of composition of the inorganic solid phase and of the interstitial water of the sediments: biological activity in an early stage, then temperature and pressure. Furthermore, organic–inorganic interaction can occur at different stages of the sediments evolution. Nature and abundance of organic matter may result in different behavior of the mineral phase, shortly after deposition; composition of minerals and structure of the rock may influence composition and distribution of organic fluid phases at depth. An excellent picture of the conditions of deposition and early history of sediments has been given by Strakhov (1962), and will be frequently used in the following paragraphs.

A general scheme of evolution of the organic matter from the time of deposition to the beginning of metamorphism, is shown in Figure II.1.1. To understand the discussion, the following stages of evolution are considered: diagenesis, catagensis, metagenesis and metamorphism. The acceptance of these terms in this book is somewhat different from that of Vassoevich et al. (1969, 1974). The two scales of evolution are shown in Figure II.1.2, with reference to the equivalent coal ranks and to the stages of petroleum generation.

1.1 Diagenesis

Sediments deposited in subaquatic environments contain large amounts of water (porosity amounts to about 80% in clay mud at 5 cm depth, i.e., water is 60% by weight of total sediment), minerals, dead organic material (contemporaneous autochthonous or allochthonous, and reworked), and numerous living microorganisms. Such a mixture results from various sedimentary processes and primary components of very different origins; it is out of equilibrium and therefore unstable, even if microorganisms are not present. *Diagenesis* is a process through which the system tends to approach equilibrium under conditions of shallow burial, and through which the sediment normally becomes consolidated. The depth interval concerned is in the order of a few hundred meters, occasionally to a few thousand meters. In the early diagenetic interval, the increase of temperature and pressure is small, and transformations occur under mild conditions.

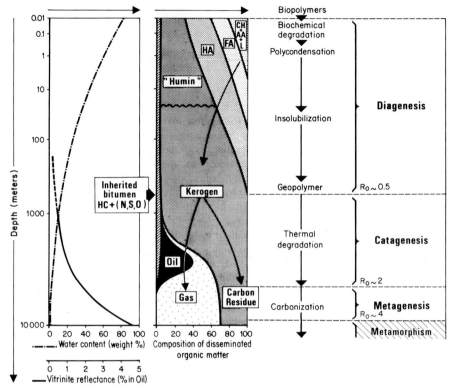

Fig. II.1.1. General scheme of evolution of the organic matter, from the freshly deposited sediment to the metamorphic zone. *CH:* carbohydrates, *AA:* amino acids, *FA:* fulvic acids, *HA:* humic acids, *L:* lipids, *HC:* hydrocarbons, *N, S, O:* N,S,O compounds (non-hydrocarbons)

During early diagenesis, one of the main agents of transformation is microbial activity. Aerobic microorganisms that live in the uppermost layer of sediments consume free oxygen. Anaerobes reduce sulfates to obtain the required oxygen. The energy is provided by decomposition of organic matter, which in the process is converted into carbon dioxide, ammonia and water. The conversion is usually carried out completely in sands and partly in muds. At the same time E_h decreases abruptly and pH increases slightly. Certain solids like the organodetrital $CaCO_3$ and SiO_2 dissolve, reach saturation and re-precipitate, together with authigenic minerals such as sulfides of iron, copper, lead and zinc, siderite, etc.

Within the sediment, organic material proceeds also towards equilibrium. Previous biogenic polymers or "biopolymers" (proteins, carbohydrates) are destroyed by microbial activity during sedimentation and early diagenesis. Then their constituents become progressively engaged in new polycondensed structures ("geopolymers") precursing *kerogen*. When deposition of organic matter derived from plants is massive compared to mineral contribution, *peat* and then *brown coals* (lignite and sub-bituminous coal) are formed. The most important hydrocarbon formed during diagenesis is *methane*. In addition organic matter

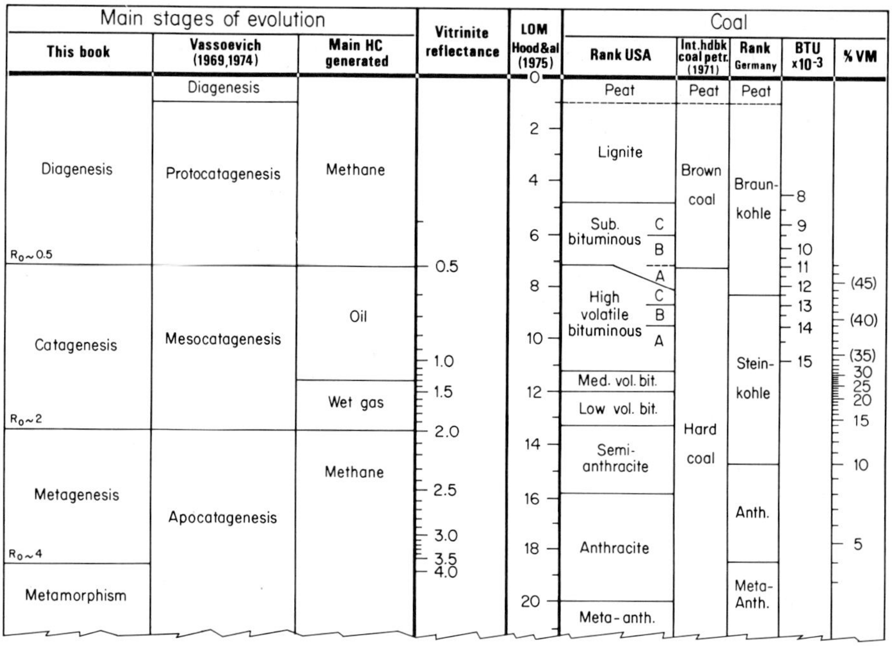

1.2 Catagenesis

Main stages of evolution			Vitrinite reflectance	LOM Hood & al (1975)	Coal				
This book	Vassoevich (1969, 1974)	Main HC generated			Rank USA	Int.hdbk coal petr. (1971)	Rank Germany	BTU x10⁻³	% VM
Diagenesis	Diagenesis				Peat	Peat	Peat		
	Protocatagenesis	Methane		2	Lignite	Brown coal	Braun- kohle	8	
				4				9	
$R_o \sim 0.5$			0.5	6	Sub. bituminous C / B / A			10 11	
Catagenesis	Mesocatagenesis	Oil	1.0	8 10	High volatile bituminous C / B / A		Stein- kohle	12 13 14 15	(45) (40) (35) 30 25
$R_o \sim 2$		Wet gas	1.5 2.0	12 14	Med. vol. bit. Low vol. bit. Semi- anthracite	Hard coal			20 15 10
Metagenesis	Apocatagenesis	Methane	2.5 3.0 3.5 4.0	16 18 20	Anthracite Meta-anth.		Anth. Meta- Anth.		5
$R_o \sim 4$ Metamorphism									

Fig. II.1.2. Main stages of evolution of the organic matter. The stages used by Vassoevich (1969, 1974) and the LOM scale of Hood et al. (1975) are shown for comparison, and also the equivalent coal ranks. (Adapted from Vassoevich 1969, 1974, International Handbook of Coal Petrology, 1971, Hood et al. 1975, Stach et al. 1975)

produces CO_2, H_2O and some heavy heteroatomic compounds during later stages of diagenesis.

The end of diagenesis of sedimentary organic matter is most conveniently placed at the level where extractable humic acids have decreased to a minor amount, and where most carboxyl groups have been removed. This is equivalent to the boundary between brown coal and hard coal, according to the coal rank classification of the International Handbook of Coal Petrology (1971). It corresponds to a vitrinite reflectance of about 0.5%.

1.2 Catagenesis

Consecutive deposition of sediments results in burial of previous beds to a depth reaching several kilometers of overburden in subsiding basins. This means a considerable increase in temperature and pressure. Tectonics may also contribute to this increase. For this stage of sedimentary evolution, we suggest the use of the word *catagenesis*, proposed by Vassoevich (1957), and also used by Strakhov (1962). Temperature may range from about 50 to 150°C and geostatic pressure due to overburden may vary from 300 to 1000 or 1500 bars. Such increase again places the system out of equilibrium and results in new changes.

Composition and texture of the mineral phases are conserved, with some changes mostly in the clay fraction. The main inorganic modification still concerns the compaction of the rock: water continues to be expelled, porosity and permeability decrease markedly; normally salinity of interstitial water increases and may come close to saturation.

Organic matter experiences major changes: through progressive evolution the kerogen produces first *liquid petroleum;* then in a later stage "wet gas" and condensate; both liquid oil and condensate are accompanied by significant amounts of *methane.* Massive organic deposits progress through the various ranks of *coal,* and also produce hydrocarbons, mostly methane.

The end of catagenesis is reached in the range where the disappearance of aliphatic carbon chains in kerogen is completed, and where the development of an ordering of basic kerogen units begins. This corresponds to vitrinite reflectance of about 2.0 which, according to various coal classifications, is approximately the beginning of the anthracite ranks.

Since these are severe changes in the organic material, and since with further evolution there is no more generation of petroleum and only limited amounts of methane, this point seems to be at a natural break. Therefore, we propose to draw here an additional borderline, one that terminates catagenesis, and to call the subsequent stage *metagenesis.*

1.3 Metagenesis and Metamorphism

The last stage of the evolution of sediments, which is known as metamorphism, is reached in deep troughs and in geosynclinal zones. Here temperature and pressure reach high values; in addition, rocks are exposed to the influence of magma and hydrothermal effects. Petroleum geology, however, is only concerned with the stage precursing metamorphism, which has been variously characterized and designated as early metamorphism, epimetamorphism, anchimetamorphism, etc. As far as organic constituents are concerned, we shall refer to this stage precursing metamorphism as *metagenesis* of organic matter.

Minerals are severely transformed under those conditions: clay minerals lose their interlayer water and gain a higher stage of crystallinity; iron oxides containing structural water (goethite) change to oxides without water (hematite) etc.; severe pressure dissolution and recrystallization occur, like the formation of quartzite, and may result in a disappearance of the original rock structure. The rock reaches temperature conditions that lead to the *metagenesis* of organic matter. At this stage the organic matter is composed only of *methane* and a *carbon residue,* where some crystalline ordering begins to develop. Coals are transformed into *anthracite.*

True conditions of *metamorphism* result in greenschist, and amphibolite facies development. Coal is transformed into metaanthracite, which has a vitrinite reflectance of more than 4%. The constituents of the residual kerogen are converted to graphitic carbon.

Summary and Conclusion

The three main stages of the evolution of organic matter in sediments are diagenesis, catagenesis and metagenesis.
1. Diagenesis begins in recently deposited sediments where microbial activity is one of the main agents of transformation. Chemical rearrangements then occur at shallow depths: polycondensation and insolubilization. At the end of diagenesis, the organic matter consists mainly of kerogen.
2. Catagenesis results from an increase in temperature during burial in sedimentary basins. Thermal degradation of kerogen is responsible for the generation of most hydrocarbons, i.e., oil and gas.
3. Metagenesis is reached only at great depth. However, this last stage of evolution of organic matter begins earlier (vitrinite reflectance approximately 2%) than metamorphism of the mineral phase (vitrinite reflectance of about 4% corresponding to the beginning of the greenschist facies).

Chapter 2
Early Transformation of Organic Matter: The Diagenetic Pathway from Organisms to Geochemical Fossils and Kerogen

2.1 Significance and Main Steps of Early Transformations

The time covering sedimentation processes and residence in the young sediment, freshly deposited, represents a very special stage in the carbon cycle. The first few meters of sediment, just below the water–sediment contact, represent the interface through which organic carbon passes from the biosphere to the geosphere. The residence time of organic compounds in this zone of the sedimentary column is long compared to the lifetime of the organisms, but very short compared to the duration of geological cycles: a 1-m section often represents 500 to 10000 years. During sedimentation processes, and later in such young sediments, organic material is subjected to alterations by varying degrees of microbial and chemical actions. As a result its composition is largely changed, and its future fate during the rest of the geological history predetermined within the framework of its subsequent temperature history.

When comparing the nature of the organic material in young sediments with that of the living organisms from which it was derived, the striking point is that most of the usual constituents of these organisms, and particularly the biogenic macromolecules, have disappeared. Proteins, carbohydrates, lipids, and lignin in higher plants amount to nearly the total dry weight, on an ash-free basis, of the biomass living in subaquatic or subaerial environments. The total amount of the same compounds that can be extracted from very young sediments is usually not more than 20% of the total organic material, and often less. This situation results from degradation of the macromolecules by bacteria into individual aminoacids, sugars, etc. As monomers, they are used for nutrition of the microorganisms, and the residue becomes polycondensed, forming large amounts of brown material, partly soluble in diluted NaOH, and resembling humic acids.

As time and sedimentation proceed, the sediment is buried to several hundreds of meters. Most of the organic material becomes progressively insoluble as a result of increasing polycondensation associated with loss of superficial hydrophilic functional groups. This completely insoluble organic matter from young sediments has received limited attention until recently. It is called "humin" by the few soil scientists who have worked on sub-aquatic soils. In ancient sediments the insoluble organic matter, called *kerogen,* is obtained by demineralization (HCl, HF) of the rock. This procedure is not degradative for ancient organic matter, as shown by Durand et al. (1977) on coals and oil shales. The situation is rather different in young sediments, where an important part of the humin can be hydrolyzed by this procedure (Huc, 1973). The hydrolyzable fraction progressively decreases with burial. Thus humin, collectively with other insoluble organic matter such as pollen, spores, etc. may be considered as a precursor of kerogen, but the terms *humin* and *kerogen* are not strictly equivalent. Petroleum geochemists consider kerogen as the main source of petroleum compounds.

The whole process is referred to here as *diagenesis* and leads from *biopolymers* synthesized by living organisms to *geopolymers* (kerogen) through fractionation, partial destruction and rearrangement of the building stones of the macromolecules. For convenience in the discussion, three steps will be considered in the following paragraphs (Fig. II.1.1):

- Biochemical degradation
- Polycondensation
- Insolubilization

It has to be realized that the first and the second step follow in immediate succession, resulting in a zone where both processes are active at the same time. This zone comprises to some extent the water and the top layer of the sediment. Nevertheless, small or minute amounts of aminoacids, lipids and sugars are still found even in very old rocks. Insolubilization may also start early for some organic material, but humic acids, which are soluble in dilute NaOH, are still found at several hundred meters depth in sediments containing abundant detrital material of continental origin.

2.2 Biochemical Degradation

The four constituents of the young sediments are minerals, organic matter from dead plants or animals, interstitial water, and living benthic organisms. The last ones usually include a wide range of forms, but microorganisms are the most active and widely distributed in shallow water sediments. They are usually less important and active under deep oceanic conditions.

a) Microbial Activity

Microorganisms — mostly bacteria, actinomycetes, fungi and algae — are abundant in sub-aerial soils, waters, and sediments deposited under moderate water depths. Their normal activity is the decomposition of organic matter, thus providing energy and/or material to build up the constituents of their cells.

In aquatic environments, bacteria seem to be especially important: they are present in both sea and lake waters. In certain sites, they are abundant in the uppermost half meter of the young sediment. As burial continues, the quantity of bacteria and other microorganisms decreases rapidly with depth (Table II.2.1). Since photosynthesis is impossible inside the sediment, the energy necessary for the synthesis of microbial constituents is obtained there by degradation of existing carbon compounds (Debyser, 1969), essentially through respiration under aerobic conditions and fermentation under anaerobic conditions. The material utilized for synthesis is also normally taken from existing organic compounds.

Table II.2.1

1. Quantity of bacteria in Northwestern Pacific bottom sediments. (After Limberg-Ruban, 1952)

Level (cm)	Total number of bacteria (mill. per g sediment)	Quantity (%)		
		Rods	Cocci	Spores
Surface of sediment	10.1	88.6	7.7	3.7
30	2.6	77.7	4.3	18.0
70	0.6	75.6	1.4	23.0

2. Quantity of bacteria in a recent sediment from San Diego Bay (Calif.). (After Zobell and Anderson, 1936)

Depth (cm)	Anaerobic bacteria (per g sediment)	Aerobic bacteria (per g sediment)	Anaerobic/aerobic ratio
0– 3	1,160,000	74,000,000	1:64
4– 6	14,000	314,000	1:21
14–16	8,900	56,000	1: 6
24–26	3,100	10,400	1: 3
44–46	5,700	28,100	1: 5
66–68	2,300	4,200	1: 2

As nutrition of bacteria operates in an osmotic way, the organic material has to be dissolved before it can be used. The necessary decomposition of the organic matter is accomplished by microorganisms through enzymatic processes, providing the material to be used as a source of energy and a base for heterotrophic processes. Proteins and carbohydrates are hydrolyzed, providing amino acids and sugars; lipids and lignin are the object of a less active degradation.

In oxidizing environments, the destruction of the organic material can go very far: aerobic bacteria use molecular oxygen to oxidize organic compounds to carbon dioxide and water, and to liberate energy (respiration). That action may last until all organic matter is destroyed (Fig. II.2.1). Under deep oceanic conditions, the situation is somewhat different, as bacterial activity seems to be reduced by hostile environment.

In fine-grained sediment, like shales, silts or fine carbonate muds, the pore space becomes quickly a closed environment, where interstitial water is separated from the overlying sea or lake water. Most of the oxygen confined in that space is exhausted by aerobic activity and anaerobic conditions are rapidly established, first on a restricted scale, around decomposing organisms, then on a wider scale (Fig. II.2.1). Under these conditions, organic matter is partly decomposed by fermentation: sulfates are reduced and also ferric and other hydroxides; oxida-

2.2 Biochemical Degradation

Preservation in fine clay or carbonate mud

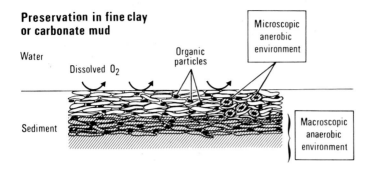

Destruction in a porous sediment deposited under aerobic conditions

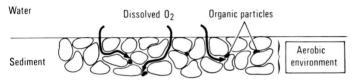

Fig. II.2.1. Preservation or destruction of the organic matter in a freshly deposited sediment. In a fine clay or carbonate mud *(top)*, pore water becomes a nearly closed microenvironment. There is no replenishment of oxygen, and anaerobic conditions are rapidly established, first on microscopic, then on a macroscopic scale. In a porous sand deposited under aerobic conditions *(bottom)*, free circulation of water containing dissolved oxygen results in the destruction of the organic matter

tion-reduction potential is lowered below zero and often drops as low as −150 to −400 mV (Strakhov, 1962; Huc, 1973).

With increasing depth, the organic matter deposited in marine, fine-grained sediments normally goes through three successive zones (Claypool and Kaplan, 1974):

- an oxidizing zone, where free oxygen is still available for oxidation of the organic matter; this zone may be absent, when the deposition environment is already anoxic;
- an anaerobic sulfate reduction zone, after all molecular oxygen has been used;
- an anaerobic methane generation zone after most of sulfate has been reduced.

Anaerobic conditions, however, are first established on a microscopic, then on a macroscopic scale (Fig. II.2.1). Then sulfate reduction and methane generation are not mutually exclusive. Thus there may be some overlap, with respect to depth, of the various zones. For instance, Doose (1980) observed in sediments of Santa Barbara basin, off California, that the sulfate reduction zone changes to the methane generation zone at about 1.5 m depth, with some 50 cm overlap.

Sulfate-reducing bacteria *(Desulfovibrio)* utilize oxygen from SO_4^{2-} and reduce the sulfur to S^{2-} under anaerobic conditions. This normally happens below the water–sediment interface where interstitial water is separated from the overlying

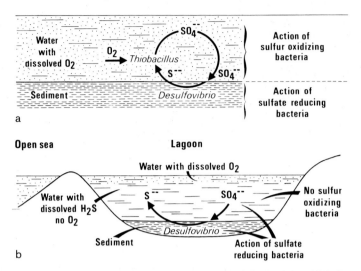

Fig. II.2.2a, b. Formation and destruction of hydrogen sulfide in sediments and bottom waters. (a) Open sea conditions: there is an equilibrium between formation and destruction of hydrogen sulfide. (b) Restricted water circulation: no dissolved oxygen in bottom waters. This results in the absence of aerobic sulfur-oxidizing bacteria. Thus, there is no destruction of hydrogen sulfide in bottom waters, which are anoxic and contain H_2S

free water: bacteria extract the limited amount of sulfate initially present in interstitial water or provided by the slow process of diffusion. The zone of sulfate reduction is normally confined within the upper level of sediment, and the water column above the sediment does not contain hydrogen sulfide. Sea or lake water usually contains dissolved oxygen, and hydrogen sulfide emanating from the sediment interface is oxidized again into sulfate, under aerobic conditions, by other types of bacteria *(Thiobacillus)*. Thus, an equilibrium is established between supply and destruction of H_2S: Orr (1974) (Fig. II.2.2). In some cases, however, the bottom water layer of the sea or lake becomes devoid of oxygen, and sulfur-oxidizing bacteria cannot work in such anaerobic conditions — except photosynthetic colored bacteria in shallow waters. Furthermore the free water becomes subject to sulfate reduction; large quantities of H_2S and free sulfur may be produced from seawater; and the benthic life disappears, except anaerobic microbes, partly due to lack of oxygen, and partly due to toxicity of H_2S (Fig. II.2.2).

The phenomenon may occur for various reasons, but it is due mainly to restrictions in water circulation and to eutrophic conditions. Restrictions in water circulation may prevent oxygen supply of water layers. For example the Black Sea, nearly separated from the Mediterranean Sea, receives more freshwater from rains and rivers than is lost by evaporation. This results in a low salinity and a low density of surface waters, thus preventing convection. Therefore the water column is stratified and bottom waters remain deprived of oxygen. Anaerobic bacteria generate abundant H_2S, but it is not destroyed by aerobic sulfur-oxidizing bacteria. Eutrophic conditions result from superabundant supplies of

2.2 Biochemical Degradation

organic matter or nutrients (phosphate, nitrate) to lake or seawaters, as is presently happening in Lake Erie due to human pollution. Populations of bacteria and other organisms that live at the expense of the organic matter and use dissolved oxygen grow considerably. They subsequently eliminate oxygen from the bottom waters. The same effect can be postulated for fossil sediments which, earlier in geological history, had received an abundant supply of organic matter, as in some lacustrine basins or in tropical swamps found today in certain coastal areas.

Sulfur is not incorporated in the bacterial cell. In clay muds, where iron is usually abundant, it is likely that sulfur readily recombines with iron to form hydrotroilite and troilite, which are slowly converted to pyrite. In carbonate muds, however, where iron is much less abundant, sulfur may recombine mainly with residual organic matter. Massive incorporation of sulfur into sediments in certain confined environments with H_2S-saturated bottom water may lead to a subsequent recombination with organic matter during late diagenesis and finally explain the origin of a sulfur-rich petroleum formed during catagenesis.

Fermentation is an anaerobic process by which bacteria, instead of using molecular oxygen (respiration), use oxidized forms of organic matter, particularly carbohydrates. Cellulose is degraded enzymatically by bacteria, such as *Clostridia* (Doose, 1980), producing the low molecular weight precursors for methane formation (acetate, bicarbonate) and the final stage is a reduction of carbon dioxide or acetate by methane-generating bacteria, such as *Methanobacteria*.

In marine sediments containing organic matter, Doose (1980) observed that methane generation becomes important once the content of sulfate has been largely diminished by anaerobic sulfate-reducing bacteria. This observation is interpreted as resulting from a competition for hydrogen (issued from organic matter) between the sulfate-reducing bacteria and the methane producers. In fact, Oremland and Taylor (1978) have demonstrated by laboratory incubations of sediments that methanogenic bacteria cannot really compete with sulfate-reducing bacteria: although the two processes are not mutually exclusive, sulfate reduction is largely predominant as long as sulfate is abundantly available.

Anaerobic fermentation has long been considered by various authors as a possible way to formation of petroleum, as methane generation was well established through this process in swamps (marsh gas). Laboratory experiments were carried out on fatty acids, carbohydrates and natural muds. The results show that methane is the single hydrocarbon produced in major quantities. Other hydrocarbons (ethane, propane, butane) show very low concentrations only, between 10^{-3} and 10^{-6}, in the gas generated in the experiments (Davis and Squires, 1953, Bokova, 1959).

When conditions are particularly favorable, the formation of methane may reach an important level. Lake Kivu, in Central Africa, is an example of methane formation by bacterial reduction of carbon dioxide on the bottom of the lake (Louis, 1967). The oxygen circulation is restricted because the lake is deep (480 m) and the valley was sealed off by recent lava flows. The total volume of water is about 580 km^3 and at depths greater than 265 m it contains dissolved gas comprising CO_2 (75%), CH_4 (24%), H_2, N_2, Ar (1%). The total amount of methane is evaluated to be 57 billion m³ under normal temperature and pressure conditions. In fossil sediments bacterial degradation of organic matter may be the source of "dry gas" fields at shallow depths, like the large gas field discovered in Upper Cretaceous sandstones of Western Canada (Medicine Hat, depth: 310 m; CH_4: 96.2%) and in the Cretaceous of Western Siberia. These examples suggest that under certain conditions, such as in the case of abundant deposition of cellulose, an important part of the organic matter may be converted to methane. Carbon isotope measurements have confirmed this assumption in several cases. The isotopic aspects of methane generation by bacteria are discussed in Section 2.5 of this chapter. Biogenic formation of gas in sedimentary basins and its significance in terms of gas reserves will be further discussed in Sect. 6.2.

It is difficult to ascertain which limiting factor will stop the degradation of organic matter by bacteria under anaerobic conditions: lack of nutrient from the remaining organic material, toxicity of products from their own metabolism, etc. In deep oceanic conditions the hostile environment (pressure, temperature) might prevent aerobic and anaerobic bacteria from being very active.

b) Free or Hydrolyzable Organic Compounds in the Young Sediment

As macromolecules, like proteins and polysaccharides, are degraded by microbial activity, individual amino acids and sugars may be found in the young sediment, in addition to fatty acids and hydrocarbons. Surprisingly, though, their total amount is rather low: even in the top layer of the sediments, most of the organic material (about 75 to 95%) is neither hydrolyzable nor extractable by organic solvents[1]. This material will be considered below (Sect. 2.3) under the generic names of "humic compounds" and "humin".

Amino acids, mostly hydrolyzable and not free, are generally the most abundant (Fig. II. 2.3) among the free or hydrolyzable organic compounds. In the top 0–15 cm layer of a reducing fjord, the Saanish Inlet in British Colombia (Brown et al., 1972), they represent 6% of the total organic carbon. In the Bering Sea (NW Pacific Ocean) the "readily hydrolyzable organic matter" reaches 25% (Bordovskiy, 1965) and in the central Black Sea 40% of the organic matter (Huc et al., 1977). However, these latter figures are likely to include also the hydrolyzable fraction of humic acids or humin, as defined below. In all cases, however, these compounds decrease at depth — rapidly in the fjord sediment, where they amount to less than 1% at 35 m, and more slowly in the Black Sea sediments, where they still amount to ca. 20% at 500 m.

Sugars show a still more abrupt decrease with depth from 3.8% of organic carbon at the water–sediment interface down to 0.02% at 44 m, in a core from Pacific Ocean (Vallentyne, 1963).

The total amount of free or combined fatty acids, hydrocarbons, terpenes, sterols and pigments is generally below 1% of the organic carbon, although some

---

1 In addition we must be careful when considering analysis of proteins, carbohydrates and lipids in recent sediments, as these are the normal constituents of the microorganisms living in the sediment (Debyser, 1969)

2.3 Polycondensation

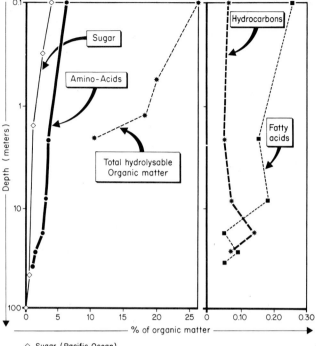

○ Sugar (Pacific Ocean)
● Amino-acids (Reducing fjord, British Columbia) Cores 4/3B
∗ Hydrolysable Organic matter (Bering Sea)
■ Fatty acids } Reducing fjord
∗ Hydrocarbons } British Columbia (Cores 4/3B)

Fig. II.2.3. Variation of amino acids, sugars and lipids in Recent sediments, as a function of depth (wt. % of the total organic matter). (Data from Vallentyne, 1963, Bordovskiy, 1965, Brown et al., 1972)

exceptions, like algal deposits, may occur. There is a decrease of the total amount of these compounds with increasing depth. This decrease may be due to consumption by bacteria or to the fact that these compounds become bound in the insoluble organic fractions. In addition to that, the free molecules, such as sterols, terpenes, etc., are subject to chemical changes in the upper layer of sediments. These changes do not affect the carbon skeleton, but improve the stability of these molecules (Chap. II.3). Finally there is evidence for early transformation by microbial decomposition and resynthesis of hydrocarbon chains (Rhead et al., 1971). This results, for instance, in changes of the length of the aliphatic chains.

2.3 Polycondensation

The major part of the organic material in young sediments is neither hydrolyzable, nor extractable by organic solvents. Most free or hydrolyzable compounds

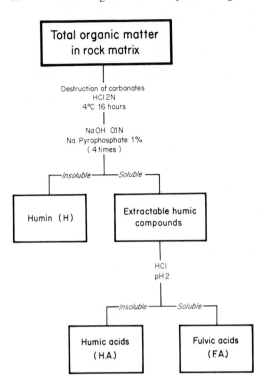

Fig. II.2.4. Flow diagram of the separation of humic material

disappear at shallow depth; and their residue, which is not directly used by microorganisms, becomes quickly incorporated into new polymeric, insoluble structures.

Techniques used by soil scientists (Fig. II.2.4) applied to Recent sediments (sub-aquatic soils) result in extraction of humic material. The extraction with a mixture of dilute sodium hydroxide (0.1 N) and sodium pyrophosphate (1%), and a subsequent separation by acid precipitation yields compounds similar to *fulvic acids* (soluble in acids) and *humic acids* (insoluble in acids). The extraction leaves a residue similar to *humin* (insoluble in NaOH). In the following, the terms humic and fulvic will be used, instead of humic-like and fulvic-like material. The sum of the humic and fulvic acids will be called humic compounds.

In subaerial soils, humic acids result from polycondensation of the organic residue of microbial (fungi, bacteria, etc.) metabolism, under more or less oxidizing conditions. Oxidative condensation of phenols is an important process (Flaig, 1966). Addition of nitrogenous compounds occurs mainly through random polymerization of free radicals like semiquinone (Flaig, 1972). The structure of soil humic acids has been described by Swain (1963): they are polycondensates made of nuclei bearing reactive groups, connected by bridges. Nuclei are simple or condensed cycles, aromatic, naphthenic or heterocyclic; bridges are mostly oxygen, sulfur, peptidic or methylene bonds; the most important reactive group is OH. Molecular weights of humic acids cover a wide range, ca. 10,000 to 100,000, and the size of particles may reach 100 Å.

2.3 Polycondensation

In subaquatic sediments, and also in the overlying water column, similar processes of polycondensation occur, under aerobic and anaerobic conditions. In such sediments humic acids may develop by polycondensation of autochthonous, mostly plankton-derived material (protein, carbohydrate, and lipid derivatives), and/or allochthonous land-derived material (mostly lignin and cellulose). Amino acids and carbohydrates may react rapidly, through the Maillard reaction (Maillard, 1912) to produce a polycondensate, linked by peptidic bond, similar to humic acids (Nissenbaum and Kaplan, 1972, Hoering and Hare, 1973). In addition to that, there may be some admixture of detrital or dissolved humic acids from continental origin.

The structure of humic acids formed in subaquatic sediments is rather different from soil humic acids. Phenolic constituents are less abundant in marine humic acids. On the contrary their H/C atomic ratio is higher with values of 1.00 to 1.50, compared with 0.50 to 1.00 in soil humic acids. This results from the abundance of aliphatic chains and alicyclic rings in marine humic acids, which is clearly shown on IR spectroscopy (Fig. II.2.5) by the importance of CH_2, CH_3 bands (around 2900, 1450 and 1375 cm^{-1}). The role of peptidic bonds is confirmed by the importance of the corresponding amide I and II infrared bands (1650 and 1540 cm^{-1}), which is greater in marine humic acids of shallow origin than in

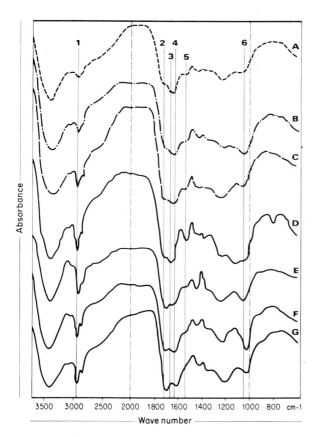

Fig. II.2.5. Comparison of IR spectra of humic acids from terrestrial soils, and humic acids from Recent marine sediments (Huc, 1973). Identification of humic acids: *Terrestrial soils: A* rendzin, *B* brown soil, acid, *C* podzol. *Marine sediments: D* and *F* France, Atlantic coast, *E* Eastern Mediterranean, *G* West Africa. *Identification of IR bands: 1* aliphatic C–H, *2:* C = O, *3* and *5* amides, *4* aromatic C = C, *6* C–O of polysaccharides

terrestrial humic acids (Huc and Durand, 1974). The sulfur content also appears to be higher in marine humic acids. Carboxyl groups are abundant, as shown by the strong absorption at the 1700 cm^{-1} IR band.

Rashid and King (1969) have studied the molecular weight distribution of the humic material from the Scotian shelf. Fulvic acids cover a wide range from below 700 to 10,000 mol. wt. Molecular weights of humic acids extend much further and some are greater than 2.10^6.

Fatty acids might be fixed on humic material, as aliphatic ester substituents $-\overset{\overset{\displaystyle O}{\|}}{C}-O-$ bonds) in a similar way, as suggested by Burlingame et al. (1969) for the kerogen of the Green River Shales. This process results in fixation of alkyl chains in the polymeric humic material. Polymerization of unsaturated fatty acids might also occur, although this has not been proven under mild temperature conditions of the young sediment.

Hydrocarbons are not affected by the polycondensation process, as they lack the necessary functional groups to be linked to the humic material. However, various types of organic compounds, including hydrocarbons, may be attached to humic material by weaker bonds: physical adsorption or hydrogen bonds.

The abundance and nature of humic and fulvic acids in Recent sediments may vary, according to the conditions of environment. In places where continental runoff is considerable, allochthonous material from higher plants is often the main source of organic material: Niger delta, Texas-Louisiana Gulf Coast, Bay of Bengal and Amazon fan in modern environments; Lower Jurassic of Luxembourg (Hagemann, 1974) in ancient sediments. In such cases fulvic and humic acids are generally abundant. They may include humic acids formed in continental soils and subsequently transported to the basin of deposition. Allochthonous material may also include reworked organic matter eroded from ancient sediments and transported by rivers to the basin of deposition. For instance, Gadel and Ragot (1974) measured the amount of anthracite material, eroded from the Carboniferous of the Alps and incorporated in the Recent sediments off the Rhone delta.

In other places, autochthonous marine organic matter is likely to be the major constituent. For instance, in the Saanich Inlet sediments of British Colombia (Brown et al., 1972) the average carbon isotopic ratio of humic acid in sediments is $-22.5‰$, compared with $-29.1‰$ in humic acid from neighboring subaerial soil, and $-19‰$ for the total organic carbon of plankton. Other examples are seen in the occurrence of humic acids in sediments far away from any coast and in sediments of the Antarctic area, where terrigenous organic supply from the continent is unlikely (Bordovskiy, 1965). However, on the average, there is more humic material produced from terrestrial organic matter than from autochthonous marine sources.

In addition to the attachment of organic compounds, humic acids may also collect and fix heavy metals, such as U, V, Ni, Cu and others.

2.4 Insolubilization

Decomposition and polycondensation of organic material, mostly occurring in the first few meters of sediment, result in macromolecules that represent more than 90% of the total organic matter in young sediments. However, a certain part of the macromolecules still consists of NaOH-soluble compounds: in the upper layer from 0 to 10 m the amount of fulvic plus humic acids ranges from 10 to 70% of the total organic matter, the highest values being recorded in deltaic or estuarine terrigenous muds (Huc and Durand, 1974). With increasing depth of burial, these wide variations are reduced, because the fulvic and humic acids are being converted to insoluble humin (Huc and Durand, 1977).

Insolubilization occurs during diagenesis of sediments on a depth and time scale much wider than the previous steps: several hundreds, or even occasionally thousands, of meters, one to several millions of years. With increasing depth, the insoluble humin predominates over fulvic and humic acids (Fig. II.2.6). Also, the fulvic acid/humic acid ratio decreases (Fig. II.2.7). The organic material becomes more condensed, as shown by a general darkening and an increase of absorption of visible light.

These facts can be interpreted as a result of an increased polycondensation, accompanied by elimination of a sufficient part of the superficial hydrophilic

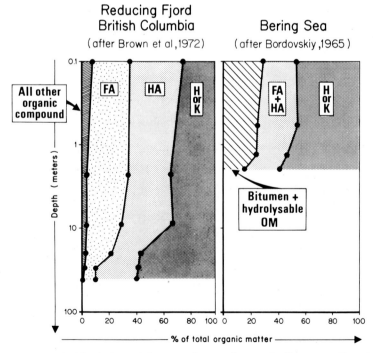

Fig. II.2.6. Evolution of the organic constituents in Recent sediments as a function of depth. *FA:* fulvic acids, *HA:* humic acids, *H or K:* humin (or kerogen). (Data from Bordovskiy, 1965; Brown et al., 1972)

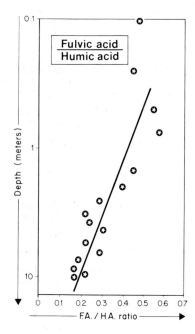

Fig. II.2.7. Variation of the fulvic acid: humic acid ratio in a core from the Bay of Benin (West Africa). (Data from Huc, 1973)

functions. Thus the organic material becomes progressively insoluble. It evolves from fulvic acid to the humic acid type, and then to the completely insoluble humin type. A similar process has been previously proposed by Swain (1963) for humic material of soils.

Several kinds of functional groups are clearly affected during the phase of insolubilization. Two stages can be tentatively separated. An initial stage covers the first tens of meters. During that time, the evolution of the humic acid fraction (Huc and Durand, 1974) shows a progressive loss of peptidic bonds, accompanied by a progressive disappearance of the $1650\,\text{cm}^{-1}$ and $1540\,\text{cm}^{-1}$ IR bands (amide I and II). Furthermore, there is a decrease of hydrolyzable nitrogen and also of the N/C ratio. The elimination of nitrogen from the solid organic phase has also been recorded by Bordovskiy (1965) and Kemp et al. (1972) on the total dry sediment (Fig. II.2.8) and Brown et al. (1972) on the humic acids. As a result of nitrogen expulsion, the ammonia nitrogen content in pore water increases. During the same interval carboxylic and aliphatic groups are little affected and may relatively increase for this reason (Huc and Durand, 1974). A second step is much less known, as it covers the interval from tens to hundreds of meters, where adequately cored intervals are the exception. Over this interval the oxygen content is reduced: O/C ratio ranges from 0.3 to 0.6 in humic acids from young sediments and decreases to 0.1 to 0.2 in kerogen of relatively shallow ancient sediments (500–1000 m depth). The oxygen decrease is mostly related to elimination of carboxylic acid groups (Fig. II.2.9).

The succession of fulvic acid to humic acid to humin seems to be the most probable evolution with respect to organic matter derived from continental plants incorporated in deltaic or estuarine muds: there the initial content of humic

2.4 Insolubilization

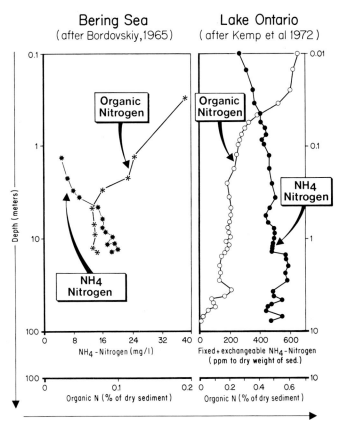

Fig. II.2.8. Organic and ammonia nitrogen in Recent sediments from the Bering Sea and Lake Ontario. (Data from Bordovskiy, 1965, Kemp et al., 1972)

compounds is high. However, a certain part of the organic matter may be incorporated in the sediment through a different pathway, which does not comprise the fulvic or humic acid stage. On the one hand, part of the insoluble fraction may be of purely detrital orgin, i.e., reworked from a previous organic deposit. On the other hand, there are geological situations where humic compounds amount to only a minor part of the organic matter in the young sediment. For instance in the central Black Sea, the upper level, which corresponds to the well-known "euxinic" environment (0–70 cm, 0–7000 years), contains only 10–15% of humic plus fulvic acids, whereas 85–90% of the organic matter is completely insoluble in alkali. A deeper sedimentary level (70–330 cm, 7000–13,800 years), which was laid down without an anoxic water column on top of it, contains less organic matter, but ca. 35% of it is composed of humic and fulvic acids, versus ca. 65% insoluble (Debyser and Pelet, 1977). These examples suggest that some autochthonous marine organic matter deposited in a very reducing environment never passed through the fulvic or humic acid stage. The same could apply to material mostly made of lipids, like some algal kerogens,

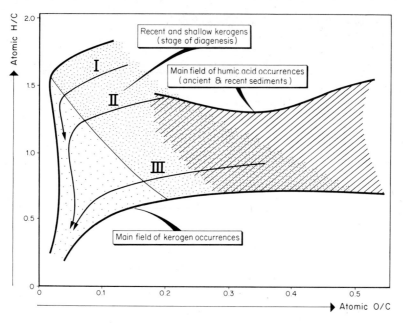

Fig. II.2.9. Elementary composition and main fields of occurrences of humic acids and kerogens from Recent or immature ancient sediments. The major types of kerogen *(I, II, III)* and their paths of evolution *(arrows)* are shown for comparison and will be described in Chapter II.4

such as the Coorongite of Australia. The respective fields of humic acids and immature kerogens, shown in Figure II.2.9, are in agreement with this interpretation.

Whatever the path of organic diagenesis may be, the result is the formation of a polycondensate. This is insoluble in alkali and it can be termed "humin", by analogy with humin from soils (Fig. II.2.4). It usually comprises most of the marine autochthonous organic matter in the depth range greater than 50 or 100 m. Huc and Durand (1977) analyzed ancient sediments from the diagenesis zone (burial depth 200–1000 m). They report that the insoluble fraction amounts to more than 90% of the marine-derived organic matter, whereas humic plus fulvic acids amount to less than 10%. The distribution is more variable in land-derived organic matter, where humic plus fulvic acids still amount from 5 to 70%. They also last down to a greater depth. For example, in the Upper Cretaceous of the Douala basin, Cameroon, extractable humic acids may still amount up to 5% of the total organic matter down to a depth of 1000 m.

The general structure of humin is already comparable to what is known from the structure of kerogen, which comprises most of the organic matter at depth. The main difference is that an important part (15 to 40%) of humin in young sediments is hydrolyzable. With increasing burial the hydrolyzable fraction decreases, as shown by Huc et al. (1977) on the Black Sea sediments. This observation indicates that some structural changes occur at depth towards a more resistant status of the organic matter. In kerogen of ancient rocks, the hydrolyzable fraction is insignificant. Therefore, humin should be considered as the precursor of the *kerogen*, as defined by petroleum geochemists.

2.5 Isotopic Composition of Organic Matter in Young Sediments

The distribution of carbon isotopes in humin depends on the original isotopic composition of the source of organic matter, and on isotope fractionation during the process which leads to humic compounds and humin. Galimov (1973, 1978) has shown that the isotope distribution in biogenic molecules is directly related to the chemical structures. For example, carbon of aliphatic chain and methoxyl group is depleted in ^{13}C, whereas carbon of carbonyl, carboxyl, phenolic and amine groups is enriched in ^{13}C, comparatively (Abelson and Hoering, 1961, Galimov, 1973, 1978, Vogler and Hayes, 1980). As a result, lipids are relatively lighter (i.e., enriched in ^{12}C) and proteins and carbohydrates heavier (i.e., enriched in ^{13}C). The variations are usually reported as

$$\delta^{13}C\text{‰} = \frac{(^{13}C/^{12}C)\text{ sample} - (^{13}C/^{12}C)\text{standard}}{(^{13}C/^{12}C)\text{standard}} \times 1000.$$

The most common standard is the "Peedee Belemnite" or PDB.

Isotopic variations are also observed with respect to hydrogen and its heavier isotope deuterium (D). The standard for hydrogen isotopes is "Standard Mean Ocean Waters" or SMOW. Similarly as for carbon, the isotope variations are expressed as

$$\delta D\text{‰} = \frac{(D/H)\text{ sample} - (D/H)\text{standard}}{(D/H)\text{standard}} \times 1000.$$

The main chemical constituents (lipids, carbohydrates, etc.) of terrestrial plants are usually isotopically lighter (i.e. depleted in ^{13}C) than those of marine plants (Silverman, 1967, Galimov 1978). A similar effect has not been observed with respect to hydrogen and deuterium (Stiller and Nissenbaum, 1980). For carbon, the global difference ranges from 5 to 10‰. This distinction reflects the isotopic composition of the carbon source for photosynthesis: marine plants utilize carbonate complexes in seawater, whereas terrestrial plants use atmospheric carbon dioxide, with a lower $\delta^{13}C$. The difference is reduced in kerogen of ancient sediment, but it still amounts to from 3 to 5‰. During the process of polycondensation and insolubilization, which leads from biogenic constituents to fulvic acid, humic acid and humin, there is a slight, but systematic, enrichment in the lighter isotope ^{12}C. This effect results from the elimination of carboxylic and other functional groups and from increasing polymerization (Abelson and Hoering, 1961, Galimov, 1973, 1978). There does not appear to be a significant fractionation of hydrogen isotopes during this process (Redding et al., 1980).

Generation of biochemical methane results in an important isotope fractionation. The $\delta^{13}C$ of biogenic methane is about 30 to 50‰ lower than the $\delta^{13}C$ of the organic matter, i.e., the methane is enriched in the light ^{12}C isotope. On the contrary, the carbon dioxide also generated by microorganisms, is enriched in ^{13}C. Thus, the isotope composition of kerogen is little affected by this process. Furthermore, it has been observed that generally carbon isotopic composition of kerogen from ancient sediments is in the same range as it is in young sediments. The generation of biochemical methane results in an important fractionation of hydrogen isotopes also. The hydrogen in the biogenic methane is depleted in deuterium by approximately 160‰ compared to the deuterium concentration of the associated waters which are the most likely source for hydrogen (Schoell, 1980).

A special remark concerns the possibility of methane oxidation in near-surface conditions, with changes of the isotopic composition which may result in misinterpretation of the origin of this gas. Methane-oxidizing bacteria are widespread in young sediments and soils. Coleman et al. (1981) have shown that cultures of these bacteria fractionated both carbon and hydrogen isotopes, resulting in the residual methane being enriched in the heavy isotopes ^{13}C and D. In particular the change in the δD value of methane is ca. 10 times greater than the change in the $\delta^{13}C$ value. Thus, if biogenic methane migrates into an aerobic environment and becomes partially oxidized by bacteria the residual gas might show a $\delta^{13}C$ comparable to that of thermogenic gas; however, comparison of $\delta^{13}C$ and δD should in many case allow identification of this situation (Coleman et al., 1981).

A comparable observation was reported by Doose (1980) in young marine sediments from California. Above the methane-generation zone (which begins at 1.5 m depth) the quantity of methane decreases to very low values in the sulfate-reducing zone and this methane is enriched in ^{13}C: the $\delta^{13}C$ value may reach -20‰ in the upper slice of sediment, as compared to -80 or -90‰ between 1.5 and 2 m. At the same time, the isotope ratio of the bicarbonate becomes enriched in ^{12}C. These data are interpreted as evidence for anaerobic methane oxidation by sulfate-reducing bacteria.

These remarks should be kept in mind when the occurrence of low quantities of methane in marine sediments, soils or ground waters is to be interpreted in terms of origin and significance for hydrocarbon prospection or inland waters protection.

2.6 Result and Balance of Diagenesis

As a result of microbial activity in water and in subaquatic soils, biogenic polymers have been degraded, then used as much as possible for the metabolism of microorganisms. Thus, even in fine muds, a part of the organic matter has been consumed and has disappeared through conversion into carbon dioxide and water. Another part has been used to synthesize the constituents of the microbial cell, and thus is reintroduced into the biological cycle. The residue unassimilable by microorganisms is now incorporated in a new polycondensate, which is insoluble: *kerogen* (Fig. II.2.10). This chemical process occurs under mild temperature and pressure conditions: thus the influence of the increase of

2.6 Result and Balance of Diagenesis

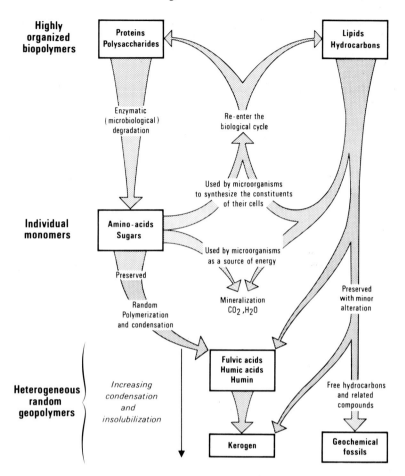

Fig. II.2.10. Fate of organic material during sedimentation and diagenesis, resulting in two main organic fractions: kerogen and geochemical fossils

temperature and pressure is likely to be subordinate, compared to the nature of the original organic constituents. This view is confirmed by the results of experimental evolution tests: heating organic matter under inert atmosphere is able to simulate the transformations occurring at greater depths — catagenesis and metagenesis — but not diagenesis.

Besides kerogen, organic matter still comprises at the end of diagenesis a minor amount of free hydrocarbons and related compounds. They have been synthesized by living organisms and incorporated in the sediment with no or minor changes. Thus, they can be considered as geochemical fossils, witnessing the depositional environment.

Several authors have studied the relationship between the original amount of organic matter deposited in a fresh sediment and the amount that was finally preserved when the sediment was fossilized (Strakhov and Zolmanzen, 1955, Bordovskiy, 1965, Hartmann et al., 1973). Their calculations were based on the determination of the expenditure of organic matter for the reduction of iron compounds. The results show that from 15 to 50% of the original organic matter may be lost by microbial decay.

It has been observed that many fine-grained sediments which are rich in organic matter, such as Eocene Green River Shale and Lower Toarcian of Western Europe, are extensively laminated down to a millimeter scale. This lamination can be preserved only if the sediments were not disturbed by benthic burrowing organisms. Burrowing organisms, when present, would intensively rework the top sedimentary layers. In addition to consuming organic matter they also allow molecular oxygen to come in contact with sedimentary or organic carbon. Thus, they favor further aerobic degradation. On the contrary, the absence of benthonic fauna would favor preservation. Therefore, anoxic conditions in the water column, as in the Black Sea today, enhance the preservation of organic matter, not only because of the reductive character, but also because those conditions are hostile for benthonic life. In this respect, an abundance of benthonic fossils in a sedimentary rock might suggest an unfavorable environment for petroleum generation, as pointed out by Hedberg (1964).

Biogenic methane is the single hydrocarbon which may be generated in abundance over the diagenetic stage of organic matter evolution.

Summary and Conclusion

Diagenesis of organic matter leads from biopolymers (proteins, lipids, carbohydrates and lignin synthesized by plants and animals) to geopolymers collectively called kerogen, which is the main organic material in ancient sediments.

Biopolymers such as proteins and carbohydrates are first degraded by microorganisms into individual amino acids and sugars, some of which are used for their nutrition. The residue not used by microbes recombines by polycondensation and polymerisation to form brown compounds comparable to fulvic and humic acids. These transformations occur mostly at or near the water sediment interface.

Insolubilization of the previous constituents occurs over the first tens or hundreds of meters, during progressive burial of sediments. Increasing polycondensation and loss of functional groups are responsible for a progressive insolubilization leading from fulvic to humic acids and finally to kerogen.

Biogenic methane is the single hydrocarbon which may be generated in abundance during diagenesis.

Chapter 3
Geochemical Fossils and Their Significance in Petroleum Formation

3.1 Diagenesis Versus Catagenesis: Two Different Sources of Hydrocarbons in the Subsurface

Diagenesis in young sediments results in two main organic fractions of very different quantitative importance: *kerogen* amounts to the bulk of organic matter, whereas some *free molecules* of lipids include hydrocarbons and related compounds. These molecules have been synthesized by living organisms and get trapped in the sediment with no or only minor change (Fig. II.3.1). They comprise specific compounds of relatively high molecular weight (Fig. II.3.2), and can be considered as fossil molecules, or *geochemical fossils*. Such molecules will be considered later in greater detail (Sect. 3.3 ff.) They represent a first source of hydrocarbons in the subsurface.

Kerogen also includes lipid material bound to the polycyclic network which will be described in the next chapter. From the kerogen, hydrocarbons are generated at depth, during catagenesis, as a result of the thermal degradation of kerogen (Fig. II.3.1). To a certain extent some of these hydrocarbons also resemble the original biogenic structure, but most of the newly formed hydrocarbons are of a more simple structure. They are metastable under subsurface conditions and comprise alkanes and 1- or 2-ring cycloalkanes or aromatics of low to medium molecular weight (Fig. II.3.2). They represent a second source of hydrocarbons in the subsurface.

The question that immediately arises is whether both sources of hydrocarbons contribute to petroleum accumulations, and which is the main source. Some contribution of the molecules inherited from living material is well established by the occurrence of characteristic molecules, like steroids, triterpenoids, etc. in Recent and ancient sediments and in crude oils. Some authors have expressed the view that the first source of hydrocarbons is the main one, and that most or all hydrocarbons of a crude oil were formed at an early stage of sedimentary history. That opinion prevailed for some time among researchers engaged in petroleum exploration, stimulating many disputes about the conditions of petroleum generation, its depth and time. The alternatives of this discussion were:

- petroleum generation at shallow depth: *diagenesis* (first tens or hundreds of meters) versus *catagenesis,* at a depth greater than 1000 m,
- *early* generation immediately following deposition of sediment versus *late* generation at a time when a lot of the overlying reservoir and cover rocks are deposited,

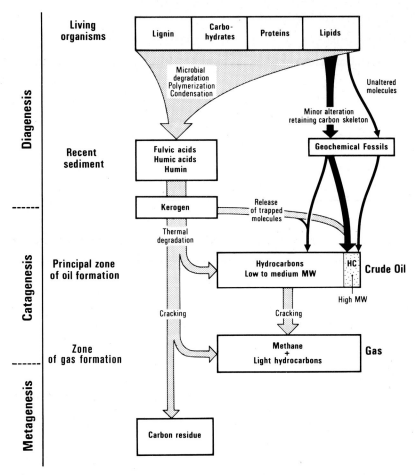

Fig. II.3.1. Sources of hydrocarbons in geological situations, with regard to the evolution of organic matter. Geochemical fossils represent a first source of hydrocarbons in the subsurface *(black solid arrows)*. Degradation of kerogen represents a second source of hydrocarbons *(grey dotted arrows)*

– *bacterial* origin of petroleum, as microorganisms are very active in the shallow freshly deposited sediment, versus *thermal* or *thermocatalytic* degradation of kerogen at depths, where microorganisms are no more active.

Quantitative and qualitative considerations have now led to the conclusion that hydrocarbons which are more or less directly inherited from living organisms are definitely part of crude oils, but they represent only a minor part. The bulk is generated later at a depth where thermal degradation of kerogen becomes important (Fig. II.3.3):

a) *Quantitative evaluations* on hydrocarbons in Recent sediments have been made by Smith (1954), Hunt (1961, 1972) and Blumer (in: Philippi, 1965), and are shown in Table II.3.1. All the values for coastal basins, continental shelf and open

3.1 Diagenesis Versus Catagenesis: Two Different Sources of Hydrocarbons

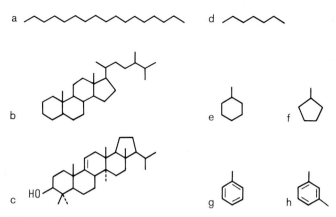

Fig. II.3.2a–h. Examples of petroleum molecules from two different sources in the subsurface: *Left:* (a–c) examples of hydrocarbons and related molecules of biological origin: (a) *n*-heptadecane: *n*-alkane occurring in algae, (b) ergostane: sterane deriving from a precursor sterol present in yeast, (c) isoarborinol: triterpenoid alcohol from a tropical plant. *Right:* (d–h) examples of low molecular weight hydrocarbons, without biogenic structure, generated at depth from the degradation of kerogen: (d) *n*-heptane, (e) methylcyclohexane, (f) methylcyclopentane, (g) toluene, (h) xylene

marine basins are in the range of 20 to 100 ppm (on dry weight basis of rock). The only exceptions are encountered in peats and in some reducing basins or trenches close to the coast line: 100 ppm off the California coast, 100 to 140 ppm in the Cariaco Trench (Venezuela). In deep oceanic cores from the Deep Sea Drilling Project, hydrocarbon yields range from 10 to 40 ppm (Hunt, 1972).

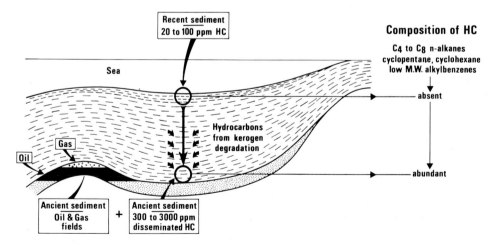

Fig. II.3.3. Differences of the hydrocarbon content in Recent and ancient sediments. The low hydrocarbon content of the Recent sediments is insufficient to account for the disseminated hydrocarbons in ancient sediments plus the pooled oil and gas

Table. II.3.1. Average values of hydrocarbons, non-hydrocarbon bitumen, and organic carbon in Recent sediments

Origin	Type of sediment	Number of samples	Extractable bitumen (ppm)		Organic carbon (%)	Average HC / Average Org.C (mg/g)	Author
			HC	non HC			
Soils (Texas)		–	60	–	1.2	5	Philippi (1965)
Peats (Florida)		–	350	–	37.0	0.9	Philippi (1965)
Freshwater lake (Louisiana)	Continental	–	41	–	0.3	14	Philippi (1965)
Mud flat (Sabine Pass, Texas)		–	23	–	0.9	3	Philippi (1965)
Laguna Madre, Texas	Coastal	–	20	–	0.4	5	Philippi (1965)
Louisiana		2	30	735	–	–	Smith (1954)
California (estuary pond)		1	800	6800	–	–	Smith (1954)
Lake Maracaibo		8	68	1060	2.2	3	Hunt (1961)
Mississippi		–	80	–	1.2	7	Philippi (1965)
Mississippi		–	80	–	0.8	10	Philippi (1965)
Mississippi		4	65	860	–	–	Smith (1954)
Louisiana	Deltaic	–	20	–	0.8	2	Philippi (1965)
Pelican Is (0–100 ft)		3	38	210	0.5	8	Smith (1954)
Orinoco		3	55	709	4.9	1	Smith (1954)
Orinoco		10	60	555	0.9	7	Hunt (1961)
Grande Ile (18–53 ft)	Open Marine	–	65	255	–	–	Smith (1954)
Gulf of Mexico		10	32	275	0.5	6	Hunt (1961)
Gulf of Mexico		–	30	–	0.9	3	Philippi (1965)
Mediterranean sea		1	29	461	0.7	4	Hunt (1961)
Cuba	Carbonate muds	10	40	575	1.4	3	Hunt (1961)
Florida		–	20	–	1.2	2	Philippi (1965)
DSDP (Various locations)	Deep marine	–	10–40	–	0.1–0.3	–	Hunt (1972)
Chubasco Trench		–	23	–	1.7	1	Philippi (1965)
Off W. Africa		4	31	376	1.5	2	Smith (1954)
Cariaco Trench		16	105	1250	2.0	5	Hunt (1961)
Cariaco Trench		–	140	–	3.4	4	Philippi (1965)
Off California		–	100	–	2.1	5	Philippi (1965)

3.1 Diagenesis Versus Catagenesis: Two Different Sources of Hydrocarbons

Table II.3.2. Average values of hydrocarbons, non-hydrocarbon bitumen, and organic carbon in ancient nonreservoir sediments

Origin	Type of rock	Number of samples	Extractable bitumen (ppm) HC	Extractable bitumen (ppm) non HC	Organic carbon (%)	$\dfrac{\text{Average HC}}{\text{Average Org.C}}$ (mg/g)	Author
200 formations from 60 sedimentary basins	shales	791	300	600	1.65	18	Hunt (1961)
	carbonates	281	340	400	0.18	151	Hunt (1961)
Rocks of	shales	very large	180		0.9	20	Vassoevich (1967)
various	silts		90		0.45	20	Vassoevich (1967)
origins	carbonates		100		0.2	50	Vassoevich (1967)
Source rocks	all types	668	860	780	1.82	47	Institut Français du Pétrole (unpublished)
from 18 sedimentary basins	shales and silts	418	930	810	2.16	43	
	calcareous shales	97	1260	1220	1.90	66	
	carbonates	118	335	440	0.67	50	
9 source rocks formations	not specified	not specified	267 to 2360		0.53 to 3.67	34 to 101	Philippi (1965)

Disseminated hydrocarbon contents in ancient nonreservoir sediments are shown in Table II.3.2. They average 300 ppm in shales and 340 ppm in carbonates, according to Hunt (1961). In petroliferous series, values observed by Philippi (1965) in nonreservoir rocks are in the range of 300 to 3000 ppm. From our evaluations, based on 668 samples from source rocks of 18 petroliferous basins, the average hydrocarbon content is above 800 ppm.

The hydrocarbon yield of Recent and ancient petroliferous sediments is compared in Figure II.3.3. It is evident that the hydrocarbon content of Recent sediments is by no means able to explain the presence of oil and gas fields plus the hydrocarbons of disseminated bitumen in nonreservoir rock of petroliferous sequences. Thermal degradation of kerogen during catagenesis, resulting from burial of sediments to a depth level of one to several thousand meters, has generated the greater part of crude oil hydrocarbons (perhaps 90%), and probably also a large part of those hydrocarbons constituting natural gas.

An alternative hypothesis would be the generation of hydrocarbons at depth from the nonhydrocarbon fraction extractable from young sediments, e. g., acids. However, these compounds are also biogenic, and their distribution is roughly comparable to that of the hydrocarbons of Recent sediments. Thus, the qualitative aspects, discussed below, apply as well to this hypothesis.

b) *Qualitative composition* of hydrocarbon in Recent sediments compared to crude oils, provides a confirmation of the fact that petroleum hydrocarbons are mainly generated later and not directly inherited from organisms. Low molecular weight alkanes (C_4–C_8) and cycloalkanes (cyclopentane and cyclohexane) are almost absent from recent clay and carbonate sediments (Dunton and Hunt, 1962; Erdman, 1965), although very small amounts have locally been detected in marine core samples of shallow depth (Hunt, 1974). The same applies to low molecular weight alkyl-benzenes and alkyl-naphthalenes (Erdman, 1958, 1961, 1965). All these molecules are important constituents of crude oils and of ancient rock bitumen, even in Upper Tertiary beds where sufficiently buried. A few polynuclear aromatics such as perylene, chrysene and fluoranthene are present in Recent sediments, with little or no substitution, whilst crude oils show a wide variety of substituted polyaromatics (Aizenshtat, 1973; Tissier and Oudin, 1973).

Hydrocarbons already present in Recent sediments are the source of a very minor fraction of crude oils. Nevertheless, they are of great interest to us, as they represent the heritage from living organisms and may be regarded as geochemical fossils, being helpful in defining biological environments. Hydrocarbons generated later, at depth, through thermal degradation of kerogen, are the source of the bulk of crude oils; however, they are likely to provide only limited information on the original organic material.

3.2 Hydrocarbons Inherited from Living Organisms, Directly or Through an Early Diagenesis: Geochemical Fossils (Biological Markers)

The occurrence of hydrocarbons and closely related substances is known from a wide range of living organisms: many of them have been found in Recent and ancient sediments and in crude oils. Treibs (1934), in his pioneering work, identified porphyrins in crude oils and suggested that they may originate from chlorophyll of plants. Smith (1952, 1954, 1955) showed that some hydrocarbons are incorporated in the Recent sediments of the Gulf of Mexico. Since that time, numerous types of hydrocarbons or substances with a similar carbon skeleton have been traced from the living organisms into Recent sediments: for example Erdman (1958 and 1965) studied lipids, sterols and carotenoids in the Mississippi delta, in marshes and in offshore sediments of Southern California; Kidwell and Hunt (1968) found saturated and aromatic hydrocarbons in Recent sediments of the Pedernales area in Orinoco delta; Meinschein (1959, 1961) reported saturated hydrocarbons, including steranes and triterpanes, and polycyclic aromatics from sediments of the Gulf of Mexico; Emery (1960) investigated gaseous and liquid hydrocarbons in sediments of Southern California; Blumer et al. (1963, 1964) isolated pristane from Recent marine sediments, and traced it back to the phytol chain of chlorophyll through zooplankton fed on phytoplankton. Many others, since that time, have found various types of hydrocarbons or closely related molecules in young sediments.

Such molecules may be derived from terrestrial (mostly plants), marine pelagic (mostly plankton), marine benthonic (algae, bacteria and other microbes), or

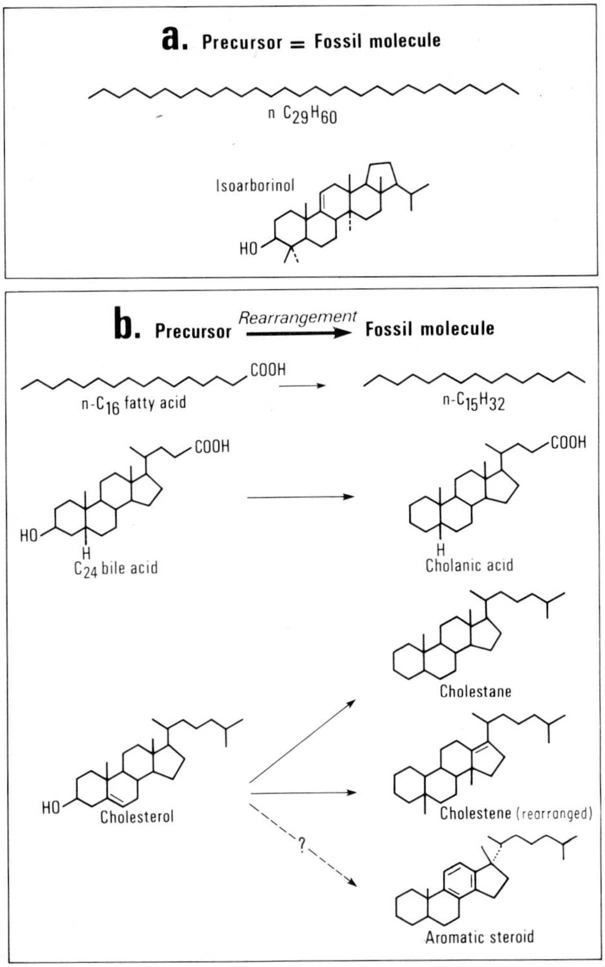

3.2 Hydrocarbons Inherited from Living Organisms: Geochemical Fossils

Fig. II.3.4a and b. Examples of geochemical fossils (a) molecules of biologic origin, without chemical alteration; (b) molecules inherited with minor changes. (Occurrences reported by Albrecht and Ourisson, 1969; Seifert, 1973; Rubinstein et al., 1975; Spyckerelle, 1975)

limnic life. The carbon skeleton of the molecules is preserved and they can easily be linked to some major structural types: steroids, cyclic or acyclic terpenoids, etc. They stem either directly from living organisms, i.e., without any chemical alteration, or indirectly with only minor changes which occur mostly during diagenesis in the young sediment: loss of functional groups, stabilization of the molecule by hydrogenation, aromatization, etc. (Fig. II.3.4). We propose to use the expression "geochemical fossil" to designate a molecule synthesized by a plant or an animal: the molecule being unchanged or having suffered only minor subsequent changes, with preservation of the carbon skeleton. These molecules are also called "biological markers".

At depth, the biogenic molecules suffer not only thermal degradation, but also dilution with newly formed hydrocarbons resulting from kerogen degradation. However, identification of the most stable inherited molecules remains possible for a long time and depth span, i.e., into the ancient sediments and crude oils on the following grounds. Silverman (1967) and Welte (1969b, 1972) have demonstrated by means of isotope measurements and optical rotation that the fraction of crude oils boiling between 400 and 500°C, which corresponds to molecular weights between 350 and 450, has the highest concentration of these fossil molecules. It includes, in particular, tetracyclic and pentacyclic structures with 26 to 35 carbon atoms and *n*-alkanes in the same range of molecular weight. The geochemical fossils are usually restricted to a small number of predominant molecules of each type. On the contrary, molecules issued from kerogen degradation are mostly abundant in the low to medium molecular weight range (below 25 carbon atoms) and cover a wide spectrum of each structural type.

In some cases certain classes of compounds of biogenic origin, e.g., straight-chain compounds, isoprenoids, terpenes, sterols or porphyrins, may be linked to the young kerogen by relatively weak bonds, and later released during the initial stage of kerogen degradation (see Chaps. II.3.13 and II.5.4).

The occurrence of biological markers in sediments and crude oils and the comparison with molecules synthesized by living organisms has been discussed many times and particularly by Eglinton and Calvin (1967), Maxwell et al. (1971), Blumer (1973) and Whitehead (1973). The laboratory and field simulations may help to determine the fate of biogenic compounds in geological situations (Eglinton, 1973). The main classes of compound will be considered successively, with respect to origin and occurrence of specific molecules.

3.3 *n*-Alkanes and *n*-Fatty Acids

3.3.0 Various sources of linear aliphatic chains in organisms have already been discussed (Chap. I.4.4). The distribution of those molecules bears the imprint of their biochemical synthesis, i.e., predominance of certain medium to high molecular weight compounds with specific carbon numbers, like fatty acids with an even number of carbon atoms, or *n*-alkanes with an odd number of carbon atoms. The preservation of such molecular features in ancient sediments is commonly observed, although it is progessively obliterated with increasing depth of burial and age.

A statistical study on the occurrence of odd and even predominance of higher *n*-alkanes ($> C_{20}$) based on 1302 ancient sediment and crude oil samples is presented in Table II.3.3, according to Tissot et al. (1977). From the table it can be seen that the existence of an odd-carbon-number predominance is rather common. An even predominance is less frequent, but by no means exceptional.

3.3.1 Odd-carbon-numbered n-alkanes of high molecular weight ($n\text{-}C_{25}\text{--}n\text{-}C_{33}$) are frequently reported from Recent detrital sediments with an important contribution of continental runoff and comprising both clay minerals or silts and

3.3 n-Alkanes and n-Fatty Acids

Table II.3.3. Odd and even-carbon-number predominances of higher n-alkanes in sediments and crude oils. (Data from Tissot et al., 1977)

Age		Paleozoic and Triassic	Jurassic	Cretaceous	Tertiary	All ages[a]
Number of samples		231	185	589	289	1302
Odd predominance	Moderate	38	46	200	124	408
	Strong	2	11	61	65	139
Even predominance	Moderate	4	7	15	24	50
	Strong	1	1	2	11	15

[a] 8 samples: age unknown.

organic material from plants (Fig. II.3.5). They have been found by Bray and Evans (1961) in various Recent sediments from offshore Southern California, bays and continental shelf of the Gulf of Mexico and freshwater lakes. The predominance of molecules with an odd number of carbon atoms can be measured by the "Carbon Preference Index" (CPI), i.e., the ratio, by weight, of odd to even molecules[2]. The ratio varies from 2 to 5.5 in the various samples studied by these authors. Other studies of Recent sediments from offshore Louisiana (Meinschein, 1961), San Francisco Bay (Kvenvolden, 1962) and in numerous other locations have confirmed these figures.

---

2 Bray and Evans (1961) computed the CPI over the C_{24}–C_{33} range, i.e.,

$$\text{CPI} = \frac{1}{2}\left[\frac{C_{25} + C_{27} + C_{29} + C_{31} + C_{33}}{C_{24} + C_{26} + C_{28} + C_{30} + C_{32}} + \frac{C_{25} + C_{27} + C_{29} + C_{31} + C_{33}}{C_{26} + C_{28} + C_{30} + C_{32} + C_{34}}\right];$$

Philippi (1965) preferred to use the ratio

$$\frac{2C_{29}}{C_{28} + C_{30}};$$

Scalan and Smith (1970) introduced another coefficient, which is a sort of moving average, centered on C_{i+2}, and showing the variation of the odd–even preference with the increasing molecular weight

$$\text{OEP} = \left[\frac{C_i + 6C_{i+2} + C_{i+4}}{4C_{i+1} + 4C_{i+3}}\right](-1)^{i+1}.$$

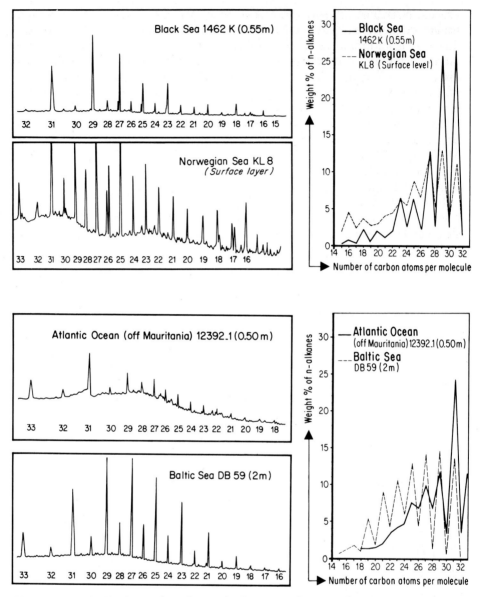

Fig. II.3.5. Distributions of *n*-alkanes in Recent sediments of various origin. Gas chromatograms of saturated hydrocarbons *(left)* showing the peaks of *n*-alkanes, with their carbon numbers. *n*-Alkane distribution curves derived from gas chromatograms *(right)*. (After Dastillung, 1976, Debyser et al., 1977)

These hydrocarbons are derived from cuticular waxes of the continental higher plants. Odd-carbon-numbered *n*-alkanes may be either directly synthesized by the plants or derived through an early diagenesis (defunctionalization) from the even-numbered acids, alcohols or esters. When both marine and continental

3.3 n-Alkanes and n-Fatty Acids

organic matter are incorporated in a sediment, the continental contribution usually defines the n-alkane fingerprint, especially in the C_{25}–C_{33} range. This is due to a higher proportion of the n-alkanes in continental organic matter than in marine planktonic matter. This explains why so many Recent marine sediments display a "continental" fingerprint (Fig. II.3.5). However, the observation is not necessarily valid for areas of pure carbonate deposition without terrestrial contribution, as no odd-carbon-number predominance (with very rare exception) is found in the alkanes higher than n-C_{18} from contemporaneous marine organisms (Koons et al., 1965; Clark and Blumer, 1967). As a result, Powell and McKirdy (1973) have shown the absence of odd-carbon preference in that range of molecular weight, even at shallow depth, in Tertiary carbonate series with little or no contribution from continental plants deposited around Australia.

In ancient detrital sediment (Fig. II.3.6), the important predominance of the original odd n-alkanes of high molecular weight is preserved in relatively shallow shales and silts, where degradation of kerogen has not yet started, and where the hydrocarbon content remains low: in a Lower Cretaceous (Aptian) shale from

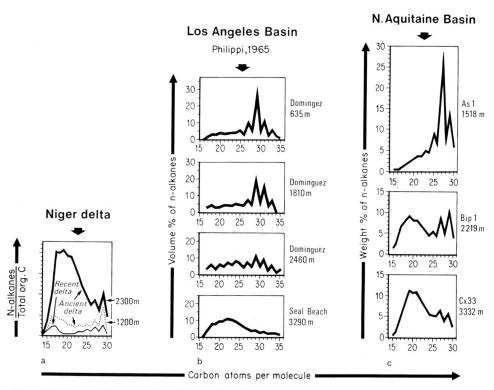

Fig. II.3.6a–c. Change of n-alkane distributions with depth (a) Recent and fossil sediments from Niger delta. Samples noted 1200 and 2300 m are from the ancient Tertiary delta. Values are related to organic carbon (Pelet, 1974). (b) Upper Miocene to Pliocene, Los Angeles Basin, California. Values are expressed as percent of n-alkanes (Philippi, 1965). (c) Lower Cretaceous, North Aquitaine Basin, France. Values are expressed as percent of n-alkanes (Tissot et al., 1972)

1518 m depth in North Aquitaine basin, n-C_{27} is still five times more abundant than n-C_{26} or n-C_{28} (Tissot et al., 1972); in an Upper Cretaceous shale from Gabon, West Africa, at 1155 m depth, n-C_{29} is four times more abundant than n-C_{28} or n-C_{30}; in a fossil *Equisetum* of Triassic age, Knoche and Ourisson (1967) found that n-C_{23}, n-C_{25}, n-C_{27} (predominant) and n-C_{29} were the four main components of the hydrocarbons in almost the same proportions as they are in contemporaneous *Equisetum*. However, such an important predominance of several successive odd n-alkanes seems infrequent in rocks of that age. It is more usually encountered in Lower Cretaceous and younger beds, on the basis of our observations on 1302 samples from 18 sedimentary basins in Table II.3.3 (Tissot et al., 1977). This fact may be due to an increasing maturity of older rocks, but may also derive from an increase in diversity and abundance of higher plants associated with the development of angiosperms in the Lower Cretaceous time. Also, it is perhaps favored by the occurrence of important detrital series in continental margins during Cretaceous and Tertiary.

Furthermore, odd n-alkanes derived from higher plants occur also in oil shales like the Green River shales (Robinson et al., 1965), the very shallow Messel and Bouxwiller shales in the Rhine valley, or the Ménat shales in Central France with CPI from 2.5 to 30 (Albrecht, 1969; Arpino, 1973). An odd-carbon preference of the n-alkanes is also known from lignites (Arpino, 1973) and coals (Welte, 1967; Leythaeuser and Welte, 1969; Brooks, 1970).

In crude oils, the high molecular weight n-alkanes inherited from terrestrial plants are normally diluted by hydrocarbons from kerogen degradation, and the CPI is around 1.0. However, some oils, probably derived mainly or solely from

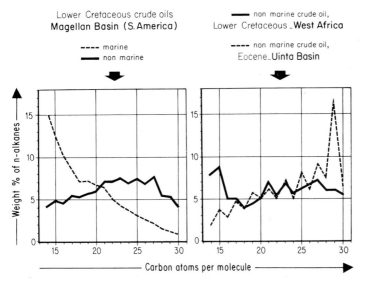

Fig. II.3.7. n-Alkanes distribution curves of three crude oils derived from continental plant material, probably extensively reworked by bacteria. On the *left*, a common type of crude oil, derived from marine organic matter, is compared to the nonmarine crude oil occurring in the equivalent formation of the same basin

3.3 n-Alkanes and n-Fatty Acids

terrestrial organic material which has been reworked by bacteria and other microbes, still show a large amount of high molecular weight n-alkanes ($> C_{20}$) with a moderate odd predominance. Figure II.3.7 shows some examples of such oils found in Lower Tertiary of the Uinta basin, USA, Lower Cretaceous from parts of the Magellan basin, South America, and Lower Cretaceous from West Africa.

3.3.2 Odd-carbon-numbered n-alkanes of medium molecular weight, mostly n-C_{15} and n-C_{17}, may also represent in some cases a direct heritage from the hydrocarbons present in algae and from the related acids. In ancient rocks the distinction is possible only when n-C_{15} and/or n-C_{17} are largely predominant compared to n-C_{14}, n-C_{16} and n-C_{18} (Fig. II.3.8). This situation has been found in Lower Cretaceous rocks from the coastal basin of West Africa (Spyckerelle, 1973), in some Devonian beds and in the Colorado Group of the northeastern flank of the western Canadian basin (Alberta), and in the Upper Jurassic from the northeastern shelf of the Aquitaine basin. Most of these locations are consistent with the hypothesis of a shelf that is covered with benthonic algae. In upper beds of the Green River shales (Mahogany beds), n-C_{17} is the main n-alkane constituent in the medium molecular weight range.

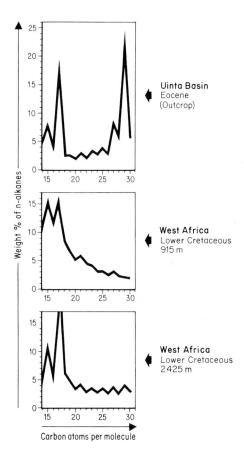

Fig. II.3.8. n-Alkane distribution curves observed in Eocene sediments of the Uinta Basin and Lower Cretaceous of West Africa. The predominance of n-C_{15} and n-C_{17} is derived from algal contribution. The organic matter from Uinta Basin shows a mixture of n-alkanes from algae (C_{15}, C_{17}) and heavier n-alkanes from higher plants (C_{27}, CC_{29})

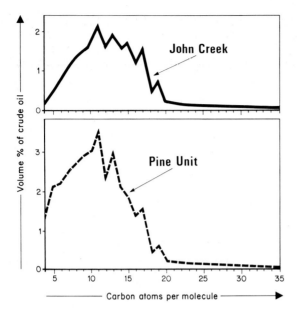

Fig. II.3.9. Predominance of odd numbered n-alkanes of medium molecular weight in two crude oils. (After Martin et al., 1963)

In several oils of lower or middle Paleozoic age from the United States, Martin et al. (1963) found an unusual distribution with some predominance of the odd n-alkanes from C_{11} to C_{19} (Fig. II.3.9).

3.3.3 Several cases of even-carbon-number predominance are reported from carbonate or evaporite sediments and occasionally from crude oils (Fig. II.3.10). Paleocene and Eocene carbonates from South Tunisia show a systematic and important predominance of even n-alkanes from n-C_{16} to n-C_{30}. Similar observations have been made in some other sediments: Zechstein from NW Germany (Welte and Waples, 1973), a few samples of the Upper Jurassic from north Aquitaine basin (n-C_{20} to n-C_{30}), Corbicula shales of Oligocene age from the Rhine valley (Welte and Waples, 1973). A few Tertiary crude oils from the Mediterranean area show the same character, but this is rather infrequent in crude oils: Amposta, Spain (Albaiges and Torradas, 1974) and Greece (Tissot et al., 1977). Douglas and Grantham (1974) reported a similar distribution in Wurtzilite, a native bitumen from the Uinta basin.

This particular type of distribution is usually associated with a high phytane: pristane ratio, whereas an equal abundance of phytane and pristane or a pristane predominance is much more frequent in other sediments. Welte and Waples (1973) suggest that in very reducing environments, reduction of n-fatty acids, alcohols from waxes, and phytanic acid or phytol is prevalent over decarboxylation, resulting in predominance of even-carbon-numbered n-alkane molecules over odd molecules (CPI < 1) and a predominance of phytane over pristane. In less reducing environments, decarboxylation results in a majority of odd n-alkanes (CPI > 1) and a predominance of pristane over phytane. A typical example of reduction is given by Knoche et al. (1968), who found in a fossil plant *Voltzia* (Coniferales) a hydrocarbon fraction consisting almost wholly of one

3.3 n-Alkanes and n-Fatty Acids

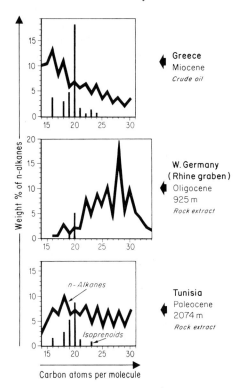

Fig. II.3.10. Predominance of even n-alkanes in rock extracts from the Lower Tertiary of Tunisia and the Rhine Valley, and in a Miocene crude oil from Greece. *Vertical bars* are isoprenoid hydrocarbon distributions

component: n-$C_{28}H_{58}$. In waxes of closely related living plants, n-$C_{28}H_{58}$ is never an important component, but the n-C_{28} alcohol is present and could be the precursor of the n-C_{28} alkane.

Shimoyama and Johns (1972) proposed an alternative interpretation of the even n-alkane predominance. They carried out experimental degradation of n-fatty acids in the presence of montmorillonite or calcium carbonate. The two minerals seem to have different catalytic effect: decarboxylation with loss of one carbon atom is favored by montmorillonite, while beta cleavage with loss of two carbon atoms is favored by calcium carbonate. Therefore, an original even-carbon-numbered fatty acid would generate mostly an odd n-alkane in shales and an even n-alkane in carbonates.

Actually, the two interpretations proposed are not conflicting, as the mechanism proposed by Welte would occur during diagenesis in young sediment, whilst the interpretation of Shimoyama and Johns refers mostly to the catagenetic transformations occuring at depth. However, two remarks suggest that reduction is a more frequent process than beta cleavage:

- the comparison between the fossil plant and the present equivalent, made by Knoche et al. (1968),
- the association of even-numbered n-alkanes with the C_{20} isoprenoid (phytane) derived from the C_{20} alcohol (phytol), and not with the C_{18} isoprenoid.

In Recent sediments a certain even predominance limited to C_{16} and C_{18} n-alkanes has been observed (Simoneit, 1975; see also Fig. II.3.5, Norwegian Sea). The predominance does not extend to higher molecular weights, as heavier odd n-alkanes, inherited as such from plants are prevalent over those issued from an elimination of the functional group of acids or alcohols in young sediments.

3.3.4 Long-chain n-alkanes without odd or even predominance are an important part of organic matter in some ancient sediments, ranging from lower Paleozoic to Tertiary (Tissot et al., 1977). They also constitute an important fraction of many high-wax crude oils (Hedberg, 1968). The distribution of n-alkanes often extends up to C_{40} or C_{50}. They are probably derived from bacterial and other microbial waxes, possibly from reworking higher plant waxes. They are associated with a series of iso- and anteiso-alkanes which are typical of a bacterial origin (see Sect. 3.4): Figure II.3.11. In contemporaneous non-photosynthetic bacteria and yeast, Han et al. (1968) found that long-chain hydrocarbons with more than 20 carbon atoms amount to more than 50% of the nonaromatic hydrocarbons. Important accumulation of bacterial biomass possibly resulted from degradation of large quantities of higher plants in paralic or intracontinental basins, such as the Tertiary of Indonesia or the Eocene of the Uinta Basin.

3.3.5 n-fatty acids have been identified in numerous Recent and ancient sediments. When large proportions of terrestrial material accumulated in a restricted limnic environment, n-fatty acids with a range from C_{24} to C_{32} and with an even-carbon-number predominance are found. This is identical with the range of plant waxes. Messel and Bouxwiller shales (Arpino, 1973) are examples for this kind of facies. Where continental runoff is only subordinate, the main fatty acids range from C_{14} to C_{22} and are probably derived from fats of marine organisms. In that range, they show again an even-carbon-number predominance, e. g., C_{16}, C_{18} and sometimes C_{22} in Recent calcareous muds from the Persian Gulf (Welte and Ebhardt, 1968), C_{16} in muds from the Tanner basin, off southern California (Kvenvolden, 1970). A single fatty acid may even be the major constituent, such as C_{16} in Lower Cretaceous from Gabon (Spyckerelle, 1973). Usually both marine and continental sources contribute to the n-fatty acids distribution. This results in two ranges, C_{16}–C_{18} (from marine or continental source) and C_{24}–C_{30} (from continental source only). This is the case in Recent muds from the San Nicolas basin off southern California, from the Mississippi delta, in carbonate muds from the Florida bay (Kvenvolden, 1970), and in many oceanic sediments (Simoneit, 1975).

This situation is also frequent in ancient sediments, where the even predominance normally decreases with catagenetic evolution. In rocks older than Lower Cretaceous, the second mode of n-fatty acids (C_{24}–C_{32}) is poorly represented. This is the case even in samples where C_{14}, C_{16}, C_{18} are abundant, with an even: odd ratio of 3 or 4, indicating that thermal maturation has been only moderate. This situation occurs, for instance, in the Permian Bone Spring limestone and Lamar limestone from Texas (Kvenvolden, 1970). As a similar situation occurs in respect to n-alkanes, we can extend our previous observation: the accumulation of not only long-chain C_{25}–C_{33} n-alkanes with an odd or even predominance, but also even-numbered n-fatty acids in the same carbon range, occurs only occasionally in immature sediments older than Cretaceous. This is again related to the evolution of higher plants. On the contrary C_{14} to C_{20} n-alkanes and n-fatty acids, which are mostly derived from plankton and benthonic algae, are presumed to have been abundantly available since Precambrian time. Indeed their predominance is found in sediments ranging in age from lower Paleozoic to Recent.

3.3.6 Comparison between n-alkane and n-fatty acid distributions. On comparison, the distribution of n-alkanes, free fatty acids, and fatty acids bound to kerogen in young sediments are generally rather different. The distribution of n-alkanes is usually dominated by odd-numbered C_{25}–C_{33} alkanes. The fatty acids bound to kerogen contain mostly C_{16}–C_{18}, whereas free fatty acids may include both ranges of molecular weight, but they are comparatively less abundant. Obviously, a certain amount of n-alkanes is directly inherited from organisms, because they are known to occur as such in cuticular waxes, for example. The balance is likely to be derived from acids and/or alcohols, bound somehow to kerogen. The alkanes released from kerogen may be of various chain lengths and include abundant alkanes of medium molecular weight ($< C_{20}$).

3.4 Iso- and Anteiso-Alkanes

Iso- (2-methyl-) and anteiso- (3-methyl)alkanes are associated with n-alkanes in plant waxes, where they comprise a comparable number of carbon atoms (about 25–31) with an odd predominance. A distribution of isoalkanes comparable to their occurrence in plants has been reported by Arpino (1973) in an Eocene sediment from the Rhine valley. The iso- and anteisoacids are also known in bacterial waxes, where they cover a wider range of molecular weight. In the lower part of the Green River shales from the Uinta basin, 2-methyl- and 3-methyl-alkanes form a homologous series, without odd or even predominance, in the C_{20}–C_{30} range, sometimes extending up to C_{45}. They are probably of microbial origin. The same compounds commonly occur in the high-wax crude oils (Fig. II.3.11).

Iso- and anteiso-fatty acids of lower molecular weight represent a large fraction of bacterial acids, whereas they are much less abundant in higher organisms. They have been reported in the lipids of marine organisms and also in Recent sediments. The related isoalkanes were found by Johns et al. (1966) in ancient oils and sediments, and their occurrence may be rather frequent.

3.5 C_{10}-branched Alkanes

Several C_{10}-isoalkanes which are relatively abundant in some ancient sediments and crude oils, such as 2,6-dimethyloctane and 2-methyl-3-ethylheptane, may be derived from the terpene constituents of essential oils from higher plants, as suggested by Mair (1964) and Mair and Ronen (1966, 1967); (Fig. II.3.12). Although they are particularly abundant in the Ponca City crude oil (0.55 and

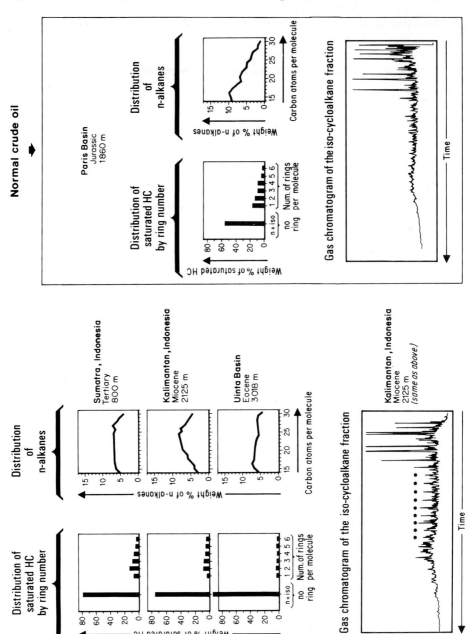

Fig. II.3.11. Examples of high-wax crude oils probably derived from continental organic matter extensively reworked by microorganisms. High molecular weight *n*-alkanes are abundant, but they do not show a marked odd or even predominance. Note the homologous series of 2-methyl and 3-methyl alkanes identified by *solid circles* on the gas chromatogram of the iso-cycloalkane fraction *(bottom)*. A normal crude oil from Paris Basin is shown for comparison

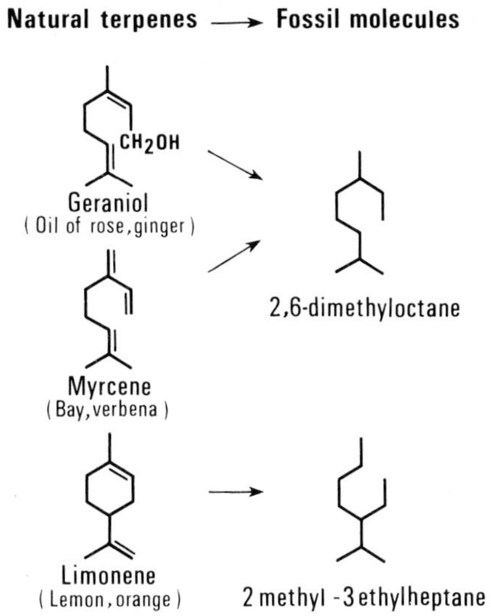

3.6 Acyclic Isoprenoids

Natural terpenes ⟶ Fossil molecules

Geraniol (Oil of rose, ginger) — CH₂OH

Myrcene (Bay, verbena)

⟶ 2,6-dimethyloctane

Limonene (Lemon, orange) ⟶ 2 methyl-3 ethylheptane

Fig. II.3.12. Possible origins of C_{10} branched alkanes from terpenes of essential oils

0.64% volume respectively) they have not been sought in a sufficient number of different crudes to appraise their value as geochemical fossils.

3.6 Acyclic Isoprenoids

A wide range of linear or cyclic compounds, formally built up from several isoprene units, is known in living plants and bacteria and to a smaller extent in animals. They occur as hydrocarbons, alcohols or other derivatives or in a

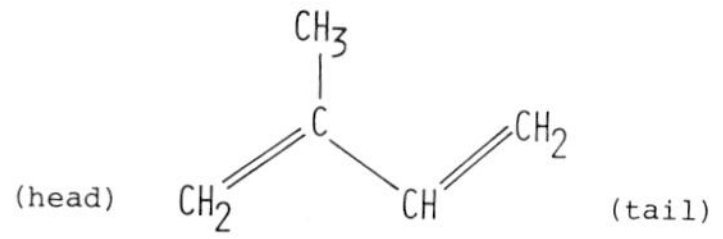

Isoprene unit (head) CH₂=C(CH₃)–CH=CH₂ (tail)

combined state, such as ester in the phytyl chain of chlorophyll (see Chap. I.4.2) or glycerol ethers in bacterial membrane. In sediments and crude oils, similar structural arrangements have been frequently reported, but they occur mostly as fully saturated aliphatic or alicyclic molecules, or they may include one or more aromatic cycles. Polycyclic structures will be dealt with in Sections 3.7 and 3.8.

Generally, in geochemistry the term "regular isoprenoid" is restricted to an acyclic, branched, saturated molecule with a methyl group on every fourth carbon atom, irrespective of the total number of carbon atoms: such arrangement implies a "head to tail" linkage of the isoprene groups. The best known and most common isoprenoids are pristane (2, 6, 10, 14 tetramethylpentadecane) and phytane (2, 6, 10, 14 tetramethylhexadecane). In addition a wide spectrum is already known from C_9 to C_{25}, and others might be detected in the future (Fig. II.3.13). Aside from the regular isoprenoids, the term "irregular isoprenoid" is used for some

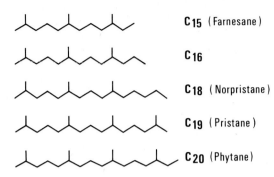

Fig. II.3.13. Principal molecules of regular isoprenoid hydrocarbons present in ancient sediments and crude oils

molecules with "head-to-head" or "tail-to-tail" linkage, such as squalane and lycopane, which are known to occur in certain geochemical environments, where they are probably derived from squalene and lycopene (Chap. I.4.2, Figs. I.4.9 and I.4.12).

In living organisms, the most widespread sources of isoprenoid chains seem to be the C_{20} phytol, a diterpene bound through an ester linkage in the chlorophyll molecule, and C_{15}, C_{20}, C_{30} and C_{40} isoprenoids from bacterial membranes. They will be successively discussed below (a) and (b).

Non-polar lipids of methanogenic bacteria also include a C_{25} irregular isoprenoid, the C_{30} squalene and its hydrogenation products up to squalane (Holzer et al., 1979). Other acyclic terpenes and sesquiterpenes (hydrocarbons, alcohols, aldehydes, ketones) are also widely known in plants and flowers. Sesterterpenes are rather rare (Spyckerelle, 1973), but squalane is an acyclic triterpene from shark liver oil, and lycopene is an acyclic tetraterpene from tomatoes and other fruits. In addition to these olefins, Clark and Blumer (1967) found pristane (C_{19} isoprenoid alkane) at a low concentration in marine benthonic algae, belonging to the Laminariales, and in planktonic algae. Zooplankton contains an assemblage of pristane, phytane (C_{19} and C_{20} isoprenoid alkanes) and various related alkenes, like zamenes and phytadienes, probably derived from phytol of the chlorophyll through feeding on phytoplankton (Blumer and Thomas, 1965, 1971).

In Recent marine sediments, pristane and phytane are present, their abundance depending on the local environment. Isoprenoid acids and an isoprenoid alcohol, dihydrophytol, have also been identified. In ancient sediments, crude oils and coals, isoprenoids from C_9 to C_{25} have been identified in various concentrations. They generally seem to be more abundant in marine sediments comprising autochthonous organic matter than in nonmarine sediments where organic matter is of terrestrial origin only (Lower Cretaceous of Canada, eastern South America and West Africa). The lower carbon number C_9–C_{14} isoprenoids have been scarcely reported (Göhring et al., 1967), possibly because that range of molecular weight is not frequently analyzed on rock extracts. The medium C_{15} to C_{25} isoprenoids are present in most sediments, including Precambrian, and oils: Bendoraitis and Hepner (1962), Welte (1967), Han and Calvin (1969), Spyckerelle et al. (1972), Waples et al. (1974). The pristane C_{19}, phytane C_{20}, and sometimes the C_{16} isoprenoid, are the most abundant, whereas C_{17}, C_{22} and C_{24} are scarce. Pristane and phytane are also suspected to occur as mixtures of

3.6 Acyclic Isoprenoids

stereoisomers in several rocks and crude oils (Maxwell et al., 1972). Isoprenoid acids have also been identified in few sediments such as Green River shales or Serpiano shale.

a) Isoprenoids from a Phytol Source

The phytol side chain of chlorophyll is a widespread source of isoprenoid structures in the biosphere. The stereochemistry of isoprenoids $< C_{20}$ and related acids in sediments is compatible with an origin from the phytyl chain of chlorophyll (Maxwell et al., 1972, 1973; Cox et al., 1970, 1972). Animals, particularly zooplankton, feed on plants, and seem able to produce several

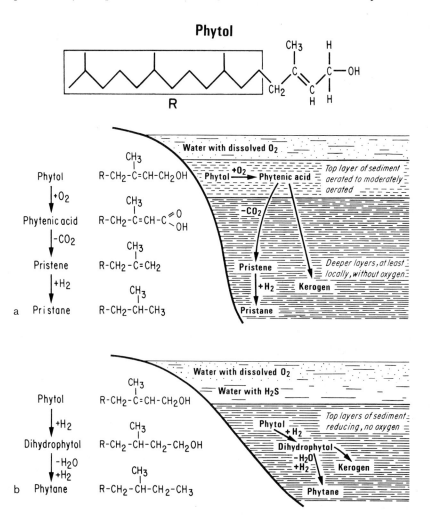

Fig. II.3.14a and b. Reductive and oxidative pathways from phytol to C_{19} and C_{20} isoprenoids. Phytol diagenesis initiated in the presence of oxygen (a), in the absence of oxygen (b)

saturated or unsaturated isoprenoid hydrocarbons from the phytyl chain, as proved by the occurrence of these molecules in zooplankton (Blumer and Thomas, 1965). This path may be a way of direct introduction of isoprenoids with 20 or less carbon atoms in sediments.

Another possibility, however, is the incorporation of phytol from chlorophyll in sediments: in a very reducing environment, phytol is quickly hydrogenated into dihydrophytol, followed by a subsequent reduction into hydrocarbon (phytane). In a normal, oxygenated environment, oxidation to phytanic acid is followed by decarboxylation (pristane) or fragmentation (isoprenoids $< C_{19}$) processes. According to the more or less reducing character of the local environment, phytane or pristane may be predominant, as suggested by Welte and Waples (1973) (Fig. II.3.14). The first step of the transformation of phytol in fresh sediments is illustrated by in situ experiments of Brooks and Maxwell (1974), who injected ^{14}C-labeled phytol into sealed sediment cores from Recent lacustrine environments. After two to eight weeks of incubation they identified various compounds, including dihydrophytol, a C_{18} isoprenoid ketone, phytyl and dihydrophytyl esters and phytadiene. This result suggests that microbiological processes have already been active in a few weeks.

On the other hand, laboratory experiments on hydrogenolysis of chlorophyll at high pressure and high temperature (Bayliss, 1968), have confirmed that saturated isoprenoids, pristane and phytane, can be produced from the phytyl chain. Direct heating of pure phytol (Albrecht, 1969) has resulted in C_{20}, C_{19}, C_{18}, C_{16}, C_{15} and possibly C_{14}-saturated isoprenoids plus some olefins. Using ^{14}C-labeled phytol in contact with different clays, De Leeuw et al. (1974, 1977) obtained isoprenoid hydrocarbons, alcohols, aldehydes, ketones, acids and other products.

Low molecular weight isoprenoids ($< C_{15}$) may also be derived from natural molecules with 15 carbon atoms, i.e., acyclic sesquiterpenes such as farnesol and associated hydrocarbons. The problem is different with isoprenoids from C_{21} to C_{25}, which are not likely to originate from phytol. Furthermore, the structure of C_{21} and C_{25} isoprenoids allowed Spyckerelle (1973) to eliminate the possibility of an origin from squalene or lycopene. He suggests that a comparable process may lead from phytol to C_{15}–C_{20} isoprenoids and from natural molecules with 25 carbon atoms (sesterterpenes) to C_{21}–C_{25} isoprenoids. Some sesterterpenoid alcohols are known from fungi, leaves and insect wax, and the same structure has been found in furan and ubiquinone derivatives.

b) Isoprenoids from Bacterial Membranes (Archaebacteria)

Archaebacteria have been recently discovered to be a third major line of descent, besides eubacteria ("true bacteria") and eukaryotes: Fox et al. (1980). They are characterized by a fundamental genetic difference (nucleic acids) and also by a different composition of their membrane lipids.

Those are mainly composed of glycerol ethers in archaebacteria (de Rosa et al., 1977, 1980; Tornabene and Langworthy, 1979; Kushwaha et al., 1981), and not esters as they are in other prokaryotes and eukaryotes. Furthermore hydrocarbon

3.6 Acyclic Isoprenoids

chains bound to glycerol are generally branched. Archaebacteria may represent ancestral life forms and they are presently confined in specific ecologic niches corresponding to extreme environments: thermoacidophilic, halophilic, methanogenic (anaerobic, methane-producing) bacteria. The latter, and especially those living in the young sediment, are a most probable source of fossil lipids.

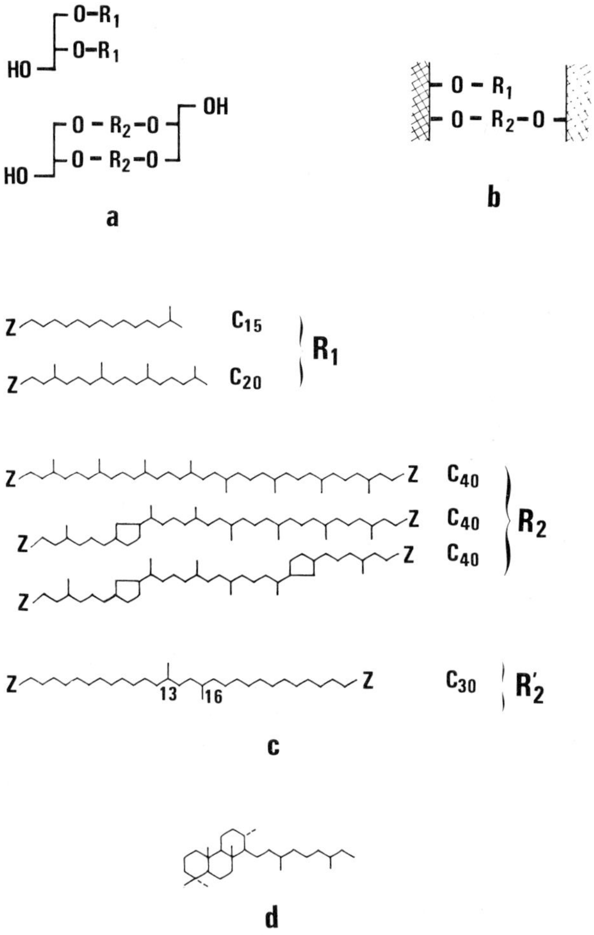

Fig. II.3.15 a–d. Branched and isoprenoid alkanes probably derived from membranes of microorganisms. (Adapted from Chappe et al., 1982, and Aquino Neto et al., 1982.) (a) Glycerol di- and tetraethers in membrane lipids of archaebacteria. (b) Ethers, derived from (a), bound to the kerogen or asphaltene matrix. (c) Branched and isoprenoid alkanes obtained from (b) by selective cleavage of ether bonds in kerogen or asphaltenes:
R_1: C_{15} isoalkane and C_{20} isoprenoid alkane from monoethers
R_2: C_{40} isoprenoid alkanes from diethers: they are typical of archaebacteria
R'_2: C_{30} branched alkane from diethers, not yet found in living organisms.
(d) Tricyclohexaprenane, a possible precursor of the C_{19} to C_{30} tricyclic terpane series, not yet found in living organisms

In geological environments, fossil molecules have been recently identified and related to membrane lipids of archaebacteria: they occur in kerogen and in the polar fraction of Recent and ancient sediments, and crude oils: Michaelis and Albrecht (1979), Chappe et al. (1979, 1980, 1982). Among them, most remarkable are glycerol tetraethers comprising C_{40}-isoprenoid chains, the structure of which is characterized by a "head to head" junction of the C_{20} sub-units (Fig. II.3.15). This macrocycle probably plays the same role in archaebacterial membranes as sterols in eukaryotic ones, i.e., to improve the mechanical properties. Among the C_{40} isoprenoid units, the acyclic C_{40} bi-phytane usually predominates, although the structures including one or two cyclopentane rings are also found.

Other ethers comprise the C_{20}-phytane chain, the C_{15} (2-methyl) alkane, and a C_{30} chain resulting from the "head to head" junction of the previous one.

The related isoprenoid hydrocarbons up to C_{40}, with the "head to head" junction have been found in crude oils: Moldowan and Seifert (1979), Albaiges (1980). They are probably issued from degradation of the ethers during catagenesis. Several other hydrocarbons known in archaebacteria have been identified in marine sediments and/or crude oils: C_{30} squalane (Gardner and Whitehead, 1972; Brassell et al., 1981), C_{25} irregular isoprenoids and various branched alkanes (Brassell et al., 1981). C_{40} lycopane, identified by Kimble et al. (1974) may have a comparable origin.

Apart from isoprenoids directly inherited from living organisms or through an early diagenesis in sediments, there is a definite possibility that the isoprenoid chains may be bound to the kerogen matrix, for instance by an ester or an ether bond, and subsequently released at depth during catagenetic degradation of kerogen. This hypothesis will be discussed in Chapter II.5.

3.7 Tricyclic Diterpenoids

Tricyclic diterpenoids have been found in sediments of the Deep Sea Drilling Project by Simoneit (1975, 1977). They have been interpreted as being derived from diterpenes such as abietic acid, and they are thought to represent evidence for the contribution of higher plant material. Abietic acid is also the most likely precursor for saturated fichtelite and triaromatic retene found in peat and lignite (Fig. II.3.16).

A series of tricyclic terpanes extending from C_{19} to C_{30} was reported by Aquino Neto et al. (1982) and Ekweozor and Strausz (1982) in some ancient sediments and crude oils. Three of them, with 19 and 20 carbon atoms, were identified as probable degradation products of a C_{30} tricyclohexaprenol (made of a tricyclic nucleus, plus a long isoprenoid chain) (Aquino Neto et al., 1982). This last compound is as yet unknown, but could be formed anaerobically from regular hexaprenol, which is a universal cell constituent. In this case, the biogenic precursor might be again a constituent of bacterial membranes (Fig. II.3.15d).

Tricyclic molecules, saturated or aromatic, are widely present in ancient sediments and crude oils, but they are generally subordinate in the saturated

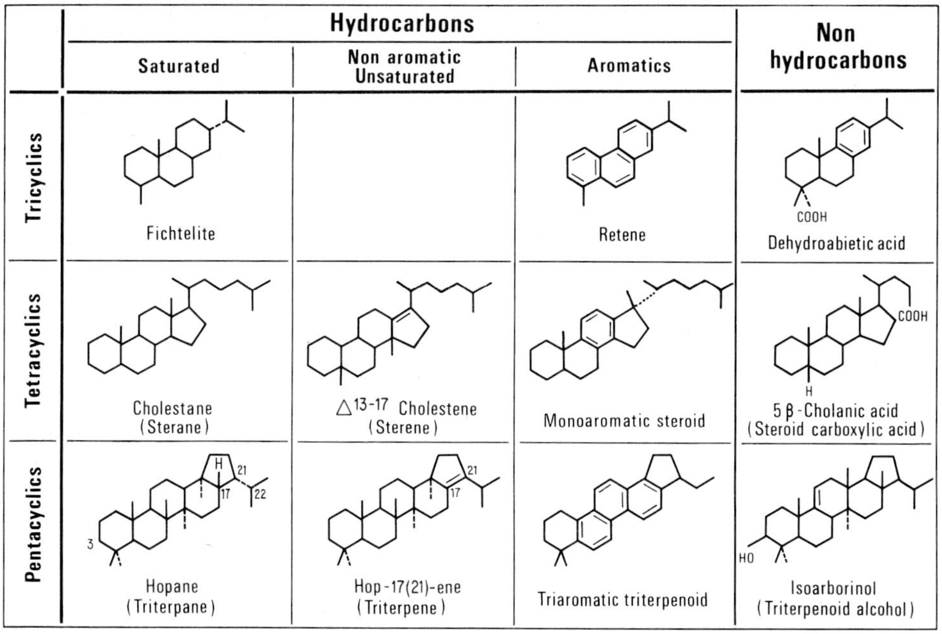

3.8 Steroids and Petacyclic Triterpenoids: Occurrence in Sediments   117

	Hydrocarbons			Non hydrocarbons
	Saturated	Non aromatic Unsaturated	Aromatics	
Tricyclics	Fichtelite		Retene	Dehydroabietic acid (COOH)
Tetracyclics	Cholestane (Sterane)	Δ¹³⁻¹⁷ Cholestene (Sterene)	Monoaromatic steroid	5β-Cholanic acid (Steroid carboxylic acid)
Pentacyclics	Hopane (Triterpane)	Hop-17(21)-ene (Triterpene)	Triaromatic triterpenoid	Isoarborinol (Triterpenoid alcohol)

Fig. II.3.16. Examples of polycyclic molecules present in ancient sediments and crude oils (Maxwell et al., 1971; Rubinstein et al., 1975; Tissot et al., 1972; Seifert, 1975; Albrecht and Ourisson, 1969; Spyckerelle, 1975; Ludwig, 1982)

fraction. In some shales from the Upper Devonian of Northern Africa, the Lower Cretaceous of the Northern Aquitaine Basin, France, the Cretaceous and Tertiary of West Africa, etc., the distribution of cycloalkanes by ring number shows a definite maximum corresponding to three-ring molecules, which may represent up to 40% of the total cycloalkanes and 20% of the saturated hydrocarbons. In some cases, like shales of the Lower Carboniferous-Upper Devonian of Eastern Sahara, the aromatic fraction shows a high content of tricyclic diaromatics of the tetrahydrophenanthrene group with a distinct predominance of $C_{19}H_{24}$.

Other tricyclic and also dicyclic aromatic molecules found in ancient sediments may be derived from triterpenes by fragmentation and dehydrogenation. They are C_{12} or C_{13} (di- or tri-methyl) naphthalene and C_{16} or C_{17} (di- or tri-methyl) phenanthrene.

3.8 Steroids and Pentacyclic Triterpenoids: Occurrence in Recent and Ancient Sediments

In this section, we shall review the first and also the most recent reports of polycyclic molecules belonging to the steroid or triterpenoid classes, or obviously derived from them. Their occurrences will be successively reported in Recent and ancient sediments and in crude oils. Then their mutual relationship, and the progressive changes of the various molecular types in subsurface will be discussed in Section II.3.9.

Steroids form a group of compounds widely distributed in living organisms (Fig. I.4.11), where they play an important biological role: in particular some of them are a constituent of the membrane lipids of eukaryotes. All naturally occurring steroids comprise a nucleus of perhydro 1,2 cyclopentanophenanthrene bearing methyl groups, one alkyl chain, and functional groups. Pentacyclic *triterpenoids* are made of 6-membered rings only or of four 6-membered and one 5-membered ring. They normally comprise 30 carbon atoms, and have been reported to occur frequently in plants, and also in prokaryotic organisms. In particular, triterpenes of the hopane series (Fig. II.3.16) are a widespread constituent of the membrane lipids of prokaryotes (eubacteria, blue-green algae). Various tetracyclic triterpenes also occur in animals and plants.

a) Recent Sediments

In Recent sediments, Schwendinger and Erdman (1964) described the occurrence of sterols that may amount to 0.01% of the total organic carbon (Erdman, 1965), Stanols, a reduced form without double bonds, are also present and may amount to as much as 10–50% of the sterols (Peake and Hodgson, 1973). The presence of the corresponding hydrocarbons, steranes, together with triterpanes, was reported by Meinschein (1961), who analyzed 4- and 5-ring cycloalkanes in a Recent sediment from the Gulf of Mexico. More recently, Gagosian et al. (1980) analyzed the occurrence and transformation of the various steroid compounds in Recent sediments of three different marine environments (Walvis Bay, Black Sea and Western North Atlantic). The occurrence of a series of triterpanes with the pentacyclic hopane skeleton in various Recent sediments was definitely established by Kimble (1972) and van Dorsselaer (1975). Spyckerelle (1975) found tetra- and pentacyclic polyaromatic, probably derived from triterpenes, in lacustrine and marine Recent sediments. Later tetracyclic di- and triaromatic molecules derived from the loss of one ring in pentacyclic triterpenes were characterized by Spyckerelle et al. (1977) in comparable situations.

b) Ancient Sediments

Steranes or pentacyclic triterpanes have been reported to occur frequently in ancient immature sediments buried to shallow depths (Fig. II.3.16). Both have been identified in Green River shales, by Burlingame et al. (1965), Hills et al. (1966), Henderson et al. (1968) and Anderson et al. (1969). The Messel and Bouxwiller shales from the Rhine valley contain several triterpanes, but only small amounts of steranes: Albrecht (1969) and Arpino (1973). The triterpanes mostly belong to the hopane series, which include a succession of molecules from C_{27} to C_{32} and sometimes to C_{35}, with two main stereoisomeric forms (Ensminger et al. 1974). Furthermore, these hopane derivatives show an ubiquitous occur-

3.8 Steroids and Petacyclic Triterpenoids: Occurrence in Sediments

rence in various types of sediment of any age (van Dorsselaer et al., 1974). The shallow samples of the Lower Toarcian shales from the Paris basin contain C_{27} to C_{30} steranes and triterpanes (Califet-Debyser and Oudin, 1969). From those same shales, Rubinstein et al. (1975) and Huc (1976) identified sterenes and triterpenes; Tissot et al. (1971, 1972) have also identified C_{27} to C_{30} monoaromatic and diaromatic steroids and reported some possible C_{27} to C_{30} monoaromatic triterpenoids. Similarly, Albrecht (1969) reported the occurrence of mono-, di-, tri- and tetra-aromatic molecules related to pentacyclic triterpenes. These compounds have been identified by Spyckerelle (1975) and Greiner (1976), and result from the progressive aromatization of triterpenes of the hopane series (Greiner et al. 1976). Since that time it has been observed that mono- and triaromatic steroids are rather common in ancient sediments and crude oils. Monoaromatic steroids are sometimes aromatized in the A-ring (Hussler et al., 1981), and frequently in the C-ring (Ludwig et al., 1981 a). Triaromatic steroids are also widespread and they usually become prevalent over monoaromatics with

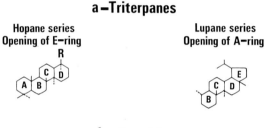

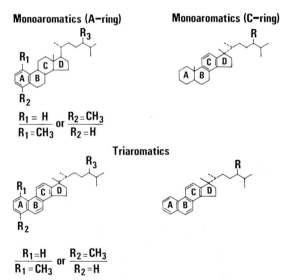

Fig. II.3.17a and b. Examples of altered triterpenes and steroids found in sediments and crude oils. (a) Degraded triterpanes by opening the E-ring (hopane series) or the A-ring (lupane series). (b) Monoaromatic (in A-ring or C-ring) and triaromatic steroids

increasing thermal maturation (Ludwig et al., 1981b; Mackenzie et al., 1981); Figure II.3.17.

Tetracyclic molecules derived by degradation from pentacyclic triterpenes have also been identified in ancient sediments: a C_{24} to C_{27} naphthene series (17, 21-secohopanes) probably derived from the loss of E-ring in hopanes (Trendel et al., 1982); saturated, unsaturated and aromatized tetracyclics derived from the loss of A-ring in lupane, friedelin and amyrins series are widespread in deltaic or lacustrine ancient sediments and in the related crude oils (Spyckerelle et al., 1977; Corbet et al., 1980; Schmitter et al., 1981); Figure II.3.17.

Pentacyclic triterpanes have also been reported from lignite (Jarolim et al., 1963, 1965) and coal (van Dorsselaer, 1975). The corresponding aromatized hydrocarbons are also present in lignite from Czechoslovakia (Jarolim et al., 1965; Streibl and Herout, 1969).

c) Crude Oils

In crude oils, steranes and pentacyclic triterpanes have been identified. These molecules are principally responsible for the optical rotation observed for natural hydrocarbon mixtures: Hills and Whitehead (1966), Louis (1964, 1968), Hills et al. (1968, 1970). The monoaromatic steranes were first reported by O'Neal and Hood (1956). Since that, the mono- and di-aromatic steroids have been found in several crude oils, including Lower Cretaceous oils from South America (Tissot et al., 1972) and Lower Paleozoic oils from the Eastern Sahara (Tissot et al., 1974). Later, mono- and triaromatic steroids were recognized to be widespread constituents of crude oils, as mentioned above. Cyclopentenophenanthrene and its methyl derivative have been identified by Carruthers and Douglas (1957) and Mair and Martinez-Pico (1962).

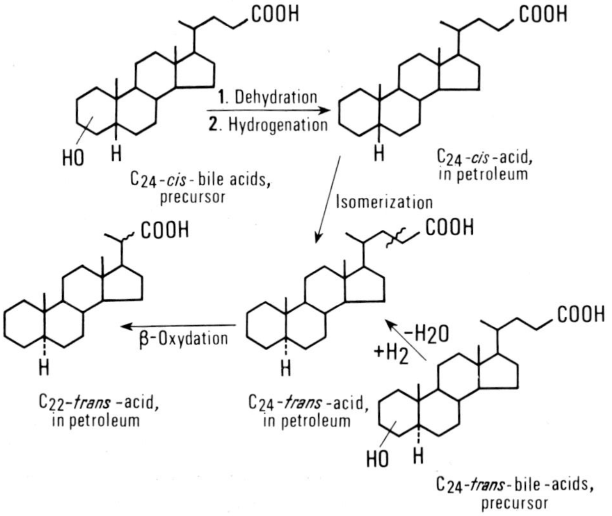

Fig. II.3.18. Possible origin of steroid acids from animal bile acids (Seifert, 1973)

Rearranged steranes (diasteranes) have been observed in several crude oils and ancient sediments which have already gone through partial diagenesis (Ensminger et al., 1978). In the Lower Toarcian shales of the Paris basin, they are supposed to derive from the diasterenes present in immature samples. This hypothesis is confirmed by simulation on clay minerals (Sieskind et al., 1979).

d) Other Related Substances

Closely related substances, including alcohols, acids and ketones, have been reported from sediments and fossil fuels (Fig. II.3.16). In particular, Albrecht (1969) found in the Messel shales of the Rhine valley the association of a pentacyclic triterpenoid alcohol, isoarborinol, and the corresponding ketone, arborinone. In the nearby Bouxwiller shale, Arpino (1973) found triterpenoid ketones and acids, but no alcohol. A series of C_{31} to C_{33} acids containing the pentacyclic hopane skeleton has been identified by van Dorsselaer (1975) in various types of sediments. Oxygenated triterpenes (alcohols, ketones) have also been reported from peat and lignite (Ikan and McLean, 1960; Jarolim et al. 1961, 1963). Tetracyclic acids derived from triterpenes (lupane, friedelin and amyrin series) by opening the A-ring were identified in deltaic sediments by Corbet et al. (1980).

Stanols and stanones are also reported by Mattern et al. (1970) in Messel shales, but few or no sterols seem to be present.

Seifert et al. (1971, 1972) and Seifert (1975) have determined numerous different carboxylic acids in a young Californian crude oil. A prominent feature is the relative abundance of the steroid acids, which amount in this particular oil to more than 0.03% of the total crude oil. The relative abundance of the isomers of C_{22} and C_{24} steroid acids are interpreted by Seifert (1973) as deriving from animal bile acids (Fig. II.3.18).

3.9 Fate of Steroids and Triterpenoids During Diagenesis and Catagenesis

The origin of these numerous steroids or pentacyclic triterpenoids found in sediments and crude oils is obviously the naturally occurring sterols and triterpenes. The various molecular types identified will be discussed later with respect to their significance in terms of environment of deposition or stage of thermal evolution (Chap. IV.3.) However, several aspects of the occurrence and distribution of these compounds in sediments and crude oils require an explanation at this point, namely:

- the relative abundances of steroids and triterpenoids,
- the fate of the biological markers in relation to the original configuration of functional groups and C=C bonds,
- the changes in stereochemistry due to isomerization during diagenesis or catagenesis,

– the occurrence of extended or dealkylated molecules, as compared to the usual C_{27}–C_{30} sterols and C_{30} triterpenes.

a) Relative Abundances of Steroids and Triterpenoids

The relative abundances of steroids and triterpenoids may vary. They sometimes occur simultaneously, but in other cases one group may be predominant. Steroids are more abundant than triterpenoids in the Lower Toarcian shales of the Paris basin or in the Cretaceous source rocks of the Persian Gulf. In the contrary, Messel and Bouxwiller shales, in the Rhine valley, contain abundant triterpanes, but only a small proportion of steranes. Both types are fairly abundant in the Green River shales (Kimble et al., 1974).

It is considered that pentacyclic triterpanes of sediments may originate in part from plants. However, the main sources are prokaryotic organisms: in particular microbial life in the upper layers of young sediments is mainly responsible for the ubiquitous hopanes, as proposed by Ensminger et al. (1972) and by van Dorsselaer et al. (1974). Steranes may be derived partly from animals and partly from plants, as already proposed by Seifert (1973) regarding steroid acids. As terrestrial animals are usually insignificant in the contributing biomass, C_{27} and C_{28} steroids found in sediments may be derived from the autochthonous aquatic life, i.e., zoo- and phytoplankton, whilst C_{29} steroids may be derived either from continental plants or from brown algae and phytoplankton. In some cases sterols can be traced back to the contributing organisms, especially algae. For instance, Boon et al. (1979) identified the major sterol of the Black Sea sapropel layer as the C_{30}, 4-methyl steroidal alcohol, dinosterol, which is the major sterol in some dinoflagellates only. The quasi-absence of steranes in some sediments where triterpanes are present might in some instances result from sterols being selectively taken out by certain microorganisms, which are unable to synthesize these compounds. However, this absence generally results from the original composition of the organic matter.

Steroid-rich sediments are usually marine or lacustrine, and their kerogen belongs to types I or II (for definition, see Section 4.8) which are mainly derived from autochthonous organic matter. In this respect, plankton may be the major source of sterols in geological conditions. Thus the ratio of steranes to ubiquitous hopanes in crude oils may indicate the origin of the main source material: planktonic when the sterane/hopane ratio is high, terrestrial when it is low.

b) Fate of Biological Markers in Relation to Their Original Structure

The fate of the functional groups, including C=C bonds, which occur in the biogenic molecules, is of particular interest and may influence the subsequent evolution of the biological markers toward the various hydrocarbon (saturated, mono- and poly-aromatics) and non-hydrocarbon types. In this respect steroids and triterpenoids behave differently.

1. Sterols are rather easily converted to saturated stanols and to the corresponding hydrocarbons, unsaturated sterenes and saturated steranes, all of which

3.9 Fate of Steroids and Triterpenoids During Diagenesis and Catagenesis

are known in Recent or ancient sediments. A high content of steranes is often observed in marine sediments deposited in a reducing environment, with a high concentration of sulfur compounds. This observation may suggest that a very reducing and hydrogenating environment leads to sterane, whereas less reducing conditions of deposition lead to sterene, which may be subsequently converted to either aromatics or saturates.

The process of aromatization of steroids also takes place in sediments, where it may start from ring A or, more frequently, from ring C and terminate with triaromatic molecules of the cyclopenteno-phenanthrene type (nomenclature of rings is shown in Figure II.3.17). Aromatization in ring C may start in ancient sediments from sterene precursors. Such monoaromatic steroids have been identified by Ludwig et al. (1981a). The related triaromatic steroids correspond to the rearrangement or loss of a methyl group and offer three alternative structures with 1-methyl, or 4-methyl, or no methyl on ring A (Ludwig et al., 1981b; Mackenzie et al., 1981). Mono- and triaromatic steroids of this group are widespread in ancient sediments and crude oils.

Less frequent are monoaromatic (cycle A) steroids, which have been reported by Hussler et al. (1981) in Cretaceous black shales cored in the Atlantic during the Deep Sea Drilling Project (DSDP). These shales contain a rich organic matter of type II (up to 30% organic carbon): they were deposited in anoxic, very reducing conditions and they still belong to the diagenesis stage.

A suggestion was made that aromatization of steroids starting in ring A is restricted to confined reducing environments, whereas aromatization starting in ring C would occur in the more common open environments (Hussler et al., 1981). In all cases aromatization might subsequently proceed (C to A or A to C) and generate triaromatic steroids. For instance, Mackenzie et al. (1981) have shown that the ratio triaromatic/monoaromatic steroids increases with thermal maturation in the Lower Toarcian shales of the Paris Basin. This aspect will be further emphasized in Chapters IV.3.4 and V.1.3.

2. The fate of triterpenoids during diagenesis has been summarized by Spyckerelle (1975).

– Pentacyclic triterpenes with an oxygenated functional group at C-3, are known from higher plants: lupane, friedelin and amyrins series. The corresponding molecules are known in sediments as alcohols, ketones and aromatic or naphthenoaromatic hydrocarbons, but only rarely as saturated hydrocarbons. These observations suggest that conversion of triterpene alcohols to triterpanes is more difficult than the sterol to sterane conversion. In fact, tetracyclic products resulting from the opening or the loss of ring A seem to be rather common in the alkane, alkene or acidic fractions of deltaic or lacustrine sediments, and also in crude oil derived from terrestrial organic matter (Corbet et al., 1980; Schmitter et al., 1981). This photomimetic transformation may be controlled by microorganisms during diagenesis. A further step would be a progressive aromatization of rings B to D (nomenclature of rings shown in Fig. I.4.10) generating in particular the tetracyclic diaromatic and triaromatic derivatives from the amyrins group characterized in sediments and crude oils by Spyckerelle et al. (1977). Another possible way of conversion of C-3

oxygenated triterpenes would be to yield unsaturated compounds by loss of the functional group and then undergo aromatization starting with the A-ring: in this case pentacyclic aromatized triterpenes would be formed, such as the succession of mono- to pentaaromatic derivatives reported in lignite by Jarolim et al. (1965).
- 3-desoxytriterpenes of the hopane series, without a functional group in C-3 position, are known from ferns, mosses, bacteria and blue algae. They have been reported in Recent and ancient sediments as acids, saturated hydrocarbons, unsaturated hopene and aromatic hydrocarbons. In fact, the pentacyclic-saturated fraction (triterpanes) of sediments and crude oils mainly consists of the hopane series. Aromatization would not start from the A-ring, but from the D-ring, where the unsaturated bond is located in hopene II.

Furthermore, tetracyclic terpanes with 24 to 27 carbon atoms, possibly derived from hopanes, are rather widespread constituents of sediments and crude oils. Trendel et al. (1982) identified their structure and showed that they could be derived from hopanes by opening the E cycle, either by microbial action at an early stage of diagenesis, or by thermal degradation during catagenesis.

c) Changes in Stereochemistry Due to Isomerization

Isomerization at chiral centres, either part of the ring system of steroids and triterpenoids or located in the side chains, is now well documented. This action results in a change of the stereoisomeric structure. The first reports concerned the D/E ring junction in hopanes which changes during diagenesis. This junction is usually cis in biogenic molecules and in young immature sediments ($\beta\beta$-hopanes), whilst maturation results in the trans stereochemistry[3] in ancient sediments and crude oils ($\alpha\beta$-hopanes): Ensminger et al. (1974), van Dorsselaer (1975), Ensminger (1974). A similar process could affect the ring junction of the biogenic oleananes and ursanes (Whitehead, 1973).

Comparable observations have been made at a later stage of evolution in the ring system of steranes (Mackenzie et al., 1980). Furthermore, isomerization also occurs in the side chains of steranes and hopanes during diagenesis and part of catagenesis (ibid.). These changes have been proposed as indices for measuring maturation of source rocks and crude oils and they are discussed in greater detail in Chapters IV.3.3, IV.3.4 and V.1.3.

d) Occurrence of Extended or Dealkylated Molecules

Natural pentacyclic triterpenoids normally contain 30 and sometimes 29 carbon atoms. There is more variation among steroids (from 18 carbon atoms in estrone and estradiol and 19 in androsterone and adrenosterone to 29 in stigmasterol and

---

3 The notation $\beta\beta$-hopane is used for 17β-H, 21β-H hopane. The notation $\alpha\beta$-hopane is used for 17α-H, 21β-H hopane

sitosterol and 30 in dinosterol). The distribution is much wider in ancient sediments and crude oils: triterpanes or oxygenated triterpenes occur from 27 to 32 carbon atoms, and sometimes up to 35 carbon atoms; steranes and aromatic steroids occur with 27 to 30 carbon atoms. Other tetracyclic molecules, where one or several cycles are aromatized, occur with 17 carbon atoms, and are frequently abundant with 19 to 23.

Several points need explanation:
- the occurrence of triterpanes beyond C_{30}
- structure and origin of the C_{17}–C_{23} aromatized tetracyclics, ie., comprising mostly a partly aromatized cyclic carbon skeleton with some methyl substituents.

The discovery of tetrahydroxybacteriohopane, a C_{35} pentacyclic polyalcohol (Förster et al., 1973), and the probable existence of other similar compounds among natural substances provide an explanation for the occurrence of triterpanes with more than 30 carbon atoms. Rohmer and Ourisson (1976) have shown that this class of compounds is rather ubiquitous in prokaryotes (bacteria and blue-green algae), where bacteriohopane polyols are an important constituent of the membrane lipids. According to the conditions of deposition and diagenesis, the C_{35} precursors may yield a series of alkanes and acids extending to C_{35} in confined anoxic environments, and C_{32} in the more common open environment. There is, however, little doubt that in all cases the hopane series with extended side chain found in sediments and crude oils is of prokaryotic origin (Ourisson et al., 1979).

Among the aromatized tetracyclics, Tissot et al. (1974) reported the importance of C_{17}–C_{21} aromatic steroids in very mature Lower Paleozoic crude oils from the Sahara, Algeria. The same constituents were identified by Mackenzie et al. (1981) in the deep samples from the Paris Basin as mono- and triaromatic steroids. Their occurrence may be explained by various hypotheses. A concentration of particular types of steroids in that range of molecular weight might be considered as several steroids are known with 18 to 21 carbon atoms. However, these steroids today belong mainly to the group of sex hormones or corticosteroids, and these can make only minor contributions to sediments. Bile acids, comprising a steroid structure with 24 carbon atoms, may also be considered. Microbiological dealkylation processes are not excluded. However, an abiotic, purely thermal disproportionation is the most likely way of derivation from C_{27}–C_{30} steroids: the elimination of methyl groups and of the main alkyl chain, along with subsequent formation of alkanes, would be balanced by aromatization of one or more cycles. In fact, Mackenzie et al. (1981) showed a correlation between the increasing catagenesis in the Paris basin, marked by burial depth, and the proportion of C_{20} molecules in the triaromatic steroids.

Other aromatized tetracyclics have a different structure and origin. Spyckerelle (1977) identified in the Messel shales some tetracyclic mono-, di- and triaromatic C_{23} to C_{21} compounds derived through degradation of the A-ring from pentacyclic triterpenes containing an oxygenated functional group in C-3, during diagenesis. This scheme of degradation, possibly initiated by microorganisms, may constitute another path to the C_{17}–C_{23} aromatized tetracyclics.

In conclusion, steroids and triterpenoids are widely distributed compounds of biogenic origin which carry a wealth of information. The relative abundance of steranes and triterpanes, or the specific occurrence of certain compounds may give a clue to the origin of the organic matter. Subsequent rearrangements, such as changes in stereochemistry and progressive aromatization, provide information on diagenesis and catagenesis processes. The use of steroids and triterpenoids for these various purposes will be discussed in Part IV, Chapter 3, whereas their use for correlation will be presented in Part V, Chapter 2.

3.10 Other Polyterpenes

Carotenoids are pigments comprising eight isoprene units. They are C_{40} compounds, occurring in continental plants and in algae (Fig. I.4.12). They have been identified in Recent sediments, both lacustrine (Lyubimenko, 1923) and marine (Trask and Wu, 1930). Schwendinger and Erdman (1963) and Erdman (1965) found the carotenoid content to be around 5.10^{-5} (as a fraction of organic carbon) in swamp sediments, and 3 to 80.10^{-5} in marine sediments. The corresponding saturated hydrocarbons are probably widely present in ancient sediments. For example, lycopane has been suspected in several cases and is reported in Messel shales by Kimble et al. (1974). Murphy et al. (1967) found perhydro-β-carotene, and Gallegos (1971) identified two isomers of this compound in the branched cyclic hydrocarbon extract from the Green River shales, where they are abundant in particular horizons (Mahogany beds).

3.11 Aromatics

The abundance of aromatic hydrocarbons in Recent sediments is low. Furthermore, Erdman (1961) has shown the absence of low molecular weight aromatic hydrocarbons in Recent sediments from various origins. This is in contrast to their importance in ancient sediments and crude oils.

A few polycyclic aromatic hydrocarbons (PAH) were first reported by Meinschein (1959) in marine sediments of the Gulf of Mexico. Thomas and Blumer (1964) identified pyrene and fluoranthene in manganese nodules. Later Tissier and Oudin (1973) identified by mass spectrometry and spectrofluorimetry fluoranthene, chrysene, triphenylene, perylene, benzo-8, 9-fluoranthene and traces of 3,4-benzopyrene in nonpolluted sediment of the Bay of Veys (France): Figure II.3.19. With the improvement of analytical techniques it became obvious that PAH from Recent sediments represent a complex assemblage of the parent compounds and their alkyl homologs (Youngblood and Blumer, 1975).

In the deeper and older layers of Recent sediments the concentrations of PAH are generally very low. A drastic increase in PAH concentrations with approach to the Earth's surface is indicative of an enhanced contribution of anthropogenic

3.12 Oxygen and Nitrogen Compounds

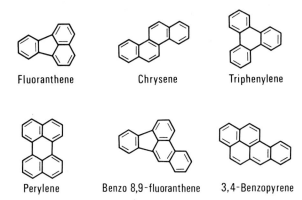

Fig. II.3.19. Polyaromatic hydrocarbons found in an nonpolluted Recent sediment of the bay of Veys, France (Tissier, 1974)

PAH within the past 75–100 years (Müller et al., 1977; Wakeham et al., 1980). Depth profiles of individual PAH may be used to discover their source, e. g., it has been shown that perylene concentrations increase with depth, suggesting an in-situ formation mechanism (Orr and Grady, 1967; Brown et al., 1972; Aizenshtat, 1973; Wakeham, 1977; Wakeham et al., 1980). Perylene may be derived from polyaromatic precursors synthesized by living organisms, e.g., fungi. For instance, Aizenshtat (1973) suggests that perylene is derived from land organisms and is incorporated in sediment with the detrital mineral fraction. Recently substantial amounts of perylene have been obtained from Namibian Shelf sediments (Wakeham et al., 1980), where input of terrestrial material is thought to be minimal. Thus, the precursor is not necessarily of terrestrial origin.

Other PAH may have been formed in natural fires. Their distribution varies little over a wide range of depositional environments, suggesting a common mechanism of generation by combustion of land plants and dispersion by air transport (Blumer and Youngblood, 1975; Youngblood and Blumer, 1975).

Finally, aromatized diterpenoids, steroids and triterpenoids with one to four aromatic rings are widespread (Spyckerelle, 1975; Simoneit, 1977; Laflamme and Hites, 1978) and have been discussed already (Sects. 3.7, 3.8 and 3.9).

3.12 Oxygen and Nitrogen Compounds

They are likely to include many geochemical fossils, but their study is still very incomplete. Linear and branched acids, ketones and steroid acids have been identified in a limited number of sediments or crude oils (Seifert et al., 1971, 1972; Seifert, 1973, 1975). Nitrogen compounds (pyridines, quinolines, carbazoles) are probably skeletal fragments of plant alkaloids (Whitehead, 1973).

Pigments containing the porphin skeleton (four pyrrole rings linked together by four methine $-CH=$ bridges) are very important in the plant and animal kingdom. They include mainly chlorophyll, the green photosynthetic pigment of plants (Fig. I.1.2), and hemin, the red pigment of animal blood. Chlorophyll and hemin are magnesium and iron chelates, respectively.

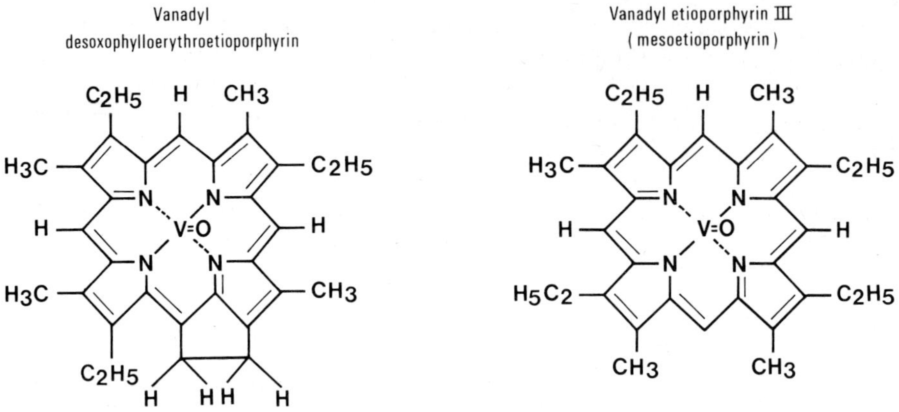

Fig. II.3.20. Examples of porphyrins found in ancient sediments and crude oils

Petroleum porphyrins also contain the porphin skeleton. They have been extensively studied since Treibs (1934) proved their biological origin and related them to the chlorophyll or hemin molecules. They have been identified in a sufficient number of sediments and crude oils to establish a wide distribution of these geochemical fossils: Hodgson et al. (1967), Baker (1969), Blumer (1965, 1973), Baker and Palmer (1978). Two recent reviews of porphyrin geochemistry have been presented by Maxwell et al. (1980) and Baker (1980).

The major type of fossil porphyrins are desoxophylloerythroetioporphyrin (DPEP), which contains an isocyclic ring in addition to the four pyrrole rings (Fig. II.3.20), and mesoetioporphyrin or etioporphyrin III without the isocyclic ring.

In addition, three minor types are observed: di-DPEP, rhodo-etio and rhodo-DPEP (Barwise and Whitehead, 1980). Fossil porphyrins are either metal-free or mostly chelated with nickel or vanadyl (VO) in ancient sediments and crude oils. Cu-porphyrins have also been reported in deep sea sediments, where they may be an indicator of oxidized terrestrial organic matter: Palmer and Baker (1978). Ga-complexes and Al-porphyrins have been found in bituminous coal: Bonnett and Czechowski (1980). These authors also isolated manganese (III) and iron (III) porphyrins from a lignite.

The transformation of chlorophyll, or hemin, is likely to start during sedimentation or early diagenesis. Magnesium or iron would be lost at an early stage and a re-chelation with vanadyl or nickel may stabilize the molecule and insure its preservation. Bergaya and Van Damme (1982) have studied the stability of metalloporphyrins adsorbed on clay minerals: Mg-porphyrin is particularly unstable, whereas Ni- and VO-porphyrin are among the most stable in this environment. The hydrolysis of the phytol may also occur at that stage. Later, reduction of the vinyl and carbonyl groups, and decarboxylation of the carboxylic acid groups lead to the formation of petroporphyrins. Evidence of thermal conversion from DPEP to etio type has been presented by Didyk et al. (1975), and also by Maxwell et al. (1980) in the Toarcian shales of the Paris Basin.

An alternative biological source of fossil porphyrins can also be proposed according to some recent discoveries of microbiologist R. K. Thauer et al. They identified in a methanogenic Archaebacteria *(Methanobacterium thermoautotrophicum)* the first natural nickel complex with a porphinoid skeleton (Pfalz et al., 1982). This possible origin might explain the widespread occurrence of Ni-porphyrins, as these bacteria commonly live in young sediments. More work is obviously needed to find out whether vanadyl complex may also occur.

The mass spectra of fossil porphyrins show a distribution of molecular weights suggesting a series of alkyl-substituted porphyrins. For instance, Barwise and Whitehead (1980) observed porphyrins ranging from C_{25} to more than C_{50}, with a maximum at C_{32}, in a concentrate of vanadyl porphyrins from Boscan crude oil. A transalkylation between pairs of porphyrins, or between porphyrins and other compounds has been proposed to explain this distribution (Baker, 1966). Alternatively, Blumer and Snyder (1967) suggested that porphyrin structure could be integrated into kerogen and later regenerated with various chain lengths. Later Mackenzie et al. (1980) showed that C_{35}–C_{39} porphyrins are generated below 2000 m in the Toarcian shales of the Paris Basin and considered them to arise by thermal cracking of the kerogen. Quirke et al. (1980) studied porphyrins with extended side chains and concluded that these porphyrins were formed by binding chlorin precursors via the vinyl substituent onto kerogen; subsequently the porphyrin products were generated as a result of cracking kerogen during catagenesis. This is likely to be the major pathway to porphyrins found in petroleum. However, Casagrande and Hodgson (1976) found homologous series also in near-surface sediment from the Black Sea; this fact suggests that some of these homologs are generated during the very early stages of diagenesis.

The recent developments in analytical methods (high performance liquid chromatography, gas chromatography, mass spectrometry, nuclear magnetic resonance) make possible a precise structural identification of porphyrins (Eglinton et al., 1980; Maxwell et al., 1980). Thus new prospects are open for the geochemistry of porphyrins and also for their use as correlation or maturation indicators.

Other pigments, named fringelites, have been found by Blumer (1960, 1962) in a fossil crinoid of Jurassic age. They are polyhydroxyquinones, which can be linked to similar pigments (aphin, hypercin) in presently living organisms.

3.13 Kerogen, the Polar Fraction of Sediments, and Asphaltenes of Crude Oils as Possible Sources of Fossil Molecules

Most of the geochemical fossils reviewed in this chapter have been traced from organisms through Recent sediments into ancient sediments and crude oils. However, there is evidence that certain molecules with a biological fingerprint are produced during the thermal evolution of kerogen. On the one hand, the abundance (related to organic carbon) of molecules like odd-carbon-numbered *n*-alkanes, isoprenoids, iso- and anteiso-alkanes, increases with depth in several geological situations. On the other hand, experimental evolution of some previously extracted kerogens generates compounds with a biological fingerprint, e. g., odd-numbered *n*-alkanes of high molecular weight (Welte, 1966, on Messel shales), isoprenoids (Albrecht, 1969, on Upper Cretaceous shales of the Douala basin), and steroids and triterpenoids (van Dorsselaer et al., 1977, on Yallourn lignite; Gallegos, 1975, on Green River shales).

A comparable observation has been made by pyrolysis of asphaltenes from normal crude oils, heavy oils and tar sands: asphaltene yields squalene and also steranes and hopanes, similar to, although not identical with, those naturally present in crude oil (Rubinstein et al., 1979; Samman et al., 1981). Selective cleavage of ether bonds, when applied to kerogen, asphaltene from crude oils, and the polar fraction of sediments, yields branched alkanes and isoprenoids which can be traced back to archaebacteria (Chappe et al., 1982).

Thus, it seems reasonable to consider that biogenic molecules with a functional group may be fixed on the kerogen or asphaltene matrix by chemical bonds, e.g., esters or ethers, and subsequently released by thermal evolution. The variable position of the cleavage might explain some of the alkylation or dealkylation observed, compared to actual biogenic molecules (Blumer, 1973). This process could generate some of the homologous series of acyclic isoprenoids, steroids, triterpenoids, porphyrins, etc. that occur in ancient sediments and crude oils, but are not presently known in living organisms. Furthermore, there may be less conversion of biological to nonbiological stereochemical types of steroids and triterpenoids in kerogen or asphaltenes (Huc, 1978; Rubinstein et al., 1979; Samman et al., 1981). Thus trapping in kerogen or asphaltene matrix may result in a better preservation of the stereochemistry.

A final remark concerns the prominence among fossil molecules of the constituents designed to ensure mechanical, thermal or chemical preservation of the cells: membrane lipids of archaebacteria (C_{40} isoprenoids), other prokaryotes (hopanes) and eukaryotes (sterols); cuticular waxes from higher plants (n-alkanes of high molecular weight); spores exine, etc. These constituents seem to be more resistant to microbiological and chemical alteration occurring during early diagenesis. Furthermore, some of them with a functional group may be rather easily trapped in the kerogen or asphaltene network, thus insuring a better preservation, including that of the original stereochemistry.

Summary and Conclusion

Geochemical fossils are molecules synthesized by plants or animals and incorporated in sediments with only minor changes. In particular, the carbon skeleton of hydrocarbons or other lipids is preserved. These molecules represent only a minor fraction of crude oils. However, they are of great interest to geologists and geochemists, as they provide information on the original organic material.

Alkanes, fatty acids, terpenes, steroids, and porphyrins are the major groups of geochemical fossils. They can be traced from Recent to ancient sediments where they progressively suffer thermal degradation and/or dilution by other hydrocarbons generated at greater depths.

Kerogen is a possible additional source of geochemical fossils. Some lipids may be trapped in the kerogen network, or alternatively bound to kerogen by chemical bonds. These molecules are released from kerogen with increase in depth and temperature.

Chapter 4
Kerogen: Composition and Classification

4.1 Definition and Importance of Kerogen

The term *kerogen* will be used here to designate the organic constituent of the sedimentary rocks that is neither soluble in aqueous alkaline solvents nor in the common organic solvents. This is the most frequent acceptance of the term kerogen, and results from a direct generalization to other rock types of the definition by Breger (1961) in carbonaceous shales and oil shales. However, it should be kept in mind that some authors still restrict the name kerogen to the insoluble organic matter of oil shales only, because kerogen originally was applied to the organic material found in Scottish shales, which yielded oil upon a destructive distillation. Such a distinction seems very artificial from a geochemical point of view, as the definition of "oil shale" is itself mostly an economic concept (a rock able to provide commercial oil products by heating) and subject to variations, according to the progress of technology and the fluctuation of petroleum economy.

A few authors seem to use the term kerogen for the total organic matter of sedimentary rocks. It is here understood that the fraction extractable with organic solvents is called *bitumen* and that the term "kerogen" does not include soluble bitumen (Fig. II.4.1).

As pointed out before, the early form (precursor) of kerogen in young sediments is the insoluble material that is also called "humin" by soil scientists, although its composition is different from compounds present in continental soils. The main difference between humin of young sediments and kerogen of ancient sediments is the existence of an important hydrolyzable fraction in humin; this fraction progressively disappears at depth. In geologic situations, there is usually an information gap, with respect to kerogen evolution, at relatively shallow depths of burial. Observations on core samples taken by oceanologists often cover the depths from 0 to 10 m (0 to 30 ft). By contrast, core samples taken by the oil industry usually start at 500 or 1000 m depth. Some wells drilled on behalf of the JOIDES program may help to fill the gap, although many sections cored in deep oceanic basins show little or no organic content. The observations reported in Chapter II.4 refer to the compositions and properties of kerogen to the extent that they can be analyzed below the previously mentioned depth gap. Furthermore, most of the considerations are related to the amorphous fraction of kerogen, that usually represents the bulk of the kerogen.

It should be noted that the definition used by Durand (1980) "the fraction of sedimentary organic matter which is insoluble in the usual organic solvents" is not

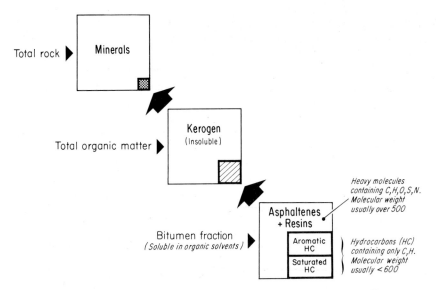

Fig. II.4.1. Composition of disseminated organic matter in ancient sedimentary rocks

strictly equivalent, except in ancient sediments, once the early diagenesis is completed. In Recent sediments extractable humic compounds are included in the kerogen defined by Durand and excluded from our definition. In ancient sediments usually studied in the petroleum industry (i. e., below the sampling gap mentioned above), both definitions become equivalent: Fig. II.4.1.

Kerogen is the most important form of organic carbon on earth. It is 1000 times more abundant than coal plus petroleum in reservoirs and is 50 times more abundant than bitumen and other dispersed petroleum in nonreservoir rocks (Hunt, 1972). In ancient nonreservoir rocks, e. g., shales or fine-grained limestones, kerogen represents usually from 80 to 99% of the organic matter, the rest being bitumen.

4.2 Isolation of Kerogen

The first and probably main hindrance when studying kerogen is to isolate kerogen quantitatively without noteable alteration of the general structure. This preliminary isolation from the inorganic material is required for most physical or chemical analyses, which are constrained when minerals — generally far more abundant — are present.

Physical methods of separation based on difference of specific gravity (sink–float method) or differential wetting of the kerogen and minerals by two immiscible liquids, such as oil and water (Quass method), have been reviewed by Robinson (1969). The advantage of these methods is the absence of chemical alteration of kerogen, but recovery is normally incomplete and thus a fractionation of kerogen may occur, generally favoring the larger organic fragments. These methods, however, are successfully used by some petrographers, to reduce pyrite content without appreciable alteration of ancient kerogens (Durand and Nicaise, 1980).

Destruction of the inorganic material by hydrochloric and hydrofluoric acids has been widely used and can hardly be avoided, if the aim is a quantitative recovery of kerogen. However, the acid treatment has to be performed under mild conditions in a nitrogen atmosphere (Durand et al., 1972). Chemical procedures should be restricted to that step in order to avoid, as much as possible, an alteration of the chemical structure of kerogen. In particular, decantation in various liquids such as CCl_4 or $CHCl_3$ has to be preferred to the elimination of pyrite by oxydation (HNO_3) or reduction ($LiAlH_4$, $NaBH_4$, etc.) treatments, which oxidize or reduce also kerogen. Ferric sulfate has been used successfully in very acid conditions to eliminate pyrite from ancient kerogens (Durand and Nicaise, 1980). The procedure cannot be applied, however, to kerogens from Recent sediments or from the early diagenesis interval.

Pyrite is by far the most frequent mineral remaining after treatment, followed by very fine-grained rutile and zircon; however, the latter do not hinder the kerogen analysis. A part of the pyrite is in intimate association with kerogen, e.g., the framboidal type of pyrite described below. Thus, it cannot be removed without alteration of kerogen.

During HF treatment, some complex fluorides may be formed, which can be prevented by adequate operating procedure, or subsequently dissolved by concentrated HCl (Durand and Nicaise, 1980).

A complete operating procedure for chemical isolation of kerogen has been provided by Durand and Nicaise (1980), with a description of the apparatus used for routine kerogen isolation.

Alpern (1980) compared the original content of the principal constituents of the organic matter in five Toarcien shales of the Paris basin with their content in two organic concentrates prepared by the chemical procedure of Durand and Nicaise, and the physical method used by coal petrographers, respectively. He found that a large proportion of the organic particles, particularly the finer fraction (e.g., the finely disseminated fluorescent material which amounts to ca. 80% of this particular organic matter), is lost by the physical method of preparation. On the contrary, there is a preferential concentration of larger fragments, particularly identifiable microorganisms.

4.3 Microscopic Constituents of Kerogen

Techniques of microscopic examination (natural light reflectance and transmittance, or UV fluorescence) can be applied to both kerogen-containing rock and previously isolated kerogen. Microscopic observations have been in use for a long time in coal petrology to identify the various coal macerals and evaluate their rank along the carbonization path that leads from peat to anthracite. More recently, these microscopic techniques have also been applied to study the finely disseminated kerogens in other sedimentary rocks in order to determine their grade of evolution (reflectance measurement). Much work has been devoted to that aspect, which will be dealt with later. Identification and diagnosis of kerogen particles has, however, received attention only more recently. First, isolated work has been done, mostly as a result of palynological and coal petrographic studies: Combaz (1964), Staplin (1969), Alpern (1970), Hagemann (1974), Teichmueller (1975b), Raynaud and Robert (1976). Later, several attempts were made to provide a systematic view of the microscopic approach to the identification of organic constituents: Rogers (1979), Combaz (1980). The principal concepts are briefly reviewed here, whereas the practical application will be discussed in Chapter V.1.

4.3.1 Locating Kerogen in Rocks

Locating the organic matter in common sedimentary rocks is difficult, even by microscopic techniques, because most of it is in a finely dispersed state without clear contours. Defined organic remnants (such as algae, spores, pollen and vegetal tissues) usually amount to a minor part of the organic matter with a range of a few percent of the total kerogen. However, in certain rare types of rocks like torbanite, tasmanite, etc. (Fig. II.4.2) they occur in great quantities. In average sedimentary rocks, the combined application of UV fluorescence and electron microprobe has shown that most of the amorphous organic material is fluorescent at shallow depth and can be located through this technique. Some minerals, such as calcite, also exhibit fluorescence, but their crystalline forms, or the Ca diagram obtained with electron microprobe makes the distinction easy.

Besides such studies of defined organic remnants, these techniques allow the observation of a second far more abundant type of organic matter, namely organic matter which is finely disseminated, often laminated or bedded between mineral particles, and especially clay minerals. In Figure II.4.3, fluorescence shows the size and the texture of this material in the Lower Toarcian shales of the Paris basin. It is obvious that physical techniques such as those used to purify coal macerals (fine grinding and density separation) are of little assistance in separating this type of kerogen from the minerals. In fact, these techniques, applied to shales of the Paris basin, result in a concentration and recovery of only a small percentage of the total organic matter (Sect. 4.2).

Furthermore, the association of reflection microscopy and fluorescence makes evident the intimate relationship between pyrite and organic matter. For

---

Fig. II.4.2 a and b. Examples of microscopic aspects of kerogen (Photo A. Combaz). (a) ▶ Thin sections observed in transmitted light. *Left:* boghead, Lower Permian, Autun, France; each elongated yellow body is a colony of Botryococcus (Chlorophyceae algae). *Right:* laminated carbonate source rock, Devonian, Alberta; the *dark brown bands* are laminae rich in organic matter. (b) Kerogen concentrates. *Left:* "sapropelic" or amorphous organic matter, Turonian, Senegal. *Right:* "humic" organic matter, Maestrichtian, Senegal

4.3 Microscopic Constituents of Kerogen

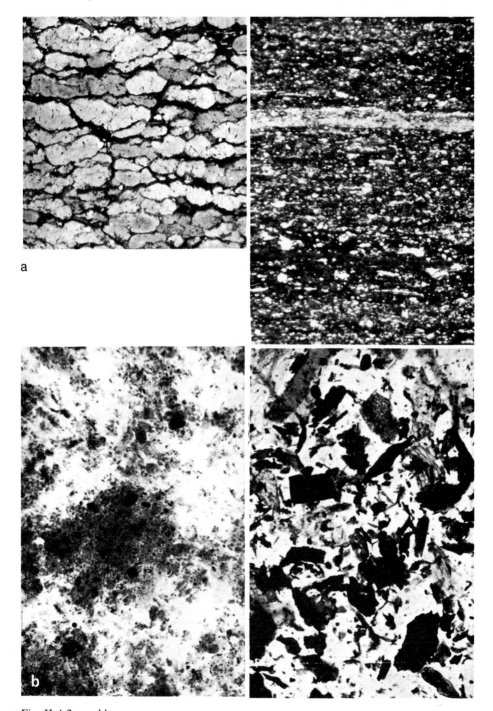

Fig. II.4.2a and b

example, pyrite commonly occurs in the form of microcrystals coated with organic matter and associated to form spherical aggregates (framboids). The shape of the aggregates is clearly visualized by the electron scanning microscope (Combaz, 1970; Fig. II.4.4). Such assemblages behave as traps for trace elements, as proved by electron microprobe. The association of organic material and pyrite is so intimate that it is not surprising that no physicochemical technique is able fully to separate one from the other.

4.3.2 Identification

Examination of kerogen concentrates isolated from the minerals shows that, except in certain types of sediments formed by massive accumulation of a single type of organism, heterogeneity of kerogen is common in sedimentary rocks; the size, shape, color and structure of the organic remnants vary. Kerogen appears to be an association of various kinds of organic debris. The tiny fragments of organic matter occurring in finely laminated shales or carbonates are not susceptible to an exact diagnosis. Larger fragments may show some structure, directly or by UV fluorescence, or after artificial etching (acidic, oxidative treatments), and may be identified as spores, pollens and plant tissues, algae, etc. Gelified tissues are frequent; other fragments include resinous excretions, sclerotes, or various microorganisms.

A general classification based on optical examination has been proposed by Combaz (1980), who first characterizes the individual organic constituents and groups them in major classes: remnants of terrestrial plants (spores and pollen grains, vegetal fragments); algae and related microflora (Acritarchs, Leiosphaeridaceae and Tasmanaceae, Dinoflagellates, colonial and benthic algae); microfauna (Chitinozoa, Gigantostraceae, Graptolites, Scolecodontes and microforaminifera); and the amorphous fraction (aggregates, granular, subcolloidal, pellicular gelified). Based on their occurrence and association, it becomes possible to define *palynofacies*, which can be compared with the types of organic matter defined upon chemical analysis of kerogen (Combaz, 1980).

More general terms are often used for kerogen by petroleum geologists. Unfortunately there is sometimes confusion along this line between description and interpretation. For example "humic" and "sapropelic" organic matter are terms associated with the respective types of organic matter in poor and good oil

---

Fig. II.4.3 a–c. Observations of kerogen in the Lower Toarcian shales of the Paris Basin by ▶ UV fluorescence microscopy (Photo B. Alpern). (a) Fluorescence of kerogen in rocks (polished section). *Left:* abundant fluorescence allows one to locate the finely disseminated organic matter in a shallow immature shale (Fécocourt). *Right:* fluorescence has almost disappeared in a deep shale buried at 2500 m (Bouchy). (b) Kerogen concentrate isolated from shale by chemical techniques. (c) Kerogen concentrate isolated from shale by physical techniques. Larger fragments of algae (Tasmanaceae) and pollen grains are proportionally more abundant than in the original rock

4.3 Microscopic Constituents of Kerogen

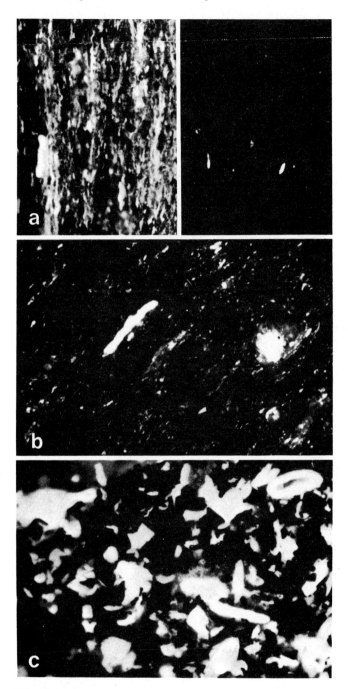

Fig. II.4.3 a–c

sources. They are also based on palynological type preparation, examined in transmitted light (Fig. II.4.2). Humic organic matter refers to kerogen where plant tissues are considered to be a major contributor. Sapropelic organic matter is commonly used to designate a finely disseminated kerogen, exhibiting an amorphous, cloudy, or vaporous shape after acidic destruction of minerals. However, physicochemical analysis of kerogen shows that the sapropelic optical type may include some kerogen whose chemical composition is comparable to the humic type (Section 4.8).

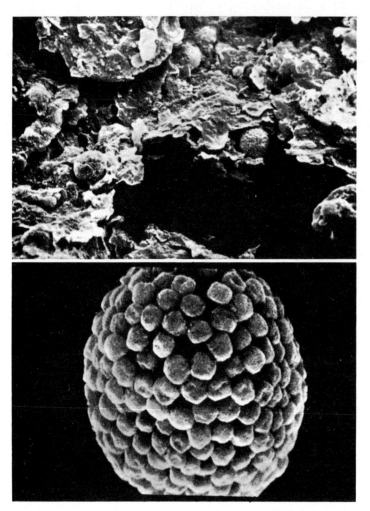

Fig. II.4.4. Kerogen containing spherical aggregates of pyritic microcrystals (framboids): Silurian of Libya. The photograph was taken by electron scanning microscope, and the horizontal dimension of the picture is ca. 70 μm. *Below:* the detail of a single framboid is shown. Each grain is a pyrite crystal coated with a thin film of organic matter. The diameter of the framboids is ca. 10 μm, and the size of the individual grains is 0.3 to 0.6 μm (Combaz, 1970)

Observations made in reflected light are usually described in terms of *liptinite, vitrinite* and *inertinite*. This order corresponds to increasing values of reflectance in the same piece of rock. Liptinite macerals are considered to be derived mostly from algae (alginite) or spores (sporinites), with occasional contributions of cutin, resins, waxes, etc. Vitrinite and inertinite are usually thought to have issued from higher plant tissues, the structure of which is still recognizable in the oxidized inertinite, whereas vitrinite has passed through a gelification stage. Other constituents are more difficult to relate to a specific origin: bitumites (Alpern, 1970), bitumens (Robert, 1971), bituminite and exudatinite (Teichmueller, 1974). They are probably not strictly equivalent, despite the similarity of names.

The combination of reflectance and fluorescence microscopy sometimes makes possible the distinction between reworked and contemporaneous organic matter in sedimentary sequences. The organic concentrates of rocks containing "sapropelic" kerogen, taken at shallow or moderate depth, normally show a strong fluorescence. However, some fragments differ by the absence of fluorescence and by strong reflectances. These fragments frequently represent reworked organic material that once had been buried to great depth, and therefore has gained high reflectance and has lost fluorescence. Obviously, this organic material comes from the erosion of older rocks. As highly carbonized organic material (like anthracite or other high-rank coal) is fairly resistant to weathering, it can be transported to the new depositional basin and mixed with young organic debris. Hagemann has shown the occurrence of "trimacerals", which are a typical reworked detrital component of coal. There are also types of organic particles, which show a high reflectance e.g., fusinite, although they have never been buried deeply. The high reflectance of these particles is probably caused by oxidation processes at the surface, either microbially or by abiological means. Even peats contain such high reflecting particles.

4.4 Chemical and Physical Determination of Kerogen Structure

Different physical and chemical techniques of analysis are able to give valuable information on the structure of kerogen. However, as shown in the previous discussion, kerogen is composed of different basic constituents in varying proportions. Some of them can be identified as macerals, the rest being frequently an amorphous mass. Therefore, an analysis of a ground rock sample results usually in a global figure, possibly integrating a wide range of individual compositions from different constituents. The only exception is the electron diffraction method, which gives a true image of the particles tested, even in the size range of several angstroms. The problem of integrating the various properties of different constituents is particularly acute when reworked material, already deeply transformed, is reincorporated in a new sediment. For this reason, a preliminary microscopic examination is always advisable.

From a quantitative point of view, amorphous kerogen is usually far predominant over recognizable organic debris. Therefore, all chemical and physical results reviewed below can be considered as mostly related to amorphous kerogen unless otherwise stated.

The principal techniques of chemical and physical constitutional analysis have been reviewed by Robinson (1969) for kerogen of the Green River formation, and by Durand et al. (1972) and Espitalié et al. (1973) for the Lower Toarcian shales of the Paris basin. More recently a group of scientists, working on several evolutive sequences of organic matter from different geological origin, provided a complete review of the principal analytical techniques and their contribution to the understanding of kerogen structure: Durand and Monin (1980), Durand-Souron (1980), Rouxhet et al. (1980), Oberlin et al. (1980), Marchand et al. (1980), Vandenbroucke (1980).

4.5 Chemical Analysis

a) *Elemental analysis* shows that carbon and hydrogen are the main atomic constituents in all types of kerogen (Table II.4.1). For 1000 carbon atoms, hydrogen varies between 500 and 1800 atoms, depending on the type and the evolution of the organic matter. The next abundant atom, oxygen, amounts to only 25 to 300 atoms. Usually, nitrogen and sulfur atoms are less abundant, 5–30 sulfur and 10–35 nitrogen atoms per 1000 carbon atoms.

b) *Bitumen extraction.* Bitumen can be separated from kerogen by extraction with organic solvents. It is reasonable to expect different kerogens to yield different bitumens. Among the constituents of bitumen, hydrocarbons have relatively low to medium molecular weights (generally below 500). Therefore, they are not likely to provide valuable information on the structure of the insoluble kerogen. On the contrary, larger structures, like asphaltenes (the part of bitumen insoluble in pentane or hexane, containing heteroatoms, with a molecular weight of several thousands), may provide more information. Analyses that are feasible on both asphaltenes and kerogen (elemental analysis, IR spectroscopy, etc.) show a broad similarity between these materials, and also a parallel variation with burial depth. Asphaltenes, being soluble in several solvents, are easier to analyze. A general description of asphaltenes from petroleum and of their structure is given by Yen (1972). He notes that asphaltenes are composed of aromatic sheets, heterocycles, and aliphatic chains. They are similar to small entities of kerogen. Because of their lower molecular weight, they are partially soluble and thus more easily analyzed than kerogen. Composition and structure of the asphaltenes are discussed in Chapter IV.1.8.

c) *Oxidation, hydrogenolysis and pyrolysis* are three methods of degrading kerogen to form lower molecular weight compounds, which are easier to analyze.

Robinson (1969) and co-workers have shown that alkaline potassium permanganate oxidizes the kerogen of the Green River formation into oxalic and volatile acids and CO_2. Using chromic acid, Burlingame et al. (1969) were able to determine part of the chain substituents on kerogen. Philp and Calvin (1977) applied the same technique to kerogen-like material in Recent algal mats from Baja California.

4.5 Chemical Analysis

Table II.4.1. Elementary composition of some typical kerogens

Kerogen type	Basin or country	Formation and/or age	Evolution stage[a]	Weight %					Atomic %				
				C	H	O	N	S	C	H	O	N	S
I	Piceance basin	Green River shales, Eocene	D	76.5	10.0	10.3	0.6	2.6	37.2	58.3	3.7	0.3	0.5
	Uinta basin	Green River shales, Eocene	D	75.9	9.1	8.4	3.9	2.6	38.7	55.9	3.2	1.7	0.5
	Uinta basin	Green River shales, Eocene	C	80.9	8.6	4.4	3.8	2.3	42.4	54.0	1.8	1.0	0.8
	Uinta basin	Green River shales, Eocene	C	82.7	4.1	8.3	1.7	3.2	58.8	34.9	4.4	1.0	0.9
	Australia	Coorongite, Recent	D	77.5	10.8	9.3	0.4	2.0	35.9	60.4	3.2	0.2	0.3
	Australia	Tasmanite, Permian	D	75.9	9.4	8.8	2.1	3.8	38.4	56.7	3.3	0.9	0.7
	France	Autun boghead, Permian	C	82.2	9.9	4.1	1.3	2.5	39.8	57.7	1.5	0.5	0.5
II	Paris basin	L. Toarcian	D	72.6	7.9	12.4	2.1	4.9	40.3	52.5	5.2	1.0	1.0
	Paris basin	L. Toarcian	C	85.4	7.1	5.0	2.3	0.2	48.4	48.4	2.1	1.1	0.0
	N. Sahara basin	Silurian	C	80.6	5.9	6.4	3.4	3.8	50.3	44.1	3.0	0.9	1.8
	N. Sahara basin	Silurian	M	85.4	3.5	5.6	2.1	3.3	63.6	31.0	3.1	0.9	1.4
	Esthonia (USSR)	Kukersite, Ordovic.	D	73.5	8.3	15.6	0.4	2.2	39.5	53.6	6.3	0.2	0.4
	Rhine Graben	Messel shales, Eocene	D	69.3	8.3	18.0	2.6	1.8	37.4	53.7	7.3	1.2	0.4
III	Douala basin	U. Cretaceous	D	72.7	6.0	19.0	2.3	0.0	45.2	44.8	8.8	1.2	0.0
	Douala basin	U. Cretaceous	C	83.3	4.6	9.5	2.1	0.2	55.1	38.9	4.7	1.2	0.0
	Douala basin	U. Cretaceous	M	91.6	3.2	2.9	2.0	0.3	68.1	29.0	1.6	1.3	0.0
Coal	Yugoslavia	Lignite, Pliocene[b] Rm = 0.43%	D	68.6	5.1	21.2	2.6	2.5	46.0	41.2	10.7	1.5	0.6
	Germany	Coal, Carboniferous[b] Rm = 1.22%	C	88.3	5.0	3.9	2.0	0.8	57.5	39.3	1.9	1.1	0.2
	Germany	Anthracite, Carboniferous[b] Rm = 2.75%	M	91.3	3.2	3.2	1.8	0.5	68.0	28.9	1.8	1.1	0.2

[a] Evolution stages: *D* diagenesis; *C* catagenesis; *M* metagenesis.
[b] *Rm* reflectance of vitrinite.

Vitorović et al. (1974, 1977) used a method of stepwise oxidation of kerogen from Aleksinac shale (Miocene, Yugoslavia) by alkaline permanganate. Oxidation products of each step were analyzed separately by gas chromatography–mass spectrometry. They proved this kerogen to consist of two types of organic matter: one is more aromatic; the other, mainly of aliphatic type, is more abundant. A general review of the main oxidative studies of kerogen is presented in Vitorovic (1980).

Pyrolysis with a subsequent analysis of the pyrolysis products by gas chromatography was also used with the same objective. In particular, Espitalié et al. (1977) developed a pyrolysis method for classification of the various types of kerogen, based on the relative contents of hydrocarbons and carbon dioxide generated during pyrolysis. Horsfield and Douglas (1980) pyrolyzed various kerogen constituents and coal macerals. Béhar et al. (1984) have studied kerogens and asphaltenes by pyrolysis-gas chromatography showing a great similarity in composition. Furthermore, the pyrograms are diagnostic of the different kerogen types. Hydrogenolysis has been extensively used by coal chemists.

d) *Functional analysis* has been developed to evaluate oxygen functional groups. For the Green River formation, Fester and Robinson (1966) determined the distribution of oxygen between the various groups: carboxyl 15%, ester 25%, hydroxyl, carbonyl, amide 6.5%, unreactive oxygen (probably ether) 52.5%. Later, Robin and Rouxhet (1976) combined reactions selective of functional groups with IR spectrophotometry to evaluate the respective amounts of these groups, and especially those including oxygen. Their respective amounts vary greatly with increasing depth. A selective degradation of ether bonds enabled Michaelis and Albrecht (1979) to identify acyclic isoprenoid structures present in kerogen, and to relate them to membrane lipids of archaebacteria.

4.6 Physical Analysis

a) Electron Diffraction and X-ray

Electron diffraction has been used by Oberlin et al. (1974) to show the degree of organization of carbon structures in amorphous kerogen at different scales from 5 to 500 Å and their evolution during burial. The method shows the existence of stacks of two to four more or less parallel aromatic sheets in kerogen. The diameter of an aromatic sheet is between 5 and 10 Å, corresponding to less than ten condensed aromatic cycles. In addition the aromatic sheet can bear aliphatic or functional groups, but these groups are not seen by this technique. The most frequent number of sheets per stack is two, but sometimes three or four occur. The interlayer distance ranges from 3.4 Å to 6 or 7 Å in shallow kerogens but it is relatively narrow in deep ones, with a maximum frequency from 3.4 to 4.0 Å. In shallow kerogens the orientation of the building blocks is completely random. With increasing depth, larger aggregates or clusters tend to emerge; the aromatic sheets keep their size (5–10 Å) and individuality, but become progressively parallel. The extension of such aggregates may reach 80–500 Å in very deep samples. However the orientation of the aggregates are in mutual disorder.

The electron micro-diffraction technique has been used to study the artificial carbonization of industrial substances. It allows one to locate carbon diffracting structures down to 10 Å or less. The dark field image technique is particularly interesting for kerogen studies: a diaphragm placed in the focal image plane gives way only to the beams at a preselected 2 Θ angle, corresponding to (002) of the carbon lattice. Thus an image on a black background is formed, where the only illuminated areas are those where a carbonaceous structure is present. The use of this technique in combination with normal electron microscopy on the same sample makes possible the location of these structures (Fig. II.4.5).

The lattice fringe image technique, using interferences of normal and diffracted beams in combination with a very high magnification ($\times$ 5,000,000–8,000,000), allows a visualization of the edge of the individual aromatic carbon sheets. Thus, it becomes possible to measure the distance and the diameter of the associated aromatic sheets (Fig. II.4.6).

A review of electron diffraction techniques and their application to understanding kerogen structure is provided by Oberlin et al. (1980).

More classic X-ray diffraction has been used successfully in coals and in petroleum asphaltenes. A carbon crystalline organization is known to emerge progressively in coals with increasing rank from lignite to anthracite. On the other hand, Yen et al. (1961) have used X-ray diffraction and other techniques in order to propose a structure of the asphaltenes.

b) Infrared Spectroscopy (Absorption)

This technique allows an evaluation of the relative importance of carbonyl and/or carboxyl groups versus aliphatic chains plus saturated rings. However, it is not possible to discriminate between the latter two. The progressive disappearance of carbonyl and aliphatic groups is also indicative for the diagenetic and catagenetic evolution of kerogens.

Infrared spectra of kerogen closely resemble the spectra obtained from coal or asphaltenes. The most likely assignment of the various bands is shown in Figure II.4.7, based mainly on publications of King et al. (1963), Erdman (1965), Espitalié et al. (1973) and Robin (1975, 1976). The main zones of interest seem to be:

- a wide asymetric band centered at 3430 cm^{-1}, related to OH groups (phenolic, alcoholic, carboxylic OH);
- a strong absorption resulting from several fine bands but showing mostly two maxima (2920 and 2855 cm^{-1}), related to CH_2 and CH_3 aliphatic groups;
- a wide band with a maximum around 1710 cm^{-1}, attributed to various C=O groups (ketones, acids, esters);
- a wide band with a maximum around 1630 cm^{-1}, mostly related to aromatic C=C, although there is definitely some contribution of other structures: olefinic C=C, particular types of C=O groups (bridged quinonic carbonyl), and of free water: Robin and Rouxhet (1976);

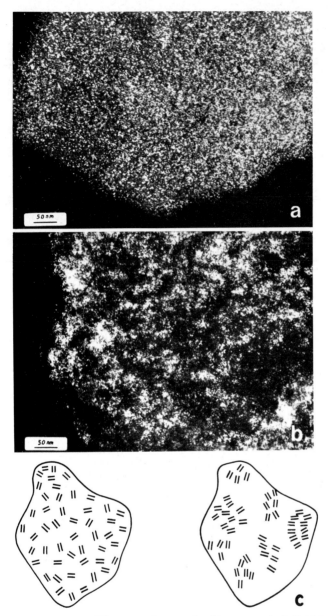

Fig. II.4.5a–c. Electron micrograph of kerogen: dark field technique. *Illuminated areas* are related to carbonaceous structures. (a) Immature kerogen, where the stacks of aromatic sheets are still orientated at random. The sample corresponds to the diagenesis–categenesis boundary, and catagenetic rearrangements have not yet started. (b) Very mature kerogen, where aggregates or clusters appear. Within these clusters, aromatic sheets show a parallel orientation. The sample corresponds to the catagenesis–metagenesis boundary: catagenetic rearrangements are completed. (c) Schematic representation of the orientation of the aromatic sheets corresponding respectively to (a) *(left)* and (b) *(right)* (A. Oberlin and J. Boulmier)

4.6 Physical Analysis

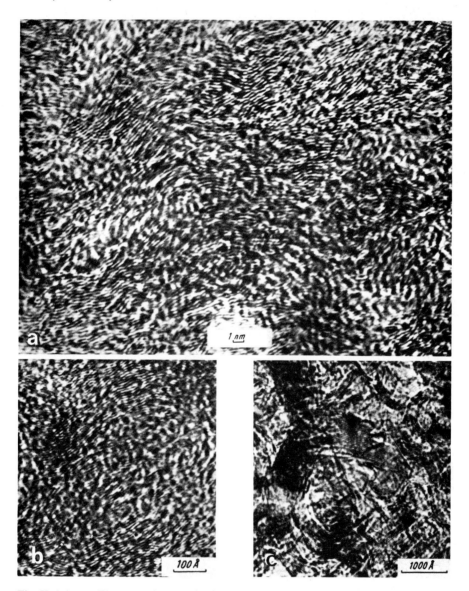

Fig. II.4.6a–c. Electron micrograph of kerogen: lattice fringe technique. This technique, using interferences of normal and diffracted beams, allows a visualization of the aromatic sheets. These are marked in the micrograph by a succession of several parallel black and white stripes. (a) *Top:* very mature kerogen corresponding to the catagenesis–metagenesis boundary (vitrinite reflectance 2.4%). *Bottom:* development of the aromatic sheets in anthracite is shown for comparison: (b) natural anthracite; (c) anthracite heated to 1000°C (A. Oberlin and J. Boulmier)

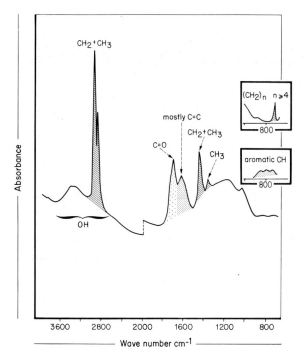

Fig. II.4.7. Typical IR spectrum of kerogen (type II). The *two insets* show specific absorption bands of two other kerogens: (a) 720 cm^{-1} related to long aliphatic chains mostly occurring in type-I kerogen; (b) 930 to 700 cm^{-1} related to aromatic CH (bending out of plane) mostly occurring in very mature kerogen. (Adapted from Robin, 1975)

- an absorption band at 1455 cm^{-1}, due to CH$_3$ and to linear and cyclic CH$_2$ groups, and another band at 1375 cm^{-1}, related to CH$_3$ only;
- a very wide band from 1400 to 1040 cm^{-1}, that includes C–O stretching and OH bending;
- a succession of weak bands from 930 to 700 cm^{-1}, related to various aromatic CH (bending out of plane) and dependent on the number of adajcent protons;
- a 720 cm^{-1} band, due to aliphatic chains of 4 or more carbon atoms.

c) Nuclear Magnetic Resonance

Nuclear magnetic resonance (NMR) has been applied to dissolved asphaltenes in liquid phase for characterization of their degree of aromaticity. Recently, some new possibilities of applying NMR to kerogen were offered by the Cross Polarization, Magic Angle Spinning (CP/MAS)^{13}C NMR. This technique operates on the solid phase and allows an evaluation of the aliphatic/aromatic carbon in kerogen. Dennis et al. (1982) applied it to characterize the thermal alteration of Cretaceous black shales in the Eastern Atlantic by basaltic intrusions. Miknis et al. (1982) used the same technique to evaluate the genetic potential of oil shales.

4.7 General Structure of Kerogen

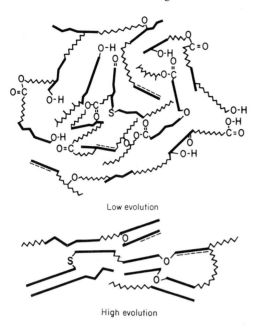

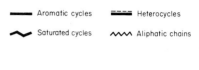

Fig. II.4.8. Structural model of macromolecular arrangements occurring in kerogen (type II). *Top:* immature kerogen, corresponding to a low evolution stage. *Bottom:* the same kerogen after reaching a high evolution stage through catagenesis. (After Tissot and Espitalié, 1975)

There are, however, limitations to the interpretation of NMR spectra due to the complexity of kerogen (Miknis et al., 1982).

d) Thermal Analysis

Differential thermal analysis (DTA) has been applied to kerogen by von Gaertner and Schmitz (1963). More recently Espitalié et al. (1973) used both DTA and thermo-gravimetric analysis (TGA) for studying kerogen. These techniques may give some indication of the structure of kerogen, but their main contribution is to follow the evolution of kerogen (Chap. II.5).

When combined with a technique for characterizing the products from pyrolysis, however, TGA provides valuable information on the kerogen structural type. For example, Souron et al. (1974, 1980) using TGA and mass spectrometry, have shown the relative importance of aliphatic, naphthenic and aromatic structures in several types of kerogen, and also their changes during catagenetic evolution of the lower Toarcian shales from the Paris basin. Comparable techniques have been used by Juentgen on coal, and by Marchand et al. (1974) on spores and pollens.

4.7 General Structure of Kerogen (Fig. II.4.8)

A general structure of kerogen may be proposed as a result of the data collected from various analytical techniques. The proposed structure applies mainly to amorphous kerogen because of the large predominance of amorphous over structured material in the kerogens studied. The average amorphous kerogen is a three-dimensional macromolecule. It consists very probably of nuclei that are crosslinked by chain-like bridges. Both the nuclei and the bridges may bear functional groups. In addition, lipid molecules can be entrapped in the kerogen matrix, as in a molecular sieve.

a) *Nuclei* are formed by stacks, comprising two to four more or less parallel aromatic sheets. Each of the sheets or layers contains a relatively small number (less than 10) of condensed aromatic rings, including occasional heterocycles containing nitrogen, sulfur and possibly oxygen. The diameter of such cyclic assemblages is smaller than 10 Å. The number of layers is frequently two per stack, and the interlayer distance greater than 3.4 Å, with a wide range in shallow kerogen, and a narrow maximum around 3.7 Å in deeply buried kerogens. The stacks are the basic building blocks of kerogen, whose increasing organization with depth determines the evolutionary trend of the organic matter in sediments. These nuclei bear alkyl chains (linear or with few short branches) substituted on the rings, naphthenic rings, and various functional groups.

b) Bridges may consist of structural arrangements such as:
- linear or branched aliphatic chains: $-(CH_2)_n-$, attached to the nuclei as substituents
- oxygen or sulfur functional bonds: ketones: $-\underset{\underset{O}{\|}}{C}-$

  ester: $-\underset{\underset{O}{\|}}{C}-O-$, ether: $-O-$, sulfide: $-S-$, or disulfide: $-S-S-$

- combination of an aliphatic chain R with a functional group, e.g., aliphatic ester: $-\underset{\underset{O}{\|}}{C}-O-R$

c) Superficial functions, substituted on nuclei or chains, may include various groups such as hydroxyl: $-OH$, carboxyl: $-\underset{\underset{O}{\|}}{C}-O-H$ methoxy: $-OCH_3$ etc.

d) There are indications that kerogens have molecular sieve properties similar to those observed in coal. A hint in this direction is the fact that samples which have first been extracted exhaustively, and then treated by acid for kerogen isolation released additional hydrocarbon molecules upon a new extraction.

Oxygen seems to play a particular role in the structure of kerogen. It may amount to from 10 to 25% of kerogen by weight in shallow, immature sediments. The highest values occur in detrital formations, where the input of terrestrial plants is important, e.g., kerogen of the Upper Cretaceous of the Douala basin, type III, see Section II.4.8. An evaluation of the proportion of carbonyl and carboxyl $C=O$ groups, versus other oxygen combinations can be made by comparing the 1710 cm^{-1} $C=O$ band on IR spectra, with the total oxygen content. In the Lower Toarcian shales of the Paris basin (Fig. II.4.9), oxygen amounts to an average of 10% by weight in shallow immature samples. With increasing depth the $C=O$ bands is reduced to zero, and an average amount of 5% oxygen still

4.7 General Structure of Kerogen

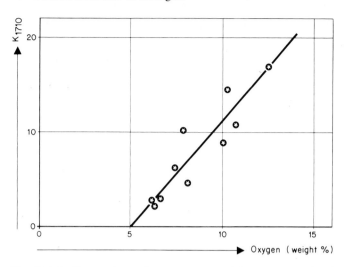

Fig. II.4.9. Comparison of oxygen content (elementary analysis) and IR absorption band C=O at 1710 cm^{-1} (K 1710): kerogen of the Lower Toarcian shales, Paris Basin. (Data from Robin, 1975)

remains in kerogen (Espitalié et al., 1973). Thus, about half of the original oxygen was present in carbonyl or carboxyl groups. Computed as C=O, these labile groups account for up to 10% by weight of the total immature kerogen of the Paris basin. The proportion would be ca. 20 to 30% in kerogen of detrital series, with abundant contribution of higher plants (type III, Douala basin). The rest of the oxygen is likely to be present in ether bonds, in phenols, or in heterocyclic structures.

Other observations, also made on the Lower Toarcian shales of the Paris basin, provide some information on the location of the C=O groups. At a depth of 2500 m, where bitumen is abundant, the 1710 cm^{-1} C=O band was found to be more pronounced in the total organic matter — especially the extractable bitumen fraction — than in the kerogen. This fact suggests that carbonyl and carboxyl groups are more frequently related to aliphatic chains and small cyclic structures than to polycondensed aromatic and heterocyclic nuclei. Furthermore, the inverse relationship (Fig. II.4.10) between the amount of bitumen generated and the intensity of the C=O infrared band in kerogen is an indirect proof that in this particular shale many petroleum constituents are bound to kerogen by ester bondings.

Nitrogen of peptidic bonds and functional groups seem to have been removed during shallow diagenesis of the organic matter (Chap. II.2). The remaining nitrogen in kerogen generally amounts to 2 to 3% by weight and may be essentially present in heterocyclic condensed structures, because it is released only to a limited extent during catagenesis but more abundantly during high-temperature pyrolysis experiments, e.g., Fisher assays and oil-shale retorting. However, a certain proportion of nitrogen is still present in the condensed polyaromatic nuclei of the most evolved kerogens obtained upon pyrolysis.

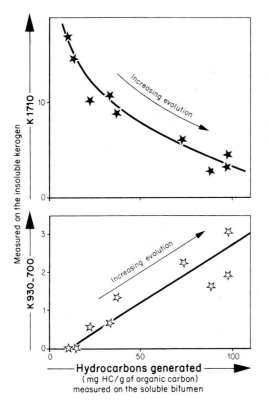

Fig. II.4.10. Relationship between the amount of hydrocarbons extracted from the Lower Toarcian shales of the Paris Basin and the intensity of specific IR absorption bands of the insoluble kerogen: *Top:* C=O band at 1710 cm^{-1} (K 1710). *Bottom:* aromatic CH bands from 930 to 700 cm^{-1} (K 930–700). These are due to desubstitution of aliphatic chains from aromatic nuclei. (Data from Tissot et al., 1971; Robin, 1975)

Sulfur is probably present in both forms, as an atomic constituent of heterocycles, and possibly as sulfide or disulfide bonds. During catagenesis it is released as benzothiophene derivatives, and later as hydrogen sulfide.

There is definitely some relationship between the respective structures of kerogens, asphaltenes from bitumen, and asphaltenes from crude oils. Kerogens and asphaltenes from bitumen are especially similar, judging from elemental analysis, IR spectroscopy, etc. Therefore, this type of asphaltene has to be thought of as small fragments of the kerogen structure. Although they have an overall composition similar to that of kerogen, the molecular weight of asphaltenes from bitumen is smaller: therefore, they are soluble in the usual organic solvents.

Asphaltenes precipitated from crude oils have been studied by numerous physical and chemical techniques, and their structure has been described by Yen (1972). The overall arrangement is somewhat similar to those of kerogens and of asphaltenes from bitumen. The oxygen content is rather different, however, as shown by elementary analysis (Table IV.1.8, Chap. IV. 1). Oxygen atoms represent 1.5–10% of the total number of atoms constituting kerogen or bitumen asphaltenes, but 1.5% or less in crude oil asphaltenes. This difference can be explained by the elimination of heteroatomic bonds during catagenetic evolution and formation of petroleum, and also by a low migration efficiency of the most oxygenated asphaltic fraction.

4.8 Depositional Environment and Composition of Kerogen: the Evolution Paths

The chemical structure of a shallow immature kerogen (relative importance of the different nuclei, bonds and functions) varies with the original mixture of organisms and with the physical, chemical and biochemical conditions of deposition. These differences are attested by microscopic examination of the constituents, elementary analysis. IR spectroscopy, degradative oxidation and various experiments of pyrolysis. The variations inherited from the young sediment remain substantial for shallow ancient sediments, but become progressively weaker with increasing burial depth. However, the differences can still be recognized over a long time and depth span, i.e., until the beginning of the zone where gas is generated by thermal cracking.

The global atomic composition of the three major elements (C, H, O) is shown in Figure II.4.11, which is a graph of atomic H/C versus O/C ratios. This figure provides a useful approach to classification of kerogens (Forsman, 1963; McIver, 1967; Welte, 1969; Durand et al., 1972). Kerogens taken at various depths from the same formation normally group along a curve, called here an *evolution path*. Closely related environments of deposition result in the same path (Tissot et al., 1974). These different curves start with different hydrogen: oxygen ratios, according to the original organic material and conditions of deposition. The curves come together for very deep samples as the kerogen approaches 100% carbon. This plot was first used by van Krevelen (1961) to characterize coals and their coalification path (Fig. II.4.12). We propose to call this type of diagram a "van Krevelen diagram".

The extreme types of disseminated organic matter correspond on one hand (type I) to kerogen, which is rich in aliphatic structures and consequently in hydrogen, as in some algal deposits; on the other hand (type III) to kerogen, which is rich in polyaromatic nuclei and oxygen groups, like organic matter made of plants from terrestrial origin (Fig. II.4.11).

a) *Type I* refers to kerogen with a high initial H/C atomic (ca. 1.5 or more) and a low initial O/C ratio (generally smaller than 0.1). Such kerogen comprises much lipid material, particularly aliphatic chains. The content of polyaromatic nuclei and heteroatomic bonds is low, compared with the other types of organic matter. The small amount of oxygen present is mainly found in ester bonds. When subjected to pyrolysis up to 550 or 600°C, the kerogen produces a larger yield of volatile and/or extractable compounds than any other type of kerogen (up to 80% by weight from shallow immature samples) and likewise a higher yield of oil. Paraffinic hydrocarbons are predominant over cyclic structures.

The high proportion of lipids may result either (1) from a selective accumulation of algal material or (2) from a severe biodegradation of the organic matter (other than lipids and microbial waxes) during deposition. The first source includes organic rich sediments made up mostly of algae, particularly those derived from lacustrine *Botryococcus* and associated forms (Marchand et al., 1969), e.g., Autun and Campine bogheads, torbanite from Scotland, coorongite from South Australia, and their marine equivalents, such as tasmanite from

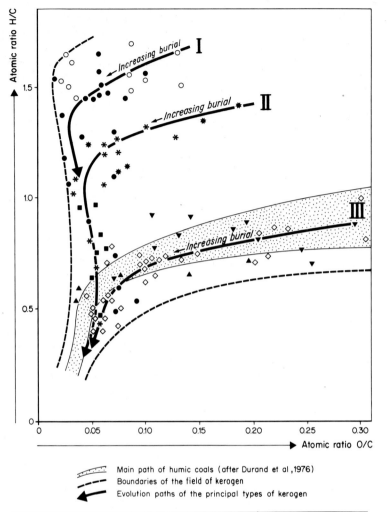

Fig. II.4.11. Principal types and evolution paths of kerogen: types I, II and III are most frequent. Kerogens of intermediate composition also occur. Evolution of kerogen composition with increasing burial is marked by an *arrow* along each evolution path I, II and III. We propose to name this type of diagram after van Krevelen

4.8 Depositional Environment and Composition of Kerogen: Evolution Paths

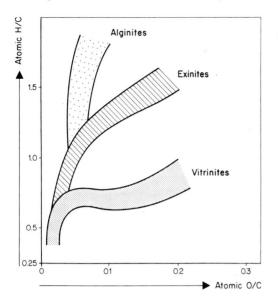

Fig. II.4.12. Evolution paths of maceral groups in coals. (After van Krevelen, 1961)

Tasmania. The second source includes some rich oil shales, where disseminated organic matter has been extensively reworked by microorganisms, so that the kerogen is derived mostly from a biomass of reworked and other microbial lipids. This situation seems to occur particularly in lacustrine environments. The prolific oil shales of Colorado, Utah and Wyoming (Green River shales) seem to result from a combination of both algal and microbial lipids. The occurrence of type-I kerogen is relatively rare compared to the other types.

b) *Type II* is particularly frequent in many petroleum source rocks and oil shales, with relatively high H/C and low O/C ratios. The polyaromatic nuclei and the heteroatomic ketone and carboxylic acid groups are more important than they are in type I, but less than in type III. Ester bonds are rather abundant (Fig. II.4.13). Saturated material comprises abundant naphthenic rings and aliphatic chains of moderate length. Sulfur is also present in substantial amounts, located in heterocycles and probably also as sulfide bond. In the associated bitumen, the cyclic structures (naphthenic, aromatic or thiophenic) are abundant and the sulfur content is higher than in the bitumen associated with the other types.

Type-II kerogen is usually related to marine sediments where an autochthonous organic matter, derived from a mixture of phytoplankton, zooplankton and microorganisms (bacteria), has been deposited in a reducing environment.

The yield from pyrolysis of this type of kerogen is lower than type I, but it may still provide commercial oil shales. For example, on pyrolysis, the Lower Toarcian shales of the Paris basin produce ca. 60% of the initial weight of organic matter. In dispersed state, these kerogens are the source material of a great number of oil or gas fields: Devonian and Colorado group of Cretaceous age from Western Canada, Paleozoic from North Africa, some Cretaceous and Tertiary source beds of West Africa, Jurassic of Western Europe and Saudi Arabia, etc.

c) *Type III* refers to kerogen with a relatively low initial H/C ratio (usually less than 1.0) and a high initial O/C atomic ratio (as high as 0.2 or 0.3). This type of

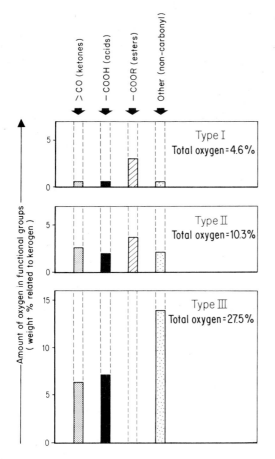

Fig. II.4.13. Amount of oxygen engaged in various functional groups in shallow immature samples of different kerogens: Eocene, Green River shales. Uinta Basin (type I); Lower Toarcian, Paris Basin (type II); U. Cretaceous, Douala Basin (type III). Type I contains little oxygen, which is mostly engaged in ester groups. Type III contains much oxygen in C=O bonds: acids, ketones. Ester bonds have not been detected. It also comprises a large amount of noncarbonyl oxygen, probably engaged in heterocycles, in quinones, or in ether bonds, methoxyl groups, etc. (After Robin, 1975)

kerogen comprises an important proportion of polyaromatic nuclei and heteroatomic ketone and carboxylic acid groups, but no ester groups (Robin, 1975; Fig. II.4.13). Noncarbonyl oxygen is possibly included in ether bonds. Aliphatic groups are subordinate constituents in varying proportions of the organic matter. They consist of a few long chains (more than 25 carbon atoms), originating from higher plant waxes, some chains of medium length (15 to 20 carbon atoms) from vegetable fats, methyl groups and other short chains. This type of kerogen is comparatively less favorable for oil generation than are types I or II, although it may provide convenient gas source rocks, if buried at sufficient depth. It is also less productive on pyrolysis. It does not include any oil shales.

Type-III kerogen is derived essentially from continental plants and contains much identifiable vegetal debris. Plant organic matter is incorporated in sediment either directly or through its alteration products in soil humic acids. Microbial degradation in the basin of deposition is usually limited due to important sedimentation and rapid burial, because this type is rather frequent in thick detrital sedimentation along continental margins. Examples of type-III kerogen are found in the Upper Cretaceous of the Douala basin (Cameroon) and in the

4.8 Depositional Environment and Composition of Kerogen: Evolution Paths   155

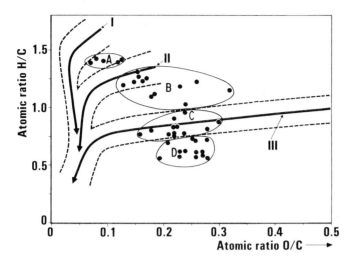

Fig. II.4.14. Elemental analysis of some kerogens from the Cretaceous black shales of the North Atlantic cored during the Deep Sea Drilling Program. The main evolution paths of type I, II and III kerogens are shown. *Group A* refers to marine organic matter; *group B* to a mixture of marine and terrestrial organic matter, some of the samples being possibly degraded *(right part): group C* to terrestrial moderately degraded organic matter; *group D* to residual organic matter. All samples belong to the diagenesis stage of evolution. (Adapted from Tissot et al., 1979)

lower Mannville shales of Alberta, and also in some Tertiary deltas, such as the Mahakam delta in Indonesia.

d) In addition to these three major types of kerogen, and their admixtures, there is also a *residual type* of organic matter, which is characterized by abnormally low H/C ratios associated with high O/C ratios. This situation was, for instance, met in several cores taken in the North Atlantic Ocean as part of the Deep Sea Drilling Project (DSDP): kerogens isolated from many Cretaceous sediments of the Bay of Biscay (leg 48) and some of the Blake-Bahama basin (leg 44) show a H/C ratio as low as 0.5 or 0.6, associated with a high O/C ratio (0.25 to 0.30) at shallow depths of burial (200 to 1000 m): Figure II.4.14, group D; Tissot et al. (1979, 1980). Such low values of H/C ratio are normally found only in sediments which have been buried at great depths and, in such case, they are associated with low O/C ratios (less than 0.10). Infrared spectroscopy (Castex, 1979) has confirmed the abundance of aromatic nuclei and oxygen-containing groups and the absence of aliphatic chains in this type of kerogen: Figure II.4.15. Furthermore, optical examination of the same samples mainly shows coaly fragments of oxidized organic matter or charcoal, in transmitted light, and inertinite in reflected light.

This residual type of kerogen exhibiting a very low H/C ratio and a relatively high O/C ratio cannot generate any hydrocarbons and is considered therefore to be one form of "dead carbon" in the sense of petroleum generation. The term "inertinite" is originally referring to a comparable property of this kind of organic

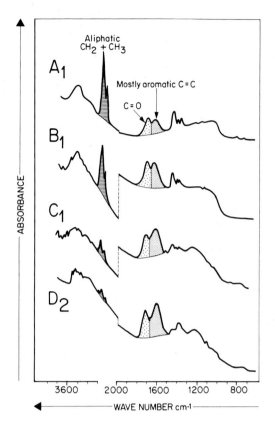

Fig. II.4.15. Infrared spectrometry of four typical kerogens. The symbols refer to samples obtained from the Cretaceous black shales of the North Atlantic and shown in Figure II.4.14 (Tissot et al., 1979)

material. It is derived from coal science, where it was observed that in the coking process certain particles in the coking coal did not react and were hence termed inertinitic. Obviously, there are several origins of this kind of nonreactive organic material with a low H/C ratio in sedimentary rocks: it may be reworked, oxidized organic material; it may be inertinitic material stemming from subaerial weathering or biological oxidation in swamps and soils. Examples of this kind are found among the above-cited DSDP samples and on microscopic examination of high-reflectance particles in Recent peats. In all cases, this type of kerogen has no potential for hydrocarbon generation.

From a practical point of view, another form of "dead carbon" is that organic material that has lost its hydrogen through a natural hydrocarbon generation process and can be observed as micrinite (see also p. 243) in petroleum source rock. In such cases both H/C and O/C ratios are very low. This concept has also been adopted by Leythaeuser et al. (1979) when giving a mass balance of the organic carbon in source rocks mainly based on Rock Eval pyrolysis.

e) *General remarks.* Although type-II and -III kerogens represent rather frequent situations, the composition of kerogen may cover other areas in Figure II.4.11 between the evolution paths I and III. There may be different reasons for that situation: variability of contributing organisms, including mixture of marine and terrestrial inputs, and also different conditions of preservation.

A first source of variability may result from the predominant character of some contributing organisms. For example, a marine sediment, deposited in a reducing environment may contain abundant diatoms under temperate or cold water conditions, and dinoflagellates or coccolithophorids in warm seas. The relative abundance of zooplankton, such as copepods, may introduce another cause for variability. The main constituents of these various groups of organisms, shown in Table I.4.1, explain that the original composition of the related kerogens may vary, although they all belong to type II.

Furthermore, some kerogens, although marine, show a relatively high O/C ratio, probably due to an important contribution of plant debris from the continent, like the Viking shales, Cretaceous from Alberta (McIver, 1967), or some Cretaceous black shales of the North Atlantic. They occupy an area between types II and III on the van Krevelen diagram: part of group B, Figure II.4.14.

Finally, another cause of variability is suspected, although insufficiently documented: biochemical and chemical degradation of the organic matter prior to or during deposition may result in a decrease of the H/C ratio and an increase of the O/C ratio. An extreme case is obviously the residual type of organic matter presented above, which has suffered a strong subaerial oxidation, and no longer presents potential for hydrocarbon generation. Moderately oxidated marine or marine/terrestrial organic matter may eventually result in an increase of the O/C ratio (≥ 0.2) with a lowered H/C ratio (ca. 1.0) still sufficient to generate some hydrocarbons and particularly gas. Among the black shales of the North Atlantic (Fig. II.4.14) the samples located on the right hand side of group B (O/C 0.20 to 0.30) might result from such situation.

Facies variations in sediment may cause a change of the kerogen from one type to another within a given horizon or formation. For instance, Breger and Brown (1962) showed in the Chattanooga shales (Kentucky, Tennessee and Alabama) a progressive change in kerogen composition. In the area close to the shore line at the time of deposition, kerogen shows low hydrogen values. This is attributed to material of terrestrial origin, deposited in a relatively oxidizing environment. To the north, i.e., basinward, the hydrogen content increases due to a higher proportion of marine phytoplankton. Upon pyrolysis, the oil yield varies directly with the H/C ratio of the kerogen.

When comparing disseminated organic matter to coal macerals by using the van Krevelen diagram (Figs. II.4.11 and II.4.12), it can be seen that the main types I, II and III broadly fall into the same areas of the diagram as do alginite, exinite and vitrinite, respectively. This comparison is quite reasonable for alginite and some of the type I kerogens, which are made up mostly of algae, like *Botryococcus* shales. It is also valid for vitrinite and type-III kerogen. In terms of coal chemistry, these kerogens have H/C, O/C distributions comparable to the gross composition of peat, lignite and coal, and especially to the vitrinite maceral. This is understandable, as peat is primarily made of the same vegetal debris, deposited in a confined environment, where biodegradation is also limited for other reasons (setting up of antiseptic conditions).

However, the comparison, based on elementary composition, has no meaning with respect to type-II kerogen and exinites. This kerogen results from a complex mixture of organic debris, which have undergone biochemical and chemical reorganization, whereas exines have preserved their shape and individuality. Furthermore, this constituent is likely to account for a minor part only of the organic matter deposited in a marine environment.

The relationship between palynofacies observed in transmitted light (amorphous or sapropelic, humic facies) and chemical composition of kerogen is rather complex. Humic material, mainly consisting of land-derived higher plant debris, normally falls into the area of type-III kerogen. Amorphous facies results in fact from severe alteration of the original constituents, which have lost their identity. Thus, if many amorphous kerogens belong to type II, some others undoubtedly belong to type III. Floculated humic acids, for instance, stemming from a colloidal solution will undoubtedly have an amorphous appearance under the microscope. At the same time, however, their elemental analysis will group them along a low evolution path. This amorphous material is often mistaken as indicative of a "sapropelic" facies, with good hydrocarbon potential.

An example of the variability of amorphous organic matter with respect to kerogen types is shown in Figure II.4.16. The spread of the H/C and O/C ratios is covering the entire field and hence clearly shows that there is a variety of origins for amorphous organic material. Powell et al. (1982), using 58 samples of the Beaufort-Mackenzie Basin and the Canadian Arctic Islands, showed a low level of correlation between the content of amorphous material and the chemical analysis of kerogen. They attributed this situation mainly to a failure to distinguish consistently between hydrogen-poor and hydrogen-rich amorphous organic matter.

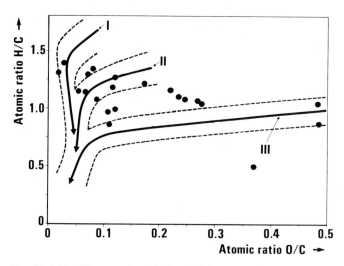

Fig. II.4.16. Elemental analysis of 21 kerogens from "amorphous" organic matter. The samples are from various origins; they all belong to the diagenesis or early catagenesis stages. The main evolution paths of type I, II and III kerogens are shown for comparison. (After Durand and Monin, 1980 with additions)

4.9 Conclusion

All physical and chemical methods for kerogen analysis converge to provide a coherent picture of the organic matter and its evolution. In particular, elemental analysis provides the framework within which other methods can be used (Durand and Monin, 1980). There are, however, limitations to its use, especially for the advanced stages of thermal evolution.

The composition of a shallow, immature kerogen depends on the nature of the organic matter incorporated in the sediments, and on the extent of microbial degradation. In turn, the composition of kerogen, particularly in respect to hydrogen (aliphatic chains) and oxygen (functional groups), determines the genetic potential of a sediment, i.e., the amount of hydrocarbons that can be generated during burial.

The most classic source rock, containing "marine organic matter, deposited in a reducing environment", broadly corresponds to type II and has a high genetic potential. On the contrary, continental organic matter deposited in environments where biodegradation is limited by quick burial, corresponds to type III and has only a moderate potential as far as oil is concerned. Most kerogens, when plotted on the van Krevelen diagram, fall on or between these types. Type I is much less frequent and refers either to selectively biodegraded organic matter, enriched in lipids, or to rocks consisting almost entirely of organic matter such as alginites.

Summary and Conclusion

Kerogen is the fraction of the organic matter in sedimentary rocks which is insoluble in organic solvents, whereas bitumen is the soluble part. Kerogen is a macromolecule made of condensed cyclic nuclei linked by heteroatomic bonds or aliphatic chains.

Different types of kerogen can be recognized by optical examination and physicochemical analyses. Three main types seem to account for most existing kerogens. They are characterized by their respective evolution path in the van Krevelen (H/C, O/C) diagram.

Type-I kerogen contains many aliphatic chains, and few aromatic nuclei. The H/C ratio is originally high, and the potential for oil and gas generation is also high. This type of kerogen is either mainly derived from algal lipids or from organic matter enriched in lipids by microbial activity.

Type-II kerogen contains more aromatic and naphthenic rings. The H/C ratio and the oil and gas potential are lower than observed for type-I kerogen but still very important. Type-II kerogen is usually related to marine organic matter deposited in a reducing environment, with medium to high sulfur content.

Type-III kerogen contains mostly condensed polyaromatics and oxygenated functional groups, with minor aliphatic chains. The H/C ratio is low, and oil potential is only moderate, although this kerogen may still generate abundant gas at greater depths. The O/C ratio is comparatively higher than in the other two types of kerogen. The organic matter is mostly derived from terrestrial higher plants.

Residual kerogen may consist of reworked, oxidized organic material, or inertinitic material stemming from subaerial weathering or biological oxidation. It is one form of "dead carbon" and has no potential for oil or gas.

Chapter 5
From Kerogen to Petroleum

As sedimentation and subsidence continue, temperature and pressure increase. In this changing physical environment, the structure of the immature kerogen is no longer in equilibrium with its surroundings. Rearrangements will progressively take place to reach a higher, and thus more stable, degree of ordering. The steric hindrances for higher ordering have to be eliminated. They are, for instance, nonplanar cycles (e.g., saturated cycles) and linkages with or without heteroatoms, preventing the cyclic nuclei from a parallel arrangement.

This constant adjustment of kerogen to increasing temperature and pressure results in a progressive elimination of functional groups and of the linkages between nuclei (including carbon chains). A wide range of compounds is formed, including medium to low molecular weight hydrocarbons, carbon dioxide, water, hydrogen sulfide, etc. Therefore, the petroleum generation seems to be a necessary consequence of the drive of kerogen to adjust to its new surroundings by gaining a higher degree of order with increasing overburden.

In unmetamorphosed sedimentary basins this process of rearrangement is not carried to the end. The level of regional metamorphism is needed to reach the graphite stage, which is the stable configuration under high temperature and pressure.

5.1 Diagenesis, Catagenesis and Metagenesis of Kerogen

Kerogen is a polycondensed structure formed under the mild temperature and pressure conditions of young sediments and metastable under these conditions. Therefore, its characteristics seem to remain rather constant, even in ancient sediments, as long as they are not buried deeply. A typical example is provided by the lower Carboniferous lignites of Moscow. Although they are about 300 million years old, their carbonization rank is still very low, because they have never been buried deeper than 200 m (Karweil, 1956). In most cases, however, as sedimentation and subsidence proceed, kerogen is subjected to a progressive increase of temperature and pressure. It is no longer stable under the new conditions. Rearrangements occur during the successive stages of diagenesis, catagenesis, and metagenesis toward thermodynamic equilibrium (Fig. II.5.1).

Such evolution has been studied in detail in various rock sequences, and particularly in the Devonian and Cretaceous of Western Canada (McIver, 1967), the Tertiary of Louisiana (Laplante, 1974), the Lower Toarcian shales of the Paris

5.1 Diagenesis, Catagenesis and Metagenesis of Kerogen

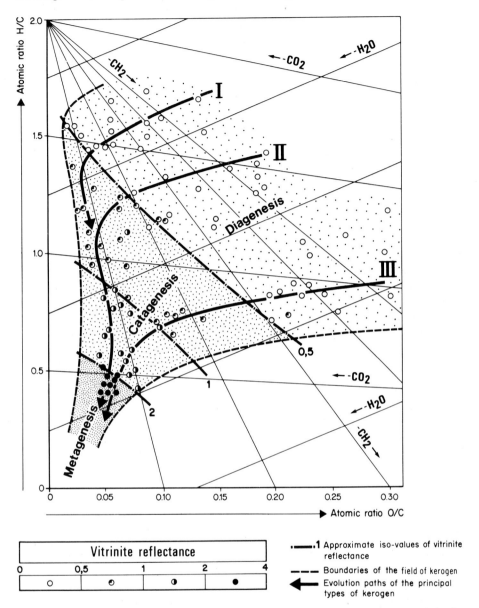

Fig. II.5.1. General scheme of kerogen evolution from diagenesis to metagenesis in the van Krevelen diagram. Approximate values of vitrinite reflectance are shown for comparison

Basin, the Silurian of the Sahara (North Africa), the Upper Cretaceous of the Douala Basin (West Africa) and the Eocene Green River shales of the Uinta Basin. In these studies, observations on the natural transformation in a geological situation were compared with experimental evolution through laboratory simulation. As the main steps of kerogen evolution are the same for all types of kerogen, we shall describe in detail degradation of type II, whose elementary composition is intermediate between other types. By using Figure II.5.1, these evolution steps can be extended to the other types of kerogen.

Kerogen is transformed when subjected to higher temperature and pressure. An ideal situation to study this transformation is in a sedimentary sequence where all geological and geochemical parameters except burial remain constant. The Lower Toarcian shales of the Paris Basin (Fig. II.5.2) are rich in organic matter (organic carbon varies from 3 to 15%) and seem to meet these requirements (Louis and Tissot, 1967). The shales represent a well-defined Ammonite zone in the upper part of the lower Jurassic sediments, dated around 180.10^6 years. Their mineralogical composition, including clay minerals, is reasonably constant over most of the basin, as is their fossil organic contents as well. As the structural history of the basin is simple, it is possible to reconstruct the burial history and compute the maximum burial depth of the formation at every location in the basin. The availability of a set of cores from various depths in the basin enables a study by microscopic, chemical and physical methods of the progressive transformation of kerogen as a function of maximum burial depth. The detailed results are reported in Tissot et al. (1971), Durand et al. (1972), Espitalié et al. (1973), Tissot et al. (1974).

In order to extend the results to deeper geological conditions (the maximum depth of the Lower Toarcian shales is slightly beyond 2500 m), some additional data from the same shales cored in Western Germany are included. The Lower Silurian shales of the Sahara (North Africa) will also be considered. They represent a well-defined and rather homogeneous stratigraphic unit which covers a wide area. The depositional environment is rather similar to that of the Lower Toarcian shales (black shales, containing Tasmanites) and, according to elementary and IR analysis, the kerogen exhibits characteristics of the same evolution path (type II, see Fig. II.4.11; Tissot et al., 1974).

The degradation of kerogen as a function of burial has been observed by the physical and chemical techniques mentioned in Chapter II.4. The main trend is a continuous increase of the carbon content of kerogen. However, three successive steps are distinguished.

a) *Diagenesis* of kerogen is marked by an important decrease of oxygen and a correlative increase of carbon content with increasing depth. When referring to the van Krevelen diagram (Figs. II.5.1 and II.5.3), this stage of evolution results in a slight decrease of H/C and a marked decrease of O/C ratios. Infrared spectroscopy has demonstrated that the diminution of oxygen is due essentially to the progressive elimination of C=O group (1710 cm^{-1} band). By combination of the analyses for functional groups and IR techniques, Robin et al. (1977) have shown the respective behavior of the various groups containing C=O, namely that acids and ketones are more affected than esters.

Kerogen reflectance is rather scattered, depending on the type of organic debris: ca. 0.5% or below on vitrinite, less on exinite. Free radicals remain scarce as paramagnetic susceptibility, χ_p, is usually low: $\chi_p \leq 5.10^{-9}$ u.e.m. CGS/g of organic carbon. The χ_p value may be somewhat higher in type-III kerogen.

In the Paris Basin the final stage of diagenesis is reached in location corresponding to maximum depth of burial from 800 to 1200 m (sample A in Fig.

II.5.3). In terms of petroleum exploration, this stage corresponds to an immature kerogen, and little hydrocarbon generation has occurred in the source rocks. However, large quantities of carbon dioxide and water and also some heavy heteroatomic (N, S, O) compounds may be produced in relation to oxygen elimination. Displacement in the van Krevelen diagram and experimental evolution (Sec. 5.2) confirm these views.

b) *Catagenesis*, the second stage of kerogen degradation, is visible in deeper samples and marked by an important decrease of the hydrogen content and of the H/C ratio (from 1.25 to about 0.5 in type-II kerogen), due to generation and release of hydrocarbons. Table II.5.1 shows the variation of atomic composition, from the stage of diagenesis (burial depth 780 m) to the various steps of catagenesis (burial depth 2070 and 2510 m in the Paris Basin, ca. 3000 m in the Sahara) and until the beginning of metagenesis (ca. 4000 m in the Sahara). Figure II.5.3 shows the corresponding evolution path from sample A (diagenesis) to B, C and D (catagenesis) and E (metagenesis). During catagenesis the main features on IR spectra are a progressive reduction of aliphatic bands, correlative to the H/C lowering (Fig. II.5.4), and the appearance of aromatic CH bands from 930 to 700 cm^{-1}. The appearance of these aromatic bands is due to a desubstitution on aromatic nuclei and possibly to an increasing aromatization of naphthenic rings. A careful study of the aliphatic bands has also proved that long chains are preferentially removed, resulting in a relative increase of methyl groups, compared to the total saturated structures (Robin, 1975; Robin et al., 1977). Infrared spectra also show the elimination of the residual C=O band to be completed rapidly; subsequently the O/C ratio remains low and constant (ca. 0.05), suggesting that the remaining part of oxygen (restricted to heterocyclic or ether groups) is not affected. The O/C ratio may even increase by relative concentration, due to a preferential elimination of hydrogen and carbon. ^{13}C NMR of solid kerogen provides comparable informations on the elimination of saturated hydrocarbon structures and the increase of aromaticity during catagenesis. Conclusions are similar to those obtained from IR spectroscopy.

Paramagnetic susceptibility increases progressively to 50 or 60.10^{-9} u.e.m. CGS/g of organic carbon. This growing abundance of free radicals has probably the same origin as the increase in aromatic CH bands, i.e., desubstitution and increasing aromatization. Vitrinite reflectance also increases, gently at first, then more rapidly from about 0.5 to 2%. The difference in reflectance between vitrinite and other macerals, e.g., exinite, vanishes beyond 1.5%. However, the mutual arrangement of the stacks of aromatic sheets remains randomly distributed, as noted from electron microdiffraction.

Some of the numerical values mentioned for type-II kerogens are different in other types of kerogen. For instance, the range of variation of the H/C ratio during catagenesis is wider in type-I kerogen (1.5 to 0.5) and narrower in type III (0.8 to 0.5) compared with type II (Fig. II.5.1). The paramagnetic susceptibility of type-III kerogen, which is more aromatic, is often higher than observed on type I or II at the same depth of burial.

In terms of petroleum exploration, the stage of catagenesis corresponds to the main zone of oil generation and also to the beginning of the cracking zone, which produces "wet gas" with a rapidly increasing proportion of methane.

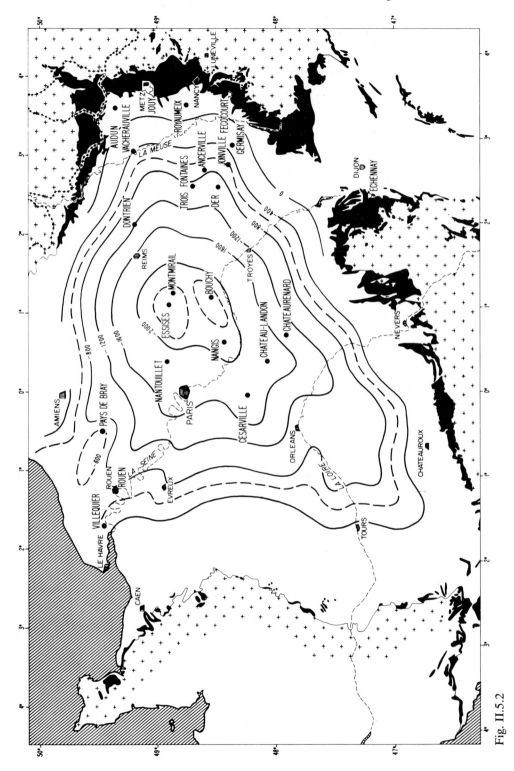

Fig. II.5.2

5.1 Diagenesis, Catagenesis and Metagenesis of Kerogen

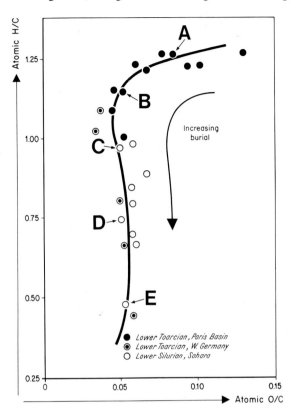

Fig. II.5.3. Evolution of the elemental composition of type-II kerogen during burial: Lower Toarcian shales from Paris Basin and W. Germany, Silurian shales from the Sahara (Algeria and Libya). Evolution stages: *A* end of diagenesis; *B–D* catagenesis; *E* metagenesis. (Modified after Tissot et al., 1974)

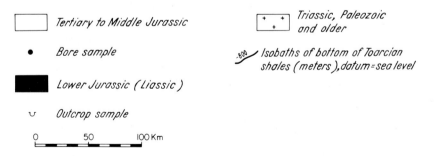

Fig. II.5.2. Map showing the outcrops of the Lower Jurassic of the Paris Basin, and the location of surface and subsurface sampling. The Lower Toarcian shales are situated near the top of the Lower Jurassic outcrops. Isobaths show the present depth to the base of the shales. (After Tissot et al., 1971)

Table II.5.1. Catagenetic evolution of the elementary composition of kerogens (type II) from the lower Toarcian shales, Paris basin, and from the Silurian shales, Northern Sahara

A. *Atomic composition*

Basin	a	Location	Maximum burial depth (m)	C	H	O	N	S	Total
	D	Vacherauville	780	40	53	5	1	1	100
Paris	C	Césarville	2070	43	53	3	1	0.08	100
	C	Bouchy	2510	48	48	2	1	0.04	100
	C	AB 1	ca. 3000	56	39	3	0.2	1.9	100
N. Sahara	M	OR 1	ca. 4000	64	31	3	0.9	1.4	100

B. *Atomic composition* (related to 100 carbon atoms)

Basin	a	Location	Maximum burial depth (m)	C	H	O	N	S	Total
	D	Vacherauville	780	100	130	13	2.5	2.5	248
Paris	C	Césarville	2070	100	123	7	2.3	0.2	233
	C	Bouchy	2510	100	100	4.5	2.3	0.1	207
	C	AB 1	ca. 3000	100	70	5	0.4	3	178.4
N. Sahara	M	OR 1	ca. 4000	100	48	5	1.4	2.2	157

[a] Stages of evolution: *D* Diagenesis (final stage), *C* Catagenesis, *M* Metagenesis.

c) *Metagenesis* of kerogen can be observed in very deep samples or in locations with a high geothermal gradient. Elimination of hydrogen is now slow and the residual kerogen usually consists of two carbon atoms or more out of three atoms (H/C $\leqq$ 0.5), as in samples from the Sahara that are buried to 4000 m (Table II.5.1, and sample E in Fig. II.5.3). In extreme cases the carbon content may reach 91% by weight and the H/C atomic ratio is only 0.4. Aliphatic and C=O bands have vanished, and aromatic C=C is the main band remaining on IR spectra (Fig. II.5.4, bottom).

The beginning of metagenesis corresponds approximately to a vitrinite reflectance of 2%: at this stage, a major rearrangement of the aromatic nuclei occurs and is clearly observed by electron microdiffraction. The stacks of aromatic layers, previously distributed at random, now gather in clusters with a preferential orientiation (Fig. II.4.5). The size of the clusters is from 100 to 200 Å in type-II kerogen. It is smaller in type III, less than 80 Å, and larger in type I, where it may reach 500 Å or more (Oberlin et al., 1974; Boulmier et al., 1977). At the same time dispersion of the interlayer spacing decreases (Oberlin et al., 1975).

This major change is accompanied by other important events, in respect to vitrinite structure and also observed by IR and electron spin resonance (ESR) characteristics. Beyond the level of 2% reflectance, anisotropy of vitrinite begins to appear. Aromatic CH bands (from 930 to 700 cm^{-1}) and paramagnetic

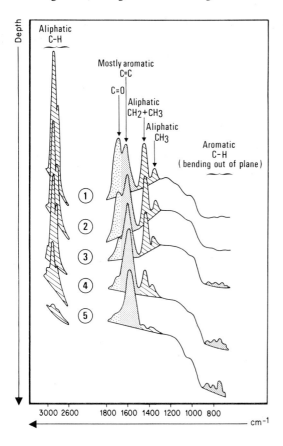

Fig. II.5.4. Infrared spectra showing evolution of type-II kerogen during burial (Lower Toarcian, Paris Basin; Silurian, Sahara)

susceptibility both reach a maximum at about 2% vitrinite reflectance, then decrease markedly (Durand et al., 1977; Robin et al., 1977).

All these data point to the progressive development of a higher degree of order by increasing polycondensation, more regular spacing of the aromatic layers and appearance of a preferential orientation over areas ranging from 100 to 500 Å, comprising hundreds or thousands of individual stacks, smaller than 10 Å.

In terms of petroleum exploration, the stage of metagenesis is entirely situated in the dry gas zone.

Results obtained by optical techniques of examination support this general scheme of kerogen degradation through diagenesis, catagenesis and metagenesis. *Microscopic examination* of kerogen from various depths shows a progressive evolution of the spores and pollen with respect to their state of preservation (shape, ornamentation etc.) and color (Correia, 1969). In the Paris Basin they change from yellow in shallow samples to orange, red, and finally brown. In deeper samples from the Sahara they lose part of their morphological character and become black. *UV fluorescence* of organic material shows a marked decrease with depth (Alpern et al., 1972). Both phenomena are interpreted as a result of increasing condensation of the kerogen. *Reflectance* of the various macerals also changes as a function of burial. The values corresponding to the three steps of

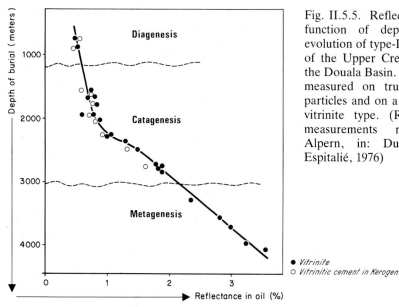

Fig. II.5.5. Reflectance as a function of depth during evolution of type-III kerogen of the Upper Cretaceous of the Douala Basin. Values are measured on true vitrinite particles and on a cement of vitrinite type. (Reflectance measurements made by Alpern, in: Durand and Espitalié, 1976)

kerogen evolution have been shown in Figure II.5.1, and an example of evolution as a function of depth is presented in Figure II.5.5: the reflectance has been measured on vitrinite (and vitrinitic cement) of the kerogen in the Cretaceous of the Douala Basin (type III).

Additional information on the evolution of kerogen is provided by *differential thermal analysis* (DTA) and *thermogravimetric analysis* (TGA). Combustion in an oxygen atmosphere shows a strong exothermic DTA peak, which shifts towards higher temperature with increasing depth of the sample (Espitalié et al., 1973). Similar observations are made on a coal sequence with increasing rank. In

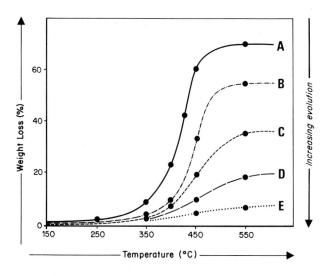

Fig. II.5.6. Thermogravimetric analysis of kerogen type II (temperature increase of $4°C\ min^{-1}$). Kerogens from the Paris Basin and from the Sahara are arranged in order of increasing evolution. The letters *A* to *E* refer to samples shown on Figure II.5.3. (After Tissot et al., 1974)

a nitrogen atmosphere, the weight loss measured by thermogravimetric analysis (TGA) (Fig. II.5.6) amounts to 70% (sample A) in shallow kerogens of the Paris Basin corresponding to the stage of diagenesis. During catagenesis, the weight loss decreases regularly to 55% (sample B), then 40% (sample C), and finally to 20% (sample D). In very deep samples from the Sahara (sample E), where metagenesis is reached, the weight loss is depressed to less than 10%. The TGA temperature corresponding to the maximum degradation rate also increases with depth.

This general evolution scheme is in agreement with the hypothesis of a progressive elimination, through bond breaking, of the noncondensed constituents of kerogen (e.g., aliphatic chains and saturated or aromatic cycles, single or associated in a small number), the residue becoming progressively more condensed and less sensitive to degradation.

5.2 Experimental Simulation of Kerogen Evolution

Much experimental work has been carried out by heating kerogen in the laboratory. Some of these experiments were conducted by industries as a part of the search for adequate raw materials to produce oil from oil shales.

Transformations occurring over geological periods are in essence not accessible to human experiments, as we shall never be able to account for the millions of years involved in the natural processes. Therefore, in the laboratory it is necessary to increase temperatures in order to accelerate the reactions and compensate for time. However, because a valid basis for comparison has not yet been established, we are not yet able to determine which time–temperature correspondence should be used. For this reason, the first objective of such laboratory experiments is to determine the extent to which they are representative of natural transformation.

In this regard, experimental assays of thermal evolution have been carried out on shallow samples from the three main types of kerogen. The shallow samples, artificially transformed by heating experiments, are then compared to the naturally transformed samples from greater depths of the same basin. These experiments are based on thermogravimetric analysis with temperature program of 4°C/min under inert atmosphere in an attempt to represent a progressive burial. At regular times corresponding to sample temperature of 300, 350 to 600°C, the elementary composition and physical properties of the kerogen (IR spectroscopy, electron diffraction, electron spin resonance, reflectance) are measured. The composition of the reaction products is evaluated by mass spectrometry (Souron et al., 1974) or by catharometer and flame ionization detectors.

Again we shall review in detail the results related to type-II kerogen, which provide us with the general scheme of artificial evolution. The shallow and immature kerogen from Fécocourt, in the Lower Toarcian shales of the Paris Basin, has been subjected to such experiments in order to compare the resulting kerogen with samples buried naturally to greater depths. Some of the resulting analyses are shown on Figure II.5.7, and they may be compared to the natural evolution upon burial by reference to Figures II.5.3 to II.5.6. In addition, Figures II.5.8 and II.5.9 show a comparison of some significant data obtained by heating all three of the main types of kerogen (I, II and III) under identical conditions.

Although the degradation of kerogen and the associated weight loss are continuous and progressive, three successive stages may be distinguished from the laboratory heating of immature Fécocourt kerogen (type II). These three stages are described in the following paragraphs.

a) Heating to ca. 350°C results in a rather small weight loss, due mostly to production of water and carbon dioxide, as shown by mass spectrometry. Optical examination of the samples shows a progressive darkening of defined organic debris with a correlative decrease of UV fluorescence. Elemental analysis and infrared spectroscopy show a loss of oxygen associated with a diminution of the C=O band (Figs. II.5.7 and II.5.8). The elemental composition H/C, O/C is comparable to natural samples buried to depths of 1000 to 1500 m (Fig. II.5.3). The degree of ordering of the carbon nuclei, as measured by electron microdiffraction, remains as low as it is in the shallow natural samples. Likewise, reflectance of kerogen is still weak.

Artificial evolution at about 350°C can be compared with the final phase of diagenesis in sedimentary basins. Thus the first artificial heating stage (350°C) can be correlated with subsurface late diagenesis.

b) From ca. 350°C to 470–500°C, kerogen encounters its major degradation stage. The weight loss is maximal per time or temperature unit (Fig. II.5.7). The products are mostly hydrocarbons and particularly aliphatics. The sample seems to swell, becomes plastic and is possibly melted, as already observed by Robinson (1969) when heating Green River shales. Elemental composition changes rapidly: the H/C ratio decreases to ca. 0.5 and aliphatic CH_2, CH_3 bands decrease markedly, whereas aromatic CH bands appear due to desubstitution (Fig. II.5.9). The C=O band progressively disappears: acids are removed first, then esters, which are more stable, whereas ketones are progressively eliminated (Fig. II.5.8). The reflectance progressively increases up to 2%, and the paramagnetic susceptibility to 50–60 · 10^{-9} u.e.m. CGS/g of organic carbon. The deepest natural sample of the Lower Toarcian shales of the Paris Basin (Bouchy 2500 m, corresponding to ca. 100°C) has a composition and a structural organization equivalent to 400°C in laboratory experiments. Other kerogens of similar composition from Silurian of the Sahara that have been deeply buried and subjected to natural temperatures over 130°C for a long period of time, as D in Figure II.5.3, have a composition and structural organization corresponding to the end of that phase (470–500°C in experiments).

This step of evolution can be correlated with the catagenesis occurring in sedimentary basins, and the laboratory assays can be considered as reasonably representative of kerogen transformation during that step. However, the situation is rather complex with regard to the bitumen products formed during the laboratory assays as compared with oil and gas generated during natural catagenesis. As temperature conditions are much higher in laboratory experiments than in sedimentary basins, disproportionation and cracking reactions of the generated bitumen may proceed differently, according to the experimental

5.2 Experimental Simulation of Kerogen Evolution

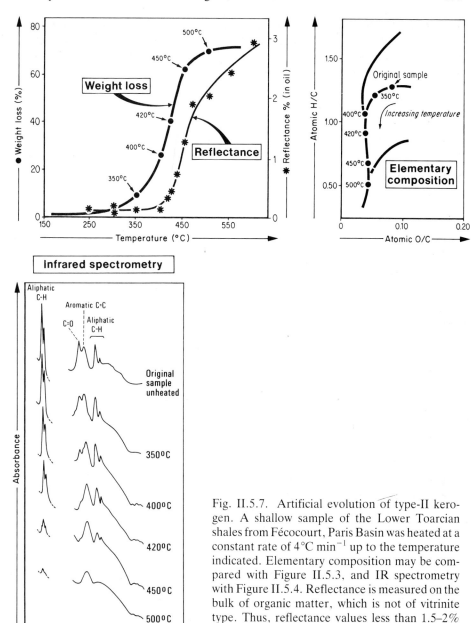

Fig. II.5.7. Artificial evolution of type-II kerogen. A shallow sample of the Lower Toarcian shales from Fécocourt, Paris Basin was heated at a constant rate of 4°C min^{-1} up to the temperature indicated. Elementary composition may be compared with Figure II.5.3, and IR spectrometry with Figure II.5.4. Reflectance is measured on the bulk of organic matter, which is not of vitrinite type. Thus, reflectance values less than 1.5–2% cannot be compared with a scale of vitrinite reflectance. (After Tissot et al., 1974)

procedure (bitumen remaining in the reaction cell or recovered in a cool trap, etc.). Therefore, the same stage of kerogen degradation may result in different yields of heavy constituents and medium to light hydrocarbons, according to the particular experimental conditions.

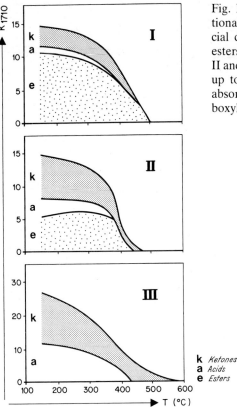

Fig. II.5.8. Elimination of the various functional groups containing oxygen during artificial degradation of kerogen: ketones, acids, esters. Samples of immature kerogen of type I, II and III were heated to various temperatures up to 600°C. K 1710 is the intensity of the absorption band related to carbonyl and carboxyl C=O groups. (After Robin, 1975)

c) Beyond 470–500°C, the weight loss to 600°C is very slow and small. However, as most aliphatic chains and functional groups have been removed previously, a major structural rearrangement occurs. Stacks of aromatic sheets gather to form clusters or aggregates within the range of 100 to 200 Å and can be seen by electron microdiffraction. At the same time, aromatic CH bands (on IR spectra) and paramagnetic susceptibility drop (Fig. II.5.9) whereas reflectance increases from 2 to 3%. All these transformations can be paired with the observations made on very deep natural samples, as E in Figure II.5.3, and this interval of artificial evolution corresponds to metagenesis.

Figures II.5.8 and II.5.9 allow a comparison of some of the results obtained by heating kerogens of types I and III with those obtained by heating type-II kerogen. Qualitative evolution is quite comparable. However, quantitative aspects vary from one type to the other due to differences in the original composition of kerogen. The total amount of products generated upon heating, as measured by the weight loss, is greatest for type I and least for type III. The proportion of hydrocarbons is higher for types I and II, and lower for type III, in accordance with the respective drop of the aliphatic IR bands and H/C ratio on Figure II.5.9. Furthermore, the light to medium hydrocarbons ratio (gas: oil ratio) is low for type I, moderate for type II, and high for type III. Contrary to

5.2 Experimental Simulation of Kerogen Evolution

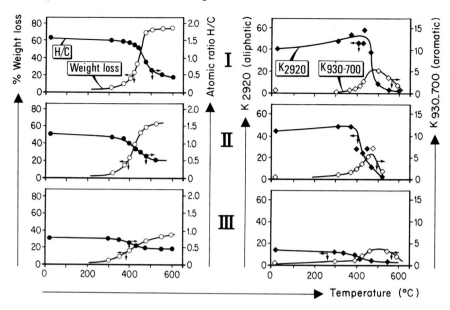

Fig. II.5.9. Changing composition of kerogen during artificial evolution in heating experiments. *Left:* weight loss and atomic H/C ratio. *Right:* IR absorption bands related to aliphatic groups (K 2920) and to aromatic CH (K 930–700). Samples of immature kerogen of types I, II, III were heated up to temperatures shown, as in Figure II.5.8. (After Robin, 1975)

hydrocarbons, the proportion of carbon dioxide generated is higher for type III kerogen in agreement with the total drop of carbonyl IR band shown in Figure II.5.8. Furthermore, the transformation of type-III kerogen is progressive, and extends over a wide temperature range, while the interval is narrow and the temperature of maximum weight loss is higher for type-I kerogen.

A very important result of the heating experiments is the step-by-step coincidence of the kerogens resulting from the rapid, high temperature experimental evolution and the slow, low-temperature natural evolution. It is possible to associate one stage of laboratory-simulated degradation to each stage of the natural evolution path. The respective results of elemental analysis, IR spectroscopy, thermogravimetric analysis and electron microdiffraction are identical and prove the identity of composition and structure of the two paired kerogens. This conclusion allows one to make an evaluation of the degradation potential of a given kerogen in the deep parts of the basin, by using shallow or outcropping samples from the same formation.

However, there are limitations to the use of artificial heating experiments. In general, satisfactory simulations are obtained from samples which have a low oxygen content, i.e., either type I kerogen or types II and III kerogens which have almost completed their diagenesis. Other trials, made on more immature samples of types II and III — e.g., a maximum burial depth ca. 200–300 m in the Messel shales (type II) or in the Douala basin shales (type III) and also the brown coals of the Mahakam delta — resulted in a systematic shift of the O/C ratio due to a higher oxygen context. This result is interpreted as being due to the preferential elimination of oxygen as water (a hydrogen-consuming process) during pyrolysis, as compared to geological situations where oxygen is mostly eliminated as carbon dioxide (Monin et al., 1980; Tissot and Vandenbroucke, 1983). Thus the elimination of oxygen is crucial for the completion of diagenesis. As diagenesis occurs at rather shallow depth and mild temperature, the temperature increase is probably not the single determining factor in diagenesis. Diagenesis cannot be adequately simulated by heating. For instance, the various reactions of oxygen elimination may have different kinetics, some of them having a higher temperature dependence than others. Furthermore, there may be an influence of the experimental conditions (open vs. confined environment). Finally experimental evolution cannot be readily applied to predict the nature (especially oil vs. gas) of hydrocarbons generated.

5.3 Structural Evolution of Kerogen

These observations, as a whole, lead to the conclusion that burial — i.e., increase in pressure and especially temperature — constitutes the determining factor in the evolution of organic matter. The temperature rise promotes the formation of bitumen, and particularly of hydrocarbons, the main constituent of petroleum, at the expense of kerogen (Louis and Tissot, 1967; Tissot et al., 1971).

Considering the general structure of kerogen proposed above (cyclic nuclei, bearing alkyl chains and functions and linked by heteroatomic bonds and aliphatic chains), the following mechanism is suggested (Fig. II.4.8). As burial, and thus temperature and pressure increase, the structure of the immature kerogen is not any more in equilibrium with the new physicochemical conditions. Rearrangements take place progressively toward increasing aromatization and development of an ordered carbon structure. The stable configuration under high temperature and pressure would be graphite, but this stage is never reached in nonmetamorphosed sediments. However, the building blocks of kerogen, each nucleus made of two or more aromatic sheets, tend to become progressively parallel over areas covering up to one or several hundreds Å. Electron microdiffraction has clearly shown that the stacks of aromatic sheets gather to form aggregates or clusters, that the number of nuclei reaching a parallel arrangement within the aggregate increases, and that the regularity of interlayer spacing increases.

The steric hindrances for development of such higher degrees of order have to be eliminated. They include, for instance, nonplanar cycles (e.g., saturated cycles), heteroatomic bonds, or aliphatic links which prevent the cyclic nuclei from forming a parallel arrangement. In a first stage, corresponding to *diagenesis*, heteroatomic bonds are broken successively and roughly in order of ascending rupture energy, starting with some labile carbonyl and carboxyl groups (like ketones and acids). Heteroatoms, especially oxygen, are partly removed as volatile products: H_2O, CO_2. The rupture of these bonds liberates smaller structural units made of one or several bound nuclei and aliphatic chains. These fragments are the basic constituents of bitumens. The larger ones are structurally

5.3 Structural Evolution of Kerogen

similar to kerogen but of lower molecular weight and therefore soluble (MAB extract[4], asphaltenes). The smaller ones are linear, branched and cyclic hydrocarbons and closely related heterocompounds, such as thiophene derivatives. During the first (diagenetic) stage, the larger fragments containing heteroatoms, especially oxygen (resins, asphaltenes, MAB extract), are predominant. Their elimination, and that of CO_2 and H_2O, cause the oxygen and sulfur content in kerogen to decrease drastically.

As temperature continues to increase, the kerogen reaches the stage of *catagenesis*. More bonds of various types are broken, like esters and also some carbon–carbon bonds, within the kerogen and within the previously generated fragments (MAB extract, asphaltenes etc.). The new fragments generated become smaller and devoid of oxygen: therefore, hydrocarbons are relatively enriched. This corresponds first to the principal phase of oil formation, as named by Vassoevich (1969), and then to the stage of "wet gas" and condensate generation. At the same time, the carbon content increases in the remaining kerogen, due to the elimination of hydrogen. Aliphatic and alicyclic groups are partly removed from kerogen. Carbonyl and carboxyl groups are completely eliminated, and most of the remaining oxygen is included in ether bonds and possibly in heterocycles.

When the sediment reaches the deepest part of the sedimentary basins, temperatures become quite high. A general cracking of C–C bonds occurs, both in kerogen and in bitumen already generated from it. Aliphatic groups that were stil present in kerogen almost disappear. Correspondingly, low molecular weight compounds, especially methane, are released. The remaining sulfur, when present in kerogen, is mostly lost, and H_2S generation may be important. This is the principal phase of dry gas formation.

Once most labile functional groups and chains are eliminated, aromatization and polycondensation of the residual kerogen increases, as shown by alteration of optical characteristics (e. g., spores and pollen become black) and by IR spectra. Parallel arrangement of the aromatic nuclei extends over wide areas from 80 to 500 Å, forming clusters. Physical properties evolve accordingly (high reflectance, electron diffraction). Such residual kerogen is unable to continue to generate hydrocarbons, as shown by the negative results of thermogravimetric assays. This stage is reached only in deep — or very old — sedimentary basins and corresponds to *metagenesis*.

These various stages are shown on Figure II.1.2 with their correspondence to vitrinite reflectance levels, coal ranks and hydrocarbons generated.

Table II.5.2 presents the atomic balance of the organic matter transformation by catagenesis from a shallow immature stage (700 m) to a deeper one (2500 m) situated in the principal zone of oil formation in the Paris Basin. The figures for kerogen, hydrocarbons, resins, asphaltenes and MAB extract, result from isolation of the constituents and elemental analysis. Water and carbon dioxide result from laboratory experiments (see above), simulating the evolution of

---

4 Heavy bitumen, extracted by methanol-acetone-benzene mixture. It can be considered a being made of heavier, or more polar, asphaltenes than those extracted by chloroform or comparable solvents.

Table II.5.2. Atomic balance of the total organic matter in the evolution from 700 m to 2500 m, corresponding to 1000 carbon atoms (lower Toarcian shales, Paris Basin)

Burial depth (location)	700 m (Vacherauville)					2500 m (Essises, Bouchy)				
	C	H	O	N	S	C	H	O	N	S
Kerogen	940	1220	120	23	24	806	876	40	18	5
Hydrocarbons	11	17	–	–	–	106	180	–	–	–
Resins	14	21	1.2	0.2	0.1	33	41	1.6	0.4	0.1
Asphaltenes	13	17	1.4	0.3	0.2	22	23	1.6	0.4	0.1
MAB extract	22	31	3.7	0.7	0.2	28	35	3.4	0.6	0.2
CO_2						5	–	10	–	–
H_2O						–	102	51	–	–
H_2S						–	36	–	–	18
N_2						–	–	–	5	–
Total	1000	1306	126.3	24.2	24.5	1000	1293	111.6	24.4	23.4

kerogen. Hydrogen sulfide and nitrogen have not been measured and are only included as estimations for balancing purposes.

As a conclusion, it can be emphasized that the main feature in kerogen evolution is the emergence of a carbon order, which progressively extends over wider areas, and becomes stronger with increasing temperature. The elimination of the steric hindrances to ordering results in the formation of a wide range of compounds, including medium to low molecular weight hydrocarbons, carbon dioxide, water, hydrogen sulfide, etc. Therefore, the petroleum generation appears to be a necessary consequence of the constant adjustment of kerogen to new temperature conditions with increasing overburden.

5.4 Formation of Hydrocarbons During Catagenesis

5.4.1 General

A better understanding of the thermal degradation of kerogen requires the knowledge of the evolution of the total organic content of the sediment. For instance, the carbon content of immature kerogen of the Paris Basin at a burial depth from 400 to 800 m amounts to 94% of the total organic carbon of the rock, while kerogen buried at 2500 m in the central part of the basin account only for 81% of the total organic carbon recovered in the cores. The balance, 6 and 19% respectively of the total organic carbon, is represented by extractable bitumen, which has been generated from kerogen (Table II.5.2).

In order to observe the formation of petroleum, we usually measure the total bitumen extractable from the rock, i.e., hydrocarbons + resins + asphaltenes

5.4 Formation of Hydrocarbons During Catagenesis

(HC + R + A) or hydrocarbons only (HC). The total organic carbon being C_0, we express:

a) The transformation ratio of organic matter into petroleum as:

$$\frac{\text{Bitumen}}{C_0} = \frac{HC + R + A}{C_0} = \text{bitumen ratio}$$

b) The transformation ratio into hydrocarbons as:

$$\frac{HC}{C_0} = \text{hydrocarbon ratio}$$

Bitumen ratio usually ranges from 0.02 to 0.20, i.e., 20 to 200 mg/g of total organic carbon. Hydrocarbon ratio ranges from 0.01 to 0.15, i.e., from 10 to 150 mg/g of total organic carbon.

Actually, the use of the ratios "carbon of bitumen/C_0" and "carbon of HC/C_0" would be better, but most data in literature lack the carbon analysis on extracts.

Such ratios can be considered as significant if numerous samples taken at random from a given formation at about the same depth (for instance over a 10–20-m cored interval) show a reasonably constant value, regardless of the total organic content. Examples from various sections in the Mowry shale of Wyoming (Schrayer and Zarrella, 1966), from the Lower Toarcian shales of Paris Basin and from Lower Miocene of West Africa (Fig. II.5.10) show this requirement to be satisfied.

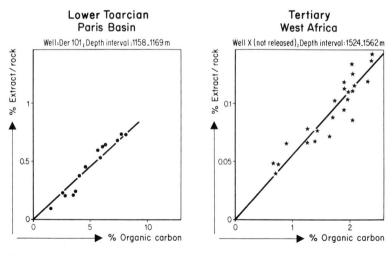

Fig. II.5.10. Significance of the ratio of bitumen to organic carbon. This ratio remains reasonably constant over a short interval, despite considerable variations of the absolute amount of organic matter. Examples are shown over cored intervals of 11 m (Lower Toarcian, Paris Basin) and 38 m (Tertiary, West Africa) respectively

5.4.2 Presentation of Some Typical Examples of Petroleum Generation

Burial of sediments is responsible for a temperature increase, which in turn determines the generation of hydrocarbons. Early observations of the formation of petroleum as a function of burial in sedimentary basins came from Larskaya and Zhabrev (1964), working on Mesozoic-Cenozoic beds from Azov-Kuban basin; Louis (1964), working on the Lower Toarcian shales of the Paris Basin, and Philippi (1965), working on Mio-Pliocene of Los Angeles and Ventura Basins. They noted an increase with depth of the bitumen ratio $\frac{HC + R + A}{C_0}$ and hydrocarbon ratio $\frac{HC}{C_0}$ accompanied by an increase in hydrocarbon content of the total extracted bitumen.

As already stated, with regard to kerogen evolution, a study of the influence of burial on petroleum generation in sediments requires a situation in which all

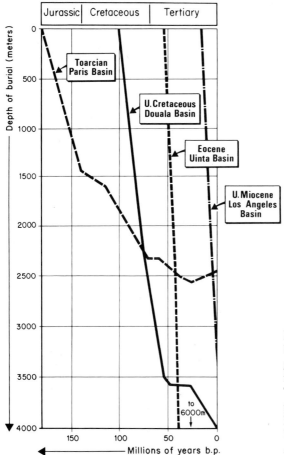

Fig. II.5.11. Reconstruction of burial histories (depth versus geological time) for deep samples from various formations: Lower Toarcian of the Paris Basin, Upper Cretaceous of the Douala Basin, Paleocene-Eocene of the Uinta Basin, Miocene of the Los Angeles Basin

5.4 Formation of Hydrocarbons During Catagenesis

geological and geochemical parameters remain constant except depth. Four examples have been the object of detailed geochemical studies comprising both quantitative and qualitative aspects. Their overall geological conditions and type of organic matter are rather different.

a) The case of Lower Toarcian shales of the Paris Basin has already been discussed in Section 5.1. The kerogen belongs to type II, which is most frequent in source rocks. The detailed results are reported by Louis and Tissot (1967) and Tissot et al. (1971).

b) The Upper Cretaceous from the Douala Basin (Cameroon) is a thick sedimentary sequence in which the nature of the sediments and conditions of deposition can be considered as reasonably constant from Turonian–Coniacian to Campanian (90–70 million years): Dunoyer de Segonzac (1969). The corresponding depth interval is 800–4200 m in the Logbaba wells, studied by Albrecht and Ourisson (1969) and Albrecht et al. (1976). The kerogen belongs to type III.

c) The Green River formation (both the Eocene and the associated Flagstaff member) of the Uinta basin, Utah, has reached maximum burial depths up to 6000 m. Kerogen composition is reasonably constant and belongs to type I. Formation of petroleum has been studied by Tissot et al. (1978).

d) The Cenozoic beds of the Los Angeles and Ventura basins (California), studied by Philippi (1965), range from Recent to upper Miocene (about 0–12 million years). The sedimentation is supposed to have been of rather constant composition. The depth varies from 0 to 3500 m in the Los Angeles basin to 4600 m in the Ventura basin.

The rocks containing organic matter consist of shales and silts in the Paris, Douala and Los Angeles basins. They range from shales to marls in the Uinta basin. The burial history in the deepest part of each basin is shown on Figure II.5.11.

5.4.3 Quantitative Aspects (Fig. II.5.12)

At shallow depths, the bitumens and hydrocarbons are fairly low. The bitumen ratio $\frac{HC + R + A}{C_0}$ is only ca. 30 mg/g in the Douala basin and 40 mg/g in the Paris Basin. Hydrocarbon ratio is 5 to 15 mg/g in the Los Angeles and Ventura basins, around 10 mg/g in the Paris and Douala basins, and somewhat higher (30 mg/g) in the Uinta basin.

Bitumen and hydrocarbon ratios change little down to depths of ca. 1500 m in Paris and Douala basins, more than 3500 m in Uinta basin, 2400 m in the Los Angeles basin and still deeper in the Ventura basin. According to local geothermal conditions, these depths correspond, respectively, to about 60°C (Paris Basin), 70°C (Douala basin), 100°C (Uinta basin) and 115°C (Los Angeles and Ventura basins). Beyond that threshold, the transformation ratios increase rather abruptly in the Douala basin, more gently in the other basins. The bitumen ratio reaches maximum values as high as 180 mg/g at 2500 m and 100 mg/g at 2100 m, respectively for the Paris and Douala basins. At the same time, the proportion of hydrocarbons increases with respect to heavy heterocompounds

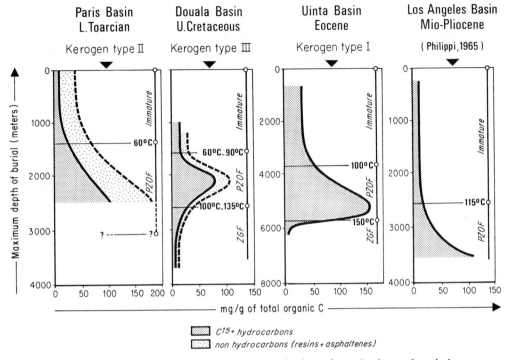

Fig. II.5.12. Formation of hydrocarbons and nonhydrocarbons (resins and asphaltenes containing N, S, O) as a function of burial depth, in different basins (cf. also Fig. II.5.11). The major steps of evolution of the organic matter are marked: "immature", "PZOF" (principal zone of oil formation), and "ZGF" (zone of gas formation by cracking). Corresponding temperatures are shown according to present geothermal gradients. In the Douala Basin, the first temperature is the present one the second temperature is a calculated paleotemperature according to Tissot and Espitalié (1975)

(resins, asphaltenes), and the hydrocarbon ratio reaches maximum values ranging from 70 mg/g at 2100 m in the Douala basin (kerogen type III) to 100 mg/g at 2500 m in the Paris Basin (kerogen type II) and to 150 mg/g at 5200 m in the Uinta basin (kerogen type I).

At greater depth in the Douala and Uinta basins, both bitumen and hydrocarbon ratios decrease progressively to very low values (below 10 mg/g). This disappearance of liquid hydrocarbons and other petroleum constituents is attributed to cracking, which produces low molecular weight hydrocarbons (gas) and a carbon residue. This is in agreement with the occurrence of gas, mostly methane, in Logbaba wells. The decrease in liquid hydrocarbons and the generation of methane at great depth have been recognized in several other formations of various sedimentary basins, like the Upper Devonian of Western Canada (Deroo et al., 1973) and Northern Africa (Tissot et al., 1974) and the Lower Cretaceous and Upper Jurassic of the Aquitaine basin (Le Tran, 1972). It has not been observed in the Paris and Los Angeles basins, as the depth of investigation is not sufficient to reach this stage.

5.4 Formation of Hydrocarbons During Catagenesis

However, the molecular weight distribution of hydrocarbons in the Paris basin shows that the light hydrocarbons increase with depth more abruptly than medium and heavy ones. Hydrocarbons $< C_{15}$, which are very scarce at shallow depth, may amount to as much as one third of the total hydrocarbons at 2500 m. This observation is in agreement with the hypothesis of an overwhelming occurrence of light hydrocarbons at great depth. In the South Aquitaine basin, Le Tran (1972) observed, after a maximum at 3500 m (about 100°C), a marked decrease of the bitumen ratio, accompanied by an abrupt increase of gaseous hydrocarbons and H_2S at 4200 m (about 120°C). In the Devonian of Northern Alberta and Northeastern British Colombia, Evans et al. (1971) noted an abrupt increase of methane content in cuttings gas at depth corresponding to 100°C. All these features point toward a cracking of organic constituents under conditions of increasing temperature.

The existence of two critical temperature thresholds in source rocks — the first one representing a lower threshold to massive generation of liquid hydrocarbons, and the second a lower threshold to cracking and gas generation — led Vassoevich (1969) to the conclusion that in between those thresholds there is the *principal zone of oil formation*. Following the early diagenesis of organic matter in young sediments, subsequent evolution will thus include the successive steps (Fig. II.1.2) of hydrocarbons formation.

a) *Diagenesis*
– an immature stage with little change: hydrocarbons are scarce. They are more or less directly inherited from living organisms and comparable to the molecules present in the young sediment.

b) *Catagenesis*
– the principal zone of oil formation,
– the zone of gas formation, where "wet gas" with increasing proportion of methane is generated in large amounts through cracking.

c) *Metagenesis*
– some additional generation of hydrocarbons (mainly methane) from kerogen; liquid hydrocarbons previously generated are also cracked and converted to gas,
– stage of structural rearrangement of the residual kerogen.

5.4.4 Qualitative Aspects: Composition of Hydrocarbons and N, S, O-Compounds

It is obvious from Figures II.5.13, II.5.14, and II.5.15, that the proportions of linear and cyclic hydrocarbons generated during the principal phase of oil formation depend on the type of kerogen.

In the Paris Basin, type-II kerogen generates abundant cyclic hydrocarbons of saturated and aromatic types, whereas normal and iso-alkanes are comparatively less important (20% of the hydrocarbons). In the Douala Basin, the hydrocarbons generated from type-III kerogen contain abundant *n*-alkanes compared to type II; the proportion of *n*- plus iso-alkanes amounts to 40% of hydrocarbons. In

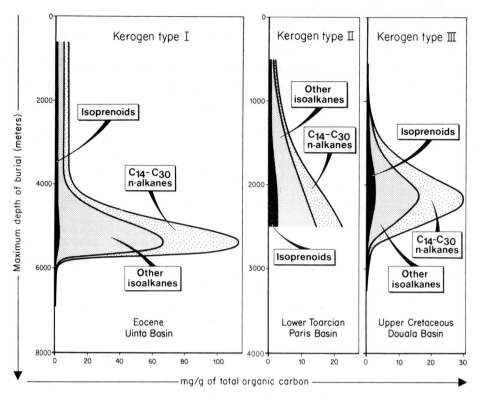

Fig. II.5.13. Formation of normal and branched alkanes in different types of kerogen as a function of burial depth, as exemplified in the Paris, Douala and Uinta Basins (same formations as in Fig. II.5.12). Values are expressed in mg per g organic carbon

the Uinta basin, type-I kerogen generates more linear plus branched alkanes than cyclic molecules: both, normal and iso-alkanes, are major constituents of the hydrocarbons (60%).

We shall now present a brief review of the variations among the main classes of hydrocarbons and their relation to the various stages of petroleum generation.

a) Alkanes

n-alkanes in shallow samples resemble the *n*-alkanes already occurring in Recent sediments. The role of high molecular weight molecules is generally important, and the distribution still reflects the depositional environment. In the Upper Tertiary of the Los Angeles and Ventura basins or the lower Cretaceous of the North Aquitaine basin, where there are contributions from terrestrial organic matter derived from higher plants, a strong odd-carbon-number predominance of the higher *n*-alkanes is observed during increasing burial to the threshold of main hydrocarbon generation. The CPI is still around 4 down to 2400 m in California, and around 5 at 1500 m in North Aquitaine basin (Fig. II.3.6). Where *n*-alkanes

5.4 Formation of Hydrocarbons During Catagenesis

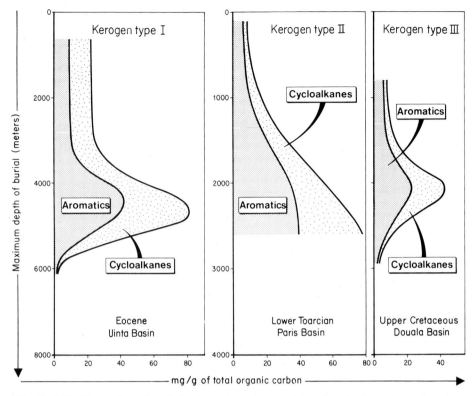

Fig. II.5.14. Formation of cyclic hydrocarbons (aromatics and cycloalkanes) as a function of burial depth, in the same basins and formations as in Figure II.5.13. Values are expressed in mg per g organic carbon

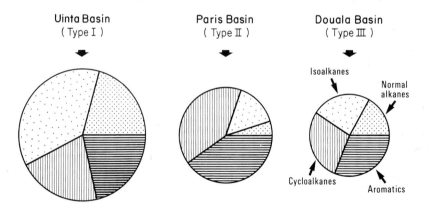

Fig. II.5.15. Compositions of hydrocarbons generated from the three main types of kerogen at the depth of maximum oil formation. Areas are proportional to the respective amount of each hydrocarbon class per g organic carbon

derived from higher plants are mixed with those from either phytoplankton or the microbial biomass living in sediments, the odd predominance is smaller but still distinct.

In the principal zone of oil formation, substantial amounts of new alkanes are generated with less or no odd preference[5], thus causing the progressive disappearance of the odd/even predominance by dilution. Thus the CPI is lowered to about 1. In addition, alkanes of low molecular weight are favored. This causes a shift of the maximum in the n-alkane distribution curve towards a smaller number of carbon atoms per molecule. These changes can be seen clearly on n-alkane distribution curves of samples from the Los Angeles (Fig. II.3.6) and Douala basins.

Where conditions are not suitable for the generation of new n-alkanes (unfavorable composition of kerogen, low geothermal gradient, or very short time of burial), however, the odd-carbon preference may last down to a great depth. In the Neocomian shales of the North Aquitaine basin, France, where the major part of the organic matter is composed of carbonaceous debris from higher plants, the preference for odd-numbered molecules remains at any depth higher than it is in good source rocks of upper Jurassic age (Table II.5.3). In the Ventura basin, California (Philippi, 1965), where the geothermal gradient is low (26.6°C/km) and the age of sediments is Miocene to Recent, the CPI is over 2 down to 4000 m.

On the contrary, the absence of a moderate to strong odd-carbon preference is not necessarily a consequence of catagenesis. Some young and shallow sediments show a distribution of n-alkanes containing many low to medium molecular weight molecules and few alkanes beyond C_{25}. The odd preference is slight or absent. This situation occurs in Pliocene beds of the Pannonian basin at depths, 400–800 m where temperature has always been low. The immature character of these sediments is confirmed by the occurrence of interbedded sedimentary

Table II.5.3. Compared values of the CPI as a function of depth in the Jurassic (source rock) and lower Cretaceous (nonsource rock) of the North Aquitaine basin

Depth	Jurassic	Lower Cretaceous
< 1500 m	1.4	2.2
1500–2500 m	1.3	1.7
> 2500 m	1.0	1.3

---

5 There may be various reasons for this fact. For example, aliphatic chains, substituted on kerogen nuclei, may be broken at various positions. Degradation of wax esters experimentally produces odd n-alkanes and even n-alkenes, which are subsequently converted to even n-alkanes (Esnault, 1973).

5.4 Formation of Hydrocarbons During Catagenesis

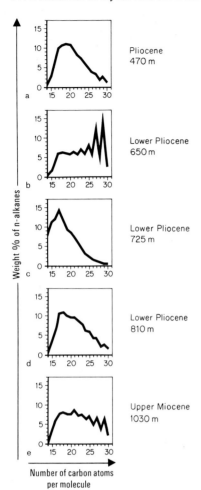

Fig. II.5.16a–e. Distribution of n-alkanes in Pliocene sediments of the Pannonian Basin. The absence of an odd/even predominance in sediments at relatively shallow depths 470, 725 and 810 m is not due to catagenetic evolution. This is proven by a strong odd predominance occurring at intermediate levels (650 m) or even deeper (1030 m) (Tissot et al., 1977)

layers, containing organic matter from plants with a strong odd-predominance of the higher n-alkanes (CPI 2 to 3; Fig. II.5.16).

At greater depth, where cracking becomes important, distribution curves are more and more dominated by lighter molecules in all types of sediments.

Some *acyclic isoprenoids* are also inherited from living organisms, either directly or through an early diagenesis. As the abiogenic formation of such peculiar structures is unlikely, one would expect the isoprenoids to be diluted by newly formed n- and iso-alkanes in relation to the abundance of alkanes generated from kerogen degradation. A plot of isoprenoids/n-alkanes as a function of increasing catagenesis shows that the isoprenoid content decreases with depth[6] and with increasing hydrocarbon generation, but in a different way from what could be expected. During burial to the temperature threshold of the

---

6 The pristane: n-C_{17} ratio may show more complex variations, due to the change of n-alkane distribution with depth.

main hydrocarbon generation, no change is noted. In the first part of the principal zone of oil formation, the ratio of isoprenoid to total saturated hydrocarbons remains nearly constant. This result indicates that some isoprenoid alkanes are being generated during catagenesis. The generation of new isoprenoids may be explained by the fixation of isoprenoid acids or alcohols as esters on the kerogen of the Recent sediment, as suggested by Burlingame et al. (1969). During the principal stage of oil formation such bonds, weaker than C–C bonds, would be more rapidly broken and isoprenoids would be liberated. An alternative interpretation of the phenomenon would be the existence of isoprenoid molecules trapped within the kerogen and subsequently released.

As oil generation continues, fewer and fewer isoprenoid chains of this type would be available, and they are no longer produced when the oil generation curve reaches its maximum. Therefore, the isoprenoid: n-alkanes ratio decreases. At greater depth, cracking of the hydrocarbons becomes important, and the distribution of isoprenoids is seriously affected.

b) Steroids and Triterpenoids

These tetra- and pentacyclic molecules of biogenic origin are mostly present in young sediments as fully saturated, unsaturated or partly aromatized hydrocarbons, or as related structures such as acids and alcohols (see Sect. 3.8). At shallow depth and down to the temperature threshold of principal oil formation, their distribution remains rather similar. The ring analysis of nonaromatic hydrocarbons by mass spectrometry shows a high proportion of 4- and 5-ring molecules, comprising mostly sterane, sterene, triterpane, and triterpene types in many shallow samples, as in shallow samples of the Paris Basin. They are distributed essentially in the carbon number range C_{27}–C_{30} (Fig. II.5.17). A similar situation occurs with respect to aromatic steroids and triterpenoids. At the end of the diagenetic stage, unsaturated polycyclics have been converted to saturates or aromatics.

During the principal stage of hydrocarbon formation, the abundance of polycyclics decreases, either by dilution with newly generated hydrocarbons or by degradation. A typical case of progressive decrease with depth is found in the Lower Toarcian shales of the Paris Basin (Tissot et al., 1971; Ensminger et al., 1977) with a correlative reduction of the optical activity. However, some tetracyclic and pentacyclic hydrocarbons, including probably steranes and triterpanes, have been reported in a Precambrian sediment (Burlingame et al., 1965). Generally, aromatized forms seem to be more stable, as they are still rather abundant in lower Paleozoic sediments from Eastern Sahara (Algeria). In these old sediments fully saturated steranes or triterpanes are not detected (Tissot et al., 1974).

c) Aromatic Hydrocarbons

The ratio of aromatic hydrocarbons to organic carbon increases with depth, but at a smaller rate than does the ratio of total hydrocarbons to organic carbon. In the

5.4 Formation of Hydrocarbons During Catagenesis

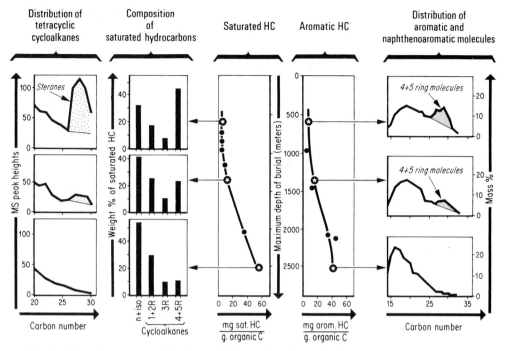

Fig. II.5.17. Occurrence of the 4- and 5-ring cycloalkanes *(left)* and of the corresponding naphthenoaromatics *(right)* in the Lower Toarcian shales, Paris Basin. The amount of saturated and naphthenoaromatic polycyclics decreases with depth. At the same time, the relative amount of chain-like *n*- and iso-alkanes increases

Paris Basin, total hydrocarbons increase from 500–700 m to 2500 m by a 10:1 ratio, saturated hydrocarbons by 15:1, and aromatics by only 7:1. In particular, the aromatic:saturated ratio decreases in the 2000–2500 m range of the Paris Basin, probably due to the cracking of side chains on aromatic nuclei (Fig. II.5.17).

Apart from the decreasing importance of steroid and triterpenoid type naphtheno-aromatics with depth, there is a general tendency to favor, at great depth, the purely aromatic types: alkyl-benzene, -naphthalene and -phenanthrene with a few chains of short length (1–3 carbon atoms in addition to the cyclic nucleus). The latter molecules may result in part from thermal cracking of steroids and triterpenoids with their methyl substituents.

d) Global Composition of Hydrocarbons

These various aspects of the catagenetic evolution result in a strong tendency for *n*- and iso-alkanes to increase their relative importance, compared with other structural types, during the principal stage of oil formation (Tissot et al., 1971; Fig. II.5.18). Polycyclic molecules, either saturated or unsaturated, are consider-

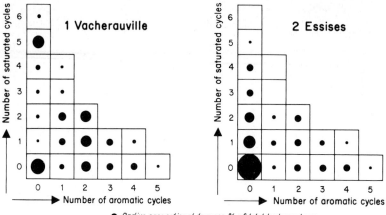

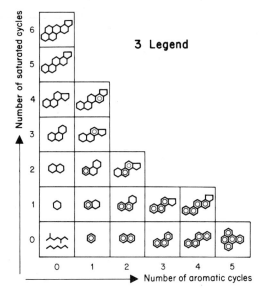

Fig. II.5.18. Abundance of the various structural types of hydrocarbons in the Lower Toarcian shales of the Paris Basin. *Left:* a shallow and immature sample; *right:* a deep sample from the principal zone of oil formation. The radius of the circles is proportional to the abundance of each molecular type. The main structural types are shown on the bottom (Tissot et al., 1971, modified)

ably less frequent in this stage of evolution than compared to the amount that was inherited from biogenic molecules of living organisms.

Beyond the principal phase of oil formation, thus when cracking becomes a dominating process, the thermodynamically more stable low molecular weight hydrocarbons prevail. Therefore, in an advanced stage of catagenesis of sediments, only the light gases and benzene survive (Leythaeuser et al., 1980). Accordingly, increasing amounts of light hydrocarbons (saturated and aromatic compounds) are generated from the kerogen and/or from the bitumen previously generated at these maturation levels. Due to the strong temperature dependence of their formation reactions, their analysis represents a sensitive indicator for the hydrocarbon generation process in the main phase of petroleum generation (Tissot et al., 1971; Philippi, 1975).

e) N, S, O Compounds

These compounds include (1) resins and asphaltenes, which are extractible with chloroform, and (2) other heavy asphaltenic compounds, which can only be extracted by a methanol-benzene or methanol-acetone-benzene (MAB) mixture. From Table IV.1.8 in Chapter IV.1, it is clear that resins and asphaltenes of bitumen have an average elementary composition which is different from that of resins and asphaltenes of crude oils. Oxygen, nitrogen and sulfur are more abundant in the resins and asphaltenes from shallow bitumen, but their abundance progressively decreases with depth and becomes more similar to those observed in crude oil constituents. When plotted on a H/C versus O/C diagram, asphaltenes from the bitumen of a given formation follow an evolution path broadly parallel to the kerogen evolution path previously described; the same is true for resins.

The following interpretation is proposed. During the first steps of kerogen degradation rather large fragments of kerogen are released, such as resins, asphaltenes and MAB-extract. They have a composition broadly similar to that of the kerogen and still include structures with various types of bonds. As temperature continues to increase, other bonds are broken in the remaining kerogen and in the break-down products. Therefore, the composition of the break-down products, which is originally different, becomes progressively more similar to the composition of crude oil such as resins and asphaltenes during catagenesis. Due to that evolution, hydrocarbons of low to medium molecular weight are likely to be produced in the same way as they are produced from kerogen. Thus, N, S, O compounds appear as possible intermediates in some of the reaction pathways from kerogen to petroleum.

5.5 Isotope Fractionation and Kerogen Evolution

The break-down of kerogen during catagenesis involves the splitting of various types of bonds and the release of smaller molecules, particularly hydrocarbons. It has been observed that the carbon of the bitumen released by this process is isotopically lighter than carbon of the related kerogen (in the range of 1 to 4‰). The same applies to pooled oils derived from the disseminated bitumen as compared to its source kerogen. This isotope fractionation may be explained by the relationship established by Galimov (1973) between isotope distribution and chemical structure (Chap. II.2.5). Bitumen, especially the hydrocarbon fraction, is derived from chemical groups, originally depleted in ^{13}C, such as aliphatic chains or saturated cycles. This isotopic character is preserved when hydrocarbon molecules are released from kerogen.

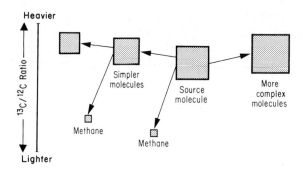

Fig. II.5.19. Schematic representation of carbon isotope ratio changes during petroleum formation and evolution (Silverman, 1967)

Table II.5.4. C-Isotope fractionation during formation of CH_4 in pyrolysis experiments (500°C). (After Sackett, 1968)

Source molecule A	$\delta^{13}C^0/_{00}$ CH_4 versus A
Propane	− 3.8
N-butane	− 9.7
N-heptane	− 9.6
N-octadecane	−23.6
Crude oil, Penn.-Okla.	−25.8

Corrected value according to least square method for a straight line function.

A similar effect, i.e., a tendency toward enrichment of the light hydrogen isotope versus deuterium is observed in the hydrocarbons released. The range is larger than in the case of the carbon isotope fractionation and there is also more variation. Values as high as a 40‰ depletion in D may be obtained (Schoell, 1980).

Methane generally shows a still lower $\delta^{13}C$, i.e., the difference between isotopic composition of methane and the source material is larger than between crude oil and its source material. Methane and other low molecular weight hydrocarbons are mostly generated by splitting of C–C bonds in kerogen or in petroleum. Silverman (1964, 1967) and Sackett (1968) pointed out that the production of methane results in an isotope fractionation. This is based on the fact that there is less energy required to break a $^{12}C-^{12}C$ bond than a $^{13}C-^{12}C$ bond. Hence in thermal cracking $^{12}C-^{12}C$ bond rupture occurs about 8% more frequently (Stevenson et al., 1948; Brodskii et al., 1959). Thus, the methane generated is enriched in ^{12}C, compared to heavier hydrocarbons and kerogen: Table II.5.4 and Figure II.5.19. Because of the production of isotopically light methane, the source molecule is relatively enriched in the heavy ^{13}C isotope. Furthermore, because of the relatively large hydrogen content of the methane molecule, the source material becomes dehydrogenated, which necessarily results in unsaturation and the creation of more reactive double bonds. This in turn would lead to polymerization reactions of the residual molecules, giving rise to components of even higher molecular weight. Silverman (1967, 1971) has followed this line of thought in explaining the observed carbon isotopic composition of narrow distillation fractions from crude oils. He points out that a minimum in the carbon isotope distribution curve is observed in the molecular weight region which contains the highest concentration of biologically produced carbon skeletons. This view is illustrated in Figure II.5.20, where the highest optical activity — witness of biogenic compounds — is observed in the fraction having the lowest $\delta^{13}C$. The general scheme of crude oil evolution and the accompanying changes in isotope ratios of the degradation products are shown in Figure II.5.19.

Sackett (1978) showed that the production of methane results in a hydrogen isotope fractionation as well. Laboratory simulation experiments indicated a significant enrichment of the light hydrogen isotope in generated methane, although the predicted fractionation under natural conditions could not be made with any reliability. Carbon isotope fractionation is associated with the cleavage of a carbon-carbon bond to form a methyl free radical which subsequently combines with a hydrogen atom to form methane. The hydrogen atom is formed by cleavage of a carbon-hydrogen bond so that one-fourth of the hydrogens in methane has participated both in a carbon-hydrogen bond cleavage and a carbon-hydrogen bond formation. The rate-determining step is presumably the bond cleavage and therefore responsible for the fractionation.

With increasing evolution of the source rock (measured by depth of burial or vitrinite reflectance), a progressive increase of the $\delta^{13}C$ and δD of methane is generally observed. In other words, the difference between the isotopic composition of methane and that of the source material (coal or kerogen) is progressively

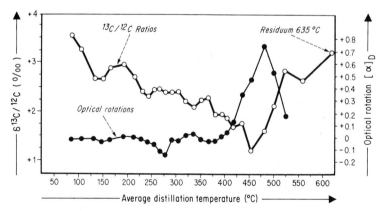

Fig. II.5.20. Changes in optical rotation and carbon isotope ratio with increasing molecular weight (boiling point) of distillation fraction of a crude oil. Highest optical activity is observed in the molecular weight range from 350 to 450, corresponding to distillation temperatures 420 to 550°C. Tetra- and pentacyclic ring structures fall in that range. Carbon isotope ratios exhibit a minimum in the same fraction. Isotope ratios are related to Petroleum Standard-NBS-22, which is 29.4‰ lower than PDB-1. (After Silverman, 1967; Welte, 1972)

lowered (Colombo et al., 1966; Galimov, 1973; Stahl, 1975; Schoell, 1980). Sackett (1968) has shown that the isotope fractionation effect in methane generation decreases with decreasing length of the carbon chain in the source molecule, and also with increasing cracking temperature up to a temperature of 500 °C (Sackett, 1978). Both factors could be responsible for the fact that methane and source material are isotopically more similar at greater depth of burial or higher rank levels. Furthermore, higher rank levels of organic matter also represent a higher degree of aromaticity. This change in chemical structure toward higher aromaticity would also result in a lower isotope fractionation between methane and its source material (Galimov, 1973) with progressing catagenesis and metagenesis. The relatively low isotope fractionation observed for gases derived from more aromatic coals as compared to gases derived from less aromatic marine kerogen might be explained similarly.

5.6 Experimental Generation of Hydrocarbons from Organic Material

Experimental heating of organic matter has been carried out by many authors in order to support the theories on thermal generation of petroleum. Results concerned with the evolution of kerogen have been presented in Section 5.2. Although experimental heating has proved to be a convenient means of simulating kerogen evolution, the simulation may not be as good with respect to liquid and gas products, which depend somewhat on the experimental procedures.

The experiments are based on different artificial or natural substances: pure chemical compounds, substances extracted from living organisms, Recent muds, and kerogen from ancient sediments. Experiments carried out on pure chemical compounds allow an investigation of the possible catalytic effect of minerals.

5.6.1 Pure Chemical Compounds

Fatty acids have been the object of many experiments since it was suggested pointed out that they could be a source of hydrocarbons. Decarboxylation of fatty acids by heating is a well-known mechanism studied by numerous authors, including Bogomolov et al. (1960, 1961, 1963), Eisma and Jurg (1964), Shimoyama and Johns (1971, 1972) and Almon (1974). Degradation of alcohols (Sakikana, 1951), esters, ketones (Frost, 1940; Demorest et al., 1951) and aldehydes (Levi and Nicholls, 1958) was also studied, since all of these groups of oxygen compounds are known to be present in the original organic matter. The form in which they are incorporated in sediments, however, is not clear. Other substances, such as chlorophyll (Bayliss, 1968), aminoacids, and polyterpenes (Erdman, 1965) have also been studied.

Heating $C_{21}H_{43}COOH$ acid at temperatures between 200 and 300 °C has shown that n-alkanes of various molecular weight are produced in presence of mineral catalysts (Eisma and Jurg, 1964; Shimoyama and Johns, 1972). With clay minerals (kaolinite, Ca-montmorillonite), the main n-alkane constituent is $C_{21}H_{44}$, resulting from direct decarboxylation. With calcium carbonate, the main n-alkane constituent is $C_{20}H_{42}$, suggesting that beta cleavage of the acid is the most likely process of n-alkane formation. Besides these predominant n-alkanes with 20 or 21 carbon atoms, the generation of other alkanes with shorter or longer carbon chains is observed in heating experiments of long duration. They probably result from cracking and polymerization. Unsaturated hydrocarbons and other fatty acids than the original $C_{21}H_{43}COOH$ have also been observed.

Almon (1974) investigated the mechanisms of fatty acid decarboxylation. Anhydrous smectite favors the formation of branched alkanes, with a distribution of isomers approaching thermodynamic equilibrium. The presence of water favors some sort of free radical reaction and results in a strong predominance of the n-alkanes; such a distribution of alkanes is comparable to the situation known in crude oils.

Thermal and catalytic cracking of n-octacosane ($nC_{28}H_{58}$ hydrocarbon) was studied by Henderson et al. (1968) in order to simulate the thermal alteration of n-alkane distribution in sediments.

Erdman (1965) has shown that thermal degradation of a polyterpene (β-carotene) and an aminoacid (phenylalanine) results in a production of low molecular weight alkyl-benzenes and alkylnaphthalenes, which are absent in Recent sediments.

5.6.2 Natural Recent Substances

Experiments on natural substances go as far back as Albrecht and Engler, who heated fish oils at the end of the last century. In more recent works, mainly marine muds were used. For instance, Erdman and Mulik (1963), by heating muds, have produced low molecular weight hydrocarbons, in particular alkylbenzenes, that are absent in Recent sediments. Harrison (1976) studied the evolution of free and bound fatty acids by using pyrolysis of a modern sediment. Ishiwatari et al. (1976) heated kerogen from a young marine sediment collected in the Tanner basin, offshore California. Kerogen produced liquid products and gas comprising initially carbon dioxide and water, then methane, other hydrocarbons and hydrogen. n-Alkanes were mostly generated by successive reactions through intermediate products (heavier, probably heteroatomic bitumen). Separate heating of the main organic constituents of the sediment showed that the bitumen (extractable with benzene-methanol) is responsible for 58% of n-alkanes generated, kerogen for 41% and humic acids for 0.4% only.

5.6.3 Kerogen of Ancient Sediments

Numerous experiments of heating ancient sediments, which have previously been subjected to an exhaustive extraction, have shown that kerogen is able to produce new hydrocarbons when heated at a higher temperature than the maximum reached during its burial history. The industrial treatment of oil shales consists of a pyrolysis to generate the shale oil. This oil, obtained from 350 to 500°C, has a rather different composition from a normal crude oil. For example, unsaturated hydrocarbons and heterocompounds (N, S, O) are abundant.

A series of experiments has been carried out on the shallow and immature Messel shales of the Rhine Valley (Welte, 1966; Albrecht, 1969; Bajor et al., 1969), on the Lower Toarcian shales of the Paris Basin (Debyser and Oudin, Durand, in: Tissot and Pelet, 1971) and on the Yallourn lignite from Australia (Connan, 1974; van Dorsselaer, 1975).

The Messel shales, rich in organic matter, were deposited in the Rhine Valley during Eocene time and have a maximum thickness of 150 m. The maximum depth of burial is 200 to 300 m, so their kerogen is still very immature (e.g., hydrocarbon ratio is only 7 mg/g of organic carbon). After exhaustive Soxhlet extraction, Welte (1966) heated the rock to various temperatures from 190 to 490°C. The bitumen content increases regularly with heating temperature; hydrocarbons, and especially saturated hydrocarbons, are favored with increasing temperature. A particularly interesting point is the distribution of n-alkanes (Fig. II.5.21). In the natural bitumen a striking odd predominance was present from nC_{23} to nC_{29}. A weaker, but still definite predominance of odd n-alkanes is noted in the new bitumen which was generated by heating to 210–300°C, but no odd predominance of n-alkanes was observed in bitumens produced above 300°C. In addition to the particular distribution of n-alkanes, Albrecht (1969) showed that heating the kerogen of Messel shales also produced C_{15}, C_{16}, C_{18}, C_{19}, and C_{20} isoprenoids.

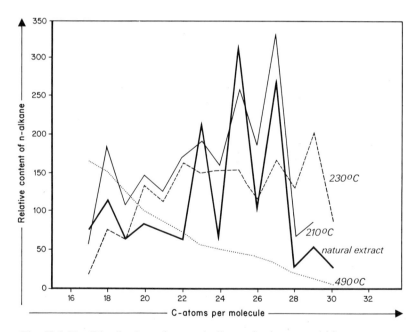

Fig. II.5.21. Distribution of normal alkanes in the natural bitumen extracted from the Messel shales, and in the bitumen generated by heating the shales at temperatures ranging from 210 to 490°C (Welte, 1966)

5.6 Experimental Generation of Hydrocarbons from Organic Material

Connan (1974) heated Yallourn lignite and showed the formation of odd-numbered *n*-alkanes and even-numbered *n*-alkenes by decomposition of wax esters. Decarboxylation of fatty acids generates the odd-numbered *n*-alkanes, whilst the alcohol produces the even *n*-alkenes. In the same experiments, van Dorsselaer et al. (1977) reported also the formation of some pentacyclic triterpanes which were probably present in the kerogen matrix. By heating the Green River shales, Gallegos (1975) showed some generation of terpanes and steranes, although they are comparatively less abundant in pyrolysis oil than in the natural bitumen.

All these results suggest that "ghost" structures, inherited from the original biogenic molecules, can be either attached to the kerogen by chemical bonds (e. g., linear or isoprenoid chains linked by an ester or ether bond) or entrapped in cavities of the kerogen matrix (as in a molecular sieve) and then released upon temperature increase. Thus molecules retaining characteristic structures (isoprenoid chains, steroid or triterpenoid polycyclic nuclei) or distribution patterns (odd *n*-alkanes) may be produced during the first stage of the catagenetic evolution. Following that line of thought, Seifert (1978) compared the hydrocarbons present in crude oils with those generated by pyrolysis of kerogen, in order to identify the source rock responsible for generation of the oil.

With the Messel shale, Bajor et al. (1969) carried out heating experiments with increasing temperatures in order to simulate a progressive burial. They showed, as a function of temperature, a progressive increase of the liquid bitumen followed by a decrease of liquid bitumen explained by cracking and massive generation of light hydrocarbons. This succession is in agreement with the progressive evolution observed in sedimentary basins as a function of burial.

Experiments on Lower Toarcian shales of the Paris Basin are of two different types. A first set was carried out on shallow immature samples in order to determine the influence of time and temperature on bitumen and hydrocarbon generation (Table II.5.5): after 90, 180, and 270 days, the amount generated was

Table II.5.5. Bitumen (hydrocarbons + resins + asphaltenes) generated by heating lower Toarcian shales (Fecocourt; Paris basin) at 180, 200 and 220°C

Time (days)	Bitumen generated (mg/g of original organic carbon)		
	at 180°C	at 200°C	at 220°C
1	5.9	6.8	9.2
3	6.6	8.7	13.1
10	6.6	9.2	14.0
30	8.8	14.8	21.4
90	10.4	12.2	30.0
180	10.7	22.2	30.4
270	12.8	21.2	36.2

about three times more at 220°C than at 180°C. In addition, generation of high molecular weight heterocompounds seems to precede the formation of hydrocarbons. The latter would mostly result from successive reactions in which heavy heterocompounds act as intermediate products (Louis and Tissot, 1967). This interpretation is confirmed by the results of Ishiwatari et al. (1976) reported above.

In a second set of experiments on Lower Toarcian shales, samples from various depths in the basin were heated under standard conditions after exhaustive Soxhlet extraction. They resulted in more bitumen generation from shallow immature samples than from deep rocks which had already undergone some maturation and had produced hydrocarbons during their natural burial. The role of pressure has been tested in experiments on the same shales of the Paris Basin with geostatic pressure varying from a few bars up to 800 bars. The hydrocarbons generated are apparently of the same order of magnitude at all pressures studied. Heterocompounds, such as resins and asphaltenes, might be slightly less abundant under high pressures. This is an indication of the relatively subordinate role of pressure in the formation of oil.

More recently, two important aspects were investigated by heating kerogen of ancient sediments in conditions closer to subsurface situations. Lewan et al. (1979) showed that in the presence of water ("hydrous pyrolysis"), the observed distribution of hydrocarbons generated is more comparable to that of crude oils than it is without water. In particular no olefins are generated, when water is present.

The role of the mineral matrix was observed in pyrolysis conditions by two series of experiments. Espitalié et al. (1980) compared the products yielded by heating the kerogen in an ancient shale, the isolated kerogen from the shale, and the same kerogen with varying proportions of minerals, such as illite, smectite, etc. They observed that some clay minerals, such as illite, are responsible for retention of heavy hydrocarbons ($> C_{15}$) and non-hydrocarbons generated during pyrolysis, thus providing an effluent rich in light hydrocarbons. In turn, the trapped heavy constituents are progressively cracked and produce light hydrocarbons and a carbon residue, if heating is continued. Horsfield and Douglas (1980) showed comparable results by pyrolyzing various kerogens and coal macerals (vitrinite, sporinite, alginite) in the presence of minerals.

5.6.4 Significance of Heating Experiments

The conclusions derived from heating experiments cannot be compared directly and without great care to subsurface conditions in sedimentary basins.

The main result of the heating experiments is to prove that kerogen is able to form petroleum constituents under temperature increase. Laboratory experiments are not able to simulate natural diagenesis, as there are probably important factors other than temperature at that stage; however, they provide a convenient simulation of catagenesis with respect to kerogen evolution. The results are in agreement with the mechanisms of petroleum generation deduced from the observations of case histories in sedimentary basins. Furthermore, experiments

5.6 Experimental Generation of Hydrocarbons from Organic Material

have also proved that biological markers can be fixed to kerogen by chemical bonds, such as ester or ether bonds, or entrapped in the kerogen matrix and subsequently released upon catagenesis.

Laboratory results have also shown that time and temperature can, to some extent, compensate for each other. The important role of the heavy heteroatomic compounds, including asphaltenes, is confirmed. Some of them may be already present at the stage of early diagenesis, whereas the major part results from the thermal breakdown of kerogen. Heavy heteroatomic compounds contribute to the generation of hydrocarbons and other constituents of low to medium molecular weight. In that respect they can be considered as intermediate products in some of the reaction pathways from kerogen to hydrocarbons.

The possible role of minerals, especially clay minerals, is not clear. Catalytic influence of clay minerals is only possible if there is an intense contact guaranteed between the reacting organic molecule and the clay surface. Such a contact between the bulk of the kerogen and clay minerals is not possible because they are both solid materials. Therefore a catalytic effect on the breakdown of kerogen can be largely excluded. Free petroleum compounds liberated by non-catalytic thermal breakdown of kerogen, however, can come in contact with clay surfaces and can be adsorbed. Thereafter a catalytic reaction is easily possible. In pyrolysis experiments this effect is observed. At present it is difficult to assess the importance of catalytic reactions under geological conditions. Finally, more experiments are probably necessary to measure the role of water in the reactions generating light or medium hydrocarbons.

It can be observed that an intermediate situation, between laboratory experiments and catagenesis due to burial in sedimentary basins, is provided by igneous intrusions in sediments. There, volcanic flows, dikes or sills provide high temperatures over a short time range. Catagenetic transformation of kerogen can be observed to a certain distance of the igneous rock and petroleum is generated. This situation has been reported by Hedberg (1964) and McIver (1967) and occurs in South Africa, Southern Colorado , Argentina, etc. and more recently in cores taken during the Deep Sea Drilling Program: Simoneit et al. (1978, 1981).

Summary and Conclusion

During burial of sediments, the increase in temperature results in a progressive rearrangement of kerogen:

1. During the last part of diagenesis, heteroatomic bonds and functional groups are eliminated. Carbon dioxide, water and some heavy N, S, O compounds are released. In terms of petroleum exploration, source rocks are considered as immature at this stage.
2. During catagenesis, hydrocarbon chains and cycles are eliminated. Thus, first mainly crude oils and then gas are successively formed. This stage corresponds to the principal stage of oil formation, and also to the principal stage of wet gas formation.
3. During metagenesis, a rearrangement of the aromatic sheets occurs. The stacks of aromatic layers, previously distributed at random in kerogen, now gather to form larger clusters. At this stage, only dry gas is generated.

The amount and the composition of hydrocarbons present in source rocks change progressively, as a function of increasing evolution. During diagenesis, molecules of biological origin predominate, including certain n-alkanes and iso-alkanes, steroids, and terpenes. During catagenesis, medium to low molecular weight hydrocarbon become predominant, and particularly n- and iso-alkanes. Where metagenesis is reached, methane is practically the only remaining hydrocarbon.

Chapter 6
Formation of Gas

6.1 Constituents and Characterization of Petroleum Gas

There is a wide variety of natural gas occurrences whose composition and modes of formation may be rather different. However, in this chapter only natural petroleum gas will be considered. While methane (CH_4) is always a major constituent of the gas, other components may be present:

– hydrocarbons heavier than methane (mostly ethane C_2H_6, propane C_3H_8, butane C_4H_{10}),
– carbon dioxide CO_2; hydrogen sulfide H_2S,
– nitrogen, hydrogen, argon, helium,
– condensate (liquid hydrocarbons dissolved in the gas: they are separated when the gas reaches surface or shallow subsurface positions).

The origin of gaseous constituents is multiple and may be not only organic, but also inorganic (atmospheric air, volcanic and geothermal gases, radioactivity).

Characterization of natural gases has to rely on a small number of parameters, as the number of their constituents is relatively small, compared to the large variety of molecules found in crude oil. Furthermore, dry gas is mainly composed of methane, with minor fractions of nonhydrocarbon gases.

The main parameters used to characterize gas are the following:
– the methane content, and also that of higher hydrocarbons, when they are present; this is conveniently expressed by the $C_1/\Sigma C_n$ ratio (methane/total hydrocarbons),
– the content, of CO_2, H_2S, N_2 and other nonhydrocarbon gases, if present,
– the carbon isotope composition of hydrocarbon gas, usually expressed as $\delta^{13}C(‰)$ (for definition see Chap. II.2.5) in particular $\delta^{13}C_1$ refers to the isotopic composition of methane,
– the hydrogen isotope composition of hydrocarbon gas, usually expressed as δD (for definition see Chap. II.2.5).

The principal characters of hydrocarbon gases generated during the successive stages of kerogen evolution are shown in Table II.6.1, according to the data recently presented by Stahl (1975), Galimov (1980), Schoell (1980), Rice and Claypool (1981), and Tissot et al. (1982). Both the first and the last stage (diagenesis and metagenesis) generate dry gas, i.e., methane with a very low content of higher hydrocarbons: $C_1/\Sigma C_n > 0.97$. However, these two stages are easily differentiated by their respective isotopic fractionation: the $\delta^{13}C_1$ ratio of

Table II.6.1. Principal characteristics of hydrocarbon gases generated during the successive stages of kerogen evolution. (After Stahl, 1975; Galimov, 1980; Schoell, 1980; Rice and Claypool, 1981)

Main stages of evolution		$C_1/\Sigma C_n$	$\delta^{13}C_1$	δD
Diagenesis	Dry gas	$\geqslant 0.97$	-90 to -55	< -180
Catagenesis	Oil-associated gas	< 0.98	-55 to -30	< -140
	Wet gas			
Metagenesis	Dry gas	$\geqslant 0.97$	-40 to -20	-150 to -130

the early, biogenic methane ranges from -90 to $-55\%_0$, whereas the metagenetic methane shows $\delta^{13}C_1$ values between -40 and $-20\%_0$. The δD of methane could provide additional valuable information, as shown in Table II.6.1. The intermediate stage of catagenesis generates wet gas, including oil-associated gas, with a variable proportion of higher hydrocarbons: $C_1/\Sigma C_n < 0.98$. The isotopic ratios $\delta^{13}C_1$ and δD show intermediate values.

Thus, a systematic study of the $\delta^{13}C_1$ measured on methane and the $C_1/\Sigma C_n$ ratio allows a first and relatively simple characterization of pooled gases, as shown by Figure II.6.1 which is compiled from the data of Stahl (1975), Galimov (1980), Schoell (1980), Rice and Claypool (1981). This illustration shows the three successive stages of gas generation during the evolution of sediments:

- biogenic gas, generated during diagenesis,
- thermal gas generated during catagenesis,

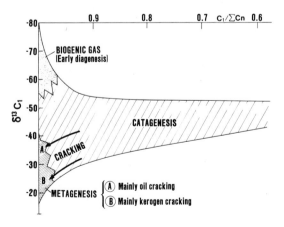

Fig. II.6.1. Relative abundance and isotopic composition of methane in gases from biogenic origin (diagenesis) and thermogenic origin (catagenesis and metagenesis). The original data used for compilation of this figure are shown, with references, in Figure II.6.7

6.2 Gas Generated During Diagenesis of the Organic Matter

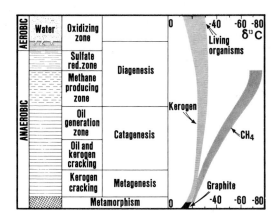

Fig. II.6.2. Successive stages of gas generation in sediments, after Claypool and Kaplan (1974), Demaison and Moore (1980), Galimov (1980) and Rice and Claypool (1981). Isotopic composition of kerogen and methane are shown after Galimov (1980), with minor changes

- thermal cracking gas generated during the late part of catagenesis and metagenesis.

The principal events responsible for gas generation in sediments are shown in Figure II.6.2, according to Claypool and Kaplan (1974), Demaison and Moore (1980), Galimov (1980), Rice and Claypool (1981), together with a comparison of the isotopic composition of kerogen and the related methane (Galimov, 1980). These aspects will be reviewed in the following Sections 6.2 and 6.3.

It should be noted that the isotopic characterization of methane presented in Table II.6.1 and Figure II.6.1 is based on pooled gases. The greatest care should be taken when applying these results to small amounts of gas found near the surface. The isotopic composition of low quantities of methane occurring in soils, groundwaters and young sediments may be altered by action of microorganisms. In particular, oxidation of methane by methane-oxidizing bacteria utilizes preferentially ^{12}C, as compared to ^{13}C. Thus the residual methane may be enriched in ^{13}C (see above, Chap. II.2).

6.2 Gas Generated During Diagenesis of the Organic Matter

6.2.1 Hydrocarbons

The upper slice of a young sediment containing organic matter and deposited in a normal, open and oxygenated sea, may well be aerobic (Fig. II.6.2). However, the sediment becomes anaerobic at very shallow depth, first at a microscopic, then at a macroscopic scale, and bacterial reduction of sulfate is initiated, generating S^{2-}. The available amount of sulfate is quickly lowered under normal conditions, and the sediment enters the zone of methane generation by bacterial degradation of the organic matter whose ultimate step is carbon dioxide or acetate reduction by methanogenic bacteria. This biogenic methane shows a very low isotopic ratio $\delta^{13}C_1 = -90$ to $-55‰$. The succession of these events is discussed in Chapter II.2.2.

A comparison of the isotopic ratio of the kerogen and of the related methane shows that such biogenic methane is highly depleted in the heavier isotope of carbon, as its $\delta^{13}C_1$ is ca. 30 to 50‰ lower than the $\delta^{13}C$ of the original kerogen. However, the isotopic composition of the kerogen is only slightly changed, as the global system simultaneously generates CH_4 (^{13}C-depleted) and CO_2 (^{13}C-enriched), providing some internal compensation (Galimov, 1980).

Favorable conditions for generation of abundant biogenic methane are the following, according to Rice and Claypool (1981):

- elimination of oxygen,
- low sulfate content,
- moderate temperatures: 75°C seems to be the maximum acceptable temperature, even to thermophilic bacteria; this assumption is rather in agreement with the observations made on subsurface oil degradation by other types of bacteria (see below, Chap. IV.5.3),
- sufficient space available to the bacterial bodies (1–10 μm), i.e., a moderate stage of compaction.

From these considerations, and also from observations made in sedimentary basins, it is assumed that most biogenic gas is generated at depth shallower than 1000 m. However, there are situations where active generation could have persisted at greater depth.

The proportion of higher hydrocarbons (ethane, propane, etc.) generated by bacterial activity is very low, as compared to methane. However, a systematic examination of the ethane/methane ratio in the cores taken as part of the Deep Sea Drilling Project has shown that the ratio increases with depth from 10^{-4} to 10^{-2} at depths ranging between 500 and 1500 m: Figure II.6.3 (Rice and Claypool, 1981). This progressive change is interpreted as a result of mixing biogenic methane with some early products of the thermal evolution of kerogen.

The composition of the major accumulations of biogenic gas is shown in Table II.6.2 and in Figure II.6.4. They are found in relatively young sediments, ranging from Cretaceous to Pleistocene, which never experienced high temperatures. Present reservoir depths are shallower than 1300 m in the major fields of Western Siberia and in the Great Plains of Canada and United States. Biogenic gas may

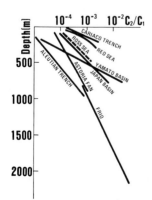

Fig. II.6.3. Ethane to methane ratio of natural gas from some wells of the Deep Sea Drilling Project, and from the Frio sands in Gulf Coast (Rice and Claypool, 1981)

6.2 Gas Generated During Diagenesis of the Organic Matter

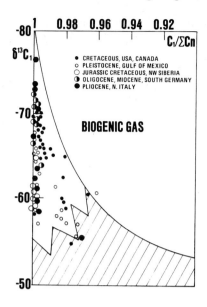

Fig. II.6.4. Relative abundance and isotopic composition of methane in various gases from biogenic origin. Stratigraphic ages given denote the age of the reservoir rock. (Data from Schoell, 1980; Rice and Claypool, 1981)

Table II.6.2. Composition of the major accumulations of biogenic gas. (Adapted from Rice and Claypool, 1981; with additional data from Schoell, 1980)

Country	U.S.S.R.	Canada	U.S.A.	Italy	U.S.A.
Basin or field	W. Siberia	S.E. Alberta	N. Great Plains	Po Valley	Gulf of Mexico
Reservoir age	Cretaceous	Cretaceous	Cretaceous	Pliocene	Pleistocene
Depth (m)	700–1300	300–1000	120–840	1200–3000	460–2800
$C_1/\Sigma C_n$	>0.99	>0.99	>0.97	>0.97	>0.97
$\delta^{13}C_1$ (‰)	−68 to −58	−68 to −60	−72 to −55	−76 to −55	−69 to −55
Estimated reserves ($10^{12} \times m^3$)	11	0.4	?	0.4	0.3

also be found at greater depths in younger sediments, such as the Pliocene of North Italy and the Pleistocene of the Gulf of Mexico, where a very young age and a low geothermal gradient make the sediments still immature down to 2800 m.

6.2.2 Nonhydrocarbons

Nonhydrocarbon gases may also be generated during the stage of diagenesis, particularly carbon dioxide and nitrogen. Carbon dioxide may result from bacterial activity at shallow depth degrading the organic matter in sediments. Marsh gas may contain 1 to 10% CO_2 (Sokolov, 1974). However, at this stage carbon dioxide is likely to be largely dissolved in subsurface waters. Later, when thermal degradation of kerogen is initiated, at depths such as 500 to 1500 m, the main feature of diagenesis is the elimination of functional groups, including carbonyl and carboxyl groups, which generate abundant CO_2: this corresponds to the elimination of oxygen from kerogen and the associated decrease of the O/C ratio. The amount of CO_2 generated is directly related to the original composition of kerogen and increases from type I to type III, which is able to generate large quantities of carbon dioxide. In fact the labile C=O groups may amount to 30% by weight of some type III kerogens (see above, Sect. 4.7). Karweil (1969) reported that 1 kg of coal is able to generate up to 75 l of CO_2.

Nitrogen is sometimes found in association with shallow biogenic gas (for instance in Alberta: Hitchon, 1963). Although nitrogen may be a remnant of air trapped in interstitial waters of sediments, another possibility is that diagenetic ammonia may be oxidized to N_2 by meteoric waters carrying oxygen.

Hydrogen sulfide is also produced by sulfate-reducing bacteria in young sediments (Chap. II.2.2). However, it is not preserved and usually recombines very quickly: either with sulfate to produce free sulfur, or with iron to generate sulfides (mainly in detrital sediments), or with organic matter (mainly in carbonate-evaporite sediments).

6.2.3 Importance of Biogenic Gas

The world reserves of biogenic methane have been estimated by Rice and Claypool (1981) to 17×10^{12} m^3 (17,000 billion cubic meters). In addition to this, new reserves could be discovered in places such as the Great Plains of Canada and the USA.

The present evaluation amounts to ca. 20% of the total world gas reserves estimated to be 77.5×10^{12} m^3. However, this figure is heavily influenced by the huge accumulations of biogenic gas in Western Siberia (ca. 11×10^{12} m^3), and the figure could be quite different on a regional basis: e.g., in Western Europe the fraction of biogenic gas is approximately 7% versus 93% of thermal gas generated during catagenesis and metagenesis.

6.3 Gas Generated During Catagenesis and Metagenesis of Organic Matter

6.3.1 Hydrocarbon Gas

Catagenesis is primarily the stage of oil generation. However, formation of liquid hydrocarbons is always accompanied by some generation of hydrocarbon gas, especially C_2–C_4, which were virtually absent during diagenesis. The occurrence of C_2–C_4 hydrocarbons is even a characteristic feature diagnostic of the catagenesis zone: an example from Western Canada, where this method has been particularly used, will be discussed in Chapter V.1.3.4.

Methane is also generated along with crude oil and C_2–C_4 hydrocarbons. Early methane is still depleted in heavy isotope ^{13}C, as compared to the kerogen which is

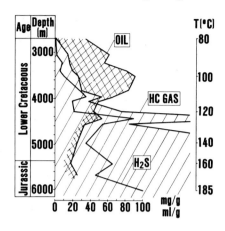

Fig. II.6.5. Abundance of liquid and gaseous hydrocarbons, and hydrogen sulfide, in cuttings from a deep well in the Aquitaine basin, France (Le Tran, 1972)

the source for it, but the difference is progressively reduced when passing from diagenesis to catagenesis and metagenesis stages: Figure II.6.2. Thus, with increasing evolution of the sediments, the proportion of thermal methane becomes greater and the $\delta^{13}C_1$ becomes less negative.

When the advanced stage of catagenesis (R_o = 1.3 to 2.0) is reached, cracking of C–C bonds becomes important in crude oil already generated and also in the remaining kerogen. The amount of liquid hydrocarbons decreases, whereas gas increases abruptly, as has been observed by Le Tran (1972) in a deep well of the Aquitaine basin: Figure II.6.5. Gas generated by cracking usually shows a lower $\delta^{13}C$ than the related oils, due to a greater probability for breaking $^{12}C-^{12}C$ bonds than $^{13}C-^{12}C$ bonds (Silverman, 1964, 1967; see above Sect. 5.5). Thermal cracking experiments carried out by Sackett (1968, 1978) have confirmed this view and showed that the resulting methane is 4‰ to 25‰ lower than the parent material (see above, Sect. 5.5).

During metagenesis, some of the kerogen constituents are still able to generate methane, as shown by the experiments of Jüntgen and Karweil (1966). In particular, the short alkyl chains, which are still present and would prevent coalescence of the polyaromatic nuclei, are broken at that stage. The resulting methane shows a $\delta^{13}C_1$ close to that of kerogen: Figure II.6.2.

From observation of the amount and isotopic composition of methane generated by thermal cracking during late catagenesis and metagenesis in geological conditions we can suggest that actual situations fall between two extreme cases (Fig. II.6.1):

– methane is mainly generated by the cracking of oil; this situation could occur frequently when kerogen belongs to type I or II, i. e., if it is very oil-prone: A in Fig. II.6.1;
– methane is mainly generated by cracking of the remaining kerogen; this situation could occur frequently when kerogen is of type III, which is less oil-prone, or when oil has been largely expelled out of the source rock before metagenesis is reached: B in Fig. II.6.1.

Both situations may be found in gases analyzed by Stahl (1975): an example of case A would be provided by the gases of the Permian Basin, Texas, whereas an example of case B would be provided by gases of the NW European basin, generated from Upper Carboniferous coals and associated shales.

6.3.2 Nonhydrocarbons

Nonhydrocarbon gases — mainly hydrogen sulfide and nitrogen — may also be generated during late catagenesis and metagenesis and they are associated with methane and light hydrocarbons in certain basins, according to the type of sediments and the geological history.

a) Hydrogen sulfide may be generated in large amounts by thermal cracking from kerogen and from liquid sulfur-containing compounds present in crude oils: the deep wells drilled in the South Aquitaine basin have clearly shown that hydrogen sulfide is generated together with methane beyond 120°C: Figure II.6.5 (Le Tran, 1972). Hydrogen sulfide is particularly abundant when the organic matter itself (type II) is rich in sulfur, which may be the case with carbonate and carbonate-evaporite sequences. On the other hand, type III kerogen is usually a poor source for sulfur compounds.

b) Free sulfur, if present, may also react with hydrocarbons to produce H_2S. Sulfate reduction (if present in sediments), and the related oxidation of hydrocarbons, may also occur during metagenesis — beyond 150°C — resulting in generation of hydrogen sulfide and carbon dioxide.

As a result of both processes (a) or (b), it has been observed that the average abundance of H_2S in natural gases reaches a maximum at depths beyond 3000 m (see below Chap. IV.4.3.2).

The existence of H_2S together with SO_4^{2-}-containing rocks at great depth and high temperatures often results in formation of free sulfur. Free sulfur and H_2S together may form polysulfides which in turn destroy methane and generate H_2S (Welte, 1976).

c) Nitrogen may also be generated from certain types of kerogen during metagenesis. Lutz et al. (1975) have shown on cuttings of Westphalian coal measures, in a deep well close to the huge Groningen gas field, that the nitrogen content increases markedly during metagenesis and may reach 60% of the total gas generated at that stage. Experimental evolution of various coals seems to confirm the view that the nitrogen present in organic matter is little affected until an advanced stage of thermal evolution is reached and the activation energies related to N_2 generation range from 40 to 70 kcal mol^{-1} (Klein and Jüntgen, 1972). Furthermore, because of the smaller molecular diameter N_2 (ca. 3.4 Å) should migrate more easily than CH_4 (ca. 3.8 Å), if diffusion processes are involved. The preferential migration of nitrogen may also result in an isotopic fractionation, with an enrichment of ^{15}N.

d) Hydrogen is often observed during high-temperature pyrolysis of oil shales; it may also be generated from kerogen in a geological situation where advanced stages of thermal maturation are reached. Diffusion of hydrogen molecules is, however, relatively easy, thus lowering concentrations.

6.4 Gas Originating from Inorganic Sources

Some constituents of natural gas may be produced by purely inorganic processes, i.e., without participation of the organic matter of the sediments.

a) The atmospheric air dissolved in shallow waters may be trapped with interstitial waters in the pore space of young sediments. Oxygen is consumed by aerobic microbial activity. Nitrogen and argon might be preserved in certain cases. However, the argon/nitrogen ratio is frequently different from the ratio in atmospheric air (1.2×10^{-2}), suggesting that other sources of nitrogen and argon are probably more important.

b) Gas originating from magmatic rocks, such as volcanic and geothermal gases may be a source of CO_2, H_2 and smaller amounts of CH_4, H_2S, N_2, He and Ar. In the Caucasus area, Sokolov (1974) reported carbon dioxide concentrations in free and dissolved gases reaching 90 to 99%. In some gas accumulations, CO_2 is a major component and is probably of magmatic origin. This is the case, for example, in certain areas of the Rocky Mountains (up to 98% CO_2 in Colorado, New Mexico and W. Texas), the Panuco-Ebano area of Mexico (96% CO_2) and part of the Pannonian basin in Central Europe (Mihai field 95% CO_2). Sokolov (1974) interpreted these occurrences as resulting from magmatism with igneous intrusions and/or volcanic activity. Fluid inclusions observed in magmatic or metamorphic rocks also contain methane and other light hydrocarbons probably generated by inorganic synthesis.

c) Radioactivity is likely to be the source for helium and most argon. Helium is produced by disintegration of the radioisotopes of the uranium and thorium families, ^{235}U, ^{238}U and ^{232}Th, respectively. Argon results from disintegration of potassium ^{40}K. Natural uranium is composed of 99.3% ^{238}U and 0.7% ^{235}U. ^{232}Th amounts to 100% natural thorium, and ^{40}K amounts to 0.12% natural potassium. With regard to the average content of natural uranium, thorium and potassium in the earth's crust, and to the respective half-life of these radio-isotopes, one can evaluate the number of disintegration per year in 1 g of average rock (Table II.6.3), which is directly related to the amount of gases generated.

Therefore, with increasing age of the rock, the cumulative amount of gases generated by radioactive decay processes increases. This age-dependence has been effectively observed, and is shown for helium in Figure II.6.6 based on data from 1970 gas fields. It is clear that the distribution of the helium concentration in gas fields changes with the age of the reservoir rock. The mode of the distribution in Paleozoic gas fields exceeds by more than two orders of magnitude the mode in Tertiary gas fields.

Table II.6.3. Number of disintegration g^{-1} of average rock per million years

Potassium	294×10^{12}
Uranium	1.42×10^{12}
Thorium	1.56×10^{12}

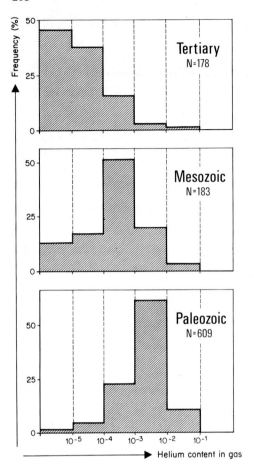

Fig. II.6.6. Distribution of the helium content in gases from Paleozoic, Mesozoic, and Tertiary reservoirs (Tissot, 1969)

6.5 Occurrence and Composition of Gas in Sedimentary Basins: Example of Western Europe

Two considerations make the distribution of gas accumulations more difficult to explain than the distribution of oil fields. Firstly, gas constituents may derive from different sources (organic, inorganic) through different processes of generation (biogenic, thermogenic, or even radioactive). Furthermore, they migrate more easily than oils through the sedimentary column. In this connection it has to be remembered that for gaseous constituents, because of their small molecular diameters and their relatively high water solubility, diffusion processes may play a special role with respect to migration and also dispersion of gas and destruction of existing fields. Unfortunately there are not yet enough data available to assess the importance and influence of diffusion processes upon the distribution and occurrence of gases. In general some rules can be given concerning the composition of the hydrocarbon fraction in relation to the geological situation of the field, but the distribution of the non-hydrocarbon constituents may change from one basin to another.

Various basins of Western Europe, ranging in age from Carboniferous to Mio-Pliocene, offer a wide distribution of natural gas composition and origin. Tables II.6.4 to II.6.6 show the composition and origin of gas accumulations in the principal gas provinces of Western Europe; they are plotted in Figures II.6.7 and II.6.8. The various gas fields can be grouped as follows.

a) *The alpine realm* (Table II.6.4; Schoell 1977, 1980; Colombo, 1966; Mattavelli et al., 1983) comprises mainly Tertiary and Pleistocene sediments, where biogenic gas was generated during early diagenesis and preserved in the Po Basin (mainly Pliocene) and the Eastern Molasse basin of South Germany (Oligocene and Miocene). Methane amounts to 97–99% of the gas, higher hydrocarbons are almost absent. The high $C_1/\Sigma C_n$ ratio and the low $\delta^{13}C_1$ are in agreement with the general scheme. In the Po Basin, the immature character of the Late Tertiary beds in attested by low vitrinite reflectances ($\leq 0.5\%$ R_o at 4000 to 6000 m) and results from a low geothermal gradient and a short duration of burial.

In older beds of early Tertiary and Mesozoic age, thermogenic gas has been generated during catagenesis in the Western Molasse basin and in Malossa, Italy, where depth is beyond 5000 m. The proportion of ethane and higher hydrocarbons is quite variable, depending on the local degree of catagenesis, whereas

Table II.6.4. Composition and origin of the main gas fields in the alpine regions of Western Europe

Country	N. Italy	S. Germany		N. Italy
Basin or field	Po Basin	E. Molasse	W. Molasse	Malossa
Reservoir Age	Pliocene (Pleistocene, Miocene)	Oligo-Miocene	Eocene	Mesozoic and Tertiary	Triassic
Depth (m)	1200–3000	900–2000	1800–3000	800–2000	5400
C_1 (%)	97–99.4	97.5–99.3	88–97	48–97	79
$\geq C_2$ (%)	0.04–0.3	0.05–0.5	1.5–8	2–21	19.8
$C_1/\Sigma C_n$	0.970–0.999	> 0.99	0.90–0.98	0.7–0.9	0.8
CO_2 (%)		< 0.4	0.3–4	0.2–4	0.4
N_2 (%)	≤ 0.5	0.1–7	0.1–10	1.5–10	0.7
$\delta^{13}C_1$ (‰)	−76 to −55	−72 to −60	−60 to −50	−47 to −35	−36.2
Origin	Biogenic	Biogenic	Mixed	Thermogenic	Thermogenic
Stage of evolution	Diagenesis	Diagenesis	Diagenesis + Catagenesis	Catagenesis	Catagenesis

Table II.6.5. Composition and origin of some gas fields derived from Jurassic source rocks of W. and N.W. Europe

Country or province	W. Germany	North Sea		France
Basin of field	N.W. Germany	Oil-Associated gas fields	Frigg	Lacq
Reservoir {Age	Jurassic E. Cretaceous	—	Eocene	E. Cretaceous L. Jurassic
Depth (m)	800–2000	—	1850	3200
C_1 (%)	65–91	54–84	95	69
$\geqslant C_2$ (%)	8–30	13–38	4	4.9
$C_1/\Sigma C_n$	0.7–0.96	0.58–0.86	0.96	0.93
CO_2 (%)			0.8	9.6
N_2 (%)	< 1			
H_2S (%)				15.4
$\delta^{13}C_1$ (‰)	−54 to −44	−51 to −45	−43.3	−44.2
Origin	Thermogenic	Thermogenic	Thermogenic	Thermogenic
Stage of evolution	Catagenesis	Catagenesis	Late Catagenesis	Late Catagenesis

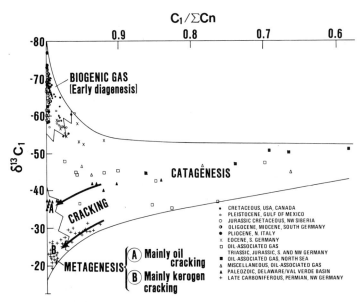

Fig. II.6.7. Relative abundance and isotopic composition of methane in gases from W. Europe, compared with other locations in North America and Siberia. Stratigraphic ages given denote the age of the reservoir rock. (Data from Boigk et al., 1976; Stahl, 1978; Schoell, 1980; Menendez, 1973; Héritier et al., 1979)

Table II.6.6. Composition and origin of some gas fields derived from Westphalian source rocks (coal measures) of N.W. Europe

Basin or field		Groningen	N.W. Germany	
Reservoir	Age	Permian (Rotliegend)	Carboniferous	Permian-E. Triassic
	Depth (m)	2450	2000–4000	1700–2500
C_1 (%)		81	82–97	20–95
$\geq C_2$ (%)		3	0.5–6	
$C_1/\Sigma C_n$		0.96	0.94–0.996	0.94–0.996
CO_2 (%)		1	1–10	0.5–7
N_2 (%)		14.8	1–10	1–80
$\delta^{13}C_1$ (‰)		−36.6	−30 to −22	−32 to −19
Origin			Thermogenic	
Stage of evolution			Late catagenesis to metagenesis	

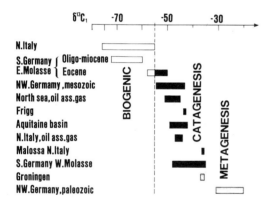

Fig. II.6.8. Origin and isotopic composition of the principal gases of Western Europe (Tissot and Bessereau, 1982)

$\delta^{13}C_1$ is typical of that evolution stage. An intermediate situation occurs in the Eocene beds of the Eastern Molasse basin and in some Miocene fields of the Po Basin, where a mixture of dry gas generated during diagenesis and wet gas from early catagenesis results in a certain proportion of higher hydrocarbons, and a $\delta^{13}C_1$ ranging from −60 to −50‰.

b) *The Jurassic source beds*, which are the major source for crude oil in Western Europe (particularly in the North Sea), have also generated large amounts of gas in the North Sea, the Aquitaine Basin, and smaller but sizable amounts in North Germany (Table II.6.5; Le Tran, 1972; Menendez, 1973; Stahl, 1978; Héritier, 1979; Schoell, 1980).

In the North Sea, the oil-associated gas is accumulated in reservoirs ranging from Jurassic to Paleocene age. It has been generated during catagenesis; the proportion of C_2 and higher hydrocarbons is important ($C_1/\Sigma C_n$ = 0.6 to 0.9) and the $\delta^{13}C_1$ is in the range of -50 to $-40‰$. The Frigg gas field, where an important gas column (170 m) overlies a thin oil ring (10 m) probably originated during the late catagenesis stage of Mio-Pliocene, when the Jurassic source beds were buried to 4000–5000 m. It contains less C_2 and higher hydrocarbons ($C_1/\Sigma C_n$ = 0.96). The stage of evolution of the source rock is attested by vitrinite reflectivities from 1.5 to 1.8% (Héritier et al., 1979).

A comparable situation occurs in the Aquitaine basin, where the Lacq and Meillon gasfields were formed during late catagenesis and occur presently at 3200–4500 m and 4000–5000 m depth, respectively. The $C_1/\Sigma C_n$ ratio (0.93) and the $\delta^{13}C_1$ ($-44‰$) are comparable (Menendez, 1973), and the stage of evolution of the source beds is attested by vitrinite reflectivities in the range of 1 to 2% at Lacq (Robert, 1980). Hydrogen sulfide and carbon dioxide are also present. H_2S may result from cracking of both kerogen and crude oil previously generated (many crude oils in the Aquitaine basin contain a significant amount of sulfur); however, destruction of methane due to polysulfides, or reduction of hydrocarbons by evaporite could also be considered as a process of H_2S generation. The latter reaction would also contribute to CO_2 generation.

c) *The Westphalian coal measures*, and the associated dark shales, are the oldest and major source for gas in N.W. Europe: southern part of the *North Sea*, *Netherlands* and *northern Germany* (Table II.6.6; Hedemann, 1963; Patijn, 1964a, b; Stahl, 1968; Lutz et al., 1975; Boigk et al., 1976; Stahl, 1978; Schoell, 1980). The main reservoirs are — in order of decreasing importance of reserves — Permian (Rotliegend), Early Triassic, and Late Carboniferous. The major feature of gas trapping is the occurrence of a wide belt of Permian salt (Zechstein) providing an excellent seal.

There were two major phases of burial: the first one ended in early Mesozoic time (Triassic to Early Jurassic), the second one reached its maximum in Cenozoic (3000 to 7000 m on top of Carboniferous). During this last burial, the Westphalian coal measures reached the stage of late catagenesis to metagenesis, depending on their geographic and stratigraphic location. Teichmüller et al. (1979) have shown that vitrinite reflectance in the top Carboniferous beds range from 1.0 to 2.6%. The bulk of gas now found in these fields is thought to have been generated during the latter phase of burial. It cannot be excluded, however, that some of the gas is even now being generated where temperatures are high enough to reach necessary activation energies and where the coal and organic matter in shales is not yet too mature.

Gas composition shows a strong predominance of methane, as $C_1/\Sigma C_n$ ranges from 0.94 to 0.996‰, whereas $\delta^{13}C_1$ ranges from -36 to $-19‰$. Thermal cracking of coal and coaly kerogen present in the associated dark shales is probably responsible for the high values of $\delta^{13}C_1$: in the vicinity of Groningen, where $C_1/\Sigma C_n$ is 0.96 and $\delta^{13}C_1$ of methane gas is only $-36.6‰$, reflectance value on top of Carboniferous is only 1.0 to 1.7% (late catagenesis); in Wüstrow and Ebstorf, where gas is almost dry ($C_1/\Sigma C_n$ = 0.99) and $\delta^{13}C_1$ reaches $-21‰$, reflectance values are in the range of 2.5% (metagenesis) (Boigk et al., 1976; Teichmüller et al., 1979).

Other constituents of gas are 1 to 10% CO_2 and N_2, which is found in variable amounts from 1 to 90%. In Late Carboniferous reservoirs, where the extension of vertical migration is limited, the highest values (10% N_2) seem to be associated with the stage of metagenesis. In Permian reservoirs, the highest figures are observed in the Wüstrow field, with a varying content of N_2 reaching 80%, and in the German part of the North Sea with a maximum 90% N_2 content. Boigk et al. (1976) have interpreted this fact as an enrichment of N_2 due to extensive migration. However, there may be other sources for nitrogen gas.

6.6 Distribution of Gases in Sedimentary Basins

The composition of hydrocarbon gases in Western Europe, and also the carbon isotope ratio vary in general agreement with the principles expressed in Sections 6.2 and 6.3, and also in Table II.6.1. From these examples, it appears that the distribution of quite a number of hydrocarbon gases, including their isotopic composition, can be reasonably interpreted in terms of stages of maturation: diagenesis, catagenesis, metagenesis (Fig. II.6.7). The distribution of nonhydrocarbon gases is more complex, as the origin may be different from one basin to another, and furthermore it can also be affected by migration. For instance, Hitchon (1963) described in the Alberta basin a succession of gas zones with high concentrations of N_2, CO_2, and H_2S, respectively, from the shallow accumulations of the eastern flank to the deeper accumulations close to the Rocky Mountains. In Western Europe the situation is comparable as far as H_2S and accumulation depths are concerned. N_2, however, is found preferentially in accumulations derived from advanced maturation stages, and moreover from important vertical migration. CO_2 is found at relatively shallow depth (late diagenesis to early catagenesis) in South Germany, but also in deep accumulations derived from late catagenesis (Lacq) or even metagenesis (North Germany). Thus the interpretation of the distribution of nonhydrocarbons has to be made in the frame of the regional geological conditions.

A special question is the possible existence of a lower limit — a floor — to exploration for hydrocarbon gas in sedimentary basins. The deepest gas fields presently known range from 7 to 8 km approximately. Beyond these depths, temperatures may range from ca. 160° to 350°C in sedimentary basins. At this point, most source beds have reached an advanced stage of metagenesis and there is little chance for additional methane generation, except in some young sediments where fast burial is associated with a low geothermal gradient. Thus, in general, the existing fields at such depths would have been generated in a shallower situation, and subsequently buried at greater depths. This scheme requires a good seal and a smooth structural history to ensure physical preservation of the accumulation.

Thermal stability of methane, at these temperatures and even higher (Hunt, 1975), is such that present and foreseeable drilling depths do not reach zones where methane can be destroyed because of temperature. Destruction of methane, however, is possible by chemical reactions involving free sulfur or sulfates at elevated temperature and resulting in hydrogen sulfide and carbon dioxide generation. This may particularly happen in carbonate or carbonate-evaporite series where sulfur has not been trapped by iron to form iron sulfides.

Finally, there may be another limitation due to the loss of reservoir characteristics, especially porosity. At great depth, dissolution and recrystallization of minerals under high pressure conditions become important when approaching metamorphism and they can obliterate the existing porosity. There are, however, situations where a hydrocarbon saturation was reported to prevent cementation in the pores where hydrocarbons are present.

Summary and Conclusion

There ist a wide variety of natural gas occurrences, whose composition and modes of formation may differ considerably. Characterization of natural gases has to rely on a small number of parameters, as there are not many constituents in gases. Prominent features are the ratio of methane to total hydrocarbon $C_1/\Sigma\, C_n$ and the isotopic composition: carbon and hydrogen isotopes.

Biogenic methane wich is only generated at low temperature levels (below 75 °C) can be recognized by very low isotopic ratios ($\delta^{13}C_1$ from $-90‰$ to $-55‰$). Pure biogenic gas does not contain higher hydrocarbons in appreciable proportion. Biogenic methane is of economic importance in certain regions of the world (e.g., Western Siberia, Canada).

There is a continuous generation of wet gas parallel to oil generation. Beyond the main phase of oil generation, due to increasing temperature, cracking of oil and kerogen becomes a predominant process. Therefore more and more wet and subsequently dry gas is being generated.

The thermal stability of methane is such that a lower limit — or gas floor — is not determined by high temperatures as such, but by the depletion of hydrogen in the source material. Foreseeable drilling depths, therefore, do not reach the stage where methane becomes thermally unstable. However, methane can be destroyed chemically, for instance by conversion to H_2S in the presence of sulfur.

Nonhydrocarbon constituents (e.g., CO_2, H_2S) of natural gases may have an inorganic origin.

Chapter 7
Formation of Petroleum in Relation to Geological Processes. Timing of Oil and Gas Generation

7.1 General Scheme of Petroleum Formation

The history of petroleum formation is summarized as a function of increasing burial of the source rock in Figure II.7.1, which represents the abundance and composition of the hydrocarbons generated, and in Figure II.7.2, which shows the correlative evolution of kerogens. The depth scale represented in Figure II.7.1 is based on examples of Mesozoic and Paleozoic source rocks. It is only approximate and may vary according to the nature of the original organic matter,

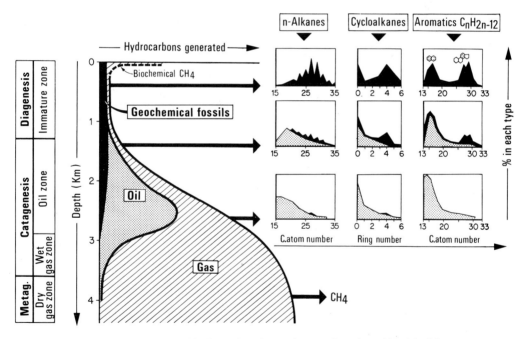

Fig. II.7.1. General scheme of hydrocarbon formation as a function of burial of the source rock. The evolution of the hydrocarbon composition is shown in insets for three structural types. Depths are only indicative and correspond to an average on Mesozoic and Paleozoic source rocks. Actual depths vary according to the particular geological conditions: type of kerogen, burial history, geothermal gradient. (Modified after Tissot et al., 1974). This figure can be compared with other diagrams proposed by Sokolov (in: Kartsev et al., 1971) and Hedberg (1974)

its burial history and the geothermal gradient. Based on the fundamental knowledge presented in the previous chapters, the general scheme of oil and gas formation may be summarized as follows.

7.1.1 Diagenesis

a) Young Sediments

At shallow depth, small amounts of hydrocarbons are present. They are inherited from living organisms, directly of with minor changes during the early diagenesis of the young sediment. They have characteristic structures resulting from their biogenic origin, and may be considered as geochemical fossils. At this stage, the composition of the kerogen is controlled mainly by the initial input of organic matter and by the nature and extent of microbial activity in the upper sedimentary layers of sediment.

The only new hydrocarbon generated at that stage is methane. In special cases, microbial activity may result in abundant methane generation (biogenic gas).

b) Immature Stage

During an appreciable time and depth span, little transformation occurs, as both hydrocarbons and kerogen are metastable under near-surface conditions and need a sufficient increase in temperature and time for rearrangements to be initiated.

When depth and temperature have increased to a sufficient level, heteroatomic bonds in kerogen are progressively broken. Oxygen elimination from kerogen is particularly important during this phase and results in CO_2 and H_2O formation.

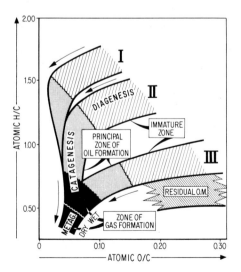

Fig. II.7.2. General scheme of kerogen evolution presented on van Krevelen's diagram. The successive evolution stages are indicated and the principal products generated during that time. Residual organic matter does not exhibit an evolution path. (Modified after Tissot, 1973)

7.1 General Scheme of Petroleum Formation

The first petroleum products liberated by this transformation include mostly heteroatomic (N, S, O)-compounds of high molecular weight particularly asphaltenes and resins. A certain amount of gas might also be generated, especially from organic matter of type III.

7.1.2 Catagenesis

a) Principal Stage of Oil Formation

As temperature continues to increase, more and more bonds are broken, e.g., ester and some C–C bonds. Hydrocarbon molecules, and particularly aliphatic chains, are produced from kerogen and from the previously generated N, S, O-compounds. Some of the hydrocarbons released are C_{15} to C_{30} biogenic molecules comparable to the geochemical fossils which were formerly entrapped in the kerogen matrix or linked by various bonding such as ester, ether, etc. However, most of the new hydrocarbons generated during the main zone of oil generation have a medium to low molecular weight. They have no characteristic structure or specific distribution, contrary to the geochemical fossils which are progressively diluted by these new hydrocarbons.

This is the principal stage of oil formation, as described by Vassoevich et al. (1969). However, liquid oil generation is accompanied by formation of significant amounts of gas.

b) Stage of Condensate and Wet Gas Formation by Cracking

As burial and temperature continue to increase, breaking of carbon–carbon bonds occurs more and more frequently, and affects both the source rock hydrocarbons already formed and the remaining kerogen. Light hydrocarbons are generated through this cracking and their proportion increases rapidly in the source rock hydrocarbons and petroleum. Due to the kinetics of formation and the advanced structure of kerogen, methane becomes quickly the dominating compound liberated.

The overall transformation occurring during catagenesis is equivalent to a process of disproportionation. On the one hand, hydrocarbons of increasing hydrogen content are generated. The average atomic ratio H/C is 1.5 to 2.0 in curde oil and is 4.0 in pure methane. On the other hand, the residual kerogen becomes depleted in hydrogen with an atomic H/C ratio of about 0.5 by the end of the stage of catagenesis.

7.1.3 Metagenesis: Dry Gas Zone

After most labile material has been eliminated through catagenesis, a structural reorganization occurs in kerogen, with the development of a higher degree of ordering. However, in this stage (metagenesis) no significant amounts of hydrocarbons are generated from kerogen except for some methane. Large amounts of methane may result from cracking of source rock hydrocarbons and of reservoired liquid petroleum.

The floor of the possible existence of methane in the subsurface is contrary to all other hydrocarbons not controlled by its thermal stability (Hunt, 1975). The stability of methane, even at higher temperatures (up to about 550 °C) is such that present and foreseeable drilling depths do not reach zones where methane can be destroyed because of temperature. Methane, however, can be destroyed chemically due to the presence of sulfur or sulfate (see above Chapter II.6).

In fact there may be an economic limit for commercial occurrence of methane, due to degradation of reservoir characteristics, especially due to lack of porosity.

7.2 Genetic Potential and Transformation Ratio

The succession of the main steps of evolution outlined above (Sect. 7.1) is a common fact, but the temperature corresponding to the thresholds and the amount of petroleum generated depends on the nature of the organic matter and also on the temperature history, namely temperature versus time relationship. The influence of pressure and potential catalysts is less clear.

In an effort to distinguish between the respective influence of the kerogen composition and of the catagenesis intensity, we can define and make use of two terms: the genetic potential and the transformation ratio.

a) *The genetic potential* of a given formation represents the amount of petroleum — oil and gas — that the kerogen is able to generate, if it is subjected to an adequate temperature during a sufficient interval of time. This potential depends on the nature and abundance of kerogen, which in turn are related to the original organic input at the time of sediment deposition, and to the conditions of microbial degradation and rearrangement of the organic matter in the young sediment. Thus, the genetic potential depends on external factors, such as topography, climate, oceanology, existence of biological associations etc., as pointed out by Debyser (1969). It may be noted that the genetic potential of a formation may be conveniently characterized by the type of kerogen (or by the corresponding evolution path) and the abundance of kerogen. A quantitative evaluation of the genetic potential can be made on the basis of a standard pyrolysis technique, which is discussed in Part V (Espitalié et al., 1977).

The genetic potential of a petroleum source rock is not essentially different from the total oil and gas yield of an oil shale upon pyrolysis, as there is no fundamental difference between a shallow immature source rock and an oil shale — provided the organic richness is sufficient. This point will be discussed further in Chapter II.9.

b) *The transformation ratio* is the ratio of the petroleum (oil plus gas) actually formed by the kerogen to the genetic potential, i.e., to the total amount of petroleum that the kerogen is capable of generating. The ratio measures the extent to which the genetic potential has been effectively realized. An evaluation of the transformation ratio is provided by the standard technique of pyrolysis, described in Part V, together with the genetic potential. The ratio can also be approximated, down through the main stage of oil generation (thus before reaching the zone of wet gas formation), by using the bitumen ratio.

The transformation ratio depends on the nature of the organic material and on the subsequent geological history (mostly the temperature versus time history). Therefore, the transformation ratio depends mostly on internal factors: geothermal gradient, subsidence, and tectonics. The temperature corresponding to the beginning of the principal stage of oil formation is also dependent on the nature of kerogen, but the beginning of the stage of gas formation is less dependent on this factor, in accord with the kinetics of cracking reactions.

7.3 Nature of the Organic Matter. Gas Provinces Versus Oil Provinces

1. The nature of the original organic material is an important factor that cannot be disregarded. Larskaya and Zhabrev (1964) pointed out that the various types of organic matter involved in their Azov-Kuban survey reacted differently to burial. When gathering data from many sedimentary basins, it is clear that kerogen made of marine or limnic autochthonous organic matter, including the microbial biomass living in sediment (type I or II, rich in aliphatic groups), generates abundant bitumen: 180 mg/g of organic carbon in the Paris basin, 200 mg/g in the Uinta basin. On the contrary, kerogen comprising a large amount of continental plant debris (Type III, rich in aromatic groups and oxygenated functions) generated a smaller amount of bitumen: 100 mg/g of total organic carbon in the Douala basin.

An illustration of the importance of the composition of the organic matter is shown in Figure II.7.3, where two shallow kerogens, which have not yet reached the stage of catagenesis, are compared: one belongs to type II, i.e., a "high" evolution path in the van Krevelen diagram; the other, from a "low" evolution path, belongs to type III. Upon pyrolysis under inert atmosphere, the first one is more than twice as productive as the second. Furthermore, the relative abundance of aliphatic structures and carbonyl or other groups, as illustrated by IR spectra, results in a different composition of the degradation products. For example, $CO_2 + H_2O$ amounts to about 50% of the products yielded by type-III kerogen, as compared with only 25% for type II. Thus, the genetic potential of the hydrogen-rich type II is about three times the potential of the hydrogen-poor.

2. The frequently discussed question of oil versus gas provinces is linked to the genetic characteristics of the organic matter and also to the degree of thermal maturation.

a) In general, the organic matter incorporated in marine or lacustrine sediments is derived from natural associations of flora and fauna, and it has been mixed and homogenized by natural processes of sedimentation (in rivers and

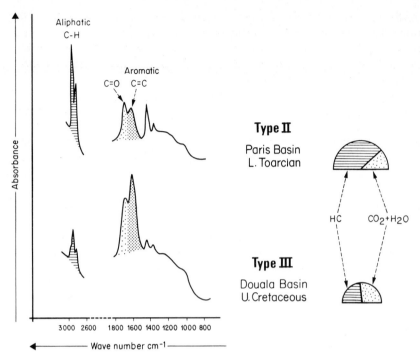

Fig. II.7.3. Comparison of the petroleum potential of two different kerogens: type II, Paris Basin; type III, Douala Basin. *Left:* IR spectra recorded on shallow samples show that type II contains more aliphatic material, and type III more aromatic and carbonyl groups. *Right:* the areas of the *half circles* represent the amounts of products yielded upon experimental heating. The respective proportions of hydrocarbons (HC) and carbon dioxide plus water ($CO_2 + H_2O$) are also shown (Tissot et al., 1974)

marine currents) and by microbial activity. This situation corresponds to the "high evolution paths" on the van Krevelen diagram (kerogen type I or II).

The progressive evolution of this type of organic matter during diagenesis and catagenesis may result in both oil and gas generation, depending on the thermal history. In shallow buried sediments, for example, gas is normally "dry", i.e., methane is almost the single hydrocarbon present. Where the sediments are buried to one or several kilometers, "wet" gas, containing methane, ethane, propane etc., is found associated or not with oil fields. At great depth, gas accumulations are again "dry", as methane is largely predominant.

b) On the other hand, intracontinental lowlands or basins bordering continents may sometimes receive abundant organic matter provided mostly by higher plants. This organic matter consists primarily of lignin and terrestrial humic substances with varying subordinate proportions of waxes, resins, etc. ... Depending on the relative organic and mineral contents, it may result in dispersed carbonaceous matter or coal beds. They correspond to the "low evolution paths" on the van Krevelen diagram (type III).

7.3 Nature of the Organic Matter. Gas Provinces Versus Oil Provinces

The structure of such organic matter is in essence an assemblage of aromatic cycles with short carbon chains. For instance, the structure of lignin can be thought of as a polymer of substances like coniferyl alcohol, shown in Figure I.4.15. Subsequent degradation produces mainly light hydrocarbons (particularly methane, by cracking the short chains).

Therefore, this type of organic matter mostly generates "dry" gas, which contains CH_4, CO_2 and N_2, and is rather similar to gases from coal mines, especially in advanced stages of catagenesis or metagenesis. The gas found in the Netherlands and in the southwestern part of the North Sea is thought to originate from this type of source i.e., coal and associated carbonaceous shales of Carboniferous age (Patijn, 1964).

Besides the predominantly aromatic structure derived from lignin and humic substances, there may be various proportions of lipid, chain-like or cyclic, material possibly inherited from higher plants (waxes, resins), microbial biomass (membranes, waxes) or even algal input. These lipids are able to generate liquid hydrocarbons at depth. They are usually subordinate, but they may reach a significant proportion of the kerogen in certain deltaic environments.

In this case, source rocks containing type III kerogen and the associated coal are responsible for commercial oil and gas accumulations, such as Handil and Bekapai fields, in the Mahakam delta, Indonesia (Durand et al., 1979). A comparable situation might be found in some other deltaic environments.

c) In a more general way, it can be said that some organic matters are less able than others to generate oil in commercial amounts, but any organic matter may generate gas, provided it is buried to a sufficient depth during a long enough time interval.

The relation between oil and gas occurrences from the same organic material, depending on the temperature history, is illustrated by the comparison of the southeastern and southwestern Sahara in Algeria: i.e., the Illizi Basin and Ahnet-Mouydir Basin, respectively. The Illizi Basin produces large amounts of oil and associated gas from Silurian and Devonian source rocks, while the Ahnet-Mouydir Basin produces only methane. From geological reconstitutions, the maximum burial depth to the top of Silurian rocks is at least 1000 m deeper in the Ahnet-Mouydir Basin, and furthermore the geothermal gradient is somewhat higher. The corresponding temperatures are more than 130°C for the Ahnet-Mouydir Basin and 100°C for the Illizi Basin. If we assume an activation energy for cracking reactions of approximately 50 kcal mol^{-1}, the reaction rate is multiplied by more than 100 between these two temperatures[7]. Therefore, the cracking may be negligible in the one case, and completed in the other. Furthermore, this interpretation is in agreement with the observations on the carbonization of spores and pollen.

---

[7] The classical rule of doubling of rate every 10°C is acceptable for reactions with an activation energy in the range of 10–20 kcal mol^{-1}. It is not applicable for cracking reactions, where the activation energy is much higher.

7.4 Temperature, Time and Pressure

Burial, and thus temperature increase are obviously of primary importance for hydrocarbon generation. However, the exact roles of temperature, time, and pressure have to be distinguished, although they are to some extent interdependent.

Oil and gas are produced from the kerogen of the source rocks through a succession of chemical reactions. These reactions are governed by the usual kinetics of the chemical reactions, and this has been verified on examples taken in sedimentary basins (e. g., the Paris basin, Tissot, 1969). Therefore, the transformation ratio depends on temperature and time. This conclusion is intuitively understood, as one would expect a different amount of petroleum to be generated from a source rock heated to a given temperature during 1 as compared with 100 million of years.

The reaction time is implicitly accounted for in the various indices of catagenesis (vitrinite reflectance, spores and pollen coloration, etc.), because the changes of these parameters also obey kinetic laws, and thus these indices integrate the effects of temperature and time. However, geological time has seldom been explicitly discussed, although it is of primary importance to know the timing of oil and gas generation and to compare it with the time of the formation of traps (deposition of impervious cover, folding, faulting, etc.).

The respective influence of temperature and time has been observed from laboratory experiments (Chap. II.5, Table II.5.5) and is confirmed by compari-

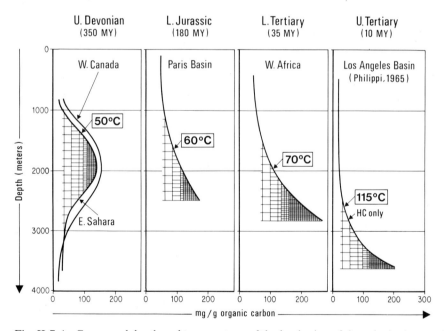

Fig. II.7.4. Compared depth and temperature of the beginning of the principal zone of oil formation in several source rocks of different ages (Tissot et al., 1975)

son of the actual evolution of source rocks of different ages in sedimentary basins. The transformation ratio of the organic matter is plotted in Figure II.7.4 as a function of depth for several source rocks of ages ranging from Devonian to Miocene, in places where the geothermal gradient is broadly similar. It is clear that the oil-generation threshold, situated at the top of the principal zone of oil formation, varies with the ages of the source rock: about 50°C in the Silurian and Devonian of the eastern Sahara, 60°C in the Lower Jurassic of the Paris basin, 70°C in the Upper Cretaceous and Lower Tertiary of West Africa, 115°C in the Mio-Pliocene of the Los Angeles basin.

From these observations it can be assumed that there is a time–temperature relationship, each one of the two parameters being able to compensate for the other. The kinetic equations show that the transformation ratio of the organic matter is influenced more by temperature than by time: i.e., the influence of time is linear, while that of temperature is exponential. For this reason, very old organic sediments such as the Moscow lignites of lower Carboniferous age, buried to depths less than 200 m, have never reached a more advanced stage of coalification (Karweil, 1956).

The exact role of pressure is difficult to elucidate, because temperature and pressure are not independent factors, both being linked to burial depth. The consideration of two neighboring areas, like the Los Angeles and Ventura basins, that comprise the same sedimentary sequence, but have a different geothermal gradient, provides some information. The threshold of oil generation is 2400 m in the Los Angeles basin, and 3600 m in the Ventura basin. Both depths correspond to the same temperature, 115°C, despite the very different pressures (assumed to be approximately proportional to the respective depths). Therefore, the influence of pressure is probably subordinate, compared to that of temperature. Experimental work has confirmed this opinion (Chap. II.5.6.3).

7.5 Timing of Oil and Gas Generation

Determination of the timing of oil and gas generation can be done in two ways, both using the graph of subsidence in the various parts of a sedimentary basin, i.e., the depth versus time curve in a certain number of selected representative locations.

a) If a sufficient number of samples is available from the source rock formation, the bitumen ratio or hydrocarbon ratio is plotted against maximum burial depth to construct the curve of petroleum generation. The main thresholds are determined — top of the principal zone of oil formation — top of the cracking and gas formation zone. Then they are drawn on the various graphs of subsidence to find the time when the respective thresholds are crossed by the depth versus time curve. Oil (or gas) is assumed to be generated in significant amounts as from that time.

b) A more general way has been proposed by Tissot (1969) and Tissot et al. (1975), using a mathematical model to simulate the thermal degradation of kerogen and the formation of petroleum (oil and gas). The principles of the simulation are discussed in Chapter V.4.

The model is calibrated for a given formation by using laboratory experiments on kerogen pyrolysis. Data from extraction of cores and cuttings samples (bitumen ratios and hydrocarbon ratios) are also used if they are available, but they are not strictly necessary. Then, the amount of oil and gas generated by the formation is computed as a function of geological time for any location in the basin where the depth versus time curve and the geothermal gradient (or reflectance measurements) are provided. Thus, a dating of the phenomena becomes possible with an evaluation of the time when oil generation began in significant amounts, the duration of this period, and the time of cracking and gas generation, if this is the case.

The results of the timing determination, by either method, can be checked against geological information on existing oil or gas fields (time of sealing-cover deposition, time of folding, faulting, etc.).

Examination of the results obtained in various geological situations shows that oil generation is never an abrupt phenomenon. It results from the kinetics of the successive chemical reactions involved and lasts an appreciable period of time. A quick oil generation extends over 5 or 10 million years, whereas a slow generation may cover over 100 million years or more.

Examples of quick and early generation of oil are found in sedimentary areas where the rate of subsidence has been very high and the geothermal gradient abnormally important in relation to the global tectonic phenomena. In the Pannonian basin of Central Europe, Pliocene source rocks have produced commercial oil over a few million years. The thickness of the Pliocene beds may reach up to 3000 m (a subsidence rate over 500 m per million years), and the geothermal gradient reaches in places more than $50\,°C\ km^{-1}$. Other examples of rather quick oil generation are found in Mio-Pliocene beds of the circum-Pacific area, such as Indonesia, Sakhalin or California; again high rates of subsidence are involved, and geothermal activity is generally high. Similar situations may occur in rift opening areas, like the Red Sea-Gulf of Suez at present time, or the south Atlantic rift in Cretaceous time.

Examples of petroleum generation over a short period of time (10–25 million years), but not necessarily early, are found in basins, troughs or grabens where a thick sedimentation occurred in a short period of time, burying previously deposited sediments of any age to great depth. In the Ahnet-Mouydir basin and in the southwestern part of the Illizi basin (Algeria), source rocks of various ages, ranging from Silurian to Upper Devonian, were deeply buried in lower Carboniferous time and generated oil and gas almost at the same time. In many parts of the world a thick sedimentation occurred in Cretaceous time, and determined burial and catagenesis transformation of the source rocks previously deposited in platform conditions. In western Canada, both Devonian source rocks of the Leduc area and lower Cretaceous source rocks produced oil and gas in late Cretaceous and early Tertiary time. Such situations are usually associated with normal geothermal gradients.

Generation of oil over a very long period of time usually occurs in platform areas where subsidence rate was always moderate. This is the case of the

Paleozoic source rocks of northern Sahara (Hassi Messaoud area), progressively buried in Mesozoic and Tertiary times at a rate ca. 25 m per million years. Jurassic source rocks of the Paris basin were progressively buried during Jurassic, Cretaceous and lower Tertiary at various rates from 5 to 15 m per million years. In both of these examples oil generation extended over ca. 100 million years.

7.6 Comparison Between the Time of Source Rock Deposition and the Time of Petroleum Generation

Besides variations in the duration of the main stage of oil generation, the elapsed time from sedimentation until the rock enters the principal zone of oil formation may vary considerably from a few million years to more than 300 million years.

It has been mentioned that many source rocks of southern Alberta in the western Canadian basin, whether Devonian or Lower to Middle Cretaceous, produced most of their petroleum in late Cretaceous and lower Tertiary time. This means that the principal stage of oil formation began ca. 300 million years after deposition of the Devonian source rocks and ca. 40 million years after deposition of the Cretaceous source rocks from the Colorado Group. Other previously mentioned examples, Pliocene source rocks (Pannonian basin), reach the main stage of oil generation within a few million years after deposition.

An interesting example, with geological control of the timing, is provided by the Hassi Messaoud area in the northern Sahara basin (Algeria; Fig. II.7.5). In this basin two sedimentation cycles occurred. Paleozoic shales and sandstones range from Cambrian to Carboniferous, and they include the prolific source rocks

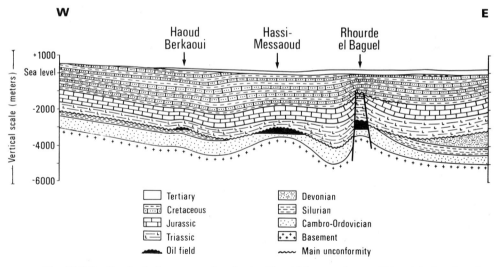

Fig. II.7.5. Geological cross-section in the Hassi Messaoud area (Algeria) showing the Paleozoic reservoir (Cambro-Ordovician) and source rock (Silurian), unconformably overlain by Mesozoic series (Poulet and Roucaché, 1969)

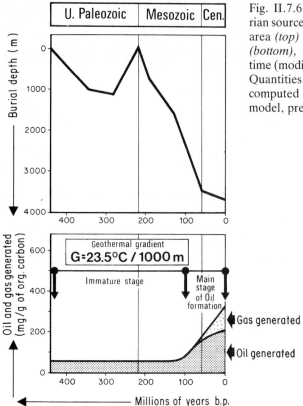

Fig. II.7.6. Burial history of the Silurian source rocks in the Hassi Messaoud area *(top)* and hydrocarbons generated *(bottom)*, as a function of geological time (modified after Tissot et al., 1975). Quantities of oil and gas produced are computed by using the mathematical model, presented in Part V

of Lower Silurian age. The area was gently folded in late Carboniferous time, then deeply eroded. In particular, the Cambrian and Ordovician reservoirs were exposed to the atmosphere during Permian. The second sedimentation cycle started in Triassic time with a thick salt deposit, followed by 4000 m of various sediments. The oil fields are located just below the unconformity in the Cambro-Ordovician sandstone and sealed by the salt deposit. Should oil have been generated during Paleozoic time and trapped in the anticlines, it would have been weathered during all Permian time, because the reservoir was exposed to the atmosphere for several tens of million years. Therefore, it may be hypothesized that oil was generated in both the Paleozoic and the Mesozoic-Tertiary cycles, but that the oil trapped during the first cycle was lost by weathering, and only the oil trapped during the second cycle makes up the present accumulations. An alternative hypothesis is that oil was generated only during Mesozoic or Tertiary, once the thick salt deposited provided an adequate seal.

Consideration of the burial curve, as a function of time (Fig. II.7.6), shows that the source rocks remained at moderate depth (less than 1500 m) during Paleozoic; then folding and erosion were followed by a thick sedimentation during Mesozoic time, resulting in a far more important burial (ca. 4000 m). Use of the mathematical model showed that only a small amount of oil was generated during the first 300 million years. The principal stage of oil generation was reached only during Cretaceous, a time when the salt cover was already available. The results of the mathematical simulation can be checked against geological data on the nearby Rhourde el Baguel oil field, where the reservoir is also the Cambro-Ordovician sandstone below the unconformity. The trap is provided there by a horst, bounded with north–south faults, whose age has been determined by sedimentation studies. The faults occurred in Lower Cretaceous time, and no structural trap of any kind was available before faulting. Therefore, the oil field has necessarily been formed after Lower Cretaceous, in complete agreement with the results of the mathematical simulation. Other instances of late formation of petroleum have been reported by Hedberg (1964) in Venezuela, Canada and Argentina: again, a source rock located below an unconformity yielded oil only after deposition of a reservoir and/or a sealing formation above the unconformity.

A more general question concerns the time of formation of existing commercial oil or gas accumulations. The answer would require the study of the main oil-producing basins by using the methods proposed here. However, the evaluation of the timing of petroleum generation has been made for several basins in Europe, Africa and North America which are considered to be typical of the main geotectonic situations.

In this regard, a large proportion of the presently existing oil and gas fields has been formed during the last 100 million years, thus in Cretaceous or Tertiary time, and they may amount to more than 90% of the petroleum reserves. This category obviously includes oil and gas fields derived from Cretaceous and Tertiary source rocks: among them the important detrital series deposited in major deltaic systems and along the continental margins are of particular importance. In addition to this, the category includes oil or gas fields, formed in the last 100 million years from older source rocks and, particularly, from Paleozoic formations previously deposited in platform basins. As subsidence was often moderate in this type of basin, the source rocks were not buried to a sufficient depth to reach the principal stage of oil formation during Paleozoic: two examples of this situation have already been mentioned, the Silurian of the northern Sahara and the Devonian of the Leduc area. A subsequent deeper burial was necessary, and this frequently occurred in Cretaceous or Tertiary time, as a part of the major phenomena of global tectonics which took place since Lower Cretaceous.

On the contrary, Paleozoic source rocks deposited in more mobile areas, where subsidence was important, have often been involved in Paleozoic orogenies and submitted to intense folding and erosion, which caused escape or weathering of petroleum. This situation concerns most of Paleozoic beds from western Europe. However, in certain places of the world, hydrocarbons definitely generated in Paleozoic time have been preserved, e.g., in oil and gas fields located along the western border of the Appalachian and Ural mountains, in the Amadeus basin of Central Australia, and in gas fields of the Ahnet basin (western Sahara) in Algeria.

Summary and Conclusion

The succession of main steps of evolution of organic matter (diagenesis, catagenesis and metagenesis) is common to all types of sediment. However, the amount of hydrocarbons, their composition, and the depth of oil and gas generation may change. The most important parameters are the nature of the organic matter and the temperature versus time relationship.

The occurrence of gas versus oil provinces depends on the nature of the organic matter and/or the thermal history. In particular, kerogen from terrestrial plant debris (type III) generates comparatively less oil than the other types I or II. However, it could provide a good source rock for hydrocarbon gas at greater depths.

The threshold corresponding to the beginning of significant oil generation varies with the geothermal gradient, the depth and the duration of burial. The related temperature ranges from 50°C in basins of Paleozoic age to 115°C in Mio-Pliocene sediments.

The timing of oil generation may also vary. The principal phase of oil formation may follow source rock deposition after a delay ranging from a few million to 300 million years. In particular, a large number of the present oil fields have been generated during the Cretaceous and Tertiary, regardless of the age of the source rock.

Chapter 8
Coal and its Relation to Oil and Gas

8.1 General Aspects of Coal Formation

When discussing the formation and occurrence of petroleum, it is necessary to include a brief review on relevant aspects of coal formation. Both petroleum and coal originate predominantly from organisms of the plant kingdom and both are subjected to the same geological processes of bacterial action, burial, compaction, and geothermal heating that constitute diagenesis and catagenesis. There are, however, also some essential differences between the modes of coal and petroleum formation. Basically, these differences center around the fact that coal is found at its site of deposition as a solid and relatively pure massive organic substance, whereas petroleum is liquid and migrates readily from its place of origin into porous reservoir rocks. Kerogen is the main precursor material of petroleum compounds. It is finely dispersed and intimately mixed with the mineral matrix in petroleum source beds. Most coals are remnants of terrestrial higher plants, whereas the kerogen of petroleum source beds is generally dominated by phytoplankton and bacteria. Most acknowledged petroleum source beds were deposited in marine environments and most coals formed under nonmarine conditions.

In the following discussion on coal, especially on general aspects of coal formation, the International Handbook of Coal Petrography (1963, 1971, 1975) has been consulted frequently.

Coal contains a variety of plant tissues in different states of preservation. Tissues of distinct origin are microscopically identifiable and can frequently be related to certain parts of the plant, such as cuticles, woody structures, spores, etc. Together with particles of less certain origin they are termed macerals, and are the petrographic components of coal. During and after deposition in sedimentary basins, plant remains undergo a sequence of physical, biochemical and chemical changes (diagenesis and catagenesis), which result in a series of coals of increasing rank. The series begins with practically unaltered plant material and peat, and continues with increasing rank through brown coal, bituminous coal and finally to anthracite. Peat is considered to be the first member in this series of coals.

A distinction is made between *humic coals* and *sapropelic coals*. Humic coals, typically, pass through a peat stage with accompanying processes of humification after accumulation at the site where the plants grew. The major organic component of most humic coals is a lustrous dark brown to black material, visible to the naked eye and mainly derived from the humification of woody tissues. In lower rank coals, this material is represented by a group of macerals called huminite, and in bituminous and anthracite coals by a group of macerals called vitrinite. Humic coals typically are stratified.

Sapropelic coals on the other hand are not stratified macroscopically and have a dull appearance. They are formed from relatively fine-grained organic muds in quiet, oxygen-deficient bodies of shallow water such as ponds, lakes and lagoons. They normally do not pass through a peat-bog stage, but follow the diagenetic path of organic-rich sediments. Like these sediments, sapropelic coals contain varying amounts of allochthonous (transported) organic and mineral matter. The organic fraction consists of autochthonous (local) algal remains and varying amounts of degradation products from nearby existing peat swamps or spores from more distant plants. Microscopically, sapropelic coals can be subdivided into boghead and cannel coals. Boghead coals contain larger amounts of algal remains. Cannel coals are characterized by higher concentrations of spores. There are many transitions between the two basic types of sapropelic coal. Sapropelic coals are relatively rare and, therefore, need not be discussed further. There are, however obvious similarities between humic and sapropelic coals on one hand and source rock-type sediments on the other and, therefore, the accumulation, diagenesis and catagenesis of *coals* are pertinent to the understanding of petroleum source beds.

8.2 The Formation of Peat

8.2.1 Geological Aspects

Large amounts of plant material must be accumulated and preserved to facilitate the formation of peat. The accumulation of large masses of organic matter depends on the efficiency of photosynthesis. High primary productivity on land became possible only after the advent of higher plants with self-supporting skeletons which represent a high level of plant evolution. Plant communities with trees, shrubs and reed-type vegetation evolved during Devonian, and for the first time allowed the production of the necessary biomass for the formation of peatbogs, and hence coal. The floras that produce peat deposits are generally of a rich coastal type. In the Paleozoic they consisted mainly of a Pteridophytic flora in the Carboniferous coals of Europe and North America and a Glossopteris flora in the Permian Gondwana coals (Given, 1972). Late Mesozoic and Tertiary peat swamps supported an angiosperm flora like that growing in peat-forming environments today.

Besides the evolutionary state of plants, there is a number of other important factors in peat formation. Of primary importance are the climate and the tectonic conditions of the area (Teichmüller, 1962). There is a direct relation between rates of plant growth and climatic conditions. With increasing temperature and humidity, plant growth increases, resulting in a mounting production of plant biomass. Tectonic events, such as the formation of depressions, graben structures or otherwise subsiding areas are favorable for the formation of peat.

8.2 The Formation of Peat

Highest growth rates are reached in tropical swamp forests such as are found together with their associated forest peats along the northeastern coast of Sumatra and the southern coast of Borneo and New Guinea. However, degradation processes that affect dead plant material are also more intense and faster in these humid and hot climates. Therefore, it may very well be that the abundance of vegetable matter in tropical swamps is generally more than offset by the more rapid decay and destruction there. Peat and peat-like accumulations are at present frequently found in cold countries such as Scandinavia, Scotland, Alaska, Canada, etc. Thus, optimum conditions for peat formation and preservation may also be found in cold countries where the vegetation can accumulate and be preserved before decay (Hedberg, 1977). In general for the accumulation and preservation of thick deposits of peat an equilibrium between the accumulation rate of organic matter and subsidence has to be maintained over long periods of time.

Yearly accumulation rates of peats in present-day peat areas average about 1 mm (Teichmüller, 1962; Given and Dickinson, 1973). Peat deposits up to several hundred meters thick are known, and brown coal seams more than 300 m in thickness in the State of Victoria, Australia, and over 400 m in Western Canada attest to uninterrupted peat formation over periods of about 1 million years. This illustrates a second essential requirement for the accumulation of great quantities of organic debris and peat. It is a special tectonic condition providing persistant, moderate subsidence without the development of strong relief. To complete the peat-forming process, it is necessary to have coverage of the accumulating plant debris by stagnant water to prevent its oxidation and destruction. The efficiency of peat-forming processes is relatively low. Less than 10% of plant production is accumulated as peat. The larger part is decomposed either during peat formation or after burial (Given and Dickinson, 1973). Nevertheless, the efficiency of peat formation is higher by about a factor of 10 than organic accumulation in average fine-grained marine sediments which may ultimately become source beds.

Most important coal occurrences were originally laid down in basins of long-lasting subsidence in coastal or paralic (coastal swamp) environments. Examples are the peat areas of Indonesia, the Tertiary brown coals of western Germany and the Carboniferous coals of eastern USA and northwestern Europe, and the Donbas in USSR. Important coal occurrences also developed around huge freshwater lacustrine basins, such as the Carboniferous coals of Bohemia, the Saar district in Germany, Massif Central in France, and Spain. The major eras of coal formation are summarized in Table II.8.1. As a consequence of their deltaic and coastal depositional environments, recent peat bogs, as well as their ancient counterparts, are found associated with a variety of sedimentary types (Spackmann et al., 1966; Hemingway, 1968). These associated sediments are generally fine-grained clastics (sandstone, mudstone, shale), but may develop coarser lithologies (conglomerates) locally. Carbonates are generally minor components. Repetitive sequences with several lithologies, typically sandstone, shale, coal and some limestone, are a feature of most coal-bearing strata (Westoll, 1968). Study of these rhythmic or cyclic sedimentation sequences has indicated that tectonic or climatic events or both produce a relative rise and fall of sea level which is an important geological factor controlling the process for coal formation.

The sedimentary cycles mentioned above represent deposition under sub-aerial, fluviatile, intertidal and shallow marine conditions. Because of the unstable sedimentary setting of such a region, deposition followed by erosion and redeposition is common. Rivers may migrate laterally and erode a partially buried peat seam resulting in "wash-outs" which are the products of such erosion. The peat material that was washed away might be transported to quieter water and admixed in minor quantities to autochthonous algal material, and thus contribute to the formation of a sapropelic coal. It might also be deposited with a higher mineral content and form part of the kerogen of a sedimentary rock unit. Another part of the eroded peat material may be oxidized and permanently lost.

Table II.8.1. Major eras of coal formation. (Modified after Given, 1972)

Cenozoic	N. America	Europe	Far East	Southern Hemisphere
Pliocene	+ (Alaska)	+	−	−
Miocene	−	++	−	+ (Australia)
Eocene	++	+	++	+ (Australia)
Mesozoic				
Cretaceous	++	−	++ (Japan, China)	++
Jurassic	−	+	++	++ (Australia)
Triassic	−	+	−	+ (Australia)
Paleozoic				
Permian	−	−	++ (China)	++ (All Gondwana-land)
Carboniferous	++	++	−	−

++ coals very abundant, + abundant, − absent

8.2.2 Biochemical and Geochemical Aspects

Almost instantaneously following accumulation of dead plant debris on the ground, especially biochemical processes are initiated. Mechanical breakage and compaction and geochemical processes also occur. At the surface these changes take place under oxidizing conditions and after coverage by additional plant debris, sealment and stagnant water, reducing conditions prevail. The pH is typically neutral or mildly acid at the surface, and becomes increasingly acidic with depth of burial (M. Teichmüller and R. Teichmüller, 1975). These various chemical environments support different bacterial and fungal populations (Table II.8.2). Microbial activity decreases rapidly with depth, but it is uncertain whether the relatively slow and limited decomposition of plant tissue in peats is caused by

8.2 The Formation of Peat

Table II.8.2. Microbial populations in peat profiles, as recorded on dilution plates. Counts of bacteria are in millions, of other organisms in thousands. (After Given, 1972)

\multicolumn{4}{Joe River}				\multicolumn{4}{Little Shark River}			
Depth, cm	Bacteria	Strepto-mycetes	Fungi	Depth, cm	Bacteria	Strepto-mycetes	Fungi
0– 8	25.0	17.4	239	0.8	6.21	0	4.4
38– 46	0.226	2.1	2.7	61– 69	1.01	<0.1	1.8
213–221	0.013	<0.2	1.1	190–198	0.0014	<0.2	<0.2
				318–326	0.0018	0	0
\multicolumn{4}{Rookery Branch}				\multicolumn{4}{North Harney River}			
0– 8	32.1	78.7	57	0– 8	70.9	0.7	13.9
116–124	0.0142	<0.1	<0.2	61– 69	0.857	1.6	13.4
178–186	0.0110	<0.2	<0.2	235–243	<10^{-4}	0	7.4

chemical inhibition or by nutritional factors (Given and Dickinson, 1973). Woody tissues are attacked almost exclusively by fungi (Moore, 1969), particular groups of which preferentially decompose lignin and cellulose components. It appears that lignin degradation occurs only under aerobic conditions, or at least under conditions when oxygen can be transferred or dehydrogenation can take place (Flaig, 1972). Cellulose is more easily removed by hydrolysis, and thus preferentially lost relative to lignin. Peat still contains free cellulose which is absent in soft brown coal (lignite-B).

The composition of the bacterial flora changes markedly between oxidizing and reducing environments. Bacteria preferentially attack the carbohydrate and proteinaceous portion of plant cells. Because these macromolecules are easily hydrolyzed, they serve as the major substrate for microbial activity. It is also important that the biomass of the sedimentary bacteria and fungi form a portion of the final organic deposit. Their metabolites are not generally lost from the sediment, except for the gaseous products such as CO_2, NH_3 and CH_4. The presence in deep peats of microbially derived compounds such as certain lipid components, carbohydrates, amino sugars and amino acids, provides indirect evidence for microbial contributions (Given and Dickinson, 1973).

Humic substances are the most characteristic products of peatification. They are defined as that part of the peat that is soluble in a basic solvent, such as an aqueous solution of sodium or potassium hydroxide. By definition, humic substances can be extracted in this way inclusively until the stage of hard brown coal (sub-bituminous). Typically, they occur as brown, hydrated gels with no internal structure. The formation of humic substances is a two-stage process. In the first stage, plant material is mechanically disintegrated and with the help of microorganisms depolymerized into aromatic, phenolic and carboxylic moieties. In the second stage, the polymers of humic substances are built up by random repolymerization and polycondensation of the molecular types (Flaig, 1972). The changes which occur during the humification of plant material are complex, and involve not only the modification of the carbon structures of molecules, but also alterations of the composition of the hetero-atomic groups attached to the carbon skeletons by sulfur-, nitrogen- and other specific bacteria. The role of lipids during the formation of peat is partly unknown. Higher plants are relatively poor in lipids, but certain parts, such as leaves, spores, pollen, fruits and especially resin-bearing tissues (e. g., of conifers) are rich in lipids and lipid-like substances. Because lipids are insoluble in water, and not so easily utilized by microorganisms, they are preferentially concentrated during the formation of peat.

In summary, two stages of peat formation can be recognized (Kurbatov, 1963): a primary phase at and immediately below the surface, which is characterized by rapid oxidation, and a secondary phase in which slower conversions occur under reducing conditions. The extent of decomposition and humification is largely a function of the severity of the primary phase. Factors such as rate of burial, water table and climate control the duration and conditions of the primary phase, and thus control the composition of the organic matter that is available for the coalification process. There are obvious similarities between the diagenesis of organic matter in soils and sediments as described in Chapter II.2, and the diagenetic steps during the formation of peat and brown coals. Special reference is made to the nature and role of humic substances in coal and its relationship to the maceral groups huminite (in brown coal) and vitrinite (in bituminous hard coals) and the importance of humic substances in the formation of kerogen in sedimentary rocks.

8.3 Coalification Process

Coalification or carbonification is the process of chemical and physical change by biochemical interferences (e. g., bacteria), temperature, pressure and time, imposed on the organic components that survived the peat formation process. Coalification is conveniently subdivided into a biochemical phase, as long as organisms such as bacteria and fungi are actively engaged, and into a subsequent geochemical phase when biochemical processes have ceased to play a role for further coalification increase. The well-known terms peat, brown coal, bituminous coal and anthracite represent different stages of the coalification series. These stages of the coalification process are termed levels of rank which indicate the maturity of the coal. Some of the parameters used to define rank are shown in Figure II.8.1.

Coal rank can be determined by general parameters such as moisture and volatile matter content, reflectance, or by essentially chemical parameters, such as carbon or hydrogen content and calorific value. No one rank indicator is ideal for all rank ranges: moisture and calorific value are typically used for brown coals; volatile matter and vitrinite reflectance are used over the bituminous hard coal range, and carbon or hydrogen content or volatile matter determinations are best

8.3 Coalification Process

Rank stages		% reflectance of vitrinite	Important microscopic characteristics	% C in vitrinite	Volatile matter % d.a.f. in vitrinite	% H_2O in situ	Calorific value of vitrite (a.f.)	Applicability of the different parameter for the determination of rank
Brown coal	Peat		Large pores Details of initial plant material still recognizable Free cellulose	50		~75		H. (d.a.f.) Volatile matter (d.a.f.) Carbon (d.a.f.) Reflectance of the vitrinites Calorific value (a.f.) or moisture in situ (moisture-holding capacity) X-ray diffraction (graphite lattice)
Brown coal	Soft brown coal	ca. 0.3	No free cellulose Plant structures still recognizable (cell cavities frequently empty) *Marked gelification and compaction takes place*	60	ca. 53	~35	7,200 B.t.u./lb (4,000 kcal/kg)	
Brown coal (Hard brown coal)	Dull brown coal		Plant structures still partly recognizable (cell cavities filled with collinite)		ca. 49	~25	9,900 B.t.u./lb (5,500 kcal/kg)	
Brown coal (Hard brown coal)	Bright brown coal	ca. 0.5		75	ca. 45	8–10~	12,00 B.t.u./lb (7,000 kcal/kg)	
Hard coal	Bituminous hard coal	ca. 2.5	Exinite becomes *markedly* lighter in color ("Coalification jump")	90	30		15,500 B.t.u./lb (8,650 kcal/kg)	
Hard coal	Anthracite	11.0	Exinite no longer distinguishable from vitrinite in reflected light Reflectance anisotropy	100	10 0			
	Graphite							

Rank increasing →

Fig. II.8.1. Rank stages, important microscopic and chemical characteristics to describe and define coals. Exinite and liptinite are synonyms. (Adapted from International Handbook of Coal Petrography, 1963)

236   Coal and its Relation to Oil and Gas

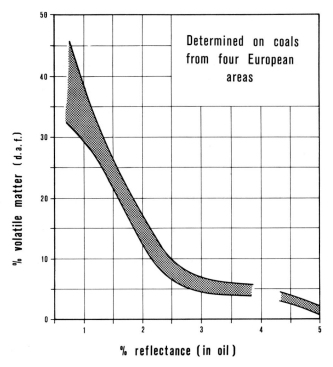

Fig. II.8.2. Correlation between reflectance and volatile matter of European coals. (After McCartney and Teichmüller, 1972)

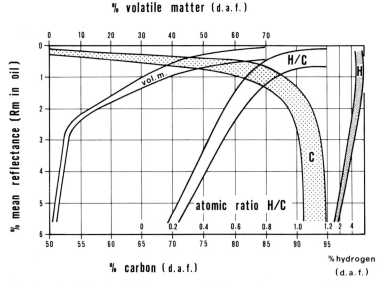

Fig. II.8.3. Relationship between vitrinite reflectance and different chemical rank parameters. (After Teichmüller, 1971, taken from Stach et al., 1975)

8.3 Coalification Process

applied in low volatile bituminous hard coals and anthracites. Attempts have been made to standardize international coal-rank terms. A correlation between reflectance and volatile matter content of coals from different areas is given in Figure II.8.2. The relationship between vitrinite reflectance and several chemical rank parameters is shown in Figure II.8.3.

Just as the peat-forming process acts differentially on the various plant tissues, for example on lipid-rich parts such as cuticles and resins versus woody tissues rich in cellulose and lignin, coalification affects different components in different ways. The number of readily distinguishable plant remains decreases with increasing coalification. By analogy with the term mineral applied to rocks, each microscopically identifiable component of a coal is termed a maceral according to the classification Stopes/Heerlen (Anonymous: International Handbook of Coal Petrography 1963, 1971, 1975). The three main groups of macerals are: the remains of woody and humic components variously termed huminite or vitrinite in low and high rank coals respectively; the remains of lipid-rich relics of plants, such as resins, waxes, spores, cuticles and algal bodies, termed liptinite (or exinite); and harder carbon-rich, brittle particles, called inertinite. The inertinite group contains a number of macerals that are present in all rank ranges and which represent the more aromatic and oxidized components. Some macerals of the inertinite group are probably remnants of biological oxidation, forest fires or even reworked material from older sediments. Huminite in brown coals and vitrinite in bituminous hard coals are the most common components.

Among the major physical changes coalification brings about, are a reduction in the bed moisture content, an increase in density, a decrease in porosity and an increase in refractive index in later stages. Chemical changes which occur are condensation, polymerization, aromatization, and loss of functional groups, i.e., of functions containing O, S, and N linked to the molecular structure of coal. Polymerization and condensation occur as an extension of the humification process that takes place in lower-rank coals. The net result of these changes is a continuous but nonlinear enrichment of carbon with increasing rank.

Van Krevelen (1961) reports that the number average molecular weight of pyridine extracts of coals rise from about 600 at 80% C (on a water- and ash-free basis, i.e., w.a.f. or dry and ash-free, d.a.f.) to 1000–1200 in the 86–88% C range. The general nature of the coal molecule has been suggested both from chemical (Given, 1960, 1961; Cooper and Murchison, 1969) and from X-ray evidence (Cartz and Hirsch, 1960). Models derived from these studies are illustrated in Figure II.8.4. Chemical identification of functional groups and general information on the carbon skeleton are established, but the structural examples given must be seen as attempts only to give the general framework of the coal structure. They are mainly intended to describe structures of hard brown coal to bituminous coal. The aromatic centers are generally linked by hydroaromatic and methylene bridges and fringed with methyl, hydroxy, carboxy, carbonyl, amino and other functional groups.

The aromaticity of vitrinite varies with rank, and has been measured by such diverse techniques as X-ray diffraction, optical and spectroscopic examination, and chemical studies. Vitrinites typically contain about 70% aromatic carbon atoms in hard brown coals and over 90% in anthracites. Liptinites initially have a

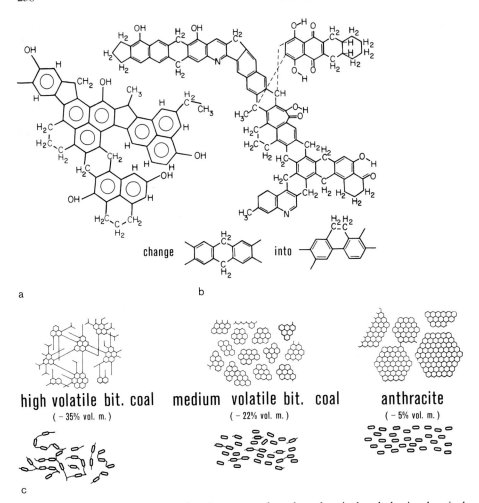

Fig. II.8.4a–c. General nature of coal structures based on chemical and physicochemical evidence. (a) Vitrinite 84.5% C, mainly from X-ray data. (After Cartz and Hirsch, 1960). (b) Vitrinite 82.0% C, from chemical data. (After Given, 1960, 1961). (c) Vitrinite at various rank levels. (After Teichmüller et al., 1968)

very low aromatic content, but in low-rank bituminous coal, 50% carbon atoms are found in aromatic rings. Thereafter the aromaticity rises rapidly to join that of vitrinite in anthracites. The inertinite maceral group has 90–100% aromatic carbon at all ranks.

The hetero-atom concentration, as well as the types of functional groups, change with rank. Hydrogen, oxygen, sulfur, and nitrogen are the major noncarbon atoms in coals at all ranks. The types and concentrations of the oxygen functional groups are shown in Figure II.8.5, plotted against rank. Nitrogen occurs as primary, secondary and tertiary amines and heterocyclic pyridyl rings and sulfur as thiols, sulfides and heterocyclic thionaphthyl- and thiophenic rings.

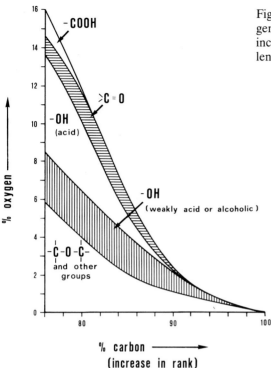

Fig. II.8.5. Relationship between oxygen functional groups of vitrinites with increasing rank level. (After van Krevelen, 1963)

Inorganic sulfur, often in the form of pyrite, is an almost ubiquitous component of coals. Similarities can be observed between the structural models derived for coal (Fig. II.8.4) and for sedimentary rock kerogen (Fig. II.4.8). Both models comprise aromatic centers or nuclei, crosslinked by bridges and fringed by functional groups. A major difference is found in the nature of bridges. In kerogen, especially at lower to medium levels of evolution, the bridges contain more and, on the average, longer aliphatic chains than can be observed in coal. Bridges in coal are more often cyclic and hydro-aromatic in nature, possibly alternating with short chain-like linkages including heteroatomic functions. In general, the overall aromaticity in coal is higher than in kerogen. With increasing rank, the average stacking number of aromatic sheets increases in coal as well as in kerogen.

A vast amount of chemical and geochemical data has been accumulated for coal (van Krevelen, 1961; M. Teichmüller and R. Teichmüller, 1967, 1968; Stach's Textbook of Coal Petrology, 1975), but one of the most useful and succinct presentations is the van Krevelen-diagram (Fig. II.8.6). This diagram has been used throughout this part of the book to describe the chemical evolution of kerogen (Figs. II.2.9, II.4.11 and II.4.12). It incorporates information on the three most abundant elements in coal, and also displays the essential differences between the maceral groups, as in Figure II.4.12. In addition the diagram (Fig. II.8.6) shows the coalification trends of the various plant components, such as woody tissues, spores and cuticles as they evolve through the peat, brown coal,

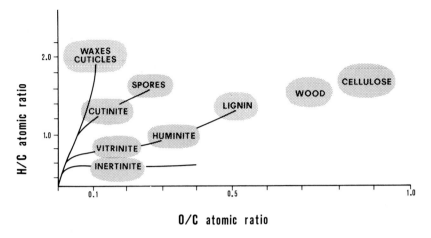

Fig. II.8.6. Selected plant and coal materials and their respective position in the H/C-, O/C-diagram (van Krevelen-diagram)

bituminous hard coal, and anthracite stages. The original diagram by van Krevelen (1961) showed all simple reaction processes like dehydration, decarboxylation and demethanation (Fig. II.8.7). It was easy to read because all these reaction processes are represented by straight lines. In the highly aliphatic hydrogen-rich waxes and exines, the predominant trend is loss of hydrogen. The considerable oxygen content of humic substances (humic acids, huminite, vitrinite) is markedly reduced with increasing rank. The relatively small changes experienced by the inertinite group of macerals indicates their high initial carbon content and chemical inertness due mainly to their higher initial aromaticity, and possibly due to the fact that they may be reworked material. The final product of

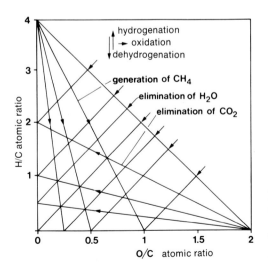

Fig. II.8.7. Original H/C versus O/C diagram proposed by van Krevelen (1961) showing the trends for elimination of water, carbon dioxide and generation of methane from coal

8.4 Coal Petrography

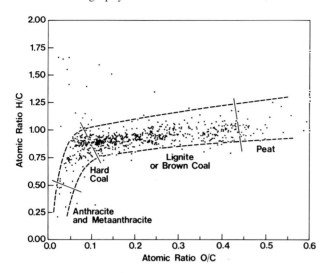

Fig. II.8.8. Evolution of the elemental composition of coal during burial. The path of humic coals starts with peat and reaches finally anthracite and metaanthracite. A few cannel coals and boghead coals are also shown with higher H/C ratios (Durand et al., 1983)

the coalification process is a meta-anthracite (high rank anthracite) which has a highly aromatic carbon structure with similarities to a poorly ordered graphite.

The chemical changes in coal during its evolution through the different rank stages can be compared with the evolution of various kerogen types. It can be deducted from a van Krevelen diagram (Fig. II.4.11) that humic coals and type-III kerogen, rich in terrestrial plant material, are similar to each other. This is, of course, no surprise, as the bulk of the starting material and, to a certain extent, the environment of deposition and early diagenesis are alike. Coal and kerogen follow the same general trend throughout the evolutionary process (Fig. II.8.8), typified by an increase in carbon (carbonification) and a loss in functional groups as documented in the loss of oxygen functions (Figs. II.8.5 and II.5.8). With respect to oxygen functions, coal (vitrinites), and type-III kerogen show a similarity by the absence of ester groups. Type-I and type-II kerogen, however, contain ester groups.

8.4 Coal Petrography

Coal petrography is the study of the macroscopically, and more importantly, of the microscopically recognizable components of coal. It describes coal in terms of its maceral composition and its rank level, which is most conveniently determined by reflected light microscopy. A petrographic microcsope is used to observe a highly polished surface of a coal block or grains in reflected light. In reflectance measurements, the amount of incident light which is reflected from the surface of a maceral is compared to the light reflected from a standard of known reflectance. Measurements are performed with monochromatic green light at a wavelength of 546 nm. Oil immersion objective lenses are typically used to enhance the contrast of petrographic components. Mean random reflectance values in oil $\overline{R}_0$ (%) are generally given at lower ranks, i.e., below 1.2% reflectance. Because of the increasing anisotropy of vitrinite with rank, mean maximum reflectance ($\overline{R}_0$ max. %), obtained using a linearly polarized light beam (rather than mean random values ($\overline{R}_0$ %), employing no polarizer) should be given at least beyond 1.2% mean reflectance. While this procedure is practical when studying coals, it is also useful in kerogen studies. Due to small particle size, however, it may be a very time-consuming procedure. The reflectance of particles as small as 3 μm in diameter can be measured routinely.

8.4.1 Micropetrographic Components: the Macerals

Macerals of coal are microscopically differentiated on the basis of their reflectance, their shape and structure and sometimes by such additional methods as etching, electron microscopy, fluorescence, and luminescence. In a typical coal, three different maceral groups, i.e., liptinite, vitrinite and inertinite (Fig. II.8.9) can readily be distinguished on the basis of reflectance values. Liptinites have a relatively low medium reflectance and inertinites have the highest reflectance. Difference in reflectance between maceral groups are greater in brown coal and lignite than in bituminous coal and gradually decrease toward anthracite. The

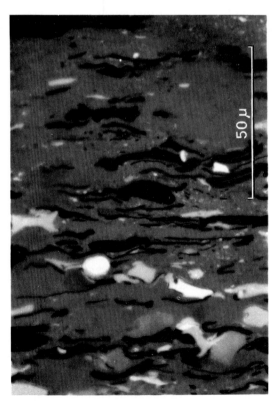

Fig. II.8.9. Polished surface of bituminous coal, reflected light, oil immersion. The macerals liptinite *(dark)*, vitrinite *(grey)* and inertinite *(light)* can be seen. (Photo Hagemann)

8.4 Coal Petrography

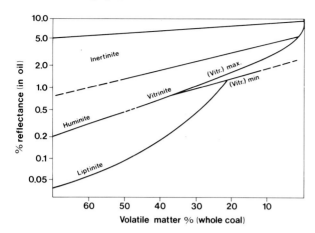

Fig. II.8.10. Reflectance changes of main groups of macerals as related to degree of coalification. (Modified after Alpern and Lemos de Sousa, 1970)

curve of vitrinite and liptinite reflectance merges around 1.3–1.6% $\overline{R}_0$ max. (Fig. II.8.10). Inertinites as compared to vitrinites remain of higher reflectance up to the meta-anthracite level (3.5% $\overline{R}_0$ max.). The liptinite components, however, retain some of their original morphology. Spores, leaf cuticles, algal bodies, resins and suberins are identifiable.

Fluorescence is a useful technique for identification of liptinite material. An excitation wavelength (emission peaks between 366 and 435 nm of a mercury lamp) in the UV or blue range of the spectrum is used to produce a fluorescence, which is generally restricted to the liptinitic components. Using this technique, liptinitic components are able to emit light from blue to red, this evidence is called fluorescence (Alpern et al., 1972). A color change from blue to green to yellow to orange to red with increasing rank is the basis of a rank parameter that may be especially useful for lower rank coals and corresponding sporinite-type sedimentary organic matter (Ottenjann et al., 1974). The color of spores in transmitted light has also been used as a rank parameter, especially in kerogens (Staplin, 1969; Jones and Edison, 1979).

The inertinitic components of coals are generally characterized by a high reflectance. It has recently been suggested (Teichmüller, 1974) that highly reflecting micrinites may be formed from liptinitic (or huminitic) substances at about the rank level near the maximum of petroleum generation. It has been generally accepted in the past that inertinitic components are the product of oxidation by forest fires or by bacterial or atmospheric oxidation. Reworked material may also be included among the inertinites.

The macerals of the huminite or vitrinite group are physically and chemically intermediate between the liptinite and inertinite groups. In the brown coal stage huminite is thought to be formed from massive cellular tissues, such as wood and bark and from the precipitation of, and gelification by, humic substances. With increasing rank, the cellular structures are changed. Cell walls swell, and cavities and pores become more and more impregnated and filled with structureless humic substances. In this way, the cellular character is increasingly obscured. Beginning in hard brown coals, around 0.4% mean reflectance, the huminitic component is from here on and throughout higher ranks termed vitrinite. At this stage cell structures may still be recognizable, then the vitrinite components are called telinite. By contrast, massive structureles vitrinite is called collinite. In reflected light under oil immersion, vitrinite appears as medium gray intermdiate between the brighter inertinite and darker liptinite (Fig. II.8.9). Macerals become progressively more alike in coal of higher ranks.

8.4.2 Application of Reflectance Measurements for Source Rock-Type Sediments

The rank or maturity of a sedimentary rock containing organic matter can be determined by measuring the reflectance of finely dispersed small huminite or vitrinite particles. In particular, this parameter allows a sediment to be evaluated with respect to whether oil or gas generation has taken place (Vassoevich et al., 1969; Teichmüller, 1971; Dow, 1977). This method is further discussed in Chapter V.1.

A number of peculiar points are associated with measuring reflectance in sedimentary rocks which differ from measuring coal samples. Proper identifica-

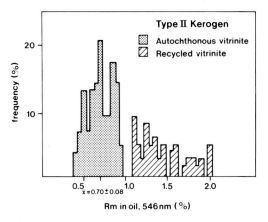

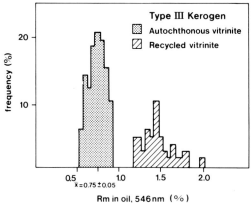

Fig. II.8.11. Frequency histogram showing vitrinite composition of a type II and type III kerogen of acknowledged source rocks

tion of vitrinite, which also relies on its specific range of reflectance, may be a problem, since there are generally no adjacent macerals for comparison. This may be especially difficult because there is a continuous reflectance and morphological sequence between vitrinite, and common particles of the inertinite group, i.e., semifusinite and fusinite. Identification of vitrinite, and distinction from fusinite in polished sections of sediments is, therefore, at least partly a matter of personal experience (Jones and Edison, 1978). Thus, although the relative rank of sediments can be determined, it is not always possible to asign unequivocally an absolute rank on the basis of vitrinite reflectance to a particular sample. A commonly used method to minimize operator bias is to plot a histrogram of all reflectance measurements from a large number of maceral grains (Fig. II.8.11). In sediments poor in organic particles, this is often difficult and time-consuming. Therefore, it is recommended (Hagemann, 1974) to prepare concentrates of organic particles by removing most of the rock matrix with HCl and/or HF, sometimes followed by heavy liquid treatment. This facilitates proper identification of vitrinite and allows a statistically sufficient number of particles to be measured within a reasonable time frame. It is also suggested to use a sequence of samples in a well to determine the maturation gradient in a sedimentary rock section rather than absolute rank of a particular sample (Dow, 1977).

It is frequently observed that low rank vitrinite and huminite show a considerable spread of reflectance values. This spread probably reflects the diversity of plant material from which the vitrinite is formed and possibly a different diagenetic history. Anisotropy also causes an increasing spread of reflectance values as rank is increased above about 1.5 $\overline{R}_0$ (%).

8.5 Petroleum Generation

8.5.1 Generation of Low Molecular Weight Volatiles

Physical and chemical changes in coal, due to increasing temperature and pressure with burial, result in coalification and produce a sequence of coals of increasing rank. Following microbial and physical changes during the diagenesis of peat and brown coal, geochemical changes, such as a decrease in oxygen and hydrogen and a relative increase in carbon, take place during the catagenesis of bituminous and anthracite coal (Fig. II.8.8). This results in an elimination of functional groups, increasing aromaticity and a decrease in the chain-like structures in coal. Simple material balance considerations indicate that the loss of hydrogen, oxygen, and also carbon results in liberation of hydrogen- and oxygen-rich carbon containing molecules during coalification processes. In general, two types of reaction are responsible for the liberation of smaller molecules. They are condensation reactions on one hand and the splitting of carbon–carbon bonds, as in cracking, on the other. There are also defunctionalization reactions, such as decarboxylation. In condensation reactions, two reactive molecules are

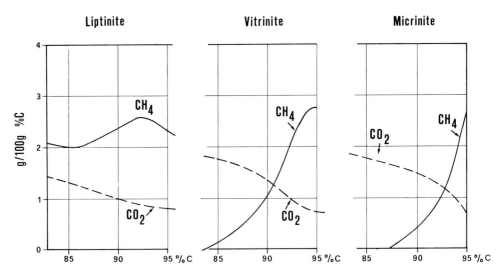

Fig. II.8.12. Amount of gas liberated from different macerals during increasing coalification. Amount of gas was calculated on the basis of elemental analysis of macerals. (After Jüntgen and Karweil, 1966)

linked to form a product of higher molecular weight than either of the reactants. At the same time, low molecular weight compounds such as CH_4 and H_2O are liberated. For example, the condensation of aromatic systems like benzene and toluene or benzene and phenol would produce biphenyls and CH_4 or H_2O respectively. The generation of low molecular weight hydrocarbons, especially methane and other volatile nonhydrocarbon compounds, such as CO_2, is a necessary by-product of increasing coalification, and is determined by type of organic matter present, temperature, and time.

Many authors have contributed to the understanding of these processes of coal maturation. Fundamental papers on the principles of gas formation in coal were published by Jüntgen and Karweil (1966) and Jüntgen and Klein (1975). It is important that the different maceral groups liptinite (exinite), vitrinite and inertinite exhibit very different degassing curves during the course of coalification according to their different chemical make-up (Fig. II.8.12). Unfortunately, their data cover only coals in the low rank bituminous to anthracite range which contain between 85 and 95% C and about 30 to 5% volatile matter. This corresponds to vitrinite reflectance data between 1.0 and 3.0 $\overline{R}_0$ (%). Some methane is liberated from liptinite at relatively low rank, e.g., under 0.85% mean reflectance. This is especially noteworthy in the context of source rock-type sediments. Data on the generation of volatiles in peat and the lower rank coals are very scarce. The first volatile products liberated in both coal and kerogen at temperatures below 100°C are mainly H_2O and CO_2 with small amounts of CH_4 (van Heek et al., 1971).

A calculated model is presented in Figure II.8.13 (Jüntgen and Klein, 1975), where the integral amount of volatile products during coalification is shown as a

8.5 Petroleum Generation

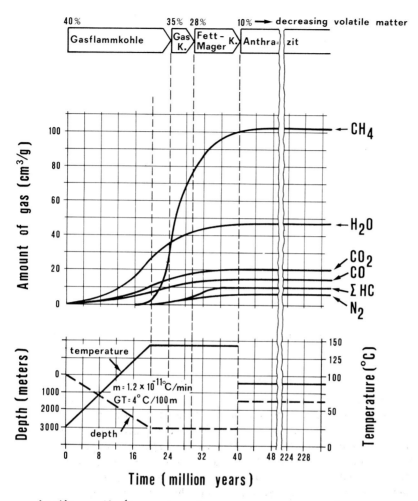

Fig. II.8.13. Integral amount of volatile compounds generated during coalification shown as a function of geological subsidence, i.e., temperature and time. (After Jüntgen and Klein, 1975)

function of rate of subsidence (temperature and time). The model is calculated according to actual geological and geochemical data from the Carboniferous in the Ruhr area of Germany. A subsidence rate of 3000 m over 20 million years, and an increase in temperature from 20 °C, at the surface to 140 °C at maximum depth of burial (3000 m) results in a heating rate of 6 °C/10^6 years or about 1.2×10^{-11} °C/min. For extrapolation of natural degassing rates, experimental data with heating rates in the range from 10^{-3} to 10^3 °C/min are available. According to this model, the evolution of volatiles begins with H_2O, CO_2 and CO before large amounts of CH_4 are released. This is in agreement with observations in nature

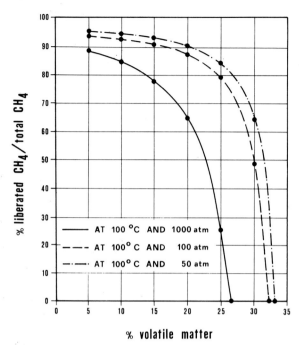

Fig. II.8.14. Percentage of the total generated gas that cannot be stored in a coal, either in an adsorbed state or in free form. Storage conditions are 100°C at pressures of 1000, 100 and 50 atmospheres, respectively. (After Jüntgen and Karweil, 1966)

and with theoretical considerations such as bond energies of functional groups and carbon–carbon bonds. The bulk of methane generation in coal begins generally in the range of medium volatile bituminous coal (about 1.3 to 1.4% reflectance). As shown in Figure II.8.12, this is more characteristic for vitrinite- and inertinite-type materials than for liptinites. With respect to this phenomenon, more data are needed in lower-rank bituminous coals, especially in the range of 48 to 30% volatile matter or 0.50 to 0.85 $\overline{R}_0$ (%), and at still lower ranks. This is also true with respect to N_2 generation, which has not yet been sufficiently documented.

The generation of volatile compounds during coalification greatly exceeds the storage capacity for these products, either in an adsorbed state or in free form. This is illustrated in Figure II.8.14, where the percentage of CH_4 that cannot be retained in coal is calculated as a function of increasing degree of coalification.

8.5.2 Generation of Heavier Hydrocarbons and Nonhydrocarbons

Aside from volatiles produced with increasing coalification, higher molecular weight substances, similar to those found in petroleum, are also generated in coal. These heavier hydrocarbons and nonhydrocarbons should be viewed in connection with the scheme of oil and gas formation developed in Chapters II.5 and II.6 and with the comparison of coalification stages and petroleum generation published by Vassoevich et al. (1969). It was observed that during coalification, extraction yields from coal increased to a maximum in high volatile bituminous

8.5 Petroleum Generation

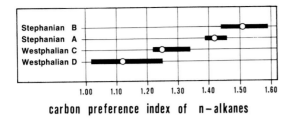

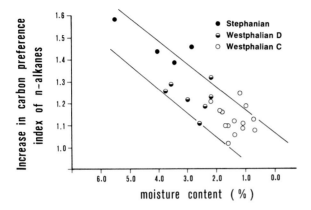

Fig. II.8.15. Ratios of the odd–even predominance of the higher n-alkanes (CPI, n-C_{23} to n-C_{29}) in Carboniferous coals of the Saar area, SW Germany, in relation to rank of coal. (After Leythaeuser and Welte, 1969)

coals at a level of maturity of 0.8–1.0 $\overline{R}_0\%$ vitrinite reflectance (Leythaeuser, 1968; Leythaeuser and Welte, 1969). With further increase in rank, a continuous decrease in the amount of extracts was observed. The composition of extracts also varied regularly with increasing coalification. Very striking was a decrease in the odd–even predominance (CPI) of the higher n-alkanes (Fig. II.8.15) and an increase in lower molecular weight compounds and aromatics with increasing

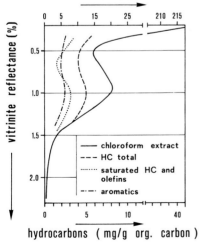

Fig. II.8.16. Quantity of extract and hydrocarbons as a function of vitrinite reflectance in a series of coals from peat to anthracite. (After Durand et al., 1977)

rank. These data were later confirmed by Hood and Gutjahr (1972), Hood et al. (1975) and Mukhopadhyay et al. (1979).

The general scheme of hydrocarbon generation from coal parallels that from kerogen in petroleum source rocks. Durand et al. (1977) investigated a series of coals of different ranks with the same geochemical methods used to study organic matter in source rock-type sediments. He found a close similarity between the physicochemical properties of coal and kerogen and in their structural and chemical evolution brought about by catagenesis. As previously noted, the greatest chemical and evolutionary similarity is observed between coal and type-III kerogen (Fig. II.4.11). The catagenetic changes in coal and in kerogen are basically a progressive elimination of steric hindrances, such as heteroatomic links, aliphatic chains and saturated cycles, in order to attain a more stable molecular configuration under higher temperature and pressure conditions. The result is the well-documented formation of CO_2, H_2O, H_2S, resins, asphaltenes, and hydrocarbons.

There are detailed studies of C_{15+}-hydrocarbons extracted from a series of coals between peat and anthracite (Hagemann and Hollerbach, 1980; Hollerbach and Hagemann, 1981). It serves as an example for the similarity in hydrocarbon generation in coal and in sediments. The relationship between the quantity of extract and the types of heavier hydrocarbons from coals of different ranks is given in Figure II.8.16. The maximum yield of hydrocarbons coincides with $0.9\overline{R}_0$ (%) which is about at the same rank level as observed by Leythaeuser and Welte (1969). At low ranks (0.2–$0.5\% \overline{R}_0$), high molecular weight n-alkanes with 25 to 33 C atoms, having a very strong odd predominance, clearly dominate the distribution curve. From about 0.5–$1.3\% \overline{R}_0$ the maximum on the n-alkane curve shifts toward C numbers around 20 and the odd predominance is gradually lost (Fig. II.8.17). Thus the generation of hydrocarbons and nonhydrocarbons in coal, as well as in petroleum source rocks, is a normal process that accompanies the diagenesis and catagenesis of organic matter.

Analyses of the saturates by combined gas chromatography and mass spectrometry reveals that in coals of the peat and brown coal stages only cyclic diterpanes occur, and that they disappear in bituminous hard coals around 0.8% vitrinite reflectance (Hollerbach and Hagemann, 1981). Chain-like diterpanes (pristane and phytane) appear around 0.4%, peak at about 1.2% mean reflectance and decrease toward anthracite. The maximum amount of cyclic and acyclic diterpanes in each group of compounds is about 1% (10^4 ppm) of the toal extract. The behavior of pentacyclic triterpenes with a double bond in ring C (Fig. I.4.10) is very similar to that of the cyclic diterpanes. They exhibit a maximum around 0.4% mean reflectance, comprise about 1% of the extract and disappear in low rank bituminous hard coals. Saturated pentacyclic triterpanes and steranes exhibit a behavior similar to chain-like diterpanes. They appear around 0.4% mean reflectance and make up about 0.1–1% of the extract. Unlike acyclic diterpanes, they disappear completely in high-rank bituminous hard coal (around $1.6\% \overline{R}_0$ max.) before the anthracite stage ist attained. This behavior of acyclic and cyclic C_{15+}-isoprenoids and steroids is tentatively interpreted as an attachment to and subsequent release from insoluble organic matter (coal or kerogen) during diagenesis and catagenesis. An important controlling mechanism seems to be the

8.5 Petroleum Generation

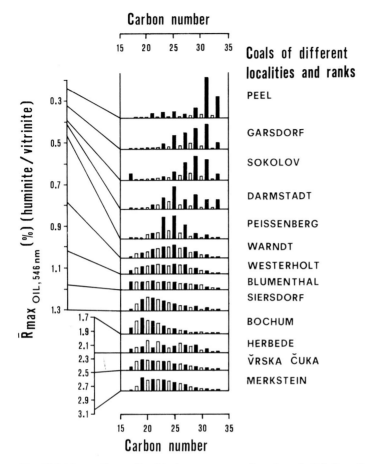

Fig. II.8.17. n-alkane distribution curves as a function of vitrinite reflectance in a series of coals from peat to anthracite. (After Hollerbach and Hagemann, 1981)

location and distribution of reactive sites (especially double bonds and OH-groups) on the original linear isoprenoids and the terpenoid and steroid molecules.

8.5.3 Coal as a Source Rock

The importance of coal as a source for gas initially recognized by the geochemical work of Karweil (1956, 1969) and the geological considerations of Patijn (1964a, b). In the latter paper the role of the Carboniferous coals in the formation of gas fields in northwestern Europe was defined. In later years additional evidence was accumulated to verify that Carboniferous coals are a primary source of gas in this region (Stahl, 1968; Bartenstein and Teichmüller, 1974; Lutz et al., 1975). Hence it is well established that coals are able to generate and release sufficient gas to form large commercial gas accumulations. However, evidence for commercial oil accumulations derived from coal is not so abundant. It is known that some liquid hydrocarbons are generated in coal during catagenesis. There are numerous reports of oil shows, small oil seeps and oil-impregnated sand lenses etc. being closely associated with coals from all over the world.

Crude oils associated with coals often belong to the high-wax type, as the "Frankenholz" oil from the Saar Region in Germany; "La Machine" oil from Central France, and oils from the Midland coal measures in Great Britain (Hedberg, 1968). Crude oils of the Mahakam Delta area of Indonesia (Durand and Oudin, 1979) also have to be mentioned here. The close association of oil and coal in the Officina area of Venezuela (Banks, 1959) must be ranked as another indication for a liquid hydrocarbon potential of coals.

When considering coal, one usually thinks in terms of rank and forgets that coal may consist of rather different material with respect to its basic structure. In this sense the liptinite content of different coals may vary considerably. Likewise the availability of chainlike molecular structures fixed to aromatic nuclei may vary from coal to coal. Depending on the amount of liptinite in coals and chainlike molecular structures there might be a potential for liquid hydrocarbons even in coals. An example for this is the Mahakam Delta, Kalimantan, Indonesia (Durand and Oudin, 1979; Durand et al., 1983). Other authors stressed specifically the role of resinite and exinite macerals for the generation of light oils and condensates (Snowdon, 1980; Thomas, 1981; Powell et al., 1982). It must be concluded therefore, that the generative potential for higher hydrocarbons is definitely present in certain types of coal. The generative potential of coal has a great similarity to type-III kerogen which yields gas rather than oil, but may generate commercial amounts of crude oil, depending on the liptinite content. For instance, it has been observed in Australia that Mesozoic and Tertiary coals have generated oil accumulations, whereas Permian coals are only a source for gas (Thomas, 1981). Likewise, the Miocene coals of the Mahakam Delta generated oil and gas, whereas Westphalian coals of Western Europe only provide gas, even in the areas where vitrinite reflectivity is 1% or lower. As far as migration mechanisms are understood there seems to be a lower expulsion efficiency in coals especially for heavier hydrocarbons. The limited primary migration of heavier hydrocarbons out of coal is probably due to the high absorption capacity of coal, and the fact that coal generally occurs as a massive, continuous, solid organic phase.

Summary and Conclusion

Coal consists mainly of detritus from higher (terrestrial) plants. Most coals are formed under nonmarine conditions.

Almost immediately following accumulation of dead plant debris on the ground, first biochemical, and later, after burial, geochemical processes are initiated. With continued burial, these processes progressively cause coalification of the organic matter to form the maturational sequence of peat, brown coal, bituminous (hard) coal and finally anthracite. These stages of the coalification process are termed levels of rank. Coal rank can be determined by physical parameters such as reflectance and moisture content or by essentially chemical parameters, such as volatile matter content, carbon or hydrogen content and calorific value. Chemical changes in coal during its evolution through the different rank stages can be compared with the evolution of various kerogen types. The greatest chemical and evolutionary similarities are observed between coal and type III kerogen.

During coalification, low molecular weight hydrocarbons, especially methane, and other volatile nonhydrocarbon compounds, such as carbon dioxide and water, are generated.

The main phase of methane generation in coal begins in the range of medium volatile bituminous coal (about 1.3 to 1.4% reflectance). The formation of volatile compounds during coalification greatly exceeds the storage (absorption) capacity for these products in coal. In addition heavier, nonvolatile hydrocarbons are formed. However, in contrast to the large amounts of methane generated, the potential of coals to produce higher hydrocarbons is limited. The generative potential of coal is in this respect similar to type III kerogen, which yields mainly gas, but may generate commercial oil accumulations, depending on the liptinite content.

Chapter 9
Oil Shales: A Kerogen-Rich Sediment with Potential Economic Value

9.1 Historical

The first mention of the oil shale industry goes back to the seventeenth century, when a British Patent no. 330 was issued in 1694 to Martin Eale, who "found out a way to extract and make great quantities of pitch, tar and oil out of a sort of rock" (Cane, 1967). The first industrial plant was developed in Autun, France in 1838, followed by another exploitation in Scotland, 1850. Since then many countries developed an oil shale industry: Australia (1865), Brazil (1881), New Zealand (1900), Switzerland (1915), Sweden (1921), Estonia (now USSR) (1921), Spain (1922), China (1929), South Africa (1935). The highest point of the development was reached during or immediately after World War II.

However, all plants built during or immediately after the war had a small capacity, roughly 50 to 200 t of oil per day, and their cost of production was high due to the importance of labor for mining and crushing the rock. Around the middle of the century, the low cost of the oil shipped from the Middle East made the shale oil uneconomical, and the exponentially increasing consumption made the amount of shale-oil production insignificant. Therefore, all oil shale plants closed down progressively between 1952 (Australia) and 1966 (Spain) with the exception of two countries: USSR (SSR of Estonia) and China (Manchuria), where the mining conditions were adequate to justify a continued exploitation. More recently, several countries made a survey on the occurrence of large oil shale deposits, with sufficient reserves to justify important investments and the potential to contribute significantly to the demand in oil. Brazil and the United States developed pilot plants on the worlds largest oil shale deposits: the Green River Shales in Colorado, Utah, and Wyoming, and the Irati Shales in Brazil. With energy becoming short, other countries are again investigating their oil shale resources.

The shale oil produced in the year 1972 was estimated at 10 million t in China, 3 million t in USSR (Estonia) and 50,000 t in Brazil (Combaz, 1974).

9.2 Definition of Oil Shales.
Oil Shale Versus Petroleum Source Rock

There is actually no geological or chemical definition of an oil shale. In fact, any shallow rock yielding oil in commercial amount upon pyrolysis is considered to be an oil shale.

9.2 Definition of Oil Shales. Oil Shale Versus Petroleum Source Rock

The mineral constituents of oil shales vary largely according to types. Some are true shales, where clay minerals are predominant, but others, like the well-known Green River Shales, are carbonates with subordinate amounts of quartz, feldspar, clay minerals, etc. In fact, the various oil shales mined around the world during the last century range from shale to marls and carbonates. The only excluded lithology is sandstone, as this particular type of sedimentary deposit is not compatible with the accumulation and preservation of organic matter.

The organic matter contained in the oil shales is mainly an insoluble solid material, kerogen. There is no oil and little extractable bitumen naturally present in the rock. Shale oil is generated during pyrolysis, a treatment that consists of heating the rock to ca. 500°C. The kerogen of oil shale is not distinct from the kerogen of petroleum source rocks defined in Part II. To some extent the pyrolysis leading to shale oil formation from kerogen is comparable to the burial of the source rocks at depth, that generates oil by the resulting elevation of temperature.

However, there are different requirements for petroleum source rocks and oil shales in respect to organic richness and rank of evolution of the rock. Many investigators have concluded that any rock containing about 0.5% of organic carbon may produce oil or gas, provided it is buried to a sufficient depth. Hydrocarbons are mobile, so that migration may result in commercial accumulations of oil or gas, even if starting from low concentrations of organic matter in a source rock. On the contrary, an oil shale must have a large amount of organic material to be of economic interest. At least, one would require that the oil shale yields more energy as shale oil than it requires to process the rock. The average pyrolysis temperature is 500°C. The energy to be provided for heating to that temperature is approximately 250 calories per g of rock, and the calorific value of kerogen is 10,000 calories per g. Therefore, if the kerogen content of the shale is 2.5% by weight, the total calorific value is used for heating the rock. Below the threshold of 2.5%, the rock cannot be a source of energy (Burger, 1973). In fact, a lower limit of 5% organic content is frequently used, and it corresponds approximately to an oil yield of 25 l per metric t of rock, or 6 US gallons per short t[8]. It should be remembered that in the present situation such low ratios are still uneconomical. The US literature frequently uses 10 US gallons per short t (42 l per metric t) as a lower limit for oil shales.

The requirements regarding the evolution stage are quite different for a petroleum source rock and an oil shale. A petroleum source rock requires a sufficient burial history and the stage of catagenesis in order to degrade a substantial part of kerogen and thus generate oil. On the contrary, if one looks for an oil shale with a large yield upon pyrolysis, the best situation is a shallow burial, i.e., an immature stage of kerogen evolution before catagenesis begins. The rocks containing organic matter which have been buried to great depth do not represent promising oil shales, even if a subsequent folding and erosion has brought the rock back to the surface.

It can be said that the equivalent of an oil shale, sufficiently buried, constitutes a petroleum source rock. However, the reverse (i.e., that the equivalent of a

---

8 1 US gallon = 3.785 l; 1 short ton = 0.907 metric t; 1 US gallon/short t = 4.17 l/metric t.

petroleum source rock, shallowly buried, constitutes an oil shale) is not necessarily true, due to the requirement of richness.

9.3 Composition of Organic Matter

Optical examination of oil shale kerogens shows that some of them are almost entirely made of algal remains, whereas some others are an admixture of amorphous organic matter with a variable content of identifiable organic remnants. The main algal types are Botryococci and Tasmanaceae.

Botryococcus is a fresh or brackish-water alga belonging to the Chlorophyceae and forming colonies. It has a wide extension from Precambrian to the present time: several beds with *Botryococcus* are known as early as the Precambrian of Bohemia. Kerogens from Autun boghead (Permian, France) and torbanite (Carboniferous, Scotland; Permian, Australia and South Africa) are considered to be almost entirely made of *Botryococcus* colonies, and the Recent coorongite from Australia represents a similar situation. The present *Botryococcus braunii*, source of the coorongite, shows two different states: a green growth stage and an orange resting stage (Cane, 1976).

Tasmanites are considered as marine algae close to the present *Pachysphaera*: the kerogen of some oil shales like tasmanite (Permian, Australia; Jurassic-Cretaceous, Alaska) consists almost entirely of these algal remains (cysts). They are present in many other oil shales, as in the Lower Toarcian shales of the Paris Basin, France, or in the Lower Silurian shales, also containing Graptolites, from Algeria.

Rocks consisting mainly of a single type of organism are anomalous from a geological point of view. The organic matter of these anomalous rocks cannot be readily compared with the classification of kerogens used so far in this book, which has been defined on amorphous kerogen. For instance, Maxwell et al. (1968) have studied such anomalous organic matter (i.e., the resting stage of *Botryococcus braunii*) comparable to the source material of coorongite and torbanite, and they found that it contained up to 76% hydrocarbons. Furthermore, the bulk of the hydrocarbons was made solely of two unsaturated branched alkenes ($C_{34}H_{58}$).

In many instances, only a minor part of the kerogen from oil shales is made of recognizable organic remnants. The balance is amorphous, probably due to microbial alteration during sedimentation. Amorphous organic material, commonly referred to as sapropelic matter, may be associated with minerals and, in this form, constitutes thick accumulations of oil shale, such as the Irati shales (Permian, Brazil) or the Green River shales (Eocene, Colorado, Utah and Wyoming). The organic material may be derived from planktonic organisms (microscopic Algae, Copepods, Ostracods) and from microorganisms normally living in the fresh sediment (Bacteria, Algae, etc.).

The chemical composition of organic matter from oil shales may vary to a large extent (Fig. II.9.1). Elementary composition always shows a high hydrogen content. The atomic H/C ratio ranges from 1.25 to 1.75. The oxygen content is

9.3 Composition of Organic Matter

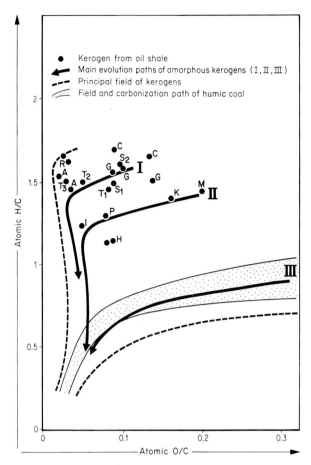

Fig. II.9.1. Van Krevelen diagram showing the elemental composition of oil shale kerogens. The evolution paths of the three main kerogen types and of the humic coals are shown for comparison. *A:* Autun boghead, Permian, France; *C:* Coorongite, Recent, Australia; *G:* Green River shales, Eocene, Colorado, USA; *H:* Marahunite, Tertiary, Brazil; *I:* Irati shales, Permian, Brazil; *K:* Kukersite, Lower Paleozoic, USSR; *M:* Messel shales, Eocene, W. Germany; *N:* Cannel coal, Carboniferous, W. Germany; *P:* Paris basin, Lower Toarcian, France; *R:* Kerosene shales, Permian, Australia; S_1: Tasmanite, Permian, Tasmania, Australia; S_2: Tasmanite, Jurassic, Alaska, USA; T_1: Torbanite, Permian-Carboniferous, South Africa; T_2: Torbanite, Carboniferous, Scotland; T_3: Torbanite, Permian, Australia

relatively variable. The atomic O/C ratio ranges from 0.02 to 0.20. Nitrogen is much less abundant and also varies a great deal. The atomic N/C ratio ranges from 0.5×10^{-2} to 5.8×10^{-2}.

A comparison of the elemental composition of kerogen from oil shales with organic matter of petroleum source rocks shows that, according to the classification previously established, the amorphous sapropelic organic matter of oil shales usually belongs to either type I or type II. There is no oil shale related to type III.

Concentrations of type-III organic material, derived from higher plants, are usually classified as humic coal, or carbonaceous shales accompanying coal. Furthermore, this type of material generally yields a large amount of gas upon pyrolysis, but only a limited amount of oil.

Infrared spectrophotometry suggests that type-I kerogen present in the Kerosene shales, Green River shales, torbanite, and bogheads, contains large amounts of long aliphatic chains and little polyaromatic material. This is confirmed by analysis of bitumen extracted from Green River samples, which contain a large proportion of normal and isoalkanes, particularly those of high molecular weight. On the other hand, type-II kerogen present in Irati shales, kukersite, and various other oil shales (Messel shale in the Rhine Valley, Lower Silurian shales from Algeria, Lower Toarcian shales from western Europe) contains more cyclic material, aromatic and naphthenic. This, in turn, is confirmed by the composition of the natural bitumen that includes a large amount of polycyclic naphthenic and naphthenoaromatic molecules.

The reason for the different composition and structure of oil shale kerogen may be found in the different organisms contributing to the shales. However, the extent of degradation versus preservation at the time of deposition may also be of importance. Selective microbial degradation would eliminate most of the easily hydrolyzable compounds and preserve mainly lipids. These lipids, enriched by bacterial lipids, would result in kerogen comparable to type I.

9.4 Conditions of Deposition

The main geological environments in which oil shales are deposited have been reviewed by Duncan (1967) and are listed below:

a) *Large lake basins,* particularly those of tectonic origin formed during orogenic mountain building. Mineralogically, these oil shales are marls or even argillaceous limestones. The associated sediments may include volcanic tuffs and saline minerals, as in the Green River formation. The main oil shales deposited in such an environment are the Green River shales of Eocene age in Colorado, Utah and Wyoming (up to 600 m thick), the Triassic beds in the Stanleyville basin, Zaire and the Albert Shales of Mississippian age, New Brunswick, Canada.

b) *The shallow seas* represent in the first place large stable platforms with rather thin deposits of oil shales (a few m to a few tens of m thick) extending sometimes over hundreds of thousands of square kilometers. The mineral phase is mostly silica and clay minerals, but carbonate may also be present. There are widespread deposits of black shales of this type in several geological systems: Cambrian of northern Siberia and northern Europe, Silurian of North Africa, Permian of southern Brazil, Uruguay and Argentina, Jurassic (Toarcian and Kimmeridgian) of western Europe, etc.

Shallow sea deposits also include those formed in subsiding geosynclinal basins. However, such deposits usually have been subsequently buried to greater depth, and may even have been folded. As a result, their oil potential has been partly consumed through a natural evolution, and the remaining organic material results only in a lower grade oil shale. Only the geologically younger ones, such as the Miocene shales of Sicily, Italy, and California, have kept a sufficient potential to be classified among the richer oil shales.

c) *Small lakes, bogs and lagoons* associated with swamps result in oil shale deposits associated with coal measures. This situation is met in Permian beds of western Europe (St. Hilaire, France) and in Tertiary beds of China (Fushun, Manchuria), where oil shale deposits overlie coal beds.

A typical characteristic of the various types of oil shales is a very fine lamination of thin (less than 1 mm) alternating layers of organic matter and minerals. This lamination witnesses quiet sedimentation, where the minerals are either precipitated from solution (carbonates) or transported as very fine detritus (clay minerals, silt). It suggests a succession of seasonal or other periodic events. Lamination also proves the absence of benthos. A situation of this kind represents a confined physicochemical environment, where the decay of part of the organic matter uses up oxygen to produce CO_2. Thus organic matter is preserved from reworking by benthic fauna and from aerobic microbial degradation.

9.5 Oil Shale Density

Density of the organic material is low compared with density of the mineral grains. The specific gravity of concentrated kerogen is usually ca. 0.95–1.05. Therefore, oil shale richness may be evaluated by using density, provided the mineral composition does not change significantly within the section considered. Stanfield et al. (1960), from the US Bureau of Mines, have shown a good correlation between increasing oil yield and decreasing formation density in the Green River shales.

Well logs permit in situ evaluation of oil shales (Tixier and Curtis, 1967). A density log run in the Piceance basin, Colorado, allows prediction of oil yield from the modified Fisher assay, in the 0–40 gallons per t range, with a maximum error of 2–4 gallons. Sonic logging might also be of some interest, as an increasing kerogen content is reflected by an increasing transit time.

9.6 Pyrolysis of Oil Shales

The technology of oil shale processing is based upon pyrolysis of the rock at a temperature around 500 °C. The fundamental aspects of kerogen degradation upon heating in industrial processes are the same as those described in part II in respect to experimental evolution of kerogen in the laboratory, by using thermogravimetric and related techniques. The various bonds of the kerogen macromolecule are broken, liberating small molecules of liquid and gaseous hydrocarbons, and also nitrogen, sulfur and oxygen compounds. However, due to the shale quickly reaching a high temperature, the kinetics of the experimental and industrial reactions are somewhat different from those under natural subsurface conditions. In particular the industrial products include usually a large proportion of olefins, sulfur and nitrogen compounds in the liquid phase; gases may contain abundant hydrogen sulfide and ammonia. The standard procedure for laboratory assays of oil shales is the modified Fischer assay.

Various retorting processes have been used (Burger, 1973):

a) External heating through the wall of the retort. The Pumpherston process, used in Scotland since 1860, was the origin of most techniques for a century; it was widely used in Europe with various amendments (Scotland, France, Spain, Sweden). The capacities of the retorting units were low, and energetic balances were poor.

b) Combustion inside the retorting unit results in a better energetic balance, but the disadvantage is the low calorific value of the gas diluted by nitrogen and combustion products. Valorization of hydrogen sulfide and ammonia is difficult. This technique is widely used in the USSR (Pintsch process) and China (Fushun). In the USA, the experimental Gas Combustion process of the US Bureau of Mines is the origin of several processing units now under investigation.

c) Another method used is the circulation of externally heated gas through the shale. The energetic balance is satisfactory, especially if the residual carbon of the retorted shale is burnt to recover also these residual calories. The gas produced is of high calorific value. This technique has been used in France (Grande Paroisse) and is now used in Brazil (Petrosix).

d) Heating the shale by hot solids is a new type of process that insures a good energetic balance and a high calorific value of the gas. The UTT process in the USSR uses hot shale ashes as a calorific vehicle, whereas the Tosco process in the USA uses ceramic balls externally heated. However, the technology is more complex than for the other processes.

e) Finally, powdered oil shales are used in the USSR (Estonia) to burn and provide calories in electric plants.

9.7 Oil Yield; Composition of Shale Oil

The oil yield upon pyrolysis depends on the abundance of kerogen in oil shale (usually determined by the organic carbon content of the rock), on the nature of kerogen, and on the natural evolution of the rock (due to geological history, in particular to depth of burial). The values of oil yields and the ratio of organic matter converted to oil (computed on the basis of the amount of carbon converted) are shown in Table II.9.1.

Oil shales containing type-I kerogen, with H/C atomic ratio over 1.5 usually show the highest conversion of organic matter into shale oil: Green River shales (Colorado) 70%; Kerosene shales (Australia) 66%. Oil shales containing type II kerogen, with a comparatively lower H/C atomic ratio, extend over a wide range of organic matter conversion ratio: from only 26% in Sweden to 66% in USSR kukersite. The natural evolution of organic matter during geological history of the sediment may influence the oil yield of the shales. Burial due to subsidence, and the associated elevation of temperature, may have already converted part of the organic matter, even if subsequent folding and erosion has brought the shale up to surface again. This is probably the cause of the comparatively low conversion rate of Permian and Carboniferous oil shales of western Europe: Autun and St. Hilaire 45–55%; Scotland 56%; and lower Paleozoic oil shales of Sweden 26%.

Shale oil densities are usually higher than those of the average crude oil: they range from 0.88 to 0.98, and the most frequent values are 0.91–0.94. Oils derived from type-I kerogen, which has a high content of aliphatic chains, are somewhat lighter (oil from Kerosene shales 0.89, torbanite 0.88) than oil derived from type-II kerogen, which has a higher proportion of cyclic structures (oil from kukersite 0.97, from the lower Toarcian shales of western Europe 0.93–0.94). Viscosity and cloud points are of industrial importance in respect to pipe-line transportation of shale oil. The cloud point is frequently related to the content of high molecular weight alkanes (paraffin). It is particularly high for shale oil derived from several type-I kerogens, like the Green River shales (cloud point 25–30°C, viscosity ca. 25 centistokes at 37.8°C). It is comparatively low for oil derived from type-II kerogen, which contains less heavy paraffins, for example the lower Toarcian shales of western Europe (cloud point -15°C, viscosity 5 centistokes at 37.8°C).

Distillation curves of shale oil show a large proportion of medium fractions that distil below 300°C: Green River shales 30 to 40%, lower Toarcian of western Europe 50%. This medium distillate contains about 30–50% olefins, which are virtually absent in natural crude oil. Another important difference is the abundance of nitrogen-containing cyclic compounds. The nitrogen content of shale oil ranges from 0.5% in Kerosene shale oil to 1.8 to 2.1% in Green River shale oil. These figures are considerably higher than those measured in natural crude oils, where the average value is 0.09%. An exception is the kukersite oil that contains only 0.1% of nitrogen. Sulfur compounds are also abundant in many shale oils, but the total sulfur content is more comparable to the values measured in natural crude oils from many producing countries, particularly from the Middle East.

9.8 Oil Shale Distribution and Reserves
(Tables II.9.1 and II.9.2)

Lower Paleozoic oil shales were mostly deposited in the shallow seas bordering the Scandinavian-East Canadian precambrian shield: central-eastern United States and Canada, Sweden, USSR (Estonia). They are only moderately rich, as many of them have been subjected to a moderate thermal evolution for a very long period of time. However, kukersite and associated deposits, of Ordovician age, are among the richest oil shales. They are situated between Tallinn and Leningrad (USSR) and consist of relatively thin (2.5–3 m) calcareous oil shale at shallow depth (5–100 m). Kukersite contains up to 40% organic matter with a

Table II.9.1. Some properties of oil shales and related shale oil. (Proparte after Robinson and Dinneen, 1967; Burger, 1973)

Country	Locality, type or age	Oil shale Organic carbon %	Kerogen (atomic ratio) H/C	O/C	N/C × 10^2	Retorting Oil yield (%)	Conversion ratio (1) of organic matter (%)	Density (15°C)	Shale oil H/C (atomic)	N (%)	S (%)
Australia	Glen Davis Kerosene shales Permian-Carbon.	40	1.6	0.03	0.8–1.6	31	66	0.89	1.7	0.5	0.6
Australia	Tamania Tasmanite Permian	81	1.5	0.09	0.9	75	78				
Brazil	Irati Permian		1.2	0.05	2.6	7.4		0.94	1.6	0.8	1.0–1.7
Brazil	Tremembe-Taubate Tertiary	13–16.5	1.6		2.7	6.8–11.5	45–59	0.92	1.7	1.1	0.7
Canada	Nova Scotia Carboniferous	8–26	1.2		5.8	3.6–19	40–60	0.88			
China	Fushun Tertiary	7.9				3	33	0.92	1.5		
France	Autun, St. Hilaire Permian	8–22	1.4–1.5	0.03	1.3–2.1	5–10	45–55	0.89–0.93	1.6	0.6–0.9	0.5–0.6
France	Crevenay, Séverac Toarcian	5–10	1.3	0.08–0.10	2.2–2.5	4–5	60	0.91–0.95	1.4–1.5	0.5–1.0	3.0–3.5
W. Germany	Messel Eocene	25–45	1.4–1.5	0.20	3.2	5–19	30		1.6	1.0	0.6
Great Britain	Scotland Torbanite Carboniferous	12	1.5	0.05	1.6–3.2	8	56	0.88		0.8	0.4

Table II.9.1 (continued)

Country	Locality, type or age	Oil shale Organic carbon %	Kerogen (atomic ratio) H/C	O/C	N/C × 10²	Retorting Oil yield (%)	Conversion ratio (1) of organic matter %	Density (15°C)	Shale oil H/C (atomic)	N (%)	S (%)
South Africa	Ermelo Permian-Carbon.	44–52	1.35		1.4	18–35	34–60	0.93	1.6		0.6
Spain	Puertollano Permian-Carbon.	26	1.4		1.8	18	57	0.90		0.7	0.4
Sweden	Kvarntrop Lower Paleozoic	19				6	26	0.98	1.3	0.7	1.7
USA	Alaska Tasmanite Jurassic?	25–55	1.6	0.10	0.4–0.5	28–57	80				
USA	Colorado Green River sh. Eocene	11–16	1.55	0.05–0.10	2.8–3.0	9–13	70	0.90–0.94	1.65	1.8–2.1	0.6–0.8
USSR	Estonia Kukersite Ordovician	77	1.4–1.5	0.16–0.20	0.5	22	66	0.97	1.4	0.1	1.1
Yugoslavia	Aleksinac sh. Tertiary	21				10	48	0.90	1.55		

All percentages are by weight. (1) Conversion to oil, based on organic carbon.

conversion ratio of 66% into oil plus gas. They have been used to produce shale oil and chemicals, domestic and industrial gas, and electricity by burning the leaner beds in electric power plants. Total reserves of the area would be 21 billion t of oil shale, or approximately 3.5 billion t of shale oil. Other oil shale deposits are located in the Siberian part of the USSR.

Table II.9.2. Approximate estimation of the main oil shale reserves. (After Burger, 1973, World Energy Conference, 1980, and others)

North America		271,000
USA	264,000	
Canada	7,000	
South America: Brazil		127,000
Northern and Western Europe		1,000
Italy		5,600
USSR (including Siberia)		56,000
Jordan		7,800
Marocco		7,400
Zaire		16,000
China		4,400
Thailand		2,000
Other countries	ca.	2,000
	Total ca.	500,000

Million m^3 potential oil

Middle Paleozoic (Silurian and Devonian) oil shales were deposited in shallow seas on the large shelves covering North Africa (Libya and Algeria) and the eastern and central part of the United States. In Africa, the richest beds are the black shales located in the transgressive lower Silurian. In the United States, the Devonian black shales represent most of the reserves. The main extension of the deposits is over Indiana, Kentucky and Ohio, but Tennessee, Alabama, and other states contribute to a total of 10 billion t of potential shale oil.

Upper Paleozoic oil shales occur widely over the Gondwana area. The world's second largest deposit is the Irati shale of Permian age in southern Brazil. In the main area, near Sao Mateus do Sul, Parana, the deposits comprise two horizons (respectively 3.20 m thick with an oil yield of 9%, and 6.50 m thick with an oil yield of 6.4%), separated by 8.6 m of barren sediments. The total reserves are estimated to be 120 billion t of oil. Stratigraphic equivalents occur in Uruguay and Argentina.

Other rich Permian-Carboniferous oil shales are located in Australia (Kerosene shales, tasmanite) and South Africa (Ermelo). Tasmanite and Ermelo shale contain mostly yellow bodies of organic material, considered to be remains of algal colonies.

In the northern hemisphere many small deposits of oil shale, often associated with coal, occur in tectonic or post-orogenic basins of Carboniferous or Permian age, along the hercynian belt of western Europe and the Appalachians: Scotland, France, Spain, Eastern Canada (Albert shale), Eastern United States. Other deposits are located in the northwestern United States.

Triassic lacustrine oil shales are extensively reported in Central Africa (Zaire), where they are associated with limestones and volcanics. Their oil yield is high, and the total reserves are estimated to be 16 billion t of potential oil. Some Triassic oil shales are also known from Switzerland, Austria and Italy.

Jurassic black shales are widespread in western Europe. They include several beds of various richness from Lower to Upper Jurassic. The most frequent is the Lower Toarcian (Posidonia shales, or Liassic ε), which occurs over a wide area in France, Luxemburg and Germany, and to a smaller extent in Spain and Great Britain. The oil content is rather low (4–6%), but the large areal distribution makes it an important potential reserve.

Other Jurassic to Cretaceous oil shales are known in northern and eastern Asia, where they are associated with coal, and in Alaska, where marine deposits include tasmanite (with Tasmanites algae), cannel coal, and a shaly coal named "whale blubber" rock.

Low-grade Cretaceous oil shales occur in South Central Canada and western United States. An association of marine black shales with phosphorite and chert extends over several countries of the Middle East (Syria, Israel, Jordan).

The main Tertiary oil shales are the lacustrine Green River shales, which constitute the biggest reserve of that type in the world, estimated to be more than 250 billion t of potential oil. The deposit of Eocene age extends over four main basins: the Piceance basin, Colorado, which is considered to contain two thirds of the reserves; the Uinta basin, Utah, where the Green River shales are outcropping along the borders, and in the Green River basin and the Washakie basin, Wyoming. In the Uinta basin, the Green River shales are buried to a great depth (up to 6000 m) in the northern part of the basin, where the shales act as source rock for Altamont and other oil fields.

Lacustrine deposits of Tertiary age also include the Paraiba Valley (Taubate–Tremembe) in Brazil, and the Aleksinac shales in Yugoslavia. Other nonmarine oil shales are associated with coal beds in China (Fushun, Manchuria). Several small deposits occur in local basins or grabens caused by Tertiary orogenic movements in Europe, South America and Western United States.

Marine deposits of Tertiary black shales are known from Sicily, Italy, where the reserves are evaluated at ca. 5 billion t of potential oil, and from California and Southern USSR.

Summary and Conclusion

Oil shales contain large quantities of insoluble organic matter, kerogen, which yields oil upon pyrolysis at temperatures of about 500°C. The rock contains no producible oil, and usually little extractable bitumen. The mineral fraction of the rock may include clay minerals, carbonates and silica.

The origin of the organic matter is frequently marine or freshwater algae, but other planktonic organisms and also bacterial biomass may contribute significantly. When compared with the classification established for petroleum source rocks, kerogen of the oil shales belongs to type I or II.

Oil shales have been deposited in lake basins, shallow seas, bogs and lagoons from late Precambrian to Tertiary.

Shale oil contains a significant proportion of unsaturated olefins which are absent in natural crude oils. Sulfur and nitrogen are also abundant in shale oil.

References to Part II

Abelson, P. H., Hoering, T. C.: Carbon isotope fractionation in formation of amino acids by photosynthetic organisms. Proc. Nat. Acad. Sci. US. **47**, 623–632 (1961)

Aizenshtat, Z.: Perylene and its geochemical siginificance. Geochim. Cosmochim. Acta **37**, 559–568 (1973)

Albaigés, J.: Identification and geochemical significance of long chain acyclic isoprenoid hydrocarbons in crude oils. In: Advances in Organic Geochemistry 1979. Douglas, A. G., Maxwell, J. R. (eds.). Oxford: Pergamon Press, 1980, pp. 19–28

Albaigés, J., Torradas, J. M.: Significance of the even-carbon n-paraffin preference of a Spanish crude oil. Nature (London) **250**, 567–568 (1974)

Albrecht, P.: Constituants organiques des roches sédimentaires. Thesis, Univ. Strasbourg, 1969

Albrecht, P., Ourisson, G.: Diagénèse des hydrocarbures saturés dans une série sédimentaire épaisse (Douala, Cameroun). Geochim. Cosmochim. Act. **33**, 138–142 (1969)

Albrecht, P., Vandenbroucke, M., Mandengué, M.: Geochemical studies on the organic matter from the Douala Basin (Cameroon). I. Evolution of the extractable organic matter and the formation of petroleum. Geochim. Cosmochim. Acta **40**, 791–799 (1976)

Almon, W. R.: Petroleum-forming reactions: clay catalyzed fatty acid decarboxylation. Thesis, Univ. Missouri-Columbia, 1974

Alpern, B.: Classification pétrographique des constituants organiques fossiles des roches sédimentaires. Rev. Inst. Fr. Pétr. **25**, 1233–1267 (1970)

Alpern, B.: Pétrographie du kérogène. In: Kerogen. Durand, B. (ed.). Paris: Technip, 1980, pp. 339–371

Alpern, B., Lemos de Sousa: Sur le pouvoir réflecteur de la vitrinite et de la fusinite des houilles. C. R. Acad. Sci. **271**, Sér. D. 12, 956–959 (1970)

Alpern, B., Durand, B., Espitalié, J., Tissot, B.: Localisation, caractérisation et classification pétrographique des substances organiques sédimentaires fossiles. In: Advances in Organic Geochemistry 1971. von Gaertner, H. R., Wehner, H. (eds.). Oxford-Braunschweig: Pergamon Press, 1972, pp. 1–28

Anderson, P. C., Gardner, P. M., Whitehead, E. V., Anders, D. E., Robinson, W. E.: The isolation of steranes from Green River oil shale. Geochim. Cosmochim. Acta **33**, 1304–1307 (1969)

Anonymous: International Handbook of Coal Petrography, International Committee for Coal Petrology, 1972. Suppl. 2nd ed. (1976)

Aquino Neto, F. R., Restle, A., Connan, J., Albrecht, P., Ourisson, G.: Novel tricyclic terpanes (C_{19}, C_{20}) in sediments and petroleum. Tetrahedron Lett. **23**, 2027–2030 (1982)

Arpino, P.: Les lipides de sédiments lacustres éocènes. Thesis, Univ. Strasbourg, 1973

Bajor, M., Roquebert, M. H., van der Weide, B. M.: Transformation de la matière organique sédimentaire sous l'influence de la température. Bull. Centre Rech. Pau **3**, 113–124 (1969)

Baker, E. W.: Mass spectrometric characterization of petroporphyrins. J. Am. Chem. Soc. **88**, 2311–2315 (1966)

Baker, E. W.: Porphyrins. In: Organic Geochemistry. Eglinton, G., Murphy, M. T. J. (eds.). Berlin-Heidelberg-New York: Springer, 1969, pp. 464–497

Baker, E. W.: Evolution of the tetrapyrrole diagenesis model. In: Internationales Alfred-Treibs-Symposium, Munich 1979. Prashnowsky, A. A. (ed.). Universität Würzburg, 1980, pp. 1–11

Baker, E. W., Palmer, S. E.: Geochemistry of porphyrins. In: Porphyrins, vol. 1. London-New York: Academic Press, 1978, pp. 485–551

Banks, L. M.: Oil-coal association in Central Anzoategui, Venezuela. Am. Assoc. Petrol. Geol. Bull. **43**, 1998–2003 (1959)

Bartenstein, H., Teichmüller, R.: Inkohlungsuntersuchungen; ein Schlüssel zur Prospektierung von paläozoischen Kohlenwasserstoff-Lagerstätten? Fortschr. Geol. Rheinld. Westfalen **24**, 129–160 (1974)

Barwise, A. J. G., Whitehead E. V.: Separation and structure of petroporphyrins. In: Advances in Organic Geochemistry 1979. Douglas, A. G., Maxwell, J. R. (eds.). Oxford: Pergamon Press, 1980, pp. 181–192

Bayliss, G. S.: The formation of pristane, phytane and related isoprenoid hydrocarbons by the thermal degradation of chlorophyll. 155th Am. Chem. Soc. Nat. Meet. Div. Petr. Chem. Preprint F 117–131 (1968)

Bendoraitis, J. G., Hepner, L. S.: Isoprenoid hydrocarbons in petroleum. Anal. Chem. **34**, 49 (1962)

Bergaya, F., Van Damme, H.: Stability of metalloporphyrins adsorbed on clays: a comparative study. Geochim. Cosmochim. Acta **46**, 349–360 (1982)

Blumer, M.: Pigments of a fossil echinoderm. Nature (London) **188**, 1100–1101 (1960)

Blumer, M.: The organic chemistry of a fossil. I. The structure of fringelite pigments. II. Some rare polynuclear hydrocarbons. Geochim. Cosmochim. Acta **26**, 225–230 (1962)

Blumer, M.: Organic pigments: their long-term fate. Science **149**, 722–726 (1965)

Blumer, M.: Chemical fossils: trends in organic geochemistry. Pure Appl. Chem. **34**, 591–609 (1973)

Blumer, M., Guillard, R. R. L., Chase, T.: Hydrocarbons of marine phytoplankton. Mar. Biol. **8**, 183–189 (1971)

Blumer, M., Mullin, M. M., Thomas, D. W.: Pristane in zooplankton. Science **140**, 974 (1963)

Blumer, M., Mullin, M. M., Thomas, D. W.: Pristane in the marine environment. Helgol. Wiss. Meeresunters. **10**, 187–201 (1964)

Blumer, M., Snyder, W. D.: Porphyrins of high molecular weight in a Triassic oil shale. Evidence by gel permeation chromatography. Chem. Geol. **2**, 35–45 (1967)

Blumer, M., Thomas, D. W.: "Zamene" isomeric C_{19} monoolefins from marine zooplankton, fishes and mammals. Science **148**, 370–371 (1965a)

Blumer, M., Thomas, D. W.: Phytadienes in zooplankton. Science **149**, 1148–1149 (1965b)

Blumer, M., Youngblood, W. W.: Polycyclic aromatic hydrocarbons in soils and Recent sediments. Science **188**, 53–55 (1975)

Bogomolov, A. I., Panina, K. I., Khotinseva, L. I.: Catalytic transformation of organic compounds at low temperature on clays (in Russian). In: Trudy VNIGRI: Publ. 155, Geokh. Sbornik **6**, 163–194 (1960); Publ. 174, Geokh. Sbornik **7**, 17–34 (1961); Publ. 212, Geokh. Sbornik **8**, 66–94 (1963)

Boigk, H., Hagemann, W. W., Stahl, W., Wollanke, G.: Isotopenphysikalische Untersuchungen zur Herkunft und Migration des Stickstoffs nordwestdeutscher Erdgase aus Oberkarbon und Rotliegend. Erdöl u. Kohle, Erdgas, Petrochem. **29**, 103–112 (1976)

Bokova, E. N.: Formation of heavy gaseous hydrocarbons during anaerobic decomposition of organic matter. Geol. Nefti i Gaza **3**, 44–47 (1959)

Bonnet, R., Czechowski, F.: Gallium porphyrins in bituminous coal. Nature (London) **283**, 465–467 (1980)

Boon, J. J., Pijpstra, W. I. C., De Lange, F., De Leeuw, J. W., Yoshioka, M., Shimizu, Y.: Black sea sterol — a molecular fossil for dinoflagellate blooms. Nature (London) **277**, 125–127 (1979)

Bordovskiy, O. K.: Accumulation and transformation of organic substances in marine sediments. Mar. Geol. **3**, 3–114 (1965)

Boulmier, J. L., Oberlin, A., Durand, B.: Etude structurale de quelques séries de kérogènes par microscopie électronique. Relations avec la carbonisation. In: Advances in Organic Geochemistry 1975. Campos, R., Goni, J. (eds.). Madrid, 1977, pp. 781–796

Brassell, S. C., Wardroper, A. M. K., Thomson, I. D., Maxwell, J. R., Eglinton, G.: Specific acyclic isoprenoids as biological markers of methanogenic bacteria in marine sediments. Nature (London) **290**, 693–696 (1981)

Bray, E. E., Evans, E. D.: Distribution of n-paraffins as a clue to recognition of source beds. Geochim. Cosmochim. Acta **22**, 2–15 (1961)

Breger, I. A.: Kerogen. In: McGraw Hill Encyclopedia of Science and Technology. New York: McGraw Hill, 1961

Breger, I. A., Brown, A.: Kerogen in the Chattanooga Shale. Science **137**, 221–224 (1962)

Brodskii, A. M., Kalinenko, R. A., Lavroskii, K. P.: On the kinetic isotope effect in cracking. Int. J. Appl. Radiat. Isotopes **7**, 118 (1959)

Brooks, J. D.: The use coals as indicators of the occurrence of oil and gas. Austr. Petr. Explor. Assoc. J. **10**, 35–40 (1970)

Brooks, P. W., Maxwell, J. R.: Early stage fate of phytol in a recently deposited lacustrine sediment. In: Advances in Organic Geochemistry 1973. Tissot, B., Bienner, F. (eds.). Paris: Technip, 1974, pp. 977–991

Brown, F. S., Baedecker, M. J., Nissenbaum, A., Kaplan, I. R.: Early diagenesis in a reducing fjord. Saanich Inlet, British Columbia — III. Changes in organic constituents of sediment. Geochim. Cosmochim. Acta **36**, 1185–1203 (1972)

Bruyevich, S. V., Zaytseva, Ye. D.: On the chemistry of Bearing Sea sediments. Tr. Inst. Okeanol. Akad. Nauk. S.S.S.R. **17** (in Russian) (1958)

Burger, J.: L'exploitation des pyroschistes ou schistes bitumineux. Rev. Inst. Fr. Pétr. **28**, 315–372 (1973)

Burlingame, A. L., Haug, P., Belsky, T., Calvin, M.: Occurrence of biogenic steranes and pentacylic triterpanes in an Eocene shale (52 million years) and in an Early Precambrian shale (2.7 billion years): a preliminary report. Proc. Nat. Acad. Sci. US **54**, 1406–1412 (1965)

Burlingame, A. L., Haug, P. A., Schnoes, H. K., Simoneit, B. R.: Fatty acids derived from the Green River Formation oil shales by extraction and oxidation, a review. In: Advances in Organic Geochemistry 1968. Schenck, P. A., Havenaar, I. (eds.). Oxford: Pergamon Press, 1969, pp. 85–128

Califet-Debyser, Y., Oudin, J. L.: Influence of depth and temperature on the structure and distribution of alkanes and aromatic hydrocarbons from rock extracts. Am. Chem. Soc. Petr. Chem. Div. Preprint 14, 4, E 16–21 (1969)

Cane, R. F.: The constitution and synthesis of oil shale. Proc. 7th World Petr. Congr. **3**, 681–689 (1967)

Carruthers, W., Douglas, A. G.: The constituents of high-boiling petroleum distillates. Part IV: Some polycyclic aromatic hydrocarbons in a Kuwait oil. J. Chem. Soc. (London) 278–281 (1957)

Cartz, L., Hirsch, P. B.: A contribution to the structure of coal from X-ray diffraction studies. Phil. Trans. Roy. Soc. (Lond.) Ser. A **252**, 557–599 (1960)

Casagrande, D. J., Hodgson, G. W.: Geochemistry of porphyrins: the observation of homologous tetrapyrroles in a sediment sample from the Black Sea, Geochim. Cosmochim. Acta **40**, 479–482 (1976)

Castex, H.: Spectroscopie infrarouge et analyse élémentaire dc quelques kérogènes. Campagne JOIDES-IPOD; Oceanologica Acta **2**, 33–40 (1979)

Chappe, B., Michaelis, W., Albrecht, P., Ourisson, G.: Fossil evidence for a novel series of archaebacterial lipids. Naturwissenschaften **66**, 522–523 (1979)

Chappe, B., Michaelis, W., Albrecht, P.: Molecular fossils of Archaebacteria as selective degradation products of kerogen. In: Advances in Organic Geochemistry 1979. Douglas, A. G., Maxwell, J. R. (eds.). Oxford: Pergamon Press, 1980, pp. 265–274

Chappe, B., Albrecht, P., Michaelis, W.: Polar lipids of Archaebacteria in sediments and petroleums. Science **217**, 65–66 (1982)

Clark, R. C., Blumer, M.: Distribution of n-paraffins in marine organisms and sediment. Limnol. Oceanogr. **12**, 79–87 (1967)

Claypool, G. E., Kaplan, I. R.: The origin and distribution of methane in marine sediments. In: Natural Gases in Marine Sediments. Kaplan, I. R. (ed.). New York: Plenum Publ. Corp, 1974, pp. 99–139

Coleman, D. D., Risatti, J. B., Schoell, M.: Fractionation of carbon and hydrogen isotopes by methane-oxidizing bacteria. Geochim. Cosmochim. Acta **45**, 1033–1037 (1981)

Colombo, U. Gazzarrini, F., Gonfiantini, R., Sironi, G., Tongiorgi, E.: Measurements of $^{13}C/^{12}C$-isotope ratios on Italian natural gases and their geochemical interpretation. In: Advances in Organic Geomechistry 1964. Hobson, G. D., Louis, M. C. (eds.). Oxford: Pergamon Press, 1966, pp. 279–292

Combaz, A.: Les palynofacies. Rev. Micropaleontol. **7**, 205–218 (1964)

Combaz, A.: Microsphérules muriformes dans les roches mères du pétrole, hypothèse sur leur origine. C. R. Acd. Sci. (Paris) Sér. D **270**, 2240–2243 (1970)

Combaz, A.: L'énergie des roches: les schistes bitumineux. Revue 2000 (1974)

Combaz, A.: Les kérogènes vus au microscope. In: Kerogen. Durand, B. (ed.). Paris: Technip, 1980, p. 55–111

Connan, J.: Diagenèse naturelle et diagenèse artificielle de la matière organique à éléments végétaux prédominants. In: Advances in Organic Geochemistry 1973. Tissot, B., Bienner, F. (eds.). Paris: Technip, 1974, pp. 73–95

Cooper, B. S., Murchison, D. G.: Organic geochemistry of coal. In: Organic Geochemistry. Eglinton, G., Murphy, M. T. J. (eds.). Berlin-Heidelberg-New York: Springer, 1969, pp. 699–726

Corbet, B., Albrecht, P., Ourisson, G.: Photochemical or photomimetic fossil triterpenoids in sediments and petroleum. J. Amer. Chem. Soc. **102**, 1171–1173 (1980)

Correia, M.: Relations possibles entre l'état de conservation des éléments figurés de la matière organique et l'existence de gisements d'hydrocarbures. Rev. Inst. Fr. Pétr. **22**, 1285–1306 (1967)

Correia, M.: Contribution à la recherche des zones favorables à la genèse du pétrole par l'observation microscopique de la matière organique figurée. Rev. Inst. Fr. Pétr. **24**, 1417–1454 (1969)

Cox, R. E., Maxwell, J. R., Ackman, R. G., Hooper, S. N.: The isolation of acyclic isoprenoid alcohols from an Ancient sediment: appraoches to a study of the diagenesis and maturation of phytol. In: Advances in Organic Geochemistry 1971. von Gaertner, H. R., Wehner, H. (eds.). Oxford-Braunschweig: Pergamon Press, 1972, pp. 263–275

Cox, R. E., Maxwell, J. R., Eglinton, G., Pillinger, C. T., Ackman, R. G., Hooper, S. N.: The geological fate of chlorophyll: the absolute stereochemistry of acyclic isoprenoid acid in a 50-million year old lacustrine sediment. Chem. Comm. 1639–1641 (1970)

Dastillung, M.: Lipides de sédiments récents. Thesis Univ. Strasbourg, 1976

Davis, J. B., Squires, R. M.: Detection of microbially produced gaseous hydrocarbons other than methane. Science **119**, 381–383 (1953)

Debyser, J.: Faits d'observation sur la genèse du pétrole. I. Facteurs contrôlant la répartition de la matière organique dans les sédiments. Rev. Inst. Fr. Pétr. **24**, 21–48 (1969)

Debyser, Y., Pelet, R., Dastillung, M.: Géochimie organique de sédiments marins récents: Mer Noire, Baltique, Atlantique (Mauritanie). In: Advances in Organic Geochemistry 1975. Campos, R., Goni, J. (eds.). Madrid, 1977, pp. 289–320

Degens, E. T., Mopper, K.: Factors controlling the distribution and early diagenesis of organic material in marine sediments. In: Chemical Oceanography. Riley, I. R., Chester, R. (eds.). London-New York: Academic Press, 1975

Demaison, G. J., Moore, G. T.: Anoxic environments and oil source bed genesis. Am. Assoc. Pet. Geol. Bull. **64**, 1179–1209 (1980)

Demorest, M., Mooberry, D., Danforth, D.: Decomposition of ketones and fatty acids by silica-alumina composites. Industr. Eng. Chem. **43**, 2569–2572 (1951)

Dennis, L. W., Maciel, G. E., Hatcher, P. G., Simoneit, B. R. T.: ^{13}C Nuclear magnetic resonance studies of kerogen from Cretaceous black shales thermally altered by basaltic intrusions and laboratory simulations. Geochim. Cosmochim. Acta **46**, 901–907 (1982)

Deroo, G., Powell, T. G., Tissot, B., McCrossan, R. G.: The origin and migration of petroleum in the Western Canadian sedimentary basin. Alberta. Can. Geol. Survey Bull. **262** (1977)

Deroo, G., Roucaché, J., Tissot, B.: Etude géochimique du Canada Occidental, Alberta. Inst. fr. Pétr. Geol. Surv. Canada. Open file **171** (1973)

De Rosa, M., De Rosa, S., Gambacorta, A., Minale, L., Bu'Lock, J. D.: Chemical structure of the ether lipids of thermophilic acidophilic bacteria of the Calderiella group. Phytochemistry **16**, 1961–1965 (1977a)

De Rosa, M., De Rosa, S., Gambacorta, A., Bu'Lock, J. D.: Lipid structures in the Caldariella group of extreme thermoacidophile bacteria. J. C. S. Chem. Commun., 514–515 (1977b)

De Rosa, M., De Rosa, S., Gambacorta, A., Bu'Lock, J. D.: Structure of calditol, a new branched chain nonitol, and of the derived tetraether lipids in thermoacidophile Archaebacteria of the Caldariella group. Phytochemistry **19**, 249–254 (1980)

Didyk, B. M., Alturki, Y. I. A., Pillinger, C. T., Eglinton, G.: Petroporphyrins as indicators of geothermal maturation. Nature (London) **256**, 563–565 (1975)

Doose, P. R.: The bacterial production of methane in marine sediments. Ph. D. Thesis, Univ. of California, Los Angeles (1980)

Dorsselaer, A. van: Triterpènes de sédiments. Thesis Univ. Strasbourg, 1975

Dorsselaer, A. van, Albrecht, P., Connan, J.: Changes in composition of polycyclic alkanes by thermal maturation (Yallourn lignite, Australia). In: Advances in Organic Geochemistry 1975. Campos, R., Goni, J. (eds.). Madrid, 1977, pp. 53–60

Dorsselaer, A. van, Ensminger, A., Spyckerelle, C., Dastillung, M., Sieskind, O., Arpino, P., Albrecht, P., Ourisson, G., Brooks, P. W., Gaskell, S. J., Kimble, B. J., Philp, R. P., Maxwell, J. R., Eglinton, G.: Degraded and extended hopane derivatives (C_{27} to C_{35}) as ubiquitous geochemical markers. Tetrahedron Lett. **14**, 1349–1352 (1974)

Douglas, A. G., Grantham, P. J.: Fingerprint gas chromatography in the analysis of some native bitumens, asphalts and related substances. In: Advances in Organic Geochemistry 1973. Tissot, B., Bienner, F. (eds.). Paris: Technip, 1974, pp. 261–276

Douglas, A. G., Mair, B. J.: Sulfur: role in genesis of petroleum. Science **147**, 499–501 (1965)

Dow, W. G.: Kerogen studies and geological interpretations. J. Geochem. Explor. **7**, 79–99 (1977)

Dow, G. G., Pearson, D. B.: Organic matter in Gulf Coast sediments. Offshore Technol. Conf. Dallas, paper OTC 2343 (1975)

Duncan, D. C.: Geologic setting of oil shale deposits and world prospects. Proc. 7th World Petr. Cong. **3,** 659–667 (1967)

Dunoyer De Segonzac, G.: Les minéraux argileux dans la diagénèse. Passage au métamorphisme. Thesis Univ. Strasbourg, 1969

Dunton, M. L., Hunt, J. M.: Distribution of low molecular-weight hydrocarbons in recent and ancient sediments. Am. Assoc. Petr. Geol. Butt. **46,** 2224–2258 (1962)

Durand, B.: Sedimentary organic matter and kerogen. Definition and quantitative importance of kerogen. In: Kerogen. Durand, B. (ed.). Paris: Technip, 1980, pp. 13–14

Durand, B.: Present trends in organic geochemistry in reseach on migration of hydrocarbons. In: Advances in Organic Geochemistry 1981. Bjorøy, M., et al. (eds.). New York: Wiley, 1983, pp. 117–128

Durand, B., Espitalié, J.: Evolution de la matière organique au cours de l'enfouissement des sédiments. C. R. Acad. Sci. (Paris) **276,** 2253–2256 (1973)

Durand, B., Espitalié, J.: Geochemical studies on the organic matter from the Douala Basin (Cameroon). II. Evolution of kerogen. Geochim. Cosmochim. Acta **40,** 801–808 (1976)

Durand, B., Monin, J. C.: Elemental analysis of kerogens. In: Kerogen. Durand, B. (ed.). Paris: Technip, 1980, pp. 113–142

Durand, B., Nicaise, G.: Procedures for kerogen isolation. In: Kerogen. Durand, B. (ed.). Paris: Technip, 1980, pp. 35–53

Durand, B., Oudin, J. L.: Exemple de migration des hydrocarbures dans une série deltaïque: le delta de la Mahakam, Indonésie. Proceedings of 10th World Petroleum Congress, Bucharest, vol. 2, pp. 3–11, 1979

Durand, B., Espitalié, J., Nicaise, G., Combaz, A.: Etude de la matière organique insoluble (kérogène) des argiles du Toarcien du Bassin de Paris: I, Etude par les procédés optiques, analyse élémentaire, étude en microscopie et diffraction électroniques. Rev. Inst. Fr. Pétr. **27,** 865–884 (1972)

Durand, B., Marchand, A., Combaz, A.: Etude de kerogènes par résonance paramagnétique électronique. In: Advances in Organic Geochemistry 1975. Campos, R., Goni, J. (eds.). Madrid, 1977, 753–780

Durand, B., Nicaise, G., Roucaché, J., Vandenbroucke, M., Hagemann, H. W.: Etude géochimique d'une série de charbons. In: Advances in Organic Geochemistry 1975. Campos, R., Goni, J. (eds.). Madrid, 1977, pp. 601–632

Durand, B., Parratte, M., Bertrand, P.: Le potentiel en huile des charbons: une approche géochimique. Rev. lust. Fr. Pétr. **38,** 709–721 (1983)

Durand-Souron, C.: Thermogravimetric analysis and associated techniques applied to kerogens. In: Kerogen. Durand, B. (ed.). Paris: Technip, 1980, pp. 143–161

Eglinton, G.: Hydrocarbons and fatty acids in living organisms and recent and ancient sediments. In: Advances in Organic Geochemistry 1968. Schenck, P., Havenaar, I. (eds.). London: Pergamon Press, 1968, pp. 1–24

Eglinton, G.: Organic geochemistry. The organic chemists approach. In: Organic Geochemistry: Methods and Results. Eglinton, G., Murphy, M. T. J. (eds.). Berlin-Heidelberg-New York: Springer, 1969, pp. 20–73

Eglinton, G.: Laboratory simulation of geochemical processes. In: Advances in Organic Geochemistry 1971. von Gaertner, H. R., Wehner, H. (eds.). Oxford: Pergamon Press, 1972, pp. 29–48

Eglinton, G.: Fossil chemistry. Pure Appl. Chem. **34,** 611–632 (1973)

Eglinton, G., Calvin, M.: Chemical fossils. Sci. Am. **216,** 32–43 (1967)

Eglinton, G., Hamilton, R. J.: The distribution of alkanes. In: Chemical Plant Taxonomy. Swain, T. (ed.). London-New York: Academic Press, 1963, pp. 187–217

Eglinton, G., Scott, P. M., Belsky, T., Burlingame, A. L., Calvin, M.: Hydrocarbons of biological origin from a one-billion-year-old sediment. Science **145,** 263–264 (1964)

Eglinton, G., Hajibrahim, S. K., Maxwell, J. R., Quirke, J. M. E.: Petroporphyrins: structural elucidation and the application of HPLC fingerprinting to geochemical problems. In: Advances in Organic Geochemistry 1979. Douglas, A. G., Maxwell, J. R. (eds.). Oxford: Pergamon Press, 1980, pp. 193–203

Ekweozor, C. M., Strausz, O. P.: 18,19 — Bisnor — 13βH, 14αH-Cheilanthane: A novel degraded tricyclic sesterterpenoid-type hydrocarbon from the Athabasca oil sands. Tetrahedron Lett. **23,** 2711–2714 (1982)

Emery, K. O.: The Sea of Southern California. New York: Wiley, 1960, p. 366

Emery, K. O., Rittenberg, S. C.: Early diagenesis of California Basin sediments in relation to origin of oil. Am. Assoc. Petr. Geol. Bull. **36,** 735–806 (1952)

Ensminger, A.: Triterpenoïdes du schiste de Messel. Thesis (3e cycle), Univ. Strasbourg, 1974

Ensminger, A., Albrecht, P., Ourisson, G., Kimble, B. J., Maxwell, J. R., Eglinton, G.: Homohopane in Messel oil shale: first identification of a C_{31} pentacyclic triterpane in nature. Bacterial origin of some triterpane in ancient sediments? Tetrahedron Lett. **36,** 3861–3864 (1972)

Ensminger, A., Albrecht, P., Ourisson, G., Tissot, B.: Evolution of polycyclic hydrocarbons under the effect of burial (Early Toarcian shales, Paris Basin). In: Advances in Organic Geochemistry 1975. Campos, R., Goni, J. (eds.). Madrid, 1977, pp. 45–52

Ensminger, A., van Dorsselaer, A., Spyckerelle, C., Albrecht, P., Ourisson, G.: Pentacyclic triterpenes of the hopane type as ubiquitous geochemical markers: origin and significance. In: Advances in Organic Geochemistry 1973. Tissot, B., Bienner, F. (eds.). Paris: Technip, 1974, pp. 245–260

Ensminger, A., Joly, G., Albrecht, P.: Rearranged steranes in sediments and crude oils. Tetrahedron Lett. 1575–1578 (1978)

Erdman, J. G.: Some chemical aspects of petroleum genesis as related to the problem of source bed recognition. Geochim. Cosmochim. Acta **22,** 16–36 (1961)

Erdman, J. G.: Petroleum — its origin in the earth. In: Fluids in Subsurface Environments. Young, A., Galley, J. E. (eds.). Am. Assoc. Petr. Geol. Memoir **4,** 20–52 (1965a)

Erdman, J. G.: The molecular complex comprising heavy petroleum fractions. In: Hydrocarbon Analysis. ASTM Spec. Tech. Publ. **389,** 259–300 (1965b)

Erdman, J. G., Marlett, E. M., Hanson, W. E.: The occurrence and distribution of low-molecular-weight aromatic hydrocarbons in Recent and ancient sediments. Am. Chem. Soc. Div., Petr. Chem. (preprints) **3,** 4, C39–C49 (1958)

Erdman, J. G., Mulik, J. D.: Genesis of low molecular weight hydrocarbons in aquatic sediments. Science **141,** 806 (1963)

Esnault, C.: Evolution thermique des acides gras combinés: génèse des paraffines normales dans les sédiments. Thesis, Univ. Pau, 1973

Espitalié, J., Durand, B., Roussel, J. C., Souron, C.: Etude de la matière organique insoluble (kerogène) des argiles du Toarcien du bassin de Paris: II, Etudes en spectroscopie infrarouge, en analyse thermique différentielle et en analyse thermogravimétrique. Rev. Inst. Fr. Pétr. **28,** 37–66 (1973)

Espitalié, J., Laporte, J. L., Madec, M., Marquis, F., Leplat, P., Paulet, J., Boutefeu, A.: Méthode rapide de caractérisation des roches mères de leur potentiel pétrolier et de leur degré d'évolution. Rev. Inst. Fr. Pétr. **32,** 23–42 (1977)

Espitalié, J., Madec, M., Tissot, B.: Role of mineral matrix in kerogen pyrolysis: influence on petroleum generation and migration. Am. Assoc. Pet. Geol. Bull. **64,** 59–66 (1980)

Evans, C. R., Rogers, M. A., Bailey, N. J. L.: Evolution and alteration of petroleum in western Canada. Chem. Geol. **8,** 147–170 (1971)

Fester, J. I., Robinson, W. E.: Oxygen functional groups in Green River oil-shale kerogen and trona acids. In: Coal Science. Gould, R. F. (ed.). Am. Chem. Soc. 1966

Flaig, W.: Chemistry of humic substances. In: The Use of Isotopes in Soil Organic Matter Studies. Oxford: Pergamon Press, 1966, pp. 103–127

Flaig, W.: Some physical and chemical properties of humic substances as a basis of their characterization. In: Advances in Organic Geochemistry 1971. v. Gaertner, H. R., Wehner, H. (eds.). Oxford-Braunschweig: Pergamon Press, 1972, pp. 49–67

Forsman, J. P.: Geochemistry of kerogen. In: Organic Geochemistry. Breger, I. A. (ed.). New York: Pergamon Press, 1963, pp. 148–182

Forsman, J. P., Hunt, J. M.: Insoluble organic matter (kerogen) in sedimentary rocks of marine origin. In: Habitat of Oil. Weeks, L. G. (ed.). Tulsa: Am. Assoc. Petr. Geol., 1958, pp. 747–778

Forster, H., Biemann, K., Haigh, W. G., Tattrie, N. H., Coluin, J. R.: The structure of novel C_{35} pentacyclic terpenes from *Acetobacter xylinum*. Biochem. J. **135,** 133–143 (1973)

Fox, G. E., Stackebrandt, E., Hespel, R. B. et al.: The phylogeny of Prokaryotes. Science **209,** 457–463 (1980)

Frost, A. V.: Reactions of hydrocarbons with activated aluminosilicates (in Russian). Žurnal Fižic. Khimii. (Moskva) **4,** No 9–10 (1940)

Gadel, F., Ragot, J. P.: Sur l'allochtonie de la fraction organique particulaire des dépôts quaternaires récents du Golfe du Lion. In: Advances in Organic Geochemistry 1973. Tissot, B., Bienner, F. (eds.). Paris: Technip, 1974, pp. 619–628

Gaertner, H. R., von, Schmitz, H. H.: Organic matter in Posidonia shales as an indication of residual oil deposit. Proc. 6th World Petr. Congr. I, 355–363 (1963)

Gagosian, R. B., Smith, S. O., Lee, C., Farrington, J. W., Frew N. M.: Steroid transformations in Recent marine sediments. In: Advances in Organic Geochemistry 1979. Douglas, A. G., Maxwell, J. R. (eds.). Oxford: Pergamon Press, 1980, pp. 407–419

Galimov, E. M.: Carbon isotopes in oil and gas geology (in Russian). Moscow: Nedra, English translation: Washington: NASA TT F-682, 1973

Galimov, E. M.: $^{13}C/^{12}C$ in kerogen. In: Kerogen. Durand, B. (ed.). Paris: Technip, 1980, pp. 271–299

Gallegos, E. J.: Identification of new steranes, terpane and branched paraffins in Green River shale by combined capillary gas chromatography and mass spectrometry. Anal. Chem. **43,** 1151–1160 (1971)

Gallegos, E. J.: Terpane-sterane release from kerogen by pyrolysis gas chromatography-mass spectrometry. Anal. Chem. **47,** 1524–1528 (1975)

Galway, A. K.: Heterogeneous reactions in petroleum genesis and maturation. Nature (London) **223,** 1257–1260 (1969)

Gardner, P. M., Whitehead, E. V.: The isolation of squalane from a Nigerian petroleum. Geochim. Cosmochim. Acta **36,** 259–263 (1972)

Given, P. H.: The distribution of hydrogen in coals and its relation to coal structure, Fuel **39,** 147–153 (1960)

Given, P. H.: Dehydrogenation of coals and its relation to coal structure. Fuel **40,** 427–431 (1961)

Given, P. H.: Biological aspects of the geochemistry of coal. In: Advances in Organic Geochemistry 1971. v. Gaertner, H. R., Wehner, H. (eds.). Oxford-Braunschweig: Pergamon Press, 1972, pp. 69–92

Given, P. H., Dickinson, C. H.: Biochemistry and microbiology of peats. In: Soil Biochemistry. Paul, E. A., McLaren, A. D. (eds.). New York: Dekker, 1973, pp. 123–212

Göhring, K. E. H., Schenck, P. A., Engelhardt, E. D.: A new series of isoprenoid isoalkanes in crude oil and Cretaceous bituminous shales. Nature (London) **215,** 503–505 (1967)

Gransch, J. A., Posthuma, J.: On the origin of sulfur in crudes. In: Advances in Organic Geochemistry 1973. Tissot, B., Bienner, F. (eds.). Paris: Technip, 1974, pp. 727–739

Greiner, A.: Synthèse d'hydrocarbures aromatiques fossiles. Thesis, Univ. Louis Pasteur, Strasbourg, 1976

Greiner, A., Spyckerelle, C., Albrecht, P.: Aromatic hydrocarbons from geological sources. I. New naturally occurring phenanthrene and chrysene derivatives. Tetrahedron **32,** 257–260 (1976)

Gutjahr, C. C. M.: Carbonization of pollen grains and spores and their application. Leidse Geologische Mededelingen **38,** 1–30 (1966)

Hagemann, H. W.: Petrographische und palynologische Untersuchung der organischen Substanz (Kerogen) in den liassischen Sedimenten Luxemburgs. In: Advances in Organic Geochemistry 1973. Tissot, B., Bienner, F. (eds.). Paris: Technip, 1974, pp. 29–37

Hagemann, H. W., Hollerbach, A.: Relationship between the macropetrographic and organic geochemical composition of lignites. In: Advances in Organic Geochemistry 1979. Douglas, A. G., Maxwell, J. R. (eds.). Oxford: Pergamon Press, 1980, pp. 631–638

Han, J., Calvin, M.: Occurrence of C_{22}–C_{25} isoprenoids in Bell Creek crude oil. Geochim. Cosmochim. Acta **33,** 733–742 (1969)

Han, J., McCarthy, E. D., van Hoeven, W., Calvin, M., Bradley, W. H.: Organic geochemical studies. II: the distribution of aliphatic hydrocarbons in algae, bacteria and in a Recent lake sediment. Proc. Nat. Acad. Sci. US **59,** 29–33 (1968)

Harrison, W. E.: Thermally induced diagenesis of aliphatic fatty acids in Holocene sediment. Am. Assoc. Petr. Geol. Bull. **60,** 452–457 (1976)

Hartmann, M., Müller, P., Suess, E., Vanderweiden, C. H.: Oxidation of organic matter in Recent marine sediments. Meteor. Forsch.-Ergebn. Ser. C, **12,** 74–86 (1973)

Hedberg, H. D.: Geological aspects of the origin of petroleum. Am. Assoc. Petr. Geol. Bull. **48,** 1755–1803 (1964)

Hedberg, H. D.: Significance of high-wax oils with respect to genesis of petroleum. Am. Assoc. Petr. Geol. Bull. **52,** 736–750 (1968)

Hedberg, H. D.: Relation of methane generation to undercompacted shales, shales diapirs and mud volcanoes. Am. Assoc. Petr. Geol. Bull. **58,** 661–673 (1974)

Hedberg, H. D.: Personal communication, 1977

Hedemann, H. A.: Zur Frage der Kohlenwasserstoffgase im Oberkarbon: Erdöl u. Kohle, Erdgas, Petrochem. **8,** 833–841 (1963)

Heek, K. H. van, Jüntgen, H., Luft, K. F., Teichmüller, M.: Aussagen zur Gasbildung in frühen Inkohlungsstadien aufgrund von Pyrolyseversuchen. Erdöl u. Kohle, Erdgas, Petrochem. **24,** 566–572 (1971)

Hemingway, J. E.: Sedimentology of coal-bearing strata. In: Coal and Coal-Bearing Strata. Murchison, D., Westoll, T. S. (eds.). Edinburgh-London: Oliver and Boyd, 1968, pp. 43–69

Henderson, W., Eglinton, G., Simmonds, P., Lovelock, J. E.: Thermal alteration as a contributory process to the genesis of petroleum. Nature (London) **219,** 1012–1016 (1968)

Henderson, W., Wollrab, V., Eglinton, G.: Identification of steroids and triterpenes from a geological source by capillary gas-liquid chromatography and mass spectrometry. Chem. Commun. pp. 710–712 (1968)

Héritier, F. E.: Frigg field. Large submarine fan trap in lower Eocene rocks of North Sea Viking graben. Am. Assoc. Pet. Geol. Bull. **63**, 1999–2020 (1979)

Hills, I. R., Smith, G. W., Whitehead, E. V.: Optically active spirotriterpane in petroleum distillates. Nature (London) **219**, 243–246 (1968)

Hills, I. R., Smith, G. W., Whitehead, E. V.: Hydrocarbons from fossil fuels and their relationship with living organisms. J. Inst. Petr. **57**, 127–137 (1970)

Hills, J. R., Whitehead, E. V.: Triterpanes in optically active petroleum distillates. Nature (London) **209**, 977–979 (1966)

Hills, I. R., Whitehead, E. V., Anders, D. E., Cummins, J. J., Robinson, W. E.: An optically active triterpane, gammacerane in Green River, Colorado, oil-shale bitumen. Chem. Commun. pp. 752–754 (1966)

Hitchon, B.: Geochemical studies of natural gas. J. Can. Pet. Tech. **2**, 100–116, 165–174 (1963)

Hodgson, G. W., Baker, B. L., Peake, E.: Geochemistry of porphyrins. In: Fundamental Aspects of Petroleum Geochemistry. Nagy, B., Colombo, U. (eds.). Amsterdam: Elsevier, 1967, pp. 177–259

Hoering, T. C., Hare, P. E.: Comparison of natural humic acids with amino acid-glucose reaction products. Am. Assoc. Petr. Geol. Bull. **57**, 784 (1973)

Hollerbach, A., Hagemann, H. W.: Organic geochemical and petrological investigations into a series of coals with increasing rank. Proceedings International Conference on Coal Science, Düsseldorf, 1981. Essen: Glückauf GmbH, 1981, pp. 80–85

Holzer, G., Oro, J., Tornabene, T. G.: Gas chromatographic-mass spectrometric analysis of neutral lipids from methanogenic and thermoacidophilic bacteria. J. Chromatogr. **186**, 795–809 (1979)

Hood, A., Castaño, J. R.: Organic methamorphism: Its relationship to petroleum generation and application to studies of authigenic minerals. United Nations ESCAP, CCOP Tech. Bull, **8**, 85–118 (1974)

Hood, A., Gutjahr, C. C. M.: Organic metamorphism and the generation of petroleum. Paper presented: Ann. Meet. Geol. Soc. Am., Nov. 1972

Hood, A., Gutjahr, C. C. M., Heacock, R. L.: Organic metamorphism and the generation of petroleum. Am. Assoc. Petr. Geol. Bull. **59**, 986–996 (1975)

Horsfield, B., Douglas, A. G.: The influence of minerals on the pyrolysis of kerogen. Geochim. Cosmochim. Acta, **44**, 1119–1131 (1980)

Huc, A. Y.: Contribution à l'étude de l'humus marin et de ses relations avec les kérogenes. Thesis, Univ. Nancy, 1973

Huc, A. Y.: Mise en évidence de provinces géochimiques dans les schistes bitumieux du Toarcien de l'Est du Bassin de Paris, Etude de la fraction organique soluble. Rev. Inst. Fr. Pétr. **31**, 933–953 (1976)

Huc, A. Y.: Géochimie organique des schistes bitumineux du Toarcien du Bassin de Paris. Thesis, University Strasbourg (1978)

Huc, A. Y., Durand, B. M.: Etude des acides humiques et de l'humine des sédiments récents considérés comme précurseurs des kérogenes. In: Advances in Organic Geochemistry 1973. Tissot, B., Bienner, F. (eds.). Paris: Technip, 1974, pp. 53–72

Huc, A. Y., Durand, B. M.: Occurrence and significance of humic acids in ancient sediments. Fuel **56**, 73–80 (1977)

Huc, A. Y., Durand, B. M., Monin, J. C.: Humic compounds and kerogen in cores from Black Sea sediments. In: DSDP Initial Reports, leg 42B (in press)

Hunt, J. M.: Distribution of hydrocarbons in sedimentary rocks. Geochim. Cosmochim. Acta **22**, 37–49 (1961)

Hunt, J. M.: Distribution of carbon in crust of earth. Am. Assoc. Petr. Geol. Bull. **56**, 2273–2277 (1972)

Hunt, J. M.: Organic geochemistry of the marine environment. In: Advances in Organic Geochemistry 1973. Tissot, B., Bienner, F. (eds.). Paris: Technip, 1974, pp. 593–605

Hunt, J. M.: Is there a geochemical depth limit for hydrocarbons? Petr. Eng. **47**, 112–127 (1975)

Hussler, G., Chappe, B., Wehrung, P., Albrecht, P.: C_{27}–C_{29} ring A monoaromatic steroids in Cretaceous black shales. Nature (London) **294**, 556–558 (1981)

Ikan, R., McLean, J.: Triterpenoids form lignite. J. Chem. Soc., 893–894 (1960)

International Committee for Coal Petrology: International Handbook of Coal Petrography, 2nd ed. 1963, 1st and 2nd supplement 1971, 1975. Centre National de la Recherche Scientifique, Paris, 1963

Ishiwatari, R., Ishiwatari, M., Kaplan, I. R., Rohrbach, B. G.: Thermal alteration of young kerogen in relation to petroleum genesis. Nature (London) **264**, 347–349 (1976)

Jarolim, V., Hejno, K., Hemmert, F., Šorm, F.: Über einige aromatische Kohlenwasserstoffe des Harzanteils des Montanwachses. Coll. Czech. Chem. Comm. **30**, 873–879 (1965)

Jarolim, V., Hejno, K., Šorm, F.: Struktur einiger aus Montanwachs isolierter triterpenischer Verbindungen. Coll. Czech. Chem. Comm. **28**, 2443–2454 (1963)

Jarolim, V., Streibl, M., Hejno, K., Šorm, F.: Über einige Inhaltsstoffe des Montanwachses. Coll. Czech. Chem. Comm. **26**, 451–458 (1961)

Johns, R. B., Belsky, T., McCarthy, E. D., Burlingame, A. L., Haug, P., Schnoes, H. K., Richter, W., Calvin, M.: The organic geochemistry of ancient sediments, Part II. Geochim. Cosmochim. Acta **30**, 1191–1222 (1966)

Jones, R. W., Edison, T. A.: Microscopic observations of kerogen related to geochemical parameters with emphasis on thermal maturation. Repr. from: A symposium in Geochemistry: Low temperature metamorphism of kerogen and clay minerals. Oltz, D. F. (ed.), 1979

Jüntgen, H., Karweil, J.: Gasbildung und Gasspeicherung in Steinkohlenflözen, Part I and II. Erdöl u. Kohle, Erdgas, Petrochem. **19**, 251–258 and 339–344 (1966)

Jüntgen, H., Klein, J.: Entstehung von Erdgas aus kohligen Sedimenten. Erdöl u. Kohle, Erdgas, Petrochem., Ergänzungsband, **1**, 74/75, 52–69. Industrieverlag v. Hernhausen, Leinfelden bei Stuttgart (1975)

Jurg, J. W., Eisma, E.: Petroleum hydrocarbons: generation from a fatty acid. Science **144**, 1451–1452 (1964)

Kartsev, A. A., Vassoevich, N. B., Geodekian, A. A., Neruchev, S. G., Sokolov, V. A.: The principal stage in the formation of petroleum, 8th World Petr. Congr., Moscow **2**, 3–11 (1971)

Karweil, J.: Die Metamorphose der Kohlen vom Standpunkt der physikalischen Chemie. Dtsch. geol. Gesell. Z. **107**, 132–139 (1956)

Karweil, J.: Aktuelle Probleme der Geochemie der Kohle. In: Advances in Organic Geochemistry 1968. Schenck, P. A., Havenaar, I. (eds.). Oxford: Pergamon Press, 1969, pp. 59–84

Kemp, A. L. W.: Preliminary information on the nature of organic matter in the surface sediments of Lake Huron Erie and Ontario. In: Proc. Symp. Hydrogeochem. Biogeochem. Tokyo Sept. 1970. Ingerson, E. (ed.). Washington: Clarke Co., 1973, Vol. II, pp. 40–48

Kemp, A. L. W., Mudrochova, A.: Distribution and forms of nitrogen in a Lake Ontario sediment core. Limnol. Oceanogr. **17**, 855–867 (1972)

Khan, S. U., Schnitzer, M.: The retention of hydrophobic organic compounds by humic acid. Geochim. Cosmochim. Acta **36**, 745–754 (1972)

Kidwell, A. L., Hunt, J. M.: Migration of oil in Recent sediments of Pedernales, Venezuela. In: Habitat of Oil. Weeks, L. G. (ed.). Tulsa: Am. Assoc. Petr. Geol., 1958, pp. 790–817

Kimble, B. J.: The Geochemistry of Triterpenoid Hydrocarbons. Ph. D. thesis, Univ. Bristol, U. K., 1972

Kimble, B. J., Maxwell, J. R., Philp, R. P., Eglinton, G., Albrecht, P., Ensminger, A., Arpino, P., Ourisson, G.: Tri- and Tetraterpenoid hydrocarbons in the Messel Oil Shale, Geochim. Cosmochim. Acta **38**, 1165–1181 (1974)

King, L. H., Goodspeed, F. E., Montgomery, D. S.: Study of sedimented organic matter and its natural derivatives. Mines Branch Report R 114, Dept. Mines Techn. Surv. Ottawa, 1963

Klein, J., Jüntgen, H.: Studies on the emission of elemental nitrogen from coals of different rank and its release under geochemical conditions. In: Advances in Organic Geochemistry 1971. v. Gaertner, H. R., Wehner, H. (eds.), 1972, pp. 647–656

Knoche, H., Albrecht, P., Ourisson, G.: Organic compounds in fossil plants (Voltzia brongniarti, Coniferales). Angew. Chem. **7**, 631 (1968)

Knoche, H., Ourisson, G.: Organic compounds in fossil plants (Equisetum; horsetails). Angew. Chem. **6**, 1085 (1967)

Koons, C. B., Jamieson, G. W., Ciereszko, L. S.: Normal alkane distribution in marine organisms; possible significance to petroleum origin. Am. Assoc. Petr. Geol. Bull. **49**, 301–304 (1965)

Krevelen, D. W. van: Coal. Amsterdam: Elsevier, 1961

Krevelen, D. van: Geochemistry of Coal. In: Organic Geochemistry. Breger, I. A. (ed.). Oxford-London-New York-Paris: Pergamon Press, 1963, pp. 183–247

Kurbatov, J. M.: Zur Frage über die Genesis des Torfes und der Torfhuminsäuren (preprint). Int. Peat Congr. Leningrad, 1963

Kuswaha, S. C., Kates, M., Sprott, G. D., Smith, I. C. P.: Novel complex polar lipids from the methanogenic Archaebacterium Methanospirillum hungatei. Science **211**, 1163–1164 (1981)

Kvenvolden, K. A.: Normal paraffin hydrocarbons in sediments from San Francisco Bay, California. Am. Assoc. Petr. Geol. Bull. **46**, 1643–1652 (1962)

Kvenvolden, K. A.: Evidence for transformations of normal fatty acids in sediments. In: Advances in Organic Geochemistry 1966. Hobson, G. D., Speers, G. C. (eds.). Oxford: Pergamon Press, 1970, pp. 335–366

Laflamme, R. E., Hites, R. A.: The global distribution of polycyclic aromatic hydrocarbons in recent sediments. Geochim. Cosmochim. Acta **42**, 289–303 (1978)

Laplante, R. E. Hydrocarbon generation in Gulf Coast Tertiary sediments. Am. Assoc. Petr. Geol. Bull. **58**, 1281–1289 (1974)

Larskaya, Ye. S., Zhabrev, D. V.: Effect of stratal temperature and pressure on the composition of dispersed organic matter. Akad. Nauk SSSR Doklady **157**, 4, 897–900 (1964) (in Russian), English translation: **157**, 1–6, 135–138 (1965)

Leeuw, I. W. de, Correia, V. A., Schenck, P. A.: On the decomposition of phytol under simulated geological conditions and in the top layer of natural sediments. In: Advances in Organic Geochemistry 1973. Tissot, B., Bienner, F. (eds.). Paris: Technip, 1974, pp. 993–1004

Leeuw, J. W. de, Simoneit, B. R., Boon, J. J., Rijpstra, W. I. C., De Lange, F., v. d. Leeden, J. C. W., Correia, V. A., Burlingame, A. L., Schenck, P. A.: Phytol derived compounds in the geosphere. In: Advances in Organic Geochemistry 1975. Campos, R., Goni, J. (eds.). Madrid, 1977, pp. 61–80

Le Tran, K.: Geochemical study of hydrogen sulfide sorbed in sediments. In: Advances in Organic Geochemistry 1971. von Gaertner, H. R., Wehner, H. (eds.). Oxford-London-New York: Pergamon Press 1972, pp. 717–726

Levi, L., Nicholls, R. V.: Formation of styrenes. Ind. Eng. Chem., 1005 (1958)

Lewan, M. D., Winters, J. C., MacDonald, J. H.: Generation of oil-like pyrolyzates from organic rich shales. Science **203**, 897–899 (1979)

Leythaeuser, D.: Die Verteilung höherer n-Paraffine und anderer schwerflüchtiger Kohlenwasserstoffe in Kohlen und Gesteinen des saarländischen Karbons und Devons in Abhängigkeit von den geologischen Verhältnissen. Dissertation Univ. Würzburg, 1968

Leythaeuser, D., Welte, D. H.: Relation between distribution of heavy n-paraffins and coalification in Carboniferous coals from the Saar District, Germany. In: Advances in Organic Geochemistry 1968. Schenck, P. A., Havenaar, I. (eds.). Oxford: Pergamon Press, 1969, pp. 429–442

Leythaeuser, D., Hagemann, H. W., Hollerbach, A., Schaefer, R. G.: Hydrocarbon generation in source beds as a function of type and maturation of their organic matter: a mass balance approach. 10th World Petrol. Congress, Panel 1, Paper 4 (1979)

Leythaeuser, D., Altebäumer, F. J., Schaefer, R. G.: Effect of an igneous intrusion on maturation of organic matter in Lower Jurassic Shales from NW-Germany. In: Advances in Organic Geochemistry 1979. Douglas, A. G., Maxwell, J. R. (eds.). Oxford: Pergamon Press, 1980, pp. 133–139

Limberg-Ruban, Ye. L.: The quantity of bacteria in the water and in bottom material in the northwestern Pacific, Issled. Dal'nevost. Morey SSSR. **3**, (in Russian) (1952)

Louis, M.: Sur le pouvoir rotatoire des hydrocarbures contenus dans les roches sédimentaires anciennes. C. R. Acad. Sci. (Paris) **259**, 1889–1890 (1964)

Louis, M.: Etudes géochimiques sur les schistes cartons du Toarcien du bassin de Paris. In: Advances in Organic Geochemistry 1964. Hobson, G. D., Louis, M. C. (eds.). Oxford: Pergamon Press, 1966, pp. 85–94

Louis, M.: Cours de géochimie du pétrole. Paris: Technip, 1967

Louis, M.: Les corps optiquement actifs et l'origine du pétrole. Rev. Inst. Fr. Pétr. **23**, 299–314 (1968)

Louis, M., Tissot, B.: Influence de la température et de la pression sur la formation des hydrocarbures dans les argiles a kérogene. 7th World Petr. Congr. Proc. **2**, 47–60 (1967)

Ludwig, B.: Steroides aromatiques de sédiments et de pétroles. Thesis, University Louis Pasteur, Strasbourg, France (1982)

Ludwig, B., Hussler, G., Wehrung, P., Albrecht, P.: Identification of mono- and triaromatic steroid derivatives in ancient shales and petroleum. Tenth International Meeting on Organic Geochemistry, Bergen (Norway), oral presentation (1981a)

Ludwig, B., Hussler, G., Wehrung, P., Albrecht, P.: $C_{26}-C_{29}$ triaromatic steroid derivatives in sediments and petroleum. Tetrahedron Lett. **22**, 3313–3316 (1981b)

Lutz, M., Kaasschieter, J. P. H., van Wijke, D. H.: Geological factors controlling Rotliegend gas accumulations in the Mid-European Basin. Proc. 9th World Petr. Congr. **2**, Geology PD 2. London: Applied Science Publishers Ltd., 1975, pp. 93–103

Lyubimenko, V. N.: Chlorophyll in deposits of lake silts. Zh. Russk. Bot. Obshch. **6**, 97–105 (1923)

Mackenzie, A. S., Quirke, J. M. E., Maxwell, J. R.: Molecular parameters of maturation in the Toarcian shales, Paris Basin, France. II. Evolution of metalloporphyrins. In: Advances in Organic Geochemistry 1979. Douglas, A. G., Maxwell, J. R. (eds.). Oxford: Pergamon Press, 1980a, pp. 239–248

Mackenzie, A. S., Patience, R. L., Maxwell, J. R., Vandenbroucke, M., Durand, B.: Molecular parameters of maturation in the Toarcian shales, Paris Basin, France. I.

Changes in the configuration of acyclic isoprenoid alkanes, steranes and triterpanes. Geochim. Cosmochim. Acta **44,** 1709–1721 (1980b)

Mackenzie, A. S., Hoffmann, C. F., Maxwell, J. R.: Molecular parameters of maturation in the Toarcian shales, Paris Basin, France. III. Changes in the aromatic steroid hydrocarbons. Geochim. Cosmochim. Acta **45,** 1345–1355 (1981)

Maillard, L. C.: Action des acides aminés sur les sucres. C. R. Acad. Sci. (Paris) **154,** 66–68 (1912)

Mair, B. J.: Terpenoids, fatty acids and alcohols as source materials for petroleum hydrocarbons. Geochim. Cosmochim. Acta **28,** 1303–1321 (1964)

Mair, B. J., Martinez-Pico, J. L.: Composition of the trinuclear aromatic portion of the heavy gas oil and light lubricating distillate. Proc. API, Sec. 3 Refining (New York) **42,** 173–185 (1962)

Mair, B. J., Ronen, Z.: Composition of the branched paraffin cycloparaffin portion of petroleum, 140–180°C. J. Chem. Eng. Data **12,** 432–436 (1967)

Mair, B. J., Ronen, Z., Eisenbraun, E. J., Horodysky, A. G.: Terpenoid precursors of hydrocarbons from the gasoline range of petroleum. Science **154,** 1339–1341 (1966)

Marchand, A., Conard, J.: Electron paramagnetic resonance in kerogen studies. In: Kerogen. Durand, B. (ed.). Paris: Technip, 1980, pp. 243–270

Marchand, A., Libert, P., Achard, M. F., Combaz, A.: Etude de la pyrolyse de quelques précurseurs possibles du kérogène, I. Evolution chimique de deux sporopollénines. In: Advances in Organic Geochemistry 1973. Tissot, B., Bienner, F. (eds.). Paris: Technip, 1974, pp. 117–135

Marchand, A., Libert, P., Combaz, A.: Essai de caracterisation physico-chimique de la diagénèse de quelques roches biologiquement homogènes. Rev. Inst. Fr. Pétr. **24,** 3–20 (1969)

Martin, R. L., Winters, J. C., Williams, J. A.: Distributions of n-paraffins in crude oils and their implications to origin of petroleum. Nature (London) **199,** 1190–1193 (1963a)

Martin, R. L., Winters, J. C., Williams, J. A.: Composition of crude oils by gas chromatography: geological significance of hydrocarbon distribution. 6th World Petr. Congr. Frankfurt, Sec. V, paper 13 (1963b)

Mattavelli, L., Ricchiuto, T., Grignani, D., Schoell, M.: Geochemistry and habitat of natural gases in Po Basin, Northern Italy. Am. Assoc. Pet. Geol. Bull. **67,** 2239–2254 (1983)

Mattern, G., Albrecht, P., Ourisson, G.: 4-methyl sterols and stanols in Messel shale (Eocene). Chem. Comm., 1570–1571 (1970)

Maxwell, J. R., Cox, R. E., Ackman, R. G., Hooper, S. N.: The diagenesis and maturation of phytol. The stereochemistry of 2, 6, 10, 14-tetramethyl pentadecane from an ancient sediment. In: Advances in Organic Geochemistry 1971. von Gaertner, H. R., Wehner, H. (eds.). Pergamon Press, 1972, pp. 177–291

Maxwell, J. R., Cox, R. E., Eglinton, G., Pillinger, C. T., Ackman, R. G., Hooper, S. N.: Stereochemical studies of acyclic isoprenoid compounds: II. Role of chlorophyll in derivation of isoprenoid type acids in a lacustrine sediment. Geochim. Cosmochim. Acta **37,** 297–314 (1973)

Maxwell, J. R., Douglas, A. G., Eglinton, G., McCormick, A.: The Botryococcenes-hydrocarbons of novel structure from the algae *Botryococcus braunii,* Kutzing. Phytochemistry **7,** 2157–2171 (1968)

Maxwell, J. R., Pillinger, C. T., Eglinton, G.: Organic Geochemistry. Quar. Rev. **25,** 571–628 (1971)

Maxwell, J. R., Quirke, J. M. E., Eglinton, G.: Aspects of modern porphyrin geochemistry and the Treibs hypothesis. In: Internationales Alfred-Treibs-Symposium München, 1979. Prashnowsky, A. A. (ed.). Universität Würzburg, 1980, pp. 37–55

McCartney, J. T., Teichmüller, M.: Classification of coals according to degree of coalification by reflectance of the vitrinite component. Fuel **51**, 64–68 (1972)

McIver, R. D.: Composition of kerogen — clue to its role in the origin of petroleum. 7th World Petr. Congr. Proc. **2**, 25–36 (1967)

Meinschein, W. G.: Origin of petroleum. Am. Assoc. Petr. Geol. Bull. **43**, 925–943 (1959)

Meinschein, W. G.: Significance of hydrocarbons in sediments and petroleum. Geochim. Cosmochim. Acta **22**, 58–64 (1961)

Menendez, R.: Composition isotopique du carbone dans les gaz provenant des sondages d'Aquitaine. Bulletin du Centre de Recherches de Pau **7**, 69–81 (1973)

Michaelis, W., Albrecht, P.: Molecular fossils of Archaebacteria in kerogen. Naturwissenschaften **66**, 420–421 (1979)

Miknis, F. P., Netzel, D. A., Smith, J. W., Mast, M. A., Maciel, G. E.: ^{13}C NMR measurements of the genetic potential of oil shales. Geochim. Cosmochim. Acta **46**, 977–984 (1982)

Moldowan, J. M., Seifert, W. K.: Head-to-head linked isoprenoid hydrocarbons in petroleum. Science **204**, 169–171 (1979)

Monin, J. C., Durand, B., Vandenbroucke, M., Huc, A. Y.: Experimental simulation of the natural transformation of kerogen. In: Advances in Organic Geochemistry 1979. Douglas, A. G., Maxwell, J. R. (eds.). Oxford: Pergamon Press, 1980, pp. 517–530

Moore, L. R.: Geomicrobiology and geomicrobiological attack on sedimented organic matter. In: Organic Geochemistry. Eglinton, G., Murphy, M. T. J. (eds.). Berlin-Heidelberg-New York: Springer, 1969, pp. 265–303

Mukhopadhyay, P. K., Hagemann, H. W., Hollerbach, A., Welte, D. H.: The relation between organic geochemical and petrological parameters of coal in Indian coal basins. Energy Sources **4**, 313–328 (1979)

Mulheirn, L. J., Ryback, G.: Stereochemistry of some steranes from geological sources. Nature (London) **256**, 301–302 (1975)

Müller, G., Grimmer, G., Boehnke, H.: Sedimentary record of heavy metals and polycyclic aromatic hydrocarbons in Lake Constance. Naturwissenschaften **64**, 429–431 (1977)

Murphy, M. T. J., McCormick, A., Eglinton, G.: Perhydro-β-carotene in the Green River shale. Science **157**, 1040–1042 (1967)

Nissenbaum, A., Kaplan, I. R.: Chemical and isotopic evidence for the in situ-origin of marine humic substances. Limnol. Oceanogr. **17**, 570–582 (1972)

Nissenbaum, A., Presley, B. J., Kaplan, I. R.: Early diagenesis in a reducing fjord, Saanich Inlet, British Columbia. I. Chemical and isotopic changes in major components of interstital water. Geochim. Cosmochim. Acta **36**, 1007–1027 (1972)

Oberlin, A., Boulmier, J. L., Durand, B.: Electron microscope investigation of the structure of naturally and artifically metamorphosed kerogen. Geochim. Cosmochim. Acta **38**, 647–649 (1974a)

Oberlin, A., Boulmier, J. L., Durand, B.: Electron microscope investigation of the structure of naturally and artificially metamorphosed kerogen. Geochim. Cosmochim. Acta **38**, 647–649 (1974b)

Oberlin, A., Boulmier, J. L., Durand, B.: Natural evolution (catagenesis) related to carbonization (as studied by high resolution electron microscopy. In: Advances in Organic Geochemistry 1973. Tissot, B., Bienner, F. (eds.). Paris: Technip, 1974c, pp. 15–27

Oberlin, A., Boulmier, J. L., Durand, B.: Nouveaux critères d'évolution de quelques carbonisats naturels déterminés par microscopie électronique en fond noir. C. R. Acad. Sci. (Paris) sér. C, **280**, 501–504 (1975)

Oberlin, A., Boulmier, J. L., Villey, M.: Electron microscopic study of kerogen microtexture. Selected criteria for determining the evolution path and evolution stage of kerogen. In: Kerogen. Durand, B. (ed.). Paris: Technip, 1980, pp. 191–241

O'Neal, M. J., Hood, A.: Mass spectrometric analysis of polycyclic hydrocarbons. Am. Chem. Soc. Mtg., Div. Petr. Chem. (preprint) 127–135 (1956)

Oremland, R. S., Taylor, B. F.: Sulfate reduction and methanogenesis in marine sediments. Geochim. Cosmochim. Acta **42**, 209–214 (1978)

Orr, W. L.: Biochemistry of sulfur. In: Handbook of Geochemistry, II/4. Berlin-Heidelberg-New York: Springer, 1974, Sec. 16-L, pp. 1–19

Orr, W. L., Grady, J. R.: Perylene in basin sediments off Southern California. Geochim. Cosmochim. Acta **31**, 1201–1209 (1967)

Ottenjann, K., Teichmüller, M., Wolf, M.: Spektrale Fluoreszenz-Messung und Sporiniten mit Auflicht-Anregung, eine mikroskopische Methode zur Bestimmung des Inkohlungsgrades gering inkohlter Kohlen. Fortschr. Geol. Rheinld. Westfal. **24**, 1–36 (1974)

Ourisson, G., Albrecht, P., Rohmer, M.: The hopanoids. Palaeochemistry and biochemistry of a group of natural products. Pure Appl. Chem. **51**, 709–729 (1979)

Palmer, S. E., Baker, E. W.: Copper porphyrins in Deep-Sea sediments: a possible indicator of oxidized terrestrial organic matter. Science **201**, 49–51 (1978)

Patijn, R. J. H.: Die Entstehung von Erdgas infolge der Nachinkohlung im Nordosten der Niederlande. Erdöl u. Kohle, Erdgas, Petrochemie **17**, 2–9 (1964a)

Patijn, R. J. H.: La formation de gas due à des réhouillifications dans le Nord-Est des Pays-Bas. 5e Cong. Int. Stratigr. Géol. Carbonifère, II, 631, Louis Jean, Gap (1964b)

Peake, E., Hodgson, G. W.: Origin of petroleum-Steranes as products of early diagenesis in recent marine and freshwater sediments. Am. Assoc. Petr. Geol. Bull. **57**, 799 (abstract) (1973)

Pelet, R.: Introduction à la géochimie organique des faciès deltaïques. Rev. Assoc. Fr. Tech. Pétr. **225**, 21–27 (1974)

Pfaltz, A., Jaun, B., Fässler, A., et al.: Zur Kenntnis des Faktors F 430 aus methanogenen Bakterien. Struktur des porphinoiden Ligandsystems. Helvetica Chimica Acta **65**, 828–865 (1982)

Philippi, G. T.: On the depth, time and mechanism of petroleum generation. Geochim. Cosmochim. Acta **29**, 1021–1049 (1965)

Philippi, G. T.: The deep subsurface temperature controlled origin of the gaseous and gasoline-range hydrocarbons of petroleum. Geochim. Cosmochim. Acta **39**, 1353–1373 (1975)

Philp, R. P., Calvin, M.: Kerogenous material in Recent algal mats at Laguna Mormona, Baja California. In: Advances in Organic Geochemistry 1975. Campos, R., Goni, J. (eds.). Madrid, 1977, pp. 735–752

Poulet, M., Roucaché, J.: Etude géochimique des gisements du Nord-Sahara (Algérie). Rev. Inst. Fr. Pétr. **24**, 615–644 (1969)

Powell, T. G., McKirdy, D. M.: The effect of source material, rock type and diagenesis on the n-alkane content of sediments. Geochim. Cosmochim. Acta **37**, 623–633 (1973)

Powell, T. C., Creaney, S., Snodown, L. R.: Limitations of use of organic petrographic techniques for identification of petroleum source rocks. Am. Assoc. Pet. Geol. Bull. **66**, 430–435 (1982)

Prakash, A., Rashid, M. A., Jensen, A., Subba Rao, D. V.: Influence of humic substances on the growth of marine phytoplankton. Diatoms. Limnol. Oceanol. **18**, 516–524 (1973)

Quirke, J. M. E., Shaw, G. J., Soper, P. D., Maxwell, J. R.: Petroporphyrins: II. The presence of porphyrins with extended alkyl side chains. Tetrahedron **36**, 3261–3267 (1980)

Rashid, M. A., King, L. H.: Molecular weight distribution measurements on humic and fulvic acid fractions from marine clays on the Scotian Shelf. Geochim. Cosmochim. Acta **33,** 147–151 (1969)

Raynaud, J. F., Robert, P.: Les méthodes d'étude optique de la matière organique. Bull. Centre Rech. Pau **10,** 109–127 (1976)

Redding, C. E., Schoell, M., Monin, K. C., Durand, B.: Hydrogen and carbon isotopic composition of coals and kerogens. In: Advances in Organic Geochemistry 1979. Douglas, A. G., Maxwell, J. R. (eds.), 1980, pp. 711–723

Rhead, M. M., Eglinton, G., Draffan, G. H., England, P. J.: Conversion of oleic acid to saturated fatty acids in Severn estuary sediments. Nature (London) **232,** 327–330 (1971)

Rice, D. D., Claypool, G. E.: Generation, accumulation, and resource potential of biogenic gas. Am. Assoc. Pet. Geol. Bull. **65,** 5–25 (1981)

Robert, P.: Etude pétrographique des matières organiques insolubles par la mesure de leur pouvoir réflecteur — Contribution à l'exploration pétrolière et à la connaissance des bassins sédimentaires. Rev. Inst. Fr. Pétr. **24,** 105–136 (1971)

Robert, P.: The optical evolution of kerogen and geothermal histories applied to oil and gas exploration. In: Kerogen. Durand, B. (ed.). Paris: Technip, 1980, pp. 385–414

Robin, P. L.: Caractérisation des kérogènes et de leur évolution par spectroscopie infrarouge. Thesis, Univ. Louvain, 1975

Robin, P. L., Rouxhet, P. G.: Contribution des différentes fonctions chimiques dans les bandes d'absorption infrarouge des kérogènes situées à 1710, 1630 et 3430 cm^{-1}. Rev. Inst. Fr. Pétr. **31,** 955–977 (1976)

Robin, P. L., Rouxhet, P. G., Durand, B.: Caracterisation des kérogènes et de leur évolution par spectroscopie infrarouge. In: Advances in Organic Geochemistry 1975. Campos, R., Goni, J. (eds.). Madrid, 1977, pp. 693–716

Robinson, W. E.: Isolation procedures for kerogens and associated soluble organic materials. In: Organic Geochemistry. Eglinton, G., Murphy, M. T. J. (eds.). Berlin-Heidelberg-New York: Springer, 1969a, pp. 181–195

Robinson, W. E.: Kerogen of the Green River formation. In: Organic Geochemistry. Eglinton, G., Murphy, M. T. J. (eds.). Berlin-Heidelberg-New York: Springer, 1969b, pp. 619–637

Robinson, W. E., Cummins, J. J., Dinneen, G. U.: Changes in Green River oil-shale paraffins with depth. Geochim. Cosmochim. Acta **29,** 249–258 (1965)

Robinson, W. E., Dinneen, G. U.: Constitutional aspects of oil shale kerogen. Proc. 7th World Petr. Congr. **3,** 669–680 (1967)

Rogers, M. A.: Application of organic facies concept to hydrocarbon source evaluation. Proceedings of the 10th World Petroleum Congress, Vol. 2, pp. 23–30, 1979

Rohmer, M., Ourisson, G.: a) Structure des Bactériohopanetétrols d'Acetobacter Xylinum. b) Dérivés du Bactériohopane: variations structurales et répartition. c) Méthyl-hopanes d'Acetobacter Xylinum et d'Acetobacter rancens: une nouvelle famille de composés triterpéniques. Tetrahedron Lett. **40,** 3633–3644 (1976)

Rouxhet, P. G., Robin, P. L., Nicaise, G.: Characterization of kerogens and of their evolution by infrared spectroscopy. In: Kerogen. Durand, B. (ed.). Paris: Technip, 1980, pp. 163–190

Rubinstein, I., Sieskind, O., Albrecht, P.: Rearranged Sterenes in a shale: occurrence and simulated formation. J. Chem. Soc., Perkin Transactions I, 1833–1836 (1975)

Rubinstein, I., Spyckerelle, C., Strausz, O. P.: Pyrolysis of asphaltenes: a source of geochemical information. Geochim. Cosmochim. Acta **43,** 1–6 (1979)

Sackett, W. M.: Carbon isotope composition of natural methane occurrences. Am. Assoc. Petr. Geol. Bull. **52,** 853–857 (1968)

Sackett, W. M.: Carbon and hydrogen isotope effects during the thermocatalytic production of hydrocarbons in laboratory simulation experiments. Geochim. Cosmochim. Acta **42**, 571–580 (1978)

Sakikana, N.: Organic geochemical reactions, IX. Reaction of formation of liquid hydrocarbons from acetone and ethanol under high pressure with active soils. J. Chem. Soc. Jpn., Pure Chem. Sect. **72**, 280–285 (1951)

Samman, N., Ignasiak, T., Chen, C. J., Strausz, O. P., Montgomery D. S.: Squalene in petroleum asphaltenes. Science **213**, 1381–1383 (1981)

Scalan, R. S., Smith, J. E.: An improved measure of the odd-even predominance in the normal alkanes of sediment extracts and petroleum. Geochim. Cosmochim. Acta **34**, 611–620 (1970)

Schmitter, J. M., Arpino, P. J., Guiochon, G.: Isolation of degraded pentacyclic triterpenoids acids in a Nigerian crude oil and their identification as tetracyclic carboxylic acids resulting from ring A cleavage. Geochim. Cosmochim. Acta **45**, 1951–1955 (1981)

Schoell, M.: Die Erdgase der süddeutschen Molasse. Erdoel Erdgas Zeitschr. **93**, 311–322 (1977)

Schoell, M.: The hydrogen and carbon isotopic composition of methane from natural gases of various regions. Geochim. Cosmochim. Acta **44**, 649–661 (1980)

Schrayer, G. J., Zarrella, W. M.: Organic geochemistry of shales: II. Distribution of extractable organic matter in the siliceous Mowry shale of Wyoming. Geochim. Cosmochim. Acta **30**, 415–434 (1966)

Schwendiger, R. B., Erdman, J. G.: Carotenoids in sediments as a function of environment. Science **141**, 808–810 (1963)

Schwendiger, R. B., Erdman, J. G.: Sterols in recent aquatic sediments. Science **144**, 1575–1576 (1964)

Seifert, W. K.: Steroid acids in petroleum, Animal contribution to the origin of Petroleum. Pure Appl. Chem. **34**, 633–640 (1973)

Seifert, W. K.: Carboxylic acids in petroleum and sediments. Progr. Chem. Org. Nat. Prod. **32**, 1–49 (1975)

Seifert, W. K.: Steranes and terpanes in kerogen pyrolysis for correlation of oils and source rocks. Geochim. Cosmochim. Acta **42**, 473–484 (1978)

Seifert, W. K., Gallegos, E. J., Teeter, R. M.: First identification of a steroid carboxylic acid in petroleum. Angew. Chem., Int. Ed. Engl. **10**, 747–748 (1971)

Seifert, W. K., Gallegos, E. J., Teeter, R. M.: Proof of structure of steroid carboxylic acids in a California petroleum by deuterium labeling, synthesis, and mass spectrometry. J. Am. Chem. Soc. **94**, 5880–5887 (1972)

Shimoyama, A., Johns, W. D.: Catalytic conversion of fatty acids to petroleum-like paraffins and their maturation. Nature/(London) **232**, 140–144 (1971)

Shimoyama, A., Johns, W. D.: Formation of alkanes from fatty acids in the presence of $CaCO_3$. Geochim. Cosmochim. Acta **36**, 87–91 (1972)

Sieskind, O., Joly, G., Albrecht, P.: Simulation of the geochemical transformation of sterols: superacid effect of clay minerals. Geochim. Cosmochim. Acta **43**, 1675–1679 (1979)

Silverman, S. R.: Investigation of petroleum origin and evolution mechanisms by carbon isotope studies. In: Isotopic and Cosmic Chemistry. Craig, H., Miller, S. L., Wasserburg, G. J. (eds.). Amsterdam: North Holland Pub. Company, 1964

Silverman, S. R.: Carbon isotopic evidence for the role of lipids in petroleum. J. Am. Oil Chem. Soc. **44**, 691–695 (1967)

Silverman, S. R.: Influence of petroleum origin and transformation on its distribution and redistribution in sedimentary rocks. Proc. 8th World Petr. Congr. **2**, 47–54 (1971)

Simoneit, B. R. T.: Sources of Organic Matter in Oceanic Sediments. Thesis, Bristol Univ., Dec. 1975

Simoneit, B. R. T.: Diterpenoid compounds and other lipids in deep-sea sediments and their geochemical significance. Geochim. Cosmochim. Acta **41**, 463–476 (1977)

Simoneit, B. R. T., Brenner, S., Peters, K. E., Kaplan, I. R.: Thermal alteration of Cretaceous black shale by basaltic intrusions in the Eastern Atlantic. Nature (London) **273**, 501–504 (1978)

Simoneit, B. R. T., Rohrback, B. G., Peters, K. E., Kaplan, I. R.: Thermal alteration of Cretaceous black shale by diabase intrusions in the Eastern Atlantic: II. Effects on bitumen and kerogen. Geochim. Cosmochim. Acta **45**, 1581–1602 (1981)

Smith, P. V.: The occurrence of hydrocarbons in recent sediments from the Gulf of Mexico. Science **116**, 437–439 (1952)

Smith, P. V.: Studies on origin of petroleum: occurrence of hydrocarbons in Recent Sediments. Am. Assoc. Petr. Geol. Bull. **38**, 377–404 (1954)

Smith, P. V.: Status of our present information on the origin and accumulation of oil. Proc. 4th World Petr. Congr. I, 359–376 (1955)

Snodown, L. R.: Resinite — a potential petroleum source in the Upper Cretaceous/Tertiary of the Beaufort — Mackenzie sedimentary basin. In: Facts and principles of World petroleum occurrence: A. D. Miall (ed.). Canadian Soc. Pet. Geol. Mem. **6**, 509–521 (1980)

Sokolov, V.: Géochimie des gaz naturels. Moscow: Mir, 1974, 365 p

Souron, C., Boulet, R., Espitalié, J.: Etude par spectrométrie de masse de la décomposition thermique sous vide de kérogénes appartenant à deux lignées évolutives distinctes. Rev. Inst. Fr. Pétr. **24**, 661–678 (1974)

Spackmann, W., Dolsen, C. P., Riegel, W.: Phytogenic organic sediments and sedimentary environments in the Everglades-mangrove complex. Paleontographica **117B**, 135–152 (1966)

Spyckerelle, C.: Constituants organiques d'un sédiment crétacé. Thesis (3è cycle), Univ. Strasbourg, 1973

Spyckerelle, C.: Constituants aromatiques de sédiments. Thesis Univ. Strasbourg, 1975

Spyckerelle, C., Arpino, P., Ourisson, G.: Identification de séries de composés isoprénoides isolées de source géologique. Hydrocarbures acycliques de C_{21} à C_{25} (sesterterpanes). Tetrahedron **28**, 5703–5713 (1972)

Spyckerelle, C., Greiner, A., Albrecht, P., Ourisson, G.: Aromatic hydrocarbons from geological sources. III. A tetrahydrochrysene derived from triterpenes in recent and old sediments: 3,3,7-Trimethyl-1,2,3,4-tetrahydrochrysene. J. Chem. Res. (S), 330–331; (M), 3746–3777 (1977a)

Spyckerelle, C., Greiner, A., Albrecht, P., Ourisson, G.: Aromatic hydrocarbons from geological sources. IV. A octohydrochrysene derived from triterpenes in oil shale: 3,3,7,13-Tetramethyl-1,2,3,4,11,12,13,14-octahydrochrysene. J. Chem. (S), 332–333; (M), 3801–3828 (1977b)

Stach, E., Mackowsky, M. Th., Teichmüller, M., Taylor, G. H., Chandra, D., Teichmüller, R.: Stach's Textbook of Coal Petrology. Berlin: Gebrüder Borntraeger, 1975

Stahl, W. J.: Zur Herkunft nordwestdeutscher Erdgase. Erdöl u. Kohle, Erdgas, Petrochem. **21**, 514–518 (1968)

Stahl, W. J.: Kohlenstoff-Isotopenverhältnisse von Erdgasen, Erdöl u. Kohle, Erdgas, Petrochem. **28**, 188–191 (1975)

Stahl, W.: Reifeabhängigkeit der Kohlenstoff-Isotopenverhältnisse des Methans von Erdölgasen aus Norddeutschland. Erdöl u. Kohle, Erdgas, Petrochem. **31**, 515–518 (1978)

Stanfield, K. E., Smith, J. W., Smith, H. N., Robb, W. A.: Oil yields of sections of Green River oil shale in Colorado 1954–1957. U. S. Bureau of Mines Rept. of Investigations **5614**, 186 (1960)

Staplin, F. L.: Sedimentary organic matter, organic metamorphism, and oil and gas occurrence. Bull. Can. Petr. Geol. **17,** 47–66 (1969)

Stevenson, D. P., Wagner, C. D., Beeck, O., Otvos, J. W.: Isotope Effect in the Thermal Cracking of Propane-1-C^{13}. J. Chem. Phys. **16,** 993 (1948)

Stevenson, F. J., Butler, J. H. A.: Chemistry of humic acids and related pigments. In: Organic Geochemistry. Eglinton, G., Murphy, M. T. J. (eds.). Berlin-Heidelberg-New York: Springer, 1969, pp. 534–557

Stiller, M., Nissenbaum, A.: Variations of stable hydrogen isotopes in plankton from a fresh water lake. Geochim. Cosmochim. Acta **44,** 1099–1101 (1980)

Strakhov, N. M.: Principles of Lithogenesis, Izdat. Akad. Nauk SSSR, Moscow, 1962 (in Russian). Translation by Consultants Bureau and Oliver and Boyd, Edinburgh and London, I, 1967; II, 1969

Strakhov, N. M., Zalmanzon, E. S.: The distribution of authigenous mineralogical forms of iron in sedimentary rock and its bearing on lithology. Izv. Akad. Nauk SSSR. Ser. Geol. **1** (in Russian) (1955)

Streibl, M., Herout, V.: Terpenoids — especially oxygenated mono-, sesqui-, di-, and triterpenes. In: Organic Geochemistry. Eglinton, G., Murphy, M. T. J. (eds.). Berlin-Heidelberg-New York: Springer, 1969, pp. 401–424

Swain, F. M.: Geochemistry of humus. In: Organic Geochemistry. Breger, I. A. (ed.). Oxford: Pergamon Press, 1963, pp. 87–147

Teichmüller, M.: Die Genese der Kohle. Reprint from: 4e Cong. strat. et geol. Carbonifère, Heerlen 15–20, Sept. 1958, III, pp. 699–721 (1962)

Teichmüller, M.: Anwendung kohlenpetrographischer Methoden bei der Erdöl- und Erdgasprospektion. Erdöl u. Kohle, Erdgas, Petrochem. **24,** 69–76 (1971)

Teichmüller, M.: Über neue Macerale der Liptinit-Gruppe und die Entstehung von Micrinit. Fortschr. Geol. Rheinld. Westfal. **24,** 37–64 (1974)

Teichmüller, M.: (a) Origin of the petrographic constituents of coal (b) Application of coal petrological methods in geology including oil and natural gas prospecting. In: Stach, E., Mackowsky, M. Th., Teichmüller, M., Taylor, G. H., Chandra, D., Teichmüller, R. Textbook of Coal Petrology. Berlin: Gebrüder Borntraeger, 1975, pp. 176–238 and 316–331

Teichmüller, M., Teichmüller, R.: Diagenesis of coal (coalification). In: Diagenesis in Sediments. Larsen, G., Chilingar, G. V. (eds.). Amsterdam-London-New York: Elsevier, 1967, pp. 391–415

Teichmüller, M., Teichmüller, R.: Geological aspects of coal metamorphism. In: Coal and Coal-Bearing Strata. Murchison, D. G., Westoll, T. S. (eds.). Edinburgh: Oliver and Boyd, 1968, pp. 233–267

Teichmüller, M., Teichmüller, R.: Fundamentals of coal petrology. In: Stach's Textbook of Coal Petrology. Berlin-Stuttgart: Gebrüder Borntraeger, 1975, pp. 5–53

Teichmüller, M., Teichmüller, R., Bartenstein, H.: Inkohlung und Erdgas in Nordwestdeutschland. Eine Inkohlungskarte der Oberfläche des Oberkarbons. Fortschr. Geol. Rheinld. u. Westf. 27, 137–170 (1979)

Thomas, B. M.: Land-plant source rocks for oil and their significance in Australian Basins. Austr. Petr. Explor. Assoc. J. **22,** 164–178 (1981)

Thomas, D. W., Blumer, M.: Pyrene and fluoranthene in manganese rodules. Science **143,** 39 (1964)

Tissier, M. J.: La géochimie organique des sédiments actuels. Contribution à l'étude de la baie de Seine et à la recherche de la pollution pétrolière. Thesis (3e cycle). Parsis VI (1974)

Tissier, M. Oudin, J. L.: Characteristics of naturally occurring and pollutant hydrocarbons in marine sediments. In: Proc. 1973 Joint Conf. Prevention and Control of Oil Spills. Am. Petr. Inst. 205–214, 1973

Tissot, B.: Premières données sur les mécanismes et la cinétique de la formation du pétrole dans les sédiments. Simulation d'un schéma réactionnel sur ordinateur. Rev. Inst. Fr. Pétr. **24,** 470–501 (1969)

Tissot, B.: Vers l'évaluation quantitative du pétrole formé dans les bassins sédimentaires. Revue Assoc. Fr. Tech. Pétr. **222,** 27–31 (1973)

Tissot, B., Bessereau, G.: Geochimie des gaz naturels et origine des gisements de gaz en Europe occidentale. Rev. Inst. Fr. Pétr. **37,** 63–77 (1982)

Tissot, B., Vandenbroucke, M.: Geochemistry and pyrolysis of oil shales. In: Geochemistry and Chemistry of Oil Shales. Miknis, F. P., McKay, J. F. (eds.). Am. Chem. Soc. Symposium Series **230,** 1983, pp. 1–11

Tissot, B., Califet-Debyser, Y., Deroo, G., Oudin, J. L.: Origin and evolution of hydrocarbons in Early Toarcian shales. Am. Assoc. Petr. Geol. Bull. **55,** 12, 2177–2193 (1971)

Tissot, B., Deroo, G., Espitalié, J.: Etude comparée de l'époque de formation et d'expulsion du pétrole dans diverses provinces géologiques. Proc. 9th World Petr. Congr. **2,** 159–169 (1975)

Tissot, B., Deroo, G., Herbin, J. P.: Organic matter in Cretaceous sediments of the North Atlantic: contribution to sedimentology and paleogeography. In: Deep Drilling Results in the Atlantic Ocean: Continental Margins and Paleoenvironment. Talwian, M., Hay, W., Ryan, W. B. F. Maurice Ewing Series, 3. Am. Geoph. Union, 1979, pp. 362–374

Tissot, B., Deroo, G., Hood, A.: Geochemical study of the Uinta Basin: formation of petroleum from the Green River formation. Geochim. Cosmochim. Acta **42,** 1469–1485 (1978)

Tissot, B., Durand, B., Espitalié, J., Combaz, A.: Influence of the nature and diagenesis of organic matter in formation of petroleum. Am. Assoc. Petr. Geol. Bull. **58,** 499–506 (1974)

Tissot, B., Espitalié, J.: L'évolution thermique de la matière organique des sédiments: application d'une simulation mathématique. Rev. Inst. Fr. Pétr. **30,** 743–777 (1975)

Tissot, B., Espitalié, J., Deroo, G., Tempere, C., Jonathan, D.: Origin and migration of hydrocarbons in the Eastern Sahara (Algeria). In: Advances in Organic Geochemistry 1973. Tissot, B., Bienner, F. (eds.). Paris: Technip, 1974, pp. 315–334

Tissot, B., Oudin, J. L., Pelet, R.: Critères d'origine et d'évolution des pétroles. Application à l'étude géochimique des bassins sédimentaires. In: Advances in Organic Geochemistry 1971. von Gaertner, H. R., Wehner, H. (eds.). London-New York: Pergamon Press, 1972, pp. 113–134

Tissot, B., Pelet, R.: Nouvelles données sur les mécanismes de genèse et de migration du pétrole, Simulation mathématique et application à la prospection. Proc. 8th World Petr. Congr. **2,** 35–46 (1971)

Tissot, B., Pelet, R., Roucaché, J., Combaz, A.: Utilisation des alcanes comme fossiles géochimiques indicateurs des environnements géologiques. In: Advances in Organic Geochemistry 1975. Campos, R., Goni, J. (eds.). Madrid, 1977, pp. 117–154

Tissot, B., Demaison, G., Masson, P., Delteil, J. R., Combaz, A.: Paleoenvironment and petroleum potential of Middle Cretaceous black shales in Atlantic Basins. Am. Assoc. Pet. Geol. Bull. **64,** 2051–2063 (1980)

Tixier, M. P., Curtis, M. R.: Oil shale yield predicted from well logs. Proc. 7th World Petr. Congr. **3,** 713–715 (1967)

Tornabene, T. G., Langworthy, T. A.: Diphytanyl and dibiphytanyl glycerol ether lipids of methanogenic Archaebacteria. Science **203,** 51–53 (1979)

Trask, P. D., Wu, C. C.: Does petroleum form in sediments at time of deposition. Am. Assoc. Petr. Geol. Bull. **14**, 1451–1463 (1930)

Treibs, A.: Chlorophyll und Häminderivate in bituminösen Gesteinen, Erdölen und Erdwachsen und Asphalten. Ann. Chem. **510**, 42–62 (1934)

Trendel, J. M., Restle, A., Connan, J., Albrecht, P.: Identification of a novel series of tetracyclic terpene hydrocarbons (C_{24}–C_{27}) in sediments and petroleum. J. C. S. Chem. Commun. 304–306 (1982)

Vallentyne, J. R.: Geochemistry of carbohydrates. In: Organic Geochemistry. Breger, I. A. (ed.). Oxford: Pergamon Press, 1963, pp. 465–502

Vandenbroucke, M.: Structure of kerogens as seen by investigations on soluble extracts. In: Kerogen. Durand, B. (ed.). Paris: Technip, 1980, pp. 415–443

Vandenbroucke, M., Albrecht, P., Durand, B.: Geochemical studies on the organic matter from the Douala Basin (Cameroon). III. Comparison with the Early Toarcian shales, Paris Basin, France. Geochim. Cosmochim. Acta **40**, 1241–1249 (1976)

Vassoevich, N. B.: Terminology used for designating stages and steps of lithogenesis (in Russian). In: Geology and Geochemistry. Chaps. 1–7. Leningrad: Gostoptekhizdat, 1957

Vassoevich, N. B., Akramkhodzhaev, A. M., Geodekyan, A. A.: Principal zone of oil formation. In: Advances in Organic Geochemistry 1973. Tissot, B., Bienner, F. (eds.). Paris: Technip, 1974, pp. 309–314

Vassoevich, N. B., Korchagina, Yu. I., Lopatin, N. V., Chernyshev, V. V.: Principal phase of oil formation: Moscow: Univ. Vestnik 6, 3–37 (1969) (in Russian); English translation: Int. Geol. Rev. 1970, **12**, 11, 1276–1296 (1969)

Vassoevich, N. B., Korchagina, Yu. I., Lopatin, N. V., Chernyshev, V. V., Tschernikow, K. A.: Die Hauptphase der Erdölbildung. Z. angew. Geol. **15**, 611–622 (1969)

Vassoevich, N. B., Visotskiy, I. V., Guseva, A. N., Olenin, V. B.: Hydrocarbons in the sedimentary mantle of the earth. Proc. 7th World Petr. Congr. **2**, 37–45 (1967)

Vitorovic, D.: Structure elucidation of kerogen by chemical methods. In: Kerogen. Durand, B. (ed.). Paris: Technip, 1980, pp. 301–338

Vitorović, D., Djuricić, M. V., Ilić, B.: New structural information obtained by stepwise oxidation of kerogen from the Aleksinac (Yugoslavia) shale. In: Advances in Organic Geochemistry 1973. Tissot, B., Bienner, F. (eds.). Paris: Technip, 1974, pp. 179–189

Vitorović, D., Krsmanović, V. D., Pfend, P.: Eine Untersuchung der Struktur des Aleksinacer Ölschiefer-Kerogens mittels verschiedener chemischer Methoden. In: Advances in Organic Geochemistry 1975. Campos, R., Goni, J. (eds.). Madrid, 1977, pp. 717–734

Vogler, E. A., Hayes, J. M.: Carbon isotopic compositions of carboxyl groups of biosynthesized fatty acids. In: Advances in Organic Geochemistry 1979. Douglas, A. G., Maxwell, J. R. (eds.). 1980, pp. 697–704

Wakeham, S. G.: Synchronous fluorescence spectroscopy and its application to indigenous and petroleum-derived hydrocarbons in lacustrine sediments. Env. Sci. Technol. **11**, 272–276 (1977)

Wakeham, S. G., Schaffner, C., Giger, W.: Diagenetic polycyclic aromatic hydrocarbons in Recent Sediments: Structural information obtained by high performance liquid chromatography. In: Advances in Organic Geochemistry, 1979. Douglas, A. G., Maxwell, J. R. (eds.). Oxford: Pergamon Press, 1980, pp. 353–363

Wakeham, S. G., Schaffner, C., Giger, W.: Polycyclic aromatic hydrocarbons in recent lake sediments — I. Compounds having anthropogenic origins. Geochim. Cosmochim. Acta **44**, 403–413 (1980)

Wakeham, S. G., Schaffner, C., Giger, W.: Polycyclic aromatic hydrocarbons in recent lake sediments — II. Compounds derived from biogenic precursors during early diagenesis. Geochim. Cosmochim. Acta **44**, 415–429 (1980)

Waples, D. W., Haug, P., Welte, D. H.: Occurrence of regular C_{25} isoprenoid hydrocarbon in Tertiary sediments representing a lagoonal-type, saline environment. Geochim. Cosmochim. Acta **38**, 381–387 (1974)

Welte, D. H.: Nichtflüchtige Kohlenwasserstoffe in Kernproben des Devons und Karbons der Bohrung Münsterland 1. Forschr. Geol. Rheinl. Westfal. **12**, 559–568 (1964)

Welte, D. H.: Kohlenwasserstoffgenese in Sedimentgesteinen: Untersuchungen über den thermischen Abbau von Kerogen unter besonderer Berücksichtigung der n-Paraffinbildung. Geol. Rundsch. **55-1**, 131–144 (1966)

Welte, D. H.: Zur Entwicklungsgeschichte von Erdölen aufgrund geochemisch-geologischer Untersuchungen, Erdöl u. Kohle, Erdgas, Petrochem. **20**, 65–77 (1967)

Welte, D. H.: Organic Geochemistry of carbon. In: Handbook of Geochemistry. Berlin-Heidelberg-New York: Springer, II/1, 6L1–6L30, 1969a

Welte, D. H.: Determination of $^{13}C/^{12}C$ isotope ratios of individual higher n-paraffins from different petroleums. In: Advances in Organic Geochemistry 1968. Schenck, P. A., Havenaar, I. (eds.). Oxford: Pergamon Press, 1969b, pp. 269–277

Welte,, D. H.: Petroleum exploration and organic geochemistry. J. Geochem. Explor. **I**, 117–136 (1972)

Welte, D.: Grenzbedingungen für Kohlenwasserstoffbildung und -umbildung. Erdoel Erdgas Zeitschr. **92**, 413–415 (1976)

Welte, D. H., Ebhardt, G.: Distribution of long chain n-paraffins and fatty acids in sediments from the Persian Gulf. Geochim. Cosmochim. Acta **32**, 465–466 (1968)

Welte, D. H., Waples, D.: Über die Bevorzugung geradzahliger n-Alkane in Sedimentgesteinen. Naturwissenschaften **60**, 516–517 (1973)

Westoll, T. S.: Sedimentary rhythms in coal-bearing strata. In: Coal and Coal-Bearing Strata. Murchison, D. G., Westoll, T. S. (eds.). Edinburgh-London: Oliver and Boyd, 1968, pp. 71–103

Whitehead, E. V.: Molecular evidence for the biogenesis of petroleum and natural gas. In: Proc. Symp. Hydrogeochem. Biogeochem. Washington: The Clarke Co., 1973, Vol. II, pp. 158–211

World Energy Conference: Survey of Energy Resources, Munich, 1980, 357 p

Yen, T. F.: Present status of the structure of petroleum heavy ends and its significance to various technical applications. Am. Chem. Soc. Div. Petr. Chem. (Preprint), F. 102–114 (1972)

Yen, T. F., Erdman, G. J., Pollack, S. S.: Investigation of the structure of petroleum asphaltenes by X-ray diffraction. Analyt. Chem. **33**, 1587–1594 (1961)

Youngblood, W. W., Blumer, M.: Polycyclic aromatic hydrocarbons in the environment: homologous series in soils and recent marine sediments. Geochim. Cosmochim. Acta **39**, 1303–1314 (1975)

Zobell, C. E., Anderson, D. A.: Vertical distribution of bacteria in marine sediments. Am. Ass. Petr. Geol. Bull. **20**, 258–269 (1936)

Part III
The Migration and Accumulation of Oil and Gas

Chapter 1
An Introduction to Migration and Accumulation of Oil and Gas

Petroleum accumulations are generally found in relatively coarse-grained porous and permeable rocks that contain little or no insoluble organic matter. It is highly improbable that the huge quantities of petroleum found in these rocks could have originated in them from solid organic matter of which now no trace remains. Rather, as discussed in previous sections, it appears that fluid petroleum compounds are generated in appreciable quantities only through geothermal action on high molecular weight organic kerogen usually found in abundance only in fine-grained sedimentary rocks and that usually some insoluble organic residue remains in the rock at least through the oil-generating stage. Hence, it can be concluded that the place of origin of oil and gas is normally not identical with the locations where it is found in economically producible conditions, and that it has had to migrate to its present reservoirs from its place of origin. This migration of petroleum, and the subsequent formation of commercial accumulations, will be discussed in this part of the book.

The release of petroleum compounds from solid organic particles (kerogen) in source beds and their transport within and through the capillaries and narrow pores of a fine-grained source bed has been termed primary migration by numerous workers. The oil expelled from a source bed passes through wider pores of more permeable porous rock units. This is called secondary migration. Petroleum compounds may migrate through one ore more carrier beds with similar permeabilities and porosities as reservoir rocks before trapped by an impermeable or a very low permeability barrier. Oil and gas accumulations are thus formed. Since practically all pores in the subsurface are water-saturated, movement of petroleum compounds within the network of capillaries and pores has to take place in the presence of the aqueous pore fluid. Such movement may be due to active water flow or may occur independently of the aqueous phase, either by displacement or by diffusion. There may be a single phase (oil and gas dissolved in water) or a multiphase (separate water and hydrocarbon phases) fluid system.

The distinction between primary and secondary migration was initially not based on the recognition of different migration processes, but only on its occurrence in pores of different sizes and lithology and possibly in a different state of distribution. The loss of hydrocarbons out of a trap is frequently called dismigration.

Petroleum compounds are generated from finely disseminated organic matter in source beds, thus, the first appearance of petroleum is also in a dispersed form. Processes of primary and especially secondary migration, which finally lead to the

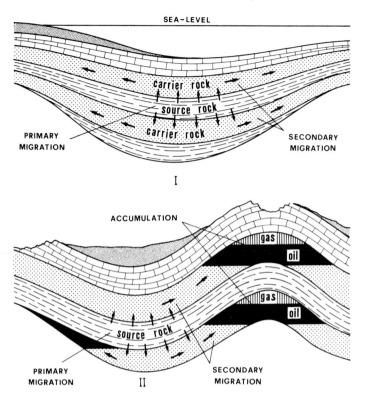

Fig. III.1.1. Formation of oil and gas acculumations: schematic presentation of primary and secondary migration in the initial and advanced stage of basin evolution. *I:* Initial phase of primary and secondary migration. *II:* Advanced stage of primary and secondary migration and formation of accumulation

formation of petroleum accumulations must, therefore, encompass mechanisms that concentrate dispersed petroleum compounds.

The specific gravities of gas and oil, the latter generally between 0.7 and 0.9 g cm^{-3}, are considerably lower that those of saline pore waters. This is why gas and oil pools are mostly found in structural highs where reservoir rocks of suitable porosity and permeability are covered by a dense, relatively impermeable cap rock such as an evaporite or a shale. A reservoir rock sealed by a cap rock in the position of a geological high, such as an anticline is known as a structural petroleum trap. Other types of traps such as sand lenses, reefs and pinch-outs of more permeable and porous rock units, are known as stratigraphic petroleum traps. In all these situations, the changes in permeability and porosity determine the location of an oil and/or a gas accumulation. A schematic presentation of primary and secondary migration and some oil trap types are given in Figure III.1.1.

Processes of primary and secondary migration are not yet understood in detail. Basic data on pore geometry, porosity and permeability relationship, and distribution of pore water in buried dense source rocks are rare. Likewise, there is little information on movement and distribution of petroleum compounds inside the pores of source rocks. Therefore, it should be realized that the following discussion on petroleum migration is largely theoretical and should not be considered definitive. The theoretical concepts should be quantified to determine their effects on the migration mechanisms for a better understanding of the migration phenomena.

Summary and Conclusion

The release of petroleum compounds from kerogen, and their transport within and through the capillaries and narrow pores of a fine-grained source rock are defined as primary migration. The movement of petroleum after expulsion from a source rock, through the wider pores of more permeable and porous carrier and reservoir rocks, is called secondary migration.

Pores in the subsurface are normally water-saturated and hence any movement of petroleum compounds takes place in the presence of aqueous pore fluid. The specific gravities of gas and oil are considerably lower than those of saline waters. This is why gas and oil are found in structural highs where reservoir rocks of suitable porosity and permeability are covered by dense and relatively impermeable cap rocks.

Chapter 2
Physicochemical Aspects of Primary Migration

2.1 Temperature and Pressure

Most petroleum and gas accumulations are found between the surface and to a depth of about 6000 to 7000 m. The physical and chemical conditions that prevail in source and reservoir rocks change with depth of burial. Most pronounced is the increase of temperature and pressure.

The increase in *temperature* with burial depth is a consequence of the transfer of thermal energy from the interior of the earth to the surface where it is dissipated. Different geothermal gradients (°C km^{-1}) are observed, depending upon the overall thermal conductivity of the rock strata, regional heat flow conditions, and subsurface water movement. A world average geothermal gradient is considered to be 25°C km^{-1} (Lee and Uyeda, 1965). Generally stable parts of the earth crust with crystalline rocks have lower geothermal gradients than mobile orogenic zones, with the exception of "arc-trench-gaps" of active margins where gradients may be as low as 10°C km^{-1}. The variation in geothermal gradients in sedimentary basins is typically in the range of 15°C km^{-1} to 50°C km^{-1}, although gradients as low as 5°C km^{-1} and as high as 77°C km^{-1}, and in special situations up to 90°C km^{-1}, have been observed. For example, a very low gradient of about 5°C km^{-1} was found in a 14,585-ft well in Andros Island in the Bahamas (Levorsen, 1954) and a gradient of 76.9°C km^{-1} was recorded in a well drilled in the upper Rhine Graben in southwest Germany near the Landau oilfield (Doebl et al., 1974). In the Walio oilfield of the Salawati Basin, Irian Jaya (Indonesia) gradients as high as 90°C km^{-1} have been reported (Redmond and Koesoemadinata, 1976).

Temperature profiles, or graphic presentations of temperature versus depth, show that geothermal gradients are not always linear, but that there are irregularities caused mainly by variations in thermal conductivity of different lithologic units, by proximity to the surface, and by subsurface water flow. Examples of such temperature profiles are given in Figures III.2.1 and III.2.2. The increase in temperature with depth is greater for higher heat flows and lower thermal conductivities. Heat flow is the amount of heat (calories) flowing through a unit area (cm^2) in a certain time (seconds). Thus, heat flow is expressed in [cal cm^{-2}s^{-1}]. Thermal conductivity (λ) is a physical quantity that defines the amount of heat that flows through a specific medium over a certain distance during a given time increment, if there is a temperature gradient. Thus, there are three main factors controlling geothermal gradients; a difference in primary heat flux related to global tectonic phenomena, differences in the thermal conductivity of the

2.1 Temperature and Pressure

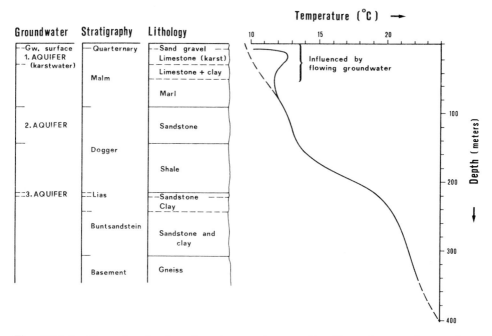

Fig. III.2.1. Example of sub-surface temperature profile. Ground water flow and differences in thermal conductivity of rock types influence the temperature-depth relationship. (Modified after Kappelmeyer, 1961)

stratigraphic column, and sub-surface fluid flow. The primary geothermal flux, commonly observed in old, stable cratonic parts of the crust tends to be low, and in young, orogenic areas and active rifting areas, tends to be high (Lee and Uyeda, 1965).

Rock units with poor thermal conductivities act as barriers to heat flow, and cause a rise in temperature when compared to average gradients. This phenomenon is well known on top of salt diapirs, especially if they are covered by shales. Rock salt is an extremely good thermal conductor ($\lambda = 11$–14) and thermal energy is easily transmitted from greater depth toward the surface. Shales are relatively poor thermal conductors ($\lambda = 3$–7) and consequently thermal energy cannot be dissipated to the surface as rapidly. This causes salt diapirs to create geothermal highs on their tops. Average thermal conductivities of a number of rock types are listed in Figure III.2.3.

Thermal conductivity is not only affected by mineralogical composition, but also depends to a great extent on porosity and permeability, and whether the pores are saturated with water or gas. Circulation of water has a very strong influence on subsurface temperatures, either as a warming or cooling agent. In graben zones, rising hydrothermal-like waters from deep-seated fracture zones may be combined with the tectonically induced primary high heat flux (Rhine Graben, Central Viking Graben of the North Sea, etc.). Heat transfer due to water circulation is now considered to be a relatively common feature in sedimentary

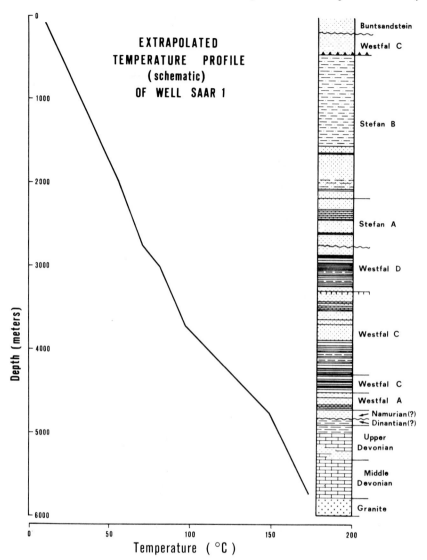

Fig. III.2.2. Extrapolated temperature profile of well Saar 1 in SW Germany. (Modified after Kappelmeyer and Haenel, 1974)

basins. Thus it would not be restricted to graben zones only. This phenomenon might explain observed heat anomalies on top of structural traps, such as anticlines.

The highest temperatures recorded in wells drilled for oil and gas are around 270° to 280°C, as observed in two very deep wells (6500 and 7800 m) in South Texas and the Gulf coast. The geothermal gradient in the 6500 m well was 42°C km^{-1} and in the 7800 m well 35°C km^{-1} (Landes, 1967).

2.1 Temperature and Pressure

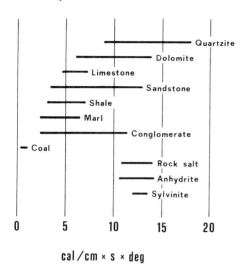

Fig. III.2.3. Average thermal conductivities of selected rock types. (After Kappelmeyer and Haenel, 1974)

Judging from oil occurrences all over the world, and from the data presented above, it appears that most oil source beds have not been exposed to temperatures much higher than 100 °C during the time of generation and migration. Light oils, condensates and gases, however, are probably generated from source beds at higher temperatures (150–180 °C), which in exceptional cases may reach an upper limit of about 250 °C. Temperatures prevailing during secondary migration in carrier rocks and reservoir rocks are similar to those encountered in generation and primary migration, especially if source rocks and reservoir rocks are in juxtaposition.

Another physical parameter increasing with depth is *pressure*. Pressure rises with increasing overburden under the gravity load of overlying sediments. As long as the pores of the sedimentary column are interconnected and ultimately open to the surface and there is sufficient permeability, the fluids inside the pore space are under normal hydrostatic pressure. The gravity load causes sediments to compact and to lose porosity. This reduction in porosity is only possible if a commensurate portion of the relatively incompressible, aqueous pore fluid is expelled and allowed to move away. If there are sedimentary layers with strongly reduced permeability, and if the gravity load continues to increase, movement of the pore fluid beneath such impermeable layers is impeded. Consequently, the pore fluid is confined in the sediment and compaction ceases or is at least retarded. Under such conditions, previously normal *hydrostatic pressures* in the pore fluid increase and the sediment becomes overpressured. Ultimately, *petrostatic pressure* (synonymous with lithostatic or geostatic pressure) is attained when the pore fluid carries the entire load of the overlying water-saturated sediment column. Interstitial fluid pressures in sedimentary rocks can, therefore, vary between normal hydrostatic and higher pressures. A plot of the normal hydrostatic pressure gradient and the petrostatic or geostatic pressure gradient is shown in Figure III.2.4.

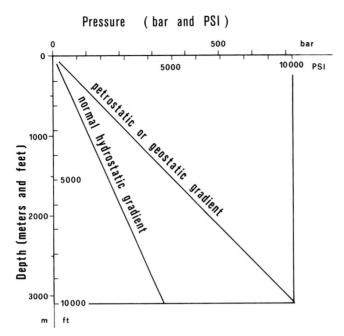

Fig. III.2.4. Plot of hydrostatic and petrostatic gradients

The hydrostatic pressure gradient is between 1 and 1.09 bar per 10 m (0.433–0.47 psi ft^{-1}), depending on the salinity of the pore water. A representative average petrostatic gradient is 2.3 bar per 10 m (1 psi ft^{-1}), however, variations in the overburden bulk specific gravity are common, and geologically important in some context (Chapman, 1973). At depths between 1000, and 6000 m, migrating petroleum and gas must move through pores saturated with water, at fluid pressures ranging between 100 and 1400 bar and at temperatures from about 50°C to about 250°C.

There is also a general increase in pore water salinity with increasing burial depth. Gradients for salinity vary between 70 mg l^{-1}m^{-1} and 250 mg l^{-1}m^{-1} (Engelhardt, 1973), but pore water salinities may also vary depending on lithology, temperature, subsurface fluid convection, and geological configuration. Pore water salinities range from fresh water to salt saturation and may reach values close to 300 g l^{-1}. Under normal pressure conditions, sandstones commonly have higher salinities than adjacent shales. Morever, pore water in shales is generally depleted in chloride and enriched in sulfate and bicarbonate as compared to sandstones. Pore water salinities are frequently abnormally low in high pressure zones and in areas of meteoric water influx.

2.2 Compaction

2.2.1 Compaction, Porosity of Clastic Sediments and Abnormally High Pressures

Compaction in sediments results in an increase in bulk density and loss of porosity with increasing effective pressure, temperature and time. The rate of compaction is largely governed by material properties of a sediment (both physical and chemical) and by the rate at which liquid pore fluid can be expelled. Because compaction causes fluid flow through sedimentary rocks, it is commonly considered to be an important factor in petroleum migration. This kind of fluid flow, however, would be restricted in carbonate-type rocks and from this point of view, carbonates and argillaceous sediments behave differently. The transport of fluid through interconnected pores of a rock is dependent on permeability, which varies considerably, although no rock is absolutely impermeable. The dynamics of fluid flow is commonly determined from Darcy's Law. Hubbert (1940) gave a correct mathematical expression of Darcy's Law,

$$Q = -\frac{K}{\mu} \varrho A \frac{\delta \Phi}{\delta l}$$

where,
- A = cross sectional area of the rock through which the flow is measured (cm^2),
- K = intrinsic permeability (darcy),
- Q = volume flux per unit time (cm^3 s^{-1}),
- μ = dynamic viscosity of the fluid (centipoise),
- ϱ = density of the fluid (g cm^{-3}),

$\dfrac{\delta \Phi}{\delta l}$ = hydrodynamic gradient along the flow path, l.

For convenience in geological context, permeability is usually expressed in millidarcies (md = 10^{-3} darcy) or even microdarcies (μd = 10^{-6} darcy). Darcy's law is valid for laminar flow, where inertial forces are negligible compared to viscous forces. During viscous flow, there is an interaction between the liquid moving through the porous rock and the surface or the inner pore space of the rock.

For clastic sediments, plots of porosity versus depth show a more or less exponential relationship. There is initially a very rapid loss in porosity at relatively shallows depths, and with further increasing overburden pressure, the rate of loss in porosity diminishes. Semilogarithmic plots of the depth-porosity relationship result in a straight line, the slope of which varies within certain limits from basin to basin. This basic fact about compaction of sediments was recognized by Athy (1930) and Hedberg (1926, 1936). They derived empirical equations for the porosity-depth relationship of sediments, where the depth of burial or overburden pressure was placed in the exponent. Figure III.2.5 shows such porosity-depth relationships, as measured on samples from three different basins: Tertiary sediments of the Gulf coast (Dickinson, 1953), Tertiary sedi-

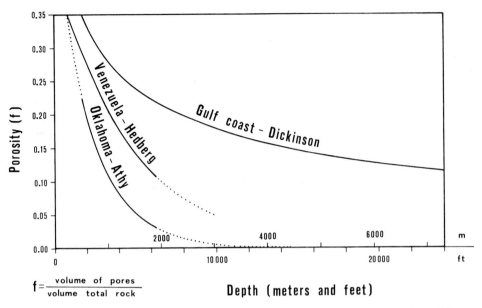

Fig. III.2.5. Depth-porosity relationship as determined for samples from three different basins: Tertiary sediments of the Gulf coast; Tertiary sediments of Eastern Venezuela; Paleozoic sediments of Oklahoma. (Modified after Levorsen, 1954)

ments of Eastern Venezuela (Hedberg, 1936) and Paleozoic sediments of Oklahoma (Athy, 1930). Hubbert and Rubey (1959) have shown on a theoretical basis that the effective compressive stress equals the maximum differential pressure, which is composed of the overburden pressure minus the internal fluid pressure.

Compaction may be delayed in thick shale sequences with little or no sand or silt intercalations (Chapman, 1972). In such sequences the escape of pore waters cannot keep pace with subsidence, and abnormally high fluid pressures may build up, especially in the central zone of thick shales. Impedance of fluid escape may also be caused by unusually effective seals such as impermeable evaporites. Another important cause of abnormally high pressures is the generation of methane and other low molecular weight hydrocarbons in organic-rich shales (Hedberg, 1974). Abnormal pressures, however, are transitory, and indicate that equilibrium has not been attained. Abnormal pressures, therefore, occur more frequently in young Tertiary sedimentary sequences than in older Mesozoic or even Paleozoic basins.

Compressive stress is widely accepted as a primary cause for porosity reduction and abnormally high pressures, but these effects have not gone undisputed. Bradley (1975) has argued that compression by overburden alone cannot cause abnormal pressures at present drilling depths. Accordingly, the extra load is transferred to the sediment below the seal and not to the interstitial pore fluid within the sealed-off portion of a water-saturated sediment. Pressures deviating from hydrostatic, according to Bradley (1975), are mainly caused by temperature

2.2 Compaction

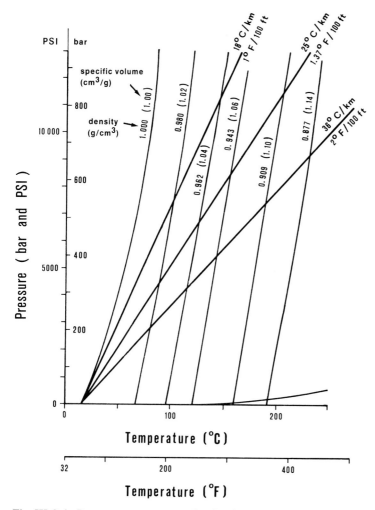

Fig. III.2.6. Pressure-temperature-density diagram for water under geological conditions. Superimposed on this diagram are geothermal gradients of 18°C km^{-1}, 25°C km^{-1} and 36°C km^{-1} for hydrostatically pressured water. (After Barker, 1972; Magara, 1974)

changes. This is in line with observations and considerations by Barker (1972) and Magara (1974), who discuss the effect of water expansion with increasing temperature, and its influence on fluid pressures in buried sediments. A pressure–temperature–density diagram for water under geological conditions, according to Barker (1972) and Magara (1974), is given in Figure III.2.6. Superimposed on this diagram are geothermal gradients of 18°C km^{-1}, 25°C km^{-1} and 36°C km^{-1} for hydrostatically pressured water. From this diagram it can be deduced that for common geothermal gradients, there is a continuous expansion of water with increasing depth. The water expansion is greater at higher geothermal gradients as the lines for these gradients intercept water

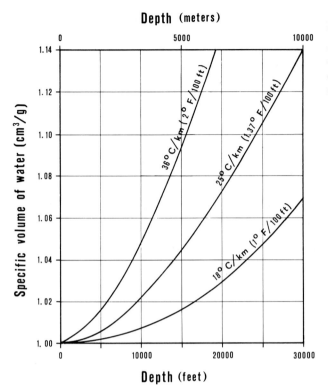

Fig. III.2.7. Relationship between specific volume of water and increasing depth of burial for different geothermal gradients. (After Magara, 1974)

isodensity lines at shorter intervals. This relationship between specific water volume (i.e., water expansion) and increasing depth of burial for different geothermal gradients is shown in Figure III.2.7. Magara (1974) pointed out that the expansion of water due to rising temperature is an additional cause for fluid migration in the sub-surface. The direction of fluid movement would be from hot to cold areas, from deep to shallow and from a basin center to the edges. Thus, temperature-induced fluid flow follows the same general directions as fluid movement caused by "normal" compaction.

It is surprising that for many years the effect and role of temperature has almost been neglected when considering compaction. The behavior of water, its specific volume, its viscosity and capacity as a solvent, are all temperature-related. Bradley (1975) remarks that porosity loss in nature is caused by a group of interrelated chemical processes dependent on reactivity, temperature, surface area, and pressure. He calls these processes "lithification". It refers to all processes whereby porosity is reduced in sandstones which become cemented, to shales, which are compacted, and to limestones, which recrystallize.

2.2.2 Compaction of Carbonates

In the context of this discussion carbonates are referred to as sedimentary rocks containing more than 50% by weight of carbonate minerals and less than 50% detrital minerals (clays and quartz). Some carbonate source rocks may be composed almost entirely out of carbonate material (less than 5% clay minerals). Carbonates are chemically more reactive than silicates and, hence, during compaction behave differently than clastic sediments. The porosity reduction in clastic sediments is more strongly influenced by mechanical and physical rearrangements of mineral grains than generally encountered in carbonates or evaporites. Chemical processes dominate the porosity and permeability reduction of carbonates.

The initial or primary porosity of fine-grained carbonate sediments corresponds to porosites observed in clastic sediments. There is experimental and empirical evidence that at lower overburden pressures, up to a depth of about 100 m to 300 m, compaction occurs as in clastic sediments. With increasing depth of burial, chemical diagenesis becomes more important. It stabilizes the rock fabric, mainly due to solution and cementation processes and is much more pronounced in carbonates than in clastic sediments. Recrystallization of carbonate-type source rocks is an important process with respect to petroleum generation and migration. With increasing burial, recrystallization converts initially fine-grained sediments into coarse-grained rocks. In this way, finely disseminated bitumen and other foreign material such as clay minerals become concentrated at grain boundaries and in intergranular spaces. This is an important step in bitumen concentration in carbonate rocks. On the other hand, it is known that in reservoir rocks, organic material may hinder chemical diagenesis of the mineral rock skeleton. It is not clear how this affects source rocks.

There is evidence for very early lithification of certain carbonate sediments. Ginsburg (1957) pointed out that in carbonate muds most of the water may be lost within the first meter of burial. Hollmann (1962) describes the underwater consolidation of limestones. Deeply buried carbonate rocks frequently do not show signs of advanced gravitational compaction, as it is demonstrated by intact macrofossils and microfossils, and sometimes by relatively high porosity values. Carbonate rocks, in addition to primary porosity, may develop a pronounced secondary porosity after deposition as a consequence of dissolution processes. Secondary porosity results from modification of primary porosity and, therefore, the two frequently cannot be clearly distinguished (Aoyagi, 1973). It should also be mentioned here that dolomitization often increases porosity because the molar volume of $CaMg(CO_3)_2$ is smaller than the volume of $CaCO_3$. For a more detailed treatment of compaction and diagenesis of carbonate sediments, the reader is referred to Engelhardt's book (1977) *The Origin of Sediments and Sedimentary Rocks*.

Finally, it should be emphasized that there are many argillaceous limestones (e.g., marls) and many calcareous shales. These hybrid sediments have a mineralogical composition intermediate between those of pure carbonates and clays. Many of these hybrid sediments have very high organic contents and appear to be petroleum source rocks (Hunt, 1967).

2.2.3 Pore Diameters and Internal Surface Areas

During sedimentary compaction and resulting porosity reduction there is also a marked regular decrease in pore diameters, especially in fine-grained clastic sediments. The sedimentary pore system on a microscopic scale is a very inhomogeneous combination of partially interconnected voids with irregular geometry and narrow pore throats leading to larger pore volumes. With increasing depth of burial, especially in clastic sediments, pores become more and more flat. Pores are incorrectly described in terms of diameters rather than by their heights and widths. Common techniques used to determine porosity, such as mercury porosimetry or gas adsorption, do not measure the true geometric configuration of pores, but only pseudo-pore diameters based on entry pressures or internal surface areas. On the basis of measurements and extrapolations (Jüntgen and Karweil, 1963; Nagumo, 1965; Seevers, 1969; Heling, 1970), a pseudo-pore diameter–depth curve can be constructed (Fig. III.2.8). According to this curve, average pseudo-pore diameters in shales are around 50–100 Å (1 nm

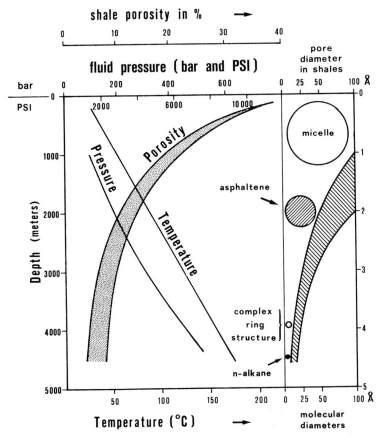

Fig. III.2.8. Interrelationship of various physical parameters with increasing depth of burial for shale-type sediments

$= 10^{-9}$ m $= 10$ Å) in a depth range of about 2000 m, and continue to decrease with increasing depth. Pore diameters of this size are of special interest with respect to petroleum migration. Because pores are usually flat at depth, it can be assumed that molecules such as asphaltenes exhibiting effective diameters of about 50 Å cannot move through a pore system with pseudo-pore diameters of 50 Å or smaller.

In clastic sediments and to a greater extent in carbonates, there is a general tendency for individual grain size to increase with increasing depth of burial. In clastic sediments, due to both these effects and the reduction in pore space, there is a distinct trend toward diminishing specific surface areas with depths. Shales have surface areas from about 10–60 $m^2 g^{-1}$ between depths of 500 and 3500 m. In sandy shales these surface areas are reduced and sandstones have even smaller specific surface areas of about 5 $m^2 g^{-1}$ (Dreier, 1968; Heling, 1970).

This concept of porosity reduction and changes of pore diameters with depth was originally developed when studying pure sandstones and shales. Source rocks, however, containing various amounts of organic matter would show a much more complex depth–porosity relationship. It is known that coal, for instance, contains a macro- and a micropore system. Whereas the macropore system is related to depth of burial, the micropore system is mainly related to the rank level of coal. In the same sense the conversion of solid kerogen particles into liquid or gaseous compounds is maturity-related and creates porosity. This newly created porosity and the accompanying changes in specific surface areas are not static, but depend on the expulsion of hydrocarbons and the reorganization of the kerogen structure. Robin (1975) for instance has measured internal surface areas in kerogen from fine-grained sediments. With increasing thermal evolution they range from less than 10 $m^2 g^{-1}$ organic matter at 1000 m to about 35 $m^2 g^{-1}$ at 4000 m depth.

2.3 Fluids

2.3.1 State and Behavior of Pore Water

Mineral surfaces have attractive van der Waals' forces that cause physical *adsorption* of liquids or gases. The extent of adsorption depends upon the nature of the minerals and of the molecules being adsorbed, and is a function of pressure and temperature. Physical adsorption may attract several monolayers of molecules to a surface in contrast to chemical adsorption (chemisorption) which accomodates only one monolayer. Several monolayers of water molecules may be physically adsorped on clay mineral surfaces. This water is not only physically adsorbed, but there are hydrogen bonds between the electronegative oxygen atoms of the mineral surfaces and the water molecules. Water molecules themselves, if not vaporized, are also hydrogen-bonded to each other. Water bound to clay surfaces has a higher degree of ordering and may be compared to ice crystals, and is immobilized. Ice has a more open structure, and consequently a lower density than liquid water. Therefore, the specific volume of bound water on clays should be higher than that of free pore water (Low, 1976). With increasing overburden and decreasing porosity, pore diameters become smaller and smaller. The surface to volume ratio of the pores is changed and free water is expelled preferentially. The ratio of free water to bound water is thus continuously reduced along with a decrease in bulk water.

At present, little is known about the state of water in source rock-type sediments at depth. Several monolayers of water may be firmly attached to capillary walls depending on pressure and temperature conditions. Clay mineralogists consider three to five monolayers (or more) of adsorbed water under surface conditions to be a reasonable assumption. With increasing temperature and pressure, the network of bound ice-like water should collapse. It is expected, however, that about two monolayers of water should be left in direct proximity to clay minerals even at depths (Low, 1976) where petroleum is generated. Dickey (1975) pointed out important consequences of the existence of structured pore water on primary oil migration in source rocks in an advanced stage of compaction. If a flat pore between clay particles is assumed to be about 30 Å in height, two monolayers of water (about 6 Å) on each side would leave little mobile water. Such pore widths are compatible with pseudo-pore openings observed in shales at about 3000 m depth.

2.3.2 Changing Chemical Composition of Petroleum Compounds

In Chapters II.4 and II.5, we discussed the composition and structure of kerogen, and its evolution with increasing depth of burial. As a consequence of this evolution under increasing temperature and pressure, a variety of petroleum compounds, i.e., bitumen with differing structures and chemical and physical properties, is generated. According to the state of evolution of the kerogen, different types of petroleum compounds in different absolute and relative quantities are available for primary migration. Furthermore, all these different compounds can be transported by migration processes, as documented by the great variety of crule oils that contain the same types of compounds as found in bitumen in situ.

Bitumen to organic carbon ratios in shallow sediments with potential source rock characteristics are around 30 to 40 mg per g. With increasing depth this ratio reaches maxima at about 150 to 180 mg per g or more. Thereafter, the ratios decrease as the principal phase of oil formation is passed. Simultaneously, throughout the phases of oil formation, the composition of the bitumen changes. The amount of saturated hydrocarbons increases and to a lesser degree so does the amount of aromatic hydrocarbons. Heterocompounds (N, S, O compounds) decrease in concentration with maturation. Low molecular weight hydrocarbons are preferentially generated with increasing depth. Beyond the principal phase of oil formation, when cracking becomes the most prominent transformation reaction, mainly gas is generated. These changes result in a decrease in average molecular weight and diameter of bitumen molecules with increasing depth and maturation. The average polarity of the molecules formed also decreases with depth and, hence, their water solubility and adsorption behavior is changed. The amounts of low boiling point aromatics increase, and these are more water-soluble than other hydrocarbons of comparable molecular weight. These differences have to be considered when discussing the principal modes of primary migration with increasing depth.

It is also important that the kerogen itself changes, becoming more aromatic with depth and more ordered, and having less and less heteroatomic functional groups.

2.4 Possible Modes of Primary Migration

The type of petroleum compounds that are transported during migration through the narrow pores and capillaries of water saturated source rocks range in molecular weight from methane (CH_4) (mol. wt. = 16) to such heavy compounds such as asphaltenes (mol. wt. up to 5000 and more). Consequently, at normal pressure and temperature, these compounds may be gaseous, liquid or solid. The effective molecular diameter of petroleum compounds range from 3.8 Å of CH_4 to about 50–100 Å in asphaltenes. A number of relevant molecular diameters are listed in Table III.2.1.

The transportation of petroleum constituents can occur in different ways, depending on the state of distribution in the source bed environment. From a theoretical point of view, there can be separate oil or gas phases, individual oil droplets or gas bubbles, colloidal and micellar solutions, or true molecular solutions. Each of these possible modes of transportation is favored by particular physical and chemical conditions within a source bed. The transfer of mass may occur either by bulk flow or by diffusion. Transport of petroleum constituents in any one of the above modes requires some kind of energy. The two most common driving forces present in source rocks to cause bulk flow are temperature and pressure. Thus the movement of petroleum compounds in bulk flow is either directly or indirectly caused by the presence of pressure gradients and probably to a lesser extent by temperature gradients. In addition to bulk flow, there may be diffusion. The flux of petroleum constituents by diffusion is proportional to their

Table III.2.1. Approximate, effective molecular diameters of selected petroleum compounds and some reference molecules in Å

Molecule	Effective diameter in Å	Molecule	Effective diameter in Å
He	2.0	CH_4	3.8
H_2	2.3	benzene	4.7
Ar	2.9	n-alkanes	4.8
H_2O	3.2	cyclo-hexane	5.4
CO_2	3.3	complex ring structures	10– 30
N_2	3.4	asphaltene molecules	50–100

1 nm = 10^{-9} m = 10 Å; see also Figure III.2.8.

concentration gradient. The direction of net transport due to diffusion is from higher to lower concentrations or from regions of higher chemical potential to regions of lower chemical potential. For example, methane in the sub-surface tends to diffuse from warm regions to cooler regions. Transport by diffusion is also controlled by the molecular size of the diffusing molecule, the smaller the molecule the easier its diffusion. Therefore the importance of transport by diffusion of hydrocarbon compounds in the subsurface is practically restricted to gaseous hydrocarbons and other low molecular species with less than 10 carbon atoms.

Two factors are very important for the transportation of petroleum and gas through the pore system of sub-surface rocks. These are whether the fluid system is monophasic or polyphasic, and whether the pore walls are oil-wet or water-wet. Whenever there are two immiscible fluids or a fluid and a gas, the boundary between these different phases have properties unlike either fluid. This boundary is called an interface and the force which acts on the contact surface, is known as interfacial tension. It is the interfacial tension between immiscible phases that largely determines the flow behavior of a polyphasic fluid system. The other important parameter controlling fluid movement is the wettability of the surrounding pore walls with respect to the different fluid phases.

Among the possible modes of primary migration there are two major categories: one which requires the active movement of pore waters, such as globules and bubbles driven by water, micellar or molecular solution; the other modes of migration can take place independently of pore water flow, such as diffusion and hydrocarbon phase movement.

2.4.1 Hydrocarbon Globules or Bubbles

The transportation of oil and gas in form of individual globules or bubbles through the narrow, water-saturated pores of compacted source rocks involves the phenomenon of capillary pressures because there are two immiscible phases and an interface. As long as globules or bubbles are of the same size or smaller than the pore diameters, there is no restriction to movement. If effective pore openings are smaller than oil or gas particles, capillary pressures must be overcome before globules or bubbles can move.

Before a globule or bubble can be forced through a smaller pore diameter, it has to be distorted. The interfacial tension between the two fluid phases, however, resists this distortion. The critical entry pressure or capillary pressure to force the bubbles or globules of the nonwetting fluid phase through the pores must be overcome. Capillary pressure is higher for small pore diameters. At pore diameters below 100 Å (0.01 μm) more than 240,000,000 dyne cm^{-2} (240 bar) would theoretically be required to move a spherical oil globule through a pore, assuming an interfacial tension between the oil and pore water of 30 dyne cm^{-1} (Hobson, 1973). However, due to the strain anisotropy in compacting shales, the oil particles should be elongated and shaped more like a spindle than a sphere. Thus, the pressure differential needed to move such a particle should be lower than calculated above.

High pressure differentials could theoretically be reached locally in source beds during advanced catagenesis and metagenesis at great depths of burial. The generation of low molecular weight hydrocarbons from the polymeric kerogen causes a great increase in molar volumes, and high pressure centers or pockets are created inside source rocks. Furthermore, a large part of the pore space is probably occupied by immobile water, fixed on mineral surfaces and the actual pore space is considerably reduced, leaving only a small volume in which generated liquid or gaseous hydrocarbons can expand. From this point of view, high pressure differentials as discussed above, could be reached. Beyond a certain pressure, however the mechanical strength of the rocks is surpassed, and fractures develop which relieve the pressure and induce migration.

It is concluded that the movement of oil globules in a water phase contributes little to primary migration of petroleum in dense-compacted source rocks without rock fracturing. Because gas has a somewhat greater capillary displacement pressure than oil, it is even more unlikely that the transport of gas in the form of bubbles plays a major role during primary migration.

2.4.2 Colloidal and Micellar Solutions

The transport of petroleum compounds through the network of pores in source rocks as colloidal and micellar solutions has been suggested by Baker (1959, 1967) and Meinschein (1959). More recently, Cordell (1972, 1973) re-evaluated colloidal solutions as a possible mechanism for primary migration. The colloidal size range is generally considered to extend from about 10 to 10,000 Å. Colloids may consist of small particles having the same individual structure or they may be formed by aggregates of smaller molecules of different types. In addition to this, single high molecular weight molecules (e. g., polymers) may be large enough to fall in the colloid size range. Colloidal particles can normally be removed from solutions by a membrane filtration process.

Polar organic molecules, having hydrophobic and hydrophilic parts, may form ordered aggregates called micelles. These micelles may contain 100 or more molecules with the hydrophobic parts on the inside and the hydrophilic parts on the outside. Crude oils and sediments are known to contain polar surface active molecules such as naphthenic acids. Louis (1967) reported more than 1% naphthenic acids in some crude oils of the USSR. Seifert and Howells (1969) reported up to 2.5% carboxylic acids in a California crude oil, being highly interfacially active in the presence of aqueous alkali. Some of these surface active compounds are definitely of primary origin (Seifert, 1975), but it is suspected that nonspecific naphthenic acids may be produced by oxidation of reservoired petroleum.

The principal reason for advocating micellar and colloidal hydrocarbon solutions as a primary migration mechanism is that it seems to be the only way to solubilize practically water-insoluble petroleum hydrocarbons in aqueous pore solutions at relatively low temperatures. Baker (1962, 1967) proposed solubilization by soap micelles as the main mechanism for transport of petroleum compounds within reservoirs and traps. According to Baker (1962), the distribution of hydrocarbons in petroleum reflects variations in the kind and size of the micelles in which hydrocarbons selectively dissolve. Ionic micelles suitable to solubilize petroleum hydrocarbons would have median diameters of about 60 Å, while neutral micelles would be about 5000 Å in diameter. A direct comparison between the postulated size of crude oil solubilizing micelles and average pore diameters in shales shows, however, that this kind of primary migration would definitely be limited down to a depth of about 2000 m (Fig. III.2.8) During this depth and time interval, surface-active polar compounds are available in source rocks in greater quantity. In calcareous source beds, argillaceous carbonates, marls or even calcareous shales, where larger pores possibly could be preserved to greater depth, there would be another severe chemical hindrance for micelle migration. Many organic acids, which are most important surface active compounds for natural hydrocarbon solubilization, have a tendency to form insoluble salts with divalent cations, such as Ca^{2+} and Mg^{2+}. Hence, in aqueous pore solutions rich in Ca^{2+} and Mg^{2+}, possibly existing micelles would be broken, and the organic acids would be precipitated as insoluble calcium or magnesium salts.

In addition to the physical barrier of very small pore diameters in shales and in addition to the chemical hindrance, there may be yet another obstacle for micelle migration. There is probably an electrical repulsion of negatively charged ionic micelles by negatively charged surfaces of clay minerals (Cordell, 1973). Finally, it should be pointed out that many times more surface active compounds (50- to 100-fold), that have not been observed, are needed in order to solubilize an equal amount of hydrocarbons. Furthermore, polar surface active compounds are preferentially adsorbed on clay minerals.

All in all, solubilization and transport of hydrocarbons in colloidal and micellar form does not seem to be a very likely process during primary migration. If, nevertheless, it were of importance, then only in relatively shallow source rocks, during the initial phase of petroleum formation. If micellar processes were responsible for petroleum accumulation, oils should also be enriched in N-, S-, O-polar compounds, which are more likely to form micellar solutions. Crude oils, however, are mostly depleted in these polar compounds as compared to source rock bitumen. Only certain shallow oils are rich in heavy and polar compounds.

2.4.3 Molecular Solution

Molecular solution is another possible transporting mechanism of petroleum compounds during primary migration. The solubility of the very different petroleum compounds of low and high molecular weight, of saturated acyclic or cyclic hydrocarbons, of aromatics and heterocompounds is very different in water. Table III.2.2 lists aqueous solubilities of some selected petroleum compounds. In general low molecular weight hydrocarbons are more water-soluble than high molecular weight hydrocarbons. Low boiling aromatics such as benzene (solubility: 1,740 ppm) and its simple derivative toluene (solubility: 554 ppm) are among the most soluble common petroleum hydrocarbons. Polar heterocompounds, such as organic acids or alcohols are more water-soluble than the corresponding hydrocarbon with the same carbon number. Yet judging from

2.4 Possible Modes of Primary Migration

the composition of rock extracts and crude oils, these compounds are generally mobilized in proportions other than their solubilities in water would suggest. Those compounds, like saturated hydrocarbons which are among the least water-soluble, are generally the most strongly enriched in crude oils, as compared to the extracts obtained from their source sediments. This is especially true for the most insoluble series of compounds on an equal-carbon-number basis, the n-alkanes, which are notoriously enriched in petroleum which has not been biodegraded.

Whereas the above argument about hydrocarbon solubilities was referring to pure compounds in distilled water, McAuliffe (1980) equilibrated whole crude oils with sea water, and other saline waters. He found the compositions of hydrocarbons dissolved in these waters to be again vastly different from those of the crude oils.

The solubility in aqueous solutions of various saturated and unsaturated hydrocarbons up to nine carbon atoms was also studied by McAuliffe (1966) and that of long-chain n-alkanes by Peake and Hodgson (1966) and Franks (1966). In these investigations it was found that a break in the solubility of n-alkanes occurs around carbon number 10. In 1980 McAuliffe confirmed and extended these findings (Fig. III.2.9).

The solubility decreases with increasing carbon number, but the long-chain hydrocarbon molecules are relatively more soluble than would be expected from an extrapolation of the solubilities of the lower members of the series. It has been suggested that this unexpected solubility of the higher members may be due to micelle formation (Peake and Hodgson, 1966). The solubilities measured ranged from about 100 to less than 0.1 ppm in the C_4–C_{10} range and below 0.01 ppm in the C_{12}–C_{30} range (McAuliffe, 1980). In case of the higher n-alkanes, millipore filtration (50 nm = 500 Å pore size) reduced the n-alkane content as much as 97% of the accomodated n-alkanes (Peake and Hodgson, 1966).

A comprehensive treatment of the molecular solution of hydrocarbons during primary migration, was presented by Price (1973, 1976). An increasing solubility of petroleum and petroleum compounds in distilled water occurs with increasing temperature. The solubilities increase gradually to approximately 100°C, where a more drastic increase is observed (Fig. III.2.10). The solubilities of oils are in the range of several up to 100 ppm below 100°C, but increase up to 100 ppm or more between 100 and 200°C. Two different solubility rates were found in the temperature range from 25 to 180°C. The more insoluble higher boiling point compounds increase their solubility with increasing temperature at a far greater rate than the more soluble lower boiling compounds.

Valuable data were also presented (Price, 1973) on the solubility differences between various classes of petroleum compounds at elevated temperature. It was found, as expected, that the presence of heteroatoms (N, S, O) on a hydrocarbon molecule drastically increases its solubility in water. The solubility of aromatic nuclei is much more increased with rising water temperature than the solubility of cyclo alkane nuclei, which are only slightly more soluble than iso-alkane or n-alkane structures. Salinity increases were found to considerably decrease the aqueous solubilities of hydrocarbons. Price (1973) concluded from his findings that only a small percentage of the world's petroleum could have undergone primary migration by molecular solution. This conclusion was reached mainly by

Table III.2.2. Aqueous solubilities of selected petroleum compounds at 25°C in ppm (wt./wt.)

Compound	Price (1973, 1976)	McAuliffe (1966)
	n-Paraffins	
Methane		24.4 ± 1
Ethane		60.4 ± 1.3
Propane		62.4 ± 2.1
n-Butane		61.4 ± 2.1
n-Pentane	39.5 ± 0.6	38.5 ± 2.0
n-Hexane	9.47 ± 0.20	9.5 ± 1.2
n-Heptane	2.24 ± 0.04	2.93 ± 0.20
n-Octane	0.431 ± 0.012	0.66 ± 0.06
n-Nonane	0.122 ± 0.007	0.220 ± 0.021
	Isoparaffins	
2,3-Dimethylbutane	19.1 ± 0.2	
2,2-Dimethylbutane	13.0 ± 0.2	
2,4-Dimethylpentane	4.41 ± 0.05	4.06 ± 0.29
2,3-Dimethylpentane	5.25 ± 0.02	
2,2,4-Trimethylpentane	1.14 ± 0.02	2.2 ± 0.12
Isobutane		48.9 ± 2.1
Isopentane	48.0 ± 1.0	47.8 ± 1.6
2-Methylhexane	2.54 ± 0.02	
3-Methylheptane	0.792 ± 0.028	
4-Methyloctane	0.115 ± 0.011	
	Bicycloparaffin	
Bicyclo[4.4.0] decane	0.889 ± 0.031	
	Napthoaromatic	
Indan	88.9 ± 2.7	
	Cycloparaffins	
Cyclopentane	160.0 ± 2.0	156.0 ± 9.0
Methylcyclopentane	41.8 ± 1.0	42.0 ± 1.6
n-Propylcyclopentane	2.04 ± 0.10	
1,1,3-Trimenthyl-cyclopentane	3.73 ± 0.17	
Cyclohexane	66.5 ± 0.8	55.0 ± 2.3
Methylcyclohexane	16.0 ± 0.2	14.0 ± 1.2
1,trans-4-Dimethyl-cyclohexane	3.84 ± 0.17	
1,1,3-Trimethyl-cyclohexane	1,77 ± 0.05	
	Aromatics	
Benzene	1740.0 ± 17.0	1780 ± 45
Toluene	554.0 ± 15.0	515 ± 17
m-Xylene	134.0 ± 2.0	
o-Xylene	167.0 ± 4.0	175 ± 8
p-Xylene	157.0 ± 1.0	
1,2,4-Trimethyl-benzene	51.9 ± 1.2	57 ± 4

2.4 Possible Modes of Primary Migration

Table III.2.2 (continued)

Compound	Price (1973, 1976)	McAuliffe (1966)
1,2,4,5-Tetramethyl-benzene	3.48 ± 0.28	
Ethylbenzene	131.0 ± 1.4	152 ± 8
Isopropylbenzene	48.3 ± 1.2	50 ± 5
Isobutylbenzene	10.1 ± 0.4	
	Sulfur and nitrogen compounds	
Thiophene	3015 ± 34	
2-Ethylthiophene	292 ± 7	
2,7-Dimethylquinoline	1795 ± 127	
Indole	3558 ± 171	
Indoline	10,800 ± 700	

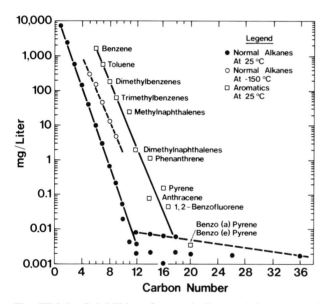

Fig. III.2.9. Solubilities of normal alkanes and aromatics in water. (After McAuliffe, 1980)

comparing n-alkane distribution patterns of different crude oils with the solubility pattern of the n-alkane series (Fig. III.2.9), as investigated by McAuliffe (1966, 1980), Franks (1966) and Price (1973). A break in the slope when plotting n-alkane abundance in petroleum versus carbon number (Fig. III.2.11) was taken as an indication for transportation by molecular solution. The position of the break with respect to carbon number should be a consequence of specific temperature and pressure conditions. Irregular n-alkane distribution patterns are

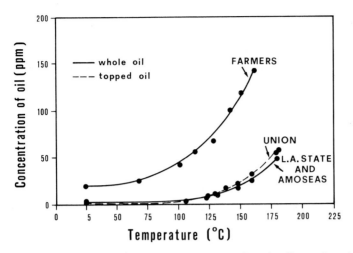

Fig. III.2.10. Solubilities of whole and topped crude oils as a function of increasing water temperature. (After Price, 1973, 1976)

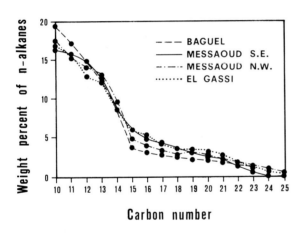

Fig. III.2.11. Distribution of n-alkanes in four different crude oils. (After Price, 1973, 1976)

considered evidence of a flush-type migration, which would have taken place at lower temperatures (below 120°C) from less mature source beds.

Bray and Foster (1980) have suggested that carbon dioxide and hydrocarbon gas which are generated together in source rocks are dissolved in pore water concurrently with oil generation. These dissolved gases would mobilize the liquid hydrocarbons so they could leave the source rock with any water expelled during compaction. However, these experiments can also be interpreted in the light gained from results of enhanced oil recovery with the help of carbon dioxide injection. In this case no increased miscibility of hydrocarbons in waters is observed. The gains in hydrocarbon production are achieved due to solubilization of carbon dioxide in oil with swelling of the hydrocarbon phase and decrease of viscosity and interfacial tension. This in turn increases the relative permeability

2.4 Possible Modes of Primary Migration

for oil and thus favors the movement of a hydrocarbon phase in an aqueous environment.

Kvenvolden and Claypool (1980) presented an example where carbon dioxide from a submarine seep in Norton Sound, Alaska, carried a minor amount of gas- and gasoline-range hydrocarbons. (The CO_2 in this case was supposed to originate from basement rocks and was believed to be of geothermal origin.) Because individual cyclic and branched-chain molecules in the gasoline range were much more abundant than straight-chain hydrocarbons Kvenvolden and Claypool deduced that the hydrocarbon mixture was immature. Thus carbon dioxide may have acted as an extraction agent when migrating through the sedimentary column.

The influence of pressure on hydrocarbon solubility under geological conditions has not been studied in detail except for methane (Bonham, 1978). It can be deduced, however, that there should be an increase in solubility among the lighter gaseous hydrocarbons with increasing pressure. The subsequent release of these hydrocarbons from solution could be facilitated by a pressure drop between source rocks and reservoir rocks.

It does not seem unreasonable to assume that molecular solution in pore waters is a potential mechanism for primary migration. This mechanism, however, seems to function best for the lighter hydrocarbon fractions, because they are the most water-soluble. The different solubility behavior of n-alkanes with more than 10 C-atoms in aqueous solutions (Fig. III.2.9) seems to be associated with the formation of n-alkane aggregates, not unlike ordered micelles. Such aggregates, however, can be filtered out by the narrow pores (100 Å and smaller) prevailing in shales below 2000 m. Finally, it has to be remembered that in a typical medium crude oil, as given by Bestougeff (1967), and as found similarly in many parts of the world, the distribution of hydrocarbon classes (Fig. III.2.12) is not at all as

Table III.2.3. Comparison in composition of light hydrocarbons between crude oil and corresponding source rock extract. (After Martin et al., 1963)

Hydrocarbon	Crude oil (vol.-%)	Source rock (vol.-%)
Pentane, n + iso	14.4	2.9
Hexane, n + iso	18.4	3.6
Heptane, n + iso	25.9	5.6
Cyclopentane	0.9	0.2
Methylcyclopentane	6.7	1.5
Ethylcyclopentane	0.6	0.7
Isomers of dimethylcyclopentane	15.3	4.4
Cyclohexane	3.8	1.6
Methylcyclohexane	12.6	11.5
Benzene	0.2	10
Toluene	1.2	58
	100	100

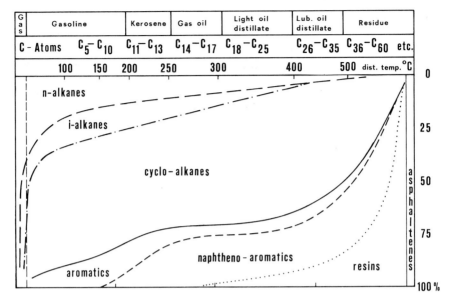

Fig. III.2.12. Distribution of various petroleum compound classes in a typical medium crude oil. (After Bestougeff, 1967)

predicted by the aqueous solubility distribution of the compounds present. According to the solubilities of hydrocarbon classes in water, there should be many more aromatics and naphtheno-aromatics and much fewer iso-alkanes and n-alkanes in crude-oil. Martin et al. (1963) presented an excellent example on the discrepancy between the actual composition of light hydrocarbons in a crude oil compared to its corresponding source rock (Table III.2.3). Contrary to the expected enrichment in water-soluble hydrocarbons such as benzene and toluene (see also Table III.2.2), the crude oil is depleted in these compounds. Transportation in aqueous solution, therefore, does not appear to be an important mechanism in primary migration of petroleum.

The foregoing discussion on molecular solution focused mainly on the concept that hydrocarbons dissolved in pore waters would be transported when a given volume of hydrocarbon containing water would be actively moving by bulk flow. Another possibility of transport of hydrocarbons in molecular solution would be diffusion.

2.4.4 Diffusion-Controlled Processes

In diffusion processes the aqueous phase would be stationary, i.e., there is no need for flow of water, whereas dissolved hydrocarbon molecules would move independently from places of higher concentration (high chemical potential) to places of lower concentration (low chemical potential). Thus, in general terms, diffusion would be a process that disseminates and destroys gas accumulations rather than contributing to accumulation processes.

2.4 Possible Modes of Primary Migration 319

Table III.2.4. Effective diffusion coefficients D ($cm^2 s^{-1}$) for diffusion of n-alkanes through the water-saturated pore space of source rock-type shales, based on in-situ measurements in shallow core holes. (After Leythaeuser et al., 1982)

				n-Alkanes					
	CH_4	C_2H_6	C_3H_8	iso-C_4H_{10}	n-C_4H_{10}	n-C_5H_{12}	n-C_6H_{14}	n-C_7H_{16}	n-$C_{10}H_{22}$
D	2.12×10^{-6}	1.11×10^{-6}	5.77×10^{-7}	3.75×10^{-7}	3.01×10^{-7}	1.57×10^{-7}	8.20×10^{-8}	4.31×10^{-8}	6.08×10^{-9}

In special geological settings, however, Leythaeuser et al. (1980, 1982) have shown that diffusion of dissolved hydrocarbon molecules in fact represents an effective process for primary migration of gas but not for oil. Based on newly determined effective diffusion coefficients for low molecular weight hydrocarbons (C_1–C_{10}), which are listed in Table III.2.4, a model was set up to calculate the amount of hydrocarbons escaping by diffusion from source rocks with geologic time. The role of diffusion was conceptualized as that of an initial process for hydrocarbon transportation within the source rock unit itself. Transportation distances are thought to be short and always following local concentration gradients. As shown schematically in Figure III.2.13, such conditions exist within the source rock near the contact to over- or underlying porous carrier rocks, around fractures or faults, or in proximity to interbedded siltstones.

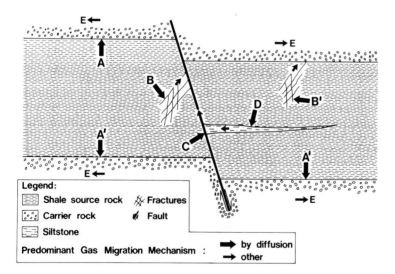

Fig. III.2.13. Schematic illustration of the most likely sites and direction of light hydrocarbon diffusion in shale source rocks:
A: toward contact with carrier rocks,
B: toward fracture zone,
C: toward fault,
D: toward interbedded siltstone lens.
Within carrier rock itself a transport mechanism other than diffusion (E) prevails. (Modified after Leythaeuser et al., 1982)

These model calculations have clearly shown that molecular diffusion by itself can account for transportation of such quantities of gas with geologic time, which are in the order of commercial-size fields. For example, it was calculated that a 1000 km^2 × 200 m volume of a high-mature gas-prone Jurassic shale source rock unit in Western Canada has yielded by diffusive transport in 540,000 years the cumulative amount of 10^9 kg (1.5×10^9 STP m^3) of methane. The same study revealed that, due to an exponential decrease of the diffusion coefficient value with increasing molecular carbon number, compositional fractionation effects occur during initial periods of diffusion of hydrocarbons through source rocks. This results in a change in reservoir gas composition with geologic time. In particular, it was shown that the composition of the gas which accumulates in a reservoir by transportation from a source rock, where diffusion was the initial process of primary migration, is controlled by three principal factors: The concentration of initially generated hydrocarbons in the source rock, their relative differences in diffusion rates, and the source rock thickness.

2.4.5 Hydrocarbon Phase Migration

When considering primary migration by molecular solution in an aqueous pore fluid or by a separate oil or gas phase, problems of pore diameters and absolute and relative rock permeabilities cannot be neglected. The limiting factors and restrictions, i.e., interfacial tension, capillary pressures and wettability, for the movement of large molecules, molecular aggregates, globules and bubbles through fine-grained source rocks with narrow pore throats have been reviewed. The main obstacle, however, which is high capillary pressures, would not occur if a continuous organic phase could be assumed to exist in the source rock at the stage of oil expulsion.

The possible movement of petroleum, as a separate, more or less continuous oil phase during primary migration, has been discussed by Dickey (1975). He pointed out that the saturation of oil in terms of total pore fluid in mature source rocks may become very large, as high as 50%, if it is assumed that most of the pore water is structured and effectively solid. The concept of relative permeability and equilibrium saturation with respect to either fluid becomes unimportant when the water is largely fixed to the pore walls and cannot be considered as a movable phase. In addition, if some of the interior surfaces are oil-wet rather than water-wet, oil can move as a continuous phase at even lower oil saturations. Fine-grained rocks rich in kerogen and with part of their clay particles coated with organic matter should be expected to contain some oil-wettable interior surfaces. Therefore, the equilibrium saturation in source rock-type shales, below which no oil can flow, may be lower than would be expected under known reservoir conditions. Sub-surface conditions that would permit primary migration of petroleum as a continuous oil phase, should be high bitumen concentrations and little mobile water in a potential source rock. Hence, deeply buried source beds at the maximum phase of oil formation and at a relatively high degree of compaction, offer the greatest opportunity for primary migration as a separate continuous oil

2.4 Possible Modes of Primary Migration

phase. The driving force for fluid movement under such conditions would be a pressure gradient.

Meissner (1978) following this line of thought, presented an example for oil phase migration from the Williston basin. He found when studying the organic-rich Bakken shale in the Williston basin that at peak hydrocarbon generation stage a continuous oil phase was in existence in that shale. This was inferred from the fact that water was never recovered during drill stem testing of the mature Bakken shale. Furthermore, the mature shale was overpressured and electrical resistivity was extremely high.

A similar alternative hypothesis visualizes kerogen as forming a three-dimensional organic network in the rock, which is wettable with oil and functions as the transport avenue (McAuliffe, 1980). Effects of this kind, combined with pressure-generating processes inside a source rock, would certainly greatly facilitate primary migration as a separate oil phase.

A comparable effect would be achieved in any kind of source rock by microfracturing due to overpressuring of the enclosed pore fluids. The microfractures would then function as conductors. The basic idea, originally introduced by Snarsky (1962) and Tissot and Pelet (1971), is that a large increase in pore pressure may be sufficient either to overcome the capillary pressure or even to exceed the mechanical strength of the rock and induce microcracking. The main causes for pressure build-up are (Durand, 1982):

- thermal expansion of water (Barker, 1972),
- specific volume increase of organic matter by generation of gaseous and liquid hydrocarbons from kerogen (Snarsky 1962; Tissot and Pelet, 1971),
- partial transfer of the geostatic stress field from the solid rock matrix to the enclosed pore fluids, resulting in an overall increase in pore pressures. This transfer is the result of conversion of part of the solid kerogen to liquid or gaseous compounds (du Rouchet, 1981).

The formation of microfractures by local centers of high fluid or gas pressure is restricted to relatively deeply buried, compacted, low permeability rocks such as shales or tight carbonates. Thus, this process is of similar importance in source rocks with different lithologies if they contain sufficient organic matter to generate liquid or gaseous low molecular weight compounds and/or are brought to sufficient temperature for thermal pressuring of water. They should also be impermeable enough to allow pressure build-up. This mechanism, unlike most others, does not necessarily demand a water drive for hydrocarbon movement.

According to Snarsky (1962), the mechanical strength of a rock is exceeded and fracturing occurs if the internal fluid pressure in a rock or in local pressure centers inside the pores is greater than by a factor of 1.42 to 2.40 over the hydrostatic pressure in the immediate surroundings. Internal pore pressures that exceed the normal hydrostatic and even the petrostatic pressure are possible, whenever massive generation of gas and oil from kerogen takes place. This process may also help to form abnormally high-pressured shales as discussed previously. Snarsky (1962) calculated that at a depth level of 3100 m, the internal fluid pressure in a rock may rise to at least 750 bar, which is higher by a factor of 2.5 than the hydrostatic pressure and higher by a factor of 1.1 than the petrostatic pressure at that depth. Tissot and Pelet (1971) report internal gas pressures of about 540 bar in a block of shale that was kept under a "petrostatic pressure" of 440 bar during an experiment. Microfracturing and a subsequent pressure release resulted in this experimental study.

Recently Ungerer et al. (1983) have calculated the overall volume expansion of organic matter during hydrocarbon genesis and also evaluated its importance for primary migration. They concluded that volume expansion of organic matter alone probably plays a minor role in the expulsion of oil. The volume expansion was estimated not to exceed 15% of the initial volume during catagenesis. Volume expansion may, however, have an important influence at the end of catagenesis and during metagenesis when dry gas is formed. Under these circumstances microfracturing of the rock matrix can occur.

It is clear that pressure build-up, microfracturing, subsequent pressure release, expansion of fluid or gas, and finally transport, is a discontinuous process. The process must be repeated many times in the source rock unit during geological time in order to accomplish meaningful gas or oil movement. After a microfracture has been opened, the enclosed fluid or gas expands and the fracture is immediately healed by the overburden pressure. A new pressure then builds up and the process begins again (Fig. III.2.14).

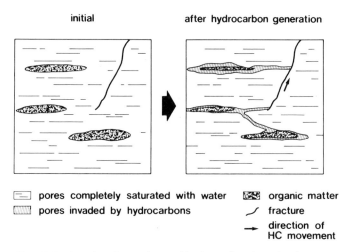

Fig. III.2.14. Microfracturing and hydrocarbon invasion in a source rock during hydrocarbon generation from kerogen. (After Ungerer et al., 1983)

A pressure build-up inside the pore network of a source rock must not necessarily cause microfracturing of the rock. If capillary pressures are not too high either because pore throats are not too narrow, or because there is a continuous organic phase and/or a three-dimensional network of kerogen, then gas and oil can slowly be driven out of the source rocks without rock fracturing. In such a case the pressure-driven movement of gas and oil would be a continuous process. From the foregoing considerations it is logical to accept pressure-driven hydrocarbon phase movement without or with microfracturing as an important mechanism of primary migration. Transportation of petroleum compounds under these conditions may take place in continuous gas or oil phases and therefore chromatographic effects have to be expected. We will see in the next chapter that geological and geochemical constraints practically single out a pressure-driven hydrocarbon phase movement as the dominating primary migration mechanism.

Summary and Conclusion

Petroleum and gas accumulations are found between the surface and depth levels of about 6000 to 7000 m and deeper. Thus, petroleum and gas must move through pores saturated with water, at fluid pressures ranging between 100 bar and 1400 bar and at temperatures from about 50°C to about 250°C.

Compaction in sediments results in an increase in bulk density and loss of porosity with increasing depth, temperature and time. At shallow depths there is rapid loss in porosity, and with increasing overburden pressure the rate of loss in porosity diminishes greatly. Because compaction causes fluid flow through sedimentary rocks it was commonly considered as an important factor in petroleum migration. This kind of fluid flow, however, would be restricted in carbonate-type rocks due to early lithification. During compaction there is also a marked regular decrease in pore diameters. At greater depths (below 3000 m) in detrital sediments they reach values around 10 to 25 Å, which is in the range of the dimensions of the molecules of certain petroleum constituents. Adsorbed, nonmobile water may make these pores even smaller.

From a theoretical point of view, primary migration can occur as discrete, more or less continuous oil or gas phases, individual oil droplets or gas bubbles, colloidal and micellar solutions, or true molecular solutions.

Physicochemical, geochemical and geological considerations make individual droplets and bubbles and colloidal and micellar solutions highly unlikely as an effective means of transport during primary migration. Molecular solution in pore waters does not seem unreasonable as potential mechanism for primary migration, especially for the lighter, most soluble hydrocarbon fractions.

Nevertheless, it does not appear to be an important mechanism in primary migration of oil. However, there are indications that primary migration in molecular solutions plays an important role for gas in the form of diffusion-controlled processes.

The most important form of primary migration during the main phases of oil (and gas) formation seems to be a pressure-driven, discrete hydrocarbon phase movement. The generation of low molecular weight hydrocarbons from polymeric kerogen causes an increase in specific volumes. Furthermore, high fluid pressures may by caused by a combination of the effects of thermal expansion of pore water, rapid burial, and partial transfer of the geostatic stress field from the solid rock matrix to the enclosed pore fluids. In this way pressure centers are created in a source rock that cause a more or less continuous pressure-driven hydrocarbon phase movement. In relatively impermeable source rocks, these pressure centers may even induce microscale rock fracturing. Pressure build-up, microfracturing, subsequent pressure release, expansion of fluid or gas, and finally transport are a discontinuous process. This process must be repeated many times in source rocks in order to accomplish meaningful gas or oil movement.

A pressure-driven, discrete hydrocarbon phase movement can function in all kinds of source rocks, independently of lithology. It is considered as the dominating primary migration mechanism.

Chapter 3
Geological and Geochemical Aspects of Primary Migration

Observed facts with respect to primary migration of petroleum can either be related to time and depth of migration, or to chemical differences, or similarities between the composition of source rock bitumen and related crude oils.

3.1 Time and Depth of Primary Migration

Petroleum geologists have long distinguished between early and late primary migration mainly on the basis of the availability of compaction waters. Early migration is considered to take place during the first 1500 m of subsidence, when sediments lose much of their porosity, yielding a large volume of water.

The question of *early migration* of oil before the onset of the main phase of hydrocarbon formation from kerogen, is vital to petroleum exploration, especially with respect to the vast offshore areas with young Tertiary and Pleistocene sediments. There is little question that the majority of the world petroleum reserves were formed in and migrated out of source rocks at temperatures of at least 50 to 70 °C in a minimum depth range of 1000 to 1500 m. In shallow sediments, however, between the surface to about 1500 m burial depth, there are only minor amounts of hydrocarbons and other petroleum-related compounds present, most of which are directly inherited from organisms, or produced as a result of early diagenesis. A major exception is very early methane production by microorganisms. The question is: can oil or gas be formed in large quantities in young, immature source sediments in the first 1500 m of burial when early sediment dewatering is most effective?

Kidwell and Hunt (1958) reported an oil-like concentration of hydrocarbons in sediments less than 10,000 years old from the delta of the Orinoco River, Venezuela. Oil and free gas were found in a lenticular sand body, at a depth of about 35 m. The hydrocarbon concentration, consisting predominantly of aromatics, was around 160 ppm. Gray clay beds enclosing the sand lens, had an average of 55 ppm hydrocarbons. Other interbedded sands were open to the surface and did not contain such high hydrocarbon concentrations. Detailed pressure studies on pore fluids indicated fluid movement from the muds towards the continuous sands. The moving fluids contained less than 16 ppm hydrocarbons. From these findings it was concluded that there was migration of hydrocarbons, either in molecular solution or in an extremely dilute colloidal dispersion and that the hydrocarbons were filtered out of the moving stream of water by capillary action in the lenticular sand. No such filtering occurred in the open sands.

Embryonic, noncommercial hydrocarbon accumulations in deep sea sediments are reported by McIver (1975). Findings from two locations, the Challenger Knoll in the Gulf of Mexico and the Shatsky Rise in the Western Pacific Ocean lend support to the idea that bitumens and petroleum hydrocarbons can begin to migrate very early. Sediments from the Challenger Knoll 136 m below the sea floor and in a water depth of 3572 m contained an immature, somewhat unusual, high-sulfur (5%) crude oil with a high specific gravity and modest volumes of gas and gasoline. Chemically this oil is similar to typical Tertiary oils, produced from cap rock reservoirs at two salt domes in Texas. Another oil seep or oil saturation in sediments from one of the Sigsbee Knolls in the Gulf of Mexico (Gealy and Davies, 1969) is also worthy of mention. When referring to these oilshows in young deep sea sediments in the Gulf of Mexico, it has to be remembered that due to sediment compaction, pore water can move past the salt and upward through the sediments and over the dome, displacing the original interstitial pore fluids (Lehner, 1969; Perry, 1970). The oil saturation reported in some young, deep sea sediments could, therefore, have been derived from deeper, more mature sediments.

The Shatsky Rise hydrocarbons were found in Pleistocene sediments, 4 m below the sea floor in 4282 m of water. Detailed geochemical analyses suggest that it is a stringer of migrated immature bitumen. About one third of the extract consisted of nitrogen–sulfur–oxygen compounds and asphaltenes. The relative enrichment of total bitumen and heavy hydrocarbons over neighboring sediments is greater than in the Orinoco delta embryonic accumulation reported by Kidwell and Hunt (1958).

At several locations in the Gulf of Mexico, the concentrations of the low molecular weight hydrocarbons, methane, ethane and propane were measured in the water column at depths between the surface and 3742 m (Frank et al., 1970). Hydrocarbon concentrations ranged from 1.2×10^{-6} to 125×10^{-5} ml l^{-1} sea water. The mol ratios of methane to ethane plus propane were low, especially at the bottom close to the sediment–water interface where values of about 10–20 $C_1:C_2 + C_3$ were measured. Such low ratios suggest a thermal cracking-derived hydrocarbon source from an actual seep, because bacterial processes typically produce $C_1:C_2 + C_3$ ratios in the thousand.

The oil and gas fields in Nagaoka Plain, Japan, produce mainly from volcanic and pyroclastic rocks enclosed in Miocene through Pleistocene mudstones and siltstones. An unconformity separates the Miocene from the underlying basement. Magara (1968) has made a very detailed study on compaction, porosities, permeabilities and pressure conditions of these sediments and on pore fluid movement. Convincing evidence is presented that hydrocarbon migration caused by differential compaction occurred between enclosing mudstone source beds and the volcanic reservoirs. In the area of the Sekihara and Katagai fields, there was apparently downward migration from the Pliocene Nishiyama mudstone source beds into the volcanic reservoir. In the Sekihara-Katagai area the presumed source beds of the oil are buried to about 1000 to 1500 m. It appears logical to see this hydrocarbon generation in connection with a high heat flow, as suggested by Klemme (1975). In this area, temperature estimates for this burial depth are up to about 75 °C. Temperature and age (Pliocene) reported for the source sediments that they are in the early stage of petroleum generation. The Sekihara and Katagai oils have a specific gravity of about $0.87\ \mathrm{g\ cm^{-3}}$. Thus, there is circumstantial geological and hydrodynamic evidence that there was primary migration of hydrocarbons out of a relatively immature high porosity (30–40%) mudstone, apparently in connection with pore water movement induced by compaction.

Young delta areas are frequently rich in oil occurrences at relatively shallow depth levels (800–1500 m) and it is still disputed whether or not this oil is of shallow origin, or derived from deeper-seated source rocks. However, there seems to be evidence that at least some of these oils, even those found in isolated sand lenses, may be derived from deeper, more mature source beds. This is the case, for instance, in the Mahakam Delta (Durand, 1979), the Niger Delta (Evamy et al., 1978; Ekweozor and Okoye, 1980) and the Gulf Coast (Milliken et al., 1981; Galloway et al., 1982). The migration avenues in such cases are thought to be fractures associated with deep-reaching growth faults (Weber and Daukoru, 1975; Price, 1976) or piercements. Pleistocene production in the area of the Gulf of Mexico has frequently been cited as evidence for early generation and migration. This area, however, is a "diapiric jungle" of shale intrusions and salt piercements associated with extensive faulting and folding (Woodbury et al., 1973), which offer many migration avenues for deeper, more mature source beds. The mere presence of oil in immature reservoirs, especially in structurally complex areas, cannot be considered compelling evidence for its origin in immature source beds.

Primary migration occurring *during the main phase of hydrocarbon generation*, generally at depths between 1500 and 3000 m to 3500 m, is documented by the majority of oil accumulations of Cenozoic, Mesozoic, and Paleozoic age around the world and should be accepted as a fact. The mechanisms of primary migration that prevail during this stage of source rock maturation, are now understood in general terms: it is conceived as a pressure-driven hydrocarbon phase movement.

The term *late primary migration,* as opposed to early primary migration, was originally used to describe migration after the first 1500 m of subsidence, which coincides with the beginning of the main phase of hydrocarbon generation. In the light of the new findings on primary migration mechanisms the term "late" primary migration has almost lost its original meaning, as this stage of migration is probably the most important, if not the only one.

With increasing depth the decrease in porosity and water content of sedimentary rocks is paralleled by a strong decrease in permeability. This is a reason why late primary migration from compacted low permeability source beds was formerly questioned by many petroleum geologists. There are, however, examples where the geological setting suggests that late primary migration from indurated low permeability source beds must have taken place. The giant oil fields of Hassi Messaoud in the eastern part of the Algerian Sahara produces from a Cambrian sandstone reservoir in a large dome-like structure (Fig. II.6.7). The seal is provided by impermeable Triassic shales and evaporites which unconformably overlie the Cambrian sandstone reservoir and Lower Paleozoic shales.

Silurian black shales have been assumed by Balducchi and Pommier (1970) for geological reasons to be the main source for the Hassi Messaoud oil. This assumption was later verified by geochemical source rock–crude oil correlation analyses (Tissot et al., 1975; Welte et al., 1975). During a first cycle of subsidence, the Silurian black shale was overlain by some 1000 to 1500 m of Paleozoic sediments in the areas of Hassi Messaoud. Since both the Silurian source and the Cambrian reservoir have been exposed to the surface during pre-Triassic uplift and erosion, any oil generated during the first, Paleozoic subsidence would have been lost to the surface. Only the oil generated during the second cycles of subsidence, which reached a depth of 3700 m during the Mesozoic, is now found in the Cambrian sandstone reservoir (Tissot et al, 1975).

The late generation and primary migration of gas was more widely accepted amoung petroleum geologists than that of oil. One example worth mentioning, however, is the huge Groningen gas field in the Netherlands. There is abundant geological and geochemical evidence which demonstrates that the gas in the Permian sandstone reservoirs of the Groningen area was generated from Late Carboniferous coal measures, including organic-rich shales and interbedded coal seams, during Mesozoic time, when these coal measures were buried between 4000 and 6000 m (Lutz et al., 1975). At these depth levels, dense, compacted shales and coals are relatively impermeable and moving compaction water cannot be considered to be an effective primary migration mechanism.

Hedberg (1964) cited several other examples of late primary migration. He pointed out that the evidence for late migration lies in cases where the supposed source must have been compacted and consolidated prior to the deposition of reservoir rocks or before fold- or fault-trapping structures were created. Some examples of late primary migration he cited are the huge Quiriquire oil feld of Eastern Venezuela and the San Pedro field of Northern Argentina. In the Quiriquire field, production occurs mainly from the Pliocene Quiriquire Formation, an alluvial fan series of continental origin which was deposited unconformably on truncated, folded beds of Miocene to Cretaceous age. The oil is thought to have been generated in these Tertiary or Cretaceous beds below the unconformity. In the San Pedro field the oil is reservoired in Permo-Carboniferous sandstones on structures which did not form until late Tertiary time. The sources of this oil, according to Reed (1964), are Devonian in age. Unfortunately, in both these cases, there is no detailed geochemical evidence for the source rock–oil correlation.

A special mechanism for primary migration during the main phase of hydrocarbon generation, was seen by some authors in the phenomenon of *clay dehydration*. Clay dehydration is the release of bound water during the alteration of smectite (montmorillonite) to illite, mainly under the influence of temperature. This phenomenon was described and analyzed by Burst (1959), Weaver (1959), Dunoyer de Segonzac (1969, 1970), and Heling and Teichmüller (1974). It was discussed with respect to petroleum generation and migration by Powers (1967) and Burst (1969). The main conclusions of Powers (1967) were that the development of a shale source rock requires a smectite-containing organic mud and its subsequent alteration to illite with deep burial, and that abnormally high fluid pressures may be caused by a volume increase of the waters desorbed from smectite during the change to illite. Burst (1969) expanded on this dehydration phenomenon of swelling clays and pointed out that in the Gulf Coast area there is a coincidence of productive horizons and the depth levels which he interpreted as a second-stage (interlayer water) clay dehydration. He suggested that large-scale water release due to dehydration and the subsequent fluid movement in a source rock-type sediment containing petroleum hydrocarbons may initiate primary migration.

Magara (1975) re-evaluating the data of Powers (1967) and Burst (1969) added new information on clay dehydration, abnormal pressures and petroleum migration. He emphasized that smectite dehydration cannot cause abnormally high pressures. The release of large amounts of bound water would result in a volume decrease, since bound water has a lower density than free water. Furthermore, Magara (1975) stated correctly that smectite dehydration alone, without good drainage, may not be sufficient to initiate primary petroleum migration. The studies of Dunoyer de Segonzac (1970) and Heling and Teichmüller (1974) show that there is a very gradual reorganization in the mixed-layer smectite–illite sequence. In particular, the pure illite stage is not reached even at 4000 m depth and a temperature of 160 °C in the Logbaba well, Cameroon. The same is true at comparably high temperatures in the Rhine Graben. Thus, clay dehydration and the release of additional water is gradual, and does not occur in a narrow depth interval. The kinetics of clay dehydration occur, therefore, in the same time and depth ranges as the phenomenon of compaction. Consequently, the mechanism of clay dehydration does not seem to be more effective for primary hydrocarbon migration during the main phase of petroleum generation than normal compaction.

The presence of expandable clay minerals has frequently been considered to be a prerequisite for an effective source rock which is able to expel oil by primary migration. In the USA, there is a coincidence between oil-producing stratigraphic horizons from Paleozoic through Cenozoic age and the distribution of expandable clays (Weaver, 1960). The presence of expandable clays and the primary migration of oil, however, are not necessarily related. For example, the lower Paleozoic black shales, source beds for the Hassi Messaoud oil in Algeria, do not contain, and almost certainly never contained, expandable clay minerals, because the clay fraction of these shales contains only kaolinite, primary illite and chlorite (Millot, 1964). The coincidence of oil occurrences and expandable clays as observed by Weaver (1960), can also be explained in terms of a better preservation of organic matter due to expandable clays or catalytic effects. Likewise, no expandable clays have been found in any of the source beds in the oil-producing Williston basin of North America (Barker, 1976).

3.2 Changes in Composition of Source Rock Bitumen Versus Crude Oil

Three different aspects are relevant with respect to changes in chemical composition between source rock bitumen to reservoired oil:

- hydrocarbon distribution in the contact zone between source rock and reservoir rock,
- gross chemical composition of crude oils versus source rock bitumen,
- the phenomenon of oil–source rock correlation.

Very little data are available on detailed investigations of *source rocks in actual contact with an oil-filled reservoir rock*. Several such examples are known and are cited here because it is this kind of information that is needed to solve the problems of primary migration.

Core samples taken at regular 1-m intervals across the transition zone reservoir rock–source rock, have been analyzed from a well in the Paris basin (Vandenbroucke, 1972). The well penetrated the lower Jurassic organic-rich Toarcian marl source rock and the calcareous Domerian reservoir rock. The mineralogical composition of the core samples of both source and reservoir rocks, and their organic content including total amount of organic carbon, extractable asphaltenes, resins, aromatic and saturated hydrocarbons, and the distribution of hydrocarbons were investigated. The source rock samples were found to be increasingly depleted in total extract as the reservoir rock is approached. The maximum depletion observed was in the order of about 10% of the total extract and was noticable over a distance of about 4 m. Less polar compounds, especially those of lower molecular weight, were preferentially lost to the reservoir rock. Thus, the movement of petroleum compounds from the source rock to the reservoir rock is apparently controlled largely by adsorption and desorption phenomena along the migration paths.

Another investigation of this kind has been presented by Tissot and Pelet (1971), who studied the transition zone between source rock and reservoir rock in a detrital shale–sand series of Devonian age in the Algerian Sahara (Table III.3.1).

Table III.3.1. Abundance and composition of extracts across the transition zone source rock–reservoir rock in a Devonian shale-sand series, Algeria, Sahara. (After Tissot and Pelet, 1971)

Proximity to reservoir in m	Extract/C_{org}[a] (mg g^{-1})	Hydrocarbons[a] % in extract	Asphaltenes[a] % in extract
2	72	54	12.2
4	86	61	11.2
7	90	63	7.5
10.5	112	63	5.7
14	118	64	5.8

[a] Each value represents the average of three or four measurements

3.2 Changes in Composition of Source Rock Bitumen Versus Crude Oil

Again a gradual depletion in extractable bitumen was observed as the sandy reservoir was approached, but in this case the zone of depletion reached about 10 m into the source rock. The depletion was strongest closest to the reservoir where the source rock had lost about 40% of its original bitumen content. The bitumen remaining in the source rock nearest the reservoir was most strongly enriched in high molecular weight heteroatomic compounds and depleted in hydrocarbons. A very similar depletion of source rock hydrocarbons in contact with reservoir rocks was reported by Neruchev and Kovacheva (1965).

Corresponding compositional fractionation effects due to expulsion have been observed, however, also on a molecular level, in two core holes penetrating sequences of interbedded mature shale source rocks and reservoir sands on Spitsbergen Island (Leythaeuser et al., 1983a, b; Mackenzie et al., 1983). Thin shales interbedded in sands and the edges of thick shale units are depleted in petroleum-range hydrocarbons to a such higher degree than the centers of thick shale units. Furthermore, it was possible in this study to obtain information on the quantities of the hydrocarbons expelled.

Figure III.3.1 shows the comparison of the C_{15+}-saturated hydrocarbon data from two source rock shales from the same sequence: In the center of a thick

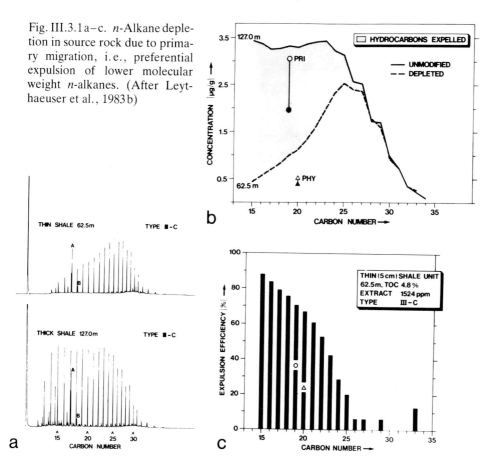

Fig. III.3.1a–c. *n*-Alkane depletion in source rock due to primary migration, i.e., preferential expulsion of lower molecular weight *n*-alkanes. (After Leythaeuser et al., 1983b)

source rock-type shale unit at 127.0 m depth the originally generated bimodal hydrocarbon distribution is presented, while the thin shale at 62.5 m shows a heavy-end biased hydrocarbon distribution (Fig. III.3.1a). Since kerogen quality and maturity are identical for both samples (Leythaeuser et al., 1983b), this compositional difference must result from expulsion, i.e., the thin shale during the course of primary migration preferentially lost the lower molecular weight hydrocarbons. In addition to that pristane was expelled to a lesser degree than n-alkanes with a comparable carbon number. Subtracting the concentration of hydrocarbons of the depleted shale from the hydrocarbon mixture of the unmodified shale, leads to information on the concentration and compositional profile of the hydrocarbons expelled from the thin source rock shale (Fig. III.3.1b). By expressing the concentration of the expelled hydrocarbons as a percentage of the originally generated hydrocarbons the expulsion efficiency is obtained (Fig. III.3.1c). It ranges for n-alkanes between about 10% and 80% depending on carbon number. These expulsion efficiencies should not be applied to source rocks in general because the example presented here is a special case where a thin source rock is sandwiched between porous reservoir strata permitting optimum hydrocarbon drainage conditions.

In the same study it could be shown that the composition of the hydrocarbons impregnating the interbedded reservoir sands is in agreement with a pronounced fractionation of hydrocarbons being expelled from the shale, i.e., low molecular weight hydrocarbons migrate preferentially from source to reservoir. Thus, in this sequence the composition of the hydrocarbon product accumulating in the reservoir appears to be controlled primarily by physical processes rather than by the type and the maturity of the organic matter in the source rock.

A comparision of the *gross chemical composition between crude oils and source rock bitumen* shows that most oils are enriched in saturated hydrocarbons and

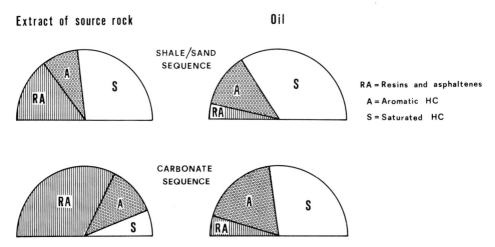

Fig. III.3.2. Comparison in terms of gross chemical composition between crude oils and source rock bitumens in a shale/sand and carbonate sequence. (After Tissot and Pelet, 1971)

depleted in polar N, S, O compounds (Fig. III.3.2). This enrichment has been frequently observed (Hunt and Jamieson, 1956; Bray and Evans, 1965; Tissot and Pelet, 1971). This is in agreement with the observed adsorption behavior of the different groups of compounds.

Any mechanism of importance in primary migration also has to account for the *geochemical correlation between crude oils and source rock bitumen.* Correlation aspects will be treated in great detail in Part V. It has been established that hydrocarbon distribution among homologous series and some individual hydrocarbon ratios remain practically unchanged during migration (Williams, 1974; Welte et al., 1975; Deroo, 1976). The numerous examples of successful oil/source rock correlations based on individual hydrocarbon ratios are a strong argument in favor of a hydrocarbon phase movement.

3.3 Evaluation of Geological and Geochemical Aspects of Primary Migration

The observation of embryonic, noncommercial hydrocarbon accumulations, such as occur in the Orinoco River delta (Kidwell and Hunt, 1958) and in young sediments of the deep sea of the Gulf of Mexico and the Western Pacific Ocean (McIver, 1975) is not in contradiction with current geochemical thinking. The hydrocarbons and bitumens found in these occurrences are enriched in aromatics and high molecular weight polar heterocompounds, or both. This would be expected if migration occurred from relatively immature source material and with compaction water as the main transportation vehicle. The majority of such immature, embryonic petroleum occurrences, however, are probably eventually lost to the surface because of the lack of suitable seals in the first few hundreds of meters of burial and the flushing action of upward moving compaction water. Autochthonous, immature bitumen apparently occurs not only in Recent and sub-Recent muds and clays, but also in sands (Palacas et al., 1972) and is never found in producible concentrations. This material should not be confused with hydrocarbons generated in more mature source rocks which subsequently migrate to shallower depth. Geochemical analyses can help distinguish between thermally immature and mature hydrocarbon assemblages.

Primary migration of petroleum hydrocarbons in young, immature, unconsolidated sediments seems to be rare, but cannot be entirely ruled out. An example of commercial quantities of oil and gas, derived from sediments buried to approximately 1000 to 1500 m, at a relatively early state of compaction (Nagaoka Plain, Japan) has been presented. In this case available information points to primary migration in connection with pore water movement induced by compaction. Commercial petroleum occurrences derived from such shallow and young source sediments apparently are rare. The Nagaoka example, however, is accepted as evidence of commercial oil pools resulting from early migration with compaction water as a main cause for primary migration. The scarcity of such young petroleum accumulations originating from relatively immature source beds is primarily due to the lack of traps and seals at this stage of burial, and by the general insolubility of petroleum compounds in low temperature compaction water. The composition of migrated petroleum at this stage should reflect the thermal immaturity of the source beds and the solubilities of the transported compounds. It should contain significant quantities of relatively unaltered biochemical molecules, such as water-soluble, polar, high molecular weight N, S, O compounds and low boiling hydrocarbons enriched in aromatics. A diagrammatic summary of these and other considerations concerning primary migration, is presented in Figure III.3.3. Early primary migration by means of compaction water is restricted to detrital, argillaceous-silty sediments. Lithification prevents compaction in most young carbonate sediments, and therefore excludes early primary migration in shallow carbonate sequences, even if sufficient soluble organic matter is present.

The occurrence of shallow accumulations of biogenic gas is a special aspect of the migration and accumulation of hydrocarbons in young sediments. Prominent

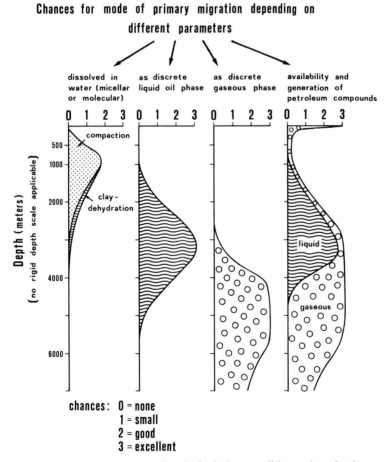

Fig. III.3.3. Diagrammatic sketch depicting possible modes of primary migration with increasing burial

3.3 Evaluation of Geological and Geochemical Aspects of Primary Migration

examples of this kind are numerous small gas fields in the Pliocene and Pleistocene sediments of the Po Valley. Other examples might be found at the eastern side of the Western Canada basin, e.g., the gas field Medicine Hat (310 m). The mode of primary migration in such cases has not been investigated.

Compaction and the resulting movement of pore water diminishes with increasing depth of burial. The chances for primary migration of petroleum compounds dissolved in and moved by compaction water are probably greatest at a depth of about 1000 m, when shales still show sufficient porosity for additional compaction, and yet are buried deep enough to reach the onset of thermal generation of hydrocarbons from kerogen. At depths below 2000 m the porosity curves for shales (Figs. III.2.5, III.2.8) tend toward the more linear region, where there is little further porosity reduction with continued increase in burial.

The overall efficiency of primary migration in molecular or micellar solution in connection with actively moving compaction waters cannot be high as compared to other possible modes of primary migration. The special role of diffusion-controlled processes during primary migration of dissolved gas will be treated separately toward the end of this chapter. The solubilities of petroleum compounds in pore waters are very low (Table III.2.2) even under the most favorable conditions. With very few exceptions, solubilities are in the range of several ppm to about 100 ppm at higher temperatures ($> 100\,°C$) and greater depths. Even the most optimistic material balance calculations show that solubilities of hydrocarbons at temperatures of less than $125\,°C$, corresponding to depths to about 3000 m, are too low to account for oil accumulations in productive petroleum provinces such as the Eastern Sahara, California or the Louisiana Gulf Coast. If a source rock volume of 100×100 km and 100 m thick is assumed, the cumulative amount of water expelled due to compaction between 1000 and 3000 m depth of burial, is about 2×10^{11} m^3. To account for oil accumulations in the range of 200 million to 2 billion t, which is a reasonable figure for oil in place in those basins, crude oil solubility in water would have to be between 1000 and 10,000 ppm. Hydrocarbon solubilities of this range are highly unlikely. At greater depths and higher temperatures, the solubilities might be higher, but in such cases the amount of compaction water is not sufficient. Between depths of 3000 and 5000 m, only a 5% pore volume reduction can be expected (Fig. III.2.8). Mass balance calculations in the Williston basin and Los Angeles basin were also used by Jones (1980) in order to point out how ineffective solution migration and how important a continuous oil phase migration must be.

Finally, the composition of most crude oils does not reflect the solubility distributions of the various classes of petroleum compounds either in micellar, or in molecular aqueous solution. Micellar solution as an effective means of primary migration is contra-indicated by additional factors, such as the lack of sufficient amounts of surface active solubilizers, micelles too large to pass through pore throats, and chemical and electrical hindrances. Thus, although there may be some early primary migration in aqueous solution in connection with compaction waters, especially with the more soluble low molecular weight hydrocarbons and some polar high molecular weight N, S, O compounds, it is exception rather than the rule.

At this point of the discussion we can now evaluate the relative importance of primary migration with or without water movement as follows:

- Compaction waters are only available in necessary quantities when hydrocarbons have not yet been generated; during the main phase of hydrocarbon generation few or no compaction waters are available.
- The relative permeability concept requires for effective movement of a hydrocarbon phase a minimum ratio of the oil/water saturation. This can only be achieved once the major part of water has been expelled.
- Any water phase transport would require an alteration of the chemical composition of reservoired petroleums in a manner contrary to that observed in nature, i.e., polar compounds should be enriched and non-polar compounds depleted; furthermore, the fact of successful oil/source rock correlation based on hydrocarbon composition also favors hydrocarbon phase movement.
- Generation and migration of hydrocarbons are comparable in compacting clastic and non-compacting carbonate source rocks.

Based on these four arguments we conclude that the majority of oil occurrences can only be explained by a hydrocarbon phase primary migration mechanism. Expulsion in an aqueous phase may play a role only under some special circumstances.

It is therefore expected that at depth levels of about 1500 m and deeper, the predominant mode of primary migration shifts from early solution migration to a form of oil phase migration. The more or less continuous oil phase is gradually superseded by a gaseous hydrocarbon phase at depths greater than about 3500 m. The beginning of late primary migration as a separate hydrocarbon phase coincides with the main phase of oil formation and continues until the generation potential for liquid oil is exhausted, and mainly gaseous hydrocarbons are formed. The onset of late primary migration is consequently controlled not only by the depth–porosity relationship, but by the generative potential of a given source rock as well. Thus, it cannot start until maturities equivalent to vitrinite reflectance values of about 0.5 to 0.6% are attained and favorable types of organic matter are present. Several factors control the formation of a discrete oil phase and its movement through the source bed. Among these are the total porosity, the ratio of fixed to mobile pore water, the average cross-section of pores, the amount and type of kerogen (organic carbon) and the presence of a three-dimensional kerogen network, and finally the amount of bitumen generated.

At depths of about 3000 m and deeper, porosities in source beds are below 10%, where a considerable part of the pore volume is occupied by immobile fixed pore water, and bitumen concentrations, as compared to free pore water, are too high to be accomodated in the aqueous pore fluid. The existence of a separate oil phase in source beds at peak of hydrocarbon generation, therefore, seems to be a necessity. Due to the continued generation of bitumen and gas with burial, this separate oil phase is pushed away from the kerogen through the mainly water-wet micropores. This type of primary migration, is at its most effective level for source rocks with high kerogen and bitumen content. The movement of a separate oil phase must overcome mounting capillary pressures until a continuous network of

3.3 Evaluation of Geological and Geochemical Aspects of Primary Migration

bitumen occupies the center pore space. At a certain point, pore diameters will become too small and oil particles too large to be forced through water-filled pore throats against extremely high capillary pressure. Continued formation of gas and other low molecular weight hydrocarbons, plus the thermal expansion of pore waters and partial transmission of geostatic stress, keep the pressure rising until the mechanical strength of the rock is exceeded. At this point microfissures open, the hydrocarbon phase expands and coalesces further, and an increment of migration is accomplished.

It is thus visualized that a pressure-driven hydrocarbon phase movement as a principal mechanism of primary migration prevails during the main phase of hydrocarbon formation. This process can probably persist without rock fracturing up to a certain depth. Beyond that depth with the onset of massive gas generation, around 3000 to 4000 m in dense rocks, microfracturing will be initiated.

Below these depth levels and at temperatures above 150°C the liquid-pressurizing phase is ultimately replaced by a gaseous phase. Zhuze et al. (1963) suggested that normal crude oils as well as low molecular weight hydrocarbons are dissolved and transported as a compressed gaseous phase. They calculated that one billion m^3 of gas (adjusted for atmospheric pressure conditions) at

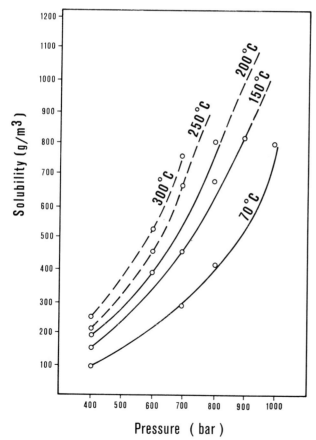

Fig. III.3.4. Solubility of petroleum in gas at various temperatures as a function of increasing pressure. (After Zhuze et al., 1963)

pressures of 400 to 800 bar and at temperatures between 70 and 200°C, could transport a quantity of petroleum in the order of 100,000 to 800,000 t. The relationship between petroleum solubility in gas at various temperatures is shown in Figure III. 3.4. Numerous fields throughout the world containing wet gases or condensates are evidence for this kind of migration mechanism. Furthermore, it can be expected that some of the deep overpressured shale source rocks, occurring in Tertiary deltas, such as the Mahakam delta, Indonesia, contain a single hydrocarbon fluid phase, due to the large proportion of gas and the high solubility of medium range hydrocarbons in gas, under such pressure and temperature.

This situation would result in a gas-phase migration, until the pressure has decreased. The remaining amount of medium hydrocarbons in gas would depend on the pressure drop. Another possibility for gas as a transport vehicle for higher hydrocarbons is the percolation of the rock column with a gas phase originating form a deeply situated source. This gas could extract heavier hydrocarbons form overlying source rocks. This may be the origin of the so-called shallow early-mature condensates. It is clear that under the conditions discussed above, asphaltenes and other heavy molecules cannot be transported.

3.4 Conclusions and Suggestions on Primary Migration

In summary, it is concluded that there is no reason to assume that one mechanism of primary migration is responsible for all petroleum accumulations (Fig. III.3.3). The predominant mechanism of primary migration may change with different subsurface conditions primarily related to increasing depth of burial. In shallow sediments down to depths of about 1000 to 1500 m, solution migration seems to be favored as compared to an oilphase migration. Solution migration may also play a role at greater depths at higher temperatures, but only involving certain light hydrocarbons. Solution migration alone does not seem to produce major deposits, except under exceptional geological conditions, as in shallow gas fields of early diagenetic origin.

The dominating and most effective form of primary migration seems to be hydrocarbon-phase migration, first in form of a liquid oilphase and at greater depth as a gaseous (or eventually supercritical) phase. Hydrocarbon-phase migration is mainly accomplished by a pressure-driven movement through preferentially oil-wet pores or a network of kerogen or by microfracturing of the rock due to an internal pressure build-up. Hydrocarbon-phase migration is probably prevalent from medium to great depths and does not necessarily exclude a minor contribution from solution migration. Hydrocarbon-phase migration is in agreement with both the empirical and factual geological and geochemical data.

In this case water flow is not needed as a driving force for migration, and in fact water hinders rather than aids movement. Microfracturing permits the release of hydrocarbons from compacted, dense, relatively impermeable source rocks. Therefore, carbonates as well as shales can expel petroleum in a similar manner. There are well-documented carbonate source rocks (Hedberg, 1964), such as the

3.4 Conclusions and Suggestions on Primary Migration

La Luna Formation of Western Venezuela which is thin-bedded carbonaceous-bituminous limestone of Cretaceous age, or the dark-colored dense limestone between reefs and calcarenites in the Devonian Swan Hills Formation of Alberta and the entire Middle East oil province. These dense, consolidated carbonate source rocks, apparently can yield oil without compaction through "hydrocarbon-phase pressure-migration". Similarly compacted, deeply buried shales can release hydrocarbons, as in the case of the Silurian source rocks of Hassi Messaoud in Algeria, the Tertiary-Cretaceous of the Quiriquire field in Venezuela and the Carboniferous coal measures of Groningen, Netherlands and the Mississippian-Devonian Bakken Shale in the Williston Basin of North Dakota (Dow, 1974; Williams, 1974).

This is in agreement with the observations that the maturation of organic matter and the generation of hydrocarbons, and also the occurrence of oil fields are comparable in siliceous, clastic, and carbonate sequences. Hydrocarbon-phase pressure-migration of oil would also be similar in clastic and carbonate sediments. All other primary migration mechanisms, however, involving larger quantities of water, would require a different kind of primary migration for carbonates only.

Another phenomenon, conflicting with primary migration of oil in aqueous solution, but in perfect agreement with hydrocarbon-phase migration, is that most crude oils are enriched in saturated hydrocarbons and depleted in polar compounds as compared to source rock bitumen. As in established chromatographic theory utilized in laboratories around the world, similar solvents preferentially extract and transport compounds of a similar chemical nature. In nonpolar (alkane) solvents, for example, polar compounds tend to be left behind. There is experimental and geological evidence for the preferential selective adsorption of different organic molecular classes on clay surfaces. The previously cited investigations by Neruchev and Kovacheva (1965), by Vandenbroucke (1972) and by Tissot and Pelet (1971) and Leythaeuser et al. (1983a, b) provide additional evidence for such chromatographic adsorption phenomena in source rocks.

The proposed model of a pressure-driven movement of an essentially hydrocarbon phase as a major mechanism of primary migration, would also explain why an effective source rock must contain a certain minimum amount of organic matter. If there is too little organic matter, the ratio of freely movable pore water to bitumen or gas may be too high and thus unfavorable for hydrocarbon-phase migration even if the porosity is low. This argument however, does not reversably mean that rocks consisting solely of organic matter, like coal, should be the most productive source rocks. Hydrocarbon-phase migration is also in agreement with the accomplishment of successful geochemical correlations between crude oils and source rocks.

Besides the previously discussed role of diffusion as an initial process for primary migration of gas, diffusion can also play a role in destroying reservoired gas accumulations by loss through the caprock. Leythaeuser et al. (1982) have also evaluated the latter process. The rate of dissipation was calculated for the Harlingen gas field in Holland, which has reserves of 1.93×10^9 STP m^3 of methane in-place. By diffusion of methane through the 400 m thick shale caprock continuing for 4.5 million years this field is reduced by one half of its original size. Based on these data about the rate of destruction of commercial-sized gas fields with geologic time the concept was proposed that gas accumulations have only a limited life in terms of the geologic time scale. Existing gas fields may either have geologically young accumulation ages, or they may be able to persist through periods of geologic time only if there was continued replenishment by additional gas generation in the source rocks. In the latter sense certain gas accumulations can be considered as dynamic systems reaching some kind of steady-state equilibrium between gas loss and gas generation from the source rocks.

Source rocks have been mainly viewed with respect to their physical and chemical properties, although they are the products of sedimentological processes, and comprise part of the sedimentary history of individual geological basins. Thus, sedimentological features and tectonic events are also of importance in petroleum migration. Source beds are often finely laminated with an organic-rich layer alternating with organic poor layers. Such geometric arrangements may provide preferential avenues for primary migration. On a larger scale some source sequences, especially clastics, may contain interbedded sand and silt layers. This would be more advantageous for drainage than thick, uniform shale sequences.

Tectonic events may cause abnormally high strain in sedimentary units, especially by horizontal compression. This may aid to increase internal fluid pressures which eventually cause rock fracturing. In active tectonic zones such as certain parts of California and the Middle East, repeated earthquakes may contribute to fracturing and hydrocarbon movement.

Summary and Conclusion

The predominant mechanism of primary migration may change with different subsurface conditions primarily related to increasing depth of burial. In shallow sediments down to depths of about 1000 to 1500 m, before the onset of the main phase of hydrocarbon formation, solution migration seems to be favored as compared to an oil-phase migration. Solution migration may also play a role at greater depths at higher temperatures. However, solution migration alone does not seem to produce major deposits.

The dominating and most effective form of primary migration is a hydrocarbon-phase migration. In relatively impermeable, dense source rocks it is accomplished by microfracturing of the rock. Hydrocarbon-phase migration is in agreement with the empirical and factual geological and geochemical data. Water flow is not needed as a driving force for migration. Well-documented examples support the conclusion that hydrocarbons can be released from such compacted, dense, relatively impermeable source rocks. The Cretaceous La Luna limestones of Western Venezuela, dense limestones between reefs in the Devonian Swan Hills Formation of Alberta, Silurian source rocks of Hassi Messaoud in Algeria and the Mississippian-Devonian Bakken Shale in the Williston Basin of North Dakota can serve here as examples.

For gases, next to a hydrocarbon-phase movement, diffusion of dissolved hydrocarbon molecules is an important process of primary migration, especially in deep basins.

It follows from the previous discussion that distances covered by primary migration are commonly in the order of meters or tens of meters.

Chapter 4
Secondary Migration and Accumulation

Secondary migration is defined as the movement of petroleum compounds through more permeable and porous carrier beds and reservoir rocks, as opposed to primary migration through dense, less permeable and porous source rocks. Secondary migration terminates in hydrocarbon pools, but tectonic events such as folding, faulting or uplifting may cause redistribution of filled oil or gas pools and thus initiate an additional phase of secondary migration. When this results in a new accumulation it is sometimes called remigration or tertiary migration.

Oil and gas accumulations generally occur in the structurally highest available part of a trap. This is because oil (spec. gr. 0.7–1.0 g cm^{-3}) and gas (spec. gr. < 0.001 g cm^{-3}) have lower densities than the surrounding aqueous pore fluid (spec. gr. 1.0–1.2 g cm^{-3}) and rise by buoyancy through the water-saturated pore space. The main driving force in secondary migration is buoyancy. Oil and gas will form a pool whenever their further, generally upward movement is retarded by a less permeable layer of rock. This empirical observation is almost as old as the oil industry, and it is exemplified in thousands of accumulations around the world. The formation of oil and gas pools requires a decrease in the pore opening size to prevent the continuation of a two- or multi-phase fluid flow. Thus, the termination or continuation of secondary migration is determined by the interplay between the driving force that causes the movement of hydrocarbon droplets or slugs, and the capillary pressures that resist this movement. As long as the aqueous pore fluid in the sub-surface is more or less stationary, the only driving force for movement of a discrete hydrocarbon phase during secondary migration is buoyancy. We know, however, that there is water flow in the sub-surface due to hydrodynamic gradients. It is important to distinguish between hydrostatic, no-flow or equilibrium state, and hydrodynamic conditions, water flow due to hydraulic head gradients. Hubbert (1940, 1953) describes the principles of hydrodynamic conditions. Water flow under hydrodynamic gradients modifies the buoyant rise of oil and gas.

Three parameters control secondary migration and the subsequent formation of oil and gas pools. These are the *buoyant rise* of oil and gas in water-saturated porous rocks, *capillary pressures* that determine multiphase flow and, as an important modifying influence, *hydrodynamic* fluid flow. Secondary migration results in the formation of hydrocarbon pools and may cause hydrocarbons to seep out at the surface.

4.1 The Buoyant Rise of Oil and Gas Versus Capillary Pressures

In the previous chapter it was concluded that our present knowledge indicates that the most effective and predominant mechanism of primary migration is most probably a hydrocarbon phase transport and that solution migration is less important. Primary migration in solution is of greater importance only in certain geological settings, before and after the main phase of petroleum generation, at relatively shallow or great depths. In any case, secondary migration in its initial stages is influenced by the predominant mode of primary migration.

As hydrocarbons leave the dense, fine-grained, low-porosity source beds as a discrete hydrocarbon phase and enter the larger pores of a carrier bed or reservoir rock, larger globules of oil or gas should immediately form, depending on its state of dispersion. Larger bodies of oil may move upward by buoyancy, but tiny droplets may not because there is more resistance to flow due to higher surface energies per unit volume in smaller bodies of oil. This phenomenon was recognized and introduced into the literature of secondary migration by Athy (1930). Hydrocarbons, however, which leave a source bed in a dissolved state cannot rise by buoyancy, but follow water flow through a carrier bed or reservoir rock. When finally released from solution by a decrease in temperature or pressure or due to increase in pore water salinity, they have a fate similar to hydrocarbons, which leave a source bed as a separate phase.

Gussow (1954, 1968), Hobson (1954) and Hobson and Tiratsoo (1975) have elaborated on the principles of buoyant rise and capillary displacement pressure for a discrete hydrocarbon phase. Refinements and practical examples of this principle were presented by Poulet (1968) and Berg (1975) and the following discussion relies heavily on the work of these authors.

The interfacial tension between two immiscible phases (liquid–liquid or gas–liquid) can be envisaged as a force that acts at the interface to produce a pressure difference across it. The pressure difference across the interface is another form, or meaning, of the term capillary pressure. The work done by the interfacial tension results, for example, in a change in the surface area of a drop of oil when immersed in water. The drop will tend to assume the smallest possible surface area, which ideally is a sphere. The interfacial tension between oil and water resists distortion of the spherical droplet and retards its passage through a pore throat with a diameter smaller than the size of the droplet. The force necessary to squeeze the droplet through the pore is also called capillary pressure, or more correctly injection pressure. Capillary pressures in rock pores are the cause for hydrocarbon entrapment. Capillary pressures are higher for smaller pore diameters, and assume specific values for a certain rock.

Under hydrostatic conditions, buoyant forces could become high enough to overcome capillary pressure, which resists secondary migration. Buoyant forces increase with the density difference between pore water and oil ($\varrho_w - \varrho_o$) and with increasing height of the oil column. Gaseous hydrocarbons behave analogously, but because they have lower densities, they develop stronger buoyant forces. Oil trapped in a reservoir under hydrostatic conditions represents an equilibrium state between buoyant forces trying to move oil and capillary pressures in the

4.1 The Buoyant Rise of Oil and Gas Versus Capillary Pressures

sealing cap rock that resist this movement. The equation [Eq. (1)] for this equilibrium situation is

$$2\gamma\left(\frac{1}{r_t} - \frac{1}{r_p}\right) = z_o \cdot g \cdot (\varrho_w - \varrho_o) \tag{1}$$

where,
γ = interfacial tension between oil and water in dyne cm^{-1},
r = radius of pores in cm,
z_o = height of oil column in cm,
g = acceleration of gravity in cm s^{-2},
ϱ_w = density of water in g cm^{-3},
ϱ_o = density of oil in g cm^{-3}.

This equation or derivatives of it can be called "buoyancy equation".

In addition to the height of the oil column z_o, the buoyant force is defined by the acceleration of gravity (g = 981 cm s^{-2}) and the density difference between water and oil ($\varrho_w - \varrho_o$) in g cm^{-3}. The capillary pressure is determined by the interfacial tension between oil and water (γ) in dyne cm^{-1} and the radius (in cm) of the pore throats in the sealing barrier rock (r_t) as compared to the reservoir rock radii (r_p).

It follows that the maximum height of an oil column which can be held in place, and which is also called critical height (z_c), is given as follows (Hobson, 1954; Berg, 1975):

$$z_c = 2\gamma\left(\frac{1}{r_t} - \frac{1}{r_p}\right) / g \cdot (\varrho_w - \varrho_o). \tag{2}$$

The secondary migration of oil through a water-wet carrier bed or reservoir rock has been described by Hobson (1954). An undistorted oil globule at rest in a rock is in equilibrium with the surrounding pore water. In such a case, the radius of the rock pore (r_p) must be equal to, or larger than, that of the oil globule. The pressure difference across the oil–water interface is energetically equatable with the capillary pressure. It causes the pressure within the oil globule to be greater than in the surrounding water. The pressure (p) (in dyne cm^{-2}) inside the globule equals twice the interfacial tension γ divided by the radius of the globule (r); thus

$$p = 2\frac{\gamma}{r}$$ (Berg, 1975). If the buoyant force is great enough to force the globule

upward through a pore throat, the globule must be distorted. The above equation shows that the excess pressure inside the globule increases as the radius of curvature decreases. The pressure at the upper end of a distorted globule in a pore throat is, therefore, greater than in the pore (Fig. III.4.1). The capillary pressure opposes the buoyant force until the radii of curvature inside the distorted oil globule are equal at the lower and the upper end. Once the globule has reached this stage and moved halfway through the throat, it can rise by buoyancy. If the curvature is smaller at the lower end than at the upper end, capillary pressure gradient and buoyant force act in the same upward direction and secondary migration is further facilitated.

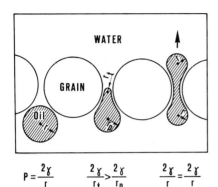

Fig. III.4.1. Transport of an oil globule through pore throats in a water-wet sub-surface environment. Capillary pressure opposes the buoyant force until the radius of curvature inside the distorted oil globule is equal at its lower and upper ends. (Modified after Berg, 1975)

This same effect may be of importance across a boundary between a fine-grained and a coarse-grained rock, i.e., between a dense source rock and a porous carrier bed or a reservoir rock. As soon as oil particles reach the boundary between a fine-grained source rock and a coarse-grained carrier bed, and are not enclosed by either type of rock, they are driven into the coarse bed by capillary forces. The transfer reduces the total surface area and the curvature of the oil particles.

The buoyant force acting on a distorted oil globule or a gas bubble is determined primarily by the density difference between the oil or the gas and the surrounding water. The density difference between normal oils and pore waters range from 0.1 to 0.3 g cm^{-3} and approach 1.0 g cm^{-3} for gas. The other important parameter influencing buoyant force, and the only variable in a given sub-surface situation, is the height of the oil or gas column [see also Eq. (2)]. All other parameters including pore and throat diameters of the rock and density and interfacial tension of the fluids under given sub-surface pressure and temperature conditions are fixed. Therefore, the equilibrium between buoyant and capillary forces can only become unbalanced through the accretion of new oil or gas material to form larger bodies with greater vertical height. When the buoyant force is greater than the capillary pressure [Eq. (1)], the oil droplet will move through the pore throat. Rearranging this equation, this would mean

$$z_o \cdot g \, (\varrho_w - \varrho_o) > 2 \gamma \left(\frac{1}{r_t} - \frac{1}{r_p} \right). \tag{3}$$

4.2 Hydrodynamics and Secondary Migration

To this point it has been assumed that essentially hydrostatic conditions prevail in carrier beds and reservoir rocks. There is, however, frequently sub-surface water flow, and the influence of hydrodynamic conditions on secondary migration cannot be neglected. Any flow of water in an aquifer, depending upon its direction, could hinder or facilitate secondary migration of hydrocarbons. Water

4.2 Hydrodynamics and Secondary Migration

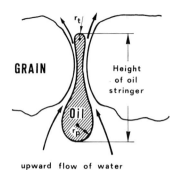

Fig. III.4.2. Transport of oil globule through a pore throat in a water-wet environment under hydrodynamic conditions. Upward flow of water helps buoyancy to overcome opposing capillary pressure. (Modified after Berg, 1975)

flow is related to hydrodynamic gradients. If the gradient is in an upward direction, it aids buoyant forces in moving petroleum. If the hydrodynamic gradient is in a downward direction, the buoyant forces have to be greater than required under hydrostatic conditions, in order to balance or overcome the opposing flow pressures. In terms of accumulation, an oil column held in place in a reservoir against a barrier rock exhibiting a certain capillary pressure can be higher if there is additional pressure from downward-flowing water which helps to balance the buoyant forces of oil.

Hydrodynamic gradients could be an important mechanism for secondary migration, especially during the initial phase. The final direction of secondary migration of oil and gas would be dependent on the relative magnitudes and directions of hydrodynamic and buoyant forces. A crucial point in the interplay between these forces is the magnitude of the hydrodynamic gradient along a stringer of oil. Consider first the theoretical situation of a vertical or a horizontal stringer of oil, and then the more probable situation of an inclined stringer. The first case refers to an oil stringer of height z_o, and density ϱ_o, in water of density ϱ_w, with a vertical positive hydrodynamic gradient, m, in the water, that results in an upward flow (Fig. III.4.2). This upward flow of water parallel to the oil stringer with the height z_o assists buoyant force. If this stringer of oil is held in place by capillary pressure, the previous equation can be rewritten as follows:

$$2\gamma \left(\frac{1}{r_t} - \frac{1}{r_p} \right) = z_o \cdot g \, (\varrho_w - \varrho_o) + z_o \cdot m. \tag{4}$$

This equation must be compared with Equation (1). It means that the capillary pressure (expression on left side of the equation) resulting from the smaller radius r_t in a pore throat at the upper end of the oil stringer, balances the buoyant force plus the force resulting from the upward flow of water (expressions on right side of the equation). If there is downward water flow, m is a negative quantity and the height of the oil stringer z_o must increase accordingly as long as the oil stringer is held in place.

Consider now a horizontal stringer of oil along the top of a horizontal carrier bed. In this case, buoyant force is negligible, and the driving mechanism can only be derived from horizontal water flow (Fig. III.4.3). The hydrodynamic gradient, m, along the length, l, of the oil stringer determines its movement under the

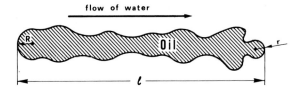

Fig. III.4.3. Horizontal transport of a stringer of oil under hydrodynamic conditions. The hydrodynamic gradient, m, along the length, l, of the oil stringer determines its horizontal disposition in the carrier bed

conditions in the carrier bed (Hobson and Tiratsoo, 1975). As buoyancy now can be neglected, the equation can be written as follows:

$$l \cdot m = 2 \gamma \left(\frac{1}{r_t} - \frac{1}{r_p} \right). \tag{5}$$

The expression at the left-hand side of the equation is the driving force, resulting from the horizontal water flow. The expression on the right-hand side of the equation is already familiar, since it is the resistance exerted by the capillary pressure to prevent the movement of the oil stringer. The smaller radius, r_t, in a pore throat, is now at the leading end of the stringer, and points in the same direction as the flow of water.

Finally, consider a stringer of oil, inclined at an angle, Θ. The flow of water occurs along the stringer, and the difference in hydrostatic heads at its lower and upper ends (Fig. III.4.4) must be considered in addition to the height of the oil column held in place only by capillary pressures (Berg, 1975). The change in the height of the oil column trapped is proportional to the hydrodynamic gradient. If the flow of water is in an up-dip direction, the oil column will be shorter. If it is in a down-dip direction, it will be longer. The basic equation [Eq. (1)] can be modified in this case as follows (Poulet, 1968):

$$l \cdot [g \cdot (\varrho_w - \varrho_o) \cdot \sin \Theta \pm m] = 2 \gamma \left(\frac{1}{r_t} - \frac{1}{r_p} \right). \tag{6}$$

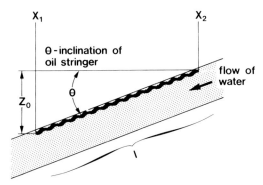

Fig. III.4.4. Transport of inclined stringer of oil under hydrodynamic conditions. The difference in hydrostatic head $(X_1 - X_2)$ at the lower and upper end of the oil stringer, the height of the oil colum (Z_o), and the flow of water determine the movement of the oil stringer

4.3 Geological and Geochemical Implications of Secondary Migration

As previously discussed, the principal factors governing secondary migration of a petroleum phase are buoyancy and hydrodynamic water flow. Capillary pressures resulting from narrow water wet-pore throats oppose this movement. The final distribution and occurrence of petroleum, once it has been generated and released from a source bed, is therefore controlled mainly by these three factors: buoyancy, hydrodynamics and capillary pressures. Whether primary migration occurs mainly in a hydrocarbon phase or less probably in aqueous solution is crucial to the initial stages of secondary migration, but less so in the final stage and the formation of pools. Ultimately oil and gas must appear as a discrete hydrocarbon phase before they can accumulate in a trap.

It is difficult to assess the exact point when hydrocarbons begin to coalesce massively into larger bodies of oil or gas. In productive basins, such as in the Eastern Sahara, Texas, or Southern California, oil stains are frequently observed in the top part of the reservoir beds even at some distance from producing fields. Oil-field studies have shown that there is always an irreducible amount of oil left behind in the reservoir rock after production is completed. This suggests that migration as a massive oil phase does not normally occur over great distances in the order of 100 km, unless special geological conditions exist, for example, along the unconformity between Cretaceous and Paleozoic beds or in the Cretaceous sands themselves in Alberta, Western Canada, and probably aide in the emplacement of the huge Athabasca heavy oil accumulation.

It is very difficult to quantify the relative importance of buoyancy or hydrodynamic forces on secondary migration. It seems clear, however, that in the initial stages of secondary migration, oil particles are on the average smaller and in a more dispersed state than during the final stage. Initially microscopic and submicroscopic oil particles should be relatively more abundant in the aqueous pore fluid than larger oil particles. In general, very finely dispersed oil droplets will not strictly follow the law of buoyancy and will be more easily moved by relatively weak hydrodynamic water flow. The overall influence of hydrodynamic water movement is, therefore, more important during the initial stages of secondary migration than during the final stages.

Hydrodynamic phenomena are crucial factors in the understanding of subsurface flow network leading to the formation of oil pools. As a consequence, an increasing number of papers are being devoted to this subject. The aim is a better understanding of hydrodynamic flow patterns in hydrocarbon-prone basins, such as in the Permian Basin of the United States (McNeal, 1965), in the Western Canada sedimentary basin (Hitchon, 1969a, 1969b; Hitchon and Horn, 1974) or in the Algerian Sahara of Northern Africa (Chiarelli, 1973). These investigations pursue a reconstruction of past and present hydrodynamic flow patterns based on actual salinity and fluid pressure data and the regional geological evolution of the basin. It is logical and necessary that such studies be linked with investigations on the regional hydrocarbon generative potential and maturity of source rocks. This means that paleo-hydrodynamic patterns need to be elaborated further.

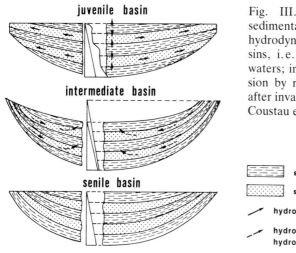

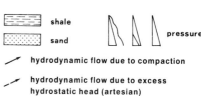

Fig. III.4.5. Three main types of sedimentary basins classified according to hydrodynamic conditions: juvenile basins, i.e., before invasion by meteoric waters; intermediate basins during invasion by meteoric waters; senile basins, after invasion by meteoric waters. (After Coustau et al., 1975)

Models have recently been developed that link the general movement of pore fluid in sedimentary basins to parameters such as type and rate of sediment deposition, porosity, pore fluid pressure and temperature. Two such models (Magara, 1976; Sharp, jr. and Domenico, 1976) on Tertiary beds in the Gulf of Mexico conclude that there exists a compaction-induced horizontal movement of pore fluid through more permeable sandy layers toward the edges of the basin or to permeable faults. Vertical movement of fluids across lithological sequences is comparatively less important. Furthermore, the temperature distribution is a sensitive indicator of fluid movement, which is largely responsible for heat transfer.

Coustau et al. (1975) classified sedimentary basins according to hydrodynamic conditions and related this classification to their petroleum potential from organic geochemical considerations. Three main types of basins are distinguished: before (1), during (2) and after (3) the invasion by meteoric (fresh) waters (Fig. III.4.5).

1. *Juvenile basins,* not necessarily young, with compaction-induced centrifugal, lateral water movement. Examples: Nigeria, Gulf of Mexico, Douala basin, North Sea, Northeast Sahara (between salt deposits) among others. Petroleum interest in the basins is very strong.
2. *Intermediate basins* with centripetal water movement, artesian properties and freshwater invasions. Examples: Persian Gulf, East Sahara (Illizi-Tinrhert basins), Paris Basin, Central Tunisia, Sahara (below the salt) and others. Petroleum interest in such basins varies from very strong to moderate.
3. *Senile basins* with hydrostatic conditions and generally invaded by meteoric waters. Examples: Northwest Aquitaine basin, parts of North Spanish basin. There is little or no petroleum interest.

The paper by Coustau et al. (1975) accounts for the fact that hydrodynamic flow of water is initially of great importance to secondary migration and formation of pools. However, if it is too strong or lasts too long, hydrodynamic

4.3 Geological and Geochemical Implications of Secondary Migration

water flow retards the formation of pools, or even destroys existing accumulations by dismigration.

As oil particles grow in size, they become more subject to buoyancy. Larger particles grow at the expense of smaller particles, and attempt to attain the smallest possible surface area with respect to volume. Upward movement of petroleum particles or gas by buoyancy starts whenever opposing capillary pressures are low and there is an insufficient counter-flow of water. Because of buoyancy, oil and gas globules always seek the highest point in a carrier bed or reservoir rock. Their movement follows the direction of lowest resistance. This may result in a very tortuous path, generally upward, more or less normal to bedding (Gussow, 1968). As they move, oil particles merge into globules, patches and stringers. This increases their buoyant force and assists their movement through narrower pore openings, enabling them to move through finer-grained rocks. Eventually a situation may be reached where at some crestal point pore diameters get too small, and capillary pressures can no longer be overcome. This could be the start of an accumulation which may finally attain economic dimensions or remain merely a small local concentration of hydrocarbons.

Poulet (1968) assumed a certain grain size (median radii of grains 0.5 mm) in an inclined sandstone (dip: 100 m over 40 km), a certain density difference between oil (0.875 g cm^{-3}) and water (1.07 g cm^{-3}) and a certain interfacial tension γ (40.4 dyne cm^{-1} at 40°C), and calculated the length of a continuous oil stringer that could be formed and held in place. These conditions are similar to those observed in the Devonian reservoir in the Illizi basin in the Algerian Sahara.

a) Under hydrostatic conditions, the critical height of the oil column (z_o) would be 35 cm, and at a greater value the stringer would move. The length L of the oil stringer is calculated from the angle of inclination (Θ) of the sandstone and the height (z_o) of the oil column according to the equation $L = \dfrac{z_o}{\sin \Theta}$. A stationary oil phase of 140 m long could be formed along the top of the inclined sandstone. With increasing depth, i.e., increasing temperature and pressure, only minor changes with respect to the height of the oil column would be expected, unless gas is added to the oil. Oil-water surface tension γ would increase with pressure, but would be approximately compensated by a decrease at higher temperatures.

b) Under hydrodynamic conditions, however, the length of the oil phase would be changed. In a horizontal reservoir, similar to that described above, a pressure gradient of 0.5 dyne cm^{-2} cm^{-1} would be required to move this oil stringer with a length of 140 m. If the reservoir had a permeability of 1 darcy and a porosity of 25%, a pressure gradient of 0.5 dyne cm^{-2} cm^{-1} would cause water of a viscosity of 1 centipoise to flow about 60 cm per year. This flow is of the same order of magnitude typically observed in artesian aquifers.

Similar calculations can be made for the movement of gas in the same reservoir (Poulet, 1968). At a shallow depth of burial, the critical height of a gas column held in place would be about 17 cm, assuming a surface tension of 50 dyne cm^{-1}. At greater depth, the density of the gas increases slowly, but the gas–water surface tension decreases strongly with increasing temperature and pressure at a rate of about 10 dynes cm^{-1} for 100 bar, and about 3 dyne cm^{-1} for 30°C. Thus, under hydrostatic conditions at 4000 m depth, the critical height of a gas

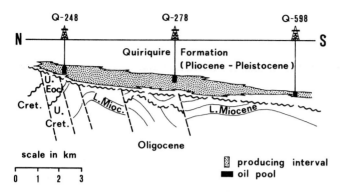

Fig. III.4.6. Geological cross section through Quiriquire oil field, Venezuela. (After Silverman, 1965)

column would be 5 cm, as compared to an oil column of 35 cm. The solution of oil in the gas would change the height of the gas column, but only to a minor extent.

From the foregoing it is deduced that a delicate balance is established when a bubble or stringer of oil or gas is held in a carrier bed. Any tectonic event such as a fold, regional tilt, uplift or subsidence, or a sudden earthquake, may change this delicate balance. Tectonic events are, therefore, also an important means of influencing secondary migration and the arrangement of flow patterns.

Moving oil particles are in steady contact with the surrounding pore water. Polar oil molecules are therefore preferentially concentrated at the oil–pore water interface. These molecules and other, more water-soluble hydrocarbons such as low boiling aromatics, are exchanged between the oil and the water phase. Although the water solubility of these compounds is very low, they will, nevertheless, be preferentially lost to the water during secondary migration. Depending on the wettability of the internal surfaces of the carrier or reservoir rock, petroleum compounds will occasionally be adsorbed and the more polar compounds will be preferentially left behind. Silverman (1965) reviewed geochemical changes observed in crude oils during secondary migration and presented new evidence by analyzing oil samples, taken in an up-dip direction from the Quiriquire oil field (Fig. III.4.6) in Venezuela (Table III.4.1). These chemical changes in crude oils, caused by secondary migration, can be summarized as follows: along the path of migration there is an increase in nonpolar hydrocarbons and a decrease in content of asphaltenes, resins, porphyrins and other nonhydrocarbons. Water washing may cause a loss in low boiling, more soluble, hydrocarbons. Furthermore, a slight decrease in the $^{13}C/^{12}C$-isotope ratio can be observed (Table III.4.1), caused primarily by the preferential loss of aromatic hydrocarbons.

The influence of secondary migration on the chemistry of gases is less understood (May et al., 1968; Stahl et al., 1975; Stahl, 1976). The changes observed to date are difficult to explain because they are apparently not always consistent. The parameters investigated are $^{13}C/^{12}C$-isotope ratios, nitrogen content, and $^{15}N/^{14}N$-isotope ratios and amount and distribution of "heavier

Table III.4.1. Quiriquire field crude-oil analyses. (After Silverman, 1965)

Well	Q-598	Q-278	Q-248
Producing interval (depth in feet)	3885–4070	2665–2855	2070–2570
Relative paraffinicity	0.78	0.81	1.02
Relative concentrations of carbonyl (C=O groups)	0.43	0.40	0.27
Sulfur content (wt. %)	1.26	0.78	0.70
Relative concentrations of vanadium porphyrins	19.6	10.6	10.3
$^{13}C/^{12}C$ (δ in ‰) PDB	−27.1	−27.3	−27.5

hydrocarbons" (C_2–C_4). The difficulty in explaining differences in chemical composition of gases along a presumed migration path may in part be due to the relatively small number of well-documented case histories. Primarily, however, the problem centers on the great variety and ambiguity of factors which can influence the chemical composition of gas. Among these factors are diffusion and effusion processes, solution and adsorption phenomena, and temperature and pressure effects and finally the great abundance and mobility of gas molecules.

4.4 Termination of Secondary Migration and Accumulation of Oil and Gas

The end of secondary migration and the final stage in the formation of oil and gas pools is the concentration of oil and gas in the highest available part of a trap. It is necessary that the total accessible pore volume in the trapping zone is of a magnitude sufficient to accommodate quantities of commercial interest. Oil or gas may be trapped in any rock of suitable porosity, regardless of lithology. The cap-rock or seal, by virtue of its general decrease in pore diameters, must exert capillary pressures which are great enough to stop the passage of oil or gas particles. Any such seal would stop hydrocarbon particles of a certain size, no matter whether the driving force of secondary migration is buoyancy or hydrodynamic flow. Under hydrostatic conditions, oil or gas would simply rise, according to the principles of buoyancy, to the highest available part of the trap without gross movement of the aqueous phase.

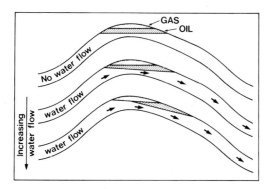

Fig. III.4.7. Displacement and separation of oil and gas in an anticlinal reservoir under the increasing action of flowing water. (Modified after Hobson and Tiratsoo, 1975)

Under hydrodynamic conditions, however, the moving water must move through the barrier rock or seal which filters out the hydrocarbons. Alternatively, water can escape through the reservoir rock, once it has passed the crestal or other part of a trap where hydrocarbons are unloaded. Under hydrodynamic conditions, it is not always necessary that oil or gas occur in the crestal part or the culmination of a trap. The force of the moving water has a direction and magnitude that may hold and trap a hydrocarbon phase at other suitable positions of the reservoir rock. A barrier effect may be caused by a change either in lithology due to finer grain size, by more intense cementation, or by a change in dip (e. g., at hinge lines). It may also be a down-dip hydrodynamic gradient that holds hydrocarbons in place against the force of buoyancy.

An example for displacement of oil and gas in an anticlinal, structural reservoir is given in Figure III.4.7. Under hydrostatic conditions and in a reservoir with uniform physical properties, gas–oil and oil–water contacts will be horizontal. At low rates of water flow the oil–water contact may be inclined, but the gas–oil contact remains horizontal. At higher flow rates the oil will be displaced further and the oil–water contact increases in steepness and, thus, oil and gas begin to become separated. A gas–oil contact remains horizontal but the gas–water contact is less steep than the oil–water contact (Hobson and Tiratsoo, 1975).

Besides hydrodynamics, capillary effects in reservoirs with inhomogeneous pore size distributions may be another cause for irregular hydrocarbon–water contacts. If the average grain size diminishes across an oil–filled sandstone reservoir, a curved oil–water contact may parallel this change. In fine-grained rock, water rises higher due to capillary attraction than in the coarse-grained part of the reservoir.

Oil and gas found in a pool reflect in composition what has been collected from the drainage area of secondary migration. A huge anticlinal structure under hydrostatic conditions, for example, collects oil and gas, that rises by buoyancy independent of their source. As oil and gas rise, they accrete and begin to form a continuous hydrocarbon phase network throughout the pore system of the reservoir rock. The water in the pore system is displaced by the moving hydrocarbons according to the principles of buoyancy and capillary pressures. As long as the mineral surfaces in a reservoir rock are water-wet, which is generally the case, a certain minimum amount of water is left behind in the reservoir rock as

4.4 Termination of Secondary Migration and Accumulation of Oil and Gas   353

a water film coating the mineral surfaces. This water is termed the irreducible minimum amount of connate water, which is found in all hydrocarbon-filled reservoirs. If the pore size distribution is inhomogeneous in a reservoir, there may be certain parts where oil or gas cannot enter, because of excessive capillary pressures. Thus, even in an oil-filled reservoir there may be isolated water-saturated pockets devoid of hydrocarbons. Such a situation occurs commonly in carbonates and in detrital reservoir rocks with secondary cementation.

The degree of oil or gas saturation of a reservoir, like secondary migration, depends largely on the dualism between those forces which move hydrocarbons, buoyancy and hydrodynamic flow of water, and the resisting force of capillary pressures. A typical case showing a cross section of the oil or water saturation with respect to total porosity in a reservoir is given in Figure III.4.8. Under hydrostatic conditions, a maximum oil saturation is reached in the upper part of the reservoir which varies according to the properties of the reservoir and the contained fluids. The relative oil saturation is usually lower in finer-grained reservoir rocks because the ratio of pore volume to enclosing surface is less favorable, and there is more water adhering to mineral surfaces. Saturation is also lower with heavier, more viscous oils. A transition zone with increasingly lower oil saturation separates the oil accumulation from the zone of complete water saturation (Fig. III.4.8). This transition zone has a lower oil saturation and it is a distinct and separate zone. Buoyant forces are not sufficient to inject oil into smaller pores which demand higher entry pressures to be filled because of a lack in oil-column height. Exact data on transition zones are rare. In many reservoirs there is apparently little indication of such a transition zone of appreciable thickness.

The gas–oil ratio at the time of emplacement depends mainly on hydrocarbon availability at the time of pool formation. If more gas than oil is available in the drainage area, the resulting accumulation is most likely to have a separate gas

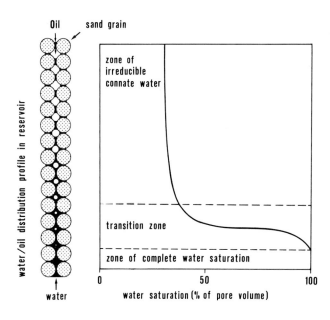

Fig. III.4.8. Cross-section through an oil-filled reservoir showing the relative distribution of oil and water approaching the oil–water contact zone. (After Mayer-Gürr, 1976)

cap. If, however, mainly oil was collected by the anticline, there will be no gas cap and the oil might even be undersaturated with gas under the reservoir conditions, i.e., at a certain pressure and temperature. The relative distribution of oil and gas in a pool, therefore, depends on many factors such as the supply of oil or gas, the reservoir pressure and temperature, the kind of oil phase in place (i.e., light oil can dissolve more gas than heavy oil) and the relative permeability of the sealing cap rock with respect to the content of the pool. Hence, the so-called bubblepoint or saturation-pressure concept which has been widely used for dating the time of oil accumulation is not valid. This concept requires that only gas-saturated oil is emplaced in a reservoir by migration. This is a rare case, and it is unjustified to assume that oil and gas would always be at a state of equilibrium at the time and under the conditions of emplacement in a reservoir. Such a case may exist, but it would be purely accidental.

4.5 Distances of Secondary Migration

Widely differing opinions have been expressed about distances of secondary migration. In some instances, a short-range migration has been inferred, such as in the isolated sand lenses in the Tertiary of the Rhine Graben and the pinnacle reefs in the Devonian of Western Canada. Long-distance secondary migration has been suggested for the Athabasca heavy oils in Canada and for many Middle East oil fields. In many instances, however, there is limited evidence and virtually no proof.

An approximate balance calculation suggests that there should be a relationship between the size of an oil accumulation and the drainage area. If it is assumed that a reservoir contains 20% of the rock volume occupied by oil, then a ratio of about 100 kg of oil per t of rock is present. If it is further assumed that this oil has been generated from a source bed of comparable thickness containing 2% organic matter and a transformation ratio of kerogen to oil of about 20%, then approximately 4 kg of oil is generated per ton of rock. If the efficiency of primary migration (expulsion efficiency) is about 25%, then 1 kg of oil is released per ton of source rock. Under these conditions, the ratio of the area of the oil pool, as compared to the drainage area of the source rock is in the order of about 1:100. In terms of distances rather than areas this would mean a ratio of 1:10. Thus, an oil field 3 km in diameter would have a drainage radius of about 15 km. The distribution of oil fields in productive basins with sufficient available traps shows that these assumptions are reasonable.

Giant oil fields (Halbouty, 1970) with over 100 million t of oil in place are known. The accumulation of such an enormous mass of oil or gas by processes of secondary migration demands a very large drainage area and a correspondingly large volume of sediments. Balance calculations for such oil fields show that, even under the most favorable conditions, secondary migration must cover distances of at least several tens of kilometers, or drainage areas of several hundred square kilometers. An oil deposit extending over several hundred square kilometers, such as the Athabasca tar sand in Canada, requires secondary migration over

4.5 Distances of Secondary Migration

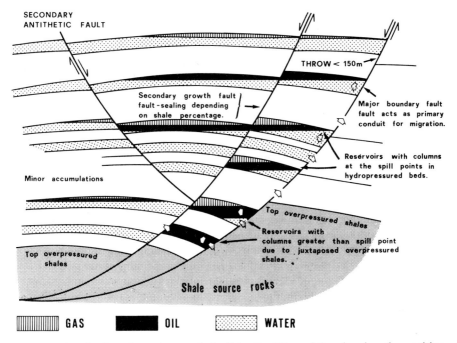

Fig. III.4.9. Section through growth fault in the Niger delta showing the position of various kinds of oil accumulations and possible avenues of migration between pools. (After Weber and Daukoru, 1975)

distances of the order of 100 km or even more. Such distances are not unreasonable, provided privileged migration avenues are available such as the major Cretaceous–Paleozoic unconformity in Western Canada.

Secondary migration over vertical distances, more than the thickness of a single reservoir rock, is only possible through faults, fracture systems, and other preferred avenues such as dikes, thrust plains, and mud volcanoes. This can lead to the superposition of several productive reservoirs, as in delta areas such as the Gulf Coast and the Niger delta, where growth faults are probably avenues of vertical migration (Fig. III.4.9). However, it is not undisputed that growth faults are conduits for vertical migration (Hedberg, 1977). Under such conditions individual accumulations in multiple-zone oil fields may represent a dynamic rather than a static system (Philippi, 1977): pools may be simultaneously receiving and losing hydrocarbons.

Another aspect of a dynamic situation with respect to petroleum accumulations is a process called separation-migration as described by Silverman (1965). It may be induced by a natural rupture of a sealing cap rock by faulting or fracturing, and causes a redistribution of the hydrocarbon content in a reservoir. Because this separation-migration process results in large differences in hydrocarbon composition, more pronounced than those caused by ordinary secondary migration, it is briefly discussed along with petroleum alterations (Part IV).

Summary and Conclusion

Secondary migration, the movement of oil and gas through carrier and reservoir rocks, and the subsequent formation of pools is controlled by three parameters. These are the buoyant rise of oil and gas in water-saturated porous rocks, capillary pressures that determine multiphase flow, and, as an important modifying influence, hydrodynamic fluid flow. As long as the aqueous pore fluids in the subsurface are stationary, i.e., under hydrostatic conditions, the only driving force for secondary migration is buoyancy. If there is water flow in the subsurface, i.e., under hydrodynamic conditions, the buoyant rise of oil and gas may be modified by this water flow. Capillary pressures in narrow rock pores are the cause for hydrocarbon entrapment.

Oil globules or gas bubbles larger than any given pore diameter have to be distorted before they can be squeezed through such a rock pore. The interfacial tension between oil, or gas, and water resists this distortion. The capillary pressure is the force necessary to squeeze the globule or bubble through that pore. Whenever capillary pressures are too high, or conversely rock pores too narrow, migrating oil or gas will be trapped. Petroleum trapped in a reservoir represents an equilibrium state between the driving forces (buoyancy or flow of water) that want to move the petroleum and capillary pressures that resist this movement.

The end of secondary migration, and the final stage in the formation of oil and gas pools is the concentration in the highest available part of a trap. The cap-rock, seal, or barrier that stops moving hydrocarbons by virtue of its general decrease in pore diameters must exert capillary pressures greater than the driving force. Distances covered by secondary migration are in the range of ten to a hundred kilometers, and occasionally even more.

Chapter 5
Reservoir Rocks and Traps, the Sites of Oil and Gas Pools

Petroleum is ultimately collected through secondary migration in permeable, porous reservoir rocks in the position of a trap. Any permeable and porous rock may act as a reservoir for oil and gas. They may be detrital or clastic rocks, generally of siliceous material, or chemically or biochemically precipitated rocks, usually carbonates. It is not uncommon that petroleum is found in fractured shales. Occasionally, igneous and metamorphic rocks are hosts for commercial quantities of petroleum, when favorably located in proximity to petroleum-bearing sedimentary sequences. The fundamental characteristic of a trap is its upward convex shape of a porous reservoir rock in combination with a more dense and relatively impermeable sealing cap rock above. The ultimate shape of the convexity may be angular, curved, or a combination of both. The only important geometric parameter is that it must be closed in vertical and horizontal planes without significant leaks to form an inverted container. The strike contours of this inverted container on a structural map must encircle closed areas comprising what is termed *closure area* or *closure* of a trap. A rare exception to this rule is true hydrodynamic trapping.

Causes for the formation of traps are numerous. They may be due to depositional features such as a sand-body embedded in and sealed by shales in a transgressive sequence, or a porous reef rock buried by dense limestones and shales. Such traps are commonly referred to as strategraphic traps. Traps formed by tectonic events such as folding or faulting are referred to as structural traps. Numerous and quite elaborate classifications of oil-field traps have envolved. In the context of oil migration and accumulation, however, such complex classifications are not necessary because most traps can be described in common geological terms, and in reality many traps bear features that make them difficult to classify.

For a detailed discussion on geological and petrological aspects of reservoir rocks and traps, such books as *Geology of Petroleum* by Levorsen (1967), *Petroleum Geology* by Chapman (1973) and *Introduction to Petroleum Geology* by Hobson and Tiratsoo (1975) may be consulted. The subsequent brief discussion on reservoir rocks and traps follows to a considerable extent these references.

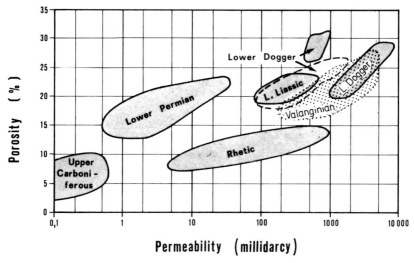

Fig. III.5.1. Relationship between porosities and permeabilities for sandstones of different geological ages in NW Germany. (After Füchtbauer, 1974)

5.1 Reservoir Rocks

The two most essential elements of a reservoir rock are porosity and permeability. The rock must contain pores or voids to store petroleum and these pores must be interconnected. The rock must be permeable to fluids and gases. A pumice rock, for example, although having very high porosity, is not a good reservoir rock because pores are not interconnected. Pore diameters also must be above a certain minimum size because the ratio of total pore volume to inner surface area must not be too low. If the pore size is too small, the capillary attraction of the mineral grains will hold the fluids in the pore space of the rock and fluids could not be produced because the absolute and relative permeabilities are too low.

Most clastic reservoir rocks have grain diameters in the range of 0.05 to 0.25 mm, resulting in average pore radii in sandstone reservoirs between 20 and 200 μ. A certain relationship can be observed between porosity and permeability in clastic rocks. An increase in porosity is paralleled by an increase in permeability. This trend is demonstrated in Figure III.5.1 for sandstones of different geological ages in northwest Germany.

Porosities in reservoir rocks usually range from 5 to 30%. The porosity of carbonate rocks is frequently somewhat less than that of sandstones, but the permeability of carbonates may be higher. A rough field appraisal of porosities in percent and permeabilities in millidarcy of the most common reservoir rocks is given in Table III.5.1 (after Levorsen, 1967).

The majority of petroleum accumulations are found in clastic or detrital reservoir rocks, including sandstones and siltstones (highly siliceous, comprised mainly of quartz), arkoses (derived from granites, mainly containing quartz,

5.1 Resevoir Rocks

Table III.5.1. Appraisal of porosities and permeabilities of common reservoir rocks (Levorsen, 1967)

Porosity %	Appraisal	Permeability in mili darcy
0–5	negligible	–
5–10	poor	–
10–15	fair	1.0– 10
15–20	good	10 – 100
20–25	very good	100 –1000

feldspar and mica), and graywackes (sandstones with particles of dark, fine-grained igneous rocks). Clastic rocks probably comprise more than 60% of all oil occurrences, and an additional 30% is found in carbonate reservoirs. It is significant, however, that more than 40% of the so-called giant oil and gas fields are found in carbonates. The enormous petroleum accumulations in the carbonates of the Persian Gulf distort such a classification. The remaining 10% of all petroleum occurrences are found in fractured shales, igneous and metamorphic rocks, etc.

Two different kinds of porosity can be identified in sedimentary rocks, primary or intergranular and secondary porosity. Sandstones commonly have a primary

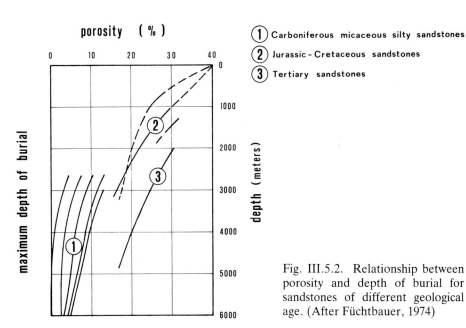

Fig. III.5.2. Relationship between porosity and depth of burial for sandstones of different geological age. (After Füchtbauer, 1974)

porosity which is largely dependent on the packing characteristics and on the variation in size and shape of the grains. Well-sorted sands with nearly spherical grains, such as certain beach or dune sands, make excellent reservoirs with high primary porosity. The Rotliegend dune sands of the giant gas field of Groningen in the Netherlands are an example for the outstanding quality of such reservoir rocks. If too much fine-grained material is present, reservoir quality drops. With increasing overburden and diagenesis, sandstones tend to lose porosity, mainly due to recrystallization and cementation. This decrease in porosity with increasing depth of burial is shown in Figure III.5.2 (after Füchtbauer, 1974). The secondary growth of quartz on sand grains is impeded when sandstones are oil impregnated. Philipp et al. (1963) used this phenomenon to estimate the time of formation of an oil pool. The degree of quartz diagenesis was determined by estimating the percentage of quartz grains with secondary euhedral quartz overgrowth in water-saturated sandstones free of hydrocarbons. This was then plotted against maximum depth of burial. Oil-impregnated reservoir sandstones were compared to this depth–quartz diagenesis plot. From the degree of quartz diagenesis of the oil-impregnated reservoir rocks the depth and time of emplacement of oil was derived.

5.2 Traps

A trap is a geological feature which enables migrating petroleum to accumulate and be preserved for a certain time interval. Traps occur in fundamentally different forms and can enclose very different volumes of pore space and hence petroleum. The maximum total holding capacity or closed colume of a trap is the volume between its highest point and the "spilling plane" or outflow level at the bottom. Traps are rarely completely full, i.e., the petroleum column rarely extends down to the point at which it would spill out of the trap. Furthermore, traps are never full in the sense that all available pore space in the reservoir rock is occupied by petroleum. There is always a certain residual amount of water, which cannot be displaced by migrating petroleum.

Traps can be formed either by tectonic activity such as faults, folds, etc. (structural traps) or from sedimentary depositional patterns (stratigraphic traps). The majority of known oil fields is found in structural traps, but structural traps are easier to locate by current geophysical and geological exploration techniques. Stratigraphic traps have become a prime target for exploration only in more recent years. Sometimes there are combined tectonic and depositional causes for the formation of a trap. Then these traps are called combination traps.

Anticlinal structures are the most common type of traps. One of the largest oil-producing anticlines is the Kirkuk field in Iraq. The productive area extends in three culminations over a length of nearly 30 miles. The "Golden Lane" fields in Mexico also occur as local "highs" along an arch extending some 50 miles. Most oil-producing anticlines, however, are only a few miles in lenght, and are often associated with some faulting.

5.2 Traps

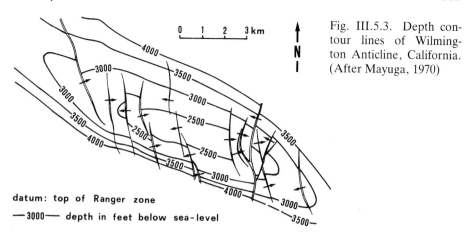

Fig. III.5.3. Depth contour lines of Wilmington Anticline, California. (After Mayuga, 1970)

The giant Wilmington oil field in the Los Angeles basin of southern California may be cited as an example of a structural anticlinal trap. The Los Angeles basin is one of the most prolific oil-producing basins in the world in terms of the ratio of pooled oil to total rock volume. The Wilmington structure (Figs. III.5.3 and III.5.4) is a broad, assymetric anticline about 12 miles long and 4 miles wide. It is dissected by a series of transverse normal faults which separate the producing sand and sandstone reservoirs into many different pools (Mayuga, 1970). Seals are provided by claystones and shales. Oil properties in the various fault blocks differ considerably, illustrating that the accumulation of hydrocarbons apparently occurred independently in each of the blocks. The different oil properties also suggest that the faults provide effective barriers to communication between adjacent fault blocks. There are seven major producing zones ranging in age from late Miocene to early Pliocene. The upper part of the Wilmington structure was eroded during the early Pliocene, and subsequent submergence resulted in the deposition of an additional 1800–2000 ft of almost horizontal Pliocene and Pleistocene sediments on top of the anticline (Fig. III.5.4). The entire Wilmington oil field is estimated to contain about 3 billion barrels of produceable oil (Mayuga, 1970).

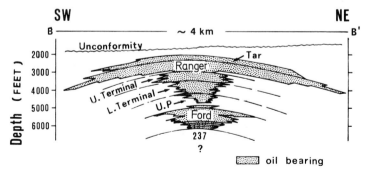

Fig. III.5.4. Geological cross section through Wilmington Anticline, California. (After Mayuga, 1970)

A different type of structural trap is associated with so-called growth faults which play an important role in such prolific areas as the Niger Delta region and the Gulf Coast, USA. The term growth fault is used because the faults remain active after their formation, and allow faster sedimentation on the down-thrown side. The enhanced deposition of sediments across the growth fault initiates a rotational movement which tilts the beds toward the fault-creating anticlinal "rollover" structures along the faults.

The subsequent discussion on oil fields and traps in the Niger Delta is mainly derived from Weber and Daukoru (1975). The majority of the approximately 150 oil fields in the Niger Delta have at least one fault which influences the accumulation in addition to the major growth fault. The fault zones can be sealing or nonsealing. The sealing capacity or conductivity of a fault depends on the throw, and on the amount of shale or sand caught up or otherwise incorporated into the fault zone. Evidence for vertical conductivity of major boundary faults is provided by the fact that in most cases the fault intersection with the upper bedding plane of a reservoir (Fig. III.4.9) acts as the spill point of the accumulation. It is possible that these spill points are also the entry points of the oil from the fault zone into the reservoir. The growth faults are in most instances sealing on the up-thrown side of the fault zone. The small dip closure of the structural traps and the fact that faults with throws of 150 m or more are considered as potential vertical leaks, is thought to be the cause for the relatively short oil columns in the Niger Delta, which rarely exceed 50 m.

There are numerous oil fields where oil is trapped in monoclines against seals provided by fault planes above the monocline. The Pechelbronn oil field in the Tertiary of the Rhine-Graben and the oil fields in Cretaceous Woodbine sandstones along the fault zone of Mexia in eastern Texas are examples of such

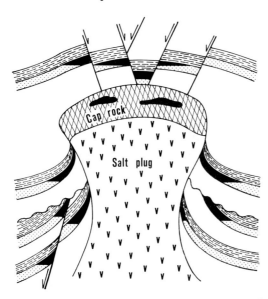

Fig. III.5.5. Some possibilities for oil pools associated with salt plugs. (Modified after Levorsen, 1967)

5.2 Traps

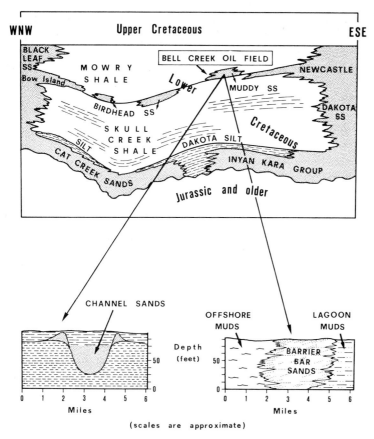

Fig. III.5.6. The Bell Creek oil field in Montana. A geological profile through the Powder River Basin, showing the oil field location and the geological setting of the channel and barrier sand which act as reservoirs. (After McGregor and Biggs, 1970)

fault traps. It should be pointed out again, however, that fault zones are not always sealing, but are sometimes nonsealing. Fault zones may even provide migration avenues, as in the Niger Delta or other areas with deep reaching fault systems.

The association of oil accumulations with salt domes is another example of structural traps formed by tectonic events. Oil accumulations in tectonic features caused by the upward movement of salt from deeply buried evaporite beds are known in several parts of the world, including northern Germany and the North Sea, Romania, West Africa and the Gulf Coast region of USA and Mexico. The relationship between the oil accumulations and the salt is only structural and there are many ways to provide suitable traps in association with salt intrusion. The traps may be formed along the folded or faulted flanks of the salt plug or on top of the plug where arching and/or faulting is produced in the overlying sediments. Some possibilities for oil pools in association with salt plugs are shown in Figure III.5.5.

The formation of *stratigraphic traps* is related to sediment deposition or erosion and is thus distinguished from formation of structural traps, which originate from tectonic events. It is clear, however, that tectonism ultimately controls many geological processes such as deposition and erosion of sediments. Stratigraphic traps can be related to facies changes, diagenesis of sediments and to unconformities in the sedimentary rock column (Rittenhouse, 1972). Most basins contain the prerequisites for the creation of stratigraphic traps. Typical examples are barrier sand bars, deltaic distributary channel sandstones and carbonate reefs of different forms and sizes. Such barrier and channel sands and carbonate reef structures, frequently embedded in fine-grained organic-rich sedimentary sequences, are ideal traps for migrating petroleum.

The giant *Bell Creek oil field* in Montana, USA (Fig. III.5.6), serves as an example for *barrier and channel sand reservoirs*. The geology of the Bell Creek field is described by McGregor and Biggs (1970). The Lower Cretaceous stratigraphic trap at the northeastern edge of the Powder River basin is not controlled by tectonic events other than regional dip. The Muddy Sandstone trap

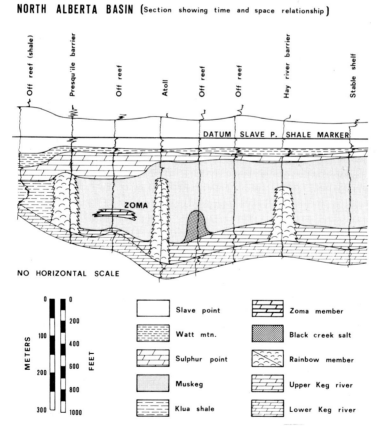

Fig. III.5.7. Geological cross-section through the North Alberta Basin, showing the temporal and spatial distribution of the Devonian reefs. (After Barss et al., 1970)

was formed near the intersection of littoral marine bars and a delta system with distributary channel sands. Profiles through such channel and barrier sands enclosed in shales are also given in Figure III.5.6.

Fossil organic reefs form very prolific stratigraphic traps in many parts of the world, for example the Middle East, Libya, southwestern USA and in Alberta, Western Canada. Reefs are constructed by various groups of organisms, such as corals, algae, echinoderms, mollusks and foraminifera. They may be interpreted as "fringing reefs", "barrier reefs", "atolls" or "pinnacle reefs", depending on their environment of growth and shape. Accordingly they vary in length, width and thickness. As an example of reef traps the giant oil fields of the *Middle Devonian reefs of the Rainbow area* in Alberta, are briefly discussed from a paper by Barss et al. (1970). The growth of the Rainbow Member reefs in the Black Creek Basin of northwestern Alberta was initiated during Upper Keg River time when banks of reef-constructing organisms started to flourish (Fig. III.5.7). Structural control of reef growth, if it did exist, was subtle. The reefs in the Rainbow sub-basin, atolls and pinnacle reefs, have vertical reliefs up to 820 ft. The reefs are enveloped by fine-grained off-reef sediments, consisting of salt, interbedded anhydrites, dolomites and shales. An evaporite cover provides an excellent seal for petroleum entrapment. Reserves from the Rainbow Member pools are estimated to be in excess of 1.3 billion barrel of oil-in-place.

Finally, it should be pointed out that migrating petroleum can be captured by all kinds of traps, by structural, stratigraphic or combination traps, only as long as they are in the path of migration.

Summary and Conclusion

Petroleum is ultimately collected through secondary migration in permeable porous reservoir rocks in the position of a trap. It is characterized by its upward convex shape in combination with a porous reservoir rock with a relatively impermeable sealing cap rock above. Generally one distinguishes between stratigraphic and structural traps. Stratigraphic traps are mainly caused by depositional features, and structural traps by tectonic events. These causes for the formation of a trap may also be combined. Typical examples for stratigraphic traps are barrier sand bars, deltaic distributory channel sandstones and carbonate reefs. Typical examples for structural traps are anticlines, fault traps, and traps associated with salt domes.

The majority of petroleum accumulations are found in clastic reservoir rocks, such as sandstones and siltstones. Next important reservoir rocks are carbonates. Fractured shales, igneous and metamorphic rocks play only a minor role. The maximum total holding capacity or closed volume of a trap is the volume between its highest point and the spilling plane or outflow level at the bottom. Porosities in reservoir rocks usually range from 5 to 30%. Traps are never full in the sense that all available pore space is occupied by petroleum. There is always a certain residual amount of water, which cannot be displaced by migrating petroleum.

References to Part III

Aoyagi, K.: Petrophysical approach to origin of porosity of carbonate rocks in Middle Carboniferous, Windsor Group, Nova Scotia, Canada. Am. Assoc. Pet. Geol. Bull. **57,** 1692–1702 (1973)

Athy, L. F.: Compaction and oil migration. Am. Assoc. Pet. Geol. Bull. **14,** 25–35 (1930)

Baker, E. G.: Origin and migration of oil. Science **129,** 871–874 (1959)

Baker, E. G.: Distribution of hydrocarbons in petroleum. Am. Assoc. Pet. Geol. Bull. **46,** 76–84 (1962)

Baker, E. G.: A geochemical evaluation of petroleum migration and accumulation. In: Fundamental Aspects of Petroleum Geochemistry. Amsterdam-London-New York: Elsevier, 1967, pp. 299–329

Balducchi, A., Pommier, G.: Cambrian oil field of Hassi Messaoud, Algeria. Geology of giant petroleum fields. Memoir 14. Am. Assoc. Pet. Geol. 477–488 (1970)

Barker, C.: Aquathermal pressuring — role of temperature in development of abnormal-pressure zones. Am. Assoc. Pet. Geol. Bull. **56,** 2068–2071 (1972)

Barker, C.: Personal communication, Univ. Tulsa (1976)

Barss, D. L., Copland, A. B., Ritchie, W. D.: Geology of Middle Devonian Reefs, Rainbow Area, Alberta, Canada. In: Geology of giant petroleum fields. Memoir 14. Am. Assoc. Pet. Geol. 19–49 (1970)

Berg, R. R.: Capillary pressures in stratigraphic traps. Am. Assoc. Pet. Geol. Bull. **59,** 939–956 (1975)

Bestougeff, M. A.: Petroleum hydrocarbons. In: Fundamental Aspects of Petroleum Geochemistry. Nagy, B., Colombo, U. (eds.). Amsterdam-London-New York: Elsevier, 1967, pp. 77–108

Bonham, L. C.: Solubility of methane in water at elevated temperatures and pressures. AAPG Bull. Vol. 62, No. 12. 2478–2488 (1978)

Bradley, J. S.: Abnormal formation pressure. Am. Assoc. Pet. Geol. Bull. **59,** 957–973 (1975)

Bray, E. E., Evans, E. D.: Hydrocarbons in non-reservoir-rock source beds. Am. Assoc. Pet. Geol. Bull. **49,** 248–257 (1965)

Bray, E. E., Foster, W. R.: A process of primary migration of petroleum. Am. Assoc. Pet. Geol. Bull. **64,** 107–114 (1980)

Burst, J. F.: Postdiagenetic clay mineral environmental relationship in the Gulf Coast Eocene. In: Clay and Clay Minerals: 6th Nat. Clays and Clay Mineral Conf. Proc. Swineford, A. (ed.). London-New York: Pergamon Press, 1959, p. 411

Burst, J. F.: Diagenesis of Gulf Coast clayey sediments and its possible relation to petroleum migration. Am. Assoc. Pet. Geol. Bull. **53,** 73–93 (1969)

Chapman, R. E.: Primary migration of petroleum from clay source rocks. Am. Assoc. Pet. Geol. Bull. **56,** 2185–2191 (1972)

Chapman, R. E.: Petroleum Geology. Amsterdam-London-New York: Elsevier, 1973

Chiarelli, A.: Etude de nappes aquifères profondes — contribution de l'hydrogéologie à la connaissance d'un bassin sédimentaire et à l'exploration pétrolière. Ph. D. thesis. Univ. Bordeaux, No. 401 (1973)

Cordell, R. J.: Depth of oil origin and primary migration, a review and critique. Am. Assoc. Pet. Geol. Bull. **56,** 2029–2067 (1972)

Cordell, R. J.: Colloidal soap as proposed primary migration medium for hydrocarbons. Am. Assoc. Pet. Geol. Bull. **57,** 1618–1643 (1973)

Coustau, H., Rumeau, J. L., Sourisse, C., Chiarelli, A., Tison, J.: Classification hydrodynamique des bassins sédimentaires, utilisation combinée avec d'autres méthodes pour rationaliser l'exploration dans des bassins non-productifs. Proc. 9th World Pet. Congr. Tokyo. London: Applied Science Publ., 1975, Vol. II, pp. 105–119

Deroo, G.: Corrélation huiles brutes — roches mères à l'échelle des bassins sédimentaires. Bull. Centre Rech. Pau-SNPA **10,** 317–335 (1976)

Dickey, P. A.: Possible primary migration of oil from source rock in oil phase. Am. Assoc. Pet. Geol. Bull. **59,** 337–345 (1975)

Dickinson, G.: Geological aspects of abnormal reservoir pressures in Gulf Coast, Louisiana. Am. Assoc. Pet. Geol. Bull. **37,** 410–432 (1953)

Doebl, F., Heling, D., Homann, W., Karweil, J., Teichmüller, M., Welte, D.: Diagenesis of Tertiary clayey sediments and included dispersed organic matter in relationship to geothermics in the Upper Rhine Graben. In: Approaches to Thaphrogenesis. Inter-Union Commission on Geodynamics, Sci. Rep. **8,** Stuttgart: Schweitzerbart'sche Verlagsbuchhandlung, 1974

Dow, W. G.: Application of oil-correlation and source-rock data to exploration in Williston Basin. Am. Assoc. Pet. Geol. Bull. **58,** 1253–1262 (1974)

Dreier, K. B.: Die Speicherung von Methan im Nebengestein von Kohlen. Brennstoff-Chem. **49,** 21–30 (1968)

Dunoyer de Segonzac, G.: Les minéraux argileux dans la diagenèse passage au métamorphisme. Mem. Serv. Carte Geol. Als. Lorr. **29,** 320 (1969)

Dunoyer de Segonzac, G.: The transformation of clay minerals during diagenesis and low grade metamorphism, a review. Sedimentology **15,** 281–286 (1970)

Durand, B.: Presents trends in organic geochemistry in research on migration of hydrocarbons. In: Advances in Organic Geochemistry 1981. Bjorøy M. et al., (eds.). Chichester: Wiley, pp. 117–128

Durand, B., Oudin, J. L.: Exemple de migration des hydrocarbures dans une série deltaïque: le delta de la Mahakam, Kalimantan, Indonesie, Proceedings 10th World Petr. Congress, Bucharest, 1979, Vol. 1, pp. 3–11

Ekweozor, C. M., Okoye, N. V.: Petroleum source-bed evaluation of Tertiary Niger Delta. Am. Assoc. Pet. Geol. Bull. **64,** 1251–1259 (1980)

Engelhardt, W. von: The origin of sediments and sedimentary rocks (Sedimentary Petrology, Part III). Stuttgart: Schweitzerbart'sche Verlagsbuchhandlung, 1977

Evamy, D. D., Harembourne, J., Kamerling, P., Knaap, A., Molloy, F. A., Rowlands, P. H.: Hydrocarbon habitat of Tertiary Niger Delta. Am. Assoc. Pet. Geol. Bull. **62,** 1–39 (1978)

Frank, D. J., Sackett, W., Hall, R., Fredericks, A.: Methane, ethane and propane concentrations in Gulf of Mexico. Am. Assoc. Pet. Geol. Bull. **54,** 1933–1938 (1970)

Franks, F.: Solute–water interactions and the solubility behavior of long-chain paraffin hydrocarbons. Nature (London) **210,** 87–88 (1966)

Füchtbauer, H.: Sediments and Sedimentary Rocks (Sedimentary Petrology, Part II) Stuttgart: Schweitzerbart'sche Verlagsbuchhandlung, 1974

Galloway, W. E., Hobday, D. K., Magara, K.: Frio formation of Texas Gulf coastal plains depositional systems, structural framework, and hydrocarbon distribution. Am. Assoc. Pet. Geol. Bull. **66,** 649–688 (1982)

Gealy, E. L., Davies, T. A.: The deep sea drilling project. Geotimes **14**, 10 (1969)
Ginsburg, R. N.: Early diagenesis and lithification of shallow-water carbonate sediments in south Florida. In: Regional Aspects of Carbonate Deposition. Le Blanc, R. J., Breeding, J. G. (eds.). Soc. Paleontol. Mineral. Spec. Publ. **5**, 80–100 (1957)
Gussow, W. C.: Differential entrapment of oil and gas — a fundamental principle. Am. Assoc. Pet. Geol. Bull. **38**, 816–853 (1954)
Gussow, W. C.: Migration of reservoir fluids. J. Pet. Technol., 353–363 (1968)
Halbouty, M. T.: Geology of giant petroleum fields. In: Am. Assoc. Pet. Geol. Memoir 14, 1–7 (1970)
Hedberg, H. D.: The effect of gravitational compaction on the structure of sedimentary rocks. Am. Assoc. Pet. Geol. Bull. **10**, 1035–1072 (1926)
Hedberg, H. D.: Gravitational compaction of clays and shales. Am. J. Sci. **31**, 241–287 (1936)
Hedberg, H. D.: Geologic aspects of origin of petroleum. Am. Assoc. Pet. Geol. Bull. **48**, 1755–1803 (1964)
Hedberg, H. D.: Relation of methane generation to undercompacted shales, shale diapirs and mud volcanoes. Am. Assoc. Pet. Geol. Bull. **58**, 661–673 (1974)
Hedberg, H. D.: Personal communication on growth faults as conduits for vertical migration (1977)
Hedemann, H. A., Wissmann, W.: Die Gebirgstemperaturen in der Bohrung Saar 1 und ihre Beziehungen zum geologischen Bau. Geol. Jb. A **27**, 433–454 (1976)
Heling, D.: Micro-fabrics of shales and their rearrangement by compaction. Sedimentology **15**, 247–260 (1970)
Heling, D., Teichmüller, M.: Die Grenze Montmorillonit/Mixed Layer-Minerale und ihre Beziehung zur Inkohlung in der Grauen Schichtenfolge des Oligozäns im Oberrheingraben. Fortschr. Geol. Rheinld. Westfal. **24**, 113–128 (1974)
Hitchon, B.: Fluid flow in the western Canada sedimentary basin. 1. Effect of topography. Water Resour. Res. **5**, 186–195 (1969a)
Hitchon, B.: Fluid flow in the western Canada sedimentary basin. 2. Effect of geology. Water Resour. Res. **5**, 460–469 (1969b)
Hitchon, B., Horn, M. K.: Petroleum indicators in formation waters from Alberta, Canada. Am. Assoc. Pet. Geol. Bull. **58**, 464–473 (1974)
Hobson, G. D.: Some Fundamentals of Petroleum Geology. London-New York-Toronto: Oxford University Press, 1954
Hobson, G. D.: The occurrence and origin of oil and gas. In: Modern Petroleum Technology, 4th ed. Hobson, G. (ed.). On behalf of Inst. Pet. G. B. London: Applied Science Publ., 1973, pp. 1–25
Hobson, G. D., Tiratsoo, E. N.: Introduction to Petroleum Geology. Beaconsfield, England: Scientific Press, 1975
Hollmann, R.: Über Subsolution und die „Knollenkalke" des Calcara Ammonitico Rosso Superiore im Monte Baldo (Malin, Norditalien). Neues Jahrb. Geol. Paläontol. Monatsh. **4**, 163–179 (1962)
Hubbert, M. K.: The theory of groundwater motion. J. Geol. **48**, 785–944 (1940)
Hubbert, M. K.: Entrapment of petroleum under hydrodynamic conditions. Am. Assoc. Pet. Geol. **37**, 1954–2026 (1953)
Hubbert, M. K., Rubey, W. W.: Role of fluid pressure in mechanics of overthrust faulting, Parts I and II. Am. Soc. Geol. **70**, 115–205 (1959)
Hunt, J. M.: The origin of petroleum in carbonate rocks. In: Carbonate Rocks (Physical and Chemical Aspects). Developments in Sedimentology 9B. Chilingar, G. V., Bissell, H. J., Fairbridge, R. W. (eds.). Amsterdam: Elsevier, 1967, pp. 225–251

Hunt, J. M., Jamieson, G. W.: Oil and organic matter in source rocks of petroleum. Am. Assoc. Pet. Geol. Bull. **40,** 477–488 (1956)
Jones, R. W.: Some mass balance and geological constraints on migration mechanisms. In: AAPG Studies in Geology No. 10, Problems of Petroleum Migration. Roberts, III, W. H., Cordell, R. J. (eds.), 1980, pp. 47–68
Jüntgen, H., Karweil, J.: Porenverteilung und innere Oberfläche der Kohlen der Bohrung Münsterland 1. Fortschr. Geol. Rheinld. Westfal. **11,** 179–196 (1963)
Kappelmeyer, O.: Geothermik. In: Lehrbuch der Angewandten Geologie, Bd. I. Bentz, A. (ed.). Stuttgart: Enke, 1961, pp. 863–888
Kappelmeyer, O., Haenel, R.: Geothermics — with Special Reference to Application. Berlin-Stuttgart: Borntraeger, 1974
Kidwell, A. L., Hunt, J. M.: Migration of oil in Recent sediments of Pedernales, Venezuela. In: Habitat of Oil. Weeks, L. G (ed.). Symposium Am. Assoc. Pet. Geol. Tulsa, 790–817 (1958)
Klemme, H. D.: Geothermal Gradients, Heat Flow and Hydrocarbon Recovery, Petroleum and Global Tectonics. Fischer, G., Judson, S. (eds.). Princeton: Princeton University Press, 1975, pp. 251–304
Kvenvolden, K. A., Claypool, G. E.: Origin of gasoline-range hydrocarbons and their migration by solution in carbon dioxide in Norton Basin, Alaska. Am. Assoc. Pet. Geol. Bull. **64,** 1078–1086 (1980)
Landes, K. K.: Eo-metamorphism and oil and gas in time and space. Am. Assoc. Pet. Geol. Bull. **51,** 828–841 (1967)
Lee, H. K., Uyeda, S.: Review of heat flow data. Terrestrial Heat Flow. Geophysics. Monogr. **8,** 87–190 (1965)
Lehner, P.: Salt tectonics and Pleistocene stratigraphy on continental slope of northern Gulf of Mexico. Am. Assoc. Pet. Geol. Bull. **53,** 2431–2479 (1969)
Levorsen, A. I.: Geology of Petroleum. San Francisco-London: Freeman, 1954
Levorsen, A. I.: Geology of Petroleum, 2nd ed. San Francisco-London: Freeman, 1967
Leythaeuser, D., Schaefer, R. G., Yükler, A.: Diffusion of light hydrocarbons through near-surface rocks. Nature (London) **284,** 522–525 (1980)
Leythaeuser, D., Schaefer, R. G., Yükler, A.: Role of diffusion in primary migration of hydrocarbons. Am. Assoc. Pet. Geol. Bull. **66,** 408–429 (1982)
Leythaeuser, D., Mackenzie, A. S., Schaefer, R. G., Altebäumer, F. J., Bjorøy, M.: Recognition of migration and its effects within two core holes in shale/sandstone sequences from Svalbard, Norway. In: Advances in Organic Geochemistry 1981. Bjorøy, M. et al. (eds.). Chichester: John Wiley, 1983a, pp. 136–146
Leythaeuser, D., Mackenzie, A. S., Schaefer, R. G., Bjorøy, M.: A novel approach for recognition and quantitation of hydrocarbon migration effects in shale/sandstone sequences, Am. Assoc. Pet. Geol. Bull. **68,** 196–219 (1983b)
Louis, M.: Cours de Géochimie du Pétrole, Société des Editions Technip et Inst. Français du Pétrole, Paris (1967)
Low, Ph.: Personal communication (1976)
Lutz, M., Kaasschieter, J. P. H., van Wijhe, D. H.: Geological factors controlling Rotliegend gas accumulations in the Mid-European Basin, Proc. 9th World Pet. Congr. Tokyo. London: Applied Science Publ., 1975, Vol. II. pp. 93–103
Mackenzie, A. S., Leythaeuser, D., Schaefer, R. G., Bjorøy, M.: Expulsion of petroleum hydrocarbons from shale source rocks: Nature (London) **301,** 506–509 (1983)
Magara, K.: Compaction and migration of fluids in Miocene mudstone, Nagaoka Plain, Japan. Am. Assoc. Pet. Geol. Bull. **52,** 2466–2501 (1968)
Magara, K.: Aquathermal fluid migration. Am. Assoc. Pet. Geol. Bull. **58,** 2513–2526 (1974)

Magara, K.: Re-evaluation of montmorillonite dehydration as cause of abnormal pressure and hydrocarbon migration. Am. Assoc. Pet. Geol. Bull. **59,** 292–302 (1975)

Magara, K.: Water expulsion from clastic sediments during compaction — Directions and volumes. Am. Assoc. Pet. Geol. Bull. **60,** 543–553 (1976)

Martin, R. L., Winters, J. C., Williams, J. A.: Composition of crude oils by gas chromatography: geological significance of hydrocarbon distribution. 6th World Pet. Congr. Sec. V, paper 13 (1963)

May, F., Freund, W., Müller, E. P.: Modellversuche über Isotopenfraktionierung von Erdgaskomponenten während der Migration. Z. angew. Geologie **14,** 376–380 (1968)

Mayer-Gürr, A.: Petroleum Engineering. In: Geology of Petroleum. Beckmann, H. (ed.). Stuttgart: Enke, 1976

Mayuga, M. N.: Geology and Development of California's Giant Wilmington Oil Field. Geology of Giant Petroleum Fields. Memoir 14, Am. Assoc. Pet. Geol., 158–184 (1970)

McAuliffe, C.: Solubility in water of paraffin, cycloparaffin, olefin, acetylene, cycloolefin and aromatic hydrocarbons. J. Phys. Chem. **70,** 1267–1275 (1966)

McAuliffe, C. D.: Oil and gas migration: chemical and physical constraints. In: AAPG Studies in Geology No. 10, Problems of Petroleum Migration, Roberts, III, W. H., Cordell, J. R. (eds.), 1980, pp. 89–107

McGregor, A. A., Biggs, C. A.: Bell Creek Field, Montana: a Rich Stratigraphic Trap. Geology of Giant Petroleum Fields. Halbouty, M. T. (ed.). Memoir 14, Am. Assoc. Pet. Geol., Tulsa, 128–146 (1970)

McIver, R. D.: Hydrocarbon Occurrences from Joides Deep Sea Drilling Projects, Proc. 9th World Pet. Congr., Tokyo. II London: Applied Science Publ., 1975, Vol. II. pp. 269–280

McNeal, R. P.: Hydrodynamics of the Permian Basin, Fluids in subsurface environments. Young, A., Galley, J. E. (eds.). Memoir 4, semicent. commem. vol. Am. Assoc. Pet. Geol., 308–326 (1965)

Meinschein, W. G.: Origin of petroleum. Am. Assoc. Pet. Geol. Bull. **43,** 925–943 (1959)

Meissner, F. F.: Petroleum geology of the Bakken Formation, Williston basin, North Dakota and Montana: Williston Basin Symposium, Montana Geological Society, 1978, pp. 207–227

Milliken, K. L., Land, L. S., Loucks, R. G.: History of burial diagenesis determinated from isotopic geochemistry, Frio Formation, Brazoria County, Texas. Am. Assoc. Pet. Geol. Bull. **65,** 1397–1413 (1981)

Millot, G.: Geology of Clays. Engl. transl. by Farrand, W. R., Paquet, H. New York-Heidelberg-Berlin: Springer, 1970; Paris: Mason et Cie., London: Chapman & Hall, 1964

Nagumo, S.: Compaction of sedimentary rocks — a consideration by the theory of porous media. Earthquake Res. Inst. (Jpn) **43,** 339–348 (1965)

Neruchev, S. G., Kovacheva, I. S.: De l'influence des conditions géologiques sur la teneur en huile des roches mères (in Russian). Dokl. Akad. Nauk, S. S. S. R. **162,** 913–914 (1965)

Palacas, J. G., Love, A. H., Gerrild, P. N.: Hydrocarbons in estuarine sediments of Choctawhatchee Bay, Florida, and their implications for genesis of petroleum. Am. Assoc. Pet. Geol. Bull. **56,** 1402–1418 (1972)

Peake, E., Hodgson, G. W.: Alkanes in aqueous systems, I. Exploratory investigations on the accommodation of C_{20}–C_{33} n-alkanes in distilled water and occurrence in natural water systems. J. Am. Oil Chem. Soc. **43,** 215–222 (1966)

Perry, D.: Early diagenesis of sediments and their interstitial fluids from the continental slope, northern Gulf of Mexico. Trans. Gulf Coast Assoc. Geol. Soc. **20,** 219–227 (1970)

Philipp, W., Drong, H. J., Füchtbauer, H., Haddenhorst, H. G., Jankowsky, W.: The history of migration in the Gifthorn trough (NW Germany). Proc. 6th World Pet. Congr. Frankfurt, Section 1, Paper 19, PD2, 457–478 (1963)

Philippi, G. T.: On the depth, time and mechanism of origin of the heavy to medium-gravity naphthenic crude oils. Geochim. Cosmochim. Acta **41**, 33–52 (1977)

Poulet, M.: Problèmes posés par la migration secondaire du pétrole et sa mise en place dans les gisements. Rev. Inst. Fr. Pét. **23**, 159–173 (1968)

Powers, M.C.: Fluid-release mechanisms in compacting marine mudrocks and their importance in oil exploration. Am. Assoc. Pet. Geol. Bull. **51**, 1240–1254 (1967)

Price, L. C.: The solubility of hydrocarbons and petroleum in water as applied to the primary migration of petroleum. Ph. D. thesis Univ. California, Riverside, 1973

Price, L. C.: Aqueous solubility of petroleum as applied to its origin and primary migration. Am. Assoc. Pet. Geol. Bull. **60**, 213–244 (1976)

Redmond, J. L., Koesoemadinata, R. P.: Walio oil field and the Miocene carbonates of Salawati basin, Irian-Jaya, Indonesia. 5th Conv. Indones. Pet. Assoc. Jakarta June, 7–9 (1976)

Reed, L. C.: San Pedro Oil Field, Province of Salta, Northern Argentina. Am. Assoc. Pet. Geol. Bull. **30**, 591–605 (1964)

Rittenhouse, G.: Stratigraphic-trap classification. In: Stratigraphic Oil and Gas Fields — Classification, Exploration Methods, and Case Histories. King, R. E. (ed.). Joint publ. Am. Assoc. Pet. Geol. and Soc. Explor. Geophys. AAPG Memoir 16, SEG Spec. Publ. No. 14, Tulsa, 14–28 (1972)

Robin, P.: Caractérisation des kérogènes et de leur évolution par spectroscopie infrarouge. Dissert. Univ. Cath. Louvain. Groupe de Physico-Chimie Min. Catalyse (1975)

Rouchet, J. du: Stress fields, a key to oil migration. Am. Assoc. Pet. Geol. Bull. **65**, 74–85 (1981)

Seevers, D. O.: Personal communication (1969)

Seifert, W. K.: Carboxylic acids in petroleum and sediments. In: Fortschritte der Chemie organischer Naturstoffe. Herz, W., Grisebach, H., Kirby, G. W. (eds.). Wien-New York: Springer, 1975, Vol. 32 pp. 1–49

Seifert, W. K., Howells, W. G.: Interfacially active acids in a California crude oil. Analyt. Chem. **41**, 554–562 (1969)

Sharp, Jr., J. M., Domenico, P. A.: Energy transport in thick sequences of compacting sediment. Geol. Soc. Am. Bull. **87**, 390–400 (1976)

Silverman, S. R.: Migration and segregation of oil and gas. In: Fluids in Sub-surface Environments. Young, A., Galley, J. E. (eds.). Memoir 4, Am. Assoc. Pet. Geol., 53–65 (1965)

Snarsky, A. N.: Die primäre Migration des Erdöls. Freiberger Forschungsh. C **123**, 63–73 (1962)

Stahl, W. J.: Carbon and Nitrogen Isotopes in Hydrocarbon Research and Exploration. 4th Eur. Coll. Geochronol. Cosmochronol. Isotope Geol., Amsterdam (1976)

Stahl, W. J., Boigk, H., Wollanke, G.: Carbon and Nitrogen Isotope Data of Upper Carboniferous and Rotliegend Natural Gases from North Germany and Their Relationship to the Maturity of the Organic Source Material. In: Advances in Organic Geochemistry 1975. Campos, R., Goñi, J. (eds.). Empresa Nacional Adaro de Investigaciones Mineras, S. A. Madrid (Spain), 1977, pp. 539–559

Tissot, B., Deroo, G., Espitalié, J.: Etude comparée de l'époque de formation et d'expulsion du pétrole dans diverses provinces géologiques, Proc. 9th World Pet. Congr., Tokyo. London: Applied Science Publ., 1975, Vol. II, pp. 159–169

Tissot, B., Pelet, R.: Nouvelles données sur les mécanismes de genèse et de migration du pétrole, simulation mathématique et application à la prospection. Proc. 8th World Pet. Cong., Moscow, PD1 (4) (1971)

Ungerer, P., Behar, E., Discamps, D.: Tentative calculation of the overall volume expansion of organic matter during hydrocarbon genesis from geochemistry data: implications for primary migration. In: Advances in Organic Geochemistry, 1981. Bjorøy, M. et al. (eds.). Chichester: John Wiley, 1983, pp. 129–135

Vandenbroucke, M.: Etude de la migration primaire: variation de composition des extraits de roche à un passage roche mère/réservoir. In: Advances in Organic Geochemistry 1972. Oxford-Braunschweig: Pergamon Press, 1972, pp. 547–565

Weaver, C. E.: The clay petrology of sediments. In: Clays, Clay Minerals. Proc. Nat. Conf. Clays Clay Min. **6,** 1959, 154–187

Weaver, C. E.: Possible uses of clay minerals in search for oil. Am. Assoc. Pet. Geol. Bull. **44,** 1505–1518 (1960)

Weber, K. J., Daukoru, E.: Petroleum Geology of the Niger Delta, Proc. 9th World Pet. Congr., Tokyo. London: Applied Science Publ., 1975, Vol. II, pp. 209–221

Welte, D. H., Hagemann, H. W., Hollerbach, A., Leythaeuser, D., Stahl, W.: Correlation between petroleum and source rock, Proc. 9th World Pet. Congr., Tokyo. London: Applied Science Publ., 1975, Vol. II, pp. 179–191

Williams, J. A.: Characterization of oil types in Williston Basin. Am. Assoc. Pet. Geol. Bull. **58,** 1243–1252 (1974)

Woodbury, H. O., Murray Jr., I. B., Pickford, P. J., Akers, W. H.: Pliocene and Pleistocene depocenters, outer continental shelf, Louisiana and Texas. Am. Assoc. Pet. Geol. Bull. **57,** 2428–2439 (1973)

Zhuze, T. P., Jushkevich, G. N., Ushakova, G. S.: Sur les lois générales du comportement des systèmes gazopétrolifères aux grandes profondeurs (in Russian). Dokl. Akad. Nauk. S. S. S. R. **152,** 713–716 (1963)

Part IV
The Composition and Classification of Crude Oils and the Influence of Geological Factors

Chapter 1
Composition of Crude Oils

1.1 Petroleum Versus Source Rock Bitumen

Petroleum originates from the bitumen of source rocks. Migration, however, and especially primary migration, is the cause of considerable changes in composition when bitumen is compared to petroleum. On the one hand, only a small amount of the total dispersed bitumen is mobilized and transferred into carrier or reservoir rocks, and an even smaller amount is accumulated in oil fields. In producing areas the ratio of reservoired oil to dispersed bitumen ranges from 1:10 to 1:10,000[9]. On the other hand, such drainage is selective, as shown by the gross comparison between the bitumen present in source rocks and the corresponding crude oil in reservoirs. The heaviest and most polar molecules, like asphaltenes, are strongly adsorbed on the source rock and can hardly be expelled into the reservoir. Therefore, the common distribution of petroleum constituents in crude oil parallels the adsorptive behavior of these constituents, i.e., the least polar saturated hydrocarbons are most frequent, then follow aromatics and benzothiophenes, and least abundant are the most polar and most easily adsorbed resins and the least soluble asphaltenes. Even within a given structural type like n-alkanes, light molecules seem to be favored compared to heavy ones.

The result of this chemical separation is that the molecular composition of crude oil hydrocarbons from different fields is more uniform than the composition of the corresponding bitumen hydrocarbons.

1.2 Analytical Procedures for Crude Oil Characterization

Adequate sampling of crude oils is essential for their characterization. A satisfactory procedure would be to obtain a single-phase sample of petroleum under pressure conditions as they are in the reservoir. This requires the use of pressure vessels and a rather sophisticated and costly sampling procedure. Therefore geochemists usually deal with samples obtained at the well head under atmospheric pressure; hence, the light hydrocarbons are completely or at least partly lost as a function of their individual boiling point. Poulet and Roucaché

---

9 These figures are averages computed over a source rock sequence, but drainage may be more efficient over certain areas or horizons close to permeable beds, and inefficient somewhere else

(1970) have demonstrated that the well-head sampling procedure is representative only for the molecules comprising 10 carbon atoms or more. In order to relate all concentrations and crude oil properties to a standard state, not influenced by sampling conditions, the crude may be distilled at temperatures ranging from 160 to 210°C according to the various laboratories.

The composition of the light distillate depends on sampling conditions, and is only regarded as an indication of the abundance and composition of the light end of the oil; the heavier fraction, which usually comprises the main part of a crude oil, can be used reliably to describe its chemical composition and to compare it with other crude oils.

However, it should be kept in mind that the most abundant individual hydrocarbons are frequently in the light fraction. For instance, in most crude oils, the *n*-alkane distribution decrease after *n*-decane. Similarly, among aromatic molecules, benzene, toluene, and other low molecular weight alkylbenzenes are very abundant constituents.

Three different approaches may be used to describe the composition of crude oils (Fig. IV.1.1):

a) Isolation and individual determination of hydrocarbons and heterocompounds was the target of the API Research Project 6. In this study the Ponca City, Oklahoma, crude oil was used to determine the hydrocarbons, and in API Research Project 48 the Wasson, Texas, crude oil was used to determine the sulfur compounds (Rossini and Mair, 1959; Smith, 1968).

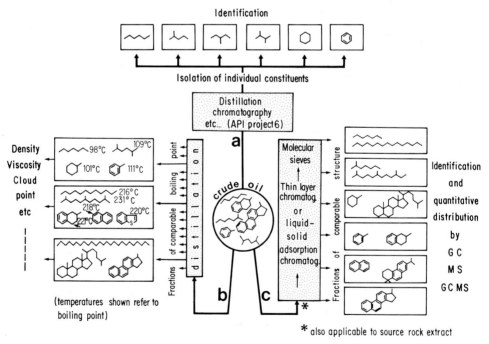

Fig. IV.1.1. Principles of the main analytical procedures for crude oil characterization

1.2 Analytical Procedures for Crude Oil Characterization

This valuable work resulted in the identification of about 350 hydrocarbons, 200 sulfur compounds and some other nonhydrocarbons. Most of them are compounds of relatively low molecular weight, comprising less than 15 carbon atoms. However, they represent a large part of usual crude oils, as 169 compounds cover 46% of the Ponca City crude, and just 10–20 compounds account for 50–70% of the light gasoline fraction from various crude oils (Bestougeff, 1967).

Identification of individual molecules has been carried out more frequently on the low molecular weight fractions. In addition to API Research Project 6, Martin et al. (1963) and Poulet and Roucaché (1970) have determined the individual C_6 and C_7 compounds, and part of the C_8 and C_9 compounds, on several crude oils of various origin. More recently, Erdman and Morris (1974) reported the individual concentrations of about 100 components up to n-decane.

b) Distillation of petroleum using wide or narrow fractions (15 in the routine U.S. Bureau of Mines procedure) is associated with some chemical or physical measurements on the fractions, e.g., density, viscosity, sulfur content, etc. This procedure is directly related to the processes used in refining crude oils, and it is designed to give valuable information about the best usage of a new crude in industrial plants.

Density, viscosity, refractive index and other physical properties depend on the relative amounts of the various groups of chemical compounds (alkanes, cycloalkanes, aromatics, thiophene derivatives, etc.) in a given fraction. Thus, it is possible to deduce from these physical properties certain aspects of the chemical composition of a crude oil. In addition, paraffin–naphthene–aromatic (P–N–A) analyses can be made from some of the distillation fractions. To take into account some regularities of crude oil composition, a correlation index (CI) has been proposed by Smith (1940) of the U.S. Bureau of Mines; the consecutive distillation fractions are represented by a number that is a function of specific gravity and boiling point. The CI value is calculated by the formula:

$$CI = \frac{48640}{K} + 473.7\, G - 456.8$$

with K = average boiling point of the fraction (Kelvin),
G = specific gravity of the fraction at 60–60 °F.

The correlation index can be plotted against the number of the successive distillation fractions, and in this way typical curves are obtained (Fig. IV.1.2). The correlation index is zero for linear alkanes, increases for cycloalkanes, and is high for aromatics. The index is quite useful for two component system — e.g., alkanes plus cycloalkanes — but less valuable for the usual three component systems (Smith, 1968).

The distillation procedure has been widely used, especially by the U.S. Bureau of Mines. A large number of crude oil analyses (around 10,000) is available in that form. Other distillation techniques used elsewhere and reported in the literature include for instance Engler, ITK (U.S.S.R.) and TBP (True Boiling Point, according to ASTM specification D 2892).

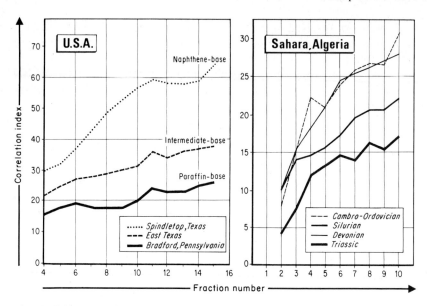

Fig. IV.1.2. Examples of correlation index as a function of distillation fraction number. *Left:* three typical U.S. crude oils; *right:* the main oil types of Paleozoic and Triassic oils from Sahara, Algeria. (After U.S. Bureau of Mines Rep. Invest. 4740; Louis, 1967)

c) Methods based on quantitative separation of the various structural types and on determination of the molecular distribution within each type, are most useful for geochemical purposes. In particular Oudin (1970), Durand et al. (1970), and Castex (1974) have proposed a complete sequence of procedures resulting in a quantitative evaluation of the chemical structure of crude oils.

Crude oils, as well as bitumens extracted from source rocks, are divided into fractions corresponding to the main structural types. Asphaltenes are precipitated by n-hexane, then saturated hydrocarbons, mono-, di-, polyaromatics and resins are separated by using liquid chromatography according to the scheme shown in Figure IV.1.3. Instead of conventional liquid chromatography on a column, medium pressure and high performance liquid chromatography or thin-layer chromatography may be used (Huc et al., 1976; Suatoni and Swab, 1976; Radke et al., 1980). Within each structural group, distribution by carbon number and identification of specific compounds are obtained by gas chromatography and mass spectrometry.

The main advantage of this procedure is to avoid any distillation which could destroy or alter some labile constituents of petroleum. Furthermore, the method is also suitable for small amounts of bitumen extracted from source rocks (100 mg bitumen if using column liquid chromatography and a few milligrams if using thin-layer chromatography), thus allowing a sound basis for oil–source rock correlation.

The results presented below are based mostly on the third type of procedure (c) described above.

1.3 Main Groups of Compounds in Crude Oils

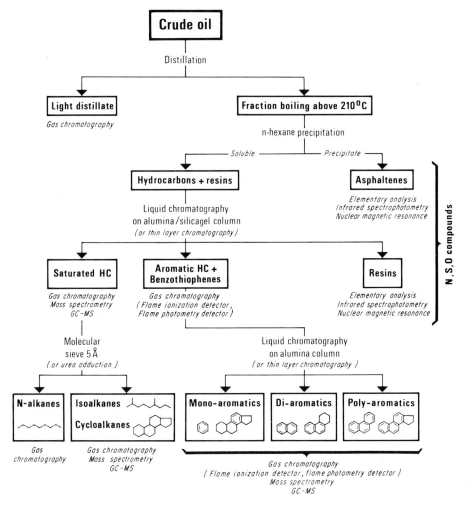

Fig. IV.1.3. Flow chart of crude oil analysis: separation of main structural types of molecules, and subsequent quantitative analysis of the fractions by gas chromatography and mass spectrometry. The sequence of analytical steps is also used for source rock bitumen, thus allowing the correlation of source rocks to crude oils

1.3 Main Groups of Compounds in Crude Oils

The gross composition of a crude oil (Fig. IV.1.4) can be defined by the content of:

- *saturated hydrocarbons,* comprising normal and branched alkanes (paraffins), and cycloalkanes (naphthenes),
- *aromatic hydrocarbons,* including pure aromatics, cycloalkanoaromatic (naphthenoaromatic) molecules, and usually cyclic sulfur compounds. The latter are

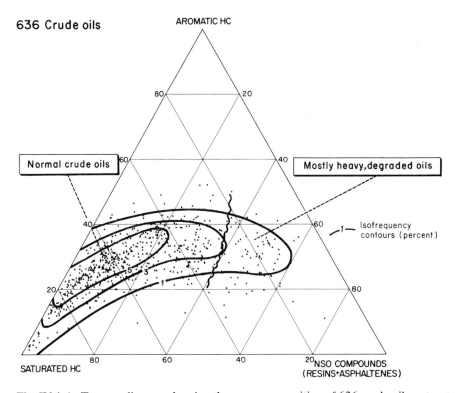

Fig. IV.1.4. Ternary diagram showing the gross composition of 636 crude oils: saturated hydrocarbons, aromatic hydrocarbons, and resins plus asphaltenes (wt. % in the fraction boiling above 210°C); for construction of isofrequency contours a triangle unit of 10:10:10 was used

most frequently benzothiophene derivatives. Their total abundance can be roughly evaluated through the sulfur content of the aromatic fraction,
- *resins and asphaltenes,* made of the high molecular weight polycyclic fraction of crude oils comprising N, S and O atoms. *Asphaltenes* are insoluble in light alkanes and thus precipitate with *n*-hexane. *Resins* are more soluble, but are likewise very polar and are strongly retained on silica gel when performing liquid chromatography, unless a polar solvent is used as the mobile phase.

These parameters are not independent, as all crude oils consist of these four groups of components. If one of these groups is missing, the other three groups, of course, amount to 100%, as saturates plus aromatics plus resins and asphaltenes are unity. This fact automatically introduces a certain degree of correlation between these groups and their subdivisions. Furthermore, the concentrations of several hydrocarbon types or N, S, O compounds show a high degree of covariance, as a result of a common origin, or common chemical affinities. This problem will be discussed later (Sect. 1.10).

In the following, various distributions with respect to compounds will be discussed, based on a statistical treatment of a large number of crude oils with the help of a computer.

1.3 Main Groups of Compounds in Crude Oils

Table IV.1.1. Gross composition of crude oils (wt. % of the fraction boiling above 210°C). (Analyses by IFP)

Average value of: (number of samples)	Normal producible crude oils (517)	All crude oils including tars (636)	Disseminated bitumen (1057)
Saturated HC	57.2	53.3	29.2
Aromatic HC	28.6	28.2	19.7
Resins + asphalt.	14.2	18.5	51.1
Aromatic sulfur (% of aromatic fraction)[a]	2.07		1.85

[a] Number of samples for aromatic sulfur: 230 and 88, respectively.

The average values of the main classes of compounds are shown in Table IV.1.1 for 636 reservoired crude oils and also for disseminated bitumens in 1057 source rocks of various ages and origins. The 636 samples of petroleum include some heavy unproducible oils and tars from tars sands. The values for 517 normally producible oils (of the 636) are also shown separately, as many of the tars are in fact degraded oils, whose composition has been altered.

Saturated hydrocarbons are usually the most important of the three main constituents listed in Table IV.1.1. The main exceptions are degraded oils, that may have partly or completely lost their alkanes, and a few immature oils rich in heavy constituents and naphthenoaromatics. The distribution of the saturated HC content shows two modes and a shoulder (Fig. IV.1.5):

- a first maximum corresponds to about 60% of saturated hydrocarbons in crude oil, and is mostly related to "paraffinic-naphthenic" oils,
- a second maximum at 40–45% is related to a more aromatic type of oil and is separated from the previous mode by a minimum corresponding to 50%,
- a shoulder around 20–25% represents the degraded heavy oils and tars that have lost part or all of their alkanes.

Aromatic and naphthenoaromatic hydrocarbons are usually the second most important group of constituents. Their content in crude oils varies from 20 to 45% by weight of the fraction boiling above 210°C, for 87% of the oils; 10% of the oils contain less than 20% aromatics, and 3% contain more than 45%.

Resins and asphaltenes usually range from 0 to 40% of non-degraded crude oils depending on genetic type and thermal maturation. Their content is usually high in shallow immature petroleum and decreases with increasing depth and subsequent cracking. In heavy oils and tars resulting from alteration by microbial activity, water washing and oxidation, their content ranges from 25 to 60%, due to an elimination or a degradation of hydrocarbons (Fig. IV.1.4).

The role of the high resins and asphaltenes content in heavy oils, and its influence on specific gravity and viscosity is discussed in Chapter IV.6.

Total number of samples: 636 Average: 53.35 Asymmetry: -0.168
Number of classes: 25 Standard deviation: 19.05 Kurtosis: -0.552

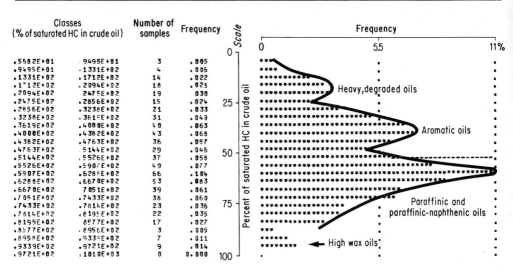

Fig. IV.1.5. Distribution of saturated hydrocarbons in crude oil (fraction boiling above 210°C), on 636 samples of normal and heavy degraded oils (analyses by IFP)

When comparing crude oils and source rock bitumens, it is evident that hydrocarbons (mostly saturated ones) are enriched in crude oils, whilst resins and asphaltenes are conversely impoverished due to their strong adsorption in source rocks and low solubility characteristics.

1.4 Principal Types of Hydrocarbons in Crude Oils

1.4.1 General

Hydrocarbons present in crude oils are normal, branched or cyclic saturated hydrocarbons, aromatic hydrocarbons and various associations of these basic types. Other unsaturated hydrocarbons, such as olefins, are essentially absent.

The composition of hydrocarbons is frequently considered in terms of alkanes (or paraffins), cycloalkanes (or naphthenes) and aromatics. However it should be remembered that association of these basic structures is a rule. Most cycloalkanes and aromatics bear normal or branched chains; molecules comprising both naphthenic and aromatic cycles plus chains account for most of the heavier hydrocarbons.

As a result of the usual procedures of separation and additional quantitative measurements, any molecule containing at least one aromatic cycle is quoted as "aromatic", even if several saturated cycles and chains are associated to the aromatic cycles. In the same manner, a saturated molecule comprising naphthenic rings and chains is counted as a cycloalkane or naphthene. Therefore, the alkane content, as computed, is restricted to molecules comprising only chains and it is underestimated in respect to the total amount of saturated chains. On the contrary, the aromatic content, as computed, is overestimated, compared to the real amount of aromatic cycles.

The respective content of alkanes, cycloalkanes and aromatics — with the above restriction — are shown on Figure IV.1.6 for 541 crude oils. The average values of these types of compound are listed in Table IV.1.2 for the three classes of crude oils and bitumen previously considered.

When looking at a great number of crude oils, the average value for each of the three different types of hydrocarbon (alkanes, cycloalkanes and aromatics) in these crude oils is similar. For all three types of hydrocarbon, it ranges from 30 to 35% (Table IV.1.2). However, the distribution of cycloalkanes is rather narrow compared to the others. Cycloalkanes vary from 20 to 40% of hydrocarbons in almost 80% of crude oils. Only 9% have less cycloalkanes and 13% have more; the latter include a large proportion of degraded oils, where a relative increase of cycloalkanes and aromatic constituents results from the selective destruction of alkanes.

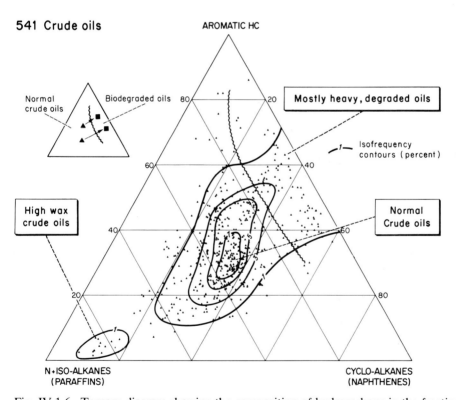

Fig. IV.1.6. Ternary diagram showing the composition of hydrocarbons in the fraction boiling above 210°C: n + iso-alkanes (or paraffins), cycloalkanes (or naphthenes), and aromatics for 541 crude oils analyzed by IFP. For construction of isofrequency contours see also Figure IV.1.4

Table IV.1.2. Average composition of hydrocarbons (wt. % of hydrocarbons). (Analyses by IFP)

Average value of: (number of samples)	Normal producible crude oils (517)	All crude oils including tars (541)	Disseminated bitumen (668)
n + isoAlkanes	33.3	31.7	27.7
Cycloalkanes	31.9	32.1	29.3
Aromatics	34.5	36.2	43.0
Saturated: aromatics[a]	2.8	2.7	1.8
Alkanes: saturated[a]	0.49	0.48	0.47

[a] Average value of the ratio.

Distribution of aromatic hydrocarbons shows two modes that in fact correspond to the bimodal distribution previously observed with respect to the concentration of saturated hydrocarbons in total crude oils. This is the result of the fact that resins and asphaltenes are usually minor constituents of crude oils:

– a first maximum around 30% is mostly related to "paraffinic-naphthenic" oils, separated by a minimum at 40% from,
– a second maximum around 45% which is related to a more aromatic type of oil, and also to degraded oils where the increase of the aromatic content is correlative to the alkane degradation.

The distribution of alkanes is also bimodal. Paraffinic-naphthenic oils result in a single broad maximum around 30–35%. A second population is clearly outlined, with an alkane content lower than 20%, and generally lower than 15%. It corresponds to degraded oils whose alkanes have been partly removed by biodegradation. The aromatic and cycloalkane content has been little altered, so that the points representing normal and degraded crude oil from the same origin join along a line issued from the alkane pole of the triangular diagram in Figure IV.1.6.

1.4.2 n-Alkanes (Figs. IV.1.7–IV.1.9)

All linear n-alkanes from C_1 to C_{40}, and a few beyond C_{40} have been identified in crude oils. They usually amount to 15–20% of crude oil, but their content can be very low, as in heavy degraded oils, or as high as 35%. The low molecular weight n-alkanes (C_5–C_7) may reach the highest values for single compounds: n-heptane accounts for 4.4% (vol %) of Fasken crude oil, Ordovician, Texas (Smith, 1968). Beyond C_{10}, the abundance decreases regularly with the number of carbon atoms in most crude oils.

High molecular weight n-alkanes ($> n\text{-}C_{20}$) are responsible for high cloud points of certain crude oils. The cloud point corresponds to the appearance of a cloud of wax crystals when the oil is chilled. Oils of this type have a high wax

1.4 Principal Types of Hydrocarbons in Crude Oil

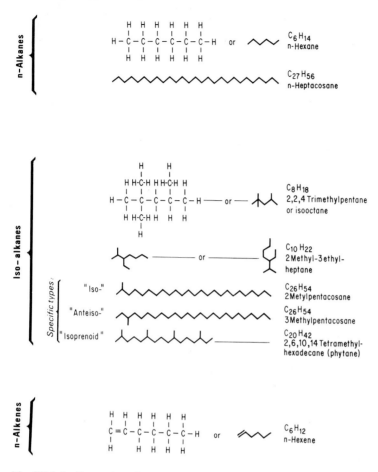

Fig. IV.1.7. Examples of normal and branched alkanes and alkenes found in crude oils

content, although the industrial "wax" or "paraffin wax" may contain also iso-alkanes and alkyl-cycloalkanes. For instance Bestougeff (1967) has found only 26 and 16% n-alkanes in paraffin waxes of Hassi Messaoud and Qatar crude oils, respectively. n-Alkanes sometimes show a slight predominance of odd- or even-numbered molecules in certain immature crude oils, such as the Uinta Basin or some Mediterranean crude oils. The significance of this particular point is discussed in Chapter II.3.

The amount of n-alkanes (Fig. IV.1.9) is largely dependent on genetic conditions, and especially on the nature of the original organic matter. High-wax crude oils and those derived from terrestrial organic matter usually contain a large proportion of n-alkanes, whereas marine or mixed organic material yields more cyclic compounds. During catagenesis of source rock organic matter, there is a general tendency towards an increase of n-alkane concentration with increasing maturation, resulting in a series of crude oils of similar origin, but of increasing n-alkane content. On the other hand, microbial degradation of reservoired

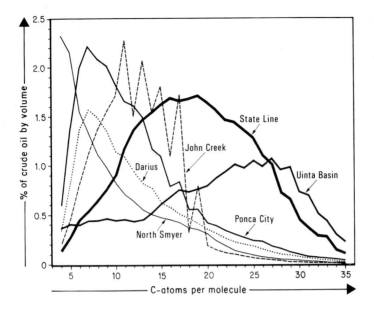

Fig. IV.1.8. Distribution of *n*-alkanes in different types of crude oils. (After Martin et al., 1963)

Field	Location	Age	Total C_4-C_{35} n-alk. in crude oil (volume %)
Uinta Basin	Utah	Eocene	20.7
State Line	Wyoming	Paleocene	31.2
Darius	Persian Gulf	Cretaceous	16.5
North Smyer	Texas	Pennsylvanian	16.0
John Creek	Kansas	Ordovician	21.2
Ponca City	Oklahoma	Ordovician	24.7

crude oils results in a removal of *n*-alkanes. Those crude oils are among those represented at the left hand side of Figure IV.1.9, with a small amount of *n*-alkanes.

1.4.3 Isoalkanes (Figs. IV.1.7 and IV.1.10)

Many branched alkanes containing 10 carbon atoms or less have been identified. Beyond C_{10}, the series of isoprenoids up to C_{25}, and only a few other isoalkanes, such as squalane, have definitely been identified. This situation results from the very large number of possible isomers, and the small concentration of higher molecular weight isoalkanes.

The highest individual concentration of isoalkanes is found in the C_6–C_8 range, namely 2-methyl- or 3-methyl-hexane and/or -heptane, and it may reach more

1.4 Principal Types of Hydrocarbons in Crude Oil

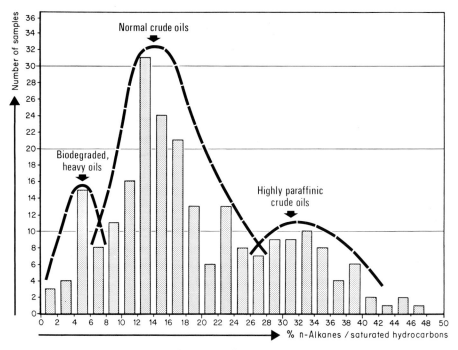

Fig. IV.1.9. Frequency distribution of *n*-alkane contents of the saturated hydrocarbon fraction: 232 crude oils and source rock bitumens of various origin (Tissot et al., 1977)

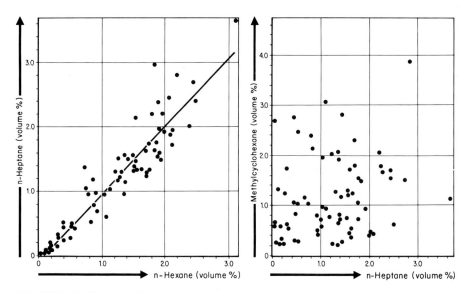

Fig. IV.1.10. Relationship between C_6 and C_7 hydrocarbons in crude oils; there is a close correlation between *n*-hexane and *n*-heptane in crude oils of different ages and geographic origins. There is, however, no correlation between methylcyclohexane and *n*-heptane (Smith, 1968)

than 1% of the crude oil. An important contribution of the understanding of the normal and isoalkane distribution has been provided by Martin et al. (1963). They carried out detailed analysis of the light fractions of 18 crude oils from various origin and ages. They noted that the order of decreasing abundance of the five hexane isomers is nearly the same in all crude oils. The same is true for the nine heptane isomers, and a comparable order of abundance was found by Poulet and Roucaché on lower Paleozoic crude oils, and others (Table IV.1.3). The most frequent configuration in the C_5–C_8 range, is (Smith, 1968) one tertiary carbon atom (2-methyl or 3-methyl); the configuration with two tertiary carbon atoms is less abundant. Other types (one quaternary carbon atom, or more than two tertiary atoms) are usually very rare.

The strong predominance of normal, iso- and anteisoalkanes (2-methyl, and 3-methyl), among C_6–C_8 alkanes, can be explained. On the one hand, the major source of long alkyl chains are alkanes, acids and alcohols from waxes of higher plants and microorganisms, which mostly comprise these structural arrangement. On the other hand, the cracking of heavier molecules preferentially results in n, iso, anteisoalkanes, and aromatics. Furthermore, the results presented on Table IV.1.3 show that isomerization equilibrium of heptanes is not reached, even in lower Paleozoic oils. The heptanes cannot be considered as an isolated system and isomerization reactions are probably slower than other reactions generating heptanes from heavier molecules. Under conditions of advanced maturation, cracking reactions removing heptanes might also be faster than isomerization.

In addition, Smith (1968) noted in 78 crude oils of various origins and ages a series of relationships between the abundances of the various C_5-, C_6- and C_7-saturated hydrocarbons: the more similar the structure, the better is the correlation. For instance n-pentane and n-hexane are strongly dependent, and the same is true for n-hexane and n-heptane. If isoalkanes are considered, molecules structurally close to the linear alkanes show a strong relationship with

Table IV.1.3. Abundance of heptane isomers in crude oils. (After Tissot, 1966)

	Isomerization equilibrium of heptanes (wt. %)				Abundance in crude oils (wt. %, liquid phase)	
	298 K		400 K		Average of 18 crudes after Martin et al. (1963)	Average of Hassi-Messaoud crudes (monophasic samples)
	Vapor phase	Liquid phase	Vapor phase	Liquid phase		
n-heptane	1.25	2.25	4.3	8.55	55.5	52.0
2-methyl hexane	9	11.2	15.4	15.65	13.8	16.0
3-methyl hexane	5.1	6.85	11.3	12.15	19.2	22.4
3-ethyl pentane	0.45	0.6	1.3	1.45	2.6	2.4
2,2 dimethyl pentane	32	24.9	16.7	13.2	0.6	0.4
2,3 dimethyl pentane	25	30	28.8	29.5	6.1	4.8
2,4 dimethyl pentane	9.9	8.15	8.4	6.8	1.7	2.0
3,3 dimethyl pentane	11.4	11.3	10.4	9.8	0.4	–
2,2,3 trimethyl butane	5.9	4.75	3.4	2.9	0.1	–
	100.0	100.0	100.0	100.0	100.0	100.0

them, e.g., methylpentane with n-hexane etc. On the contrary, when comparing two very different structures such as linear and cyclic, e.g., n-heptane and methylcyclohexane, there is no correlation (Fig. IV.1.10).

This situation can be explained by the fact that low to medium molecular weight hydrocarbons are not inherited biogenic molecules, but are generated through thermal degradation and cracking of C–C bonds of either kerogen or larger bitumen molecules already formed. The succession of chemical reactions resulting in the various light alkanes, for instance, is alike, and the more the structures of two alkanes resemble each other, the greater is the chance that the reaction paths are the same. Therefore parallel reactions and comparable kinetics result in rather constant ratios of closely related molecules. On the contrary, different structures such as cyclohexane and n-hexane are likely to derive from different parts of the kerogen macromolecule, through a different succession of reactions, and no correlation can be expected between them.

Some C_{10} isoalkanes are particularly abundant in crude oils. For example 2,6-dimethyloctane and 2-methyl-3-ethylheptane amount to 0.55 and 0.64% of Ponca City crude oil, respectively. This predominance, compared to other decanes, is probably due to their origin from monoterpenes of higher plants (Fig. II.3.12).

In the medium range of molecular weight, the most remarkable molecules belong to the series of isoprenoids (Fig. II.3.13). They are isoalkanes from C_9 to C_{25} with one methyl branch on every fourth carbon atom. Their genetic origin is discussed in Part II. They are of particular interest in respect to petroleum genesis. They frequently amount to about 1% of a crude oil. The most abundant are pristane (tetramethylpentadecane C_{19}) and phytane (tetramethylhexadecane C_{20}). Together they average up to 55% of all acyclic isoprenoids. Pristane is often more abundant than phytane, with an average ratio of 1.35. The pristane predominance may reach 4 to 10 in some high-wax paraffinic crude oils of nonmarine origin from the Gippsland, Cooper and Bowen-Surat Basins, Australia (Powell and McKirdy, 1975). Other important isoprenoids are often C_{16} and C_{18}, whereas C_{17} and C_{22} are practically absent, due to an unlikely mode of cleavage of the precursor. Some irregular isoprenoids (cf. Chap. II.3) are also present in certain crude oils, such as squalane (C_{30}) and lycopane (C_{40}).

In addition to the series of isoprenoids, gas chromatography of the branched-cyclic fraction of saturated hydrocarbons shows the existence in some crude oils of homologous series of isoalkanes in the C_{10}–C_{30} range. High molecular weight isoalkanes and anteisoalkanes (2-methyl and 3-methyl, Fig. IV.1.7) are present in paraffin wax. They are especially abundant in certain crude oils, for example the Uinta basin oils derived from nonmarine source beds, or Indonesian oils derived from a paralic environment (Tissot et al., 1977).

1.4.4 Alkenes (Fig. IV.1.7)

Unsaturated chains are relatively unstable, and thus very unusual in crude oils. However, very small quantities of n-hexene, n-heptene and n-octene were identified by Putscher (1952) in a Pennsylvania crude oil. On the other hand, polycyclic cycloalkenes, such as sterenes or hopenes, have been recently reported in ancient sediments. They have also been found in certain young crude oils (Hollerbach, 1976). However, this type of compound has not been investigated very actively until now.

1.4.5 Cycloalkanes (Fig. IV.1.11)

Cyclopentane, cyclohexane, and their derivatives of low molecular weight ($< C_{10}$) are important constituents of petroleum. In particular, methylcyclohexane is

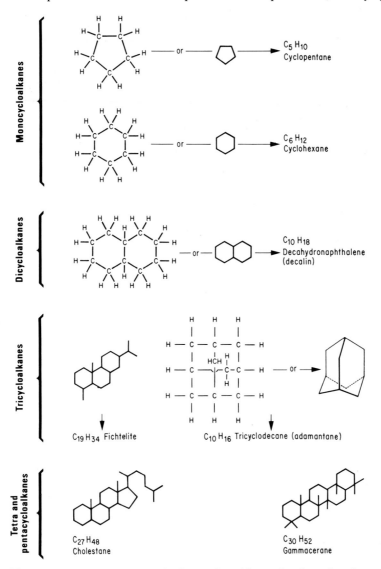

Fig. IV.1.11. Examples of cycloalkanes found in crude oils and coals

1.4 Principal Types of Hydrocarbons in Crude Oil

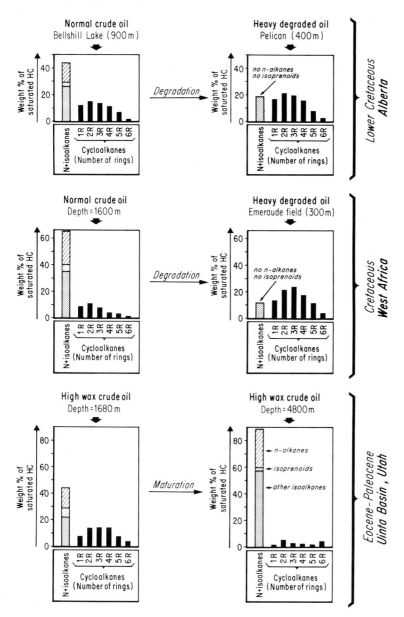

Fig. IV.1.12. Distribution of saturated hydrocarbons as found by ring analysis. Additional data on normal and isoalkanes are from separation of normal and iso-cyclic fraction with the help of 5 Å molecular sieves, followed by quantitative GLC analysis. The analysis clearly shows the effects of biodegradation and thermal alteration. (Data on Alberta oils from Deroo et al., 1977; data on West African oils from Claret et al., 1977)

Table IV.1.4. Composition of cycloalkanes (wt. %). (Analyses by IFP)

Average value of: (number of samples)	Normal producible crude oils (299)	All crude oils including tars (347)	Disseminated bitumen (448)
Mono + dicyclics	53.9	52.3	39.9
Tricyclics	20.4	20.9	21.3
Tetra + pentacyclics	24.9	25.5	35.6

frequently the most abundant of them, and it reaches 2.4% of total crude from Conroe, Texas (Smith, 1968). Methylcyclopentane is also important, and both methyl derivatives are usually more abundant than the respective parent molecules, cyclopentane and cyclohexane. Most of alkylcycloalkanes with less than 10 carbon atoms are derivatives of cyclopentane or cyclohexane, and only a few are bicyclic, e.g., bicyclooctane or bicyclononane. Peculiar structures with different ring arrangements — i.e., cyclobutane and cycloheptane — have been also reported in minute amounts.

Cycloalkanes of the medium to heavy fraction (C_{10}–C_{35}) are usually made by arrangements of one to five 5- or 6-membered rings (Table IV.1.4). The distribution by ring number of saturated hydrocarbons ("ring analysis") is usually obtained by mass spectrometry, according to the method of Hood and O'Neal (1959): the interpretation[10] is based on the fragment peaks obtained with an ionization intensity of ca. 70 eV. Examples of distribution by ring analysis in normal and degraded crude oils are shown in Figure IV.1.12.

Mono- and dicyclics generally amount to 50–55% of the total cycloalkanes >C_{10}. Those of high molecular weight usually contain one long chain (linear or slightly branched) and several short methyl or ethyl chains (Hood et al., 1959). The abundance of the various mono- and dicycloalkanes decreases regularly as a function of molecular weight (i.e., number of carbon atoms).

Tricycloalkanes average only 20% of the total naphthenes >C_{10}. Some of them are likely to present the structural arrangement of the perhydrophenanthrene. The abundance of the various tricycloalkanes decreases regularly as a function of molecular weight, beyond 20 carbon atoms.

A very peculiar compound is the tricyclodecane, or adamantane (Fig. IV.1.11); it has been identified in a very small amount in the Hodonin, Czechoslovakia, crude oil (Landa and Machacek, 1933) and from the Ponca City crude oil (Mair et al., 1959). Since then, Petrov et al. (1974) have found hydrocarbons of the adamantane series in many crude oils, where they may amount to 10–15% of the tricycloalkanes.

---

10 The method gives the number of rings per stable mass spectrometric ion, which may differ from the number of rings per molecule. On the other hand, if the method is applied to bitumen from immature source rocks, unsaturated polycyclics — e.g., sterenes — might be abundant and should be separated from saturated before the analysis is run

Tetra- and pentacycloalkanes average 25% of the total naphthenes $>C_{10}$. Their structure is directly related to the tetracyclic steroids and pentacyclic triterpenes (Fig. II.3.16). The distribution of tetracyclic steranes is mainly from 27 to 30 carbon atoms, whilst the distribution of pentacyclic triterpanes has its maximum from 27 to 32 carbon atoms, but sometimes extends up to 35 carbon atoms. Their origin and occurrence, as geochemical fossils, have been discussed in Part II. Some of them have been identified by gas chromatography and mass spectrometry, and a few have been isolated in crystalline form (Hills et al., 1970). They show a high optical rotation and are responsible for most of the optical activity of fossil hydrocarbons. In general, tetra and pentacyclic cycloalkanes are most abundant in young and immature crude oils, also rich in resins, asphaltenes and polyaromatics.

Although the total cycloalkane content of source rocks and petroleum is very similar, mono and dicyclics are more abundant in crude oils than in source rocks; the contrary is true for tetra- and pentacyclic molecules.

1.4.6 Aromatics (Fig. IV.1.13)

True aromatics are molecules containing only aromatic rings and chains. They usually include one to four or five condensed aromatic rings and a small number of short chains. Compounds belonging to this basic type have been identified: benzene (1 ring), naphthalene (2 rings), phenanthrene and anthracene (3 rings), pyrene, benzanthracene and chrysene (4 rings). A second fundamental type consists of one five-membered ring in addition to the six-membered rings: fluorene (3 rings), benzofluorene and fluoranthene (4 rings), and benzofluoranthene (5 rings). The molecular mass formula of aromatics is C_nH_{2n-p}, where p varies with the number of rings (Fig. IV.1.14). Benzene (p = 6), naphthalene (p = 12), and phenanthrene (p = 18) types are the most abundant. The major components in each type are usually not the parent compound, but the alkylated derivatives comprising one to three additional carbon atoms. For instance the major constituent of the alkyl-benzene C_nH_{2n-6} type is frequently toluene, which may reach 1.8% of crude oil, and sometimes xylene (the total o, m, p-xylenes may amount to 1.3% of crude oil), whereas benzene is usually less abundant (up to 1% of crude oil). The same situation is true for the naphthalene type C_nH_{2n-12}, the distribution of which (Fig. IV.1.15) has its maximum at C_{12} or C_{13} (di- or tri-methyl naphthalene), and for the phenanthrene type C_nH_{2n-18}, with a maximum at C_{16} or C_{17} (di- or tri-methyl phenanthrene). Beyond these maxima all distributions decrease rapidly, due to the scarcity of long chains. Higher-boiling alkylbenzenes, naphthalenes, etc. which occur at lower concentration typically contain two to three methyl groups and one long chain (Hood et al., 1959).

The preference for mono- di- and tri-methyl derivatives could be explained if naphthalene and phenanthrene series were mostly generated by fragmentation of steroids or terpenoids. However, individual methyl derivative, e.g., methylphenanthrene, are not clearly related to biological precursor molecules (Radke et al., 1982).

When several structural types are possible, as for molecules with three or more aromatic rings, one of them is favored in crude oils. Thus alkylanthracenes are

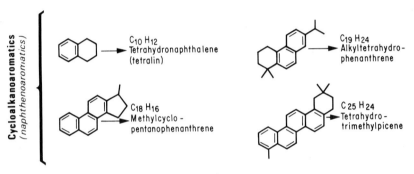

Fig. IV.1.13. Examples of aromatics and cycloalkanoaromatics (naphthenoaromatics) found in crude oils

present in small amount, and alkylphenanthrenes are largely predominant. This fact would be in agreement with the suggested origin of the phenanthrene series.

Youngblood and Blumer (1975), and McKay et al. (1975) reported traces of higher polynuclear aromatic hydrocarbons with four to eight aromatic rings in fossil fuels. Non condensed aromatic types such as diphenyl, which is composed of two benzene rings joined by a single carbon–carbon bridge, are also known in minor amounts.

1.4.7 Naphthenoaromatics (Fig. IV.1.13)

These compounds are usually the major constituents of the high boiling fraction of hydrocarbons. They include one or several condensed aromatic rings, fused with naphthenic rings and chains. The most frequent structural types, comprising up to five rings, are presented in Figure IV.1.14. The distribution of the total aromatic and naphthenoaromatic fraction by number of aromatic cycles in the

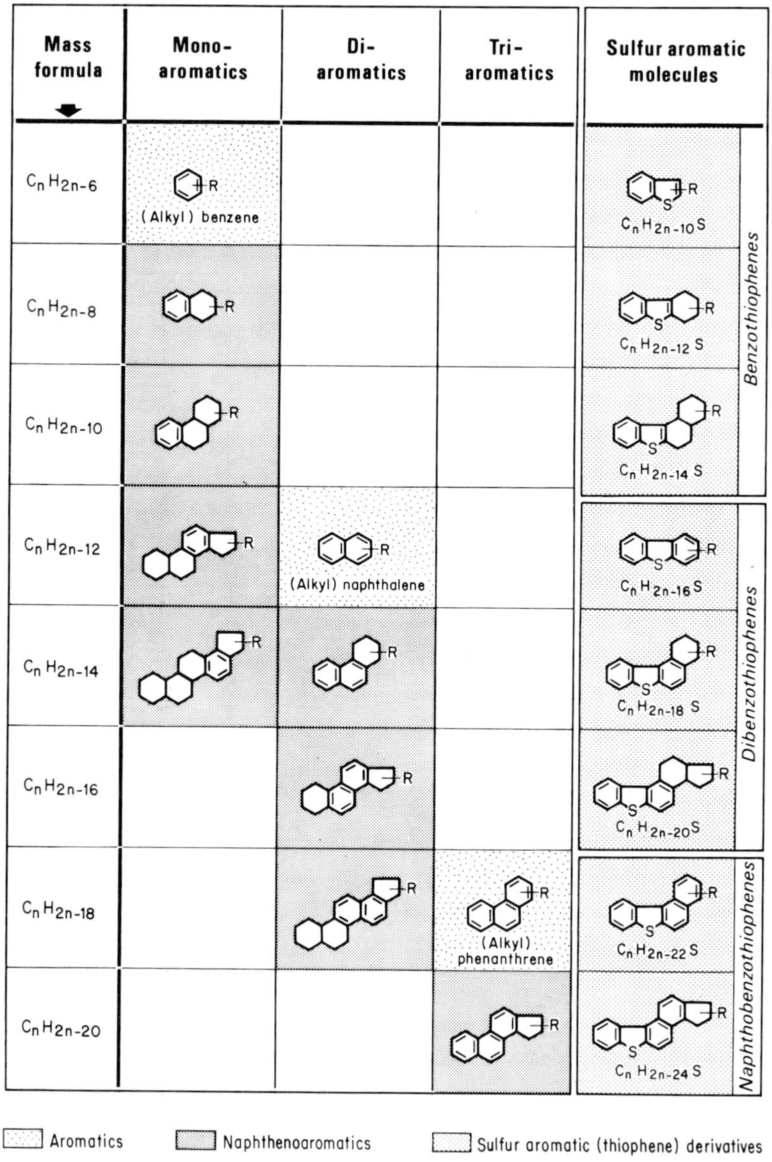

1.4 Principal Types of Hydrocarbons in Crude Oil

Mass formula	Mono-aromatics	Di-aromatics	Tri-aromatics	Sulfur aromatic molecules	
C_nH_{2n-6}	(Alkyl) benzene			$C_nH_{2n-10}S$	*Benzothiophenes*
C_nH_{2n-8}				$C_nH_{2n-12}S$	
C_nH_{2n-10}				$C_nH_{2n-14}S$	
C_nH_{2n-12}		(Alkyl) naphthalene		$C_nH_{2n-16}S$	*Dibenzothiophenes*
C_nH_{2n-14}				$C_nH_{2n-18}S$	
C_nH_{2n-16}				$C_nH_{2n-20}S$	
C_nH_{2n-18}			(Alkyl) phenanthrene	$C_nH_{2n-22}S$	*Naphthobenzothiophenes*
C_nH_{2n-20}				$C_nH_{2n-24}S$	

Aromatics   Naphthenoaromatics   Sulfur aromatic (thiophene) derivatives

Fig. IV.1.14. Examples of aromatic, naphthenoaromatic and sulfur aromatic (thiophene) derivatives corresponding to the molecular mass formula C_nH_{2n-p}. In many cases, several structures are possible and only the most frequent or most probable is shown

molecule is shown in Table IV.1.5. Mono- and di-aromatics are especially abundant, compared to polyaromatics, in paraffinic-naphthenic crude oils from detrital sedimentary series.

Naphthenoaromatics are particularly abundant, as compared to pure aromatics, in young or shallow immature crude oils. The aromatic types become predominant after an important thermal evolution.

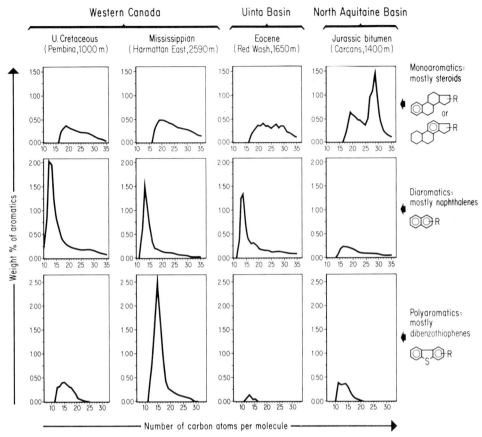

Fig. IV.1.15. Typical distributions of aromatics, naphthenoaromatics and sulfur aromatic derivatives corresponding to the mass formula C_nH_{2n-12}. Among this group of compounds, monoaromatics are tetracyclic molecules and diaromatics are alkylnaphthalenes. Polyaromatics of the same mass are in fact sulfur-containing aromatics (alkyldibenzothiophenes). Their true formula is $C_mH_{2m-16}S$, with $m = n$-2. Maximum abundances usually are recorded for the di- or tri-methyl derivatives of each type. The results are obtained from low-ionization-voltage mass spectrometry (10 eV), using parent peaks. *Left:* two crude oils from Western Canada: a low-sulfur oil from Pembina, U. Cretaceous, and high-sulfur oil from Harmattan E., Mississippian. (Data from Deroo et al., 1973.) *Right:* a high-wax crude oil from Red Wash, Utah, and a bitumen from the Jurassic of the North Aquitaine Basin, France. The latter is particularly rich in monoaromatic steroids (ca. C_{27}–C_{29})

Naphthenoaromatics may show various structural arrangements. Bicyclic (1 aromatic, 1 saturated cycle) indane, tetralin (tetrahydronaphthalene) and their methyl derivatives are usually abundant and have been determined in several crude oils. Tricyclic tetrahydrophenanthrene and derivatives are also common. Of particular importance are the tetracyclic and pentacyclic molecules which are mostly related to steroid and triterpenoid structures: besides the saturated form, discussed in Section 1.4.5, they have been identified with 1, 2, or 3 aromatic rings,

1.4 Principal Types of Hydrocarbons in Crude Oil

the other being naphthenic rings (Fig. II.3.16). The distribution of tetracyclic naphthenoaromatics, according to the number of carbon atoms, may show two distinct maxima, or only one (Fig. IV.1.15):

- a first maximum around 20 carbon atoms probably resulting from dealkylation of more complex molecules,
- a second maximum around 27 to 30 carbon atoms corresponding to the original structure of the biogenic steroids.

Their concentration may be important in naphthenic or paraffinic-naphthenic crude oils. All together the average amount of tetra plus pentacyclic molecules — naphthenic, aromatic or naphthenoaromatic — is over 10% by weight of the crude oil fraction boiling above 210°C (Table IV.1.6).

Table IV.1.5. Average composition of the aromatic fraction of crude oils and bitumen extracted from source rocks (wt. % of the total aromatic fraction boiling above 210°C). (Analyses by IFP)

Average value of: (number of samples)	Crude oils (121)	Disseminated bitumen (108)
Monoaromatics	33.0	26.3
Diaromatics	23.4	24.1
Triaromatics	12.9	18.7
Tetra- and polyaromatics	7.3	11.5
Total aromatics + naphthenoarom.	76.6	80.6
Thiophene derivatives (benzo-, dibenzothiophenes)	23.4	19.4
Total	100.0	100.0

Table IV.1.6. Tetra- and pentacyclic molecules in crude oils and bitumens (wt. % of the crude oil fraction boiling above 210°C; wt. % of bitumen extracted by $CHCl_3$). (Analyses by IFP)

Average value of tetra + pentacyclics in:	Crude oils		Disseminated bitumen	
	wt. %	Number of samples	wt. %	Number of samples
Naphthenic fraction	7.3	299	5.6	448
Aromatic + naphthenoaromatic fraction	5.4	121	4.1	108
Total	12.7		9.7	

1.5 Sulfur Compounds

The average sulfur content of crude oils, based on 9347 samples is 0.65% by weight. The distribution (Fig. IV.1.16) is bimodal with a minimum at 1% separating:

- low sulfur crude oils containing less than 1% sulfur and comprising the majority of the samples (about 7500),
- high sulfur crude oils containing more than 1% sulfur and grouping a smaller number of crudes (about 1800).

However, as pointed out by Smith (1968), the picture would be quite different if the diagram were be established on a production plus reserves basis. As crude oils from the Middle East, Venezuela and the West Texas Permian basin account for a large part of the world production plus reserves, and are generally high-sulfur crude oils, the second maximum ($>1\%$ S) may become as important as, or even more than, the first one ($<1\%$ S).

Sulfur is the third most abundant atomic constituent of crude oils, following carbon and hydrogen. It is present in medium as well as in heavy fractions of crude

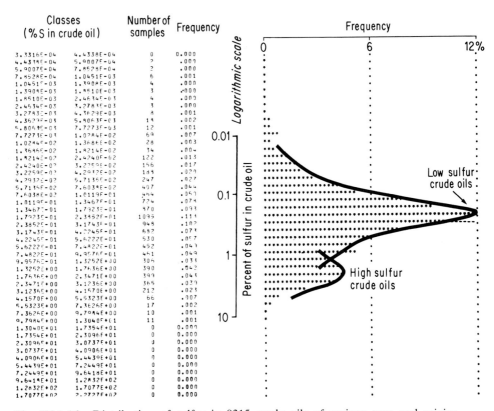

Fig. IV.1.16. Distribution of sulfur in 9315 crude oils of various ages and origins. (Analyses by U.S. Bureau of Mines and IFP)

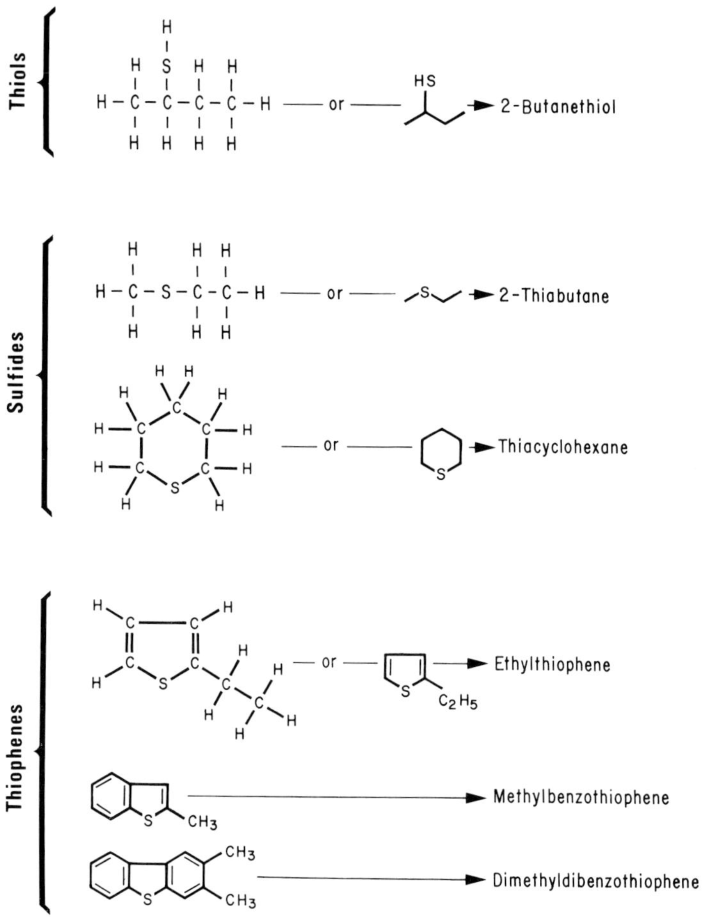

1.5 Sulfur Compounds

Fig. IV.1.17. Examples of sulfur compounds in crude oils

oils. In particular the sulfur content is not specifically related to the heaviest fractions of crude oils, as is the case for nitrogen. In the low and medium molecular weight range (up to C_{25}) sulfur is associated only with carbon and hydrogen. In the heavier fractions of crude oils, it is frequently incorporated in large polycyclic molecules comprising N, S, O (see Sec. 1.8).

Sulfur compounds identified in the light and medium fraction of crude oils (up to C_{25}) belong to four main classes of unequal importance (Fig. IV.1.17):

a) thiols, also called mercaptans
b) sulfides
c) disulfides
d) thiophene derivatives.

Nonthiophenic sulfur compounds are particularly abundant in immature crude oils, whereas thiophenic sulfur abundance increases with maturation (Ho et al., 1974).

Thiols can be thought of as derived from hydrogen sulfide, by substitution of an alkyl or cycloalkyl radical to an hydrogen atom. Normal and isoalkanethiols, cyclopentanethiol and cyclohexanethiol have been found in petroleum (Smith, 1966). Aromatic thiols have not been reported. Most thiols have a low molecular weight (less than 8 carbon atoms). The most abundant in Wasson crude oil, Texas, is ethanethiol and amounts only to 0.0053% by weight of the crude.

Sulfides can be thought of as derived from hydrogen sulfide by substitution of the two hydrogen atoms by alkyl groups. They are usually of very low abundance. The major compounds of this type in Wasson crude are the 2-thiabutane which amounts to 0.0022% by weight of the crude, and various thiacyclopentane derivatives (up to 0.0025%). Alkylaryl sulfides are substituted by one aromatic ring and one alkyl group.

Disulfides have a similar structure as sulfides, but the sulfur bridge consists of two sulfur atoms instead of one: a few of them have been identified in the low molecular range (dithiabutane to dithiahexane).

Thiophene can be described as an unsaturated 5-membered ring, comprising one sulfur and four carbon atoms. In many respects (e.g., chromatography) the thiophene ring behaves like a benzene ring. Thiophene itself is generally very scarce, but benzothiophenes, dibenzothiophenes and benzonaphthothiophenes (comprising one thiophene ring and 1, 2, or 3 benzene rings, respectively) are important constituents of all high-sulfur crude oils. They are present in small amount in low-sulfur crude oils. In Wasson, Texas, crude oil, the most abundant molecule of this type identified so far is 2-methylbenzothiophene, which amounts to 0.0025% by weight of the crude. Although dibenzothiophenes seem to be fairly abundant in certain crude oils, it is not possible at the moment to give an individual evaluation.

Ho et al. (1974) have shown that the benzothiophene: dibenzothiophene ratio normally decreases with increasing maturation of crude oil. Dibenzothiophenes are frequently the major group of thiophenic compounds in mature or altered high-sulfur crude oils. Regarding the distribution by molecular weight, the parent molecule is generally less abundant than the substituted derivatives, and the maximum content usually corresponds to molecules comprising two or three additional carbon atoms, i.e., dimethyl-, trimethyl-dibenzothiophene (Fig. IV.1.15). The same is true for benzonaphthothiophene. An exception is the distribution of benzothiophene, which sometimes covers a wide range of molecular weights. Molecules comprising a saturated cycle in addition to aromatic and thiophenic cycles are also known, particularly the dibenzothiophene with one additional saturated cycle.

Thiophene derivatives are particularly abundant in crude oils with a high content of aromatics, resins and asphaltenes. For 121 crude oils from various origins, total thiophene derivatives average more than 20% of the aromatic fraction boiling above 210°C (Table IV.1.5).

Table IV.1.7. Nitrogen content according to distillation fractions of Wilmington, California, crude oil (Smith, 1968)

Boiling range (°C)	Nitrogen (wt. %)
up to 130	0.00
130–250	0.0027
250–300	0.028
300–350	0.10
350–400	0.25
400–500	0.48
Residuum	0.97

1.6 Nitrogen Compounds (Fig. IV.1.18)

Nitrogen content is usually much lower than sulfur content in crude oils: about 90% of the crudes contain less than 0.2% of nitrogen. The average value on crude oils is 0.094% by weight. The largest contents are known in certain crude oils from California, e.g., 0.65% in Wilmington and 0.75% in Gato Ridge oils. Due to a relative minimum of the distribution close to 0.25% we can distinguish high-nitrogen crude oils (N > 0.25%) from normal, nitrogen-poor petroleum. Degraded asphalts are enriched in nitrogen, as the relative part of resins and asphaltenes is increased, compared with normal crude oils: nitrogen content reaches 1.24% in Whiterocks Canyon asphalt (Utah).

The main part of nitrogen is found in high molecular weight and high boiling point fractions, as shown by Table IV.1.7. These high molecular weight compounds are in fact polycyclic aromatic structures containing more or less at random different heteroatoms (N, S, O). However, a certain number of low to medium molecular weight specific nitrogen compounds are known from naphtha and gas oil fractions: about 25 to 30% of them usually consists of basic nitrogen compounds: pyridine, quinoline. The rest is made of nonbasic molecules, e.g., carbazoles and indoles (Seifert, 1969). More recently Schmitter et al. (1980) investigated the basic fraction of crude oils and showed that basic nitrogen compounds are mainly monoaza-arenes with a single N-atom: quinolines and benzoquinolines derivatives. In particular, the tricyclic fraction contains several dimethyl — and trimethyl — benzoquinolines, which are usually the major compounds of this group.

Porphyrins contain four pyrrole nuclei and a metal, mostly vanadium or nickel. They will be discussed in Section 1.9 as organometallic complexes.

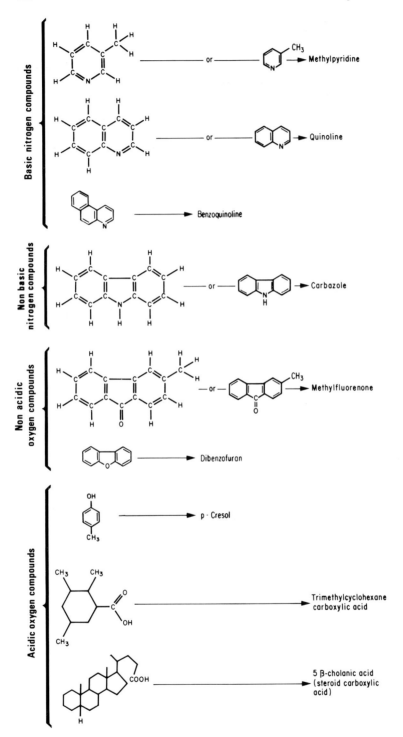

Fig. IV.1.18. Examples of nitrogen and oxygen compounds in crude oils

1.7 Oxygen Compounds (Fig. IV.1.18)

The most important groups of oxygen-containing compounds are acids, which seem to be a common constituent in young and immature crude oils. Saturated fatty acids from C_1 to C_{20}, and also isoprenoid acids have been identified. Naphthenic acids have been known for a long time to be important constituents in some naphthenic and asphaltic crude oils from USSR (Grosniy, Bibi-Eibat, etc.), Venezuela and California: cyclopentane and cyclohexane carboxylic acids C_6–C_{20}, cyclopentyl acetic acids C_8–C_{10}. Some of the naphthenic acids present in heavy crude oils are certainly original constituents of the oil (e. g., those with a hopane skeleton). Some other may result from a degradation of the original crude by chemical or biochemical oxidation.

Seifert and Howells (1969) estimated the amount of carboxylic acids up to 2.5% by weight in a young and immature California crude oil. They separated and identified a large number of carboxylic acids belonging to many hydrocarbon structural types. Naphthenic and naphthenoaromatic types are the most abundant, followed by polyaromatic and heterocyclic types (Seifert and Teeter, 1970). Of particular interest is the identification of steroid carboxylic acids (330 ppm for the fully saturated type), some of them being an indication of animal contribution to the formation of petroleum (Seifert, 1973). In spite of the considerable complexity of this class of compound, four steroid acids were definitely identified and their structure proven by synthesis.

The most ubiquitous group of oxygen compounds in crude oils is probably the group of pentacyclic acids with a hopane skeleton. They are considered by Ourisson et al. (1979) to be derived from bacterial lipids such as the C_{35} "bacteriohopane tetrol" and other constituents of bacterial membranes.

Several phenols (Seifert and Howells, 1969), and particularly cresols, are abundant in the acidic fraction of crude oils. Other oxygen compounds have been identified in minor amounts: ketones, alkyl or cyclic including fluorenones, and dibenzofurans.

1.8 High Molecular Weight N, S, O Compounds: Resins and Asphaltenes

High molecular weight constituents of crude oils usually contain N, S and O compounds. They are referred to as resins and asphaltenes. Asphaltenes and most resins are complex structural arrangements made of polycyclic aromatic or naphthenoaromatic nuclei, with chains and heteroatoms (O, N,S). They constitute the heavy ends of petroleum, and they have to be considered as the natural high molecular weight end members of the aromatic and naphtheno-aromatic series. When the molecular weight increases beyond ca. 600, the probability of the molecule containing one or more heteroatoms (O, N or S) is high. Therefore, there is practically no pure aromatic or naphthenoaromatic molecule in the heavy end of petroleum, and all constituents contain O, N or S atoms, occasionally substituting for carbon atoms in the cyclic structure. They also contain metals (Sect. 1.9).

Resins and asphaltenes are usually distinguished based on the separation procedure. However, there are several ways of preparing resins and asphaltenes, resulting in slightly different compositions. Precipitation by propane separates resins and asphaltenes from the rest of crude oil, then precipitation by n-pentane or n-heptane separates resins (soluble) from asphaltenes (insoluble). Alternatively asphaltenes are readily precipitated from crude oil by n-hexane or n-heptane, then resins are separated from hydrocarbons by liquid chromatography on alumina. In fact, molecular size and polarity are two factors which can, to some extent, compensate for each other (Long, 1979). Thus, different compounds may be precipitated with the same procedure from different crude oils, and also from synthetic oils, such as shale oil or coal liquids (Bockrath et al., 1979).

Asphaltenes plus resins generally amount to less than 10% in paraffinic oils and less than 20% in paraffinic-naphthenic oils; they may reach 10 to 40% in aromatic-intermediate oils. Among them, asphaltenes amount to only 0 to 20% of those normal, non-degraded oils. Heavy degraded oils, such as the Athabasca or Orinoco deposits, may contain 25 to 60% resins plus asphaltenes. Resins and asphaltenes are also present in source rock bitumen, where they are more abundant than in the related crude oils. Resins and asphaltenes from source rock bitumen are close to the corresponding petroleum compounds, but not identical. Table IV.1.8 shows the compositions of resins and asphaltenes from bitumen and from crude oils.

The structure of asphaltenes and resins precipitated from petroleum has been investigated by using several physical techniques, such as X-ray diffraction, nuclear magnetic resonance and infrared spectroscopy (Yen, 1961, 1972, 1979). This work resulted in a structural model of asphaltenes summarized as follows by Yen (1972) and shown in Figure IV.1.19:

a) The basic unit is a pericondensed polyaromatic *sheet*. Aromatic cycles are mostly substituted by methyl groups and a small number of long chains. Sulfur is usually found in benzothiophene groups, nitrogen in quinoline groups and oxygen in ether bonds. All of them represent a defect of the aromatic structure.

b) Individual aromatic sheets are usually stacked (ca. 5 layers) to form *particles* or *crystallites* by intermolecular or intramolecular association. Molecular association results from π–π bonds between pericondensed aromatic sheets.

Table IV.1.8. Average composition of resins and asphaltenes (wt. %)

		C	H	O	S	N	O+S+N	H/C atomic
Crude oils	Resins	83.0	10.0	2.0	4.5	0.5	7.0	1.4
	Asphaltenes	83.4	8.1	2.0	5.0	1.5	8.5	1.16
Source rock bitumen	Resins (151)	78.0	9.2	7.2	2.8	0.9	10.9	1.37
	Asphaltenes (175)	74.4	7.5	7.6	3.0	1.7	12.3	1.18

1.8 High Molecular Weight N, S, O Compounds: Resins and Asphaltenes

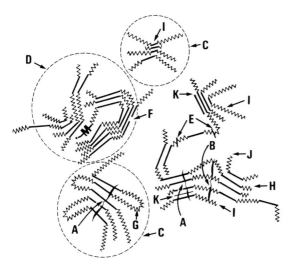

A Crystallite B Chain Bundle
C Particle D Micelle
E Weak Link F Gap and Hole
G Intracluster H Intercluster
I Resin J Single Layer
K Petroporphyrin M Metal

Fig. IV.1.19. Macrostructure of asphaltenes precipitated from crude oils. (According to Yen, 1972)

c) Several crystallites may from *aggregates* or *micelles* of different size. The size and the molecular weight vary with the analytical method applied: the molecular weight of the individual sheets is ca. 500–1000, and the size of their polyaromatic nucleus is 8–15 Å. The molecular weight of a particle resulting from the stacking is 1000–10,000, and their average thickness is 15–20 Å. Values of molecular weight reaching 50,000 or more have been measured on aggregates of particles. The size of the latter depends on the number of particles, 40–50 Å being an average value.

d) Resins include minor amounts of free acids, esters or ethers. The bulk of resins is, moreover, likely to contain molecules comparable to, but less aromatic than, asphaltenes. The number and size of aromatic nuclei would be reduced compared to asphaltenes, thus lowering the possibility of intermolecular bonds. Therefore separated resins probably comprise a single sheet and show a comparatively lower molecular weight (500 to 1200). Their lighter constituents may still comprise some true polyaromatics. However, most of the resins are rich in heteroatoms and particularly oxygen. Resins are not stable compounds especially when exposed to air and sunlight, and they probably evolve further and go on to form asphaltenes. When heat-treated, they produce, by dismutation and cracking, hydrocarbons and asphaltenes.

Some authors consider that the model developed by Yen based on precipitated asphaltenes is also a valid representation of asphaltenes in crude oils. Some others are of the opinion that aggregates or even particles do not exist and are a result of asphaltenes being separated in the laboratory from the other constituents of crude oil.

The *microstructure* of asphaltenes (e.g., the sheet and generally the particle described by Yen) is widely accepted. However, there are some discussion about the intramolecular forces and particularly the role of cross-linking by oxygen or sulfur bridges. Most of oxygen and part of sulfur are engaged in cross-linking, or in functional groups: they are removed by thermal treatment. On the contrary, nitrogen and the remaining part of sulfur are engaged in heterocycles of the polyaromatic nuclei; upon heating they are found in the residual coke (Moschopedis et al., 1978; Speight et al., 1979).

The problem of the *macrostructure* (micelles or aggregates), and the evaluation of the molecular weight arc clearly consequences of the physical state of asphaltenes in their environment. Laboratory measurements of the molecular weight of previously separated asphaltenes are dependent on the nature of the solvent, the asphaltene concentration, and the temperature (Moschopedis et al., 1976). Asphaltenes form aggregates even in dilute solution: in solvents of low polarity, such as benzene, the asphaltenes from Athabasca tar sands have a molecular weight of 4000 to 7000; in solvents of higher polarity, such as nitrobenzene, their molecular weight is lower than 2000 at 120°C (Moschopedis et al., 1976). Recently, Boduszynski et al. (1980), using chemical fractionation of a vacuum distillation residue (called "asphalt") followed by infrared analyses and field ionization mass spectrometry (FIMS), expressed the view that particles of very high molecular weight (beyond 5000) do not exist in crude oil.

In crude oils, asphaltenes are probably dispersed by the action of resins (Pfeiffer and Saal, 1940; Speight et al., 1979; Speight, 1980). Resin-asphaltene interactions, especially through hydrogen-bonding, are preferred to asphaltene–asphaltene association and thus they keep these particles in suspension. The overall structure would be of micellar type as proposed by Pfeiffer and Saal (1940): Figure IV.1.20. In this scheme, the core of the micelle is occupied by one

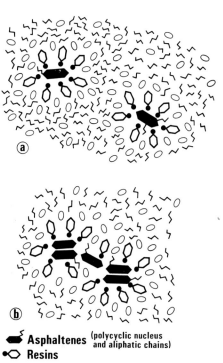

Asphaltenes (polycyclic nucleus and aliphatic chains)
Resins
Aromatic hydrocarbons
Paraffinic and/or naphthenic hydrocarbons

Fig. IV.1.20a and b. Possible micellar structure of asphaltenes and resins in crude oils. (Adapted, with changes, from Pfeiffer and Saal, 1940). (a) Asphaltenes are fully dispersed when the crude oil contains a sufficient amount of resins and aromatic hydrocarbons. (b) Asphaltene–asphaltene association occurs in crude oils where there is a shortage of resins and aromatic hydrocarbons, as compared to the abundance of asphaltenes; this is the case, for instance, in heavy degraded oils

1.8 High Molecular Weight N, S, O Compounds: Resins and Asphaltenes

or several asphaltene "molecules" and it is surrounded by interacting resins. In turn, the resins are surrounded by aromatic hydrocarbons which ensure a progressive transition to the bulk of crude oil where saturated hydrocarbons are usually predominant. This interpretation may not be conflicting with the model developed by Yen, as the core of the micelle could be a particle or crystallite.

The size of the micelles formed by asphaltenes surrounded by interacting resins in their natural environment i.e., normal crude oil or heavy oil, is likely to play an important role in the viscosity of these oils. When the oil contains a sufficient amount of resins and aromatic hydrocarbons, the asphaltenes are fully dispersed. On the contrary, if there is a shortage of these molecules, there may be interaction between asphaltenes and formation of large-sized aggregates. This situation would result in a higher viscosity of crude oil, and eventually a gel type structure. This consideration may explain the high viscosity observed in heavy degraded oils, where low molecular weight aromatics and possibly some of the resins are lost by water washing and/or biodegradation, and thus become insufficient to peptize asphaltenes.

In coal liquids, produced by hydrogenation of coal, Tewari et al. (1979) have shown that interactions by hydrogen-bonding involve phenolic OH and N-containing bases. In crude oils, there may also be interaction of phenolic OH with ketones, quinones, etc. present in the heavy ends of petroleum.

High viscosity due to the asphaltene abundance and physical structure is the main hindrance for production of heavy oils. Furthermore, asphaltenes are also considered as responsible for water in oil emulsions observed when certain crude oils are produced.

Pyrolysis of asphaltenes has provided valuable information on their structure and also on their possible role in petroleum generation. The behavior of asphaltenes during pyrolysis can be compared to that of kerogen. Up to 300–350°C, oxygen from carbonyl or carboxyle group is eliminated, whereas some light hydrocarbons are generated. Beyond that temperature the yield of methane, higher hydrocarbons and also hydrogen sulfide increases and reaches a maximum in the 410–470°C range. About 50% of Athabasca asphaltenes are converted to volatiles, whereas the residuum becomes insoluble in benzene and highly condensed: H/C = 0.45 at 600°C versus 1.24 in the original asphaltene (Moschopedis et al., 1978).

Furthermore, Rubinstein et al. (1979) have compared the products yielded by pyrolysis of Prudhoe Bay asphaltenes with those naturally present in the related crude oil. They have identified, besides olefins resulting from cracking, a number of constituents comparable to those of Prudhoe Bay crude oil: n- and iso-alkanes, polycyclic naphthenes, aromatics. Arefjev et al. (1980) made a comparable observation on asphaltenes from degraded oils of the USSR and showed that pyrolysis products can be used for reconstruction of the degraded paraffinic fraction. More recently Behar (1983) pyrolyzed the asphaltene fraction of ten different crude oils and found a remarkable similarity between the relative abundance and distribution of the main constituents in pyrolysis products and in the related crude oils.

A comparable information is provided by Chappe (1982), who submitted kerogens and asphaltenes separated from crude oils to a selective cleavage of ether bonds. In both cases, he identified glycerol ethers comprising C_{20} and C_{40} isoprenoid chains typical of the membranes of Archaebacteria.

All these data point to a comparable structure of the units constituting kerogen and asphaltenes. In fact, asphaltene constituents in source rocks should be considered as being small fragments of kerogen liberated by elimination of heteroatomic or other bonds. Thus the overall structure (nuclei bearing chains or functional groups, linked by chains or heteroatomic bonds) might be preserved from the kerogen to the heavy ends of crude oils. Progressive aromatization of the cyclic nuclei would favor the stacking of pericondensed sheets. In this respect, it can be observed that the structure proposed by Yen for asphaltenes is reasonably comparable to the one proposed in Part II of this book for kerogen.

Furthermore, the study of several typical successions of source-rocks as a function of burial suggests the existence of evolution paths of the asphaltenes in source rocks somewhat comparable to the evolution paths of kerogens (Castex in Tissot, 1981): With increasing burial, S/C, O/C and H/C ratios decrease progressively, resulting in a more aromatic and condensed asphaltene bearing fewer and fewer chains or functional groups. A consequence is likely to be the formation of some hydrocarbons, in addition to those generated directly from kerogen. This hypothesis is supported by the experimental evidences reported above.

The role of resins and asphaltenes in migration is not completely clear. Table IV.1.1 shows that the average amount of these heavy constituents in source-rock bitumen approximately equals that of hydrocarbons. In reservoired crude oils, the ratio is only 1 to 6. In particular the ratio of the average values resins + asphaltenes to aromatics decreases from 2.6 in source rocks to 0.5 in normal, non-degraded oils. Adsorption of the heavy constituents on mineral surfaces in source rocks is partially responsible for that situation. However, the previous considerations on the mechanism of asphaltene solubilization by resins suggest that each petroleum might be able to mobilize out of the source rock the amount of asphaltene which may dispersed by resins, and in turn the amount of resins which may be solubilized by aromatic hydrocarbons (Tissot, 1981). Such hypothesis would contribute to explain the strong correlation between the abundances of aromatics and heavy constituents in crude oils reported below (Sect. 1.10). Obviously this interpretation would not apply to degraded oils, which may contain a much larger amount of asphaltenes and resins, as degradation occurred in the reservoir after migration.

1.9 Organometallic Compounds

Crude oils contain metals, particularly nickel and vanadium, in variable amounts from less than 1 ppm in some Paleozoic crudes from Algeria and the USA, up to 1200 ppm vanadium and 150 ppm nickel in Boscan crude from Venezuela. The average content measured on 64 crude oils from the United States, Canada, Venezuela, North and West Africa, the Middle East, USSR and Australia are 63 ppm vanadium and 18 ppm nickel. The respective vanadium and nickel content, according to their origin, is shown in Figure IV.1.21. Both are at a low level

1.9 Organometallic Compounds

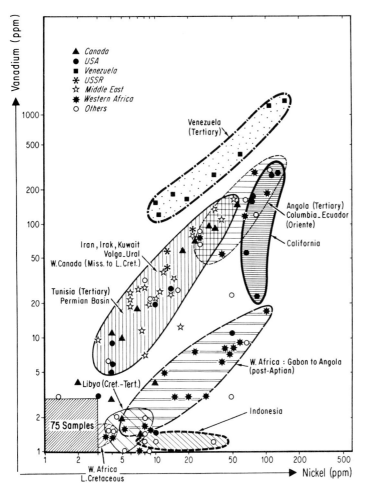

Fig. IV.1.21. Vanadium and nickel contents of 175 crude oils from various origin. The *shaded area* contains 75 crude oils of paraffinic, naphthenic, or paraffinic-naphthenic types from Devonian and Cretaceous of Alberta. Paleozoic of Pennsylvania and Oklahoma. Tertiary of Gulf Coast, Mesozoic and Paleogene of Paris and North Aquitaine basins and from the Rhine Graben and North Sea, Tertiary of Pannonian basin, Paleozoic of Algeria and Libya, Lower Cretaceous of West Africa, Tertiary of Australia, Paleozoic and Mesozoic of Bolivia, etc.

— a few ppm — in many oils from detritic sand–shales series: Paleozoic of North Africa, Cretaceous of Alberta (Colorado and post-Colorado groups) Cretaceous and Tertiary of West Africa, Tertiary of Australia. In these low-sulfur crude oils nickel is frequently prevalent over vanadium. The abundance of metal is variable, but generally higher in sulfur-rich crude oils, where vanadium is frequently prevalent over nickel: Upper Paleozoic of Ural-Volga, lowermost Cretaceous of Alberta (Mannville group and associated oils along the pre-Cretaceous unconformity), Cretaceous of Mexico, Cretaceous and Tertiary of the

Middle East, Tertiary of Venezuela. The highest values of metals are reached among crude oils from Venezuela.

There is a good correlation between metals, sulfur, and asphaltene content. This fact is quite understandable as metals are usually in the resin and asphaltene fractions. As a result of the metal–asphaltene correlation, degraded oils, which are enriched in asphaltenes, contain also more metals than the nondegraded oils of similar origin.

Other metals like iron, zinc, copper, lead, arsenic, molybdenum, cobalt, manganese, chromium, etc. have also been reported. They have not been analyzed systematically, and it is difficult to evaluate if their occurrence is common or not. However, vanadium and nickel seem definitely to be the most abundant metals. A variable part of them is incorporated in porphyrin complexes, but the structural affiliation of the remaining part is largely unknown.

Porphyrins were first identified in petroleum by Treibs (1934). They are characterized by a tetrapyrrolic nucleus, probably inherited from chlorophyll or hemin (Fig. II.3.20). Besides nitrogen, they contain oxygen and a metal which is usually vanadium (as vanadyl V=O) or nickel; iron and copper are also reported. Porphyrins vary in content from traces to 400 ppm in crude oils. They are generally associated with the resin fraction of crude oils. When purified, they are red or brown. Their mass spectra show a distribution of molecular weights suggesting a series of alkyl-substituted porphyrins. The possible origin of these series is discussed in Chapter II.3.12. Laboratory experiments have shown that porphyrins are labile at higher temperatures. This could explain why old paraffinic crude oils, with a low density and a low resin and asphaltene content, contain only traces of porphyrins. In addition, the relatively low evolutionary state of plants during early Paleozoic times might be responsible for the low porphyrin content of Cambrian and Ordovician crude oils (Louis, 1967).

Metals occur not only in porphyrin complexes. Although porphyrins account for 100% of vanadium in Baxterville crude oil, they account for only 31% in Boscan, Venezuela, crude, and for only 27% in La Luna, Venezuela, crude: Erdman and Harju (1963). The balance of metals, i.e., 0 to 70% of vanadium in crude oils, is included in structural associations other than free porphyrins.

They may include porphyrin molecules inserted between two sheets of the asphaltene particle and linked by π-vanadium bonds (Tynan and Yen, 1967). Alternatively metal may be complexed in a porphyrin nucleus, the latter bearing complex substituents.

In the heavy end of crude oils there is frequently a good correlation between the sulfur content and the vanadium content. These two elements can be eliminated simultaneously during certain steps of hydrotreatment of the heavy fraction of crude oils. It therefore seems reasonable to assume that vanadium is mainly bound in chemical structures in which sulfur plays an important role, as nitrogen does in porphyrins.

1.10 Covariance Analysis of Main Crude Oil Constituents

The concentrations of the various crude oil constituents are not independent parameters. Some of them show a high degree of covariance. The reason for one kind of dependence is obvious. It is related to the gross composition of crude oil, as:

saturated + aromatics + resins + asphaltenes = 1.

However, another kind of dependence can be observed when dealing with the concentrations of more specific classes of compounds: benzothiophenes, monocycloalkanes, etc., or with other chemical and physical parameters such as sulfur content, viscosity, density, etc.

Tables IV.1.9 and IV.1.10 show the linear correlation coefficients of important parameters, such as the abundance of the main classes of compounds and the sulfur content, on 349 crude oils from all types of reservoirs rocks deposited in various environments.

Their depths range from near surface to more than 4000 m, and the ages of the producing formation range from Cambrian to Pliocene. There are two main groups of interrelated parameters:

- the first group consists of sulfur content and aromatic constituents (asphaltenes, resins, thiophene derivatives and aromatic hydrocarbons);
- the second group consists of paraffins, mono and dicycloalkanes and alkylbenzenes.

There is a strong positive correlation among the parameters of the first group (average $\varrho = 0.8$). They are also correlated with some physical properties such as density and viscosity. They correlate, however, negatively with the concentration of saturated hydrocarbons in crude oil.

The correlation between the second group of parameters is weaker but still definite. These parameters are also correlated to the degree of maturation of the oil, namely to depth of reservoir.

Only the parameters of the first group will be discussed here. The explanation for the relationship between sulfur, the aromatic constituents and the heavy constituents, resins plus asphaltenes, may be found in the origin of sulfur in crude oils, and in the mode of asphaltene solubilization in crude oils.

Sulfur is not a major constituent of biogenic molecules, where it ranks well behind oxygen and nitrogen. Therefore, it has to be introduced into the organic matter, either into kerogen during early diagenesis in young sediments or into crude oils.

The time and conditions of sulfur introduction are discussed in Chapters II.2, III.4 and IV.5. The observed relation between sulfur and aromatic substances may be interpreted in two ways:

a) Sulfur is preferentially fixed by molecules containing an aromatic cycle. The scarcity of aromatic cycles available in Recent sediments (only some polyaromatics) would not favor this hypothesis. Nevertheless humic acids of Recent sediments already contain a fair amount of sulfur.

Table IV.1.9. Linear correlation coefficient of the main classes of compounds and sulfur content of 349 crude oils. (Analyses by IFP)

		Percent of crude oil			% sulfur	% hydrocarbons			% aromatics	
		Saturated HC	Aromatic HC	Resins +Asph.	in crude oil	Paraffins	Naphthenes	Aromatics	Total thiophene derivatives	Heavy aromatics $\geqslant C_{25}$
% crude oil	Sat.	1	−0.86	−0.91	−0.79	0.83	0.33	−0.97	−0.70	−0.62
	Arom.		1	0.56	0.63	−0.74	−0.40	0.93	0.61	0.53
	Res+As			1	0.78	−0.72	−0.20	0.80	0.56	0.50
% sulfur in crude oil					1	−0.54	−0.72	0.79	0.83	0.55
% hydrocarbons	Par.					1	−0.19	−0.81	−0.42	−0.45
	Napht.						1	−0.40	−0.59	−0.38
% aromatics	Arom.							1	0.74	0.62
	Thioph.								1	0.27
	Heavy arom. $\geqslant C_{25}$									1

Table IV.1.10. Linear correlation coefficient of some classes of compounds (349 crude oils). (Analyses by IFP)

	Saturated HC	Paraffins	1+2 ring naphthenes	Alkylbenz.	Depth
Saturated HC	1	0.85	0.59	0.38	0.45
Paraffins		1	0.59	0.49	0.60
1+2 ring naphthenes			1	0.46	0.44
Alkylbenz.				1	0.35
Depth					1

Table IV.1.11. Correlation coefficient of sulfur content, depth of producing horizon and physical properties of crude oils (2000 samples). (Data from U.S. Bureau of Mines, IFP and others)

	Depth	Specific gravity	Viscosity[a]	Pour point	Sulfur[a] (in crude oil)
Depth	1				
Specific gravity	−0.393	1			
Viscosity[a]	−0.321	0.823	1		
Pour point	−0.015	0.090	0.121	1	
Sulfur[a] (in crude oil)	−0.291	0.661	0.520	−0.070	1

[a] It should be noted that sulfur and viscosity have approximately a lognormal distribution. Therefore, for computation of the correlation coefficients, the logarithmic values of sulfur and viscosity have to be used.

b) Sulfur is incorporated in organic matter as an agent of aromatization. Sulfur would combine with unsaturated chains and promote cyclization, or with saturated or partly unsaturated cycles and promote their aromatization. This second hypothesis fits best with observed facts.

Furthermore, the way the asphaltenes are solubilized in crude oils by resins and aromatics, and thus transported during migration (see above Sect. 1.8) is probably responsible for the strong correlation between aromatics and resins plus asphaltene, irrespective of their original abundance in the bitumen of source-rocks.

Another aspect of the same problem is the dependence of some physical properties (density, viscosity) on the sulfur content of crude oils. Density, viscosity and sulfur are often the only information available on crude oils which were routinely collected many years ago. The correlation is shown in Table IV.1.11 using 2000 crude oils of various origin, age, and reservoir types.

In fact, density and viscosity of a crude oil depend mostly on the content of heavy constituents of petroleum, resins and asphaltenes. As previously seen, the abundance of resins and asphaltenes is, in turn, closely related to the sulfur content. That the relationship is indirect is expressed by the fact that the sulfur-density of sulfur-viscosity correlations (Table IV.1.11) are not as good as the sulfur–resins + asphaltenes correlation (Table IV.1.9). Furthermore, a high content of resins + asphaltenes is not the only cause of high viscosity. In high-sulfur crude oils (S ⩾ 1%), where resins and asphaltenes are generally abundant and dominate any other cause of viscosity increase, the correlation between density, viscosity, and sulfur is better.

Among the physical parameters, there is one which is completely independent from the others mentioned so far, the pour point. Pour point, therefore, is able to provide information which cannot be derived otherwise. It denotes the content of high molecular weight n-alkanes in crude oils.

Other constituents of crude oils, like tetra- and pentacyclic naphthenes, and the polyaromatics, correlate, to a certain degree, with the aforementioned two groups of parameters. As a result it can be said that generally:

- crude oils rich in aromatics contain more tetra- and pentacyclic naphthenes, polyaromatics and thiophene derivatives than the other oils, but proportionally less alkylbenzenes and other monoaromatics;
- crude oils poor in aromatics, when immature, contain a certain amount of naphthenes. In a more advanced stage of evolution, paraffins (normal or branched) are predominant, mono- and dicyclic naphthenes and alkylbenzenes are abundant.

Summary and Conclusion

The main groups of compounds in crude oils are saturated hydrocarbons, aromatic hydrocarbons, resins and asphaltenes.

Saturated hydrocarbons are usually the major group, except in degraded heavy oils. They contain normal plus isoalkanes (paraffins) and cycloalkanes (naphthenes). The relative abundance of paraffins and naphthenes is rather comparable in many crude oils, with the exception of paraffinic oils and degraded heavy oils.

Aromatic hydrocarbons comprise pure aromatics, naphthenoaromatics (condensed aromatic and saturated cycles) and benzothiophene derivatives (containing heterocycles with sulfur).

Resins and asphaltenes are high molecular weight polycyclic molecules containing N, S, and O atoms. The basic unit of their structure is a condensed polyaromatic sheet. Several sheets are usually stacked to form particles, which may in turn form aggregates or micelles.

There is a strong positive correlation between the concentrations of aromatics, resins plus asphaltenes, and the sulfur content.

Chapter 2
Classification of Crude Oils

2.1 General

Various crude oil classifications have been proposed by geochemists and petroleum refiners. The purpose of these is very different, and also the physical or chemical parameters which have been used in the classifications. Petroleum refiners are mostly interested in the amount of the successive distillation fractions (e.g., gasoline, naphtha, kerosene, gas oil, lubricating distillate) and the chemical composition or physical properties of these fractions (viscosity, cloud test, etc.). Geologists and geochemists are more interested in identifying and characterizing the crude oils, to relate them to source rocks and to measure their grade of evolution. Therefore, they rely on the chemical and structural information of crude oil constituents, especially on molecules which are supposed to convey genetic information. In that respect molecules at relatively low concentrations, such as high molecular weight n-alkanes, steroids and terpenes, may be of great interest.

The type of information on crude oil composition varies a great deal. A large number of distillation results — Hempel, Engler, ITK, and other methods — are available, often including some additional measurements on the fractions, e.g., density, viscosity, refraction index, etc. For example, about 10,000 analyses have been carried out according to the U.S. Bureau of Mines routine method. They include mostly crude oils from the United States and Canada, with several hundreds from other countries.

Analyses based on the content of the various structural types in crude oils (alkanes, cycloalkanes, aromatics, N, S, O compounds), and the distribution of molecules within each type, have been made in the last 15 years. However, they are probably less numerous than distillation data, and the results are frequently unpublished. The classification proposed below using this type of geochemical parameter is based mainly on the analyses carried out by the Institut Français du Pétrole on more than 600 crude oils of various origins and ages.

2.2 Historical

An excellent review of the main classifications proposed since the beginning of the century has been made by Smith (1968).

A well-known and up to now frequently used classification of crude oils has been set up at the U.S. Bureau of Mines by Smith (1927) and Lane and Garton (1935). This classification is based on distillation and the specific gravities of two key fractions of distillation (Table IV.2.1).

It was successfully applied to intercompare crude oils from certain provinces in the United States. However, with the progress in modern organic geochemistry, and the higher level of information derived from it, it became obvious that the classification was ambiguous. On the one hand, rather different chemical composition of oils could result in the same grouping of these oils in the classification. On the other hand, different geological situations could not be accounted for, except changes in depth. Since the U.S. Bureau of Mines classification can hardly be developed any further with respect to more compositional information on oils, a new classification based on geochemical parameters is proposed in the following Section IV.2.3.

Other classifications have been proposed by Sachanen (1950), Creanga (1961) and Radchenko (1965). In addition, in refinery practice there is a classification based on refractive index, density and molecular weight (n. d. M.). Since all these classifications have not been extensively applied in the past, they are not discussed here.

Table IV.2.1. Classification of crude oils. (According to Lane and Garton, 1935)

Specific gravity		Base	Base
Key fraction I 250–275°C (Atm. pressure)	Key fraction 2 275–300°C (40 mm Hg)	Key fraction I 250–270°C (Atm. pressure)	Key fraction 2 275–300°C (40 mm Hg)
$\leqslant 0.8251$	$\leqslant 0.8762$	paraffinic	paraffinic
$\leqslant 0.8251$	0.8767 to 0.9334	paraffinic	intermediate
$\leqslant 0.8251$	$\geqslant 0.9340$	paraffinic	naphthenic
0.8256 to 0.8597	$\leqslant 0.8762$	intermediate	paraffinic
0.8256 to 0.8597	0.8767 to 0.9334	intermediate	intermediate
0.8256 to 0.8597	$\geqslant 0.9340$	intermediate	naphthenic
$\geqslant 0.8602$	$\geqslant 0.9340$	naphthenic	naphthenic
$\geqslant 0.8602$	0.8767 to 0.9334	naphthenic	intermediate
$\geqslant 0.8602$	$\leqslant 0.8762$	naphthenic	paraffinic

2.3 Basis of Proposed Classification of Crude Oils

The newly proposed classification is based primarily on the content of the various structural types of hydrocarbons: alkanes, cycloalkanes (naphthenes), and aromatics plus N, S, O compounds (resins and asphaltenes). It also takes into account the sulfur content. It should be noted that the content of alkanes, naphthenes and aromatics is considered here as a result of analytical methods presently in use, i.e.:

- all data refer to the portion of petroleum boiling above 210°C at atmospheric pressure;
- alkane (paraffin) content includes n- and isoalkanes, but not the alkyl chains substituted on cyclic nuclei;
- cycloalkanes (naphthenes) include all molecules containing one or more saturated cycles but no aromatic cycle;
- aromatics include all molecules containing at least one aromatic cycle; the same molecules may include condensed saturated cycles and chains substituted on the nucleus.

In Figures IV.1.4 and IV.1.6, some 550 crude oils were plotted in triangular diagrams showing their relative composition with respect to the whole crude oils (saturates, aromatics, N. S, O compounds) and with respect to hydrocarbons only (n+isoalkanes, cycloalkanes, aromatics). From these plots, it is obvious that the oils cover only part of the triangles in form of elongated clouds. On the crude oil diagram (Fig. IV.1.4), the cloud starts at the pole of saturates and bends over to the N, S, O compounds. On the hydrocarbon diagram (Fig. IV.1.6) the cloud starts at the pole of n+isoalkanes and bends over to the aromatic hydrocarbons. This is, of course, another expression of the correlation matrix discussed in Section IV.1.10.

This peculiar distribution of crude oils and the geochemical consequences derived from it are in essence the basis for our proposed classification.

As a first parameter for the classification, we shall consider the concentration of saturates in crude oil, which is negatively correlated with aromatics, resins, asphaltenes, and sulfur content. The distribution of saturated hydrocarbons in crude oils has been discussed in Chap. IV.1 and is shown in Figure IV.1.5. There, two modes are clearly separated by a minimum at 50%; this cut is used to separate two different populations of crude oils, one more paraffinic or paraffinic-naphthenic, the other more aromatic or asphaltic. The 50% minimum of the saturates corresponds to the 1% minimum of the distribution of sulfur in crude oils (Fig. IV.1.16), as shown by a cross-correlation of those two parameters.

In addition to this major subdivision, another important cut may be introduced among the crude oils, based on the alkane concentration. The distribution of n+isoalkanes (paraffins) shows a minimum at about 10%, which separates another class of crude oils. Heavy oils, generally degraded, represent this class: the low content of n+isoalkanes is mainly due to biodegradation.

2.4 Classification of Crude Oils (Table IV.2.2 and Fig. IV.2.1)

a) According to the previous consideration, crude oils will be considered as "paraffinic" or "naphthenic" if the total content of saturated hydrocarbons is over 50% of a particular crude oil. For practical purposes, and future use in connection

Table IV.2.2. Proposed classification of crude oils

Concentration in crude oil > 210°C		Crude oil type	Sulfur content in crude oil (approximate)	Number of samples per class (Total = 541)
S = saturates AA = aromatics + resins + asphaltenes	P = paraffins N = naphthenes			
S > 50% AA < 50%	P > N and P > 40%	Paraffinic	< 1%	100
	P ⩽ 40% and N ⩽ 40%	Paraffinic-naphthenic		217
	N > P and N > 40%	Naphthenic		21
S ⩽ 50% AA ⩾ 50%	P > 10%	Aromatic intermediate	> 1%	126
	P ⩽ 10%, N ⩽ 25%	Aromatic asphaltic		41
	P ⩽ 10%, N ⩾ 25%	Aromatic naphthenic	generally S < 1%	36

with geological environments, additional boundaries justified by minor cuts are established at 40% alkanes and cycloalkanes. These two boundaries are shown in Figure IV.2.1. They separate "paraffinic" oils from intermediate "paraffinic-naphthenic" oils and those from "naphthenic" oils.

b) Crude oils will be considered as "aromatic" if the total content in saturated hydrocarbons is less than 50%, i.e., the total content in aromatics, resins and asphaltenes is more than 50%. According to the previous consideration on the alkane content, an additional boundary separates "aromatic-intermediate" oils, with more than 10% n+isoalkanes, and heavy degraded oils, with less than 10%. The latter class is divided for practical purpose into two subclasses:

– "aromatic-asphaltic" oils with less than 25% naphthenes,
– "aromatic-naphthenic" oils with more than 25% naphthenes.

This minor cut corresponds to a difference of the sulfur content: aromatic-asphaltic oils are high-sulfur oils, whereas aromatic-naphthenic oils usually contain less than 1% sulfur.

The subdivision into classes of crude oil, and the approximate equivalence in respect to sulfur content, is shown in Table IV.2.2 and Figure IV.2.1. Figure IV.2.2 shows that low-sulfur crude oils mostly belong to the paraffinic, paraffinic-

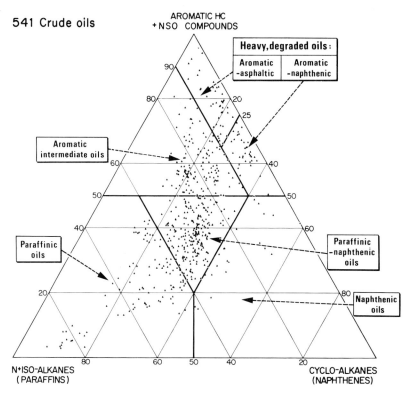

Fig. IV.2.1. Ternary diagram showing the composition of the six classes of crude oils from 541 oil fields

naphthenic or naphthenic classes. On the contrary, high-sulfur crude oils mostly belong to the aromatic intermediate class and — if they are degraded — to the aromatic-asphaltic class.

2.5 Characteristics of the Principal Classes of Crude Oils

1. The *paraffinic class* is comprised of light crude oils, some being fluid, and some high-wax, high-pour-point crude oils. The viscosity of these high-pour-point oils at room temperature is high, due to a high content of n-alkanes $>C_{20}$. At slightly elevated temperatures (35–50°C), however, the viscosity becomes normal.

The paraffinic class comprises some oils from the Paleozoic of North Africa, United States and South America, some of the lower Cretaceous oils from the basins bordering the South Atlantic Ocean (beneath the massive salt sequence in West Africa; the Magellan basin in South America, etc.), and some Tertiary oils from West Africa, Libya, Indonesia, and the Pannonian basin in Central Europe. Among them, some high-wax crude oils occur in the nonmarine Lower Cretaceous of the Magellan basin, the Cretaceous and Tertiary of West Africa, and the

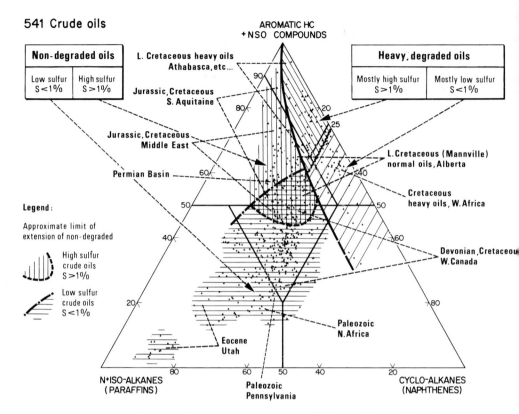

Fig. IV.2.2. Ternary diagram showing the occurrence of high- and low-sulfur oils, with respect to the crude oil classification

Tertiary of Indonesia. Most of the oils associated with the Green River and Flagstaff formations of Uinta Basin, Utah, belong to the high-wax type.

Specific gravity is usually below 0.85. The amount of resins plus asphaltenes is below 10%. Viscosity is generally low except when n-alkanes of high molecular weight are abundant. Aromatic content is subordinate and mostly composed of mono- and diaromatics, frequently including monoaromatic steroids. Benzothiophenes are very scarce, sulfur content is low to very low.

2. The *class of paraffinic-naphthenic* oils has a moderate resins-plus-asphaltenes content (usually 5 to 15%) and a low sulfur content (0 to 1%). Aromatics amount to 25 to 40% of the hydrocarbons. Benzo- and dibenzothiophenes are moderately abundant.

Density and viscosity are usually higher than in the paraffinic class but remain moderate. The paraffinic-naphthenic class includes many crude oils from Devonian and Cretaceous (mostly from Colorado and post-Colorado groups) of Alberta, from the Paleozoic of North Africa and United States — except from the Permo-Carboniferous of the Permian basin, West Texas. The class also includes Jurassic and Cretaceous oils of the Paris, North Aquitaine and North Sea basins,

2.5 Characteristics of the Principal Classes of Crude Oils

Cretaceous and Tertiary oils of West Africa, (above the massive salt sequence), Cretaceous and Lower Tertiary oils from North Africa, and some Tertiary oils from Indonesia.

3. There are only a few crude oils in the *naphthenic class*. Some immature oils can be mentioned here, such as some oils from the Jurassic and Cretaceous of South America. However, the class includes mainly degraded oils which usually contain less than 20% n+isoalkanes. They originate from biochemical alteration of paraffinic or paraffinic-naphthenic oils and they usually have a low sulfur content, although they are degraded. Examples are found in the Gulf Coast area, in the North Sea, and in USSR.

4. The *aromatic-intermediate class* is comprised of crude oils which are often heavy. Resins and asphaltenes amount to ca. 10–30% and sometimes more, and the sulfur content is above 1%.

Aromatics amount to 40 to 70% of hydrocarbons. The content of monoaromatics, and especially those of steroid type, is relatively low. Thiophene derivatives (benzo- and dibenzothiophene) are abundant (25–30% of the aromatics and more). Specific gravity is usually high (more than 0.85). This class includes most crude oils from Jurassic and Cretaceous of the Middle East (Saudi Arabia, Qatar, Kuwait, Irak, Syria, Turkey), from the Permo-Carboniferous of the Permian basin (West Texas), from the Mississippian, Jurassic and Lower Cretaceous (Mannville) of Alberta, some Jurassic-Cretaceous heavy oils of the South Aquitaine basin, some Upper Cretaceous-Tertiary oils of W. Africa, and some oils from Venezuela, California, and the Mediterranean (Spain, Sicily, Greece).

5. and 6. The *classes of aromatic-naphthenic and aromatic-asphaltic* oils are mostly represented by altered crude oils. During biodegradation, alkanes are first removed from crude oil. This results in a shift of the crude oils away from the alkane pole as shown in Figure IV.2.3. Later, a more advanced degradation may involve removal of monocycloalkanes and oxidation. Then the position of an oil in the diagram is also shifted towards the aromatic pole.

Therefore, most aromatic-naphthenic and aromatic-asphaltic oils are heavy, viscous oils resulting originally from degradation of paraffinic, paraffinic-naphthenic, or aromatic-intermediate oils. The resin-plus-asphaltene content is usually above 25% and may reach 60%. However, the relative content of resins and asphaltenes, and the amount of sulfur, may vary according to the type of the original crude oil.

a) The *aromatic-naphthenic class* is mainly derived from paraffinic or paraffinic-naphthenic oils. The Lower Cretaceous oils of West Africa, when altered, contain mostly resins, and keep a low sulfur content: the heavy oils of the Emeraude field contain ca. 25% of resins, but the sulfur content is only 0.4 to 0.8% (Claret et al., 1977). The same is true for tar-like oil in the Lower Jurassic sands of Malagasy (Bemolanga tar sands), with a sulfur content below 1% and a resin: asphaltene ratio of 2 or more.

b) The *class of aromatic-asphaltic* oils includes a few true aromatic oils, apparently nondegraded, from Venezuela and West Africa. However, it is mainly comprised of heavy, viscous, or even solid oils, resulting from alteration of aromatic-intermediate (particularly high sulfur) crude oils.

The result is usually an asphaltic oil, or tar, whose sulfur content is above 1% and may reach up to 9% in extreme cases. Their resin and asphaltene content is very high, ca. 30 to 60%, with a lower resin: asphaltene ratio than in the aromatic-naphthenic oils. Most degraded tar sands from Western Canada fall in this class: Athabasca with ca. 5% sulfur content in the MacMurray area, Peace River, and Wabasca. Heavy asphaltic oils from Venezuela and the South Aquitaine basin (France) also belong to the same group.

2.6 Concluding Remarks

The six classes of crude oils defined above are very unevenly populated. Most normal (nondegraded) crude oils belong to three classes only. They are:
- *"aromatic-intermediate"* oils, which usually contain more than 1% sulfur. They are frequently generated in marine sediments, deposited in a reducing environment;

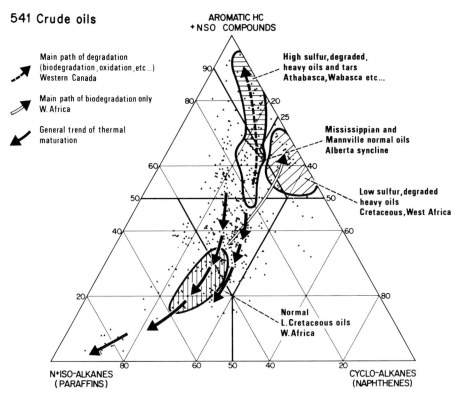

Fig. IV.2.3. Ternary diagram showing the main trends of alteration and thermal maturation of crude oils. Biodegradation is exemplified by the alteration of the paraffinic oils from the Lower Cretaceous of West Africa. A more complete degradation, including oxidation, is shown for the aromatic intermediate oils from the Mississippian and Lower Cretaceous (Mannville) oils of Western Canada. The extreme end of degradation is reached for the Athabasca heavy oils or tars

- *"paraffinic-naphthenic"* and *"paraffinic"* oils, which usually contain less than 1% sulfur. They are frequently generated in deltaic or coastal sediments of the continental margins, or in nonmarine source beds.

The above evaluation is based on the frequency of different oils investigated. However, if we now consider the total amount of known production and reserves, the relative importance of the classes is changed.

By far the most important classes, with respect to quantity, are the aromatic-asphaltic, then the aromatic-naphthenic and the aromatic-intermediate. The first two comprise the enormous reserves of heavy oils and tar sands (Athabasca, Arctic Islands of Canada, Eastern Venezuela, Malagasy etc.). The aromatic-intermediate class is represented by the large accumulations of crude oils in the Middle East.

The other extreme is the class of naphthenic oils, which is comparatively rare.

Evolution and alteration affects the composition of a crude oil (Evans et al., 1971) and thus may progressively change its classification. If we refer to the ternary diagram of Figure IV.2.3, *thermal evolution* is denoted by a diminution of aromatic and heavy constituents, and an increase of paraffins. In advanced stages, the composition approaches the paraffin pole, that would otherwise be reached only by gas (methane). Thus a paraffinic-naphthenic oil, at shallow or medium depth, may be changed into a paraffinic oil at great depth. *Alteration* is primarily denoted by a shift away from the paraffin pole, due to biochemical degradation of alkanes. Then an additional increase of resins and asphaltenes may result from oxidation.

Through alteration, normal paraffinic-naphthenic oils may be changed into aromatic-naphthenic. Aromatic intermediate oils may also be changed to the aromatic-naphthenic class or even to aromatic-asphaltic heavy oils.

Summary and Conclusion

The classification of crude oils is based on the content of n+isoalkanes (paraffins), cycloalkanes (naphthenes) and aromatic compounds (aromatic hydrocarbons, resins, asphaltenes).

The main classes of normal crude oils are the following:
- *paraffinic oils*, containing mostly normal and isoalkanes, and less than 1% S;
- *paraffinic-naphthenic oils,* containing both linear and cycloalkanes, and less than 1% S;
- *aromatic-intermediate oils,* containing less than 50% of saturated hydrocarbons, and usually more than 1% S.

Evolution and alteration change the composition of crude oils. For instance, thermal evolution of a paraffinic-naphthenic oil may result in a paraffinic oil. Alteration results in general in heavy oils of aromatic-naphthenic or aromatic-asphaltic classes:

- *paraffinic and paraffinic-naphthenic oils* are usually degraded into aromatic-naphthenic oils, with a moderate sulfur content (less than 1%);
- *aromatic-intermediate oils* are usually degraded into aromatic-asphaltic oils, with a high sulfur content (more than 1%).

Chapter 3
Geochemical Fossils in Crude Oils and Sediments as Indicators of Depositional Environment and Geological History

3.1 Significance of Fossil Molecules

Geochemical fossils are biological markers that can convey information about the types of organisms contributing to the organic matter incorporated in sediments. Thus, they can be used for characterization, correlation, and/or reconstitution of the depositional environment, in the same manner as macro- or microfossils are commonly used by geologists. However, it should be remembered that aside from geochemical fossils, also other molecules may be used for correlation, provided they are characteristic enough.

In addition, a comparison between the original biogenic molecule and the molecule found in source rocks or crude oils may give valuable information about the history experienced by the molecule: microbial activity or chemical rearrangement at the time of deposition, and subsequent chemical reactions during catagenesis.

The most common uses of geochemical fossils may be listed as follows:
a) as correlation parameters (oil–oil and oil–source rock);
b) for reconstitution of depositional environment;
c) for elucidation of chemical transformations during diagenesis and catagenesis;
d) for detection of contamination with foreign material in marine or freshwater Recent sediments.

The problem of oil–oil and oil–source rock correlations will be treated separately (Chap. V.2), as it is not specifically related to geochemical fossils but includes other markers.

The quality of information provided by geochemical fossils in terms of depositional environment depends on three factors:

– their state of conservation, which may or may not allow one to link them to their biochemical precursor molecule;
– the distribution of this biochemical precursor (parent molecule) in the present animal and/or plant kingdom;
– the assumption that the distribution was comparable in ancient organisms.

Conservation of the original characteristics depends mostly on the structure of the original biogenic molecule. Relatively inert molecules, such as n-alkanes, suffer little chemical alteration, and their relative abundances are affected especially by dilution, when new alkanes are generated during catagenesis. On the contrary, acids, alcohols, ketones, etc. may suffer several alterations,

3.1 Significance of Fossil Molecules

depending on diagenetic and catagenetic conditions: loss of functional groups, alkylation, dealkylation, reduction, aromatization, etc. As a result, the molecules found in ancient sediments may range from unchanged biogenic molecules, for example *n*-alkanes, and less frequently acids or alcohols, through compounds very close to the original molecule, such as steranes and triterpanes, to molecules keeping only the cyclic skeleton, like aromatic steroids.

Generally, one can expect geochemical fossils to be useful for correlation. For reconstitution of paleo-environments, however, the original structural details have to be preserved. For instance, the carbon skeleton of steroids may be derived from a wide range of specific steroids, known from a large number of animals or plants, so that the occurrence of molecules retaining only the tetracyclic nucleus cannot be linked to a particular environment of deposition.

The information deduced from the occurrence of fossil molecules is also a function of the distribution of the precursor in living organisms. On one hand, biogenic molecules, typical for certain organisms or for a class of organisms, provide information on the fossil biological association. For instance, odd-carbon-numbered high molecular weight *n*-alkanes and related waxes found in sediments are typical products of terrestrial higher plants. On the other hand, very common molecules occurring in many kinds of organisms provide little or no information on the paleogeography, facies, and environmental conditions. For example, C_{15}–C_{20} *n*-alkanes without odd-carbon-number predominance may be generated by all types of kerogen during catagenesis.

Therefore, the interest in geochemical fossils depends on the distribution of biochemical precursors in presently living organisms. It should be emphasized, however, that the distribution of different precursors in contemporaneous organisms is still insufficiently known. Most structural determinations of biogenic molecules carried out until now have been made for the needs of the food or drug industries.

Thus, only the organisms, or even certain parts or organisms used in industrial processes have been analyzed: tobacco leaves, grains of cereals, essential oils of higher plants, whilst the rest of the plant is practically unknown. Furthermore, the lipids, hydrocarbons, etc. synthesized or accumulated by prokaryotic organisms, and particularly bacteria, are almost unknown with regard to the numerous existing species. Nevertheless, the bacterial biomass can in some cases be very significant compared to the total organic input. An example of the still incomplete knowledge on the occurrence of biochemical precursors of geochemical fossils in living organisms is pentacyclic triterpenoids of the hopane series. They were first considered as characteristics of lower terrestrial plants, as they were mostly known from ferns, mosses, and other plants. Later they were discovered in one anaerobic bacteria, then in several other prokaryotic organisms stemming from many different environments (Ensminger et al., 1974; van Dorsselaer, 1975). In fact, since then they have been found in many ancient sediments including petroleum source rocks, oil shales, coals, and also in crude oils, proving that they are very wide-spread in geological conditions (van Dorsselaer, 1975).

A new tool for relating fossil molecules to existing biogenic compounds is stereochemistry. Ackman et al. (1972), Cox et al. (1972) and Maxwell et al. (1973) demonstrated the interest of stereochemistry of acyclic isoprenoids to clarify this relationship. Along the same line, Ensminger et al. (1974) distinguished the importance of the genetic heritage and the subsequent evolution in the stereochemistry of pentacyclic molecules of the hopane type.

3.2 Geochemical Fossils as Indicators of Geological Environments

The use of geochemical fossils for reconstitution of conditions of deposition is relatively new. However, this field of application looks quite promising, and new developments can be forecast. Fossil molecules of typical biological origin, present in the bitumen extracted from rocks, and such molecules occurring in crude oils are especially interesting. As far as petroleum is concerned, migration processes have often lowered the content of geochemical fossils which are high molecular weight and/or polycyclic molecules. The result is that characteristic molecules are less frequently identified in crude oils than in the extract of sedimentary organic matter.

The interpretation may be based, in some cases, on the occurrence of a single and characteristic compound. However, so far this practice is rather dangerous, as advances in the knowledge of the distribution of contemporaneous biogenic molecules are rapid, and may alter the present-day interpretation.

Geochemical fossils are likely to provide knowledge on the major associations of organisms contributing to the organic matter of sediments. The three main types of associations, in terms of quantitative importance of the biomass, are the following, roughly associated with kerogen types II, III and I or III, respectively:

a) *Marine organic matter:* phytoplankton is the major contributor, with subordinate zooplankton, and sometimes a significant contribution of benthic algae. The principal biogenic molecules are:

- n-alkanes and n-fatty acids of the medium molecular weight range, C_{12}–C_{20}, with a frequent predominance of C_{15} and C_{17} n-alkanes, synthesized as such by algae, and probably also derived by decarboxylation from C_{16} and C_{18} acids;
- C_{15}–C_{20} isoprenoids;
- abundant steroids (particularly C_{27} cholestane) and some carotenoids;
- ubiquitous triterpenoids particularly of the hopanes series also occur; the sterane/hopane ratio is relatively high, generally close to unity or higher.

b) *Continental organic matter,* mainly made up of the debris of higher plants, is incorporated in deltaic or other mainly land-derived sediments, or in peat bogs, sometimes without important degradation. Biogenic hydrocarbons include mostly odd n-alkanes of high molecular weight (C_{25}–C_{33})[11]; some tricyclic diterpenes in the cycloalkane and mostly in the aromatic fraction, and some ubiquitous hopanes. However, cyclic material is only subordinate. When steranes are present, C_{29} sitostane and stigmastane predominate over C_{27} cholestane. The

---

11 The alkane content is usually much higher in continental than in marine organic matter; thus, where both types of material contribute, continental matter superimposes its pattern of n-alkane distribution, especially in the C_{25}–C_{33} range

3.2 Geochemical Fossils as Indicators of Geological Environments

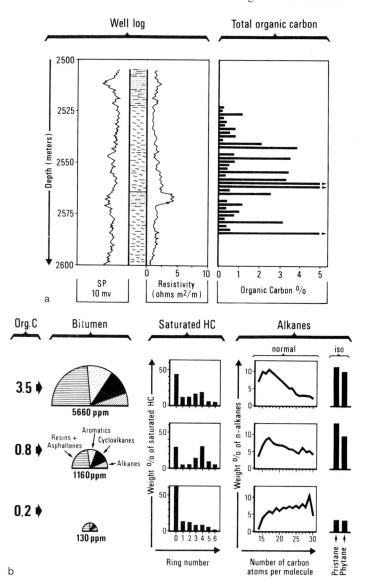

Fig. IV.3.1a and b. Comparison of bitumens from interbedded organic rich shales (2–5% organic carbon) and organic lean shales (ca. 0.5% organic carbon) in Upper Cretaceous of the coastal basin, Gabon (adapted from Claret et al., 1974): (a) the total organic content shows the alternation of organic rich and lean shales (b) the surface areas of the sectors are proportional to the amount of extractable bitumen. Organic lean shale *(bottom)* contains mostly n-alkanes of high molecular weight, and branched alkanes, but few isoprenoids. This organic matter is derived mostly from terrestrial higher plants. Organic rich shale *(middle* and *top)* contains more cyclic material, especially steranes, many isoprenoids, and lower molecular weight n-alkanes. The organic matter from the organic-rich shale is mostly marine and autochthonous. *Note:* the total amount of n-alkanes is used as reference (100%) for n-alkanes and isoprenoid distribution

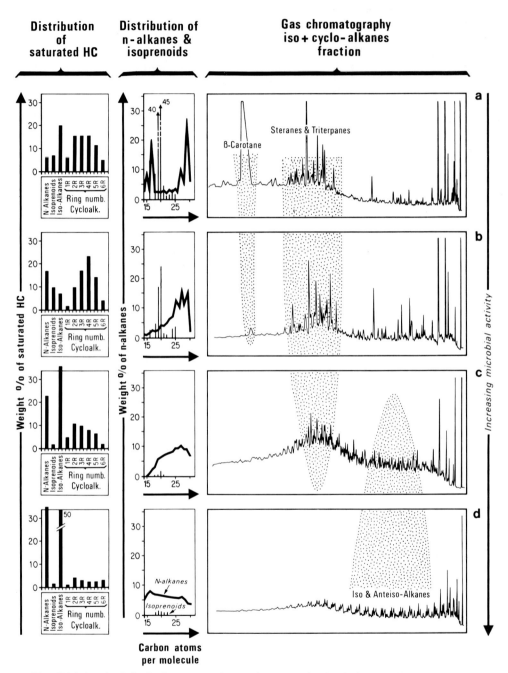

Fig. IV.3.2 a–d. Principal sources of organic matter in the Paleocene-Eocene beds of Uinta Basin (Utah). (After Tissot et al., 1977, 1978.) (a) Mostly algae (nC_{15}, nC_{17}, steranes, β-carotane) and subordinate higher plants (nC_{25} to nC_{33}); (b) higher plants, with algae and microorganisms; (c) higher plant material reworked by microorganisms; (d) mostly aliphatic chains (n, iso, and anteiso-alkanes) resulting from microbial activity; the alkanes are probably of mixed origin, partly remains of degraded plant material, partly of microorganisms. *Note:* the total amount of n-alkanes is used as reference (100%) for n-alkane and isoprenoid distribution

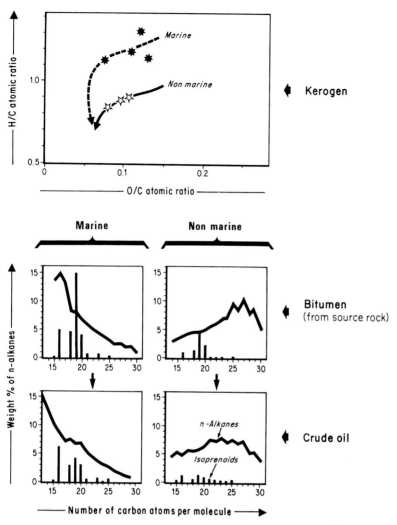

Fig. IV.3.3. *n*-Alkane and isoprenoid distribution in marine and nonmarine source rocks, and related crude oils from Magellan Basin. Nonmarine bitumens and crude oils contain more high molecular weight *n*-alkanes and less isoprenoids than do marine bitumens and crude oils. *Top:* elemental composition of kerogen and evolution path are shown on a van Krevelen diagram. *n*-Alkane and isoprenoid scale as in Figure IV.3.2

sterane/hopane ratio is usually low to very low. Isoprenoids are moderately abundant and dominated by pristane. In some cases the pristane: phytane ratio may reach 4 to 10 (Brooks, 1970; Powell and McKirdy, 1975).

c) *Microbial organic matter* is abundant in some lacustrine or paralic environments where plant material is heavily degraded during alternating periods of subaerial microbial activity and water flooding. The related organic matter is enriched in lipids, that may be either the remains of the degraded plant material, or may be reworked, or synthesized by microorganisms. Biogenetic hydrocarbons are dominated by long chain n-, iso- (2-methyl) and anteiso- (3-methyl) alkanes, sometimes extending up to C_{40} or C_{50}, normally without any appreciable predominance. Isoprenoids and cyclic hydrocarbons are scarce, except hopanes which are synthesized by prokaryotes. The geological environment of the related waxy oils has been studied by Hedberg (1968).

It is clear that geochemical fossils are a witness of the main organic contributors, whether or not they are autochthonous. Thus, for instance, organic matter of terrestrial origin may be contained in a sediment laid down in a marine environment. In particular, terrestrial organic matter deposited in deltas areas, in marine brackish or nonmarine environments, can be identified as of terrestrial origin; microbial biomass grown on land-plant debris can likewise be identified as such, whether it is transported into marine, lacustrine, or lagoonal-type sediments.

Examples of an identification of the major organic contributions are presented in Figures IV.3.1 to IV.3.3. The marine/continental discrimination is clearly shown in an example from Gabon, West Africa. The Upper Cretaceous of Ile Mandji includes a series of shales — Anguille and Azilé formations — which are the source rocks of the oil fields discovered in this area (Claret et al., 1974). The stratigraphic sequence shows a finely stratified sedimentary unit, being alternatively low and high in organic carbon (Fig. IV.3.1a). This situation results from a duplex origin of the organic material. A small amount of land-derived organic matter is present throughout the sequence; this is the only source for the horizons containing only 0.5% organic carbon, or less, and 100 to 500 ppm bitumen. The continental origin is denoted by a total high alkane content, and among these a dominance of high molecular weight n-alkanes, with odd preference, and a low content of isoprenoids (Fig. IV.3.1b). In addition to this continuous input, some horizons have incorporated autochthonous organic matter, especially of planktonic origin, due to temporary anoxic conditions. As a result, these horizons contain more organic carbon (0.5–5%) and bitumen (500–6000 ppm), with different characteristics: cycloalkanes (especially steranes) and isoprenoids are abundant; n-alkanes are preferentially of low to medium molecular weight.

Another example is provided by the lacustrine Green River formation in Uinta Basin, Utah (Tissot et al., 1978). Samples taken at comparable levels of burial depth show a progressive change with increasing age, which is tentatively interpreted to be the result of an increasing microbial degradation (Fig. IV.3.2):

- the upper beds (equivalent to Mahogany ledge) show various contributions of algal lipids: normal alkanes C_{15} and C_{17}, β-carotane; higher continental plants: n-alkanes C_{27}–C_{33}, and C_{28}, C_{29} steranes; and possibly higher animals: C_{23} steranes from bile acids (a);
- in the middle to lower part of the Green River formation, algal lipids (both C_{15}, C_{17} and carotane) are first degraded (b), then odd n-alkanes C_{27}–C_{33} from higher plants also decrease, whilst iso- and anteiso-alkanes of microbial origin start to appear (c);
- in the underlying Flagstaff member, microbial degradation is more advanced; there, the biomass mostly originates from bacteria and other microbes, or from plant lipids reworked by micro-organisms: heavy n-alkanes up to C_{40} without

any carbon-number predominance, series of iso- and anteiso-alkanes; there are no longer any steroids (d).

The associated crude oils retain the same characteristics: Redwash or Duchesne oils resemble the upper beds with algal and higher plant contribution; Altamont and other oils associated with the lower beds mostly comprise high molecular weight normal, iso- and anteiso-alkanes.

Another example is found in the Magellan basin at the extreme end of South America. The Springhill formation, of early Cretaceous age, has been deposited in a marine basin and in small adjacent continental depressions. Continental organic material (probably reworked by micro-organisms) exhibits a high alkane content, especially of high molecular weight n-alkanes with some odd predominance, and a low content of isoprenoids. On the other hand, marine organic matter shows a lower alkane content, with a preference for low to medium molecular weight, and an abundance in isoprenoids. The different associated crude oils still retain the equivalent hydrocarbon pattern (Fig. IV.3.3). Thus it is relatively easy to recognize the origin of the organic material, even where continental-derived organic matter is incorporated into a sediment deposited in a marine environment, and containing some definite marine microfossils.

However, more mature crude oils usually contain a minor amount of geochemical fossils, diluted by abundant molecules issued from kerogen upon thermal catagenesis. An exception is the Uinta basin, where the characteristic n-, iso- and anteiso-alkanes persist down to 5000 m and more. This situation is due to favorable circumstances: the composition of the plant and microbial biomass is such that the corresponding kerogen (type I) is mostly built from acids, alcohols, etc. with long aliphatic chains. Thus, hydrocarbons yielded during catagenesis resemble the original biogenic chemical structures, i.e., normal and isoalkanes.

Apart from the main types of association, specific geochemical fossils may provide a more refined interpretation in terms of plant or animal groups, climate etc. Albrecht and Ourisson (1969) found in the Eocene Messel shale of the Rhine Valley a high concentration of the pentacyclic triterpene alcohol, isoarborinol, preserved for 50 million years. Today this alcohol is known to occur in tropical plants from Indonesia, living in a climate rather comparable to the assumed climate of the Rhine-Graben area during Eocene time, according to geological reconstitutions.

Seifert (1973) studied the stereochemistry of four individual steroid acids (C_{22} and C_{24}) in the Midway Sunset crude oil from the Pliocene of California. The ratios of the cis:trans configuration of the A:B rings allowed him to infer an animal contribution to the organic matter, through bile acids. The related data have been presented above (Fig. II.3.18). Huang and Meinschein (1976, 1979) discussed the significance of sterols as environmental indicators; whereas Brassell et al. (1980) considered the various classes of lipid components as potential indicators of terrestrial, marine (non-bacterial) and bacterial inputs.

However, it should be pointed out that the most widespread biological markers are inherited from the membranes of bacteria living in the sediment, such as pentacyclic hopanes, or long isoprenoid chains from Archaebacteria. The latter are particularly observed in the asphaltene fraction of crude oils and in the kerogen of the related source rock, where they are protected by incorporation in a polymeric network (Chap. II.3.6). It is not surprising that the most frequent identifiable biological remnants are issued from the most resistant part — the membrane — of the last member of the chain of living organisms throughout the cycle of organic carbon: bacteria. In this connection, methanogenic bacteria are of particular importance, because they are active within the upper section of the sedimentary column.

The consequence of these considerations is that remnants of the organisms living in the basin of deposition may be obliterated by the effect of microbial activity. The presence of specific geochemical fossils makes possible to conclude that certain types of organisms were present in the environment of deposition. However, their absence cannot be interpreted as a proof of the absence of these organisms.

3.3 Geochemical Fossils as Indicators of Early Diagenesis

In some cases, geochemical fossils may provide some information on the physicochemical conditions prevailing in the freshly deposited sediment: reducing and hydrogenating environment, acidic conditions, etc.

Distributions showing a strong predominance of even n-alkanes are less frequent (ca. 1:10) than the odd predominance (Table II.3.3; Tissot et al., 1977).

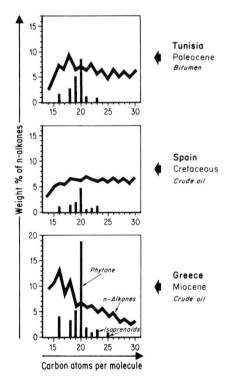

Fig. IV.3.4. Examples of the predominance of even n-alkanes and phytane isoprenoid in bitumen extracted from sediment and in crude oils. This predominance usually occurs in carbonate or evaporite series deposited in reducing and hydrogenating environment. n-Alkane and isoprenoid scale as in Figure IV.3.2

The phenomenon often concerns all n-alkanes from C_{16} to C_{30}. Living organisms usually do not synthesize even n-alkanes, but even-numbered acids and alcohols are important constituents of fats and vegetal waxes. In a normal environment, oxidation of alcohols into acids, and then a subsequent decarboxylation of the latter generates odd n-alkanes. In very reducing, hydrogenating environments, reduction of alcohols or acids produces even n-alkanes. In a similar way phytol (C_{20}) issued from the side chain of chlorophyll may transform along two different ways: in normal or oxidating environment, oxidation to phytanic acid (C_{20}) and subsequent decarboxylation generates pristane (C_{19}); in very reducing environment, reduction produces phytane (C_{20}; Fig. II.3.14). In fact the even predominance of normal alkanes is almost always associated with the predominance of phytane over pristane (Welte and Waples, 1973).

Examples of n-alkanes and isoprenoid distribution from strongly reducing paleoenvironment are presented in Figure IV.3.4. Such distributions usually occur in carbonate and/or evaporite series: Paleocene and Eocene carbonates from Tunisia, Oligocene marls and evaporites from the Rhine Graben, crude oils originated from the Miocene of the Mediterranean Sea: Amposta, Spain; Greece (Welte and Waples, 1973; Albaiges and Torradas, 1974; Dembicki et al. 1976; Tissot et al., 1977).

Sterols, like phytol, may be converted along several pathways, either to steranes or to aromatics. A suggestion is made that high sterane:aromatic steroid ratios are also correlated to very reducing paleoenvironments. Structural rearrangements involving steranes, sterenes and aromatic steroids can be induced by active acid sites present in clay minerals (Rubinstein et al., 1975), and may provide relevant information on diagenesis.

3.4 Geochemical Fossils as Indicators of Thermal Maturation

Certain molecules, although obviously having a biogenic structure, are not known as such in living organisms. For instance C_{16} and C_{18} isoprenoids have a characteristic structure — i.e., one methyl group on every fourth carbon atom — which is not likely to be generated by an inorganic process. However, they are not known in living organisms or Recent sediments, and therefore they have to be generated during diagenesis or catagenesis from existing biogenic precursor molecules, i.e., most probably from phytyl chains attached to chlorophyll. Such geochemical fossils may act as a witness of the subsequent rearrangements of the organic matter occurring during thermal maturation.

A well-known illustration of the above considerations is the distribution of heavy *n-alkanes*. Odd n-alkanes of biogenic origin are progressively diluted by n-alkanes without predominance, generated during catagenesis from kerogen. These alkanes are formed by rearrangement of long aliphatic chains originating from acids and alcohols of wax esters. The evolution of the odd predominance (CPI) is commonly used as an index of maturation (Chap. V.1).

Acyclic *isoprenoids* are also of interest in this connection, as they are rather typical and easily identifiable. Brooks et al. (1969), Connan (1974) and Powell and McKirdy (1975) worked on coals and crude oils from Australia and New Zealand. In coals and comparable sediments derived from continental organic matter, they consider that oxidation of phytol into phytanic acid occurs to a large extent during the initial aerobic stage of plant decay. Thus progressive generation of pristane by decarboxylation of phytanic acid during the late stage of diagenesis and the early stage of catagenesis would result in an increase of the pristane:phytane ratio. The highest values are recorded in the coal sequence. Brown and subbituminous coals have moderate pristane: phytane ratios (1 to 7), whereas high volatile bituminous coals have higher values (7 to 15).

When catagenesis increases, the cracking zone is reached and isoprenoids are affected: cracking reactions become prevalent over decarboxylation and C_{16}, C_{18} isoprenoids are generated. Thus, the content of the latter compounds may be used as an indicator for advanced catagenesis.

More specific characteristics such as stereochemistry of fossil molecules may also be used. The configuration of the C_{19} acyclic isoprenoid, pristane, has been studied by Patience et al. (1978). In shallow sediments, such as the immature Messel shale, the meso isomer (6R, 10S pristane) is present alone, whereas a mixture of three isomers (6R, 10S; 6R, 10R; 6S, 10S pristanes) is present in more mature shales and crude oils. The ratio of meso-pristane to other isomers becomes 1:1 toward the end of the diagenesis stage.

Triterpanes of the hopane series occur in Recent and ancient sediments, including coals, and in crude oils. They belong to different isomeric series (Fig. IV.3.5):

- (17α, 21β) H-hopane or $\alpha\beta$ hopane,
- (17β, 21α) H-hopane or $\beta\alpha$ hopane,
- (17β, 21β) H-hopane or $\beta\beta$ hopane.

However, living organisms always synthesize $\beta\beta$ hopane-type triterpenes (and sometimes subordinate $\beta\alpha$ type), but $\alpha\beta$ hopane-type triterpenes have never been detected in organisms. Thus the $\alpha\beta$ series is generated by subsequent isomerization of the $\beta\beta$, or $\beta\alpha$ series, during diagenesis and the early stages of catagenesis. In more advanced stages, another isomerization occurs in position C-22 of the C_{31} to C_{35} hopanes. Thus stereochemistry of the hopane series may be used as an indicator of the maturation of organic matter (van Dorsselaer, 1975; Ensminger, 1977; Ensminger et al., 1977; Seifert and Moldowan, 1978, 1980). These chemical fossils may also be used for detection of crude oil migration or pollution in Recent sediments.

Changes of configuration in *steranes* deserve special interest, particularly the 20S/20R ratio in C_{27}–C_{29} steranes (Mackenzie et al., 1980): Figure IV.3.6. With increasing burial depth the configuration at C-20 changes from a 20R predominance at shallow depth to equal amounts of 20R and 20S configurations around the peak of oil generation, i.e., well into the catagenesis zone. The equilibrium of the C-20 stereochemistry in steranes is reached more slowly than the stereochemistry of the asymetric centers in long-chain isoprenoids, or that of C-17 and C-22 in hopanes. Thus the 20S/20R + 20S ratio in steranes can be used to measure the evolution of source rocks over the immature zone (diagenesis) and

3.4 Geochemical Fossils as Indicators of Thermal Maturation

Structure	Series	Occurrence	Example of the Eocene shales in the Rhine Graben Locality maximum burial depth
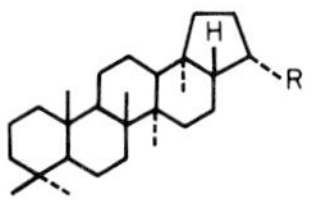	ββ-Hopane	Immature Sediments (also in living organisms)	Messel 200 m
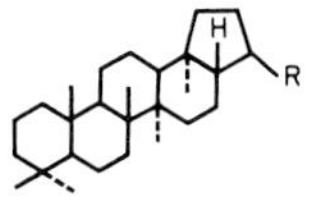	βα-Hopane	Immature Sediments (also in living organisms)	
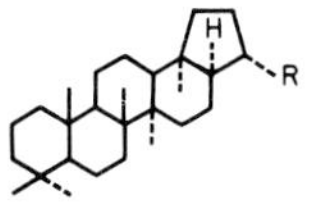	αβ-Hopane	Sediments only	Stockstadt 1800 m

Fig. IV.3.5. Stereochemistry of the hopane series. The $\beta\beta$- and $\beta\alpha$-hopanes occur in Recent and in ancient immature sediments, such as the Eocene shales of Messel, W. Germany, where the maximum burial depth is ca. 200 m. In the Rhine Graben, the same Eocene shales have been buried to 1800 m at Stockstadt and there only $\alpha\beta$-hopane occurs, being the more stable form. (After van Dorsselaer, 1975)

part of the principal zone of oil generation; in some cases, it can also be used to assess the degree of maturity of crude oils.

The conversion of monoaromatic to triaromatic steroids has also been shown by Mackenzie et al. (1981) to cover the same range of thermal maturation. In addition to this, Seifert and Moldowan (1978) suggested that the extent of cracking in the side chain of the monoaromatic steroids could be used as a maturation parameter, whereas Mackenzie et al. (1981) utilized mono- and triaromatic steroids. Cracking of the side chain seems to occur mainly in the catagenesis zone.

Porphyrins also change with increasing maturation. The conversion of DPEP to etio porphyrins has been observed for a long time, although the mechanism is not yet clear. Mackenzie et al. (1980) used the DPEP/etio ratio for vanadyl

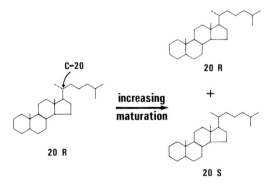

Fig. IV.3.6. Changes of configuration at C-20 in the steranes during burial: with increasing maturation the predominance of 20R is changed to equal amounts of 20R and 20S configuration

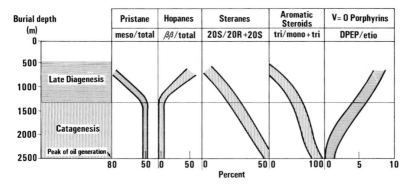

Fig. IV.3.7. Changes of some molecular parameters as a function of burial depth in the Lower Toarcian shales of the Paris Basin. (Adapted from Mackenzie et al., 1980, 1981)

porphyrins and showed it to decrease with increasing burial depth over the late diagenesis and catagenesis zones.

A comparison of the molecular parameters applied to characterize maturation is presented in Figure IV.3.7 according to the comprehensive study of their distribution in the Lower Toarcian shales of the Paris Basin by Mackenzie et al. (1980, 1981). On this example, it is clear that stereochemistry of long-chain isoprenoids and hopanes can only be used over the diagenesis zone, whereas stereochemistry of steranes at C-20, aromatization of steranes and conversion of porphyrins cover the range of late diagenesis and catagenesis, up to the peak of oil generation. However, these different chemical and stereochemical changes are likely to occur with different kinetics. Therefore, the situation in other basins, with a different time-temperature history, may be somewhat different, some conversions being possibly more sensitive to temperature than others. For instance, Mackenzie et al. (1982), studying samples from the Pannonian basin in Central Europe, suggest that aromatization of monoaromatic steroids is more accelerated by a fast elevation of temperature than isomerization at C-20 of steranes.

Thus it is difficult to provide intercalibration of the different scales until more geological examples have been studied. Also the correspondence between molecular and other parameters, such as burial depth, vitrinite reflectance, etc., might be more scattered than suggested in Figure IV.3.7, once different basins are plotted.

3.5 Present and Future Development in the Use of Geochemical Fossils

Investigation for geochemical fossils has developed recently to a considerable extent due to the association of gas-liquid chromatography (GC) and mass spectrometry (MS), and to the recent availability of computerized GC-MS. Much work is related to polycyclic terpenoids and steroids, which may occur in sediments as saturated, unsaturated, and aromatic hydrocarbons, or as acids, alcohols, ketones, etc., many of them with different stereochemistry. However, the greatest care is necessary before a specific identification is made, and before a conclusion is drawn with respect to the biological precursor. For instance, Mulheirn and Ryback (1975), working on stereochemistry of steranes, pointed out that in some cases a complete identification of stereochemistry cannot be achieved by using only GC-MS. In particular, configuration at the C-24 atom, which is of phylogenetic importance in C_{28} and C_{29} steranes, required isolation and study of optical rotatory dispersion and high resolution proton magnetic resonance spectroscopy. Since then, more powerful resolution in gas chromatography might help to solve this problem in a simpler way (Maxwell et al., 1980).

Although identification of geochemical fossils often requires a great analytical effort, their study is generally rewarding. They already provide a powerful tool to correlate various crude oils or oil shows discovered in sedimentary basins by petroleum exploration. Once the various oil types have been separated, geochemical fossils are again used to tie these crude oils to their respective source rocks (Chap. V.2).

As pointed out, geochemical fossils may also be used to characterize facies, environments of deposition, and diagenetic aspects. Even the thermal history may be understood better with their help.

Furthermore, the future use of geochemical fossils is promising, as these molecules can hopefully be taken as indicators for very characteristics bioenvironments, such as tropical forest or saline lagoon. In that respect, geochemists are heavily dependent on the progress of the natural-product chemists to accumulate data on the occurrences of characteristic molecules in living plants and animals. The introduction of stereochemistry into the determination of geochemical fossils is of great assistance in correlating living and fossil molecules.

Comparable interest should be attached to the nonhydrocarbon, polar fractions of bitumens and crude oils. They may include some geochemical fossils, retaining their original functional groups (acids, alcohols). Some of these have already been determined, particularly in the same range of molecular weight as the hydrocarbons. However, analytical procedures become defective when N, S, O content, size of the molecules, and polymerization increase.

In addition to these uses, analysis of geochemical fossils in ancient sediments and experimental evolution in recent muds are of great interest to environmental and pollution studies. They obviously provide information on the chemical and biological degradation of organic molecules in modern environments (Eglinton, 1973).

Summary and Conclusion

Geochemical fossils are biological markers that frequently convey genetic information about the types of organisms contributing to the organic matter of sediments. They are used for correlations (oil–oil and oil–source rock), for reconstitution of depositional environments, and also as indicators for diagenesis and catagenesis.

Identification of the major sources of organic material (marine autochthonous and terrestrial plants), and of the importance of reworking by microbes, can be achieved by using fossil hydrocarbons. More specific geochemical fossils such as alcohols, acids, etc. may provide a more refined interpretation in terms of plant or animal groups, climate, etc.

Normal or reducing conditions of diagenesis may also be characterized by structural rearrangement involving biogenic molecules. Other changes with respect to structure or distribution of certain types of hydrocarbons (n-alkanes, isoprenoids, triterpanes, steranes, or porphyrins) also reflect the intensity of catagenesis.

Chapter 4
Geological Control of Petroleum Type

4.1 General and Geochemical Regularities of Composition

Regularities in crude oil and gas composition have been observed for a long time. Some of those seem to have a regional significance, e.g., geologists know that crude oils from the Middle East are generally rich in sulfur, while oils from North Africa have a low sulfur content. Other regularities seem to be more general: for instance, in several places of the world, it has been observed that oil density decreases with depth; more generally, geochemists and refiners have observed that crude oil density, sulfur content, and viscosity are closely related.

The present composition of a crude oil results from many factors: nature of the original organic matter, temperature history, migration, subsequent evolution and alteration, etc. Some of these factors, such as the original organic material and the particular geological history, tend to make crude oils different; some others, for example migration or alteration processes, tend to make crude oils more similar. As a result of such complex and sometimes contradictory influences, it is often difficult to detect a single phenomenon clearly responsible for the observed distribution of crude oil constituents.

However, two main types of observations can be distinguished:

a) General Regularities

Certain regularities are well known among crude oils. The strong correlations observed between the abundance of certain hydrocarbons, or classes of hydrocarbons, elements such as sulfur, and physical properties of the crude oils (density, viscosity, etc.) are an expression of those regularities. These correlations are valid regardless of the type of deposition or geological history of the basin and should be considered as facts of general significance, resulting from purely chemical considerations. They have been discussed in Chapter IV.1.

b) Geochemical Regularities

These occur on a local, regional, or more general basis, and are expressed by changes in composition of oils according to the overall geological setting:
- the environment of deposition: marine versus nonmarine organic matter, detritic, or carbonate sedimentation,

- thermal evolution: depth history, time elapsed since source rock burial, geothermal gradients,
- alteration of petroleum in reservoir: biodegradation, oxidation, water washing.

4.2 Geochemical Regularities Related to the Environment of Deposition

Some of the oil characteristics are derived from the source of the organic material, and some others from the physicochemistry of the aquatic basin of deposition, including activity of microorganisms in the young sediment.

4.2.1 Marine Versus Nonmarine Organic Matter (Fig. IV.4.1)

It should be first pointed out that, in organic geochemistry, the marine or continental character has to be related to the origin of the predominant organic

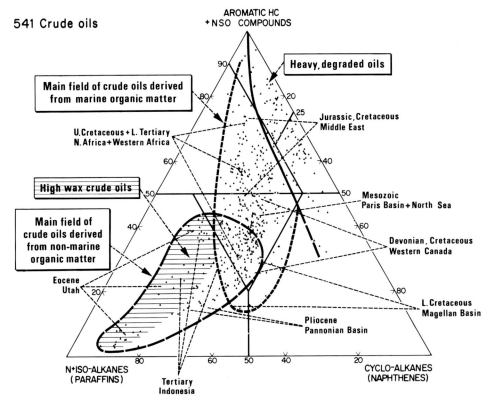

Fig. IV.4.1. Ternary diagram of crude oil composition, showing the principal field of occurrence of crude oils from marine and nonmarine origin

4.2 Geochemical Regularities Related to the Environment of Deposition

material. In all cases, this material has been deposited below aquatic media, as this is a condition for a satisfactory conservation. Organisms living in the water or on the bottom of a lake or sea generally contribute to the organic matter incorporated in the sediment. They include mostly phyto- and zooplankton and bacteria, but macroscopic algae may be, in some circumstances, important contributors. However, the land-derived organic matter can also be predominant, compared with other sources of organic material, especially in paralic basins, either intracontinental or in the sea bordering land, and in deltaic environments. The latter sometimes extend far into the open sea, as in the case of the Ganges, Niger or Amazon deltas, whose detritic material is spread over the continental shelf and slope, and even reaches the abyssal plains.

a) *Marine organic matter* usually corresponds to kerogen type II and results in crude oils of paraffinic-naphthenic or aromatic-intermediate type (Fig. IV.4.1). The amount of saturated hydrocarbons (Fig. IV.1.5) is ca. 30–70% in crude oil, and 40–75% related to hydrocarbons. Isoprenoid isoalkanes, compared with n-alkanes, are rather abundant at shallow or medium depths. Saturated hydrocarbons of sterane and triterpane type are often present in shallow and immature crude oils, and decrease with depth and thermal evolution. Aromatics amount to 25 to 60% of hydrocarbons, a value significantly higher than in crude oils derived from nonmarine organic material. The same is true for sulfur content, which may reach high values in certain types of marine crude oils: this particular aspect will be discussed in Section 4.3.3. Resins and asphaltenes are also relatively abundant in young and immature crude oils.

This type includes most crude oils from Jurassic and Cretaceous of Middle East, Devonian of Alberta, Cretaceous (above the salt series) of West Africa, Mesozoic from Paris, Aquitaine and North Sea basins, and Mesozoic of Mexico.

b) *Nonmarine organic matter* deposited in a deltaic, marginal or open marine environment usually corresponds to type-III kerogen. In rare cases (internal basins), nonmarine material may result in type-I kerogen. When it predominantes, terrestrial organic matter generates crude oils of paraffinic or sometimes paraffinic-naphthenic type. The amount of saturated hydrocarbons is ca. 60–90% in crude oil, and 70–90% in hydrocarbons. Normal and isoalkanes, and mono- and dicyclic naphthenes are abundant. Polycyclic naphthenes, and particularly those of sterane type, are rare. Total aromatics are significantly lower than in crude oils derived from marine organic matter: they amount to 10 to 30% of hydrocarbons and comprise mostly mono- and di-aromatics. Sulfur content is low to very low, usually less than 0.5%. Resin and asphaltene content is usually below 10% but may, in some cases, reach 20% in young and immature crude oils.

The nonmarine type includes most crude oils from the lower Cretaceous of West Africa, Brazil, South Argentina and Chili (Magellan basin), the lower Tertiary of the Uinta basin (Utah), the Tertiary of Pannonian basin and the Rhine Graben, and part of the oils from Nigeria, the Gulf Coast, and several Tertiary basins of Indonesia. High-wax crude oils (Hedberg, 1968) belong to this type, and are discussed in Section 4.2.2.

c) *The geochemical significance* of the difference between the more paraffinic oils of nonmarine origin and the more aromatic oils of marine origin has to be interpreted by consideration of the various organic inputs into the sediment, and the subsequent alteration of the organic matter during early diagenesis (Tissot et al., 1979).

Continental organic material is mainly composed of plants. They are rich in cellulose and lignin (higher plants) and contain subordinate amounts of other carbohydrates, proteins and lipids. The latter include particularly aliphatic hydrocarbons of high molecular weight, the closely related waxes and fatty acids of medium chain length from fats. The plant material may be either directly transported to the basin of sedimentation, or reworked and incorporated in soil humic acids, then transported as such. In many cases the environment of deposition is marine, and the organic matter is associated with abundant detrital minerals, forming a thick deltaic deposit. Alternatively, it may be transported by turbidity currents and extend much farther on the continental slope. On the one hand the fast burial, and on the other the deep water, prevent a strong reworking by microorganisms at that stage. Such situations are found in many basins of Late Cretaceous-Tertiary age located on continental margins. Under these conditions, the occurrence of anoxic-reducing environments, with sulfate reduction in the bottom water layer, is unlikely. Thus, the possibility of introduction of sulfur in the sediment, and subsequently in kerogen, is restricted here.

The resulting kerogen (type III) comprises mainly a polyaromatic network, as lignin and soil humic acids are aromatic polymers containing oxygen and some short alkyl chains. However, this material is degradable only with difficulty and constitutes a rather inert part of kerogen until an advanced stage of catagenesis is reached, when gas is generated by cracking the short alkyl chains. The subordinate lipid fraction of kerogen, derived from wax esters and fats, yields, trough decarboxylation, and rupture of C–C bonds, alkyl chains of various lengths. As a result kerogen made of continental material may generate, when subjected to thermal catagenesis, first alkanes with few cyclic molecules, then gas. Sulfur content is low.

Marine organic material contains abundant proteins, carbohydrates and lipids. The latter include polycyclic naphthenic material such as sterols from algae and marine animals, and triterpenes of the hopane series from prokaryotes (blue-green algae and bacteria). The lipids also include some molecules appropriate for cyclization, like unsaturated fatty acids from fats. Furthermore, situations where marine autochthonous organic matter is abundantly preserved in the sediment usually correspond to anoxic conditions in the bottom water layer (Tissot et al., 1979, 1980) where sulfur is extracted from sulfate of the seawater by action of anaerobic bacteria. Sulfur may be incorporated in sediment, then react with organic matter and possibly induce cyclization and aromatization.

At this stage, there is also an important contribution from anaerobic bacteria which are very active in these sediments, as algal material is rather easily degradable. In particular, membranes of methanogenic bacteria (Archebacteria) are a source of isoprenoid chains of various lengths, identified in kerogen by Michaelis and Albrecht (1979) and Chappe et al. (1979). From these various sources, type-II kerogen made of autochthonous marine and bacterial organic matter contains abundant cyclic material and isoprenoid chains. When subjected to catagenesis, it may release more biogenic polycyclic naphthenes and more

4.2 Geochemical Regularities Related to the Environment of Deposition   443

diagenetic aromatic material (aromatics, resins, asphaltenes) than will the nonmarine organic matter. It may also yield more sulfur compounds.

In terms of geology, the distribution of marine and continental organic matter can be summarized as follows. The predominance of the rather easily degradable marine organic matter occurs where conditions are favorable for its preservation and continental runoff is limited because of physiographical, climatic or other reasons. This is the case in upwelling areas, where production exceeds the potential for degradation: along the western coast of continents such as Peru or Northwest and Southwest Africa; the marine character is enhanced where tectonics and/or climate create erosional conditions improper for pedogenesis, e.g., along the Northwestern coast of Africa, bordered by the Sahara. A predominance of marine organic matter is also provided by widespread transgressions creating shallow epicontinental seas with a high biological productivity, e.g., Jurassic of Western Europe. Finally, a silled- or barred-basin geometry favoring anoxic conditions is another typical situation favoring marine organic matter, presently observed in the Black Sea. This situation was very probably more common over certain geological periods: middle Cretaceous black shales of the Atlantic Basins (Tissot et al., 1980) and carbonate platform of the Middle East.

Nonmarine organic matter predominates where a large drainage system results in an abundant supply from the continent. In turn, this situation is determined partly by the location of humid climatic belts and partly by the physiography of the continents. A highly oxygenated water column in the basin of deposition may enhance the nonmarine character by creating favorable conditions for destruction of the autochthonous plankton. The predominance of non-marine organic matter is presently associated with the existence of large river systems, such as the Mississippi, Amazon, Niger etc. This type of sedimentation represents most of the Tertiary prograding deltas, along the present stable margins. Nonmarine organic matter also occurs in some syn-rift sediments of the passive margins, e.g., early Cretaceous of W. Africa and in intracontinental lowlands and lakes (Tertiary of the Uinta Basin, some Tertiary filled basins in China and Cretaceous internal basins in Africa).

4.2.2 High-Wax Crude Oils

High-wax crude oils represent a special case of the nonmarine crude oils. They are likely to be derived from continental organic matter which was strongly reworked by microorganisms. Cellulose is rather easily degraded by bacteria. Lignin is degradable only in the presence of oxygen, by association of fungi and bacteria working subsequently. This situation can be achieved in shallow paralic or lacustrine basins where the sediments are periodically exposed to sub-aerial degradation and water flooding. Furthermore, fats are more degradable than waxes. Thus, the main components preserved from the terrestrial organic input are long-chain *n*-alkanes and wax esters from higher plants, and soil humic acids, which are a relatively inert part of kerogen. In addition, the microbial biomass amounts to an important part of the organic matter: its lipid fraction comprises

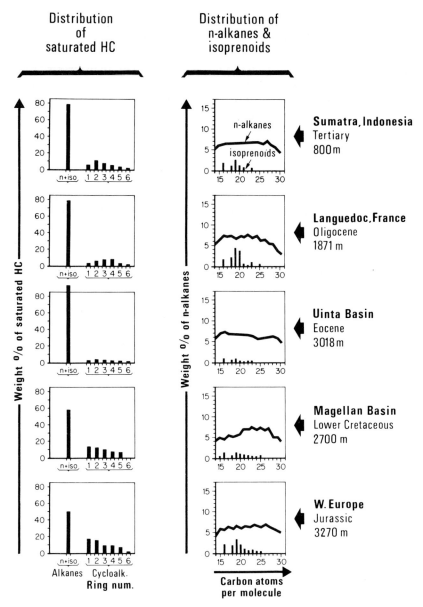

Fig. IV.4.2. Examples of high-wax crude oils. *Left:* distribution of saturated hydrocarbons with respect to structural types; *right:* distribution of n-alkanes and isoprenoids as a function of the number of carbon atoms per molecule. *Note:* the total amount of n-alkanes is used as reference (100%) for n-alkane and isoprenoid distribution

4.2 Geochemical Regularities Related to the Environment of Deposition

mostly aliphatic chains (*n*, iso, anteiso) from the microbial waxes, without odd or even predominance and also constituents (isoprenoids, hopanes etc.) from bacterial membranes.

The kerogen resulting from these environments is mainly composed of waxes and a relatively inert humic fraction, in spite of the fact that there was a high terrestrial input with cellulose and lignin. Therefore it may give rise to high-wax crude oils. The related kerogen usually belongs to type III, due to a sizeable proportion of humic material. In some cases, e. g. the Uinta Basin, kerogen is made essentially of long aliphatic chains and belongs to type I.

The related crude oils contain a high proportion of normal, iso- and anteiso-alkanes and belong to the paraffinic type (Fig. IV.4.1). Long-chain *n*-alkanes are abundant and may show either no predominance or a slight odd predominance derived from the contribution of higher plants. The high paraffin content is responsible for the high pour point and the waxy character of the oils. Some examples are presented in Figure IV.4.2.

These crude oils are commonly known as high-wax oils. They have been reviewed by Hedberg (1968): they are found mainly in shale-sandstone sequences and are frequently associated with coal or highly carbonaceous strata, the whole being deposited in a continental, paralic or nearshore marine environment. These crude oils are normally low in sulfur. They can be traced back to the following geological situations:

a) Rift valleys, during the first stage of oceanic opening, are occupied by intermittent swamps, lakes, or narrow seas surrounded by continental edges delivering terrestrial organic material. This situation favors alternate stages of aerobic biodegradation and flooding. This type of sedimentation occurred during Mesozoic, and particularly during lower Cretaceous, in the area of the South Atlantic Ocean. It is responsible for high-wax crude oils from Brazil, South Argentina, the Magellan basin (Chili) and Gamba (Gabon).

b) Paralic troughs bordering continents or cordilleras of islands occur in folded belts before or during the major orogenic stages. As a present-day example, the troughs bordering some islands of Indonesia may be cited. The organic matter accumulating in these troughs is provided mostly by terrestrial plants more or less degraded by microorganisms. There, high-wax crude oils are associated with coal beds. Examples are found along alpine folded belts: Tertiary of India, Burma, Malaysia, Indonesia, and Venezuela (Officina Basin).

c) Continental lowlands with nonmarine sedimentation may receive a similar supply of organic matter. They yield the same type of crude oil: Cretaceous of Wyoming and Colorado; Paleocene-Eocene of the Uinta Basin, Utah; Mesozoic and Tertiary of China; Cretaceous of internal basins in Africa (Sudan).

4.2.3 Clastic Series Versus Carbonates — High Sulfur Crude Oils

Pure carbonates usually contain little organic matter. However, carbonate sequences may well include source rocks which are rich in organic matter and which are usually made of marls or argillaceous limestones. In such carbonate source rocks, the clay content is frequently ca. 10–30% or more. We shall refer to oils having originated from such situations as "oil from carbonate sequences" and compare them to the crude oils generated from sand-shale or "clastic sequences".

All crude oils contain sulfur compounds, but crude oils rich in sulfur are more frequent in a carbonate sequence than in a clastic sequence (Table IV.4.1). The average sulfur content is 0.86% in carbonates, compared with 0.51% in clastics, and 0.65% for all crude oils. Oils produced from clastic sequences include the oils derived from nonmarine organic matter, whose sulfur content is low, as discussed in Section 4.2.1. However, there is still an obvious difference between the average sulfur content in marine clastics and its content in marine carbonates and marls.

The distribution of sulfur in all types of crude oils has been shown in Figure IV.1.16. It is bimodal, and the modes are separated by a minimum corresponding to ca. 1% of sulfur. When crude oils from carbonate sequences and those from clastic sequences are considered separately, the same type of bimodal distribution is found in both groups (Fig. IV.4.3). However, the second mode, corresponding to high sulfur oils, includes about 25% of the samples from carbonate sequences, whereas it includes only half of that percentage in crude oils from clastic sequences. Furthermore, in crude oils from carbonates, the sulfur content may reach up to 5% by weight in nondegraded oils, and up to 10% in heavy degraded oils or tars. These figures are seldom reached in oils from clastic sequences.

Aromatics usually amount to 40 to 60% in high sulfur crudes, including benzo- and dibenzo-thiophenes (10–25% of the crude oil). Complex molecules of high molecular weight containing N, S, O, i.e., resins and asphaltenes, are also abundant in the same crude oils. Thus, high-sulfur crude oils belong mostly to the aromatic-intermediate or to the aromatic-asphaltic oil types (Fig. IV.2.2).

Sulfur is not a major constituent of living organisms. Therefore the high sulfur content observed in crude oils has to be gained later from elsewhere: diagenesis, influenced by microbial activity at the time of deposition, chemical reaction with evaporites occurring during catagenesis, or degradation of reservoired crude oil.

In most cases, the origin of sulfur is to be found in the environmental conditions at the time of the source rock deposition. As already discussed in Part II, the fine

Table IV.4.1. Average sulfur content and specific gravity of crude oils from clastic and carbonate sequences. (Data from U.S. Bureau of Mines, IFP, and others)

Reservoirs	Sulfur (wt. %)	Specific gravity	Number of samples
Carbonate	0.86	0.844	2464
Clastic	0.51	0.847	5281
All types	0.65	0.847	9347

Fig. IV.4.3. Distribution of sulfur in crude oils from carbonate and clastic sequences (geometric classes). The second mode, above 1%, is more important in carbonate than in clastic sequences. (Data from U.S. Bureau of Mines, IFP and others)

4.2 Geochemical Regularities Related to the Environment of Deposition

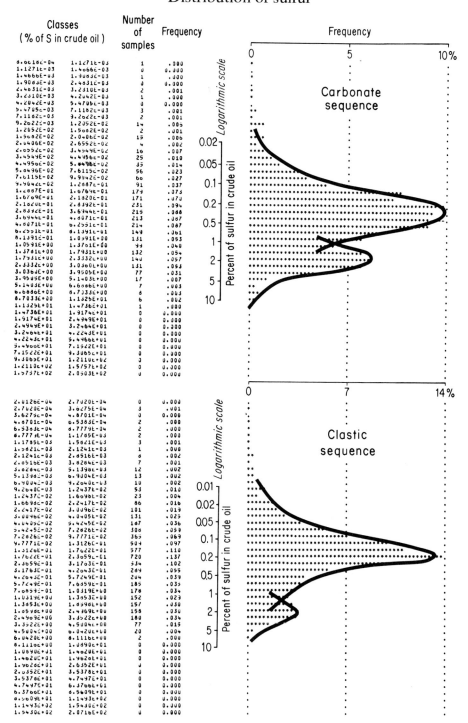

Fig. IV.4.3

muds (shales or carbonates) deposited in marine basins with a normal oxygen content quickly become a closed environment where interstitial water is separated from the overlying sea water. Oxygen confined in that space is exhausted by microbial aerobic activity, and anaerobic conditions are rapidly established. Under these conditions sulfate-reducing bacteria (e.g., *Desulfovibrio desulfuricans*) extract the sulfate available, and reduce it to H_2S. Furthermore, in confined seas or lagoons, the bottom water layer may become devoid of oxygen. In this case large quantities of H_2S may be produced directly from the sulfates of seawater.

Sulfur is not incorporated in the bacterial cell. In clay muds, where iron is usually abundant, it is likely that sulfur readily recombines to form iron sulfides. However, in carbonate muds, where iron is much less abundant, sulfur may remain as free sulfur and progressively combine with residual organic matter during diagenesis. Kerogen originating from this environment belongs to type II and may yield high-sulfur crude oils upon catagenesis. In addition, sulfur may act as an agent of aromatization, when incorporated into organic matter. This hypothesis would explain the strong correlation observed between sulfur and a high aromaticity of oils (aromatics, resins, asphaltenes).

Massive incorporation of sulfur into sediment, in confined environments of carbonate-evaporite type sedimentation, and subsequent incorporation of sulfur into organic matter during diagenesis, is probably the cause of high-sulfur petroleums generated during subsequent evolution of organic matter, such as the aromatic-intermediate crude oils of the Middle East and of the South Aquitaine basins. On the contrary, other very reducing environments associated with a clastic sedimentation result in low-sulfur paraffinic-naphthenic crude oils. For example, the crude oils derived from lower Silurian source rocks of North Africa, and the Lower to Middle Jurassic of the Paris basin are low-sulfur crudes, although the sediments were deposited under anoxic conditions.

Semi-closed seas or lagoons with carbonate-evaporite sedimentation may develop in various geological situations. However, they are particularly frequent in relation to mobile belts along active margins where frequent movements of the crust determine the formation of semi-closed seas, such as the Mediterranean sea during Neogene, or the present Black Sea. In fact, some of the major areas with high-sulfur oils occur in the neighborhood of the folded belt that originated from the old "Tethys" geosyncline: the Middle East (Arabia, Gulf States, Irak, Syria), Mediterranean (Turkey, Albania, Greece, Sicily, Spain). A comparable situation occurs in the Carribean (Maracaibo, Colombia, Cuba, Mexico).

Other types of confined environments, not related to the main folded belts, may result in similar situations as far as the fate of organic matter is concerned. For instance, in the Cretaceous and Tertiary of Egypt, in some Jurassic basins of Germany, in the Upper Jurassic of France (Aquitaine and Paris basins), in the Jurassic and Cretaceous of the South Emba area (USSR).

A somewhat different interpretation in terms of paleogeography has been proposed by Radchenko (1965), who compared the geographic distribution of high-sulfur crude oils in several geological systems (from Cambro-Silurian to Neogene) with the paleoclimateic zones, as reconstructed by Strakhov (1962). It was concluded that high-sulfur crude oils typically occur in the arid zones located

Table IV.4.2. Correlations in low- and high-sulfur crude oils. (Data from U.S. Bureau of Mines, IFP, and others)

	Specific gravity	Viscosity (log)	Pour point	Sulfur (log)
Specific gravity		0.726	0.102	0.429
Viscosity (log)	0.916		0.135	0.275
Pour point	0.352	0.413		0.058
Sulfur (log)	0.745	0.671	0.142	

⟵ High sulfur crudes

Low sulfur crudes ⟶

Low-sulfur crudes: S < 1% (1584 samples)
High-sulfur crudes: S > 1% (414 samples)

in the northern and southern tropical belts where carbonate-evaporite sedimentation is frequent.

Incorporation of sulfur into petroleum may also occur at depth, during catagenesis, although it seems to be less frequent than incorporation of sulfur into organic matter during diagenesis. Orr (1974) showed that thermal maturation of oil with sulfate present in high temperature reservoirs may involve nonmicrobial reduction of sulfate, with sulfurization of the oil and formation of hydrogen sulfide. On the contrary, thermal maturation without sulfates would result in a continuous decrease of the sulfur content, with preferential removal of non-thiophenic sulfur compounds (Ho et al., 1974).

Microbial degradation of crude oil may also result in a sulfur enrichment, as benzothiophene derivatives and high molecular weight heterocompounds are particularly resistant to bacterial degradation (see Chap. IV.5), and hence they are selectively concentrated. Furthermore, if microbial activity leads to hydrogen sulfide or elemental sulfur, these may react with altered crude oil to produce various sulfur compounds. A peculiar alteration is reported by Ho et al. (1974) in some high-gravity oils and condensates from Western Canada: hydrogen sulfide normally generated at great depth has migrated updip and reacted with mature oils and condensates to produce mercaptans.

Compared with crude oils in general, high-sulfur crudes show a stronger correlation between density, viscosity, and sulfur content (Table IV.4.2). Pour point also indicates some correlation with density and viscosity in the high-sulfur group of crudes, although none was apparent when looking at crude oils in total. These observations are probably due to the fact that the abundance of resins and asphaltenes (containing a large proportion of sulfur) is the main cause of variation of density, viscosity, and pour point in high-sulfur crude oils. Other types of compound exerting influence on viscosity or pour point (e.g., high molecular weight alkanes) are subordinate in these cases.

4.3 Geochemical Regularities in Relation to Thermal Evolution

Whatever the genetic characteristics of a source rock and of the associated crude oil may be they all experience the same general trend of evolution:

- at shallow depth, during the phase of diagenesis, the source rock is immature. The available hydrocarbons and associated molecules are in most cases more or less directly inherited from living organisms (geochemical fossils). Their abundance is generally too small to provide commercial oil accumulations. Small amounts of heteroatomic compounds are also progressively generated from kerogen, and the whole may form some accumulations of heavy oils,
- beyond a certain threshold, catagenesis begins and the principal zone of oil formation is reached; therefore the major part of pooled crude oil has to be expected within the catagenesis zone, around the depth corresponding to the maximum of oil generation,
- with increasing burial, cracking of carbon chains becomes important in source rocks; cracking may affect reservoired oil as well; light hydrocarbons are progressively favored, and the GOR increases in the pools,
- at great depth, only light hydrocarbons remain and gas accumulations are formed; gas is the only commercial deposit found at great depth.

These successive steps of evolution may be more or less productive, and occur at different burial depth, according to the genetic characteristics of organic matter, the geothermal gradient, and the time. In particular, some types of organic matter are less capable of generating oil (see Part II). However, when both oil and gas are formed, the various steps succeed each other as a function of the temperature history.

A comparison of oil and gas occurrences with temperature history should be based on the reconstitution of the temperature versus time curve for the various parts of the drainage area of each field. In fact such detailed information is available only in a limited number of cases. If we want to consider a large number of oil or gas fields, in order to obtain statistical data on a worldwide basis, we are restricted to the use of present depth, and age of the producing series. We are aware that the present depth is sometimes very different from the maximum depth of burial, especially for Paleozoic reservoirs, and the age is not a good approximation for the time elapsed since the formation has been buried. Therefore some scattering has to be expected on the distributions observed. This effect may be eliminated to a large extent if we restrict the study to Tertiary basins, where the maximum depth of burial is frequently the present one. The situation would be still better if we consider a single Tertiary basin.

4.3 Geochemical Regularities in Relation to Thermal Evolution

4.3.1 Distribution of Depths

The distribution of the producing depths of the oil fields reflects the general curve of petroleum generation (Fig. IV.4.4). At shallow depth, the number of oil fields is small; it increases downwards, reaches a maximum, then decreases again at greater depths, as most deep discoveries are gas fields. The average depth for 12,018 oil fields of any age and locality in the world is 1465 m. This is probably influenced to some extent by the large number of small and shallow pools producing in the United States, and by the Paleozoic fields, whose present depth is often shallower than the maximum burial depth. However, Table IV.4.3 shows that the average depth of 2609 Tertiary oil fields is 1552 m.

More restricted populations (classified by geological province and age) show an average depth from 1195 m (231 crude oils from the Tertiary of the Pannonian and Vienna basins) to 1959 m (1038 crude oils from the Tertiary of Gulf Coast). This range can be explained firstly by the geothermal conditions and the burial history, then in other cases by the type of organic matter or the migration processes. In particular, the geothermal gradient is high in the Vienna and the Pannonian basins (average depth 1195 m), where it may reach $50°C\ km^{-1}$ in several areas. It is still rather high in some West Africa basins (average depth 1362 m), such as the Douala basin, where a reconstitution of a paleogradient, computed by a mathematical model using vitrinite evolution, gives values ca. $50°C\ km^{-1}$; such high values are probably related to Cretaceous time. On the contrary, the geothermal gradient is comparatively low in the Gulf Coast area (average depth 1959 m): from 22 to $24°C\ km^{-1}$ in Louisiana to 29 to $33°C\ km^{-1}$ in Texas.

The observed distributions are usually far from a normal type, when grouping oil fields of various origins and ages. However, they become closer to a normal distribution if we select some populations from the same producing series in the same geological province. For instance the hypothesis of a normal distribution is acceptable on a 0.01 probability level for the Cretaceous-Tertiary fields of the Middle East (average depth: 1901 m) on the one hand, and for the Tertiary of Venezuela + Colombia + Trinidad on the other hand (average depth: 1513 m).

4.3.2 Change of Composition with Depth

The change of composition with depth was first reported by Barton (1934) on the crude oils of the Gulf Coast. He noted a progressive decrease of density and an increase in paraffinic content with increasing depth of the producing interval. The same phenomenon was observed by Hunt (1953) from the Tensleep oils of Wyoming, but it could not be generalized to all crude oils from this area, as the depth effect does not obliterate major differences resulting from different source rocks and does not account for variations of geothermal data.

Since then, comparable situations have been recorded from many places. Figure IV.4.5 shows the average density of crude oils in respect to depth rank. The change of composition with depth results mainly from the progressive cracking of carbon chains, that causes the content of light hydrocarbons to

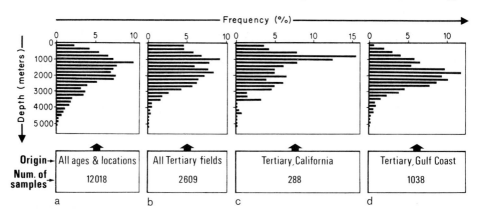

Fig. IV.4.4a–d. Depth distribution of oil-producing horizons (data from U.S. Bureau of Mines, IFP and others): (a) 12018 accumulations from Cambrian to Pliocene age; (b) 2609 accumulations of Tertiary age; (c) 288 accumulations in the Tertiary of California; (d) 1038 accumulations in the Tertiary of the Gulf Coast

Table IV.4.3. Average depths of oil fields. (Based on 12018 data from U.S. Bureau of Mines, IFP, and others)

Classes of oil fields	Number of fields	Average depth (m)	Standard deviation of depth distribution (m)
All samples	12018	1465	903
Tertiary	2609	1552	891
Tertiary of Pannonian and Vienna basins	231	1195	682
Cretaceous-Tertiary of West Africa	445	1362	766
Tertiary of California	288	1509	901
Tertiary of Venezuela Columbia, Trinidad	93	1513	810
Devonian of Alberta	241	1630	712
Cretaceous-Tertiary of Middle East	115	1901	943
Tertiary of Gulf Coast	1038	1959	820

increase. However, the correlation coefficient between depth and density, computed on the same set of crude oils (all ages and origins) is only -0.40, due to the variety of genetic types of source rocks and crude oils, and due to different geothermal conditions. It is somewhat higher ($\varrho = -0.46$) in Tertiary fields. The depth–density relation is better if we only consider individual Tertiary provinces with a simple geological history, and where present and maximum burial depth

4.3 Geochemical Regularities in Relation to Thermal Evolution

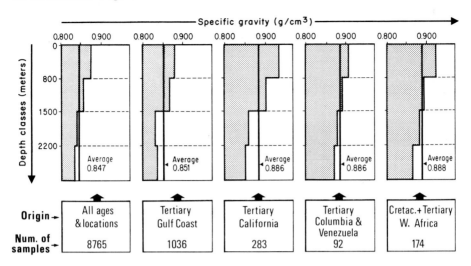

Fig. IV.4.5. Change of specific gravity of crude oils as a function of depth. Average specific gravity is calculated over four depth intervals. (Adapted from Goubin, 1975; data from U.S. Bureau of Mines, IFP and others)

are nearly the same: Tertiary + Cretaceous of West Africa ($\varrho = -0.56$); Tertiary of California ($\varrho = -0.54$).

The general trend of thermal maturation is shown in the ternary diagram of crude oils composition, Figure IV.2.3. The correlation of the $n + $ isoalkane content with depth increases from $\varrho = +0.45$ if we consider the abundance of alkanes in crude oil, to $\varrho = +0.60$ if we consider their abundance in saturated hydrocarbons. The correlation is still better for n-alkanes only. Thus, various parameters using the n-alkanes:saturated hydrocarbon ratio have been used to diagnose the level of maturity of a crude oil. We can mention, for instance, ternary diagrams (normal, branched and cycloalkanes) plotted either for the total saturated fraction, or for the light fraction comprising six or seven carbon atoms only. A related index of crude oil maturity was proposed by Jonathan et al. (1975), based on the n-hexane:methylcyclopentane ratio. They showed that the same ratio, when measured on bitumen of the source rocks, correlates with the vitrinite reflectance.

Sulfur content of crude oils generally decreases with depth, possibly due to cracking and elimination of sulfur as H_2S (Prinzler and Pape, 1964). Table IV.4.4 shows the average sulfur content in relation to depth. There is a clear decrease of the average sulfur content with depth, from 0.80% in shallow crude oils (less than 800 m) to 0.33% in deep ones (deeper than 2200 m). The different behavior of clastic and carbonate series is emphasized (Fig. IV.4.6), as carbonate series show a stronger sulfur–depth correlation ($\varrho = -0.46$) than clastic sediments ($\varrho = -0.27$). The assumption that sulfur is eliminated from the liquid phase as hydrogen sulfide is backed by the distribution of hydrogen sulfide content in natural gas, that reaches a maximum at great depth, below 3000 m (Fig. IV.4.7).

Table IV.4.4. Sulfur content in crude oils according to depth classes. (Data from U.S. Bureau of Mines, IFP and others)

Depth	Average sulfur content (%)	Number of samples
All samples	0.654	9347
0– 800 m	0.803	2085
800–1500 m	0.753	2540
1500–2200 m	0.432	1938
>2200 m	0.330	1458

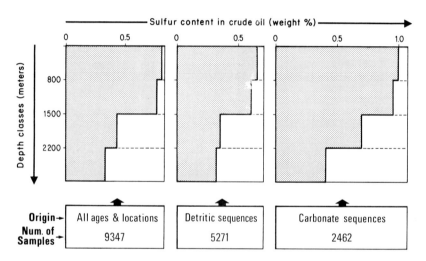

Fig. IV.4.6. Change of sulfur content in crude oils as a function of depth. Average values of sulfur are calculated over four depth intervals. (Data from U.S. Bureau of Mines, IFP and others)

In addition, Ho et al. (1974) noted that nonthiophenic compounds are preferentially removed by thermal maturation; thiophenic compounds are affected by a decrease of the benzothiophene:dibenzothiophene ratio.

4.3.3 Change of Composition with Age

The age of the reservoir rock is a rather poor indication of the time that has to be considered for the kinetics of evolution: on one hand, the age of the source rock may be different from the age of the reservoir; on the other, the time a sediment was at the appropriate temperature, i.e., deep enough, is not dependent on the stratigraphic age of the rock, in a strict sense.

4.3 Geochemical Regularities in Relation to Thermal Evolution

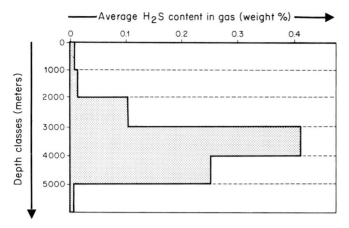

Fig. IV.4.7. Average values of hydrogen sulfide content in natural gas as a function of depth. (9340 samples analyzed by U.S. Bureau of Mines, IFP and others)

Therefore variations of density, gas:oil ratio, etc., as a function of age show only a broad tendency, when considering statistically a great number of oil fields, toward a decrease of density and an increase in light hydrocarbons with increasing age. Our statistics are not good enough if we restrict the investigation to a smaller area, e.g., a sedimentary basin. Variations due to the succession of different organic deposits may be so predominant that the time effect is not able to obliterate them.

The distribution of sulfur content with respect to age of the producing series is worth special consideration. It is shown in Figure IV.4.8. Computed on 9315 samples, the geometric mean decreases from Paleozoic (S = 0.31%), to Mesozoic (S = 0.29%) to Cenozoic oils (S = 0.19%). The shapes of the distributions are different: all three groups of crude oils — Paleozoic, Mesozoic and Cenozoic — show a principal mode around 0.2% sulfur, but Paleozoic and Mesozoic oils have bimodal distributions, with the second mode around 2%, separated by a minimum at 1%. In the latter case, high-sulfur crude oils comprise 20 to 30% of the total number of crudes. On the contrary, Tertiary oils show only a regular decrease in frequency with increasing sulfur content, and only 12% of the crudes contains 1% or more of sulfur. It is not clear whether the occurrence of confined basins with carbonate-evaporite sedimentation was actually more frequent in Upper Paleozoic (e.g., Permian basin of Texas and basins bordering Rocky Mountains, or Volga-Ural province) and in Mesozoic (e.g., Middle East) than in Tertiary. The fact that major deltaic series with low-sulfur crudes contain a large number of the producing fields in the Tertiary may be a reason for the observed distribution.

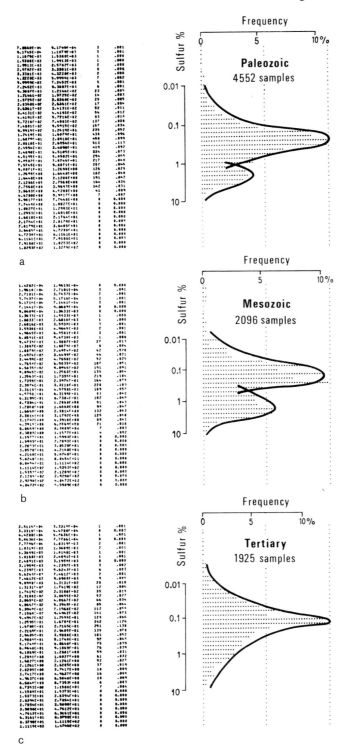

Fig. IV. 4.8a–c

4.3.4 Combined Influence of Depth and Age on Petroleum Composition

Temperature and time are the major parameters in the kinetics of petroleum generation. Consideration of both depth and age of the producing series allows one to estimate statistically, for a large number of geological situations, the effects of these parameters. However, depth and age alone are a poor evaluation of temperature and time of evolution. It is clear that depth (i.e., temperature) should be more important than age, as has been shown in Part II.

Within a given basin, use of relatively narrow age classes provides a general picture of the evolution of the various crude oils, and it shows in particular that the original variations in primary organic material remain, in many cases, predominant over the influences exerted by evolution. For instance the Mississippian to Lower Cretaceous crude oils in Alberta are statistically heavier and contain more sulfur than both younger Cretaceous and Devonian oils, whatever the depth may be.

4.4 Concluding Remarks on Crude Oil Regularities

Oil composition is primarily influenced by the nature and evolution of the source rock. The main fields of occurrence of the various types of crude oils are shown in Figure IV.4.1. However, oil composition is sometimes influenced by alterations occurring after petroleum has been pooled. Thus, the original primary composition of crude oil, bearing the imprint of source rock facies and evolution, may be obscured. These alterations of pooled petroleum are treated in the following chapter.

Summary and Conclusion

Marine organic matter usually generated paraffinic-naphthenic or aromatic-intermediate crude oils. Terrestrial organic matter, derived from plants, generates crude oils of paraffinic or sometimes paraffinic-naphthenic types. High-wax crude oils represent a particular case of the paraffinic class: the organic matter is strongly reworked by bacteria and enriched in long chain lipids.

High-sulfur crude oils contain more than 1% sulfur. They occur more frequently in carbonate-evaporite series than in clastic series. Sulfur is generally extracted from sulfate during sedimentation, and subsequently recombines either with iron in clastic series, or with organic matter in carbonate-evaporite series.

Fig. IV.4.8a–c. Distribution of sulfur in Paleozoic, Mesozoic and Tertiary crude oils (geometric classes). The absence of the second mode (high sulfur) among crude oils of Tertiary age may be caused by the predominance of clastic series over carbonates. (Data from U.S. Bureau of Mines, IFP and others)

A general trend of evolution is observed with increasing depth and age: a decrease in density and sulfur content, and an increase in light alkanes. These changes are due to thermal evolution. However, variations due to the original composition of organic matter may in certain cases be so predominant, that time or depth effect is not able to obliterate the original pattern of composition.

Chapter 5
Petroleum Alteration

Petroleum is a very complex mixture of organic compounds and has a high energy content; it is thermodynamically metastable under geological conditions. Petroleum after being pooled in a reservoir is therefore susceptible to alteration. Attention to the importance of alteration was directed by a number of papers (Williams and Winters, 1969; Evans et al., 1971; Bailey et al., 1973a), which investigated heavier crude oils of Western Canada.

The geochemical effects of alteration, however, had been recognized long before by Russian petroleum scientists (Andreev et al., 1958; Kartsev et al., 1959). In a more general way it has been known much longer, for instance in the form of the so-called tar mats at the oil–water contact of many oil fields (e.g., Burgan).

Alteration of pooled petroleum is now observed among oil accumulations around the world. The causes of alteration are numerous. They may be related to the relative instability of petroleum and/or to the fact that traps are, on the one hand open systems, and on the other may change their level of burial due to further subsidence or erosion. The composition of pooled petroleum can be altered by chemical or physical processes. An example of chemical alteration would be thermal maturation or microbial degradation of the reservoired oil, while physical alteration might be via the preferential loss of light compounds by diffusion, or the addition of new compounds to the reservoir due to further migration. The distinction between chemical and physical processes is not stressed here, because the two are often interrelated and may even occur simultaneously.

In previous chapters we have learned that variations in crude oil composition are to a certain extent inherited from different source rocks. For instance, coaly material in general yields more gaseous compounds, while high-wax crude oils are commonly associated with source material containing high proportions of lipids of terrestrial higher plants and of microbial organisms. High-sulfur crude oils frequently are related to carbonate-type source rocks. Aside from the influence of the source rock facies, the state of maturity of the source material is also of importance. However, alteration of the reservoired petroleum may affect the composition of a crude oil to a greater extent than does the character of the source material. In other words, crude oil alteration tends to obscure the original character of the oil, and therefore affects crude oil correlation studies, and furthermore influences the quality and economic value of petroleum.

A comprehensive treatment of the most important alteration processes, thermal alteration, deasphalting, biodegradation and water washing has been presented in a series of papers by Bailey, Evans, Rogers and coworkers (Evans et al., 1971; Rogers et al., 1972; Bailey et al., 1973a, b; Bailey et al., 1974; Rogers et al., 1974).

5.1 Thermal Alteration

Thermal alteration of pooled petroleum, as with the maturation of kerogen, proceeds under the influence of heat in the subsurface. For a given geothermal gradient, it increases with increasing depth of burial and time of residence at a certain temperature. Thus, thermal alteration or maturation effects can be predicted from a time–temperatur relationship. With increasing depth and rising temperature, there is a tendency for crude oils in reservoirs to become specifically lighter and contain an increasing amount of low molecular weight hydrocarbons at the expense of high molecular weight constituents. A more detailed examination of changing crude oil properties with increasing reservoir temperature in the Western Canada basin is presented in Figure IV.5.1. There is a linear increase in compounds containing less than 15 carbon atoms. This happens at the expense of C_{15+} heavy constituents. Simultaneously gases, especially methane, increase exponentially in relative abundance. At elevated temperatures, in more mature zones, only methane is found in reservoirs along with pyrobitumen generated by cracking processes (Evans et al., 1971). In reservoirs, the presence of pyrobitumen plus light hydrocarbons (below C_{15}), and the absence of the intermediate oil constituents are an indication for disproportionation reactions during thermal alteration. From oils of intermediate molecular weight, the ultimate stable end products are low molecular weight methane, and an insoluble carbon-rich residue of very high molecular weight. In this way, the system follows the inevitable thermodynamic trend by adjusting continuously toward more stable molecules.

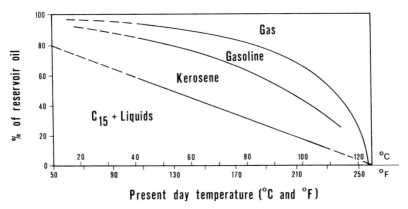

Fig. IV.5.1. Chemical changes in crude oils with increasing reservoir temperature in the Western Canada Basin. (After Evans et al., 1971)

The whole process can be viewed as a disproportionation reaction. It can be deduced that there must be a hydrogen transfer from donor to acceptor, i.e., from "aromatic-type structures", which become more condensed, to aliphatic molecules. In the NSO fraction of the oils, this process is accompanied by decarboxylation, dehydration and desulfuration yielding carbon dioxide, water and hydrogen sulfide.

It has to be realized, however, that it is difficult to distinguish between effects of thermal alteration of pooled petroleum and the effects of different stages of maturity of a source rock at the time of oil release. Sometimes the distinction is possible with the help of solid residues in the reservoir.

5.2 Deasphalting

Another relatively common alteration is deasphalting, the precipitation of asphaltenes from heavy to medium crude oils by the dissolution in the oil of large amounts of gas and/or other light hydrocarbons in the range from C_1 to C_6. The precipitation of asphaltenes out of oils by low-boiling hydrocarbons is so effective that it is routinely used in laboratories and refineries for separation of asphaltenes from other crude oil constituents. Deasphalting occurs as a natural process among heavy to medium oils whenever considerable amounts of hydrocarbon gas are either generated in larger quantities in a pool due to thermal alteration of the oil or due to gas injection from outside as a result of secondary migration. However, it is necessary for this gas to be dissolved in the oil before it can precipitate asphaltenes, being ineffective, if it accumulates in a separate gas cap. Gas deasphalting is difficult to distinguish from thermal maturation, because both processes often occur concomitantly and the net change in oil composition goes in the same direction, i.e., oils become lighter.

In the reef-bearing area of Western Canada, thermal alteration and also gas deasphalting can be observed (Evans et al., 1971; Bailey et al., 1974). Among deeply buried reservoirs in Upper Devonian reefs, it is quite common to find precipitated asphaltenes in some of the reef porosity. In this region of Western

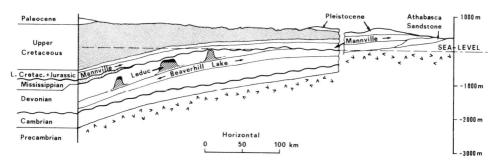

Fig. IV.5.2. Generalized geological cross-section of the Western Canada sedimentary basin. (Modified after Gussow, 1962)

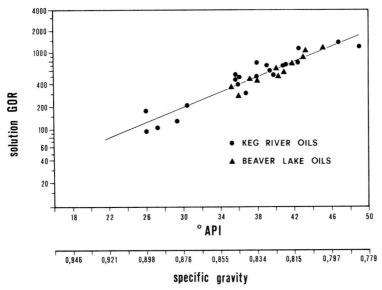

Fig. IV.5.3. Correlation between gas:oil ratio (GOR) and specific gravity of Keg River and Beaverhill Lake oils in Western Canada sedimentary basin. (After Evans et al., 1971)

Canada, it has been argued that only limited deasphalting has been caused by maturation, but extensive deasphalting has only been associated with massive gas injection. Such gas-induced deasphalting has produced oils which regionally are anomalously light, such as Devonian Beaverhill Lake and Keg River oils (Evans et al., 1971). A generalized geological West–East cross section of this region of Western Canada is shown in Figure IV.5.2. The reefs rest upon a platform, which extends downdip toward the west into an organic facies of high thermal maturity. From this higher temperature zone a regional updip flow of gas occurred through continuous porous strata, and caused the precipitation of asphaltenes in reservoirs located in the migration path of the gas. The amount of asphaltenes precipitated is correlatable with the amount of gas dissolved in the oils. The solution gas:oil ratio (GOR) for these oils is a measure of the amount of gas injected into the reservoir, while the specific gravity of the oil is a measure of its asphaltene content. Figure IV.5.3 presents the correlation between the GOR and the specific gravity of Keg River and Beaverhill Lake oils, which are undersaturated with respect to gas. Therefrom, it can be deduced that there is an inverse relationship between the gas content and the asphaltenes.

Evans et al. (1971), in connection with deasphalting, also discuss the phenomenon of the so-called gravity segregation. It is observed in oil pools with a large vertical extent, but where the temperature difference between the top and the bottom of the pool is small. In such pools the crude oil may become progressively heavier with increasing depth, which is of course the reverse of what might be expected from thermal maturation. The top oils in these reservoirs contain much more gas in solution than oil in deeper horizons, mainly because the solution of gas causes oil to increase in volume. When considering the pressure–volume relationship on a nearly isothermal basis, the following can be deduced theoretically: work done by a larger quantity of solution gas against a lower pressure roughly equals the work done by a smaller quantity of gas at a higher pressure. In other words, the quantity of gas at the top of the reservoir should be higher than the quantity of gas injectable at the bottom. Thus, the bulk density of the liquid oil will be lighter and the GOR higher at the top of the reservoir. In addition, more asphaltenes will be precipitated at the top than at the bottom. This explanation by Evans et al. (1971) of an inversed gravity relationship in large vertical oil pools is in most cases certainly more plausible than a process of gravity segregation. It is hard to believe that, if gravity segregation is involved, heavy compounds (e.g., asphaltenes) of an oil would settle selectively through the tortuous pore structure of a reservoir and would reach the bottom zone.

A distinction between gas deasphalting and thermal maturation is possible. Bailey et al. (1974) point out that thermal maturation is regional, whereas deasphalting due to gas injection is more localized, and neighboring reservoirs may produce oil rather than condensates or gas condensates. Furthermore, it is suggested that the maturation state of organic matter (kerogen) in the rocks can be measured. In pure carbonate rocks, the virtual absence of kerogen may prevent such an estimate. In these cases the nature of the bitumen-like organic matter itself can be used to determine the maturation state. Bitumen-like matter, which has been subjected to severe thermal maturation and has reached a high maturity stage (metagenetic), is less than 2% soluble in CS_2 and the H/C ratio of the insoluble fraction is below 0.53. Values less than these indicate, according to McAlary and Rogers (1971), that the bitumen-like organic matter (asphaltene precipitate) is produced by thermal metagenesis, whereas higher values suggest deasphalting. Furthermore, it has been demonstrated (Rogers et al., 1974) that "reservoir bitumens" (asphaltene precipitates) formed by deasphalting, have carbon isotope ratios like the original oil, whereas residues formed by thermal alteration i.e. pyrobitumen have significantly heavier ratios. It is argued that the cracking of carbon–carbon bonds during thermal alteration produces isotopically light methane. Simultaneously, a high molecular weight residue is produced, enriched in the heavier isotope.

5.3 Biodegradation and Water Washing

The microbial alteration of crude oil, i.e., biodegradation, and the alteration due to water washing, i.e., the removal of water-soluble compounds, are commonly observed in oil pools, located in areas invaded by surface-derived, meteoric formation waters. Both alteration processes are not necessarily coupled, but in essence they are frequently observed in combination. This is not surprising, as both processes are initiated by the action of moving sub-surface water. In case of biodegradation, meteoric waters are thought to carry dissolved oxygen and microorganisms into the reservoir and bring them in contact with the oil-water interface. Biodegradation of crude oil is a selective utilization of certain types of hydrocarbons by microorganisms. It is apparently started under aerobic conditions. In the case of water washing, formation waters undersaturated with hydrocarbons, moving along the oil–water interface, apparently extract soluble hydrocarbons selectively, and thus change the chemical composition of the remaining oil.

The different effects of *biodegradation* upon crude oils are well documented in the literature (Evans et al., 1971; Bailey et al., 1973a; Deroo et al., 1974; Connan et al., 1975). The effects of biodegradation are common and conclusive in the "tar mats" that are so widely developed at the oil–water contacts of producing oil fields, e.g., the Burgan field in Kuwait. Biodegradation by aerobic and/or anaerobic microorganisms results in partial or total removal of n-alkanes, of slightly branched alkanes and possibly of low-ring cycloalkanes and aromatics. Different intensities of biodegradation and different duration leave crude oils exhibiting a different degree of alteration. Bacteria introduced into an oil pool with oxygen-rich meteoric waters, apparently utilize this dissolved oxygen and metabolize preferentially certain types of hydrocarbons. Under anaerobic conditions, the oxygen supply of bacteria is probably derived from dissolved sulfate ions. The selective removal of hydrocarbons by bacteria seems to occur roughly in the following sequence: n-alkanes (below nC_{25}), isoprenoid alkanes, low-ring cycloalkanes and aromatics. Recently, some evidence has been presented that in cases of severe degradation even tetracyclic steranes are removed, as compared to pentacyclic triterpanes (Alimi, 1977; Reed, 1977). The general trend of crude oil biodegradation can be shown in the triangular diagram in Figure IV.2.3. For example, the chemical composition of the Cretaceous crude oils from West Africa is shifted away from the n+iso-alkane corner, showing a depletion in these compounds.

Philippi (1977), investigating crude oils of Tertiary to Cretaceous age from California, East Texas and Louisiana, USA, presented evidence that microbial degradation causes optical activity among crude oil hydrocarbons due to selective microbial digestion of optical antipodes present in the primary paraffinic crude oils. The optical activity was observed in narrow boiling point fractions of saturated hydrocarbons boiling between 80 and 325 °C. This boiling range is well below the molecular weight range of tetracyclic and pentacyclic steranes and triterpanes, which are known frequently to exhibit optical activity as a feature inherited from their primary biosynthesis. The optical activity, as described by Philippi (1977), however, is interpreted as a secondary feature caused by biodegradation of previously nonrotating, racemic mixtures of hydrocarbons. The chemical structure of the hydrocarbons responsible for the observed optical activity is not known. A very strong argument in support of this interpretation of microbial digestion of optical antipodes is the fact that the optical activity in the boiling range 80–325 °C in five different basins was only observed at reservoir temperatures below 66 °C (150 °F). Growth of bacteria may be inhibited above this temperature range.

The fate of the microbial cell material with respect to the composition of biodedraded crude oils is at present unclear. Philippi (1977) speculates that most microbial cell material formed during biodegradation is probably digested by subsequent generations of microbes and that part of the more stable oil-soluble compounds may dissolve in the crude oil. Bailey et al. (1973b), while confirming through laboratory experiments with bacterial cultures biodegradation effects as seen in the field, suggest that protein-bearing bacterial cells may augment the nitrogen content of the asphaltenes. Bailey et al. (1973b) also claim that nonhydrocarbons, particularly asphaltenes, are formed as by-products of biodegradation of hydrocarbons and increase in proportion in crude oils.

Under anaerobic conditions, certain types of bacteria reduce sulfates to satisfy their demand for oxygen. The reduced sulfur resulting from this activity may occur in different forms (Rogers et al., 1972), depending upon the geochemical environment. In some salt domes associated with biodegraded oil, elemental sulfur has been formed; in metal-rich environments sulfides of iron, zinc, lead etc. are produced in association with altered crude oil (Connan and Orgeval, 1976); in other situations, part of sulfur extracted form sulfate may react with hydrocarbons to produce sulfur-rich oils and tars.

Whereas microbial degradation of crude oils in connection with invasion of meteoric waters is amply documented in the literature, processes and effects of *water washing* are less well described. The main reason for this lack of field examples is probably that both processes, biodegradation and water washing, change crude oils in a similar direction, i.e., oils become heavier. Furthermore, water washing normally should have less severe effects upon the composition of a crude oil and, although the two processes might occur independently, they are apparently in most cases parallel to each other. Water washing results in a removal of more soluble hydrocarbons from crude oils. The individual solubilites of hydrocarbons (Table III.2.2) are a good measure for their susceptibility to water washing. In general, light hydrocarbons are more easily dissolved and selectively removed from pooled petroleum than heavier hydrocarbons. Low-boiling aromatics such as benzene, toluene, and xylene should be depleted. Biodegradation and water washing can be expected wherever reservoirs are close to the surface, and where they are accessible for surface-derived waters. Oils seeping out at the surface are always biodegraded and in addition, oxidized by inorganic oxidation. Volatile compounds are lost by evaporation. Consequently, such oils are always heavy, enriched in NSO compounds and of a tar-like appearance.

An excellent example of progressive crude oil alteration is presented by Deroo et al. (1974), based on crude oil samples from eastern Alberta in the Western Canada sedimentary basin. There, systematic changes have been observed, going in an upward direction from normal, unaltered oils in deeper pools toward heavier oils in more shallow position, and finally to severely altered heavy oils very close the surface (less than 100 m deep). The latter heavy oils, or better, the sandstone impregnated with these heavy oils is known as Athabasca tar sand. The common origin of these eastern Alberta oils and their original similarity previous to alteration had been established on the basis of cycloalkanes, aromatics and thiophenic compounds. The waters associated with this gradational alteration sequence of crude oils contain progressively less dissolved solids, indicating the increasing influence of meteoric water. Three different stages of crude oil degradation were described in this study by Deroo et al. (1974) as compared to the unaltered pooled oils. Figure IV.5.4 shows the effect of biodegradation in

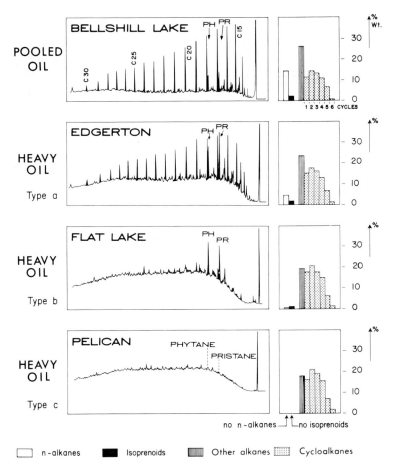

Fig. IV.5.4. Biodegradation of *n*-alkanes, isoprenoid and other branched alkanes and cyclo-alkanes in the Western Canada sedimentary basin. (After Deroo et al., 1974)

these three stages with respect to *n*-alkanes, isoprenoids and other branched alkanes, and cyclo-alkanes within the fraction of saturated hydrocarbons. The Bellshill Lake oil is a representative of unaltered oils, with *n*-alkanes clearly dominating over all other saturated hydrocarbon compounds. The Edgerton heavy oil from the Mannville Formation is characterized by a loss of *n*-alkanes, and hence a relative increase in isoprenoids, pristane and phytane is observed. In the Flat Lake oil from the Colony Formation no *n*-alkanes are left, and pristane and phytane are the most abundant chain-like alkanes. Finally, in the severely altered Pelican heavy oil from the Wabiskaw Formation, even pristane and phytane have been completely removed by microorganisms. During all this sequence of degradation, the distribution of cycloalkanes with 1 to 5 rings remains practically unchanged.

In western Canada there is, along with increasing bacterial degradation, an increase in sulfur content of crude oils. It is not yet clear whether sulfur is added to

5.3 Biodegradation and Water Washing

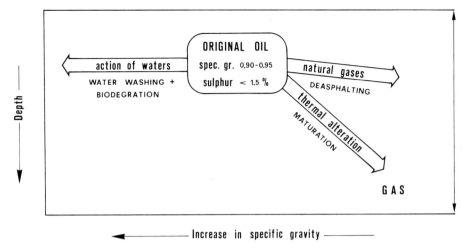

Fig. IV.5.5. General scheme of main crude oil alteration processes. (After Bailey et al., 1974)

the crude oils due to the action of bacteria, or whether the sulfur compounds in crude oils are just selectively left over, and thus relatively enriched after bacteria have eliminated certain parts of the hydrocarbon fraction.

The general trend of degradation, including biodegradation, water washing and inorganic oxidation, is shown in Figure IV.2.3 for the compositional shift of the oils from eastern Alberta, Canada. They start as normal oils of the Alberta syncline, and move toward the aromatic-NSO-corner in the triangular plot.

In summary, the ultimate composition of petroleum may be influenced strongly by alteration subsequent to the process of accumulation. The three major alteration processes are thermal maturation, deasphalting and degradation associated with the action of surface-derived formation waters. A general scheme of these alteration processes (Fig. IV.5.5.) has been presented by Bailey et al. (1974). An original crude oil of relatively low maturity would thus respond to thermal maturation or deasphalting by becoming lighter. It can be destroyed and converted into gas and a heavy insoluble residue by severe thermal maturation. If this oil is degraded by the action of formation waters, it increases in specific gravity and heavy NSO compounds; its economic value is diminished. Aside from the already described alteration, there are still other processes that may affect pooled petroleum. There may be compositional changes induced by selective losses from the hydrocarbon mixture via leakages. The leakage may be an intraformational feature of the sealing cap rock, related to its gross permeability, or it may be the consequence of a natural rupture of a sealing cap rock, due to a tectonic event such as faulting. Cases of diffusional losses and the dependence upon the permeability of seals have been described by Philipp et al. (1963) and Smith et al. (1971). However, the existence of gas fields, formed in Paleozoic time (e.g., in the Appalachian basin, USA), is proof of the low efficiency of this process in general.

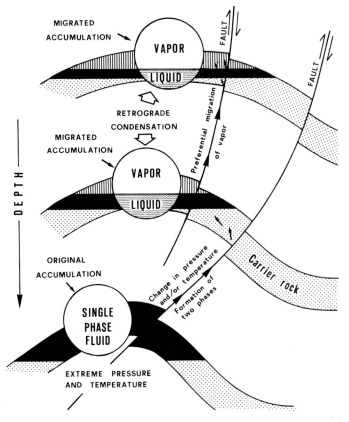

Fig. IV.5.6. Principle consecutive stages of separation-migration effects on chemical composition of pooled petrolum. In this scheme the original accumulation was a single phase fluid. (After Silverman, 1965)

Whereas losses by diffusion are gradual and increase with closeness of the reservoir to the surface, and geological duration, changes due to leakages, induced by tectonic events, may be more severe and more or less instantaneous. Profound changes in chemical composition of pooled petroleums can be expected by separation-migration, as described by Silverman (1965). For separation-migration with consecutive compositional changes, single-phase fluid systems have first to be converted into two-phase systems, for instance, by a pressure release, due to faulting. In Figure IV.5.6, this process is schematically depicted with a subsequent preferential migration of the more mobile vapor phase. The vapor phase is thus separated from the liquid, which stays behind. As the vapor migrates into shallower trapping positions, the reduced pressure and temperature conditions may cause the vapor to revert to a two-phase accumulation by retrograde condensation. This process may be continued, and thus, in an upward direction, traps with successively lighter hydrocarbon content may be found. There are no records about the frequency of separation-migration in nature.

Summary and Conclusion

The ultimate composition of petroleum may be influenced strongly by alteration after accumulation. Crude oil alteration tends to bring about changes in the character of the oil influencing its quality and economic value and adversely affecting crude oil correlation studies. The most important alteration processes are thermal maturation, deasphalting and degradation.

Thermal maturation typically occurs to pooled crude oils when increasing burial produces a rise in reservoir temperature. With increasing temperature and time of residence, crude oils become lighter, due to cracking of heavier compounds, and their gas content increases. Another relatively common alteration is deasphalting, the precipitation of asphaltenes from heavy to medium crude oils by the dissolution in the oil of large amounts of gaseous and/or other light hydrocarbons. Gas deasphalting is difficult to distinguish from thermal maturation, because oils also become lighter.

Biodegradation is the microbial alteration of crude oils. It is selective utilization of certain types of hydrocarbons by microorganisms. Meteoric waters are thought to carry the microorganisms into the reservoir. The selective removal of hydrocarbons by bacteria seems to occur roughly in the sequence, n-alkanes, isoprenoid alkanes, low ring cyclo-alkanes and aromatics. Other types of degradation occur through water washing, oxidation, and evaporation.

Chapter 6
Heavy Oils and Tar Sands

The terms "heavy oils" and "tar sands" are trivial names which are not precisely defined in a physical, chemical or geological manner. The terms heavy oils and tar sands are derived from phenomenological features as observed by exploration geologists and reservoir engineers in the field and by refiners. Although it is generally understood that the bitumen-like product in tar sands is an extra-heavy oil, it has to be realized that there is no clear-cut difference between a heavy oil and the bitumen-like product in tar sands.

There is of course a fundamental difference between the expression "heavy oil" and the term "tar sand". Heavy oils are a petroleum product that may occur in any host rock. The term tar sand denominates a sandy sedimentary rock that contains a bitumen-like, extra-heavy oil in relatively large quantities. Among refiners and chemical engineers the term tar is often used for a substance resulting from the destructive distillation of organic matter. It is clear that the term tar in this book is not used in this latter meaning, but as a naturally occurring petroleum product.

Heavy oils and extra-heavy oils of tar sands are, in most cases, the result of petroleum degradation in reservoirs: they are thus residual products that occur in porous rocks (sandstones, carbonates, etc.), where petroleum has entered by migration, accumulated and became degraded. These degraded products show little or no mobility, even at subsurface conditions, due to their high viscosity. Therefore they are frequently not producible by conventional techniques. Sometimes a small amount of heavy oil can be produced, with a very low efficiency of the primary recovery process (a few per cent of the oil in-place). Very heavy oils can only be produced by mining the reservoir rock or using sophisticated methods of enhanced recovery. Due to the importance of the known reserves of heavy oils and tar sands as compared to the reserves of conventional crude oil, great efforts are presently devoted to research and development of processes for production, transport and refining of heavy crude oils.

6.1 Definitions

As indicated before there is no formal or widely accepted definition of heavy oils and tar sands. However, it is generally considered that, beyond a viscosity of 100 centipoises at reservoir conditions, the oil is difficult to produce by common techniques. The limit for production corresponds to a density of approximately

6.1 Definitions

20° API (i.e., 0.93 specific gravity), although there is only a gross correlation between density and viscosity of heavy oils, as we shall see below (Sect. 6.3).

Some authors differentiate between heavy oil accumulations and tar sands by using the following criteria:

Heavy oil accumulations
- viscosity: 100 to 10,000 centipoises at reservoir conditions;
- specific gravity: 0.93 to 1.00 g cm^{-3} (10°–12° to 20° API) approximately.

The heavy oils are still mobile in reservoir conditions, but they may be produced by conventional techniques only with low production rate and overall efficiency.

Tar sands (extra-heavy oils)
- viscosity: > 10,000 centipoises at reservoir conditions;
- specific gravity: > 1.00 g cm^{-3} (< 10°–12° API).

The oil has no mobility at reservoir conditions, and cannot be produced by conventional techniques.

This classification, although convenient to the reservoir engineer, is not appropriate for the geologist or geochemist. Viscosity is a temperature-dependent parameter (Sect. 6.3, below), especially in extra-heavy oils, such as Athabasca oils, where viscosity is reduced by three orders of magnitude between 20° and 80°C. Thus, most of the W. Canada deposits (reservoir temperature 10°–20°C) would be classified as tar sands, and most of the Venezuela deposits (reservoir temperature 50°–80°C) would be heavy oils, although some of these oils from both countries have comparable viscosity at standard temperature conditions.

Extra-heavy oils can occur in tar sands and carbonate reservoirs as well (carbonate triangle in Alberta), although tar sands are quantitatively more important. The organic content of tar sands is called extra-heavy oil or bitumen, and also tar or asphalt. Although the term bitumen is compatible with the general definition of bitumen (the extractable fraction of total organic matter in sediments: Chap. II.4.1), it is by no means specific for tar sands, as any heavy oil or conventional oil would also satisfy the definition. Asphalt should be avoided, as this term is commonly used by refiners to designate an asphaltene-rich precipitate obtained with a C_5-cut (instead of using *n*-pentane or *n*-heptane). The term tar is also used to designate a product from distillation of organic matter, as indicated above. Thus it seems preferable to use the term "extra-heavy oils".

In the following sections the term heavy oils will be used generically, including extra-heavy oils from tar sands unless otherwise stated.

One particular type of crude oil, although viscous or solid in surface conditions should not be confused with heavy oils: these are waxy oils, which are normal, non-degraded oils (Chap. IV.4.2.2); their high viscosity is due to the great abundance of long-chain *n*-alkanes. These may be solid at surface temperature, but they are usually dissolved in other constituents of crude oil in reservoir conditions. Waxy oils usually show a low content of asphaltenes. The specific gravity of waxy oils is lower than that of heavy oils.

Heavy oils may be compared with other conventional-oil-substitutes, such as coal liquids and shale oil. All three show a high specific gravity, close to

1.0 g cm^{-3}, a large content of sulfur and nitrogen, and a hydrogen deficiency. Thus, comparable problems are met in refining these products. Coal liquids, however, are closer to heavy oils than shale oil as they contain a large proportion of asphaltene-like material (although probably not identical) and are highly viscous. Shale oil contains more alkanes than the other and also alkenes, which are specific of a high temperature cracking of the kerogen; they are frequently less viscous, except some waxy shale oils.

6.2 Composition of Heavy Oils

Like all types of petroleum, heavy oils are made of hydrocarbons, resins and asphaltenes. However, the proportions of these various constituents are different in heavy oils (Table IV.6.1) and in conventional oils (Table IV.1.1). Heavy oils contain less hydrocarbons, especially alkanes, more sulfur aromatic compounds, resins and asphaltenes. In turn, this difference controls the specific chemical and physical properties of heavy oils.

a) Saturated hydrocarbons usually amount to less than 25%, with an average value of 16%, compared to 57% in normal crude oils: heavy oils are particularly depleted in $n+$iso alkanes which are commonly less than 5%;
b) aromatic hydrocarbons and benzothiophene derivatives frequently represent 25–35% of the heavy oils, with an average of 30%: this figure is comparable to

Table IV.6.1. Gross composition of some heavy oils

Locations	Asphaltenes (wt. %)	Resins (wt. %)	Hydrocarbons (wt. %)			Number of Samples
			Total	Aromatic	Saturated	
Athabasca	23.3	28.6	48.1	32.2	15.9	15
Wabasca	21.6	30.6	47.7	32.1	15.6	7
Peace River	48.7	23.2	28.1	20.5	7.6	3
Cold Lake	20.6	28.0	51.4	30.5	20.9	7
E. Venezuela						
Tar Sands	22.1	37.6	40.3	26.0	14.3	9
Heavy oils	12.6	32.4	55.0	36.4	18.6	5
Average on 46 heavy oils	22.9	30.6	46.5	30.4	16.1	46
Conventional oils of various origin (Table IV.1.1)	14.2		85.8	28.6	57.2	517

6.2 Composition of Heavy Oils

the average value of 29% in normal crude oils; however the proportion of benzothiophene derivatives is comparatively more important in heavy oils;

c) resins plus asphaltenes range from 25 to 70%: they commonly amount to more than 40% of the heavy oils from W. Canada and E. Venezuela, where they average 54%; these figures have to be compared with an average value of 14% in conventional oils; resins are frequently in the 25–35% range, whereas asphaltenes may vary from 10 to 50% of the heavy oils. The asphaltenes/resins ratio is low in eastern Venezuela deposits (average 0.3 to 0.7), intermediate in Athabasca, Wabasca and Cold Lake deposits (0.3 to 1.0), whereas it may occasionally reach high values, such as 2 in the Peace River accumulation.

The H/C ratio of the resins and asphaltenes from heavy oils is comparable to that of the same constituents in normal crude oils. However, the large abundance of resins and asphaltenes causes a hydrogen deficiency of heavy oils, as compared to normal crude oils. In Western Canada, the atomic H/C ratio of the heavily degraded Athabasca oil is 1.45 to 1.50, to be compared with H/C = 1.7 in the Lloydminster oil, which is somewhat biodegraded, and 1.8 in normal crudes. Similar figures are measured on the heavy oils from Eastern Venezuela.

The sulfur, nitrogen and metal contents of the heavy oils depend not only on the bulk composition (saturated and aromatic hydrocarbons, resins and asphaltenes), which is controlled by the degree of degradation, but also on the abundance of the elements (S, N, V, Ni) in the original, non-degraded crude oil.

The *sulfur* content of the heavy oils is, on the average, higher than in the related normal crude oils, as the benzothiophene derivatives, resin and asphaltene fractions are more abundant in heavy and extra-heavy oils. In Western Canada, for instance, there is a strong linear correlation between the sulfur content of the heavy oils from Lower Cretaceous reservoirs and their percentage of aromatic + resin + asphaltene fractions (Fig. IV.6.1). In Eastern Venezuela, a comparable

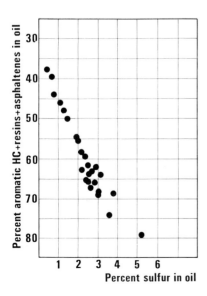

Fig. IV.6.1. Sulfur content of some heavy oils from W. Canada as a function of combined aromatic, resin and asphaltene contents (Deroo et al., 1977)

gradation is observed from normal crude oils containing 1% S or less, to heavy oils with 40–50% resins plus asphaltenes and 3–5% sulfur, and finally extra-heavy oils with 50–70% resins plus asphaltenes and 5–10% S. Sulfur concentrations of up to 5–6% are rather commonly observed in heavy oil provinces, such as Athabasca and other locations in W. Canada, or Boscan in Venezuela. Higher amounts of sulfur are less frequent, but concentrations up to 10% have been reported. It is not yet clear whether this sulfur enrichment is only a concentration process (by selective destruction of the compounds containing no sulfur), or a net uptake of sulfur from the environment.

However, some paraffinic low-sulfur crude oils, once degraded, may result in relatively low-sulfur heavy oils because benzothiophene derivatives are scarce and the sulfur content of resins and asphaltenes is low: the Cretaceous heavy oils of Emeraude, Congo with only 0.4 to 0.8% S (Claret et a., 1977); the Lower Tertiary Asphalt Ridge and Sunnyside tar sands, Utah, with 0.5% S, the Pliocene Derna heavy oil in Rumania with 0.7% S.

The *nitrogen* content of heavy oils also depends on the resin plus asphaltene content, and possibly more on the composition of the original oil: Athabasca tar sands, although heavily degraded, contain only 0.4% N, whereas some California heavy oils reach 0.8%, and the Whiterocks Canyon asphalt, Utah, 1.24%.

The occurrence of *metals* (V, Ni) is also related to the resin and asphaltene content as these fractions contain both porphyrin-bound and non-porphyrin-bound metals. Their abundance, however, is again influenced by the original type of crude oil which is probably an important factor. For instance, the highest concentration of metals (1200 ppm V, 150 ppm Ni) is reported from the Boscan, Western Venezuela, heavy crude oil which is neither the heaviest crude (11°API) nor the most viscous (400 centipoises at 80°C in reservoir conditions). But the related normal oils (20–30° API) from Western Venezuela already contain 100–300 ppm V and 10–30 ppm Ni. The Athabasca tar sands, which are extra-heavy oils, contain only 300 ppm V and 80 ppm Ni, as they are derived from normal Lower Cretaceous oils containing only 10–20 ppm V and ca. 5 ppm Ni (Fig. IV.1.21).

In terms of crude oil *classification* (see above, Chap. IV.2), heavy oils belong to the aromatic-naphthenic and aromatic-asphaltic classes: their distribution between the two classes depends on the original type of oil, and on the nature and extent of degradation (Figs. IV.2.3 and IV.6.2):
– aromatic intermediate conventional oils, when degraded, are converted to aromatic-asphaltic heavy oils: for instance the Mississippian and Mannville (Early Cretaceous) normal oils of the Alberta syncline belong to the aromatic-intermediate class, and the related heavy oils and tar sands from Athabasca, etc. belong to the aromatic-asphaltic class;
– paraffinic or paraffinic-naphthenic conventional oils are converted by biodegradation into aromatic-naphthenic heavy oils; for instance, the Early Cretaceous oils from West Africa are changed by biodegradation into the aromatic-naphthenic heavy oil of Emeraude. If degradation is more advanced and also includes inorganic oxidation processes, the same types of conventional oils can be further altered into aromatic-asphaltic heavy oils, such as the Orinoco Heavy Oil Belt in Eastern Venezuela.

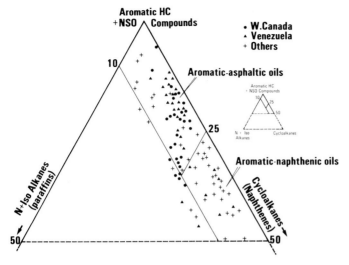

Fig. IV.6.2. Ternary diagram showing the gross composition of some heavy oils from various origins

6.3 Specific Gravity and Viscosity

Physical properties of heavy oils are largely controlled by the abundance and physical state of resins and asphaltenes.

Specific gravity of heavy oils is high and ranges from 0.93 g cm^{-3} (20° API) to 1.06 g cm^{-3} (2° API). However, the largest accumulations correspond to tar sands with a specific gravity close to 1.0 g cm^{-3} (6°–12° API): Athabasca (6°–10° API), Wabasca (10°–13° API), Cold Lake (10°–12° API) in Western Canada; Cerro Negro-Morichal-Jobo (8°–12° API) in Eastern Venezuela; Asphalt Ridge and Sunnyside (8°–12° API), in Utah, etc. Some heavy oils — not classified as tar sands — may reach comparably high specific gravities, but they are still producible because the in-situ viscosity is low, due to a high reservoir temperature. For instance, the Boscan field in Western Venezuela shows a 11° API gravity, but viscosity is only 400 centipoises at reservoir temperature of 80°C (depth: 2300 m).

Viscosity is probably the most important physical parameter of heavy oils, as far as production and transport are concerned. Like specific gravity, viscosity is influenced by the abundance of resins and asphaltenes. Viscosity, however, is also influenced by the physical state of asphaltenes in crude oils, i.e., the size and structure of the micelles formed by interaction with resins and aromatics. This aspect is discussed in Chapter IV.1.8. In heavy degraded oils, there may be a shortage of resins and aromatics and a surplus of asphaltenes due to in-situ degradation, once the oil has migrated into the reservoir. Thus asphaltenes are no more fully dispersed and they may form large-sized aggregates. In turn, this situation is responsible for an increase of viscosity.

The influence of resins and asphaltenes on viscosity is the reason for the strong viscosity–density correlation (linear correlation coefficient $\varrho = +0.823$ in Table IV.1.11). Thus a continuous trend is observed in the viscosity–density diagram (Fig. IV.6.3): at 50°C viscosity increases progressively from ca. 100 centipoises for a 20° API heavy oil to 100,000 centipoises for a 6°–8° API extra-heavy oil in a tar sand. However, there is some scatter around that general trend, as there may be additional causes of density or viscosity variations, other than the variations related to the resin plus asphaltene content. In particular the scatter seems to increase toward low temperature and high specific gravity (low API): at reservoir temperature of 10–15°C, the viscosity of W. Canada heavy oils in the range of 8°–10° API may split over two orders of magnitude, i.e., 5.10^3 to 5.10^5 centipoises.

Viscosity is highly dependent on temperature; this relationship is known for a long time by refiners who use a special chart of predicting the viscosity of petroleum products at different temperatures. Furthermore, it has been observed that crude oils, including heavy oils, broadly follow a similar trend (Fig. IV.6.4).

The main feature shown by the diagram is a strong reduction of the viscosity with an increase in temperature. Furthermore, the more viscous and the heavier the oil is, the more important will be the reduction. For instance, when going from 20°C to 75°C the viscosity of an "Arabian light" crude is reduced by factor of 3 whereas the viscosity of the Lloydminster heavy oil (15° API) is reduced by a factor of 30 and that of an Athabasca extra-heavy oil (8° API) is reduced by a factor of 1000.

The strong viscosity decrease with temperature, as seen for the heavy oils, means that a designation of the various types of heavy and extra-heavy oils based on viscosity in reservoir conditions is not applicable for geological or geochemical

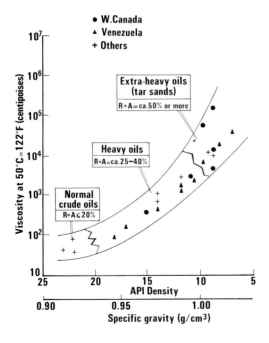

Fig. IV.6.3. Relationship between specific gravity (or API density) and viscosity of some heavy oils. R + A designates the resins plus asphaltenes content

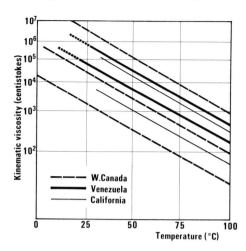

Fig. IV.6.4 Changes of viscosity of some heavy oils as a function of temperature (kinematic viscosity is the ratio absolute viscosity/specific gravity, at a given temperature)

studies: some heavy 7°–10° API crudes from Eastern Venezuela show viscosity curves intermediate between the 8° API Athabasca and the 10°–12° API Cold Lake curves. However, the two Canadian heavy oils have reservoir temperatures of 10°–15°C, whereas the Venezuelan reservoirs exhibit higher temperatures of about 50°–60°C. Therefore their in-situ viscosities are in the range of 10^5–10^7 centipoises and 10^3–10^4 centipoises respectively; thus the Canadian deposits would be called "tar sands" and the Venezuelan "heavy oils", although they are in certain respects comparable. If we now wish to decide to use a standard temperature reference for viscosity, such as 50°C (122°F), Figure IV.6.3 shows what could be a proper definition for heavy oils and extra-heavy oils (or tar sands).

The significance of the abrupt decrease in viscosity of extra-heavy oils with temperature has still to be investigated. Such a dramatic change suggests a rearrangement of the macrostructure of the high molecular weight aggregates, i.e., a change in the physical state of asphaltenes in crude oil (Tissot, 1981).

A final remark concerns the variability of physical properties, such as specific gravity and viscosity within the huge deposits of Western Canada and Eastern Venezuela. The volume of these accumulations is enormous (in the range of 40 to 800×10^9 m^3 of reservoir for each deposit), and the heavy oils or extra-heavy oils have practically no mobility. It is thus obvious that there may be substantial changes in chemical composition, density and viscosity within each accumulation. This may be the reason why so many different values of density, viscosity, etc. are cited in the literature concerning large-size accumulations, such as Athabasca, Peace River, or the Orinoco Belt.

6.4 Origin and Occurrence of Heavy Oils

Although some accumulations of immature, non-degraded, heavy oils are known, they probably amount to less than 1% of the world reserves of this category. These immature heavy oils contain abundant hetero compounds (resins and asphaltenes) resulting from an early breakdown of kerogen, with the associated high sulfur and nitrogen content. They differ from degraded heavy oils by the occurrence of *n*-alkanes which have not been bacterially destroyed. This situation is found, for instance, in some crude oils of the Mediterranean area. Favourable conditions for accumulation of these heavy immature oils may occur where a particular source rock also acts as a reservoir, or where migration conditions are particularly easy (intense fracturation, karst, etc.).

Most heavy oils originate from normal, fluid crude oils which have been subsequently degraded in the reservoir by one or several of the following processes: biodegradation, water washing, loss of volatiles, inorganic oxidation (see above, Sect. 5.3). These processes result in a decrease of the light ends of the crude oil, and also of the alkanes and low molecular weight alkylbenzenes, and an increase of the more resistant benzothiophene derivatives, polyaromatics, resins and asphaltenes. Furthermore, at certain conditions, sulfur extracted from sulfate by anaerobic bacteria may react with hydrocarbons to increase the sulfur content and the abundance of heavy constituents.

The result of this type of alteration is a heavy, highly viscous oil. The degradation is commonly associated with the invasion of surface-derived, meteoric formation waters. Thus the extent of degradation is associated with parameters such as depth, proximity to aerial contact, and salinity of formation waters. Excellent examples are provided by the two major provinces of heavy oils, i.e., Western Canada and Eastern Venezuela.

A cross-section of the Alberta basin showing the Early Cretaceous heavy oil deposits of Western Canada is presented in Figure IV.6.5. The corresponding changes of composition of the oils are shown in Figure IV.5.4. The normal crude oils of the Early Cretaceous in the Alberta syncline are represented by the Bellshill Lake, a 26° API normally producible crude oil, associated with a water salinity > 80 g l^{-1}. Closer to the outcrops, the Lloydminster area contains heavy oils (density 16° API; viscosity 2000 centipoises at 25°C) associated with water salinities of 50–80 g l^{-1}; then the Cold Lake deposit reaches 11° API and 80000

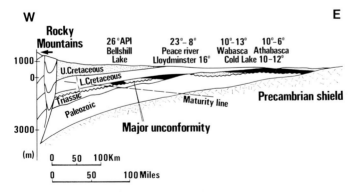

Fig. IV. 6.5. Schematic cross-section of Western Canada basin showing heavy-oil sands. *Numbers* indicate the range of API gravity of the various accumulations. (Adapted with changes from Demaison, 1977)

6.4 Origin and Occurrence of Heavy Oils

centipoises at 15 °C, associated with water salinities lower than 50 g l^{-1}. Finally, the outcropping Athabasca tar sands contain an extra-heavy oil (density 6–8° API, viscosity 2.10^6 centipoises at 10 °C) associated with low salinity waters (< 20 g l^{-1}).

Deroo et al. (1974, 1977) have shown that the main source rocks of the normal crude oil entrapped in Early Cretaceous sands are the associated shales. Due to the close relationship and the progressive change from these normal oils to heavy oils, they inferred that the same source was an important contributor to heavy oils. However, there may be some contribution also from the Paleozoic and Jurassic shales which are present basinward. Hacquebard (in Deroo et al., 1977) has shown that catagenesis of the Cretaceous source beds is reached basinward from all major heavy oil accumulations. If this general concept is accepted the distances of secondary migration have to be in the range of tens to several hundred of kilometers: Demaison (1977). This author pointed out that both stratigraphic features, such as sandstone distribution, and structural factors, such as the Peace River arch, have influenced entrapment of the oils. Thus a normal, medium-gravity oil has probably migrated into the sandstone reservoirs; later, or simultaneously, the oils came into contact with meteoric waters entering the reservoirs from the outcrops, and were subsequently degraded. This view is supported by geochemical investigations, which have proved a step by step degradation, decreasing basinward and also with the increasing salinity of formation waters (Deroo et al., 1974, 1977).

In Eastern Venezuela, the general situation is rather similar to that in Western Canada (Fig. IV.6.6); a foredeep basin, with a heavily folded and faulted northern flank and a gently dipping southern flank, contains normal and heavy oils in Cretaceous and Oligocene-Miocene (Officina) sandstones. The traps are formed by a combination of normal faults and sandstone pinchouts. The oil accumulations located at depths greater than 1500 m are normal, medium to light (25–40° API), low-sulfur crudes (≤ 1% S) with a low content of resins and asphaltenes; southwards, when approaching the border of the basin, the oils (Temblador, Merey) become progressively shallower (ca. 1200–1500 m), heavier

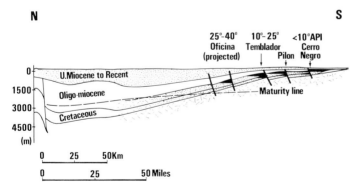

Fig. IV.6.6. Schematic cross-section of Eastern Venezuela basin showing heavy-oil sands (Orinoco Heavy Oil Belt). *Numbers* indicate the range of API gravity of the various accumulations. (Adapted with changes from Demaison, 1977)

and biodegraded (with 20–40% resins plus asphaltenes, a density of 10–25° API and 1–3% sulfur); finally they grade into the Orinoco Heavy Oil Belt (Cerro Negro, El Pao, Hamaca, etc.), a gigantic accumulation made of heavily degraded oils: more than 50% is made up of resins and asphaltenes, API density is 10° or lower, sulfur reaches 4–5% or even more.

The source rocks for normal oils are found in the Cretaceous and Oligo-Miocene beds. However, it is not yet established which one was the major source for the heavy oils. Demaison (1977) located the maturation area of the source rocks and showed, that it is basinward and completely outside the area of heavy oil occurrences. He calculated a minimum distance of migration of about 100 km. Once migrated into the reservoir, the normal, medium-gravity oil came into contact with meteoric waters entering the sandstones from the outcrops, in the Orinoco area. The oil was subsequently degraded and the locations close to the outcrops were more heavily altered. Furthermore it was observed by Hedberg (1947) and others that, in some locations, light waxy crudes in the upper part of the Oligo-Miocene beds overlie heavy oils in the lower part. Demaison (1977) suggested that this situation is explained by an easier access of meteoric waters in the more sandy lower part of the stratigraphic column.

6.5 World Reserves and Geological Setting

The present estimate of the world reserves of heavy oils is shown in Table IV.6.2. All figures are related to oil in-place as the concept of recoverable reserves is continuously changed, due to the advancement of technologies for producing these nonconventional crude oils.

The total figure — 450 to 1000 billion m^3 — is comparable to the total *in-place* discovered quantity of conventional crude oil, which can be estimated to ca. 600 billion m^3 (including the oil already exploited, but excluding non-discovered potential resources; assuming a recoverable/in-place oil ratio of 0.25).

The recoverable fraction of the heavy oils will depend on the developments of technologies for production. For instance, out of the 220 billion m^3 of heavy oils present in the Cretaceous sands of Western Canada, ca. 6 billion m^3 could probably be recovered by surface mining and ca. 40 billion m^3 by in-situ steam flooding, i.e., 25% of the total deposit. The balance, ca. 170 billion m^3, is not recoverable by existing technologies, but what fraction could be added by developing new methods is yet unknown.

The most striking feature with respect to occurrence of heavy oils and tar sands is the abnormally high quantity of in-place oil in the major deposits. This aspect was discussed by Demaison (1977). In particular, the evaluations of the in-place oil in the single Athabasca amount to 140 billion m^3, i.e., three times the in-place oil of the largest of all conventional oilfields, Ghawar, Saudi Arabia. Comparable situations probably occur with respect to the Orinoco Heavy Oil Belt, but precise definition of the individual deposits has still to be improved. The total in-place reserves of the Middle East province (ca. 300 billion m^3), however, are in the

Table IV.6.2. Principal in-place reserves of heavy oils of the world. Adapted, with changes, from Demaison, 1977; Boy de la Tour et al. 1981

Country	Area	Age of reservoir	Volume in-place (billion m^3)
Venezuela	Orinoco	Oligo-Miocene Cretaceous	150 to 500
Canada	Athabasca Cold Lake Wabasca Peace River	Early Cretaceous	220
Canada	Carbonate triangle (Alberta)	Paleozoic	0 to 200
U.S.S.R.	Melekess (Volga-Ural) Others	Permian Paleozoic	20 3.3
U.S.A.	Tar Triangle Uinta Basin Others	Permian Eocene	2.5 1.7 5.5
Others			50 to 70
Total		ca.	450 to 1000

same order of magnitude as the in-place heavy oils of Western Canada (Cretaceous only: 220 billion m^3) or Eastern Venezuela (150 to 500 billion m^3).

Thus, three questions immediately arise: why are these three provinces (Middle East, Western Canada, Eastern Venezuela) so rich? Why did the oil accumulate in a relatively small area of the basin? Why are some individual accumulations (Athabasca and probably parts of the Orinoco Belt) so huge, even when compared to the giant oil fields?

These features require in fact several conditions:
– unusual expulsion from the mature source rocks and efficient sealing,
– long distance migration,
– wide drainage area of some individual deposits.

Demaison (1977) pointed out that "foreland", or "foredeep" basins (Halbouty, 1970; Klemme, 1975; Bally, 1975) were an excellent situation in those respects. They frequently offer a large volume of mature source rocks. Then, the existence of a wide homoclinal slope, with a tremendous drainage area, and the occurrence of major unconformity surfaces as conductors (e.g., the pre-Cretaceous unconformity in Western Canada) favor a long-distance migration, especially from the deep part of the basin. In fact all three major oil provinces mentioned above

(Middle East, Western Canada, Eastern Venezuela) fit into that scheme, and also the Volga-Ural basin (USSR), where the Melekess deposits ranks immediately behind Venezuela and Canada (Table IV.6.2).

A wide drainage area for an individual deposit is an essential condition for gathering a huge amount of oil into a single accumulation. Demaison (1977) pointed out that a paleodelta system comprising far-reaching, efficiently interfingering carrier sandstones is the ideal setting in that respect. A combination of foredeep basin and paleodelta system is very probably the cause for the major heavy oil accumulations discussed here.

Finally, degradation itself seems to be an important, if not the most important, factor for trapping very large quantities of oil. The general layout of the foredeep basins is also favorable to the entry of meteoric waters, from the outcrops, along unconformity surfaces, basal sandstones, etc. In the absence of degradation, the bulk of the normal fluid oils when migrating on a wide, featureless homoclinal slope would reach the outcrops and would be lost. In-situ degradation would supress mobility of the oil and prevent it from escaping: this is insured not only by an "asphalt plug" but more efficiently by increasing viscosity of the bulk of the deposit, until it can no longer move. Thus, in a sense heavy oils provide their own seal. This particular aspect would explain why no single accumulation of conventional oil comparable in size to Athabasca or Orinoco deposits is known. Degradation and absence of mobility would be a requisite for such huge deposits.

6.6 Valorization of Heavy Oils

6.6.1 Production

Heavy oils still mobile under reservoir conditions can only be produced by conventional methods at a low rate, with a very low efficiency: the recovery factor is only a few percent. Extra heavy oils, immobile under reservoir conditions, cannot be produced by conventional techniques. Thus, ex-situ mining methods and in-situ enhanced recovery technologies are necessary to mobilize the major fraction of heavy oil reserves. The methods are usually based on the drastic viscosity reduction in response to heating up the heavy oil.

Athabasca oil sands are mined, in open pits, in two locations near Fort McMurray, Alberta. The sand is subsequently separated from the heavy oil by using an alkaline hot water process. The total production is presently ca. 8 million m^3 year^{-1}.

Heavy oils with viscosities ranging from 100 to 25000 centipoises in reservoir conditions are produced by enhanced recovery techniques, using in-situ, cyclic or continuous, steam injection, at an annual rate of ca. 25 million m^3 (mainly in California and W. Venezuela). Steam (temperature ca. 300°C, pressure ca. 100 bar) is injected through conventional wells; the heat is partly transferred to the heavy oil (including the heat of steam condensation) and the viscosity subsequently decreases. Then oil can be pumped either through separate producing wells or through the injection well, by periodically alternating injection and production. The process seems to be presently applicable at depths shallower than 1000 m, provided permeabilities are sufficient. It is not clear whether the process would be also applicable to extraheavy oils, with an in-situ viscosity beyond 100,000 centipoises. A large-scale project in the Cold Lake area has been announced; the in-situ viscosity of this deposit seems to be somewhat lower than in Athabasca, Wasbasca and Peace River deposits.

In-situ combustion is another process for reducing oil viscosity and making it producible. In this case, air is injected through conventional wells and flows toward production wells. Once ignition is started at the injection well the combustion front moves slowly in the direction of producing wells. Ahead of this front, oil is pyrolyzed and cracked: it yields a lighter oil, which is produced, and a heavy, coke-like residue which remains in place. The latter is burnt when the combustion front reaches it, and provides the energy necessary for the process. This technology is presently used in Rumania, the United States and Canada, with an annual production ca. 1.5 million m^3, and seems to be applicable down to 1500 m. The amount of heavy oil burnt as coke would be 10–25% of the in-place oil. Simultaneous injection of water is also suggested, in order to combine the advantages of combustion and steam injection methods.

Other methods, such as injection of carbon dioxide into heavy oils, or solvents, are still experimental.

6.6.2 Refining

Valorization of the heavy oils causes some difficulties in refining, due to their specific characteristics:

- high sulfur, nitrogen, and metal (V, Ni) content: they may act as poisons to heterogenous catalysts used in refining processes; furthermore, sulfur and nitrogen cause environmental problems,
- low H/C ratio, as compared to conventional oil,
- high asphaltene content, which is responsible for the high viscosity of the heavy fuels, and is also a cause of instability of some refined products.

The increase of the H/C ratio may be obtained by carbon elimination or hydrogen addition. A heavy, carbon-rich, residue can be removed by deasphalting (with the use of pure solvents, or a C_5-fraction) or coking (heating and cracking the oil). Then, hydrogenation treatments may further increase the H/C ratio, and also remove sulfur and nitrogen.

Summary and Conclusion

Heavy oils and extra-heavy oils of tar sands are, in most cases, the result of crude oil degradation in reservoirs, where petroleum has entered by migration, accumulated and subsequently become degraded. This latter phenomenon is commonly associated with the invasion of the reservoir by meteoric waters and it may include biodegradation, water washing, loss of volatiles, and oxidation.

Heavy oils contain less hydrocarbons, especially alkanes, than normal crude oils, and more asphaltenes, resins and sulfur-aromatic compounds. Asphaltenes plus resins range from 25% to 70% of heavy oil. In terms of crude oil classification, heavy oils belong to the aromatic-naphthenic and aromatic-asphaltic classes.

Physical properties of heavy oils are largely controlled by the abundance and physical state of asphaltenes and resins: specific gravity ranges from 0.93 g cm^{-3} to 1.06 g cm^{-3} (20° to 2° API). Viscosity ranges from 100 to 100,000 centipoises and it is highly dependent on temperature.

World reserves of heavy oil in-place amount to 0.5–1.0 trillion m^3, with very large deposits in Western Canada (Athabasca) and Eastern Venezuela (Orinoco belt).

References to Part IV

Ackman, R. G., Cox, R. E., Eglinton, G., Hooper, S. N., Maxwell, J. R.: Stereochemical studies of acyclic isoprenoid compounds. I Gas Chromatographic analysis of stereoisomers of a series of standard acyclic isoprenoid acids. J. Chromatogr. Sci. **10**, 392 (1972)

Albaiges, J., Torradas, J. M.: Significance of the even-carbon n-paraffin preference of a Spanish crude oil. Nature (London) **250**, 567–568 (1974)

Albrecht, P., Ourisson, G.: Triterpene alcohol isolation from oil shale. Science **163**, 1192–1193 (1969)

Alimi, H.: Untersuchungen zur Erdölgenese an Sedimenten der Unteren Odjaghgheshlagh-Formation (Obereozän) im Moghan-Becken/NW Iran. Thesis, RWTH-Aachen, 1977

Andreev, P. F., Bogomolov, A. I., Dobryanskii, A. F., Kartsev, A. A.: Transformation of Petroleum in Nature (in Russian). Leningrad: Gostoptekhizdat, 1958, 416; Engl. transl. Int. Ser. Monogr. Earth Sci. **29.** New York: Pergamon Press, 1968

Arefjev, O. A., Makushina, V.-M., Petrov, A. A.: Asphaltenes as indicators of geochemical history of crude oils (in Russian). Izv. Akad. Nauk SSR Ser. Geol. **4**, 124–130 (1980)

Arpino, P.: Les lipides de sédiments lacustres éocènes. Thesis, Univ. Strasbourg, 1973

Bailey, N. J. L., Evans, C. R., Milner, C. W. D.: Applying petroleum geochemistry to search for oil: examples from western Canada Basin. Am. Assoc. Pet. Geol. Bull. **58**, 2284–2294 (1974)

Bailey, N. J. L., Jobson, A. M., Rogers, M. A.: Bacterial degradation of crude oil: comparison of field and experimental data. Chem. Geol. **11**, 203–221 (1973b)

Bailey, N. J. L., Krouse, H. H., Evans, C. R., Rogers, M. A.: Alteration of crude oil by waters and bacteria–evidence from geochemical and isotope studies. Am. Assoc. Pet. Geol. Bull. **57**, 1276–1290 (1973a)

Baker, E. W.: Mass spectrometric characterization of petroporphyrins. J. Am. Chem. Soc. **88**, 2311–2315 (1966)

Bally, A. W.: A geodynamic scenario for hydrocarbon occurrences. 9th World Petrol. Cong., Proc. **2**, 33–44 (1975)

Barton, D. C.: Natural history of the Gulf Coast crude oil. In: Problems of Petroleum Geology. Wrather, W. E., Lahee, F. M. (eds.). Am. Assoc. Pet. Geol. Tulsa, 109–155 (1934)

Behar, F., Pelet, R., Roucaché, J.: Geochemistry of asphaltenes. In: Advances in Organic Geochemistry 1983. Schenk, P. A. (ed.), 1984 (in press)

Bestougeff, M. A.: Petroleum hydrocarbons. In: Fundamental Aspects of Petroleum Geochemistry. Nagy, B., Colombo, U. (eds.). Amsterdam: Elsevier, 1967, pp. 77–108

Blumer, M.: Chemical fossils: trends in organic geochemistry. In: Chemistry in Evolution and Systematics. Swain, T. (ed.). London: Butterworths, 1973, pp. 591–609

Blumer, M., Snyder, W. D.: Porphyrins of high molecular weight in a Triassic oil shale: Evidence by gel permeation chromatography. Chem. Geol. **2**, 35–45 (1967)

Bockrath, B. C., Schweighardt, F. K.: Structural characterization of coal-derived asphaltenes and its significance to liquefaction. Preprints, Div. Petrol. Chem., Am. Chem. Soc. **24** (4), 949–954 (1979)

Boduszynski, M. M., McKay, J. F., Latham, D. R.: Asphaltenes, where are you? In: Proceedings Assoc. of Asphalt Paving Techn. **49**, 123–143 (1980)

Boy de la Tour, X., Le Leuch, H.: Nouvelles techniques de mise en valeur des ressources d'hydrocarbures. Revue Inst. Franç. Pétrole **36**, 251–280 (1981)

Brassell, S. C., Comet, P. A., Eglinton, G. et al.: The origin and fate of lipids in the Japan Trench. In: Advances in Organic Geochemistry, 1979. Douglas, A. G., Maxwell, J. R. (eds.). Pergamon Press, 1980, pp. 375–392

Brooks, J. D.: The use of coals as indicators of the occurrence of oil and gas. Aust. Pet. Explor. Ass. J. **10**, 35–40 (1970)

Brooks, J. D., Gould, K., Smith, J. W.: Isoprenoid hydrocarbons in coal and petroleum. Nature (London) **222**, 257–259 (1969)

Calvin, M.: The Bakerian Lecture 1965: Chemical Evolution. Proc. Roy. Soc. London **288**, 441–486 (1965)

Calvin, M.: Chemical Evolution. New York: Oxford University Press, 1969

Castex, H.: Le soufre thiophénique dans les pétroles et les extraits de roche. Analyse par spectrométrie de masse et chromatographie en phase gazeuse. Rev. Inst. Fr. Pét. **29**, 3–40 (1974)

Chappe, B.: Fossiles moléculaires d'Archaebacteries. Thesis, University of Strasbourg (1982)

Chappe, B., Michaelis, W., Albrecht, P., Ourisson, G.: Fossil evidence for a novel series of archaebacterial lipids. Naturwissenschaften **66**, 522–523 (1979)

Claret, J., Roucaché, J., Gageonnet, R., Renard, B., du Rouchet, J.: Caractérisation géochimique des huiles de gisements et comparaison avec les hydrocarbures des roches non réservoirs dans le Crétacé supérieur et le Tertiaire de la région de Port Gentil (Gabon). In: Advances in Organic Geochemistry 1973. Tissot, B., Bienner, F. (eds.). Paris: Technip, 1974, pp. 335–348

Claret, J., Tchikaya, J. B., Tissot, B., Deroo, G., van Dorsselaer, A.: Un exemple d'huile biodégradée à basse teneur en soufre: le gisement d'Emeraude, Congo. In: Advances in Organic Geochemistry 1975. Campos, R., Goni, J. (eds.). Madrid, 1977, pp. 509–522

Clark, R. C., Blumer, M.: Distribution of paraffins in marine organisms and sediments. Limnol. Oceanogr. **12**, 79–85 (1967)

Connan, J.: Diagénèse naturelle et diagénèse artificelle de la matière organique à éléments végétaux prédominants. In: Advances in Organic Geochemistry 1973. Tissot, B., Bienner, F. (eds.). Paris: Technip, 1974, pp. 73–95

Connan, J., Le Tran, K., van der Weide, B.: Alteration of Petroleum in Reservoirs. Proc. 9th World Pet. Congr. Tokyo. London: Applied Science Publ., 1975, **2**, 171–178

Connan, J., Orgeval, J. J.: Un exemple de relation hydrocarbures-minéralisations: cas du filon de barytine de St.-Privat (Bassin de Lodève, France). Bull. Centre Rech. Pau-SNPA **10**, 359–374 (1976)

Cox, R. E., Maxwell, J. R., Ackman, R. G., Hooper, S. N.: Stereochemical studies of acyclic isoprenoid compounds, III. Stereochemistry of naturally occurring (marine) 2, 6, 10, 14 tetramethylpentadecane. Can. J. Biochem. **50**, 1238 (1972)

Creanga, C.: Contributü la clasificarea, chimica a titeiurilor. Clasificarea "Carpatica". Acad. Rep. Pop. Romine, Studii Cercetari Chim., Bucarest **9**, 93–108 (1961)

Demaison, G. J.: Tar sands and supergiant oil fields. Am. Assoc. Pet. Geol. Bull. **61**, 1950–1961 (1977)

Dembicki, H., Meinschein, W. G., Hattin, D. E.: Possible ecological and environmental significance of the predominance of even-carbon number C_{20}–C_{30} n-alkanes. Geochim. Cosmochim. Acta **40**, 203–208 (1976)

Deroo, G., Powell, T. G., Tissot, B., McCrossan, R. G.: The origin and migration of petroleum in the Western Canadian Sedimentary Basin, Alberta. Geol. Surv. Can. Bull. 262 (1977)

Deroo, G., Roucaché, J., Tissot, B.: Etude géochimique du Canada Occidental, Alberta. Geol. Surv. Can., open file 171 (1973)

Deroo, G., Tissot, B., McCrossan, R. G., Der, F.: Geochemistry of the heavy oils of Alberta. In: Oil Sands Fuel of the Future. Memoir 3, Can. Soc. Pet. Geol. 148–167, 184–189 (1974)

Dorsselaer, A. van: Triterpènes de sédiments. Thesis, Univ. Strasbourg 1975

Dorsselaer, A. van, Ensminger, A., Spyckerelle, C., Dastillung, M., Sieskind, O., Arpino, P., Albrecht, P., Ourisson, G.: Degraded and extended hopane derivatives (C_{27} to C_{35}) as ubiquitous geochemical markers. Tetrahedron Lett. **14**, 1349–1352 (1974)

Douglas, A. G., Mair, B. J.: Sulfur: role in genesis of petroleum. Science **147**, 499–501 (1965)

Durand, B., Espitalié, J.: Analyse géochimique de la matière organique extraite des roches sédimentaires: II. Extraction et analyse quantitative des hydrocarbures de la fraction C_1–C_{15} et des gaz permanents. Rev. Inst. Fr. Pét. **25**, 741–751 (1970)

Durand, B., Espitalié, J., Oudin, J. L.: Analyse géochimique de la matière organique extraite des roches sédimentaires: III. Accroissement de la rapidité du protocole opératoire par l'amélioration de l'appareillage. Rev. Inst. Fr. Pét. **25**, 1268–1279 (1970)

Eglinton, G.: Chemical fossils: a combined organic geochemical and environmental approach. Pure Appl. Chem. **34**, 611–632 (1973)

Ensminger, A.: Evolution de composés polycycliques sédimentaires. Thesis, University Louis Pasteur, Strasbourg, France (1977)

Ensminger, A., Albrecht, P., Ourisson, G., Tissot, B.: Evolution of polycyclic hydrocarbons under the effect of burial (Early Toarcian shales, Paris Basin). In: Advances in Organic Geochemistry, 1975. Campos, R., Goni, J. (eds.). Madrid, 1977, pp 45–52

Ensminger, A., Albrecht, P., Ourisson, G., Kimble, B. J., Maxwell, J. R., Eglinton, G.: Homohopane in Messel oil shale: first identification of a C_{31} pentacyclic triterpane in nature. Tetrahedron Lett. **36**, 3861–3864 (1972)

Ensminger, A., van Dorsselaer, A., Spyckerelle, C., Albrecht, P., Ourisson, G.: Pentacyclic triterpenes of the hopane type as ubiquitous geochemical markers: origin and significance. In: Advances in Organic Geochemistry 1973. Tissot, B., Bienner, F. (eds.). Paris: Technip, 1974, pp. 245–260

Erdman, J. G., Harju, P. H.: Capacity of petroleum asphaltenes to complex heavy metals. J. Chem. Eng. Data **8**, 252–258 (1963)

Erdman, J. G., Morris, D. A.: Geochemical correlation of petroleum. Am. Assoc. Pet. Geol. Bull. **58**, 2326–2337 (1974)

Evans, C. R., Rogers, M. A., Bailey, N. J. L.: Evolution and alteration of petroleum in Western Canada. Chem. Geol. **8**, 147–170 (1971)

Goubin, N.: Quelques exemples d'exploitation d'un fichier de géochimie organique. Rev. Inst. Fr. Pét. **30**, 213–262 (1975)

Gussow, 1962 (s. Fig. IV.5.2)

Halbouty, M. T. et al.: Factors affecting formation of giant oil and gas fields and basin classification. In: Geology of giant petroleum fields. Am. Assoc. Pet. Geol. Memoir **14**, 528–555 (1970)

Hedberg, H. D., Sass, L. C., Funckhouser, H. J.: Oil fields of Greater Oficina area, Central Anzoategui, Venezuela. Am. Assoc. Pet. Geol. Bull. **31**, 2089–2169 (1947)

Hedberg, H. D.: Significance of high wax oils with respect to genesis of petroleum. Am. Assoc. Pet. Geol. Bull. **52**, 736–750 (1968)

Hills, I. R., Smith, G. W., Whitehead, E. V.: Hydrocarbons from fossil fuels and their relationship with living organisms. J. Inst. Pet. London **56**, 127–137 (1970)

Ho, T. Y., Rogers, M. A., Drushel, H. V., Koons, C. B.: Evolution of sulfur compounds in crude oils. Am. Assoc. Pet. Geol. Bull. **58,** 2338–2348 (1974)

Hoeven, W. van: Organic geochemistry. Thesis, Univ. Calif. Berkeley, UCRL 18690 (1969)

Hollerbach, A.: Personal Communication (1976)

Hood, A., Clerc, R. J., O'Neal, M. J.: The molecular structure of heavy petroleum compounds. J. Inst. Pet. **45,** 168–173 (1959)

Hood, A., O'Neal, M. J.: Status of application of mass spectrometry to heavy oil analysis. In: Advances of Mass Spectrometry. London: Pergamon Press, 1959, pp. 175–192

Huang, W. Y., Meinschein, W. G.: Sterols as source indicators of organic materials in sediments. Geochim. Cosmochim. Acta **40,** 323–330 (1976)

Huang, W. Y., Meinschein, W. G.: Sterols as ecological indicators. Geochim. Cosmochim. Acta **43,** 739–745 (1979)

Huc, A. Y., Roucaché, J., Bernon, M., Caillet, G., da Silva, M.: Application de la chromatographie sur couche mince à l'étude quantitative et qualitative des extraits de roches et des huiles. Rev. Inst. Fr. Pét. **31,** 67–98 (1976)

Hunt, J. M.: Composition of crude oil and its relation to stratigraphy in Wyoming. Am. Assoc. Pet. Geol. Bull. **37,** 1837–1872 (1953)

Johns, R. B., Belsky, T., McCarthy, E. D., Burlingame, A. L., Haug, P., Schnoes, H. K., Richter, W., Calvin, M.: The organic geochemistry of ancient sediments, Part II. Geochim. Cosmochim. Acta **30,** 1191–1222 (1966)

Jonathan, D., L'Hote, G., du Rouchet, J.: Analyse géochimique des hydrocarbures légers par thermovaporisation. Rev. Inst. Fr. Pét. **30,** 65–88 (1975)

Kartsev, A. A., Tabasaranskii, Z. A., Subbota, M. I., Mogilevskii, G. A.: Geochemical Methods of Prospecting and Exploration for Petroleum and Natural Gas. Berkeley, Calif.: Univ. of Calif. Press, Engl. transl. Witherspoon, P. A., Romey, W. O., 1959

Klemme, H. D.: Giant oil fields related to their geologic setting; a possible guide to exploration. Bull. Can. Pet. Geol. **23,** 30–66 (1975)

Landa, S., Machacek, U.: Adamantane, a new hydrocarbon extracted from petroleum. Coll. Czech. Chem. Commun. **5,** 1–5 (1933)

Lane, E. C., Garton, E. L.: "Base" of a crude oil. U.S. Bureau of Mines, Rept. Invest. 3279 (1935)

Long, R. B.: The concept of asphaltenes. Preprints. Div. Pet. Chem., Am. Chem. Soc. **24** (4), 891–900 (1979)

Louis, M.: Cours de Géochimie du Pétrole. Paris: Technip, 1967

Mackenzie, A. S., Patience, R. L., Maxwell, J R., Vandenbroucke, M., Durand, B.: Molecular parameters of maturation in the Toarcian shales, Paris Basin, France. I. Changes in the configuration of acyclic isoprenoid alkanes, steranes and triterpanes. Geochim. Cosmochim. Acta **44,** 1709–1721 (1980)

Mackenzie, A. S., Quirke, J. M. E., Maxwell, J. R.: Molecular parameters of maturation in the Toarcian shales, Paris Basin, France. II. Evolution of metalloporphyrins. In: Advances in Organic Geochemistry, 1979. Douglas, A. G., Maxwell, J. R. (eds.). Pergamon Press, 1980, pp. 239–248

Mackenzie, A. S., Hoffmann, C. F., Maxwell, J. R.: Molecular parameters of maturation in the Toarcian shales, Paris Basin, France. III. Changes in the aromatic steroid hydrocarbons. Geochim. Cosmochim. Acta **45,** 1345–1355 (1981)

Mackenzie, A. S., Lamb, N. A., Maxwell, J. R.: Steroid hydrocarbons and the thermal history of sediments. Nature **295,** 223–226 (1982)

MacLean, I., Eglinton, G., Douraghi-Zadek, K., Ackman, R. G., Hooper, S. N.: Correlation of stereoisomerism in present day and geologically ancient isoprenoid fatty acids. Nature (London) **218,** 1019–1024 (1968)

Mair, B. J.: Terpenoids, fatty acids and alcohols as source materials for petroleum hydrocarbons. Geochim. Cosmochim. Acta **28**, 1303–1321 (1964)

Mair, B. J., Ronen, Z.: Composition of the branched paraffin-cycloparaffin portion of petroleum 140–180°C. J. Chem. Eng. Data **12**, 432–436 (1967)

Mair, B. J., Ronen, Z., Eisenbraun, E. J., Horodsky, A. G.: Terpenoid precursors of hydrocarbons from the gasoline range of petroleum. Science **154**, 1339–1341 (1966)

Mair, B. J., Shamaiengar, M., Krouskop, N. C., Rossini, F. D.: Isolation of adamantane from petroleum. Anal. Chem. **31**, 2082–2083 (1959)

Martin, R. L., Winters, J. C., Williams, J. A.: Composition of crude oils by gas chromatography: geological significance of hydrocarbon distribution. 6th World Pet. Congr. Sec. V, 231–260 (1963a)

Martin, R. L., Winters, J. C., Williams, J. A.: Distribution of n-paraffins in crude oils and their implications to origin of petroleum. Nature (London) **199**, 1190–1193 (1963b)

Maxwell, J. R., Cox, R. E., Eglinton, G., Pillinger, C. T., Ackman, R. G., Hooper, S. N.: Stereochemical studies of acyclic isoprenoid compounds, II Role of Chlorophyll in derivation of isoprenoid type acids in a lacustrine sediment. Geochim. Cosmochim. Acta **37**, 297 (1973)

Maxwell, J. R., Mackenzie, A. S., Volkman, J. K.: Configuration at C-24 in steranes and sterols. Nature **286**, 694–697 (1980)

McAlary, J. D., Rogers, M. A.: Reservoir bitumens — another aid in the geochemical evaluation of a basin (abs.). Geol. Soc. Am. Abs. Programs, **3**, 393–394 (1971)

McKay, J. F., Cogswell, T. E., Weber, J. H., Latham, D. R.: Analysis of acids in high-boiling petroleum distillates. Fuel **54**, 50–61 (1975)

Michaelis, W., Albrecht, P.: Molecular fossils of Archaebacteria in kerogen. Naturwissenschaften **66**, 420–421 (1979)

Moschopedis, S. E., Speight, J. G.: Investigation of hydrogen bonding by oxygen functions in Athabasca bitumen. Fuel **55**, 187–192 (1976)

Moschopedis, S. E., Fryer, J. F., Speight, J. G.: Investigation of asphaltene molecular weights. Fuel **55**, 227–231 (1976)

Moschopedis, S. E., Parkash, S., Speight, J. G.: Thermal decomposition of asphaltenes. Fuel **57**, 431–434 (1978)

Mulheirn, L. J., Ryback, G.: Stereochemistry of some steranes from geological sources. Nature (London) **256**, 301–302 (1975)

Orr, W. L.: Changes in sulfur content and isotopic ratios of sulfur during petroleum maturation-Study of Big Horn basin Paleozoic oils. Am. Assoc. Pet. Geol. Bull. **58**, 2295–2318 (1974)

Oudin, J. L.: Analyse géochimique de la matière organique extraite des roches sédimentaires, I. Composés extractibles au chloroforme. Rev. Inst. Fr. Pét. **25**, 3–15 (1970)

Ourisson, G., Albrecht, P., Rohmer, M.: The hopanoids, Palaeochemistry and biochemistry of a group of natural products. Pure Appl. Chem. **51**, 709–729 (1979)

Patience, R. L., Rowland, S. J., Maxwell, J. R.: The effect of maturation on the configuration of pristane in sediments and petroleum. Geochim. Cosmochim. Acta **42**, 1871–1875 (1978)

Petrov, Al. A., Arefjev, O. A., Jakobson, Z. V.: Hydrocarbons of adamantane series as indices of petroleum catagenesis process. In: Advances in Organic Geochemistry 1973. Tissot, B., Bienner, F. (eds.). Paris: Technip, 1974, pp. 518–522

Pfeiffer, J. P., Saal, R. N. J.: Asphaltic bitumen as colloid system. J. Phys. Chem. **44**, 139–149 (1940)

Philipp, W., Drong, H. J., Füchtbauer, H., Haddenhorst, H. G., Jankowsky, W.: The history of migration in the Gifhorn trough (NW-Germany). Proc. 6th World Pet. Congr. Frankfurt, Sec. I, Paper 19, PD2 1963

Philippi, G. T.: On the depth, time and mechanism of origin of the heavy to medium-gravity naphthenic crude oils. Geochim. Cosmochim. Acta **41**, 33–52 (1977)

Poulet, M., Roucaché, J.: Influence du mode d'échantillonnage sur la composition chimique des fractions légères d'une huile brute. In: Advances in Organic Geochemistry 1966. Hobson, G. D., Speers, G. C. (eds.). London: Pergamon Press, 1970, pp. 155–179

Powell, T. G., McKirdy, D. M.: Crude oil composition in Australia and Papua-New Guinea. Am. Assoc. Pet. Geol. Bull. **59**, 1176–1197 (1975)

Prinzler, H. W., Pape, D.: Zur Entstehung und zum postgenetischen Verhalten der organischen Schwefelverbindungen des Erdöls. Erdöl u. Kohle **17**, 539–545 (1964)

Putscher, R. E.: Isolation of olefins from Bradford crude oil. Anal. Chem. **24**, 1551–1558 (1952)

Radchenko, O. A.: Geochemical regularities in the distribution of the oil-bearing regions of the world (in Russian). Leningrad: Nedra, 1965. Engl. trans.: Jerusalem: Israel Program Sci. Transl. 1968

Radke, M., Willsch, H., Welte, D. H.: Preparative hydrocarbon group type determination by automated medium pressure liquid chromatography. Anal. Chem. **52**, 406–411 (1980)

Radke, M., Welte, D. H., Willsch, H.: Geochemical study on a well in the Western Canada Basin: relation of the aromatic distribution pattern to maturity of organic matter. Geochim. Cosmochim. Acta **46**, 1–10 (1982)

Reed, W. E.: Molecular compositions of weathered petroleum and comparison with its possible source. Geochim. Cosmochim. Acta **41**, 237–247 (1977)

Rogers, M. A., Bailey, N. J. L., Evans, C. R., McAlary, J. D.: An Explanatory and Predictive Model for the Alteration of Crude Oils in Reservoirs in the Western Canada Basin. Proc. Int. Geol. Montreal, Canada, Mineral Fuels Sec. **5**, 48–55 (1972)

Rogers, M. A., McAlary, J. D., Bailey, N. J. L.: Significance of reservoir bitumens to thermal-maturation studies, Western Canada Basin. Am. Assoc. Pet. Geol. Bull. **58**, 1806–1824 (1974)

Rossini, F. D., Mair, B. J.: The work of the API research project 6 on the composition of petroleum. 5th World Pet. Congr. **5**, 223–245 (1959)

Rubinstein, I., Sieskind, O., Albrecht, P.: Rearranged sterenes in a shale: occurrence and simulated formation. J. Chem. Soc. Perkin Transactions I, 1833–1836 (1975)

Rubinstein, I., Spyckerelle, C., Strausz, O. P.: Pyrolysis of asphaltenes: a source of geochemical information. Geochim. Cosmochim. Acta **43**, 1–6 (1979)

Sachanen, A. N.: Hydrocarbons in petroleum. In: The Science of Petroleum. Brooks, B. T., Dunstan, A. E. (eds.). New York: Oxford University Press, 1950, **5**, pt 1, 77

Schmitter, J. M., Vajta, Z., Arpino, P. J.: Investigation of nitrogen bases from petroleum. In: Advances in Organic Geochemistry, 1979. Douglas, A. G., Maxwell, J. R. (eds.), 1980, pp. 67–76

Seifert, W. K.: Effect of phenols on the interfacial activity of crude oil (California). Carboxylic acids and the identification of carbazoles and indoles. Anal. Chem. **41**, 562–568 (1969)

Seifert, W. K.: Steroid acids in petroleum. Animal contribution to the origin of petroleum. In: Chemistry in Evolution and Systematics. Swain, T. (ed.). London: Butterworths, 1973, pp. 633–640

Seifert, W. K., Moldowan, J. M.: Applications of steranes, terpanes and monoaromatics to the maturation, migration and source of crude oils. Geochim. Cosmochim. Acta **42**, 77–95 (1978)

Seifert, W. K., Moldowan, J. M.: The effect of thermal stress on source-rock quality as measured by hopane stereochemistry. In: Advances in Organic Geochemistry, 1979. Douglas, A. G., Maxwell, J. R. (eds.), 1980, pp. 229–237

Seifert, W. K., Gallegos, E. J., Teeter, R. M.: Proof of structure of steroid carboxylic acids in a California petroleum by deuterium labelling, synthesis and mass spectrometry. J. Am. Chem. Soc. **94,** 5880–5887 (1972)

Seifert, W. K., Howells, W. G.: Interfacially active acids in a California crude oil. Isolation of carboxylic acids and phenols. Anal. Chem. **41,** 554–562 (1969)

Seifert, W. K., Teeter, R. M.: Identification of polycyclic aromatic and heterocyclic crude oil carboxylic acids. Anal. Chem. **42,** 750–758 (1970)

Silverman, S. R.: Migration and segregation of oil and gas. In: Fluids in Subsurface Environments. Young, A., Galley, J. E. (eds.). Memoir 4, Am. Assoc. Petr. Geol. Tulsa, 53–65 (1965)

Silverman, S. R.: Carbon isotopic evidence for the role of lipids in petroleum. J. Am. Oil Chem. Soc. **44,** 691–695 (1967)

Smith, H. M.: Correlation index to aid in interpreting crude oil analyses. U.S. Bureau of Mines, Tech. Paper 610 (1940)

Smith, H. M.: Crude oil: qualitative and quantitative aspects. The Petroleum World. U.S. Bureau of Mines Inf. Circ. 8286 (1966)

Smith, H. M.: Qualitative and quantitative aspects of crude oil composition. U.S. Bureau of Mines Bull. 642 (1968)

Smith, J. E., Erdman, J. G., Morris, D. A.: Migration, Accumulation and Retention of Petroleum in the Earth. Proc. 8th World Pet. Congr. Moscow. London: Applied Science Publ., 1971, **2,** pp. 13–26

Smith, N. A. C.: The interpretation of crude oil analyses. U.S. Bureau of Mines, Rept. Invest. 2806 (1927)

Speight, J. G.: The chemistry and technology of petroleum. Marcel Dekker Inc., New York, 1980, pp. 498

Speight, J. G., Moschopedis, S. E.: Some observations on the molecular "nature" of petroleum asphaltenes. Preprints. Div. Pet. Chem., Am. Chem. Soc. **24** (4), 910–923 (1979)

Strakhov, N. M.: Principles of Lithogenesis. Moscow: Izdat. Akad. Nauk SSSR, 1962 (in Russian). Transl. Consult. Bur. and Edinburgh and London: Oliver & Boyd, I, 1967; II, 1969

Suatoni, J. C., Swab, R. E.: Preparative hydrocarbon compound type of analysis by high performance liquid chromatography. J. Chromatog. **14,** 535–537 (1976)

Tewari, K. C., Li, N. C.: On molecular interactions involving coal-derived asphaltenes. Preprints, Div. Pet. Chem., Am. Chem. Soc. **24** (4), 982–989 (1979)

Tissot, B.: Problèmes géochimiques de la genèse et de la migration du pétrole. Rev. Inst. Fr. Pét. **21,** 1621–1671 (1966)

Tissot, B.: Connaissances actuelles sur les produits lourds du pétrole. Revue Inst. Français du Pétrole **36,** 429–446 (1981)

Tissot, B., Oudin, J. L., Pelet, R.: Critères d'origine et d'evolution des pétroles. Application à l'étude géochimique des bassins sédimentaires. In: Advances in Organic Geochemistry 1971. von Gaertner, H. R., Wehner, H. (eds.). Oxford: Pergamon Press, 1972, pp. 113–134

Tissot, B., Pelet, R., Roucaché, J., Combaz, A.: Utilisation des alcanes comme fossiles géochimiques indicateurs des environnements géologiques. In: Advances in Organic Geochemistry 1975. Campos, R., Goni, J. (eds.). Madrid, 1977

Tissot, B., Deroo, G., Hood, A.: Geochemical study of the Uinta Basin: formation of petroleum from the Green River formation. Geochim. Cosmochim. Acta **42,** 1469–1485 (1978)

Tissot, B., Deroo, G., Herbin, J. P.: Organic matter in Cretaceous sediments of the North Atlantic: contribution to sedimentology and paleogeography. In: Deep Drilling Results in the Atlantic Ocean: Continental Margins and Paleoenvironment. Talwain, M., Hay, W., Ryan, W. B. F. (eds.). Maurice Ewing Series, 3. Am. Geoph. Union, 1979, pp. 362–374

Tissot, B., Demaison, G., Masson, P., Delteil, J. R., Combaz, A.: Paleoenvironment and petroleum potential of Middle Cretaceous black shales in Atlantic Basins. Am. Assoc. Pet. Geol. Bull. **64,** 2051–2063 (1980)

Treibs, A.: Chlorophyll- und Häminderivate in bituminösen Gesteinen, Erdölen, Erdwachsen und Asphalten. Ann. Chem. **510,** 42–62 (1934)

Tynan, E. C., Yen, T. F.: Am. Chem. Soc. Div. Pet. Chem. Reprints A-89 (1967)

Welte, D. H.: Petroleum exploration and organic geochemistry. J. Geochem. Explor. **1,** 117–136 (1972)

Welte, D. H., Waples, D.: Über die Bevorzugung geradzahliger n-Alkane in Sedimentgesteinen. Naturwissenschaften **60,** 516–517 (1973)

Whitehead, E. V.: Molecular evidence for the biogenesis of petroleum and natural gas. In: Proc. Symp. Hydrogeochem. Biogeochem. Tokyo, 1970. Ingerson, E. (ed.). Washington: The Clarke Co., 1973, pp. 158–211

Williams, J. A., Winters, J. C.: Microbial alteration of crude oil in the reservoir. In: Symp. Pet. Transformations in Geol. Envir. Am. Chem. Soc. Nat. Meet. Preprints 14, 22–31 (1969)

World Energy Conference. Survey of Energy Resources 1980. Munich, Sept. 8–12, 1980, 357 pp.

Yen, T. F.: Present status of the structure of petroleum heavy ends and its significance to various technical applications. Am. Chem. Soc. Meet. Reprints F 102–114 (1972)

Yen, T. F.: Structural difference between petroleum and coal-derived asphaltenes. Preprints, Div. Pet. Chem., Am. Chem. Soc. **24** (4), 901–909 (1979)

Yen, T. F., Erdman, G. J., Pollack, S. S.: Investigation of the structure of petroleum asphaltene by X-ray diffraction. Analyt. Chem. **33,** 1587–1594 (1961)

Youngblood, W. W., Blumer, M.: Polycyclic aromatic hydrocarbons in the environment: homologous series in soils and recent marine sediments. Geochim. Cosmochim. Acta **39,** 1303–1314 (1975)

Part V
Oil and Gas Exploration:
Application of the Principles of Petroleum
Generation and Migration

Chapter 1
Identification of Source Rocks

The derivation of petroleum from nonreservoir rocks is a well-established fact. Rocks that are, or may become, or have been able to generate petroleum are commonly called *source rocks*. The presence of insoluble organic matter (kerogen) is a primary requisite for an active or a potential source rock.

A crucial test in the recognition of a petroleum source bed is the determination of its content of organic matter, both soluble (bitumen) and insoluble (kerogen). A second important step in source-rock identification is the determination of the type of kerogen and the composition of the solvent extractable hydrocarbons and nonhydrocarbons. Finally, from optical and/or physicochemical properties, evolutionary stages of kerogen can be determined. This concept is commonly referred to as the "maturity of source rocks". A combination of parameters allows the identification of the content and type of kerogen, and the level of maturity of a source rock.

The term source rock is applied irrespective of whether the organic matter is mature or immature. Source rock quality is defined in terms of amount and type of kerogen and bitumen and its stage of maturity. However, while it is possible to recognize whether a source rock is immature, and thus cannot yet have yielded oil, it is yet very difficult, except in special cases, to find out whether or not oil has migrated out of a mature source rock.

1.1 Amount of Organic Matter

The kerogen content of a sediment is normally determined by combustion of the organic carbon to CO_2 in an oxygen atmosphere after carbonate carbon has been removed by acid treatment. Such organic carbon analyses are conveniently performed on finely ground rock samples in a combustion apparatus utilizing an induction furnace and a thermal conductivity cell to measure the evolved CO_2. However, this procedure really determines only organic carbon content, and not the total organic matter or kerogen. In some laboratories extractable organic matter (bitumen) is removed prior to determination of organic carbon. The fraction of organic carbon corresponding to bitumen is usually not more than 0.1 or 0.2 of the total. However, less scattering is commonly observed on organic carbon values determined on bitumen-free samples. Although organic carbon analyses are relatively simple, it has repeatedly been observed that values obtained for the same samples in different laboratories may differ by about

Table V.1.1. Conversion factors for computation of total organic matter from organic carbon content

Stage	Type of kerogen			Coal
	I	II	III	
Diagenesis	1.25	1.34	1.48	1.57
End of catagenesis	1.20	1.19	1.18	1.12

$\pm 0.3\%$ C_{org}. To compensate for other elements (H, O, N, S) present in kerogen, the value found for organic carbon has to be multiplied by a conversion factor, because the elemental composition of kerogen depends on type and level of evolution. Forsman and Hunt (1958) determined conversion factors ranging from 1.07 for metamorphosed rocks up to 1.40 for nonmetamorphosed organic matter rich in oxygen. More detailed information on conversion factors can be derived from Table V.1.1 with respect to type of organic matter and stage of maturation. Conversion factors for certain peats may be as high as 2.0.

Average values of organic carbon for shale-type source rocks are presented in Table II.3.2. They are generally in the range of 2%. Average values of organic carbon for carbonate-type source rocks are in the range of 0.6%.

Not only the amount but also the type of organic matter is critical. To take two extreme examples, a rock with 2% $C_{org.}$ composed mainly of inertinite and rederived anthracite is a much poorer source rock than one containing 0.5% $C_{org.}$ of algal material, such as is found in form of Tasmanites. Another rather frequent situation is met in rock sequences where the background of 0.5% $C_{org.}$ represents a terrestrial input, whereas particular layers contain 2 to 5% $C_{org.}$, derived mainly from autochthonous plankton. An example of this kind is shown in Figure IV.3.1.

The lower limit of organic carbon for petroleum source rocks has been established primarily in an empirical manner. Ronov (1958) investigated some 26,000 samples of different ages and environments from oil and nonoil provinces. The critical lower limit for nonreservoir rock shale-type sediments in oil provinces turned out to be 0.5% organic carbon. The average organic carbon content in oil basins of the Russian platform was found to be almost three times as high as in areas outside the oil province. This lower limit of 0.5% organic carbon is consistent with more recent findings from generally acknowledged source rock units. Some laboratories, however, prefer to use a 1% $C_{org.}$ lower limit for detrital source rocks.

A rather comprehensive study of organic matter in limestones was carried out by Gehman (1962). Analyses of about 1400 ancient rocks from many parts of the world showed that ancient limestones contain much less organic matter than ancient shales. The average organic carbon content of limestones was 0.20%, and that of shales 0.94%. The average hydrocarbon content, however, was about 100 ppm and similar for ancient limestones and shales. The observations by

Gehman (1962) and corroborative data on generally acknowledged carbonate-type source rocks lead to the conclusion that 0.3% organic carbon is the lower limit for carbonate-type source beds. Average values of organic carbon for these carbonate-type source rocks are considerably higher and range above 0.6% (see also Table II.3.2).

These minimum values of organic carbon for potential source rocks, with 0.3% for carbonates and 0.5% for shales, should be considered only as necessary background values, rather than as positive indications for a source rock. Nevertheless, minimum values are of significance, not only because hydrocarbons in source rocks are generated from insoluble kerogen, but also because a critical level of hydrocarbons has to be reached before expulsion from a source rock is possible. Prior to expulsion, the specific adsorption capacity of a source rock for hydrocarbons has to be satisfied and, moreover, sufficient hydrocarbons for movement of a pressure-driven hydrocarbon phase have to be available. Finally, it should be pointed out that this minimum organic carbon value as a source rock criterion no longer applies to rocks in a very advanced stage of maturity, i.e., in a metagenetic stage when only dry gas is generated. At this stage 0.3 or 0.5% organic carbon may merely indicate a residual amount of organic matter that initially may have been more than twice as high.

1.2 Type of Organic Matter

A distinction between various types of sedimentary kerogen is essential for proper source rock appraisal, because different types of organic matter have different hydrocarbon potentials. The differences arise from variations in the chemical structure of the organic matter.

Remains of bacteria, phytoplankton, zooplankton and higher plants have been recognized as main contributors to kerogen in sediments. Major chemical differences in gross composition exist between organisms living in an aquatic environment and those living in subaerial (nonaquatic) environment. This distinction stems from the generalization that while terrestrial plants need structural support from polymers such as lignin, aquatic plants, supported by the surrounding water, need none. Therefore, the distinction between kerogen derived from aquatic (planktonic) organisms and terrestrial higher plants is of importance. Furthermore, the environments of transport and deposition and the resulting mode of preservation also influence the chemical composition of organic matter. These factors, however, are as yet difficult to detect.

Type and quality of kerogen can be differentiated and evaluated by optical microscopic and physicochemical methods. Optical methods on one hand allow a visualization of the kerogen and thus an appreciation of details, such as mixing from different organic sources. Most optical techniques, however, recognize only a fraction of the total kerogen of a sediment, a fraction which is not necessarily representative. On the other hand, physicochemical methods monitor the total organic matter usually in form of bulk analyses without recognizing different components or sources. The various techniques commonly applied have specific

1.2.1 Optical Microscopic Methods

Optical microscopy can only be used to describe that part of the organic matter of a sediment occurring in particles large enough to be visible in an optical microscope (about 1 μm or larger). However, also some finer "amorphous" organic matter can be identified, using transmitted light or fluorescence techniques. Definitive texts on optical microscopy are available for the specialist (Ottenjann et al., 1974; Teichmüller, 1974; Alpern (ed.), 1975; Stach et al., 1975). Generally three different techniques are applied for microscopic studies of source rocks:

1. investigation in reflected light of a polished surface of whole rock pieces,
2. investigation in reflected light of a polished concentrate of organic particles, isolated from a rock (organic particles are concentrated by means of acid maceration and/or density flotation),
3. investigation in transmitted light of organic particles, isolated from a rock, and strewn on a glass side.

To undertake all three types of studies is necessarily time-consuming. In the past usually in any laboratory only one approach has been used. The application, however, of two techniques seems advisable because the different techniques, to a certain extent, yield complementary information. Results stemming from a

Table V.1.2. Approximate equivalence of various terms used in kerogen description. (After Cornford, 1977)

Provenance	Terminologies			
Aquatic	Algal	Liptinite	Amorph.	Type I
	Amorphous			
				Type II
Sub-aerial (Terrestrial)	Herbaceous (fibrous)	Vitrinite	Humic	
	Woody (plant structure)			Type III
	Coaly (angular to sub-angular fragments)	Inertinite		Residual

1.2 Type of Organic Matter

given technique are internally consistent, although interrelating the terminologies of the different techniques is still in progress. Table V.1.2 lists approximate equivalence of various terms used in kerogen description. It seems that specialists concerned with microscopic studies of kerogen are currently propagating a new genetically oriented nomenclature for sedimentary organic particles. However, for historical reasons and in order not to proliferate nomenclature, it is probably best, for the time being, to adopt coal petrographic terms based on the maceral concept to describe dispersed sedimentary organic matter. The main groups of macerals in coal, i.e., liptinite, vitrinite, and inertinite have been described in Chapter II.8. A short description of their characteristic appearance in sediments can be given as follows:

a) Liptinite (e.g., algae, resins, spores etc.): identified by characteristic shape; high transmittance and intense fluorescence at lower levels of maturity and low reflectance compared to the other maceral groups.
b) Huminite, Vitrinite: angular to subangular particles, sometimes showing cell structure. Identified by moderate transmittance, intermediate reflectance, and usually absence of fluorescence.
c) Inertinite: angular, high reflectance, often exhibiting cellular outline or granular texture. No fluorescence, opaque in transmitted light.
d) Other particles such as solid bituminites, and various microfossils.

Reflected light microscopy allows those particles to be studied that take a polish. Thus particles of huminite, vitrinite and inertinite groups are easily identified (Stach et al., 1975). Amorphous or structureless organic matter is more difficult to identify in reflected light. However, a certain part of it, which is derived from liptinitic matter, can be recognized using incident UV excitation to produce fluorescence.

In reflected light the rank of coals is determined by measurement of the reflectance of the huminitic or vitrinitic particles (see Chap. II.8). The application to the evaluation of the maturity of petroleum source rocks is discussed later (Sect. 1.3.1). Next to maturity measurement, the main use of reflected light in studying kerogen is the differentiation of humic (sometimes also called herbaceous, woody or coaly) organic matter. The two maceral groups of vitrinite and inertinite are easily identified by virtue of their reflectance, shape, and internal structure. Vitrinite is of a medium grey appearance, whereas inertinites show a much brighter reflectance. The distinction between those two types of humic matter is of importance, because inertinites do not exhibit any oil potential, and perhaps a limited potential for gas, whereas vitrinite material has some potential for oil, and a definite gas potential.

Another important observation, only possible in reflected light, is the recognition of reworked opaque organic matter, derived from older sediments and redeposited. Eroded and redeposited coal fragments are readily recognized when exhibiting typical lamination, as so called bi- or trimacerite particles in which two or three maceral groups are combined (Fig. V.1.1). Other reworked organic particles, not stemming from coal seams, may show oxidation rims on the outer side of the particle due to intense contact with oxygen during reworking. These oxidation rims in reflected light show in most cases a brighter reflectance. In most

Fig. V.1.1. A kerogen in reflected light: large redeposited coal fragment (Trimacerite) together with homogeneous vitrinite particles *(grey)* and inertinite *(white)*. Kerogen concentrate of Carboniferous sandstone, Emsland, Germany. Reflected light, oil immersion. (Photo Hagemann)

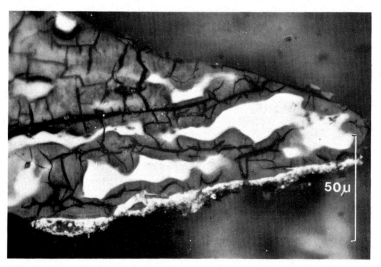

Fig. V.1.2. An altered inertinite particle, showing darkening due to weathering (oxidation). Reflected light, oil immersion. (Photo Hagemann)

cases, however, oxidation rims have a lower reflectance (Hagemann, 1977; Fig. V.1.2). Redeposited matter can also be identified by lower and redder liptinite fluorescence than expected for the rank of the vitrinite. Finally, if higher-rank matter is redeposited, a higher reflectance population of vitrinite can be found.

1.2 Type of Organic Matter

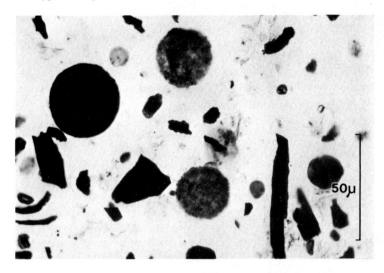

Fig. V.1.3. A kerogen in transmitted light: *large round objects* are algal bodies. *Smaller round objects* are palynomorphs. *Angular opaque particles* are vitrinitic plant remains. Kerogen concentrate of Upper Liassic shale (Pliensbachian), Luxemburg, Transmitted light. (Photo Hagemann)

Fig. V.1.4. A kerogen in transmitted light: small angular dark objects are vitrinitic plant particles. The *grey transparent object* in *upper right corner* is a fragment of cuticle; the *long dark perforated object* in *lower left corner* is fusinite. Kerogen concentrate of Upper Liassic shale (Pliensbachian), Luxemburg. Transmitted light. (Photo Hagemann)

Particles of higher reflectance are also observed in contemporaneous peats. They are obviously not reworked and could be the product of selective oxidation.

For studies in *transmitted light*, a similar microscope is employed to that used for the observation of rock thin sections or biological preparations. This approach

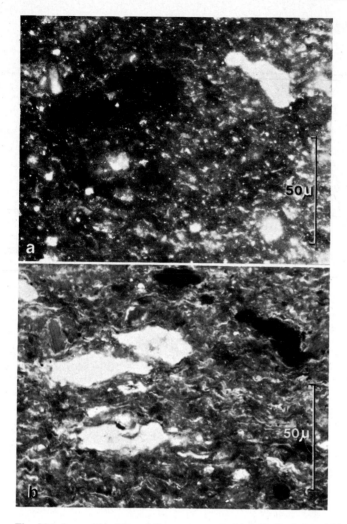

Fig. V.1.5a and b. Use of fluorescence to identify liptinites (Algae). A single spot on a polished Miocene shale from Probištip Jugoslavia, is shown (a) in reflected white light and (b) under UV excitation showing fluorescence. Three large liptinite particles *(middle, left)* not apparent under white light in (a), are identified by their bright fluorescence in (b). These liptinite particles are algal bodies. A huminite particle *(top right)* is best identified in (a) under white light. (Photos Hagemann)

to the study of kerogen using transmitted light is derived from palynological examinations (Combaz, 1964, 1980). Strewn grain concentrates of organic particles, recovered from crushed sedimentary rocks after acid maceration and/or density flotation, are used. The type of organic matter is identified, using the translucency and shape of particles, and in some cases their fluorescence. The transmitted-light technique is excellent for the identification of the structured particles of the important maceral group of liptinites, such as algal remains,

1.2 Type of Organic Matter

spores, pollen or cuticles (Fig. V.1.3 and V.1.4). "Amorphous" organic matter, without clear outlines, of no distinct shape, and without identifiable structure is also readily observed. It is moderately transparent and of a yellowish to brown, or dark-brown color. This type of organic matter is frequently termed "sapropelic". The amorphous organic matter is frequently considered as being derived from plankton and having a good petroleum potential. Although this is partly correct, it has to be kept in mind that there are other origins for amorphous organic matter which do not necessarily have a good petroleum potential. Precipitation or adsorption of dissolved or colloidal organic matter such as humic acids results in structureless, amorphous organic matter having a low petroleum potential. Furthermore, intense microbial reworking of organic debris may also destroy original morphological structures of various biological origins. This discussion is presented in Chapter II.4.8 and illustrated in Figure II.4.16. Based on these considerations, it is concluded that identification of amorphous organic matter, by using transmitted light, is by no means a proof of a good petroleum potential. Additional information has to be obtained from fluorescence studies and/or chemical analyses. In contrast to the amorphous organic matter is the opaque matter of the vitrinite and inertinite groups, which is not easily identifiable since only outlines are generally visible. The opaque organic matter is classified usually as "humic", although fibrous plant remains have been classified as "herbaceous" (Table V.1.2).

The results of transmitted light studies mainly give information on the type of organic particles present: Rogers (1979), Masran et al. (1981). Information on maturity can also be obtained. This matter is discussed later in more detail (Sect. 1.3.1).

Fluorescence studies can be carried out in either a transmitted or reflected light mode. Ultraviolet or blue light is used to excite the organic matter and the fluorescence spectrum is observed. Fluorescence can be used in three ways in the study of source rocks. Firstly, since it is diagnostic of the liptinite group of macerals, it is particularly useful to identify amorphous liptinitic matter (Fig. V.1.5a, b). There is often no other good method to visualize the finely dispersed particles of algal or microbial origin. It has also been proposed that oil-like components termed "exsudatinite", in the nomenclature of microscopy, can be seen by using fluorescence (Teichmüller, 1974). The second use of fluorescence is in determining the level of maturity of kerogen by estimating the intensity and color of fluorescence (Sect. 1.3.1). The third use of fluorescence, a quantitative measurement of the fluorescence spectrum, using a photometric microscope, has not yet been fully explored. It is also used for maturity studies.

Following the previous discussion, a general scheme of preferred microscopic techniques for identification of types of kerogen in source rocks is given in Table V.1.3.

The table summarizes some common terms, used for types of organic matter, most frequently found in source rock-type sediments and relates these to various microscopic techniques (reflected, transmitted and fluorescent light).

Microscopy cannot describe particles of kerogen smaller than about 1 μm. Amorphous organic matter composed of masses of smaller particles can be recognized without further differentiation by employing fluorescence techniques.

Table V.1.3. Scheme for subdividing organic matter in source rocks by microscopy, based on preferred microscopic techniques

Reflected light technique			
Particles inspected		Identification by	
		Reflectance	
Reworked organic particles bi-, trimacerated, oxidized matter	*Liptinite:* Resinite Sporinite Cutinite Alginite Liptodetrinite	Low	Internal structure: shape morphology
	Huminite: Vitrinite: Telinite Collinite	Medium	
	Inertinite: Fusinite Semifusinite Sclerotinite Macrinite	High	

(Humic matter spans the Huminite and Inertinite rows in the middle column area.)

Fluorescent light technique		
Transmitted light	Reflected light	Identification by
Algae, resin bodies, pollen, spores, cuticles	Alginite, resinite, sporinite, cutinite	Internal structure, shape, morphology color of fluorescence
	Distinction between fluorescent (liptinitic) and nonfluorescent (humic) amorphous organic matter	No distinct outline or shape color of fluorescence

In concentrates of the particulate organic matter, the juxtaposition of particles and good flat polish allow accurate identification of the various macerals. Polished whole rock preparations give a more representative view of the composition of the organic matter and relay information on the mode of deposition of organic matter.

When comparing the scheme for subdividing organic matter (see also Table V.1.2) in transmitted and reflected light with the types of kerogen identified by

1.2 Type of Organic Matter

Table V.1.3 (continued)

Particles inspected	Transmitted light technique	
	Identification by	
	Translucency	Internal structure, shape, morphology, color
Algae, resin bodies, pollen, spores, cuticles	Translucent	Internal structure, shape, morphology, color
Structured humic matter: herbaceous, woody, coaly	Weakly translucent – opaque	Shape
Amorphous matter	Generally translucent	No distinct outline or shape

physicochemical means, the following can be concluded: *Type-I* kerogen usually consists largely of structured algal material *(Botryococcus, Tasmanites)*, easily recognizable in a microscope (Fig. V.1.6a, b); it may also consist of structured algal remains mixed with amorphous matter. The latter possibly stems from intense microbial reworking of various kinds of organic matter.

Type-II kerogen usually consists of a large proportion of sapropelic organic matter with an amorphous appearance plus identifiable particles. Those particles include algal or other planktonic remnants and some land-derived plant material, such as pollen (Fig. V.1.7). Humic material may also be a minor component of type-II kerogen. *Type-III* kerogen usually consists of humic, opaque and coaly material (Fig. V.1.8). However, some of the amorphous material also has to be grouped with type-III kerogen. *Residual* kerogen may comprise coaly, inertinitic, and also some amorphous material.

Technical details of microscopic studies in reflected light: reflectance measurements and maceral determinations can be made on HCl-treated density concentrates (separated by heavy liquids) of particulate organic matter embedded in "epoxy" resin and polished, or alternatively on whole rock surface. The concentration procedure isolates typically between 20 to 60% of the total organic carbon in the form of organic particles. Nearly complete recovery of organic matter, as compared to total organic carbon, is only possible with HCl and subsequent HF treatment on finely ground rock. This latter method is used for kerogen isolation. The reflected-light technique for observing kerogen uses a coal petrographic microscope. The microscope is equipped with oil immersion objective lenses for reflected light, a photometer and devices for fluorescence and transmitted-light studies.

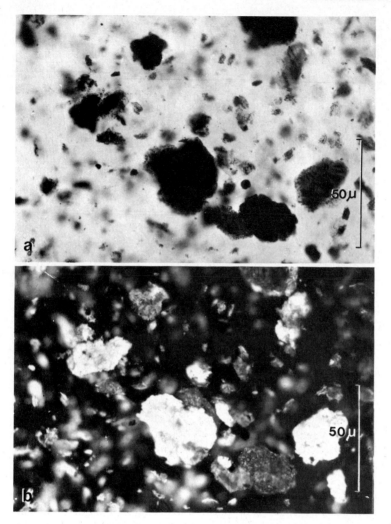

Fig. V.1.6a and b. Organic matter of the Probištip oil shale of Miocene age from Jugoslavia represents type I kerogen. In (a) a kerogen concentrate is shown in white transmitted light. The large dark objects are colonies of the algae Botryococcus. The smaller light particles represent unstructured amorphous organic matter, partly of algal origin. In (b) the same kerogen preparation is shown in reflected light using ultraviolet excitation to produce fluorescence. Strong bright fluorescence verifies the liptinitic nature of the algal material shown in (a). Some internal structure is now visible. The non-fluorescent elongated particle, at the bottom right of the figure is pyritized algal material. The amorphous organic matter shows generally less fluorescence than the structured algal material. (Photos Hagemann)

Reflectance is measured with polarized light of 546 nm wavelength, using microscopes with a photometer head. Standards are used, which give a specific reflectance in oil. The effective measuring aperture is usually 3 μm or greater. The effective magnification should be greater than 400. Fluorescence observation is

1.2 Type of Organic Matter

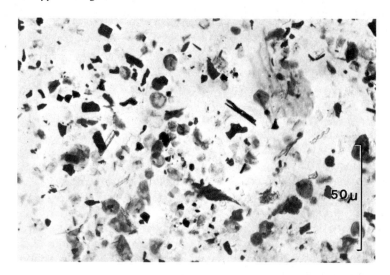

Fig. V.1.7. Type-II kerogen, composed of sapropelic organic matter (amorphous) mixed with small algae and spores, and some huminitic particles *(angular, dark fragments)*. Sapropelic matter appears to be very faint due to the low degree of coalification of this sample. Kerogen concentrate of Sinemurian shale, Luxemburg. Transmitted light. (Photo Hagemann)

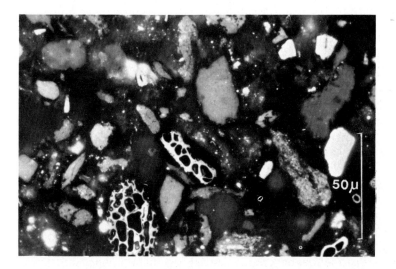

Fig. V.1.8. Type-III kerogen composed of terrestrial plant particles of huminitic and inertinitic nature. Some particles show distinct cellular structure. Kerogen concentrate of Pliocene shale from the Black Sea. White reflected light oil immersion. (Photo Hagemann)

generally carried out with a 100-watt mercury lamp (wavelength of light used for excitation: emission peaks 365–435 nm) with a diachronic illuminator and appropriate barrier filters.

1.2.2 Chemical Methods Based on Kerogen Analysis

Chemical methods for characterizing the main types of organic matter are mostly based on kerogen, either separated from the mineral fraction, or not separated. Additionally, the study of extractable bitumen can also provide information on the type of organic matter. Most chemical parameters (e. g., elementary analysis of kerogen, IR absorbance, or naphthene abundance) are influenced by the type and also by the maturation of the organic matter. Thus, it is necessary to distinguish the respective influence of the original organic type versus degree of maturation. This cannot be achieved by methods which measure a single parameter, like the carbon ratio ($C_R:C_T$). The distinction can be based only on methods using several independent parameters, such as elemental analysis, or based on a crossplot of different methods.

The fundamental properties of kerogen described in Chapter II.4 are the basis for subdividing organic matter. Their study requires a preliminary elimination of the mineral fraction by acid (HCl, HF) treatment. The efficiency of this treatment varies considerably from one rock to another. Pyrite is the major mineral present in kerogen concentrates and its residual abundance may range from 0 to more than 50%. It is generally considered that a concentrate cannot be used for analysis if it contains more than 30% pyrite or other minerals. A typical difficulty is the occurrence of framboids, where pyrite microcrystals are coated by a film of organic matter (Sect. II.4.3).

Elemental analysis of kerogen, when plotted on a van Krevelen diagram (Fig. II.4.11) provides an immediate reference to the three main types of kerogen as described in this book. However, the characterization becomes difficult in the wet and dry gas zones (beyond the 1.3–1.5% level of vitrinite reflectance), where many properties of the different kerogens become alike.

Infrared spectrophotometry provides information on the occurrence and abundance of the various functional groups in kerogen and also paraffinicity or aromaticity (Fig. II.4.7). In particular the absorption bands provide a comparative evaluation of the petroleum potential of different source rocks. This evaluation is based on the respective intensity of the absorption bands related to aliphatic CH_2, CH_3 groups (source of hydrocarbons) and to polyaromatic nuclei (inert part of kerogen; Fig. II.7.3).

Electron spin resonance (ESR) applied to kerogen may also provide some information about the type of organic matter. However the main influence on ESR results from thermal evolution, and therefore this technique is treated later (Sect. 1.3.2).

Carbon isotope distributions in kerogen also depend on the origin and chemical structure of organic matter. It has been observed that terrestrial organic matter is enriched in the lighter isotope, ^{12}C, as compared to marine plankton; the difference is 5 to 10‰ (see Sect. I.2.5). In ancient sediments the $\delta^{13}C$ of land-derived kerogen is usually 3 to 5‰ lower than the $\delta^{13}C$ of marine-derived kerogen.

In very advanced stages of maturation, e. g., beyond 1.5% vitrinite reflectance, identification of the original type of organic matter becomes difficult, because the chemical compositions of the different kerogens become progressively similar.

1.2 Type of Organic Matter

However, *electron microdiffraction* is still able to differentiate between the original types I, II or III of the organic matter. The discriminative parameter is the size of aggregates produced in kerogen by heating at 1000°C. These aggregates appear as clouds where individual aromatic platelets and stacks become parallel. The size of the aggregates is 80 Å or less in type III, ca. 200 Å in type II, and 500 or even 1000 Å in type-I kerogen (Oberlin et al., 1974).

1.2.3 Pyrolysis Method Based on Whole Rock Samples

The main difficulty in applying chemical methods on kerogen to routine work is the rather lengthy and costly character due to the preliminary step of kerogen isolation. This represents a restriction on the number of kerogen concentrates available for analysis. A quick method is needed to provide the same type of information from the original core or cuttings, and requiring no special preparation other than grinding the rock. The data obtained from the study of many kerogens have enabled Espitalié et al. (1977) to develop a standard pyrolysis method of source rock characterization and evaluation (Rock-Eval). The method uses a special pyrolysis device sketched in Figure V.1.9. The sample (ca. 100 mg) is progressively heated to 550°C under an inert atmosphere, using a special temperature program (Fig. V.1.10). During the assay, the hydrocarbons already

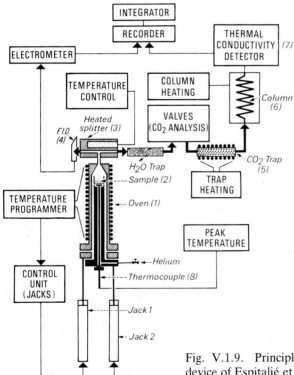

Fig. V.1.9. Principle of the Rock-Eval pyrolysis device of Espitalié et al. (1977)

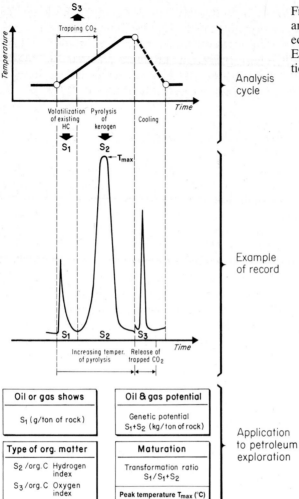

Fig. V.1.10. Cycle of analysis and example of record obtained by the pyrolysis method of Espitalié et al. (1977). Application to petroleum exploration

present in the rock in a free or adsorbed state are first volatilized at a moderate temperature. The amount of these hydrocarbons (S_1) is measured by a flame ionization detector (FID). Then, pyrolysis of kerogen results in generating hydrocarbons and hydrocarbon-like compounds (S_2) and oxygen-containing volatiles, i.e., carbon dioxide (S_3) and water. The volatile compounds generated are split into two streams passing respectively through a FID (detector) measuring S_2 and a thermal conductivity detector measuring S_3. An adequate temperature program allows a good separation of S_1 and S_2 peaks on the FID detector. However, the measurement of S_3 is limited to a convenient temperature window in order to include the main stage of CO_2 generation from kerogen, and to avoid other sources of CO_2 (such as decomposition of carbonates, and particularly of siderite, which is the most labile carbonate). A fourth parameter is the temperature T_{max} corresponding to the maximum of hydrocarbon generation during

1.2 Type of Organic Matter

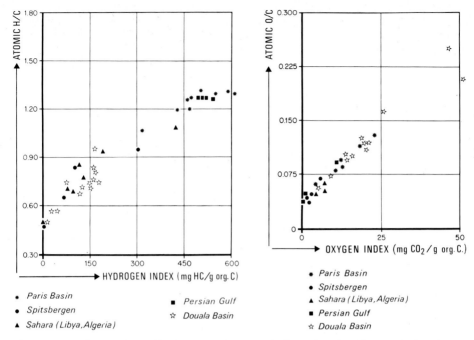

Fig. V.1.11. Correlation of hydrogen and oxygen indices measured by pyrolysis of rock with H/C and O/C ratios, respectively, measured by elemental analysis of kerogen (Espitalié et al., 1977)

pyrolysis. However, it is used mostly for evaluation of the maturation stage, which is treated later in Section 1.3.2. Figure V.1.10 shows a typical record with indication of S_1, S_2, S_3 and T_{max}.

The *type* of the *kerogen* is characterized by two indices: the *hydrogen index* (S_2/organic carbon) and the *oxygen index* (S_3/organic carbon). The indices are independent of the abundance of organic matter and are strongly related to the elemental composition of kerogen. In particular there is a good correlation between hydrogen index and H/C ratio, and between oxygen index and O/C ratio, respectively (Fig. V.1.11). Thus the two indices can be plotted in place of the normal van Krevelen diagram, and interpreted in the same way. Rocks containing typical kerogens of types I, II and III have been subjected to pyrolysis, and their hydrogen and oxygen indices are plotted in Figure V.1.12. The three main evolution paths are generally distinct, and can be readily recognized.

A semi-quantitative evaluation of the genetic potential can be achieved by using pyrolysis. The quantity S_1 represents the fraction of the original genetic potential which has been effectively transformed into hydrocarbons. The quantity S_2 represents the other fraction of the genetic potential, i.e., the residual potential which has not yet been used to generate hydrocarbons. Thus $S_1 + S_2$, expressed in kg hydrocarbons per ton of rock, is an evaluation of the *genetic potential*[12].

---

12 This evaluation accounts for two aspects, abundance and type of the organic matter

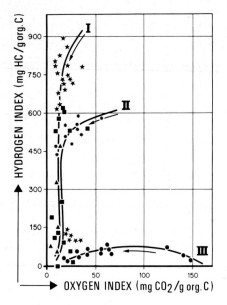

Fig. V.1.12. Classification of the source rock types by using hydrogen and oxygen indices. This diagram is readily comparable to the van Krevelen diagram plotted from elemental analysis of kerogen (Espitalié et al., 1977)

★ Green River shales
• Lower Toarcian, Paris Basin
▲ Silurian-Devonian, Algeria-Libya
● Upper Cretaceous, Douala Basin
■ Others

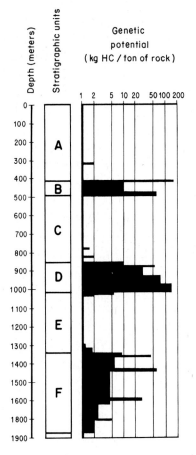

Fig. V.1.13. Example of a well log of the genetic potential. Genetic potential is plotted as a function of depth and allows a quantitative evaluation of the source rocks. Units A, C, and E have no hydrocarbon potential. Units B and D are good to excellent source rocks. Unit F shows a fair potential for hydrocarbons, decreasing to low potential at the bottom of unit F. Some individual layers have a much higher potential

1.2 Type of Organic Matter

The following classification of the source rocks is suggested:
- lower than 2 kg t^{-1} (2000 ppm): no oil source rock, some potential for gas,
- from 2 to 6 kg t^{-1} (2000 to 6000 ppm): moderate source rock,
- above 6 kg t^{-1} (6000 ppm): good source rock.

Exceptionally, some values as high as 100 or 200 kg t^{-1} may be observed. Such rocks contain abundant organic matter of type I or II and may provide either an excellent source rock, if the burial depth is sufficient, or an oil shale, if the burial depth is shallow. Figure V.1.13 shows a well log of genetic potential plotted versus depth.

Other techniques of pyrolysis have been proposed to distinguish the types of organic matter and to evaluate the source rock potential. However, their main limitation ist the difficulty to distinguish an immature rock with a low genetic potential from a more mature rock with an originally high genetic potential. The previously described method of Espitalié et al. (1977) easily allows this distinction by plotting oxygen index versus hydrogen index. Without the oxygen index, the interpretation remains dubious, unless an independent evaluation of the maturation is carried out (e.g., vitrinite reflectance). An example is the "carbon ratio" (C_R:C_T), which was originally proposed by Gransch and Eisma (1970) to evaluate the degree of maturation of a source rock; it is discussed in Section V.1.3.2. Later it was used as a tool for characterizing the types of organic matter. The residual carbon C_R, after pyrolysis at 900 °C under inert atmosphere is related to the abundance of polyaromatic nuclei. The volatile fraction C_T minus C_R (C_T being the total organic carbon) is related to the abundance of aliphatic chains and oxygen containing functional groups which are lost during pyrolysis. Thus the low C_R:C_T values truly correspond to immature source rocks containing aromatic-poor kerogen of type I or II. Their genetic potential is high. In contrast the high C_R:C_T values may be obtained either from the same good source rock after maturation, or from rocks containing aromatic-rich type-III kerogen, even at low stages of maturation. Thus the interpretation remains ambiguous, unless it is completed by another method.

Pyrolysis-fluorescence is a quick test which can be easily performed at a well site (Heacock and Hood, 1970). A weighed amount of core or cuttings is heated in a test tube. Bitumen which has condensed along the cold parts of the test tube is then dissolved in a few milliliters of solvent. The fluorescence of the solution is measured by using a simple, filter-type UV fluorescence equipment. The fluorescence intensity PF shows a good correlation with the parameter S_2 of the Rock-Eval pyrolysis technique. This is quite understandable as the same volatile fraction (hydrocarbons and related compounds) is involved in both assays. Thus, PF can be used for a quick evaluation of the residual genetic potential. Again, the lack of another independent parameter makes the true distinction of the kerogen types difficult. A cross plot of PF versus total organic carbon may provide some information in this respect.

1.2.4 Chemical Indicators Based on Bitumen

Extractable bitumen includes geochemical fossils which carry information about the origin of the organic matter. They can be used to identify the type of organic matter (e. g., type II mostly marine and type III mostly terrestrial) and, in turn, to make some appraisal of the source rock potential. However, there are limitations to this reasoning. First, the free molecules of bitumen amount to only a small part of the total organic matter. Furthermore, the content of the various classes of compounds (e.g., *n*-alkanes, acids, sterols, etc.) may vary widely from one source of organic matter to another. For example, long-chain *n*-alkanes with an odd number of carbon atoms are important only in higher plants, whereas they are negligible in plankton. Thus a marine sediment containing 90% marine plankton plus 10% higher plant debris brought from the continent usually bears a typical terrestrial fingerprint, if high molecular weight *n*-alkanes are considered. Another source of difficulty is the mobility of bitumen, i. e., it may migrate also in source rock-type sediments and not only in typical reservoir rocks. Such accumulation is marked by abnormally high bitumen to organic carbon ratios (above 200 mg:g of organic carbon). Under such circumstances the original pattern of fossil molecules may be changed by an input of bitumen with a different distribution.

From these considerations it is generally better to base the evaluation of the type of organic matter and of the source rock potential on data obtained from kerogen. Kerogen usually amounts to 80% or more of the total organic matter, and it is unable to migrate. Thus, information based on kerogen encompasses the major part of the organic matter, and is definitely representative of the rock under consideration.

The nature and occurrence of geochemical fossils is described in Chapter II.3. Their significance in terms of geological environments is discussed in Chapter IV.3. For example, the sources of organic material are clearly outlined in the Gabon and Magellan basins by the distribution of saturated hydrocarbons. In the coastal basin of Gabon (Fig. IV.3.1), there is a background of organic matter, at a level of about 0.5% organic carbon, permanently incorporated in sediments. This organic matter is of terrestrial origin and its petroleum potential is low. However, some layers contain a larger proportion of organic matter (from 1 to 5%) with a high petroleum potential. In fact, these layers are the "effective" source rocks of the oil fields, although they may amount to less than half of the total thickness of the formation. In the Magellan Basin of southern most America (Fig. IV.3.3), the Springhill formation has been deposited in several small basins, separated by structural highs. The environment changes from marine to nonmarine, depending on the local conditions. In some areas, where the sedimentation is marine, the organic matter is mainly derived from plankton. There the petroleum potential is high, and the crude oils belong to the normal paraffinic or naphthenic types. Elsewhere, sedimentation is nonmarine, and the organic matter is mainly of terrestrial origin, with a high proportion of long paraffinic carbon chains derived from plant waxes. The related petroleum potential is comparatively lower and crude oils belong to the high-wax type.

1.3 Maturation of the Organic matter

Thermal evolution of the source rock, during diagenesis, catagenesis and metagenesis, changes many physical or chemical properties of the organic matter. These properties may be considered as indicators for maturation. The parameters most commonly used in petroleum exploration are optical examination of kerogen, physicochemical analysis of kerogen, and chemical analysis of extractable bitumen. Hood and Castaño (1974) compared the various techniques of measuring the thermal evolution and proposed a numerical scale for it called LOM (Level of Organic Metamorphism; Hood et al., 1975).

1.3.1 Optical Indicators of Maturity

Examination in transmitted or reflected light, with or without fluorescence, provides different types of information on the thermal evolution of organic matter (see also Sect. V.1.2.1). The main advantage of optical techniques is to discriminate and locate organic matter of different origins and to measure their respective rank of evolution. For example, a contribution of older kerogen fragments, reworked from ancient rocks, can be identified by its higher level of reflectance or carbonization. The proportion of reworked kerogen may not be accounted for by purely chemical techniques, which measure bulk properties of the mixture of contemporaneous and reworked organic matter. Furthermore, there might be some advantage in locating the measured particles.

a) Carbonization of Palynomorphs

Observation of palynological concentrates in *transmitted light* is the basis for several scales of maturation. They use the progressive changes of color and/or structure of the spores, pollen and other microfossils. The color is originally yellow, then becomes orange or brown–yellow (diagenesis), brown (catagenesis), and finally black (metagenesis). Alteration of the structural features occurs mostly during catagenesis and metagenesis. For instance, spores and pollen at the level of metagenesis are no longer suitable for stratigraphic investigations. The original observations of spores and pollen carbonization were made fifty years ago in coal, then in other types of rocks. A review and a systematic investigation of the carbonization of spores and pollen grains, based on observations and experimental work, was made by Gutjahr (1966). He used a color scale, and also a standard procedure to measure the absorption of light. The author emphasized that different types of spore or pollen grains show different absorption values at low levels of maturation. This fact is due to the original difference in wall thickness, and possibly to differences in chemical composition. With increasing levels of maturation, differences become less pronounced and disappear in the wet gas and dry gas zones.

Correia (1967) and Staplin (1969) proposed two scales in which both color and structure alteration were used; they are respectively named state of preservation

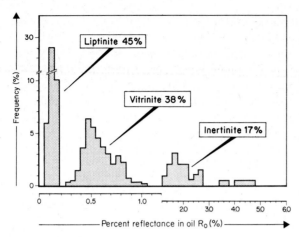

Fig. V.1.14. Histogram showing the distribution of the reflectance of organic matter, measured by reflected light on a polished section. The various constituents — liptinite, vitrinite and inertinite — have different ranges of reflectance. (Middle Liassic shale, Luxemburg; modified from Hagemann, 1974)

of palynomorphs, and thermal alteration index. Five stages are defined by color and alteration of shape and ornamentation. Stage 1 refers to fresh yellow material, whereas stage 5 corresponds to barely identifiable remnants of the zone of metamorphism (Fig. V.1.32).

b) Vitrinite Reflectance

Reflectance of coal macerals measured in *reflected light* has long been used to evaluate coal ranks. A relationship has been established between huminite-vitrinite reflectance and other properties characterizing coal ranks (Alpern, 1969; Teichmüller, 1971; McCartney and Teichmüller, 1972). Therefore, *vitrinite reflectance is now considered as one of the best parameters to define coalification stages*.

Reflectance measurements have been extended to particles of disseminated organic matter (kerogen) occurring in shales and other rocks: Vassoevich et al. (1969), Lopatin (1969), Teichmüller (1971), Hood and Castaño (1974), Dow (1977). Histograms showing the frequency distribution of reflectance are established. They usually show several groups of reflectance values corresponding to the various constituents, or macerals of the kerogen concentrate. In the same sample, the reflectance increases from liptinite particles to vitrinite and finally to inertinite (Fig. V.1.14). Both liptinite and vitrinite reflectances increase with thermal evolution (Fig. V.1.15). However, only huminite or vitrinite particles are generally used for reference to the coalification scale, because vitrinite is the main maceral of humic coals. Therefore the mean random reflectance of vitrinite is preferred to other particles. However, there are situations where several groups of vitrinite particles, with different reflectance, are present. It is frequently considered that only the group of particles showing the lowest reflectance truly represents the situation of the autochthonous material. Other vitrinite particles — with higher reflectance — are considered to be reworked: Figure II.8.11. Although this practice is frequently justified, there are geological situations

1.3 Maturation of the Organic Matter

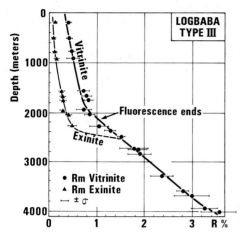

Fig. V.1.15. Compared increase of the huminite/vitrinite and exinite/liptinite macerals with burial in a Logbaba well (Douala Basin, Cameroon). The standard deviation σ is shown for each sample, and also the level where fluorescence ends (Alpern et al., 1978)

where several groups of vitrinite are present and reworking is unlikely. Thus, it can be considered that, according to the source material and conditions of early diagenesis, there may be different types of vitrinite, which may react in different ways to thermal evolution.

From correlation of huminite-vitrinite reflectance with other parameters of source rock maturation, and also with the occurrence of oil and gas fields, the following stages can be distinguished (for which $\overline{R}_o$ is the mean reflectance in oil):

a) $\overline{R}_o < 0.5$ to 0.7%: diagenesis stage, source rock is immature,
b) 0.5 to $0.7\% < \overline{R}_o < $ ca. 1.3%: catagenesis stage, main zone of oil generation; also referred to as oil window,
c) ca. $1.3\% < \overline{R}_o < 2\%$: catagenesis stage, zone of wet gas and condensate,
d) $\overline{R}_o > 2\%$: metagenesis stage; methane remains as the only hydrocarbon (dry gas zone).

There are no sharp boundaries of the oil zone, as oil represents a complex assemblage of a wide variety of components which are generated at different rates. Furthermore, kerogen has no uniform structure. Instead, the relative abundance of chemical bonds that differ in strength varies with kerogen type. A high relative abundance of weaker bonds, such as some heteroatomic bonds frequent in type-II kerogens, means that onset of oil generation occurs at a relatively early stage of the maturation process. On the contrary, a high relative abundance of stronger bonds, such as C–C bonds of the aliphatic network constituting the bulk of type-I kerogen, means that onset of oil generation occurs at a relatively late stage of the maturation process.

This effect is stronger in the lower than in the higher level of maturation. For example (Fig. V.1.16) the beginning of the oil zone corresponds to a reflectance of about 0.5% in the Paris Basin (kerogen type II) as compared with 0.7% in the Uinta Basin (kerogen type I). The maximum or peak of the oil generation curve (Fig. V.1.33) occurs at about 0.9% reflectance in the Douala Basin (kerogen type III) and about 1.1% in the Uinta Basin (kerogen type I), whereas it seems to be reached at 0.8% in the Paris Basin (kerogen type II).

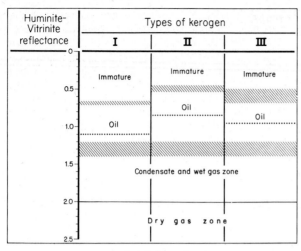

Fig. V.1.16. Approximate boundaries of the oil and gas zones in terms of vitrinite reflectance. Boundaries may change slightly according to the time-temperature relationship, and also to the mixing of various sources of organic matter

The beginning of the wet gas zone likewise has no sharp boundary. It generally occurs between 1.2% and 1.4%. However, the lower boundary at 2% has more general value, because similarities in chemical composition of kerogen become stronger at that stage. This convergence is shown by the H/C and O/C ratios of the various types of kerogen (Fig. II.5.1), and also by other physical analyses of kerogen.

Huminite-vitrinite reflectance is the most widely used optical technique for determining the maturity of a source rock. It is probably the best optical tool presently available for that purpose. However, there are several limitations to this technique, which should be kept in mind for a proper interpretation:

1. One of the parameters used for identification of syngenetic vitrinite particles is its reflectance. This reflectance, however, is subsequently measured as a maturity indicator on the vitrinite particles, which have been selected, among others, on the basis of their reflectance.

2. Vitrinite is abundant in type-III kerogen, and it can be used excellently. Vitrinite is only moderately abundant, or even scarce, in type-II kerogen, and mostly absent in type I: this situation could be an important limitation for the use of reflectance (Alpern, 1980). Furthermore, marine or lacustrine kerogens (types I and II) may also contain gelified particles resembling vitrinite although their chemical composition and the evolution of reflectance may be different (Alpern et al., 1978). For instance, Alpern (1980) found in the lower Toarcian shales of the Paris basin, besides a predominant maceral of the liptinite group, three gelified macerals with different reflectances. Two of them are definitely syngenetic, with an average reflectance of 0.25 and 0.6% respectively and their relative abundance changes with the depositional environment, from the coast line to the open sea: Figure V.1.17.

3. The reflectance intervals shown above for the oil and gas zones are only approximate. There have been many discussions about the boundaries of the various zones, in terms of reflectance: Dow (1977), Heroux et al. (1979), Alpern

1.3 Maturation of the Organic Matter

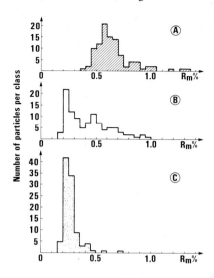

Fig. V.1.17. Reflectance of gelified macerals in the Lower Toarcian shales of the eastern Paris Basin. In addition to a predominant maceral of the liptinite group (reflectance 0.10 to 0.15% not represented here), two populations of syngenetic gelified macerals (average R_m 0.25 and 0.6% respectively) are present. The depositional environment changes from coastal *(A)* to open sea *(C)*. Locations *A, B, C* are different, but close to locations G2, G6, and G10 shown in Figure V.1.25. (After Alpern, 1980)

(1980), Robert (1980), Teichmüller (1982). In fact the observations reported above and also theoretical considerations (Tissot and Espitalié, 1975) indicate that the rate of transformation of the organic matter during burial depends on the chemical composition. Obviously the vitrinite maceral has a composition rather comparable to type-III kerogen. Its chemical reaction at increasing temperature is similar to that observed for type-III kerogen. There is no reason for different kerogen, like types I and II, to have the same rate of transformation as for type-III kerogen or vitrinite. Thus, different source rocks with the same thermal history should have reached the same level of vitrinite reflectance, but not necessarily the same stage of evolution, in terms of oil or gas generation zones. Furthermore, it should be realized that different thermal histories, e.g., heating rates, may result in the same level of vitrinite reflectance but in different amounts and types of hydrocarbons generated. This is again due to the difference in the chemical structure between vitrinite and the main hydrocarbon generating kerogens, and also because of the difference in the kinetics of the hydrocarbon-generating processes.

The *rank* of the associated humic *coal* is another aspect which has been used for studying petroleum generation. It is closely related to vitrinite reflectance, as the rank is determined either by this technique or by other techniques, such as volatile matter or carbon content which can be correlated with vitrinite reflectance. Coal rank has been used by the previously cited authors using vitrinite reflectance and also by Brooks (1970), Hacquebard and Donaldson (1970) and Demaison (1975).

c) Fluorescence

Fluorescence of various liptinite constituents is induced by blue or UV light. The visible light emitted by kerogen in response to excitation may be characterized by its intensity and color spectrum (van Gijzel, 1971; Alpern et al., 1972; Ottenjann

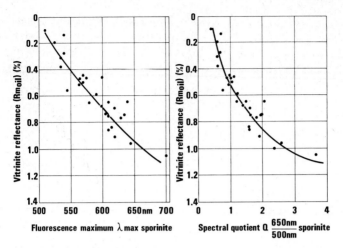

Fig. V.1.18. Observed relationship between vitrinite reflectance and two parameters characterizing fluorescence of the sporinite macerals in a series of coals (Teichmüller and Durand, 1983)

et al., 1974). Fluorescence is intense in shallow immature samples and decreases during diagenesis and most of catagenesis. It has completely disappeared at the end of the oil zone. Furthermore the color spectrum is progressively changed: the wavelength range of the fluorescence light moves from yellow toward red with increasing catagenesis. These changes vary quantitatively for the different liptinite macerals (sporinite, cutinite, fluorinite) but remain qualitatively the same (Teichmüller and Durand, 1983). Furthermore a good relationship was observed between fluorescence of coal macerals, vitrinite reflectance and pyrolysis data: Figure V.1.18. Ottenjann et al. (1974) proposed the ratio Q = intensity at 650 nm:intensity at 500 nm of sporinite fluorescence to measure this change. Fluorescence is presently getting to be a routine tool, but more work is still needed to clarify various aspects of the observations. However, reflectance and fluorescence cover different parts of the maturation scale and may be complementary. In particular, most of the decrease in fluorescence intensity occurs over the range where vitrinite reflectance is low and varies little, i.e., the beginning of the oil generation zone.

1.3.2 Pyrolysis Methods for Measuring Maturity

Many methods utilizing some type of pyrolysis have been proposed in order to characterize the rank of evolution of the organic matter. Only the most recent ones are discussed in detail, whereas the others are briefly reviewed. In general, pyrolysis is conducted under an inert atmosphere (nitrogen, helium) with a preselected rate of heating (approximately 10–50°C min^{-1}). The pyrolysis yields three main groups of compounds equivalent to S_1, S_2, and S_3 of the method described in Section 1.2.3 for characterizing the types of organic matter:

1.3 Maturation of the Organic Matter

- hydrocarbons already present in the rock (S_1) which are volatilized by moderate heating at 200 to 250°C,
- hydrocarbons and related compounds (S_2) generated at higher temperatures by pyrolysis of insoluble kerogen,
- carbon dioxide (S_3) and water.

The pyrolysis method of Espitalié et al. (1977), as presented in Section 1.2.3, utilizes the three parameters S_1, S_2 and S_3. Other methods measure only the parameters S_1 and/or S_2. In addition, the temperature T_{max} is recorded at the maximum of hydrocarbon generation during pyrolysis (Fig. V.1.10).

Barker (1974), Claypool and Reed (1976) and Espitalié et al. (1977) have shown that two indices are of great interest in characterizing the rank of evolution: the ratio $S_1:(S_1 + S_2)$ and the temperature T_{max} (Figs. V.1.10 and V.1.19). In absence of migration the $S_1:(S_1 + S_2)$ ratio is an evaluation of the *transformation ratio* defined in Chapter II.7. The continuous increase of this ratio as a function of depth (Fig. V.1.19) makes it a valuable index of maturation. In addition to measuring maturation, the values S_1 and $S_1:(S_1 + S_2)$ can be used for a *quantitative evaluation* of the hydrocarbons generated. The result is conveniently

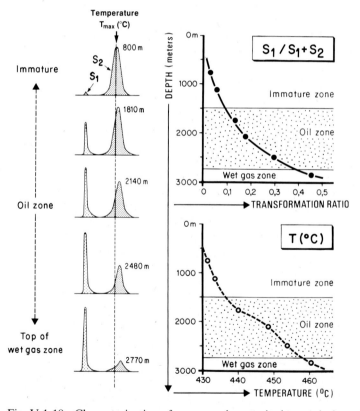

Fig. V.1.19. Characterization of source rock maturity by pyrolysis methods. Transformation ratio and/or peak temperature T_{max} may be used as indicators of thermal evolution. These parameters are defined on Figure V.1.10. (Modified after Espitalié et al., 1977)

expressed in grams hydrocarbons per ton rock and in grams hydrocarbons per kilogram organic matter, respectively. However, such evaluation cannot be made from isolated measurements, due to the possibility of minute accumulations (microreservoirs) occurring in source rocks, which would alter the figures. Therefore, it is necessary to plot the variation of the transformation ratio $S_1:(S_1 + S_2)$ versus depth in order to identify such accumulations by their anomalously high values compared with the average curve.

The temperature T_{max} also increases progressively, but the numerical scale changes with the rate of heating. Thus the T_{max} scale has to be calibrated according to the individual laboratory procedure. This temperature index is not affected by migration phenomena. Furthermore, there is a good correlation of the two indices $S_1:(S_1 + S_2)$ and T_{max} in a given series. The temperature index T_{max} can also be correlated with other maturation parameters, such as vitrinite reflectance. Figure V.1.20, for instance, shows the very good relationship between T_{max} and vitrinite reflectance for type-III kerogen and humic coals. The temperature index T_{max} is considered to be among the most reliable ones for characterizing thermal evolution. It is particularly valuable in the case of marine or lacustrine kerogen (types I and II) where vitrinite is often scarce or absent.

The questions whether these maturation indices $S_1:(S_1 + S_2)$ and T_{max} are influenced by the type of kerogen may be solved by use of the Rock-Eval pyrolysis method of Espitalié et al. (1977). Due to simultaneous availability of hydrogen and oxygen indices, the identification of the kerogen types is carried out at the same time. From such studies it has been found that the ratio $S_1:(S_1 + S_2)$ is fairly independent of the type of organic matter during catagenesis and metagenesis. The temperature T_{max} is influenced by the type of organic matter during the diagenetic stage and the beginning of catagenesis. It is lower in the terrestrial kerogen of type III and higher in the marine or lacustrine types I and II. However, the T_{max} values are almost equivalent for the different kerogen types in the peak zone of oil generation and later in the gas zone.

Older methods for determining maturity, such as the carbon ratio, are affected by the kerogen type. For example, Gransch and Eisma (1970) compared the total organic carbon C_T with the residual carbon C_R after pyrolysis at 900° (C_R is

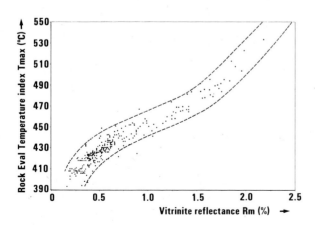

Fig. V.1.20. Relationship between the temperature index T_{max} and vitrinite reflectance for type-III kerogen and humic coals. (After Teichmüller and Durand, 1983)

approximately equivalent to C_T minus the carbon of S_2). They used the ratio $C_R:C_T$ as an indicator of increasing maturation. However, the change of sedimentation from one bed containing a more aliphatic kerogen (type I or II) to another bed containing a more aromatic kerogen (type III) is also marked by an increase of the $C_R:C_T$ ratio and can be misinterpreted as an increase of maturation. Bordenave et al. (1970) used different conditions of pyrolysis and calculated the ratio of the volatile carbon (i.e., carbon of S_2) to the total organic carbon. The problems of interpretation are rather comparable. Leplat and Noel (1974) use gas chromatography to analyze the various hydrocarbons comprising the peak S_2 and then they compute ratios of different hydrocarbons.

Another method of pyrolysis is based on the detailed analysis by gas chromatography of the hydrocarbons comprising the peak S_1 (i.e., hydrocarbons naturally present in the rock; Jonathan et al., 1975). Although pyrolysis is used as a tool, the aim is the compositional analysis of the bitumen and the method will be discussed in Section 1.3.4.

1.3.3 Chemical Indicators of Maturity Based on Kerogen

Most kerogen properties evolve during the rearrangement of the kerogen structure with increasing burial. The maturation range is different for the various physical or chemical parameters, and it corresponds to part or all of the thermal history. For instance, oxygen content is mostly affected during diagenesis whereas hydrogen content changes mostly in the catagenesis zone and crystalline organization (as seen by electron microdiffraction) appears in the metagenesis zone only. However, many of these techniques have to be considered more as research rather than routine tools, due to the time and care needed in their application.

The *elemental analysis* of kerogen, plotted on a van Krevelen diagram, gives a fairly good indication on the maturation stage: McIver (1967), Durand and Espitalié (1973), Laplante (1974), Tissot et al. (1974). The relationship of elemental analysis to a classical maturation index — huminite–vitrinite reflectance — is shown in Figure II.5.1. From this graph, it is obvious that a complete elemental analysis, or at least C, H, O, is needed. Due to the oblique trend of the iso-maturity lines, the knowledge of the carbon content alone, or even H/C is not sufficient. For example, an atomic ratio H/C = 0.8 is observed at the end of the principal zone of oil formation in type-II kerogen whereas it is also observed in immature samples of type III which are approaching the beginning of the oil generation. The approximate boundaries for the three main types of kerogen are shown in Table V.1.4. From this table and also from Figure II.5.1, it appears that the H/C ratio is a better indicator of the maturation of kerogen types I or II and the O/C ratio in the case of kerogen type III. However, the two ratios should be interpreted together.

Among physical analysis of kerogen, *electron spin resonance* (ESR) has been proposed as a routine method for measuring organic maturity or even paleotemperatures (Pusey, 1973). The observation of an ESR signal in kerogen results from the occurrence of unpaired electrons — i.e., free radicals — in kerogen.

Table V.1.4. Average value of the atomic composition of the basic kerogen types I, II, III at the beginning of the main zones of hydrocarbon generation. Intermediate composition of kerogen results in intermediate values

Beginning of the following zones of hydrocarbon generation:	Average value of the atomic ratios					
	H/C			O/C		
Oil	1.45	1.25	0.8	0.05	0.08	0.18
Wet gas	0.7	0.7	0.6	0.05	0.05	0.08
Dry gas	0.5	0.5	0.5	0.05	0.05	0.06
Kerogen type	I	II	III	I	II	III

Free radicals appear in kerogen as a result of splitting bonds. For instance, cracking of an alkyl chain substituted on a polyaromatic nucleus results in two fragments which are free radicals, each with an unpaired electron. The free radicals exist until they recombine, either with a hydrogen atom or with another alkyl chain. This happens more easily for the alkyl fragment than for the kerogen fragment. Marchand et al. (1969) studied the ESR signal of kerogen: they have shown that the paramagnetic susceptibility χ_P is proportional to the number of free radicals, and that χ_P increases with thermal evolution. Pusey (1973) proposed using the "spin density" (number of free radicals per g kerogen) to measure organic maturity and paleotemperature.

Durand et al. (1977) investigated systematically the ESR signal on different types of kerogen and coal ranging from the stage of diagenesis into the metagenesis. The general trend of variation is the same for all types of kerogen. The signal first increases with depth and temperature; it reaches a maximum at a level corresponding to approximately 2% vitrinite reflectance or more and at greater depths it decreases again (Fig. V.1.21). The initial increase of χ_P is related to the elimination of the alkyl chains substituted on polyaromatic nuclei and it is correlative with oil and wet gas generation. The subsequent decrease of the signal may be due to coalescence of the polyaromatic nuclei to form larger sheets. In terms of hydrocarbon potential, it corresponds to the dry gas zone.

The occurrence of a maximum in the χ_P curve (Fig. V.1.21) is a first source of uncertainty in appraising the maturity of a source rock by ESR. In fact, a given ESR value may correspond to two different stages of evolution (one above the maximum and the other below it). A second limitation stems from the different behavior of the kerogen type. At a certain stage of thermal evolution, the value of the signal is larger in type-III kerogen, compared to type I or II (Fig. V.1.21). This fact may be explained in terms of stability of the free radicals: free radicals are more stable on rigid polyaromatic nuclei than they are on alkyl chains. Thus, they are comparatively more abundant in aromatic-rich type-III kerogen than they are in aromatic-poor types I or II. Therefore, certain changes of organic composition

1.3 Maturation of the Organic Matter

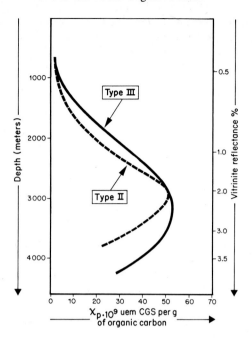

Fig. V.1.21. Influence of kerogen type and thermal evolution on the ESR signal. Paramagnetic susceptibility χ_p of kerogen type III (Douala Basin) is plotted as a function of depth. Evolution of paramagnetic susceptibility of kerogen type II is plotted for comparison on the basis of vitrinite reflectance. (Adapted from Durand et al. 1977)

may be misinterpreted in terms of increase or decrease of paleotemperature. A third source of uncertainty is the time effect on recombination of free radicals. At higher age there is a greater chance for free radicals to recombine.

Finally, ESR may be considered as a valid method for measuring organic maturity during the catagenesis stage in basins where two conditions are verified: (a) the type of organic sedimentation remains approximately constant (e.g., Gulf Coast) and (b) the stage of metagenesis (i.e., 2% vitrinite reflectance) is not reached. It can be pointed out that on certain shales, like those of the Douala basin, a direct use of natural rock gives satisfactory results (Durand et al., 1977). However, a preliminary concentration of kerogen is generally required.

Other physical techniques like *infrared spectrophotometry, thermal analysis, nuclear magnetic resonance* and *electron microdiffraction* may be applied to characterize kerogen maturation. Nevertheless, they are considered more as research tools that can be used in some instances to solve specific exploration problems.

1.3.4 Chemical Indicators of Maturity Based on Bitumen

Abundance and chemical composition of petroleum compounds (hydrocarbons, resins and asphaltenes) present in source rocks depend on the nature of the original organic matter, and the degree of thermal maturation. Thus, numerous methods have been developed using the amount or the composition of the extractable bitumen to characterize the stage of thermal evolution. However, it should be pointed out that migration of hydrocarbons may affect their abundance and composition. Certain shale beds, silt or carbonate source rocks may acquire a slight porosity and permeability (due to microfractures, recrystallization, etc.). Then they can be affected by a short-range migration with preferential accumulation of hydrocarbons, especially those of low molecular weight. This situation can usually be detected by abnormally high transformation ratio (obtained by the pyrolysis method) or bitumen:organic carbon ratio.

a) Abundance of Bitumen

The first step of the analytical process is usually solvent extraction followed by hydrocarbon separation and weighing. The *abundance of bitumen* may be expressed as bitumen ratio, i.e., total extract:total organic carbon, or as hydrocarbon ratio, i.e., hydrocarbons:total organic carbon. The general trend of variation of these ratios with increasing thermal evolution has been discussed in Chapter II.7 and is shown in Figure II.7.1. If a sufficient number of samples is available from a source rock formation at different depths, the values of bitumen or hydrocarbon ratios are plotted versus burial depth, and the generation curve can be drawn. This type of graph has been established for the Uinta, Paris and Douala basins and is presented in Figure II.5.12. On this plot, the immature zone, the principal zone of oil formation, and the zone of gas formation can be traced and subsequently converted in terms of burial depth.

Equivalent procedures may be used to obtain the abundance of hydrocarbons without proper separation, solvent evaporation and weighing, which require time and care. One such procedure uses quick extraction followed by IR spectroscopy of the solvent containing the hydrocarbon fraction, utilizing the CH-band at about 2930 cm^{-1} (Welte et al., 1981). The abundance of hydrocarbons is directly evaluated from the absorbance measurement. Another procedure is based on the volatilization of hydrocarbons by moderate heating (for instance between 200–250 °C) of the rock. In particular the first signal S_1 provided by the pyrolysis method described above shows a very good correlation with the extractable hydrocarbons, although the range of the hydrocarbons volatilized is limited to about C_{20}.

There are several limitations to determining the evolution stage from the amount of bitumen or hydrocarbon. First, a series of samples from various depths is not always available. Then, there may be changes of kerogen type or geothermal gradient occurring in the basin. In turn, these changes affect the amount of bitumen and make it difficult to trace a regular curve of hydrocarbon generation. Changes of kerogen type can be accounted for if the total organic carbon is replaced by the amount of pyrolyzable organic carbon (signal S_2 of the pyrolysis). This was done by Tissot et al. (1978) in Uinta Basin studies. Then, the hydrocarbon ratio becomes rather comparable to the transformation ratio discussed in Section 1.3.2. Finally, short-range migration and minute accumulation within the source rock may also alter the amount of bitumen.

b) Composition of Hydrocarbons: General

For the above reasons, many authors use for maturity studies the *composition* rather than the abundance of hydrocarbons. The basic concept appears clearly on the right side of Figure II.7.1. Originally, the only hydrocarbons present are the geochemical fossils with specific distributions. With increasing temperature, new hydrocarbons are generated; their distribution is wide and does not favor specific compounds. The progressive change of the hydrocarbon distribution observed in source rocks reflects the dilution of the geochemical fossils by the newly generated hydrocarbons. Thus indices reflecting these compositional changes may be used to evaluate source rock maturation. Many indices have been proposed, based on various classes of hydrocarbons. Their validity obviously depends on the uniformity of distribution of the selected compounds in the various types of Recent sediments. If the composition of a certain class of hydrocarbons is the same in Recent sediments from any depositional environment, an index based on that class is particularly valuable. This is the case for the light hydrocarbons C_3–C_8, which are practically absent in Recent sediments. The distribution of long-chain *n*-alkanes is also similar in many Recent sediments, but there are exceptions.

c) Light Hydrocarbons

Amount and composition of the light hydrocarbon fraction, comprising compounds with one to about eight carbon atoms, have been found to be suitable maturity indicators of source rocks. The basic concept of the application of light hydrocarbons as a maturity indicator rests upon the fact that living organisms do not synthesize a complex spectrum of C_3 to C_8 hydrocarbons, especially not aromatics and naphthenes. Furthermore, light hydrocarbons are generated from kerogen at a much higher rate with increasing maturation as compared to the C_{15+} hydrocarbon fraction, which needs less activation energies for their formation as compared to light hydrocarbons. A great variety of techniques has been developed to extract and subsequently analyze light hydrocarbons from rock samples. Basically, all of them use gas chromatography and the methods can be classified according to the molecular range of compounds analyzed:

– gas analysis (C_1–C_4) of either headspace gas or cuttings gas,
– gasoline-range analysis (C_4–C_8) of cores and cuttings.

Essential to all of these techniques is to collect and store the samples immediately after recovery in gas-tight containers (e.g., tin cans). The simplest method is the analysis of the headspace gas accumulated at ambient temperature in the sample container bearing the water-covered rock sample. A known volume of the headspace is taken by a syringe and injected directly into a gas chromatograph. Heating of the container to just below 100 °C and simultaneous high speed grinding of the rock sample increases the yield of the hydrocarbons accumulating in the headspace (Hunt, 1975). Other extraction procedures mobilize low boiling hydrocarbons continuously from the rock and collect them in a cold trap. Mobilization can be carried out by combined heating and grinding either under reduced pressure or with the aid of an auxiliary gas (Durand and Espitalié, 1972; Le Tran, 1975; Philippi, 1975). Durand and Espitalié (1975) also applied $CFCl_3$ extraction to rock samples, however, because of the need to remove the solvent by evaporation, this method has applications only in a higher molecular range ($\geq C_6$). Le Tran (1971) isolated occluded light hydrocarbons from carbonate rocks by acid digestion of the mineral matrix. Jonathan et al. (1975) used thermal vaporization of the C_6–C_{15} hydrocarbons present in cores or cuttings. The hydrocarbons are trapped and subsequently analyzed by gas chromatography. Thus, a complete fingerprint of the fraction $< C_{15}$ is obtained.

Using a comparable technique, Schaefer et al. (1978a, b) developed the hydrogen stripping method in order to recover hydrocarbons from C_2–C_9. This method combines both mobilization of C_2–C_9 hydrocarbons from the rock and a subsequent capillary gas chromatography into a single step. The stripping gas (hydrogen) acts also as the carrier gas in the capillary column, and splitless sample introduction ensures high sensitivity. Sample requirements are very low (between 0.1 and 1.0 g) and, therefore, facilitate the analysis of hand-picked samples from drill cuttings.

Bailey et al. (1974) analyzed the cuttings gas (C_1–C_4) and the hydrocarbons of the gasoline range (C_4–C_7) in Western Canada. The variation of composition observed can be summarized as follows:

1. *diagenetic stage:* the gas is mainly composed of methane; the gasoline fraction is lean, with most components absent,
2. *catagenetic stage:* the gas contains abundant C_2 to C_4 hydrocarbons; the gasoline fraction is rich, with all components present,
3. *metagenetic stage:* the gas is mainly composed of methane; the gasoline fraction is lean.

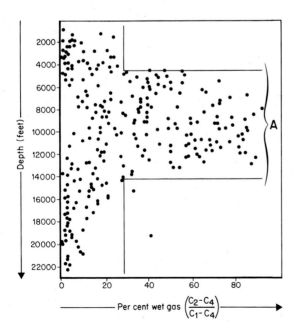

Fig. V.1.22. Percentage of wet gas (C_2–C_4 hydrocarbons) in the total C_1–C_4 cuttings gas for 14 wells located in the Sverdrup Basin, North Canada. The plot versus burial depth delineates the main zone of oil and wet gas generation. A: Main stage of HC generation (oil and wet gas). (After Snowdon and Roy, 1975)

1.3 Maturation of the Organic Matter

Another good example of the generation of light hydrocarbons during the stage of catagenesis is presented in Figure V.1.22 based on data from 14 wells in the Sverdrup basin, North Canada (Snowdon and Roy, 1975). A plot of "wet gas", defined as the proportion of C_2–C_4 hydrocarbons in the total C_1–C_4 gas mixture versus depth clearly delineates the stage of catagenesis (oil and wet gas generation) between 4,000 and 14,000 ft.

Leythaeuser et al. (1979 a, b) studied the generation of light hydrocarbons in source rocks in relation to their maturity and type of organic matter for selected exploration wells using the "hydrogen stripping" technique. They found, for instance, that over a shaly source-bed interval of Jurassic age of about 1000 m in Western Europe light hydrocarbons are progressively generated.

An overall tenfold increase of the total C_2–C_7 hydrocarbon content normalized with respect to organic carbon is observed over the interval studied. Figure V.1.23 shows the trend of concentration vs. depth for selected n-alkanes. It can be observed at an early maturation stage that the hydrocarbon concentrations increase with different rates for different hydrocarbon species by the following approximate order of magnitude: $10^{2.5}$ for ethane, 10^2 for propane, 10^1 for n-butane, etc. This means that the activation energies of the formation reactions increase with decreasing chain length of the products.

The maturity-related light hydrocarbon contents in source rocks have been investigated by Schaefer et al. (1982). As shown in Figure V.1.24 the total saturated light hydrocarbon contents, as determined by a combined hydrogen-stripping/thermovaporization technique, of selected petroleum source beds increase by more than two orders of magnitude if maturity increases from 0.4 to

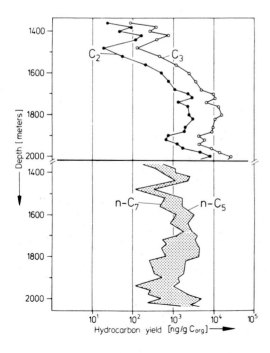

Fig. V.1.23. Depth trends for carbon-normalized yields of ethane and propane *(upper)* and pentane to heptane *(lower)* for a Jurassic shale unit of Fastnet Basin, Ireland. Hexane values fell within the *stippled* area. (After Leythaeuser et al., 1979)

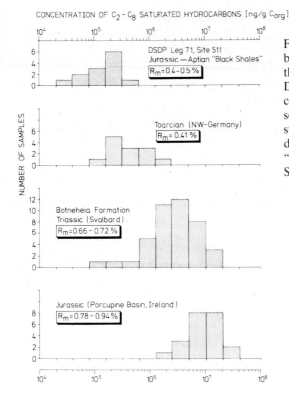

Fig. V.1.24. Saturated light hydrocarbon concentrations (in ng/g C_{org}) in the "black shales" of Site 511 of the Deep Sea Drilling Project (DSDP) compared to those of potential oil source rocks of different maturity stages. This comparison indicates additionally the low maturity of the "black shales" from Site 511. (After Schaefer et al., 1983)

about 0.9 $\overline{R}_o$. More than 10^7 ng/g C_{org} are reached in the hydrogen-rich Jurassic source bed at the maturity interval 0.78–0.94% $\overline{R}_o$ (Fig. V.1.24).

Several authors proposed using for maturation studies the distribution of the main classes of C_6 or C_7 hydrocarbons. Philippi (1975) uses the C_7 hydrocarbons and plots in a triangular diagram the abundance of paraffins (normal plus isoalkanes), naphthenes (cycloalkanes), and aromatics. With increasing maturation the proportion of paraffins usually increases. On the triangular diagram it is marked by a move in the direction of the paraffin pole. It should be emphasized that similar changes also occur in the higher range of molecular weight. For example, a similar trend on the same triangular plot (paraffins, naphthenes, aromatics) may be observed for the extractable fraction C_{15+} of source rocks with increasing maturity.

Jonathan et al. (1975) used a concentration ratio of specific compounds, e.g., *n*-hexane to methylcyclopentane, as a maturation index. The basis for interpretations is still the same as in the global C_6 or C_7 plots, i.e., an increase in the proportion of linear versus cyclic molecules. The authors have calibrated their indices against more classical parameters, such as vitrinite reflectance.

A similar approach was used by Thompson (1979) who employed a large number of light hydrocarbon data to establish so-called paraffinicity indices, e.g., the "heptane-value" which increases with increasing maturation.

d) Carbon Preference Index (CPI)

The carbon preference index has probably received the greatest attention among methods using the chemical composition of bitumen. It was originally proposed by Bray and Evans (1961), and is based on the progressive change of the distribution of long-chain n-alkanes during maturation. In Recent sediments, the major source of these n-alkanes is wax from higher plants. Therefore, it was assumed that the fingerprint of the long-chain alkanes reflects the contribution of higher plants in most marine or terrestrial environments. Under these circumstances, there is a strong predominance of molecules with an odd number of carbon atoms in Recent sediments. Thermal degradation of kerogen during catagenesis subsequently generates new alkanes without predominance. Thus the preference for odd-numbered molecules progressively disappears. This phenomenon is discussed in greater detail in Chapter II.5 and illustrated in Figure II.3.6. It is generally considered that there is little or no odd preference by the time the peak of oil generation is reached.

Several expressions of the odd preference have been proposed:
1. the original definition by Bray and Evans (1961) used the C_{24}–C_{34} interval:

$$\text{CPI} = \frac{1}{2} \left[\frac{C_{25} + C_{27} + \cdots + C_{33}}{C_{24} + C_{26} + \cdots + C_{32}} + \frac{C_{25} + C_{27} + \cdots + C_{33}}{C_{26} + C_{28} + \cdots + C_{34}} \right].$$

2. Philippi (1965) used the predominance of the C_{29} n-alkane in the C_{28}–C_{30} interval:

$$R_{29} = \frac{2\,C_{29}}{C_{28} + C_{30}}.$$

3. Scalan and Smith (1970) used an expression which is applicable to any desired five-carbon-number interval:

$$\text{OEP} = \left[\frac{C_i + 6C_{i+2} + C_{i+4}}{4C_{i+1} + 4C_{i+3}} \right]^{(-1)^{i-1}}.$$

4. Tissot et al. (1977) proposed a mathematical treatment of the distribution based on the best fit of a parabola, and a numerical expression of the deviation form that curve.

Although the carbon preference index has been widely applied, under any of these formulations there are several limitations to its use. The CPI values are influenced by the type of organic matter, and by the degree of maturity. In particular, it has been observed that sediments of the same age, brought to the same temperature due to burial, may result in different CPI values. For example, Erdman (1975) reported that Tertiary shales reaching a temperature of 110°C show an odd preference of 2 to 2.5 in the Gulf of Papua, and 1.0 in the North Sea, respectively. In addition, in the North Aquitaine basin, Lower Cretaceous shales always have a stronger odd preference than the Upper Jurassic beds at a comparable depth (Table II.5.3). Furthermore, there are situations where high and low CPI values are alternatively observed in successive beds from the same well (Fig. II.5.16).

The reason for such observations is that the alkane distribution depends on the original distribution of the *n*-alkanes and on the amount of the new alkanes generated. First, there are environments where the hypothesis on the universal shape of the distribution of long-chain *n*-alkanes in immature sediments is not valid. Huc (1976) has shown that in cores of the Lower Toarcian shales of the Paris basin, which have never been buried to more than 500 m, the CPI ranges from 1.0 to 2.2. The higher values are located close to the border of the basin, and the lower values are located basinward. Furthermore, in the same core hole, the CPI changes from 2.2 down to 1.1 over a vertical distance of 15 m (Fig. V.1.25). Where the terrestrial input is negligible, a significant proportion of the long-chain alkanes may be derived from other sources, such as planktonic algae and bacteria.

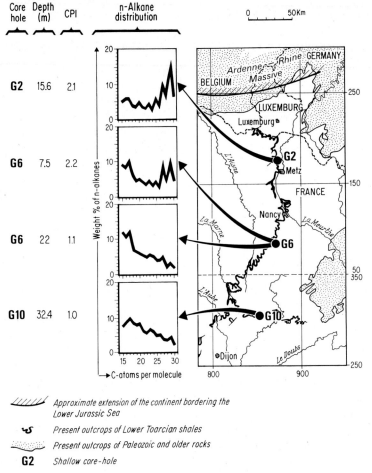

Fig. V.1.25. Changes of the Carbon Preference Index in immature shales of Lower Toarcian, Paris Basin. All samples are from shallow core holes, and their maximum depth of burial has never been more than 500 m. The changes of *CPI* are related to the importance of the terrestrial contribution from the Continent located north of the Lower Jurassic Basin. (Adapted from Huc, 1976)

These organisms do not usually generate a predominance of the odd-numbered long-chain molecules. Thus, the related CPI is near 1.0, even in immature sediments. At depth, this situation may explain the observed alternation of high and low CPI observed in the obviously immature sediments of the Pannonian basin (Fig. II.5.16).

The change of the alkane distribution with increasing depth is directly related to the amount of new alkanes generated. We have learned in Chapter II.5 that different types of kerogen generate smaller or larger amounts of hydrocarbons during catagenesis. In the North Aquitaine basin, the Lower Cretaceous beds, having a lower source rock potential than the Upper Jurassic, yield a smaller amount of hydrocarbons, and keep a higher CPI than the Upper Jurassic beds (Table II.5.3). Furthermore, the n-alkane content in the total petroleum generated varies to a large extent according to the type of kerogen (Fig. II.5.15). Thus, the same values of CPI may correspond to the formation of different quantities of petroleum, depending on the kerogen type.

The above observations represent limitations to the use of absolute CPI values for defining maturity. Despite these various limitations, the carbon-preference index can offer valuable information on the maturation of source rocks, provided it is considered as a qualitative indicator and used in association with another independent index. High CPI values (above 1.5) always refer to relatively immature samples. Low CPI values, however, do not necessarily mean higher maturity, they can also mean a lack of higher n-alkanes stemming from terrestrial input. Finally, it should be mentioned that an even predominance (CPI values of 0.8 or less) may be observed in carbonate-evaporite sediments. This has to be interpreted also as a sign of immaturity.

e) Isoprenoid Distribution

Isoprenoid abundance also varies during catagenesis (Sect. II.5.4.4) and may provide an indication of maturation. The pristane:phytane ratio has been proved to be changed by thermal evolution in coal (Brooks, 1970), and this observation may be extended to kerogens of comparable composition. However, no comparable variation has been observed in association with typical I or II kerogens. The ratio of isoprenoids to n-alkanes decreases with depth in all types of organic matter, but the rate of decrease is not directly related to the amount of hydrocarbons or alkanes generated. This is due to the fact that isoprenoid chains, linked to kerogen by weak bonds, may be released from kerogen, and thus the phenomenon is more complex than a simple dilution of isoprenoids by other alkanes. Nevertheless, the pristane:n-C_{17} or phytane: n-C_{18} alkane ratio may provide some information on thermal evolution.

f) Ring-Number Distribution of Naphthenes

Naphthenes have also been considered with respect to thermal evolution by Philippi (1965). He defined a naphthene index (NI) as the percentage of 1-ring plus 2-ring naphthenes in the 420–470 °C isoparaffin–naphthene fraction. The aim of NI is to quantify the progressive dilution of biogenic 4- and 5-ring naphthenes by more simple mono- or dicyclic naphthenes generated from kerogen. This criterion can be applied to a given source rock to compare various stages of maturation. For example, Figure II.5.17 clearly shows the progressive decrease of polycyclic naphthenes in the Lower Toarcian of the Paris basin. However, different source rocks cannot be directly compared with each other, as the original content of polycyclic naphthenes may vary considerably. This content can be high in association with type-I or -II kerogens and low in association with type-III kerogen.

g) Methylphenanthrene Index (MPI)

The Methylphenanthrene Index (MPI) has been developed recently by Radke et al. (1982a). It is based on the distribution of phenanthrene and three or four of its methyl homologs which shows a progressive change during maturation.

Phenanthrenes from petroleum and sedimentary organic matter are genetically related to steroids and triterpenoids found in biological source materials (Mair, 1964; Greiner et al., 1976). However, partial dealkylation of these potential precursor molecules would result in 1- and 2-methylphenanthrene only and, therefore, may not account for a predominance of 1- and 9-methylphenanthrene, as frequently seen for immature samples. Thus, methylation of phenanthrene has been discussed as an alternative mechanism. Coinciding depth trends of the 1- and 9-methylphenanthrene/phenanthrene ratios that have been observed in the Western Canada Basin (Radke et al., 1982a) are likely to reflect the similar reactivities of the 1- and the 9-position of the phenanthrene molecule in methylation reactions. The relative abundance increase of 2- and 3-methylphenanthrene during maturation can be explained on the basis of rearrangement reactions, which, at higher temperatures, favor these thermodynamically more stable isomers.

Based on mechanistical considerations, the Methylphenanthrene Index MPI 1 has been introduced:

$$\text{MPI 1} = \frac{1.5 \, (2\text{-methylphenanthrene} + 3\text{-methylphenanthrene})}{\text{phenanthrene} + 1\text{-methylphenanthrene} + 9\text{-methylphenanthrene}}.$$

A similar expression, the Methylphenanthrene Index MPI 2, may be used as a control means or as a substitute for the MPI 1:

$$\text{MPI 2} = \frac{3 \, (2\text{-methylphenanthrene})}{\text{phenanthrene} + 1\text{-methylphenanthrene} + 9\text{-methylphenanthrene}}.$$

The numerical value of the MPI 2 is generally somewhat higher than that of the MPI 1; the difference reflects a slight predominance of 2- over 3-methylphenanthrene which is common to the methylphenanthrene distribution.

A strong correlation of the MPI 1 and vitrinite reflectance data within the 0.6–1.3% R_m interval has been observed in a Western Canada Basin well (Fig. V.1.26). Obviously, the MPI was not significantly influenced by changes in the

1.3 Maturation of the Organic Matter

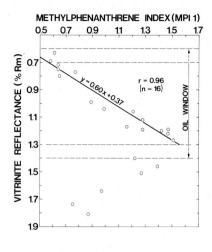

Fig. V.1.26. Correlation between the Methylphenanthrene Index and the mean vitrinite reflectance for samples from Elmworth 103571 W6M well. (After Radke et al., 1982a)

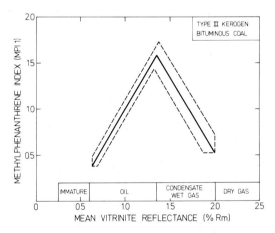

Fig. V.1.27. Relationship between Methylphenanthrene Index (MPI 1) and mean vitrinite reflectance (% R_m) as based on data from Type-III kerogen-bearing rock and bituminous coal samples. (After Radke and Welte, 1982)

organic facies, e.g., interbedded coal layers. This is understandable, because the shales in between the coals contain mainly type-III kerogen. The validity of the MPI as a maturity parameter has been further demonstrated in the Ruhr area, West Germany, where a good correlation between the MPI 1 and R_m has been shown to exist for a series of Upper Carboniferous bituminous coals (Radke et al., 1982b). Furthermore, evidence has been obtained for a general relationship between the MPI 1 and R_m (Fig. V.1.27). Based on this relationship, vitrinite reflectance values (R_c) have been calculated from the MPI 1 (Radke and Welte, 1982):

$R_c = 0.60 \text{ MPI } 1 + 0.40$ for $0.65 \leq R_m \leq 1.35$
$R_c = -0.60 \text{ MPI } 1 + 2.30$ for $1.35 < R_m \leq 2.00$.

The idea is that the MPI 1 value determined for an extract indicates the vitrinite reflectance value (R_c) of the corresponding source rock unless it has received an input of nonindigenous aromatics.

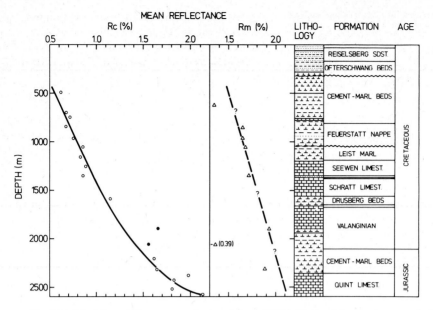

Fig. V.1.28. Mean reflectance calculated from MPI 1 (R_c) and from microscopic data (R_m) vs. depth for a well drilled in the Northern Alps. Presence of migrated C_{15+} hydrocarbons is indicated by solid signature. (After Radke and Welte, 1982)

The validity of this concept has been demonstrated for numerous wells where redistribution of heavy hydrocarbons was negligible. An impregnation of a rock sample by migrated oil generally results in an elevated R_c value; i.e., the R_c value is higher than the microscopically determined R_m value of that sample. Thus, the MPI is a good tool when relating petroleum of unknown origin to its presumed source rock.

The MPI is particularly useful to measure the maturity of organic matter in marls and limestones where vitrinite particles are frequently rare, as shown for a well drilled in the Northern Alps (Fig. V.1.28). The theoretical vitrinite reflectance gradient based on the MPI 1 differs significantly from the tentative gradient based on microscopic data. Thus, calculated vitrinite reflectance values (R_c) are a possible means to support or question actual microscopic data.

h) Isomerization and Aromatization of Biological Marker Hydrocarbons

The occurrence of geochemical fossils (biological markers) in geological environments has been discussed in Chapter II.3. The importance of these characteristic molecules as indicators of depositional environment and geological history was presented in Chapter IV.3. The purpose of this paragraph is to examine whether certain ratios of these molecules which can be measured quantitatively are useful as maturation indices. The first consideration of such an approach was the usage of carbon preference index (CPI) described in paragraph (d) or the naphthene-index described in paragraph (f) of this section.

1.3 Maturation of the Organic Matter

The assessment of the extent of thermal maturation of organic matter in sedimentary rocks by biological marker parameters has recently been reviewed by Mackenzie (1984). The best and most reliable examples of this approach involve the monitoring of the extent of isomerization and aromatization reactions by measuring the relative concentrations of the reactants (A) and products (B):

$$A \underset{k_2}{\overset{k_1}{\rightleftharpoons}} B.$$

Ideally, the concentration of B should be zero at the start of the maturation process, and the extent of the reaction is measured as percentage of

$$\frac{B}{A+B},$$

usually by combined gas chromatograph/mass spectrometry (GC/MS). When the reaction rate constant k_1 is very much larger than k_2, the reaction ideally will

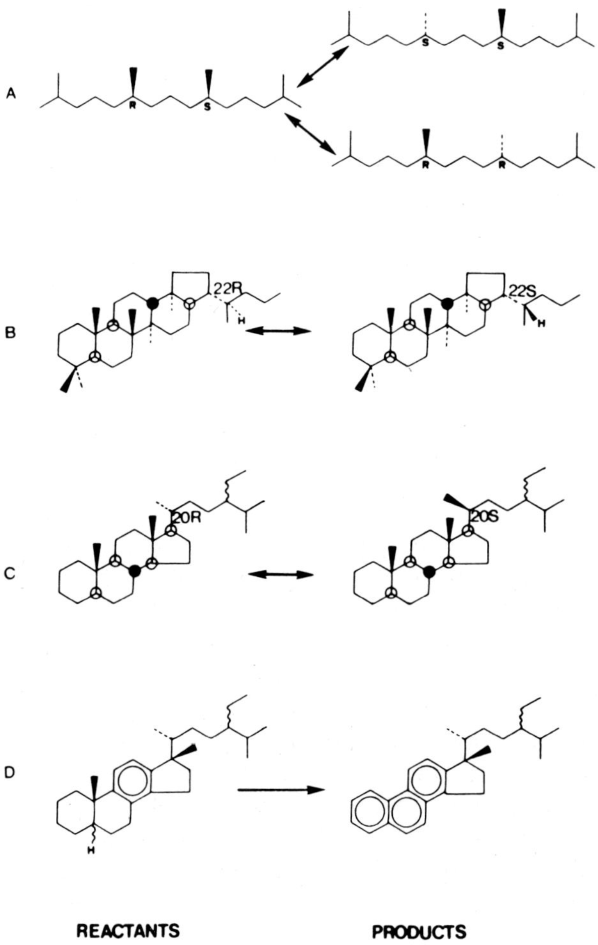

Fig. V.1.29. Isomerization and aromatization reactions used to assess the extent of thermal maturation of sedimentary organic matter (Mackenzie, 1984)

proceed from 0 to 100% with increasing maturity. If there is a noticeable back reaction, and k_1, for example, equals k_2, then a reaction equilibrium will be established at a value of 50%. When a reaction has reached completion or equilibrium it can no longer be used to assess the extent of thermal maturation.

Figure V.1.29 gives four examples where the initial concentration of the products nearly always is zero. In acyclic hydrocarbons isomerization occurs at chiral centres, like C-6 and C-10 in pristane (A in Fig. V.1.29). The biogenic mesoisomer (6R, 10S configuration) is converted to two enantiomers (6R, 10R and 6S, 10S) under geological conditions, but this reaction is apparently complete with an equilibrium value of 50% at a relatively early stage of maturity (Mackenzie et al., 1980). Similarly, isomerization at C-24 in the side chain of steranes comes to an equilibrium very early (Maxwell et al., 1980).

The isomerization at C-22 in $17\alpha(H),21\beta(H)$-hopanes (B in Fig. V.1.29) can be followed in the C_{31} to C_{35} extended hopanes by the m/z 191 mass fragmentograms showing two diastereomers (22R and 22S) for each homolog except in the most immature sediments. Either an average of all carbon numbers values (C_{31}–C_{35}) or the value for the C_{32} extended hopanes is used to measure the extent of isomerization which varies from 0 to about 60% with increasing maturity (Ensminger et al., 1974; Seifert and Moldowan, 1980), although the actual equilibrium value seems to vary slightly between carbon numbers. The isomerization at C-22 of $17\alpha(H),21\beta(H)$-hopanes requires a higher level of maturity to reach completion than at C-6/C-10 in pristane, but it is normally complete before the onset of intense hydrocarbon generation (Mackenzie and Maxwell, 1981).

Despite the fact that isomerization reactions of steranes under geological conditions are very complex (Seifert and Moldowan, 1979), the isomerization at C-20 of C_{29}-steranes (C in Fig. V.1.29) appears to be most useful in assessing the thermal maturity of both sedimentary rocks and crude oils (Mackenzie et al., 1980; Seifert and Moldowan, 1981), since it can extend well into the zone of hydrocarbon generation (Mackenzie and Maxwell, 1981). The extent of this isomerization is measured for $5\alpha(H),14\alpha(H),17\alpha(H)$-$C_{29}$ steranes using the peak areas (or hights) of the 20S and 20R isomers in the m/z 217 mass fragmentograms. The values rise from 0 to 50–60%, but it is difficult to determine the equilibrium exactly because of the complexity of the sterane distributions.

Isomerization reactions at chiral centers which are part of ring systems are more complex, but they are occasionally used as supplementary maturity parameters. Among these are the isomerization at C-17 and C-21 in hopanes (van Dorsselaer, 1975; Seifert and Moldowan, 1980) and at C-14 and C-17 in steranes (Seifert and Moldowan, 1979; Mackenzie et al., 1980; cf. also Mackenzie, 1984).

Aromatization of C-ring aromatic steroid hydrocarbons (D in Fig. V.1.29) is the only aromatization-type transformation where quantitation has been attempted to use this reaction as a maturation parameter (Mackenzie et al., 1981). The main reaction pathway is the conversion of C-ring monoaromatic steroid hydrocarbons with two nuclear methyl groups to ABC-ring triaromatic steroid hydrocarbons with one nuclear methyl substituent probably via diaromatic intermediates. The chiral centre at C-5 is lost during aromatization, and the reaction extent can be measured on the basis of two C_{29} monoaromatic reactants (m/z 253 fragmentogram) and a C_{28} triaromatic product (m/z 231 fragmentogram). The extent of aromatization increases from 0 to 100% with increasing maturity over the range of late diagenesis to the peak of oil generation, similar to that covered by the isomerization at C-20 in the steranes (reaction C in Fig. V.1.29).

Within each reaction type a sequence of changes is always observed, and Figure V.1.30 attempts a summary of the ranges covered by those reactions discussed above in relation to other reactions, like porphyrin conversion, and to vitrinite reflectance and oil generation. This interrelation is based on the analysis of Mesozoic sample suites from the Paris Basin and NW Germany (Mackenzie and Maxwell, 1981). It is only a rough guide since the relative rates of the reactions are governed by the kinetic constants of the reaction rates and vary with thermal history. Those reactions whose ranges extend into the oil generation zone have an application to the measurement of crude oil maturity as well as of source rock maturity. However, it is to be expected that the relationship between these different parameters would be different under various geological settings, as the reaction rates are strongly influenced by heating rates. For instance Mackenzie et al. (1982) suggested that in sequences of sediments whose temperature was increased at high average rates (e. g., Pannonian Basin, ca. 15 °C. Ma^{-1}) the rate of aromatization was accelerated to a greater extent than the isomerization rate, when compared with lower average heating rates (e.g., East Shetland Basin, North Sea, 1 °C. Ma^{-1}) (Fig. V.1.31). This was proposed to be partly due to the

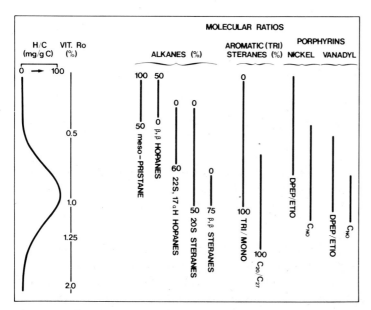

Fig. V.1.30. Ranges of individual molecular measurements of thermal maturation plotted against the hydrocarbon generation curve and vitrinite reflectance values. The data were compiled from results for Toarcian shales of the Paris Basin and for Pliensbachian shales of N.W. Germany (adepted with changes from Mackenzie and Maxwell, 1981)

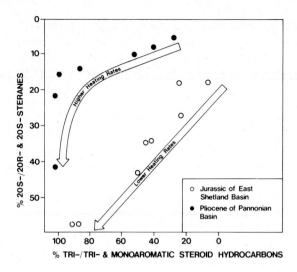

Fig. V.1.31. Comparison of the variation of the extent of aromatization of C-ring monoaromatic steroid hydrocarbons relative to the extent of isomerization at C-20 of $5\alpha(H),14\alpha(H),17\alpha(H)$-$C_{29}$ steranes between Jurassic shales of the East Shetland Basin, North Sea, and the Pliocene shales of the Pannonian Basin, SE Hungary (Mackenzie, 1984)

aromatization reaction having a higher activation energy than the isomerization reaction. These two reactions, however, may not be independent since it is conceivable that saturated steranes are converted to aromatic steroids.

1.4 Conclusions on Characterization of Potential Source Rocks

All the methods reviewed may contribute to the characterization of the source rocks. However, their degree of efficiency varies from low to excellent, and there may be some limitations to their use. These aspects are summarized in Table V.1.5. Their cost also varies considerably, depending on the degree of sophistication.

The *abundance* of organic matter is quickly and conveniently obtained through the measure of total organic carbon. It can be carried out routinely on a large number of samples. *Quality* and *maturation* of the organic matter can be evaluated separately. However, due to the strong relationship between their respective influence, it is preferable to evaluate quality and maturation simultaneously. *Pyrolysis* is probably *the best routine tool for determining type and maturation of organic matter* at the same time. Furthermore it allows a semi-quantitative evaluation of the *genetic potential* and *transformation ratio*. Pyrolysis techniques are also quick and inexpensive, and can be used routinely. Optical techniques are also of great value, particularly examination in transmitted light associated with fluorescence for determining the type of organic matter, and vitrinite reflectance to characterize maturation. These techniques are somewhat more expensive than pyrolysis; they also require more sample-preparation time, and are therefore not available as quickly. Thus, they are often applied to a limited number of samples selected on the basis of pyrolysis results. Finally, the same conclusion applies to elemental analysis of kerogen, which also requires

1.4 Conclusions on Characterization of Potential Source Rocks

Table V.1.5. Principal methods used for source rock characterization, and their degree of efficiency

Class	Type of the analysis	Abundance of organic matter	Quality = type of organic matter	Maturation of organic matter	Correlation between source-rock and petroleum
Chemical (on rock)	Organic carbon	⬤			
Optical microscopy	Transmitted light (Palynofacies, alteration)		⬤	⦁	
	Reflected light		·	⬤	
	Fluorescent light		⬤	⦁	
Pyrolysis (on rock)	Rock-Eval	⬤	⬤	⬤	
	Composition of HC in pyrolysate		⬤	⬤	⬤
Physicochemical (on kerogen)	Elemental analysis		⬤	⦁	
	Infrared spectroscopy / Thermal analysis (TGA) / Electron microdiffraction		·	·	
	ESR, NMR		·	·	
	Isotopes (C, H, S etc.)		·		·
Chemical (on bitumen or crude oil)	Amount of HC			⬤	
	Light HC		⬤	⬤	⬤
	n-Alkanes		·	⬤	·
	Isoprenoids		·	·	·
	Steroids, terpenes		·	⬤	⬤
	Aromatics		·	⬤	⬤
	Porphyrins, metals			·	⬤
Physical (on bitumen, oil or gas)	Isotopes (C, H, S etc.)		⬤	⬤	⬤

Efficiency ⬤ Good or excellent   fair   · low or limitations of use

time and care, and the technique could be restricted to calibrating the main types distinguished by pyrolysis.

As a general remark, *optical* and *chemical* techniques should not be considered as competitive alternatives, but as complementary methods. In fact, optical techniques are well designed to discriminate between autochthonous and reworked or altered organic material, whereas chemical techniques consider organic matter as a whole and measure averages. In counterpart, chemical techniques encompass the whole organic matter, whereas optical techniques often rely on a certain fraction of organic matter due to the laboratory procedures of concentration (Correia and Peniguel, 1975) or observation: this fraction is not necessarily the main oil-generative part of kerogen. Furthermore optical techniques alone do not always offer a sufficient discrimination of the quality of organic matter. Thus, the combination of two different methods, one chemical and one optical, is certainly the most rewarding.

There is obviously a need for a common expression of the results. It would be desirable for the type and maturation stage of the organic matter to be expressed in comparable terms by all laboratories, although they may be determined by different methods. The question is relatively simple with respect to the *type* of organic matter. In this book, we preferred to use numerals (type I, II and III) rather than genetic terms (sapropelic, humic, etc. ...). The reason is that denominations based on optical examination, e.g., "sapropelic", may in some cases include different chemical compositions. However, there is some hope that further work may allow the definition of a general nomenclature.

The situation is rather different with respect to the *maturation* stage of source rocks. The various scales are based on different chemical or physical (including optical) properties of the organic matter. In turn, the properties depend on different aspects of the structural evolution of the organic matter: e.g., elimination of carboxylic groups, abundance of free radicals, size of the polyaromatic nuclei, etc. The chemical transformations generating these phenomena obey different kinetic laws, with different activation energies in response to temperature increase. There are two consequences of this fact. First, there is no way to express maturation in terms of paleotemperature, since both time and temperature affect the maturation indicators. Each scale is in fact a temperature–history scale, integrating the effects of temperature and time. Furthermore, the time–temperature relationship is different for the various properties considered. Thus, the scales cannot be superimposed, as one may be affected by strong variations while the other varies only slightly, and vice-versa. For example, fluorescence and O/C ratio vary strongly during diagenesis and the beginning of catagenesis, whilst reflectance and H/C are affected to a smaller extent. In turn, reflectance and H/C ratio change strongly during late catagenesis.

For this reason, we have to consider the use of one scale of maturation as a reference, providing an approximate calibration with the other indicators of thermal evolution. Due to its wide use, the huminite-vitrinite reflectance scale has often been considered as a reference. Hood et al. (1975) proposed an independent scale, called LOM (level of organic metamorphism) designed to have a linear variation as a function of depth, and to be continuous over the whole range of generation and destruction of petroleum. More recently, Dow (1977)

1.4 Conclusions on Characterization of Potential Source Rocks

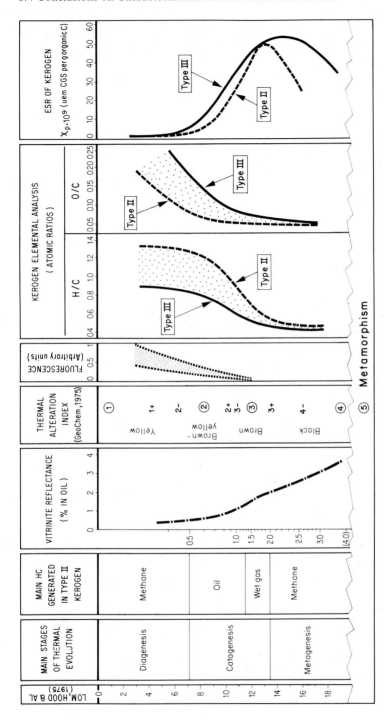

Fig. V.1.32. Comparison of various maturation indicators based on kerogen analysis

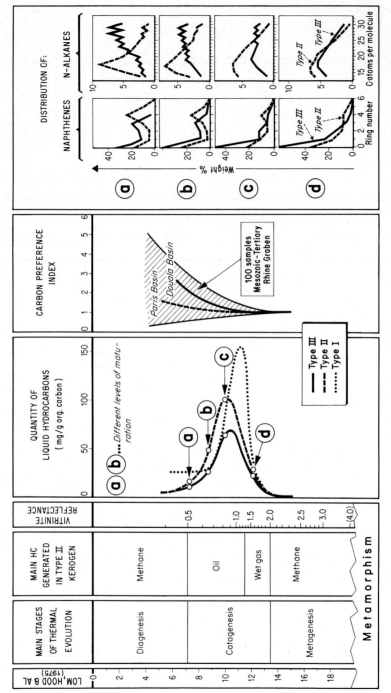

Fig. V.1.33. Comparison of various maturation indicators based on abundance and composition of the extractable bitumen. (Adapted from an original sketch by B. Durand)

1.4 Conclusions on Characterization of Potential Source Rocks

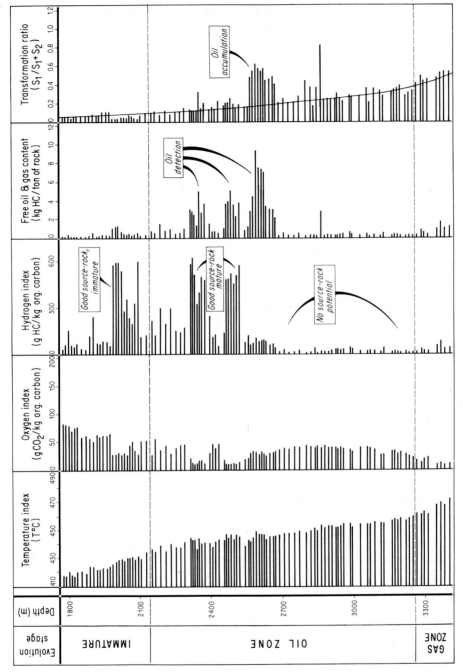

Fig. V.1.34. Example of geochemical log plotted from pyrolysis analysis performed on a well site

suggested using a logarithmic scale of huminite-vitrinite reflectance ($\log \overline{R}_o$) for the same purpose. This scale seems to be approximately linear with depth over most of the diagenesis and catagenesis stages, but not necessarily during metagenesis. The temperature index of pyrolysis T_{max} has the advantage over vitrinite reflectance of being based on a property of the bulk of organic matter, whereas vitrinite may be subordinate or even absent in type-I or -II kerogen. There is, however, an influence of the type of kerogen at low evolution stages, which may be accounted for by using a plot of T_{max} versus hydrogen index.

In terms of vitrinite reflectance, the boundary between diagenesis and catagenesis corresponds approximately to $\overline{R}_o = 0.5\%$ and the boundary between catagenesis and metagenesis to about $\overline{R}_o = 2\%$. Metamorphism (greenschist facies) appears at about $\overline{R}_o = 4\%$. The corresponding values of LOM are shown in Figure V.1.32 and V.1.33. Whatever scale is used, it should be remembered that, due to the occurrence of different types of organic matter with different kinetics of transformation, there are no exact limits for the stages of oil and gas generation. The exact boundaries of the petroleum generation zone differ slightly with the type of kerogen. A maturation scale, therefore, should be based on one of the main kerogen types. If not specified, oil generation may start at a vitrinite reflectance $\overline{R}_o$ from 0.5 to 0.7% and may reach a peak at $\overline{R}_o$ ranging from 0.8 to 1.1% according to the type of kerogen.

The main indicators of source rock evolution are presented for comparison in Figure V.1.32 (indicators based on kerogen), and in Figure V.1.33 (indicators based on composition of bitumen). The reader is referred to Figure V.1.19 for the pyrolysis indices. The course of change of a maturation parameter, e.g., its linearity versus maturation scale, defines the preferred intervals of application: (a) fluorescence and O/C ratio: diagenesis; (b) ESR and H/C ratio: catagenesis; (c) vitrinite reflectance: catagenesis and metagenesis; (d) pyrolysis and alteration of palynomorphs: from diagenesis to metagenesis.

The introduction of fast but powerful and reliable techniques of pyrolysis makes it now possible to run geochemical analysis on a large number of samples. For instance, cuttings from a bore hole can be analyzed routinely every 10 or 20 m. Furthermore, pyrolysis may even be performed on a well site. The results can be plotted against depth in the form of a log where the type of organic matter and the genetic potential are shown, together with the maturation stage and the transformation ratio (Fig. V.1.34). After evaluation of this log, selective sampling can be done in order to run a smaller number of samples for more detailed analysis.

Summary and Conclusion

There are three crucial factors for characterization of a petroleum source rock: the amount of organic matter or organic richness, the type or quality of organic matter and its state of maturity.

Summary and Conclusion

The amount of organic matter can easily be estimated from an organic carbon analysis. Quality and maturation of organic matter can be evaluated by different chemical and optical techniques. It is advisable to evaluate quality and maturation simultaneously, because of the strong interrelation of these two parameters.

Pyrolysis is probably the best routine tool for determining type and maturation of organic matter at the same time. Optical techniques are also of great value in this respect. Examination of organic matter in transmitted light is valuable for a determination of the type of organic matter. Vitrinite reflectance studies are excellent for a characterization of the state of maturity. The methylphenanthrene index is a very good means to measure the maturity of extracts and crude oils over the catagenesis stage. Optical and chemical techniques should not be considered as competitive alternatives, but as complementary methods.

Quality or type of organic matter is here divided with the help of chemical methods (pyrolysis, elemental analysis, etc.) into type-I, type-II and type-III kerogens. Divisions based on optical examinations do not per se distinguish between kerogens on the basis of different chemical composition.

Maturation stages can be determined using various physicochemical or optical techniques. The interrelation of the resulting maturation scales are currently being established. Each scale represents a temperature-history scale, integrating the effects of temperature and time. The most widely applied maturation scale is based on huminite-vitrinite reflectance data. Whatever scale is used, it should be remembered that, due to the occurrence of different types of organic matter with differing kinetics of transformation, there are no sharp boundaries for the stages of oil and gas generation.

Chapter 2
Oil and Source Rock Correlation

The objectives in crude oil correlation vary considerably, depending on exploration and production problems. In principle it is desirable to correlate different oils with each other, oils with their source rocks, gases with gases and, if possible, gases with oils and source rocks (Fig. V.2.1). Questions frequently posed to the exploration geologist are the following:

- is it possible to identify various types or families of oils, condensates or gases occurring in a sedimentary basin? Does each family have a single or multiple origin?
- is it possible to identify the source rock of a given oil, condensate or gas? Is it possible to detect intra-basinal facies changes in a given source rock that would result in the release of different and identifiable crude oils across the basin?

The solution of such questions helps to define the kind and number of exploration targets, and to develop an exploration concept. In oil production, especially in oil-field development of strongly tectonically dissected productive

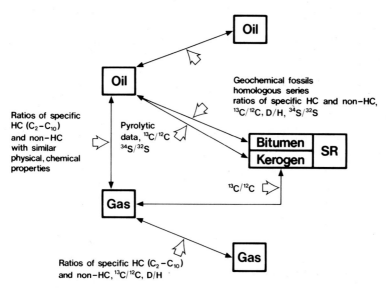

Fig. V.2.1. Schematic representation of desirable studies in oil, gas and source rock correlation. Some useful correlation parameters for certain types of correlation are listed

areas, or in multipay fields, oil–oil correlation can quickly provide valuable information on possible communication between different fault blocks or different productive horizons.

Oil or gas entrapped in a reservoir has been generated from kerogen disseminated throughout a source rock. During primary migration only a certain proportion of the petroleum compounds generated from kerogen can be released, a certain part is always retained in the source rock. Based on this concept, the three principle objects of study for correlation are the *insoluble kerogen* of a source rock, the *bitumen* extractable from this source rock and *pooled hydrocarbons (oil, condensate, or gas)*, possibly derived from this source rock.

Geochemical correlation between petroleum and source beds is based upon recognition of compositional similarities or dissimilarities. Such features are usually established on relative distribution patterns of certain compounds, although absolute concentration values may occasionally provide useful information for correlation. Generally, compositional parameters suitable for correlation should:

- not be too seriously affected by any processes acting upon source rocks and petroleum during and after their physical separation (e.g., preferential separation by migration processes, thermal and bacterial alteration);
- have sufficiently characteristic compound distributions in both rocks and oils to permit differentiation between individual source rocks and oils.

The basic idea is to recognize identical, similar, or different "fingerprint" patterns exhibited by correlation parameters for kerogens, bitumens and pooled hydrocarbons. The degree of similarity or dissimilarity then proves, suggests, or disproves a relationship between petroleums and source beds. It is advisable never to rely on one correlation parameter only, but to apply several independent correlation criteria. Any correlation based on minor constituents, e.g., biological marker distributions, should be reasonably consistent with bulk parameters like aliphatic/aromatic hydrocarbon ratios or carbon isotope values. One of the first successful correlation studies on oils was presented by Jones and Smith (1965), who compared gross chemical differences of distillation fractions of crude oils from the Permian Basin of West Texas and New Mexico. Correlation problems have been reviewed by Koons et al. (1974), Williams (1974), Welte et al. (1975), Deroo (1976), Seifert and Moldowan (1978), Seifert et al. (1980) and Seifert and Moldowan (1981).

2.1 Correlation Parameters

The parameters suitable for *oil–oil correlation* can be chosen from various classes of hydrocarbons and nonhydrocarbons normally present in crude oils. One of the best correlation tools is provided by *geochemical fossils or biological markers*. Geochemical fossils have been discussed in detail in Chapters II.3 and IV.3. Compounds of the steroid and terpenoid classes are of special interest in this respect. Furthermore, the distribution patterns of hydrocarbon molecules with a less specific origin, such as *n-* or iso-alkanes, or the various classes of naphthenes and aromatics can be successfully utilized. In addition, porphyrin ratios or even some nonbiogenic heterocompounds such as thiophene or carbazole derivatives may provide useful correlation parameters. In oil–oil correlation, it is possible to select parameters in such a way that a wide molecular weight range is covered for any given correlation parameter. Such information can be provided, for instance, by the *n*-alkane distribution curve or by the gas chromatography (GC) record of the saturated hydrocarbond fraction on a capillary column. Useful information may also be obtained by such gross parameters as isotope ratios ($^{13}C/^{12}C$, D/H, $^{34}S/^{32}S$) or aliphatic/aromatic hydrocarbon ratios. Isotope ratios may also be used in a more specific form by determining the isotopic composition of certain compound classes or even individual compounds.

The correlation between oils and condensates, and especially between oils, or condensates, and gases is a more delicate problem, because gases and to some extent condensates are related by preferential generation and accumulation mechanisms which may be different from those for oils. This fact must be taken into account. The molecular weight range of these various petroleum products is so different, that the success of a correlation analysis depends largely on the depree of overlap between the molecular weight ranges of the different products (e. g., light and gasoline range hydrocarbons for oil-condensate correlation).

Useful information for the correlation of condensates and light oils may be obtained by a detailed gas chromatographic analysis of the gasoline-range hydrocarbon fraction. In the case of gases, there are severe limitations because their composition is too simple, and they consist of rather unspecific compounds. However, carbon isotope ratios have been shown to be useful correlation parameters for grouping gases (Stahl et al., 1977; Schoell, 1980).

The correlation between *pooled hydrocarbons and source rocks* is more complicated than the oil–oil correlation. Pooled hydrocarbons have accumulated through the processes of primary and secondary migration. There are compositional differences between the bitumen left behind in a source rock and that fraction of bitumen that finally accumulated in a reservoir. Due to migration phenomena, pooled crude oils are strongly enriched in saturated hydrocarbons, moderately enriched in aromatic hydrocarbons and depleted in polar NSO compounds when compared to source rock bitumen (see also Chap. III.3; Fig. III.3.1). Saturated and aromatic hydrocarbons are therefore more suitable for oil–source rock correlation than heterocompounds. Any group of compounds selected for correlation should be easy to isolate and characterize, but the individual components should be similar in their physical properties, such as polarity, solubility and molecular weight. On these criteria again, geochemical fossils of the steroid and triterpenoid hydrocarbon group serve as most suitable correlation parameters. Likewise, other hydrocarbon-based parameters may be applied as previously mentioned for oil–oil correlation.

Re-distribution of bitumen in source beds, especially impregnation with foreign bitumen from an outside source, can obliterate oil–source rock correlation based only on source rock bitumen while neglecting the kerogen. An oil–source rock correlation, therefore, ideally should also include a study of kerogen demonstrating that the bitumen of a source rock is indigenous, and can be linked to the kerogen from which it was supposedly generated. Alternatively, the indigenous nature of the bitumen should be proven indirectly by relating the amount of extractable organic matter, the percentage of hydrocarbons in the extract or the transformation ratio (production index) from Rock-Eval pyrolysis (see Chap. V.1) to the maturity level of the kerogen. The choice of parameters for correlating bitumen directly with kerogen is limited. The most frequently used technique is a carbon isotope analysis. Bitumen derived from a certain kerogen should be isotopically lighter than the kerogen itself. The difference in δ^{13}C-values, however, should not be more than about 2–3‰. An oil connected to a certain source rock bitumen should be isotopically identical or, again, slightly lighter than the related bitumen. Gross isotope data, however, should not be overemphasized when a correlation is indicated. In contrast, noncorrelation based on isotope values, i.e., when an oil is isotopically heavier than a kerogen, is a clear indication for a misfit.

Furthermore, carbon isotope ratios of pooled gases, i.e., of methane, ethane and propane, have been applied to obtain information about their possible source rocks (Stahl and Carey, 1975). This application is based on the concept that carbon isotope ratios of gases are controlled mainly by the maturity of their source material. A refinement of this method is achieved when carbon isotope values are applied together with hydrogen isotope values and are plotted against each other. In this way, Schoell (1980) came to a very detailed classification of gases of various origins. It should be mentioned, however, that the concept that carbon isotope ratios can be used to determine the maturity of the source rock from which a gas has been derived, is not undisputed.

Another method to correlate bitumen with kerogen, employing pyrolysis of the kerogen, has been suggested by Welte (1972). The basis for this method is the idea that exhaustively extracted kerogen still exhibits a potential to generate hydrocarbons (see also Chap. II.5), and, on pyrolysis, will yield new hydrocarbons with a pattern similar to those which have been previously generated in a natural way. Recently Seifert (1978) reported the successful application of such a technique. Pre-extracted kerogen was pyrolyzed, and the hydrocarbon pattern of the pyrolyzate compared to the previously extracted bitumen using GC-MS-analyses. On the basis of a match between sterane and triterpane patterns, he was able to prove the indigenous nature of the bitumen. The same principle can be used to correlate degraded with non-degraded oils by pyrolyzing the asphaltene fraction of the degraded oil (Rubinstein et al., 1979; Arefjev et al., 1980).

2.2 Oil–Oil Correlation Examples

Oil–oil correlation is sometimes straightforward. In such cases, a simple comparison of the GC records of the saturated hydrocarbons may clearly demonstrate identical patterns for several groups of compounds (n-alkanes, branched and cyclic alkanes). Such an example is given in Figure V.2.2.

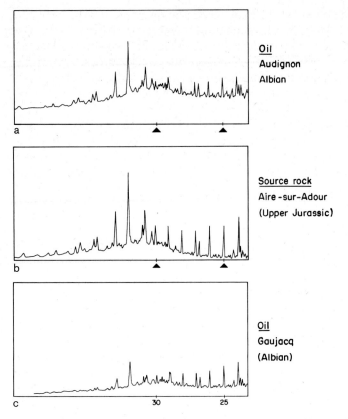

Fig. V.2.2 a–c. Comparison of the gas chromatographic tracings of the saturated hydrocarbons as a means for oil–oil and oil–source rock correlation. South Aquitaine Basin, France. (After Deroo, 1976)

Koons et al. (1974) studied petroleum characteristics of crude oils in lower Tuscaloosa Cretaceous reservoirs in the central Gulf Coast area, USA, utilizing a multiple-parameter approach. The parameters chosen for *oil–oil correlation* were the carbon isotope values of the saturates, the paraffin content of the saturated hydrocarbon fraction, the amount of steranes, and some light hydrocarbon ratios, such as those of the cyclopentanes to n-paraffins. Utilizing these four parameters, crude oils reservoired in the lower Tuscaloosa formation could be separated into two families. One family appears to be indigenous to the lower Tuscaloosa interval, which consists of sediments deposited in an intermediate-type (brackish-marine), nearshore environment receiving varying amounts of terrestrial organic matter. The carbon isotope values of the indigenous lower Tuscaloosa oils (δ^{13}C = -27.9 ± 0.6‰ PDB) are lighter by about 2‰ than those of the other oil family. They also contain more paraffins (27 vs. 13%), twice the amount of steranes and a higher ratio of cyclopentanes to n-paraffins. The relatively light carbon isotope values and the high paraffin content are in agreement with a terrestrially influenced source, as represented by the lower Tuscaloosa. The other family of

2.2 Oil-Oil Correlation Examples

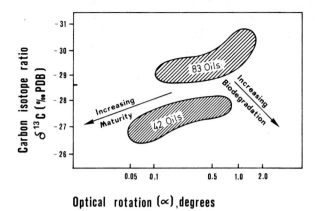

Fig. V.2.3. Relation between carbon isotope ratios and optical rotation in crude oils of the Williston Basin, USA. Two families of crude oils can be recognized. Increasing maturation of oils causes a shift to the *lower left*. (Modified after Williams, 1974). Biodegradation would have caused a shift to the *lower right corner* of the diagram

oils is probably derived from more marine organic matter and is apparently not indigenous to the lower Tuscaloosa formation.

Three different types of oil have been recognized by Williams (1974) in the Williston Basin in the northwestern United States on the basis of carbon isotope ratios, hydrocarbon composition and optical rotation measurements. A semi-log plot of carbon isotope ratio versus optical rotation was helpful in characterizing oil types (Fig. V.2.3). This method is of special interest with respect to recognition of maturation and biodegradation effects upon crude oils (see also Chap. IV.5). Thermal maturation processes decrease optical rotation (Louis, 1968) and cause a shift toward heavier carbon isotope values in the residue (Silverman, 1967). Contrary to this, biodegradation increases optical rotation (Philippi, 1977) due to a selective removal of optical enantiomers from racemic mixtures present in the original crude oil and/or addition of new optically active biogenic compounds. However, biodegradation, like thermal maturation, also causes a shift toward heavier carbon isotope values due to the preferential cleavage of the $^{12}C-^{12}C$ bond, and hence loss of the isotopically lighter carbon atoms. In Figure V.2.3, analytical data, representing 125 oil samples, show two distinct elongated clusters oriented in the graph from upper right to lower left. This elongation toward lower optical rotation and heavier carbon isotope values is the result of thermal maturation on these oils (Williams, 1974). The separation into two different oil families with different isotope values is due to their origin from different source beds. In such a plot as presented in Figure V.2.3, biodegradation would result in a shift from upper left to lower right.

Recently the identification of crude oil families and also the correlation of crude oils with their source rocks based on sulfur isotope ratios has been performed by Thode (1981) in the Wiliston Basin (USA/Canada). Sulfur isotope ratios are expressed as $\delta^{34}S$ values which are defined as follows:

$$\delta^{34}S\ (\text{‰}) = \frac{(^{34}S/^{32}S)\ \text{sample} - (^{34}S/^{32}S)\ \text{standard}}{(^{34}S/^{32}S)\ \text{standard}} \times 1000.$$

The oils of the Williston Basin were distinguished by the $\delta^{34}S$ values of their sulfur-containing compounds in the bulk oil. Three major types of crude oil were found. The $\delta^{34}S$ values of the three families were $5.8 \pm 1.2‰$, $2.8 \pm 0.8‰$ and $-4 \pm 0.7‰$ respectively. Long-distance secondary migration over some 150 km resulted in little or no change in the $\delta^{34}S$ value. However, elevated temperatures or advanced oil maturation and crude oil alteration, such as water washing and biodegradation, may change sulfur isotope ratios of the bulk oil.

Considerable efforts have been undertaken in recent years to develop the application of biological marker (geochemical fossil) parameters, in combination with others, for oil–oil correlation (e. g., Seifert and Moldowan, 1978, 1981; Shi et al., 1982; Welte et al., 1982). They all use steroid and terpenoid hydrocarbon distributions derived from capillary gas chromatography/mass spectrometry as major source-related correlation parameters, an analytical technique that makes use of a mass spectrometer as a very specific detector for molecules being eluted

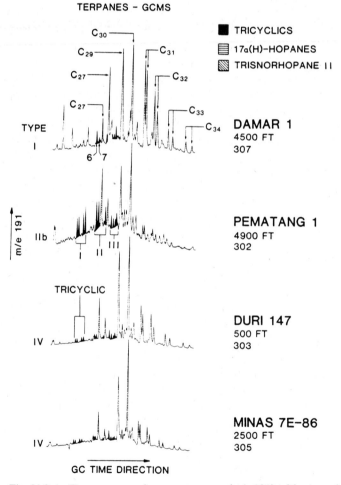

Fig. V.2.4. Terpane mass fragmentograms (m/z 191) of four crude oils from different oil fields in Sumatra indicating three oil families based on compositional characteristics. (After Seifert and Moldowan, 1981)

2.2 Oil-Oil Correlation Examples

from the capillary of the gas chromatograph. This type of analysis benefits from the fact that series of homologs or certain isomers of steranes, polycyclic terpanes and aromatic steroid hydrocarbons may be represented by the mass fragmentogram of a single mass spectrometric key fragment for each compound type. As an example, Figure V.2.4 shows the m/z 191 terpane mass fragmentograms of four crude oils from different oil fields in Sumatra. The origin of the 191 mass fragment is mainly from tricyclic and pentacyclic terpanes. From these data three oil families can be differentiated based on their characteristic triterpane compositions. It was realized early, however, that biological marker distributions of crude oils are strongly influenced by their temperature history (i.e., maturity) and

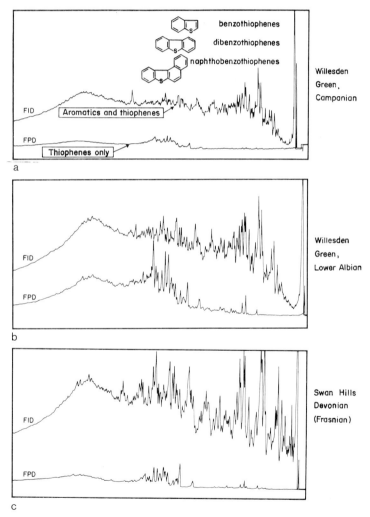

Fig. V.2.5a–c. Comparison of the gas chromatographic tracings of nonhydrocarbon compounds (benzothiophenes, dibenzothiophenes, and naphthobenzothiophenes) as a means for oil–oil correlation. (After Deroo, 1976)

possibly also by migration effects (Seifert and Moldowan, 1978). Implications related to this will be discussed in more detail in context with oil–source rock correlation.

An application of nonhydrocarbons for crude oil correlation was illustrated by Deroo (1976). He used thiophenic compounds to characterize the crude oils from the Alberta Basin of Western Canada. The thiophenic compounds used for correlation comprise mainly benzothiophenes, dibenzothiophenes and naphthobenzothiophenes (Fig. V.2.5). For crude oil characterization, a gas chromatographic record of the total aromatic fraction is used employing a conventional flame ionization detector (FID). Simultaneously, a portion of the effluent is split off after leaving the gas chromatographic column and relayed to a detector (flame photometric detector, FPD), which is only sensitive to sulfur-containing compounds. In this way, two gas chromatographic records are obtained, one of the total aromatic fraction inclusive of thiophenes, and another showing only sulfur-containing aromatic compounds (thiophenes). An example of this type of analysis is given in Figure V.2.5. Three different types of oil can be recognized. The aromatic and thiophenic gas chromatographic records allow a differentiation of the groups of oils located (a) above the unconformity in the Upper Cretaceous (Colorado group and younger), (b) at or near the main unconformity in the Alberta Basin, and (c) below the unconformity in the Devonian. The three different types of oil also differ in total sulfur content.

Crude oils become increasingly similar with maturation. In particular, the saturated hydrocarbons are progressively less characteristic; polycyclic naphthenes (geochemical fossils) are greatly diminished in concentration, and tend to be less distinctive; n-alkane and iso-alkane distribution patterns look more and more alike. Therefore, a correlation between older and more mature crude oils, based on saturated hydrocarbons, is rather difficult and might yield ambiguous results. In such cases naphthenoaromatic and aromatic hydrocarbons and thiophenic compounds have been successfully applied to oil–oil correlations (Tissot et al., 1974; Deroo, 1976; Rullkötter and Welte, 1980).

Figures V.2.6 and V.2.7 give examples of crude oil correlations based only upon the aromatics and naphthenoaromatics of the Paleozoic oils from the Illizi Basin (Algeria). A global overall analysis of these crude oils, and gas chromatography and mass spectrometry of the saturated hydrocarbons does not allow a distinction between oils reservoired in Ordovician, Lower and Upper Devonian and Carboniferous strata. However, a mass spectrometric analysis with respect to type of ring structure and carbon number of the aromatics and naphthenoaromatic hydrocarbons provides a fair differentiation between the groups of oils. The differentiation is mainly based on tetracyclic molecules very probably derived from steroids. Although the saturated tetracyclic molecules are hardly detectable any more in these oils, there are sufficient tetracyclic molecules present in the aromatic fraction. Aromatized tetracyclic molecules, such as the monoaromatic structures, containing one aromatic and three naphthenic rings, are more stable and hence are preserved in more mature oils. The total aromatics and the monoaromatic tetracyclic molecules used for oil–oil correlations in the Illizi Basin served also as correlation parameters for a successful oil–source rock correlation (Tissot et al., 1974).

2.2 Oil-Oil Correlation Examples

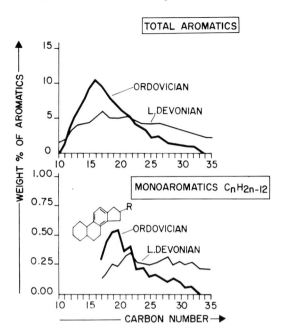

Fig. V.2.6. Oil–oil correlation based on aromatics and naphthenoaromatics. Comparison between crude oils from Lower Devonian and Ordovician reservoirs in the Tinfouyé-Tabankortoil field, Illizi Basin, Eastern Sahara, Algeria. (After Tissot et al., 1974)

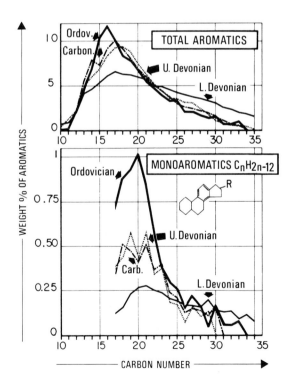

Fig. V.2.7. Oil–oil and oil–source rock correlation based on aromatics and naphthenoaromatics. The distribution of the naphthenoaromatics (monoaromatics, C_nH_{2n-12}) permits a discrimination between three different types of oil which also can be related to their source rocks. Carboniferous and Devonian oils are from the oil field Tigouentourine, the Ordovican oils from Quan Taredert in the Illizi Basin, Eastern Sahara, Algeria. (After Tissot et al., 1974)

In addition to maturation effects, biodegradation hampers or makes impossible oil–oil correlation. In slightly or moderately biodegraded crude oils, characteristic petroleum components have to be selected which are not readily eliminated or altered by microorganisms. Positive correlation was indicated by Moldowan and Seifert (1980) between a non-biodegraded and a moderately biodegraded crude oil from Minas and Duri oil fields, Sumatra, based on the presence of a rare isoprenoid biological marker, namely botryococcane, and a series of common isoprenoid alkanes in both oils (Fig. V.2.8).

Steranes and triterpanes are often found unaffected by biodegradation in moderately degraded crude oils (Welte et al., 1982) and thus in these cases are still suitable for correlation. Eventually, however, microorganisms seem to be able to attack steranes and triterpanes (see also Chap IV.5). In this case steranes are preferentially destroyed (Reed, 1977; Seifert and Moldowan, 1979), whereas triterpane structures may be specifically altered, leading to demethylation and ring opening (Rullkötter and Wendisch, 1982). Figure V.2.9 shows the capillary gas chromatograms of the saturated hydrocarbon fractions of a non-biodegraded (well VHD-2) and a biodegraded asphalt (well BML-1) from Morondava Basin, Madagascar, which demonstrate the complete removal of n- and isoprenoid alkanes in the BML-1 asphalt leaving behind the hopane-type triterpanes as only major components. The adjacent mass fragmentograms clearly show that the

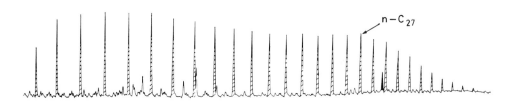

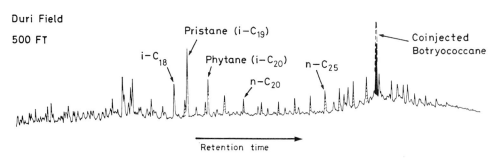

Fig. V.2.8. Capillary gas chromatograms of two Sumatra crude oils indicating positive correlation despite biodegradation by the presence of botryococcane in both oils. The biodegraded crude oil at the bottom shows a predominance of branched isoprenoid alkanes over n-alkanes. (After Moldowan and Seifert, 1980)

2.2 Oil-Oil Correlation Examples

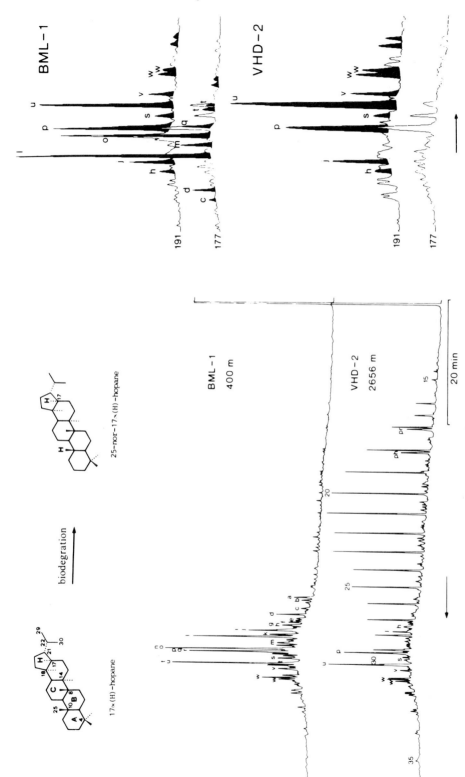

Fig. V.2.9. Capillary gas chromatograms (saturated hydrocarbon fractions) and terpane mass fragmentograms (m/z 177 and 191) of a biodegraded (well *BML-1*) and a non-biodegraded asphalt (well *VHD-2*) from the Morondava Basin, Madagascar. All *n*-alkanes and isoprenoid alkanes are removed in the biodegraded asphalt, the remaining major peaks in the gas chromatogram representing hopane-type triterpanes. The mass fragmentograms show the shift of the regular hopane series (m/z 191) to m/z 177 in the biodegraded asphalt caused by demethylation of the hopanes at C-10

regular C_{27} (h, j) to C_{31} (w) hopane series is accompanied by an equivalent series of demethylated hopanes (m/z 177) in the biodegraded asphalt (c, d corresponding to h, j. respectively, etc.).

Polyaromatic hydrocarbons and thiophenic compounds appear to be even less affected by biodegradation than are polycyclic naphthenes (steranes, terpanes). An example of correlating oils with various degrees of biodegradation is presented by Claret et al. (1977). Oils from the offshore oil field Emeraude in the Congo, West Africa, show different gross chemical properties, depending on reservoir depth and stratigraphic horizon. The oil bearing horizons of Cretaceous age in the Emeraude field are situated at relatively shallow depth (200 to 600 m). The oils are obviously affected by the products of biodegradation, and the question arises whether or not these oils might have a common origin. A detailed gas chromatographic and mass spectrometric analysis revealed the typical signs of biodegradation (e.g., loss of normal and isoprenoid alkanes). Furthermore, the oils show good correlation on the basis of their aromatic and thiophene components (Fig. V.2.10). This points to a common origin. In addition to GC records of the total aromatics, a more detailed mass spectrometric analysis of the distribution of mono-, di-, and polycyclic aromatics provided corroborative evidence for a common origin.

Similarly, Welte et al. (1982) provided evidence by low eV gas chromatography/mass spectrometry of aromatic hydrocarbon fractions that biodegraded and non-biodegraded crude oils from the Vienna Basin belong to the same oil family. These data were supported by sterane and triterpane mass fragmentograms. The oils investigated were from reservoirs between 5500 and 800 m depth. The

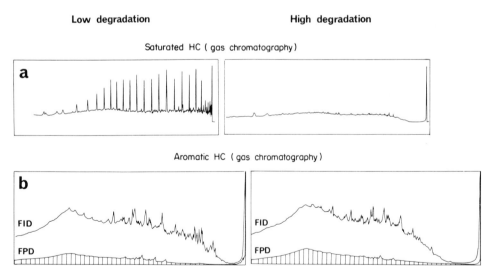

Fig. V.2.10a and b. Comparison between degraded and nondegraded crude oils in the Emeraude oil field, Congo. The gas chromatographic tracings of the saturated hydrocarbons (a) are different due to degradation. However, polycylic molecules still show an excellent correlation. (b): gas chromatographic tracings of total-aromatics *(FID)* and thiophene derivatives *(FPD)*. (After Claret et al., 1977)

compositional similarities of most of the oils in the Vienna Basin, apart from biodegradation effects, point to a common origin.

Finally, very heavily degraded oils can also be correlated by pyrolyzing the asphaltene fraction. In this way, the compositional pattern of the original crude oil can be reconstructed: this feature has been described in Chapter IV.1.8.

2.3 Oil–Source Rock Correlation Examples

In principle for oil–source rock correlation the same methods are utilized as mentioned before for oil–oil correlation. For instance, Figure V.2.2 shows that the GC record of the saturated fraction of two crude oils from the South Aquitaine Basin in France can easily be compared to the equivalent GC record of the source rock. The general approach, however, is somewhat different, in that more detailed analyses are required. This is so because a pooled oil has experienced processes of migration which may have partly changed its composition from that of the corresponding bitumen of the source rock. Oil–source rock correlation parameters, therefore, should include detailed information on hydrocarbon distribution patterns of specific molecules, such as steroid or triterpenoid hydrocarbons, which exhibit comparable physicochemical behavior. Furthermore, correlation parameters should be chosen from among at least two different molecular weight ranges, e. g., C_{12}–C_{25} saturates and aromatics and from C_{27+}-cyclic hydrocarbons. Gross chemical information alone, like that based on the amount of saturated normal or aromatic hydrocarbons, sulfur content, or even carbon isotope ratios, is generally not sufficient, but often gives valuable additional information. In oil–source rock correlation the indigenous nature of the bitumen in a given source rock should also be verified, especially if there is suspicion of oil impregnation of a rock from an outside source. Indications for this can be derived from too high hydrocarbon: total organic carbon ratios (above 150–200 mg hydrocarbons: g C_{org}) and from too high bitumen: total organic carbon ratios (above 200–300 mg bitumen: g C_{org}). By far the most convenient method is a succession of Rock-Eval measurements presenting free oil and gas contents and transformation ratios (production index): Figure V.1.22. Any zone of migrational hydrocarbon enrichment will exceed the established trend. It can also be achieved by comparing the maturity of the bitumen by a chemical maturation parameter such as the Methyl Phenanthrene Index values (Radke et al., 1982) with the maturity of the immobile kerogen (e. g., vitrinite reflectance).

Early examples for oil–source rock correlations based on several correlation parameters were presented by Welte et al. (1975). Among the correlation parameters used were the distribution patterns of steranes and triterpanes. Six molecular ion intensities of homologous series were determined by GC/MS and relative distributions established in source rock bitumens and oils (Fig. V.2.11). Other parameters used were ratios of two or more hydrocarbons, the C_{16} to C_{30} n-alkane series, the isoprenoid hydrocarbons and carbon isotope ratios. When making conclusions about a genetic relationship for an oil–source rock pair, the

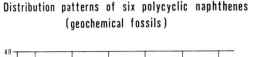
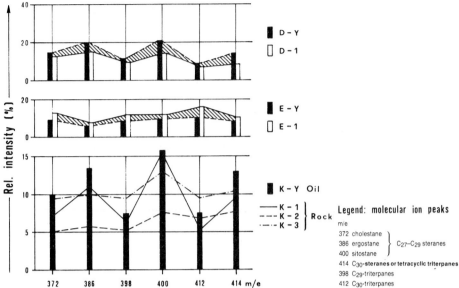

Fig. V.2.11. Distribution pattern of sterane and triterpane molecular ion intensities used as correlation parameter for oil–source rock correlation. (After Welte et al., 1975)

highest decisive value was assigned to the C_{27+}-cyclics. Figure V.2.11 shows three different oil–source rock situations from top to bottom:

- D-Y/D-1; an oil–source rock pair of Oligocene age from the Upper Rhine-Graben in SW Germany (Wild Cat well at Hüttenheim),
- E-Y/E-1, an oil–source rock pair of Eocene age from the Uinta Basin in Utah (Bluebell oil field),
- K-Y/K-1, K-2, K-3; a source rock sequence from the Upper Rhine-Graben in SW Germany. The oil K-Y was collected in the Forst oil field from a reservoir of Eocene age. The Jurassic rock samples K-1, K-2, K-3, all exhibiting source rock properties, are from the same oil field and represent the Dogger alpha (K-1), the Lias epsilon (K-2) and the Lias beta (K-3).

Based on the C_{27+}-cyclics, the correlation between oils and source rocks for the pairs D and E was considered to be good. In case of the multiple-choice source rock situation K, the closest relationship can be established between the Dogger alpha shale (K-1) and the Tertiary oil (K-Y).

In this same study Welte et al. (1975) also showed that of the more than 10 oil–source rock pairs investigated, no one showed complete agreement between all correlation parameters applied. This indicates the necessity to apply several independent correlation parameters. In a follow-up paper, Leythaeuser et al. (1977) expanded on this approach to oil–source rock correlation, and replaced

2.3 Oil-Source Rock Correlation Examples 563

the visual comparison between different distribution patterns (correlation parameters) by a statistical analysis. They also showed with the help of the C_{27+}-cyclics distribution pattern that the homogeneity of some source rock units as a function of sampling location was excellent.

An example of how to approach source rock/crude oil correlation in a complex geological system by a more sophisticated biological marker technique was presented by Seifert et al. (1980) for the area of the Prudhoe Bay oil field, Northern Alaska. The molecular structures used for fingerprinting the shale extracts and the crude oils included steranes, terpanes and monoaromatic steroid hydrocarbons. The diagrammatic east–west cross-section of the Prudhoe Bay oil field area in Figure V.2.12 shows the generalized structural and stratigraphic situation and the approximate origin of the samples analyzed by GC/MS. Based on their data, Seifert et al. (1980) concluded that the Shublik (Triassic), Kingak (Jurassic) and the deep post-Neocomian shales were the major sources of the principal oil accumulations. Other possible source rocks on a geological and bulk geochemical reasoning, like the deep Kayak (Mississippian) shale, the shallow post-Neocomian, the Neocomian and the Upper Cretaceous shales were discarded after biological marker analysis as the origin of the accumulated petroleum. This is shown for these four shales and the Sadlerochit oil (202 in Fig. V.2.12) by the sterane mass fragmentograms (m/z 217) in Figure V.2.13a. Black shaded peaks indicate for each shale sample where they differ significantly from the oil so that they can be ruled out as its source. A similar lack of correlation

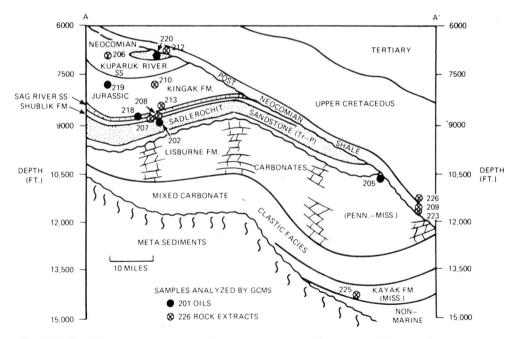

Fig. V.2.12. Diagrammatic east–west cross-section, Prudhoe Bay field area, showing generalized structural and stratigraphic relationships and the approximate location of the samples analyzed by GCMS. (After Seifert et al., 1980)

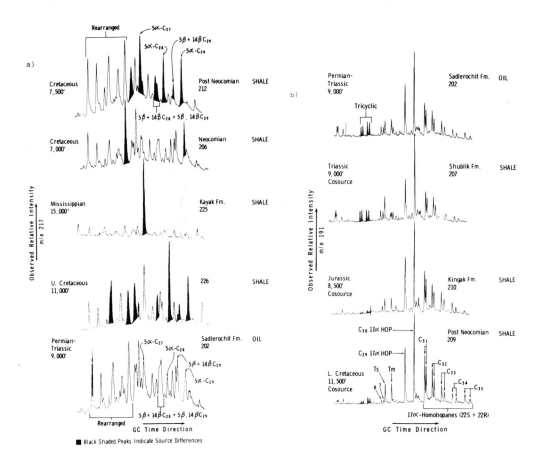

Fig. V.2.13a and b. Prudhoe Bay, Alaska, source rock–oil correlation (Seifert et al., 1980). (a) Proof of lack of correlation by sterane mass fragmentograms (m/z 217). (b) Key source rock–oil correlation for the Sadlerochit oil by terpane mass fragmentograms (m/z 191)

between the same rocks and the Sadlerochit oil was evident from the terpane mass fragmentograms (m/z 191). In contrast to this, the terpane mass fragmentograms in Figure V.2.13b provide the key for positive correlation of the Sadlerochit oil which is similar in source characteristics to all other oils from the Prudhoe Bay oil field area except the Kingak oil (219). All biological marker distributions, gross parameters and carbon isotope values confirm a one-to-one relationship between the Kingak oil (219) and the Kingak shale as its source. The Sadlerochit oil (202) however, shows hybrid values which can be explained by a combination of hydrocarbons from three different sources, namely the Shublik shale, the Kingak shale and the post-Neocomian shale, which among themselves exhibit, major compositional differences, e.g., in their triterpane mass fragmentograms (Fig.

2.3 Oil-Source Rock Correlation Examples

V.2.13), but also in their sterane and monoaromatic steroid hydrocarbon mass fragmentograms. The clue to the multiple source concept for the Sadlerochit oil (and those with similar composition) was finally derived from a quantitative assessment of the source specific biological marker parameters (Table V.2.1). This quantitation provides three aspects:

a) the consistency of data within a given source rock formation and a clear distinction between different formations;
b) the one-to-one correlation of the Kingak oil and the Kingak shale;
c) a multiple source concept as the best explanation for the composition of the Sadlerochit oil.

Source rock/crude oil correlation by biological marker techniques is a powerful tool as long as it can be reasonably well assured that the original biological marker pattern of the source has not been altered by subsequent processes such as migration, maturation or biodegradation of a crude oil. The effect of biodegradation on biological markers has already been mentioned (see Chaps. IV.5 and V.2.2). Usually more serious, however, is the influence of thermal maturation on the biological marker distribution: it can influence both the pattern and the total abundance (see Chap. V.1.3). Steroid hydrocarbons for instance undergo significant isomerization and aromatization reactions in the early phase of petroleum generation (Mackenzie et al., 1982a), and thus fingerprint correlation may fail if the source rock, although the right one, is not available at the suitable maturation level. In these cases carbon number distributions of regular steranes (C_{27}–C_{29}), often displayed in triangular diagrams, may still be used for correlation. Carbon number distributions of steranes, in contrast to isomer distributions (see Chap. V.1.3), appear to be largely unaffected by thermal maturation but

Table V.2.1. Source-specific parameters for three major source rocks and two different oils in Prudhoe Bay, Alaska. (After Seifert et al., 1980)

GCMS No.	Formation	Age	Tricyclic Terpanes, %[c]	$17\alpha(H)$- Hopanes, %[c]	$C_{28} \dfrac{5\beta + 14\beta^{a}}{5\alpha^{b}}$ Steranes
Shale 210	Kingak	Jurassic	2.3	64	3.9
Shale 213	Kingak	Jurassic	2.7	64	2.8
Shale 207	Shublik	Triassic	8.5	52	6.0
Shale 208	Shublik	Triassic	11.7	46	6.1
Shale 209	Post-Neocomian	Lower Cretaceous	4.0	65	4.4
Shale 223	Post-Neocomian	Lower Cretaceous	3.1	67	3.9
Oil 202	Sadlerochit	Permo-Triassic	8.9	55	4.1
Oil 219	Kingak	Jurassic	2.2	63	4.9

[a] C_{28} 5β, 14α, 17α(20R) + C_{28} 5α, 14β, 17β(20R + 20S) + C_{28} 5β, 14β, 17β(20R + 20S)
[b] C_{28} 5α, 14α, 17α(20R)
[c] Area % of peaks on m/z 191 mass fragmentogram

rather represent the original organic matter composition defined by the environmental conditions during sedimentation (Huang and Meinschein, 1979; Shi et al., 1982; Mackenzie et al., 1982b). Another limiting factor for the use of biological markers for correlation parameters is related to the fact that these compounds are often relatively minor components within the hydrocarbon mixture generated from a source rock and that they are thermally less stable than most of the others. Recent quantitative biological marker determinations in source rocks suggest that their concentration may drop by a factor of more than one hundred from the late diagenesis to peak oil generation stage. This effect may be caused both by the overwhelming generation of other hydrocarbons and by the early thermal breakdown of biological markers. As a consequence, mature crude oils may contain very little biological marker components. Any later addition to such a mature crude oil from a second source which is quantitatively less important, but less mature and therefore richer in biological markers, may give an imprint on a reservoired crude oil which is not consistent with the bulk, more mature, hydrocarbons. This again demonstrates the necessity of using multiple parameters in source rock/crude oil and also in oil–oil correlation.

An interesting example for an oil–source rock correlation, based on thiophenic compounds and aromatics, was given by Deroo (1976). He demonstrated that the Rousse condensate from the Arzacq Basin in SW France, reservoired in the Upper Jurassic, had a double origin, from source rocks of Albian and Oxfordian age (Fig. V.2.14). With the help of the distribution patterns of the diaromatics (type C_nH_{2n-12}) and dibenzothiophenes, even the relative contribution of each source could be estimated. This is possible as both source rocks are at about the same maturation level.

Oil–source rock correlation based on characteristic compositional features of the light hydrocarbon fraction was performed by Schaefer et al. (1978b). The availability of sufficiently accurate concentration data for the light hydrocarbons in a large number of rock samples, obtained by the "hydrogen stripping" technique, permitted the application of a statistical approach to this problem. First, a number of characteristic light hydrocarbon concentration ratios (correlation parameters such as methylpropane/n-butane, methylbutane/n-pentane, 2-methylpentane/3-methylpentane, cyclohexane/methylcyclopentane, etc.) were calculated for each sample. Based on these parameters a correlation coefficient matrix is produced by a computer program, which, upon request, may plot a dendrogram. As shown in Figure V.2.15, the compositional similarities between one crude oil and the hydrocarbons in a large number of rock samples from an exploration well (Porcupine Basin, offshore Ireland) may be demonstrated by plotting the correlation coefficients between oil and rocks against depth. On the basis of a similarity limit of 0.80 for the correlation coefficient, two intervals show significant similarities with the oil, the shallower interval exhibiting the greater similarity. From other geochemical information it was deduced that this interval is not a source rock, but a zone which is impregnated by migrated petroleum. The lower interval or the appropriate lateral equivalents were considered as the likely source rock for the oil.

Correlation of crude oils with their oil source formation, using high resolution gas chromatography of C_6–C_7 hydrocarbons, was performed by Philippi (1981).

2.3 Oil-Source Rock Correlation Examples

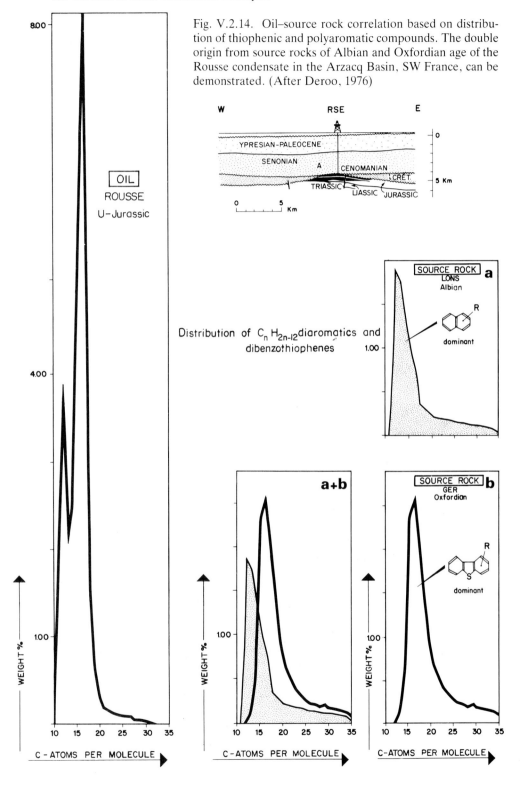

Fig. V.2.14. Oil–source rock correlation based on distribution of thiophenic and polyaromatic compounds. The double origin from source rocks of Albian and Oxfordian age of the Rousse condensate in the Arzacq Basin, SW France, can be demonstrated. (After Deroo, 1976)

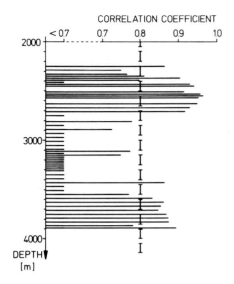

Fig. V.2.15. Compositional similarities between oil and rock samples (Porcupine Basin, offshore Ireland). Cluster analysis of one crude oil and 51 rock samples from an exploration well based on eight different parameters (concentration ratios) calculated from light hydrocarbon data. The similarity values (correlation coefficients) between the rock sample and the oil sample are plotted against depth. A similarity limit of 0.80 is represented by a *broken line* (Schaefer et al., 1978b)

The concept is similar to the previous example. Ideally, a given composition parameter should have the same value for a crude oil and the source rock which generated and expelled that crude oil. A "similarity coefficient" has been devised to define the degree of similarity. Philippi's method was tested on ten selected presumed oil–source formation pairs and it was concluded that it is functioning properly.

It has to be realized, however, that oil–source rock correlation based on characteristic compositional features of light hydrocarbons is only possible because thermodynamic equilibrium of isomer distributions is normally not reached during natural hydrocarbon generation (see also Table IV.1.3). Furthermore, this kind of correlation can only be applied when crude oil and source rock are at about the same level of maturation.

A special application of a kind of oil–source rock correlation is the prediction, and if possible, recognition, of a source rock based on certain characteristics of reservoired petroleums. Certain petroleums have been found to exhibit an even predominance of the n-alkanes in the range from C_{16} to C_{32}, which is frequently combined with a predominance of phytane over pristane. Degree and extent (with respect to C number) of these characteristics vary with type of oil and maturity. It has been shown that an even predominance of the n-alkanes is typical for highly saline, carbonate and/or evaporitic environments (Welte and Waples, 1973; Dembicki et al., 1976) and that this feature is very frequently accompanied by a phytane over pristane predominance (Waples et al., 1974; Kalbassi and Welte 1975). Thus a crude oil with these characteristics is most likely derived from a carbonate or evaporite type source rock (see also Chap. II.3 and Fig. IV.3.4). This conclusion was also verified by Tissot et al. (1977) by statistical analysis of n-alkane patterns of numerous rock extracts and crude oils.

Crude oils exhibiting an odd predominance of the higher n-alkanes, especially in the range C_{23}–C_{33}, and which simultaneously contain more n-alkanes above C_{20}

2.3 Oil-Source Rock Correlation Examples

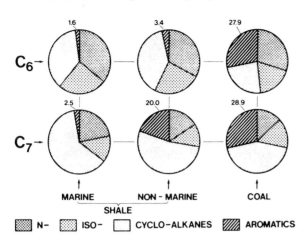

Fig. V.2.16. Hydrocarbon-type composition (%) of C_6 and C_7 compound groups for marine shale (type-II kerogen), nonmarine shale (predominantly type-III kerogen) and coals (mean values) from Upper Carboniferous (Westphalian) strata of Corehole V, Ruhr area, Federal Republic of Germany. This information should be compared to similar findings among C_{15+} hydrocarbons as presented in Figure II.5.15

than below, are most certainly derived from a source rock with a large proportion of terrestrial organic matter of higher plant origin (see also Fig. IV.3.3). Such crude oils have frequently a slight pristane over phytane predominance. However, the overall amount of isoprenoids is rather low. Oils of this kind usually originated from nonmarine source rocks, or from strata representing nearshore, brackish-marine environments.

Leythaeuser et al. (1979b) found that significant differences in the light hydrocarbon composition are associated with changes in the type of the organic matter they originate from. This dependence on the facies type is exemplified in Figure V.2.16 where the hydrocarbon-type composition of C_6 and C_7 compound groups are shown for marine shale, nonmarine shale and coals of advanced maturation level (about 1.2% R_o) from Upper Carboniferous (Westphalian) strata of a core hole in the Ruhr district. The most striking effect concerns the concentration of aromatics (benzene and toluene) with respect to the other nonaromatic hydrocarbons. In the case of the coal samples, both aromatic hydrocarbons represent a significantly higher proportion in the C_6 and C_7 groups as compared to the marine shales containing type-II kerogen, with an intermediate position for the nonmarine shales (predominantly type-III kerogen). For the cycloalkanes there is a tendency for higher relative proportions in the marine shale as compared to coal for the C_7 hydrocarbons. In summary, the C_6 and C_7 hydrocarbon mixtures generated from coaly organic matter are relatively rich in aromatic and poor in naphthenic compounds, in comparison with those found in marine shales at equivalent maturation level. The predominance of benzene and toluene appears to reflect the influence of higher plants. Similar observations were made for two adjacent source-bed type shale intervals, bearing type-III and type-II kerogen, respectively, in the main phase of petroleum generation, of Jurassic age in the Porcupine Basin (Schaefer and Leythaeuser, 1980). Comparable observations on C_{15+} medium to heavy hydrocarbons have been presented in Figure II.5.15.

Recognition of typical facies characteristics in crude oil or bitumen has not yet been extensively developed. However, more research in this direction will certainly help to characterize crude oils in more detail, which in turn will allow the prediction of the type of source rock from which it probably derived. In Chapters I.4 and IV.3, some data are presented and thoughts developed which are pertinent to recognition of depositional environments and type of biomass deposited.

Finally, it should be remarked that such features as the existence of irregularities in the n-alkane distribution pattern in oils, or an even or odd predominance etc., is always a sign of relatively low maturity. More mature crude oils show smooth n-alkane distribution curves. Such knowledge of the origin of an oil with respect to facies and level of maturation may be of great help when developing an exploration concept.

Summary and Conclusion

Most important objectives in oil and source rock correlation are the recognition of various families of pooled hydrocarbons in a basin, and the identification of the source rock of a given petroleum. Such knowledge helps to define exploration targets. Geochemical correlation is based upon recognition of compositional similarities between petroleums and source beds. Relative distribution patterns of hydrocarbons and nonhydrocarbons are typically used as correlation parameters, especially those with very specific chemical structures, such as compounds of the steroid and terpenoid classes. $^{13}C/^{12}C$-isotope ratios are also used, especially to relate oil and bitumen to a certain kerogen. It is advisable to use always more than one correlation parameter.

Oil–oil and oil–source rock correlation is hampered by both strong maturation or biodegradation of pooled petroleums. Naphthenoaromatic and aromatic hydrocarbons and thiophenic compounds have been successfully applied for oil–oil-correlation of matured oils. Aromatic hydrocarbons and thiophenic compounds have been used to correlate biodegraded oils in addition to biological markers of the sterane and triterpane type which are not always affected by biodegradation.

Chapter 3
Locating Petroleum Prospects: Application of Principle of Petroleum Generation and Migration — Geological Modeling

In the early days of petroleum exploration, wells were drilled where oil or gas seepages indicated the presence of petroleum. Later, with increasing sophistication of geological knowledge, and especially with the advent of exploration geophysics, the decision to drill a well was additionally taken on the basis of the recognition of suitable structures, such as anticlines, fault traps, unconformity traps and reefs. Frequently, the selection of a structure to be drilled was based on intuition and general experience rather than on pertinent information, because very little information was available whether or not a trap would contain hydrocarbons.

A systematic study, utilizing the new understanding of petroleum generation and migration, can help to decrease the uncertainty in predicting a petroleum-filled structure, and hence the financial risk when drilling a well. This is of special importance in offshore areas and remote, hostile exploration regions where drilling is extremely costly. The consequent application of geochemistry can also define new exploration targets in relatively well-known basins where the rate of discoveries has declined.

The purpose of this chapter is to show how petroleum exploration can benefit, firstly by collecting source rock and maturational information (acquired at relatively low cost) from application of organic geochemical studies, and secondly by relating this knowledge intelligently to the geological framework of a given exploration area. The basic idea of this concept is to identify the source rocks present, the hydrocarbon potential of each source rock in time and space, and relate this information to the geological evolution of the basin. At the end of such a study, most favorable zones for petroleum accumulation are located in the basin. This is done by relating the hydrocarbon generation of a source rock at any given time during basin evolution to the most likely paths of petroleum migration, and to the formation and age of a trap.

The determination of the most favorable zones of petroleum accumulation in a basin consists of a sequence of steps to prepare and combine the necessary geological and geochemical information, and produce maps and sections showing and rating the potential targets. This can be done "by hand" in a conventional manner, or with the help of computers. In Figure V.3.1 a scheme is presented that shows the sequence of steps and, in a general way, the kind of geological information required in order to apply organic geochemical studies to petroleum exploration. This scheme can easily be followed in practice. It is no more difficult than the introduction of facies analyses, isopach maps, or depth contour maps has previously been. In the past, all these techniques have also been assimilated by

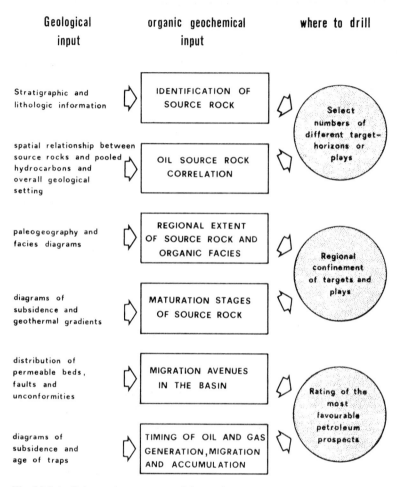

Fig. V.3.1. Schematic summary of the various steps that lead to the determination of most favorable zones of petroleum accumulation in a basin

petroleum geologists, and nowadays geochemical data should be incorporated in the philosophy of exploration.

Geological-geochemical modeling with the help of high speed computers is possibly the most effective, suitable, and certainly the most advanced method for determination of the favorable zones of petroleum accumulation. It integrates and quantifies the newly gained geochemical knowledge on petroleum formation and migration into the geological framework of basin evolution. This computer-aided numerical simulation of geological and geochemical processes has the enormous advantage that through the combination of regionally derived geological data with information about the principal processes of petroleum generation and migration, tailor-made models of a specific exploration area are available for the explorationist. In this way a ranking of exploration targets based on a network of logically consistent data can be established. This diminishes greatly the exploration risk.

The following chapter outlines the acquisition of the geochemical information (Sect. 3.1) and presents a first conceptual model of petroleum generation in a basin (Sect. 3.2). Then geological modeling is presented (Sect. 3.3), whereas geochemical modeling will be introduced in the next Chapter V.4.

3.1 Acquisition of the Geochemical Information

When a new exploration program is designed in a more or less unknown area, only a limited amount of general geological information is normally available. This, of course, applies nowadays only to certain remote areas and offshore regions. Sample material of the most important stratigraphic and lithologic units represented in the basin must be collected, outcrops at the rim of the basin being a readily accessible source. A careful study of the available information on the sedimentary filling of the basin (a facies analysis) might result in a preliminary *identification* of one or more *potential source rocks* or source rock sequences. A source rock-identification program using organic geochemical techniques to analyze available sample material should help to define potential source rocks more clearly. At the same time, neighboring nonsource rock strata should also be checked for porosity and permeability, and their role in offering possible migration avenues. If little or no sample material is available, a limited sampling program may be started to obtain surface samples and complementary unweathered core samples from a few shallow drill holes (up to 100 m depth) at strategic locations. In offshore basins, information from adjacent, comparable strata on land, and sometimes samples taken from the sea bottom by dredging or shallow coring may be helpful.

A different case is presented where basins have been abandoned some years ago, and where fresh interest arises for new exploration. The advent of the geochemical techniques and new knowledge of the principles of petroleum generation and migration frequently justifies a fresh look at old prospects. In such cases, old sample material from early wells can still be used to evaluate the source rock potential. However, in all cases, the first step in a basin study is to identify the number and kind of potential source rocks in the basin. The sample material has to be analyzed for amount and type of kerogen, and its stage of maturity.

In fact, the concept of source rock identification and determination of the main zones of oil and gas formation can be introduced into an exploration campaign at any stage. In particular, before a well is drilled, it is critical to ensure that a maximum of information can be gained with respect to both source and carrier rocks. To obtain this information from the well, an effective sampling program must be prepared. Major lithologies *and* regular depth intervals should be sampled. Cuttings should be collected at intervals of 3 to 30 m, depending on the changes in lithology. Preferably, samples should be canned in small air-tight containers (about 300–500 cc), with ample water to keep the original moist saturation, particularly if light hydrocarbon studies are intended. In addition, some core or side wall core samples should be obtained, in order to support and verify information gained from cuttings. Core sampling is already done routinely for large-scale testing of physical properties. The sample material should be subjected to geochemical analyses in such a way that basic analytical information on source rock properties etc. is available, ideally by completion of the well or, at least, with a minimum time-lag. Therefore, *rapid methods* have to be applied for source rock analyses. Such methods *should be considered as screening techniques*, rather than as a means to obtain detailed analytical results. Methods suitable for screening are the pyrolysis technique and the quantitative determination of light hydrocarbons. Both have been described in Chapter V.1.

These two techniques will provide independent *geochemical logs*, which can be readily interpreted in terms of hydrocarbon potential and level of maturity (main zones of oil and gas formation).

Even information with respect to hydrocarbon migration can be derived from these two logs. Analytical instruments for both kinds of screening methods can be installed and used directly at the well site. Under these circumstances, considerable time is saved because sample processing can be simplified (no canning, packing and transportation etc.), and results can be utilized while drilling is still in progress. The advantages of "real-time" geochemical logging are obvious, and certainly can help to cut costs and gain valuable information. For instance, well-head geochemical logging by pyrolysis can indicate when a overmature zone is reached, and no more oil is to be expected. Useless continuation of drilling, and hence the waste of money, can thus be prevented. Conversely, total amount of light hydrocarbons and certain hydrocarbon ratios may give an indication for nearby hydrocarbon accumulation, and thus stimulate further drilling and improve the success rate. Knowledge of a nearby accumulation will also stimulate necessary safety measures for further drilling.

In addition to the screening analyses, further sample material has to be analyzed in a well-equipped geochemical laboratory, to verify and complete the information gained from screening. The number of samples to be analyzed in more detail for source rock identification, type of kerogen, and maturity can be reduced considerably by utilizing the screening results to select zones of homogeneous organic facies, in particular those of most promising hydrocarbon potential. Whenever possible, a few core samples (side wall cores or others) should be included to support results obtained from cuttings and/or serve as fixed geochemical reference points in a well. Such detailed geochemical information on selected rock samples, based on various complementary physicochemical and optical microscopic techniques (Chap. V.1), serve to refine results from screening techniques and allow the improvement and updating of the preliminary conceptual model. In addition, analyses of pooled hydrocarbons possibly collected from formation tests (FT samples), can provide information on the source rock–oil relationship: identification of the source by direct correlation and the stage of maturity at the time of hydrocarbon expulsion can be attempted (Chap. V.2). In the case of severe thermal alteration occurring in the reservoir, such information

3.2 First Conceptual Model of Petroleum Generation in a Basin

cannot be obtained. In some cases, where source rocks have not yet been found, the analyses of pooled hydrocarbons may allow prediction of the type of source rock to be expected (Chap. V.2).

3.2 First Conceptual Model of Petroleum Generation in a Basin

The scientific approach for locating petroleum prospects, described below, can be introduced at any stage of exploration. In fact, it is desirable to update this determination of most favorable zones of petroleum accumulation continuously during the entire exploration campaign.

With the help of the geochemical data, a preliminary conceptual model of the basin with respect to source rocks and generation of petroleum is developed. For each source rock identified, a map is produced showing the regional extent of the source bed, its thickness, and depth level. It is also advisable to construct preliminary diagrams of the subsidence history for each source layer at strategic points in the basin (Fig. II.5.11). This probably could be done at a relatively early stage of the exploration campaign in basins situated at a passive continental margin. In the passive margin setting, the depth–time relationship is usually rather simple, as normally there has been minimal erosion. The stratigraphic identification of seismic marker horizons is the basis for the construction of diagrams of subsidence before the drilling of the first exploration wells. By interpretation of the maturity data of the rock samples investigated, the subsidence curves of source rock units help to predict approximate levels of maturity in different parts of the basin. In this way, it is possible to obtain a first idea about the depth level of the main zone of oil formation throughout the basin.

For a determination of the most favorable zones of petroleum accumulation the following series of questions has to be answered:

1. Which beds have a source rock potential? What is their regional extent and their relationship to paleography?
2. During which geological time interval, at what depth, and in which area of the basin were the above-mentioned source rocks mature enough to generate petroleum?
3. During which geological time interval, and along which preferred routes was petroleum migration possible?
4. At what geological time did the formation of traps occur, as compared to the timing of petroleum generation and migration? Where are these traps located, with respect to possible migration routes?

These questions can be answered on the basis of a set of maps and sections prepared "by hand". However, the quantitative approach using geological and geochemical models is a more effective method for answering the questions.

3.3 Numerical Simulation of the Evolution of a Sedimentary Basin — Geological Modeling

The starting point for a quantitative, mathematical-numerical handling of the formation and migration of petroleum is the simulation of the development of a sedimentary basin. The relevant factors of such an undertaking are represented schematically in Figure V.3.2. The idea is to retrace geological steps, which follow each other as a causal sequence beginning from an initial condition at time ($t = T_o$), over intermediate conditions ($t = T_1$), and to the basin configuration observed today ($t = T_x$). A particular difficulty is that only the end result at ($t = T_x$) — today's stratigraphic section, porosities, etc. — and not the initial condition at ($t = T_o$) is known for the observed sedimentary basin as empirical data available for comparison. The evolution of a particular sedimentary basin must be completely computed out, using estimated empirically derived initial data, and then the result must be compared with the real basin to see whether the present condition at time ($t = T_x$) has been reached.

The conceptual model is first set up for a real system, that is, for a real sedimentary basin, which naturally is less well known at the start of an exploration campaign than at the end. The conceptual model is the basis for a mathematical model, which then will be calculated through with estimated initial data. The first iteration will generally not achieve good correspondence with the values that describe the present-day condition of the sedimentary basin. Now the initial input data must be changed and/or the conceptual model modified as long as necessary for the calculations to give the geologically known end result. A flow diagram (Fig. V.3.3) illustrates this procedure schematically.

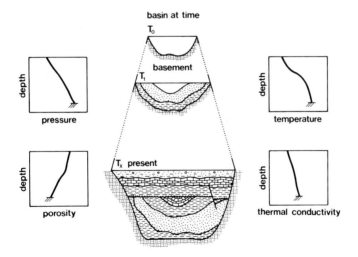

Fig. V.3.2. Evolution of a sedimentary basin from an initial condition at time ($t = T_o$) to the basin configuration observed today ($t = T_x$). Such parameters as pressure, porosity, temperature and thermal conductivity are changing continuously in each sedimentary unit with increasing depth of burial. (After Welte, 1982)

3.3 Numerical Simulation of the Evolution of a Sedimentary Basin

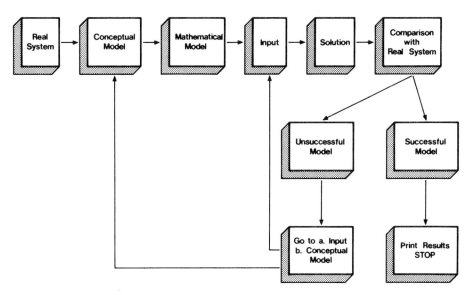

Fig. V.3.3. Flow diagram for three-dimensional quantitative basin model. (After Welte and Yükler, 1981)

The computed values are made to resemble those observed in nature within tolerance by means of successive iterations. In this way the simulation model works its way from a lower level of correspondence to another higher level until the whole simulation is ended. The fact that the most important parameters to be calculated are mutually dependent is of fundamental importance. It means that an inherent logic is followed through the iterations toward the most important parameters for the model, such as compaction and sediment thickness, or temperature and maturity. The core for the three-dimensional deterministic dynamic basin model is a pair of non-linear partial differential equations of energy and mass balance that describe the course of the respective physicochemical processes. The backbone of this approach was presented by Welte and Yükler (1981). A geological, a hydrodynamic, a thermodynamic, and a geochemical part of the program can be distinguished among the input data (Fig. V.3.4). The sedimentary basin to be investigated covering an area of many thousand km^2 and a depth of several km is divided up into a three-dimensional grid, rather, into elementary partial volumes.

The essential parameters, such as pressure, temperature, porosity, degree of maturity of the kerogen, etc., are then calculated with their variations over geological time spans for the numerous grid points, usually in the range of 100,000 of the three-dimensional grid. The necessity of repeating the programmed calculations at all grid points several times in a sequence of iterations was already pointed out. The computed values, which are mutually dependent, are adapted to each other, and they are brought into agreement with those measured in nature by this iteration process (Figs. V.3.3 and V.3.4). It is obvious that only the fastest and most capable of large computers can be used for such calculations.

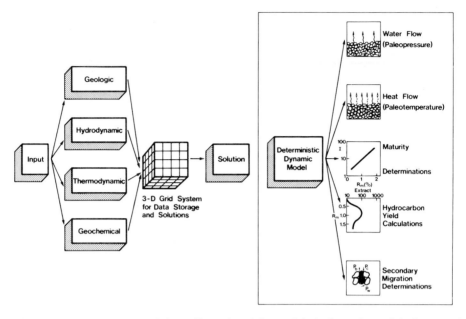

Fig. V.3.4. Development of three-dimensional deterministic dynamic model. Groups of input parameters and results after solution are schematically shown. (After Welte and Yükler, 1981)

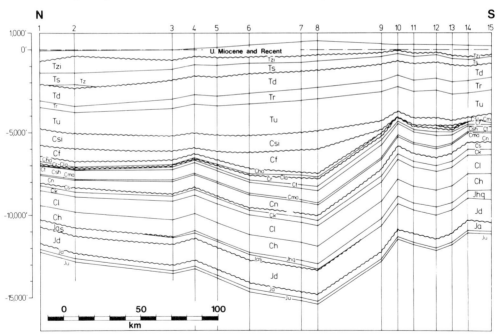

Fig. V.3.5. Computed cross-section through a real basin as based on quantitative basin modeling. Grid points in the horizontal plain are shown with numbers *1–15* (N–S). (After Welte, 1982)

3.3 Numerical Simulation of the Evolution of a Sedimentary Basin

The first important stage in the simulation of a basin's history from a geological standpoint is the calculation of cross sections (Fig. V.3.5) of the sedimentary basin that is under investigation. The comparison of the computed sedimentary thickness on the cross-sections with thickness actually measured in wells or seismic profiles is one of the first control possibilities for bringing concept and reality into agreement through purposeful iterations. How finely the cross-section is divided into discrete units in the horizontal and vertical dimensions is governed by both the complexity of the geology and by the level of geological information. The integration over time of the individual processes of the system leads to the computation of sedimentary thickness in the corresponding geological time spans.

One of the most important goals of integrated basin studies for petroleum exploration is the computation of the hydrocarbon potential of a particular source rock at every given point in time during the history of a sedimentary basin. This is possible with a relatively high level of confidence with the help of the numerical simulation of the basin's development. The maturity mapping of a source rock in space and time is an essential part when determining the hydrocarbon generating zone. In a first step, it is acceptable for an exploration campaign to establish a valid ranking of exploration targets in a given basin rather than to present a truly quantitative prediction of amounts of hydrocarbons. Whilst a truly quantitative evaluation necessitates the use of an elaborate kinetic model of kerogen degradation (presented in the next Chap. V.4), it is sufficient for a first evaluation of exploration targets to calculate in a semi-empirical manner vitrinite reflectance values (i. e., maturity). A relatively simple method to arrive at calculated vitrinite reflectance values using time–temperature relationships was, for instance, proposed by Lopatin (1971) and later modified by Waples (1980). This method is further discussed in the next chapter. Following this line of thought a present-day maturity map based on vitrinite reflectance values is presented in Figure V.3.6 for a source rock in an area of active exploration of about 165,000 km^2. These values can be converted into amounts of hydrocarbon generated per rock unit, if the quality and thickness of the source rock is known. In this example, the computed basin model typically contains about 350 data points for one slice of source bed extending over an area of 165,000 km^2. Maps of this kind can be drawn by the computer at any desired geological time for every parameter that is contained in the set of differential equations of the simulation program.

When the hydrocarbon potential of a source rock has been computed in space and time, this information can be combined in appropriate form with information about the presence of porous carrier and reservoir rocks and their geological position in the developing basin. Structural maps of such reservoir rocks or of any sealing rock can also be called up from the computer for any given point in time (Fig. V.3.7).

The numerical simulation of the geological development of a sedimentary basin makes possible an uncommon density of geological information and geological data of a completely new quality. The great advantage of the numerical simulation described here lies not only in the quantifying of geological parameters, which normally are only available through wells, but more importantly in the fact that a continuous net of geological information covering the whole area and

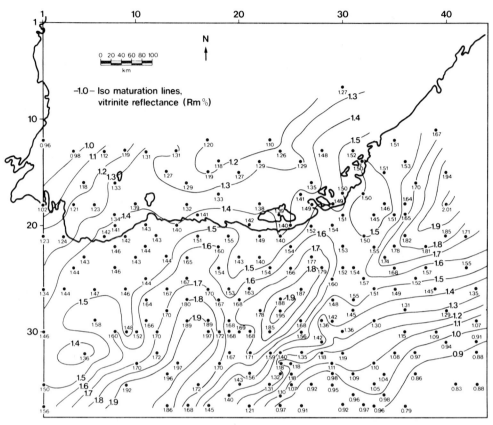

Fig. V.3.6. Computed vitrinite reflectance values for a source rock of a real basin for the present day as based on quantitative basin modeling. (After Welte, 1982)

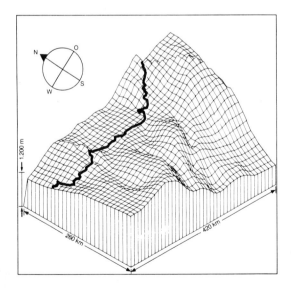

Fig. V.3.7. Three-dimensional subsurface contour map of the top of a reservoir formation in a real basin as based on quantitative basin modeling. (After Welte, 1982)

volume is made available in dependence with time. The computed maps have a considerably higher "reality content" than conventional geological subsurface maps through the mutual influence of individual parameters on each other and through the cross-checks with real geological control points.

It is quite obvious that the application of computer-aided modeling for geological and geochemical processes is only in its initial stages. Furthermore, it is realized that a truly quantitative prediction of hydrocarbon potentials and ultimate oil and gas reservees of sedimentary basins and not only a ranking of exploration targets is only possible if the most important geological and geochemical processes responsible for the formation of hydrocarbon accumulations are understood in greater detail than at present. Next to the geological modeling the geochemical modeling, especially of the kinetics of kerogen degradation and hydrocarbon generation, is of great importance in this connection. This will be discussed in the next chapter.

Summary and Conclusion

A systematic utilization of the new understanding of petroleum generation, migration, and accumulation can help to decrease the uncertainty in predicting petroleum-filled traps. This is done by identifying the source rocks, the hydrocarbon potential of each source rock in time and space, and by relating this information with the geological evolution of the basin. This determination of most favorable zones of petroleum accumulation in a basin can be done "by hand" in a conventional manner or with the help of computers.

In the initial stage of an exploration program, even with (limited) information on potential source rocks, a first conceptual model of petroleum generation in a basin is developed. In this model the relative position of potential source rocks, their presumed state of maturity and their extension throughout the basin are depicted in maps and profiles.

Such a conceptual model may provide valuable information for the selection of the site of exploratory wells. Rapid geochemical screening techniques can be applied at the well site to provide real-time geochemical logs which can be readily interpreted in terms of hydrocarbon potential of source rocks and level of maturity. Further sample material has to be analyzed in a laboratory, to verify and complete the information gained from screening.

A numerical simulation of geological and geochemical processes that lead to the formation of hydrocarbon accumulations is the prerequisite to make quantitative predictions about oil and gas reserves in a basin. Simulation of geochemical processes necessitates a kinetic model of hydrocarbon formation, and also a model of migration and accumulation mechanisms which are not yet known in sufficient detail. However, a ranking of exploration targets can be achieved by simulating the evolution of a sedimentary basin, ie., by geological modeling and a semi-empirical incorporation of relevant geochemical processes. This diminishes greatly the exploration risk.

The backbone of geological modeling is a set of non-linear partial differential equations of energy and mass balance that describe the course of respective geological processes. Main results of the model calculations are, for instance, computer maps showing porosity data, temperatures, maturities or sedimentary thicknesses after compaction for any given sedimentary unit during evolution of the basin. A typical three-dimensional grid system of an exploration area is in the range of 10,000 to 100,000 grid points. It is thus obvious that geological modeling makes possible a formerly uncommon density of geological information and geological data of a completely new quality.

Chapter 4
Geochemical Modeling: A Quantitative Approach to the Evaluation of Oil and Gas Prospects

4.1 Necessity of a Quantitative Approach to Petroleum Potential of Sedimentary Basins

The purpose of the evaluation of a sedimentary basin in terms of petroleum exploration is to know the amount of oil and gas that has been generated and accumulated and then to locate it.

In this respect, the following information is desirable:
a) Amount of oil and gas generated (per km^2) in each source rock and in every part of the basin.
b) Time of hydrocarbon formation to compare it with the age of deposition of impermeable seals and with the age of folding and faulting, i.e., with respect to formation of traps.
c) Amount of oil and gas which has been expelled out of the source rock into the porous reservoirs, amount and location of hydrocarbons accumulated in traps.
d) Evaluation of the ultimate oil and gas reserves of a sedimentary basin. This parameter will be required increasingly when exploration reaches an advanced stage. Then the problem arises whether to continue or stop exploration, depending on the possible reserves remaining to be discovered.

A quantitative approach allowing a computation of the amount of oil and gas generated and migrated in any place of the basin as a function of time is necessary to answer the above questions. As petroleum and natural gas result from kerogen degradation, various indices have been proposed to characterize the quality and evolution stage of the organic matter. These have been reviewed in Chapter V.1. They can be used as a first approach, in a semi-empirical manner, for ranking of exploration targets (see above, Chap. V.3). However, these indices do not allow a truly quantitative evaluation of hydrocarbon generation as a function of time. In particular, most of them do not account for the respective kinetics of degradation of the various types of organic matter and also for complex burial histories.

Mathematical models based on kinetics of kerogen degradation and polyphasic fluid flows and using computer simulation allow a truly quantitative approach. The problem seems to be adequately expressed in respect to formation of petroleum and gas as a function of time, but additional research is still required on migration of hydrocarbons to achieve the evaluation of the ultimate reserves.

Kinetics of kerogen degradation was first introduced through the consideration that temperature and time might, to some extent, compensate for each other. This view was known for a long time, as Maier and Zimmerley (1924) investigated the kinetics of bitumen generation from the Green River shales. Along that line, McNab et al. (1952) and Abelson (1963) investigated experimentally various degradation schemes (decarboxylation, cracking) and showed that by using an Arrhenius plot, such reactions may occur at the relatively low temperature in the sedimentary basins over geological periods of time. Another aspect is the relation between age or heating time of source rocks and temperature threshold of the main zone of oil generation. For a given type of organic matter, the threshold generally decreases with increasing age or heating time (Connan, 1974, 1976) of the source rock, according to an Arrhenius-type relationship.

On the other hand, coal researchers also considered the kinetics of the progressive carbonization, either natural or artificial during cokefaction, and the related elimination of volatile matter and increase of carbon content and reflectance. Kinetics were introduced by using nomograms, or an approximate computation (Huck and Karweil, 1955; van Krevelen, 1961; Lopatin, 1971; Lopatin and Bostick, 1973). In respect to pyrolysis of oil shales, experimental data on kinetics of kerogen degradation were obtained by Hubbard and Robinson (1950) and mathematical models were subsequently proposed by Allred (1966), Fausett (1974), Braun and Rothman (1975).

Lopatin (1971), followed by Waples (1980), made an attempt to calculate in a simple manner the combined effect of time and temperature on maturation of the organic matter, without using the actual kinetic equations, which require a numerical simulation and the use of computers. It is assumed that the rates of chemical reactions approximately double for every 10°C rise in temperature (Lopatin, 1971; Waples, 1980). Upon progressive burial they calculate an interval maturity by multiplying the time of residence ΔT in a temperature interval of 10°C by a factor 2^n, which itself changes to 2^{n+1}, 2^{n+2}, etc. every 10°C. The total maturity is expressed by TTI (time–temperature index) which is the sum of the interval maturities:

$$TTI = \sum_n 2^n(\Delta T_n).$$

Then a table is provided to calibrate TTI versus vitrinite reflectance and, through it, to interpret this index in terms of stages of oil and gas generation.

Although this calculation procedure is attractive, because it does not require the use of computers, the basic assumption that the rate of reactions doubles for every 10°C rise is only valid for reactions with an activation energy in the range of E = 10 to 25 kcal mol^{-1} (at temperatures from 20° to 160°C which include the stages of oil and gas generation). This order of magnitude is acceptable at the onset of oil generation, when relatively weak chemical bonds are broken in kerogen (Tissot, 1969; Connan, 1974).

However, it has been recognized from experiments (Weitkamp and Gutberlet, 1968) and observations in sedimentary basins (Tissot and Espitalié, 1975) that the activation energies involved in kerogen degradation progressively increase, with increasing maturation, as stronger bonds are successively broken. For instance, at the peak of oil generation the average activation energy is E = 40 to 60 kcal mol^{-1}, and in the gas zone it reaches E = 50 to 70 kcal mol^{-1}. It should be remembered that an activation energy E = 50 kcal mol^{-1} corresponds approximately to reaction rates multiplied by 5 every 10°C and E = 65 kcal mol^{-1} to rates multiplied by 10, in the temperature ranges corresponding to the peak of oil generation and the gas zone, respectively. In that interval of maturation, over a temperature increase of 30°C, the TTI will account for a rate multiplied by $2^3 = 8$, whereas the actual rate will be multiplied by 125 to 1000, if the average E = 50 to 65 kcal mol^{-1}. More information about the role and significance of activation energies may be found in Sections V.4.2 and V.4.5. In particular, there are differences in the kinetics of degradation due to differences in chemical composition of the respective kerogen types. Thus it is difficult to account with a single parameter for hydrocarbon generation from the various types of source rocks.

4.2 Mathematical Model of Kerogen Degradation and Hydrocarbon Generation

A mathematical model of petroleum generation, accounting explicitly for geological time, was first introduced by Tissot (1969) and is fully discussed in Tissot and Espitalié (1975). The model is based on kinetics of kerogen degradation and uses the general scheme of evolution, described in Part II of this book. Kerogen is a macromolecule composed of polycondensed nuclei bearing alkyl chains and functional groups, the links between nuclei being heteroatomic bonds or carbon chains. As the burial depth and temperature increase, the bonds are successively broken, roughly in the order of increasing rupture energy. The products generated are first heavy heteroatomic compounds, carbon dioxide, and water; then progressively smaller molecules; and finally hydrocarbons. At the same time, the residual kerogen becomes progressively more aromatic and evolves towards a carbon residue. All these mechanisms have been described in Part II and are summarized in Figure V.4.1. The purpose of the mathematical model is to represent the kinetics of the parallel and successive reactions shown in this figure.

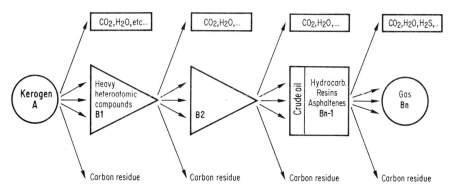

Fig. V.4.1. General scheme of kerogen degradation (Tissot and Espitalié, 1975)

Reactions are considered here as not reversible. In fact, when source rocks have been buried to a certain depth, then brought again to surface by subsequent folding and erosion, the organic content keeps the composition and physicochemical properties corresponding to the maximum burial depth. Furthermore, some of the by-products of kerogen evolution, such as water and carbon dioxide, are highly mobile in sub-surface conditions and could not be available for recombination, should it be the case.

For simulating the system of reactions, shown in Figure V.4.1, it is advisable to consider only a limited number of steps and particularly:

- A, B_1, B_{n-1}, which was the first version proposed by Tissot (1969),
- A, B_{n-1}, B_n, which appears from experience to be sufficient to account for the successive oil and gas formation (Tissot and Espitalié, 1975), and will be used here.

The last formulation corresponds to:

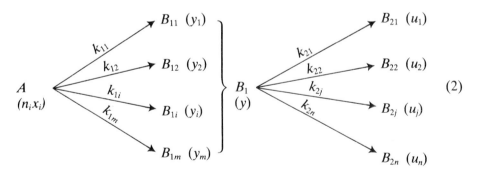

$$\text{Kerogen} \rightleftarrows \left\{ \begin{array}{c} CO_2, H_2O, \text{etc.} \\ \text{Oil} \\ \text{Carbon residue} \end{array} \right. \rightleftarrows \left\{ \begin{array}{c} CO_2, H_2O, H_2S \text{ etc.} \\ \text{Gas} \\ \text{Carbon residue} \end{array} \right. \quad (1)$$

The following symbols will be used. A is kerogen, comprising n_i bonds of a given type i, at the time t; x_i is the amount of organic matter reacting in the ith reaction (breaking of i-type bond). B_{11} to B_{1m} are the products of the first step of reactions (formation of oil), their respective amount at time t being y_1 to y_m. B_{21} to B_{2n} are the products of the second step of reactions (cracking), their respective amount at time t being u_1 to u_n.

$$A \; (n_i x_i) \xrightarrow{\substack{k_{11}\\k_{12}\\k_{1i}\\k_{1m}}} \begin{array}{c} B_{11}\,(y_1) \\ B_{12}\,(y_2) \\ B_{1i}\,(y_i) \\ B_{1m}\,(y_m) \end{array} \Bigg\} B_1\,(y) \xrightarrow{\substack{k_{21}\\k_{22}\\k_{2j}\\k_{2n}}} \begin{array}{c} B_{21}\,(u_1) \\ B_{22}\,(u_2) \\ B_{2j}\,(u_j) \\ B_{2n}\,(u_n) \end{array} \quad (2)$$

The first set of transformations represents a certain number of parallel and/or successive reactions. It is assumed that the probability of breaking a bond of type i is independent of the abundance of bonds of the other types and also independent of the time elapsed. Then the breaking of i-type bond obeys the Poisson law:

$$-\frac{dn_i}{n_i} = k_{1i} dt,$$

4.2 Mathematical Model of Kerogen Degradation and Hydrocarbon Generation

k_{1i} being a constant at a given temperature. If the structure is homogenous, i.e., the density of bonds is statistically the same in the kerogen volume, we have:

$$x_i = \mu n_i,$$

where μ is a constant, and, with regard to parallel reactions:

$$x_{i0} = x_0 P_i,$$

P_i being the frequency of i-type bond at the origin $t = 0$, and $x_0 = \Sigma x_{i0}$ being the total amount of labile — or pyrolyzable — organic matter. Then, it becomes:

$$-\frac{dx_i}{dt} = k_{1i} x_i.$$

Therefore, we obtain for each of the i reactions a kinetic equation similar to that of the first-order reactions. It has been shown that such formulation is able to account for kerogen degradation in geological conditions (Tissot, 1969).

If we make the additional hypothesis that the B_{1i} constituents of oil have the same general behavior in respect to cracking, we can consider for the second step of reactions $B_1 = \Sigma B_{1i}$. The respective amounts x_i, y_i and u_j can be obtained from the following system of differential equations [Eq. (3)]:

$$\left. \begin{array}{l} -\dfrac{dx_i}{dt} = k_{1i} x_i \\[1em] \dfrac{du_j}{dt} = k_{2j} y \\[1em] y = \sum_i y_i \\[1em] \sum_i x_{i0} + \sum_i y_{i0} + \sum_j u_{j0} = \sum_i x_i + \sum_i y_i + \sum_j u_j \end{array} \right\} \quad (3)$$

The first two equations express the kinetics of the system, whereas the last two express the mass balance. The variation of k_{1i} and k_{2j} with temperature may be described by using the Arrhenius formula:

$$k_{1i} = A_{1i} e^{-\frac{E_{1i}}{RT}}$$

which is basically valid for fast laboratory or industrial reactions, but may be extended to slow reactions occurring under geological situations (Tissot, 1969; Connan, 1974). E_{1i} is the activation energy of the ith reaction, A_{1i} is a constant and T is the absolute temperature (Kelvin). In geological situations, temperature is a function of time through subsidence and related burial. As burial is reconstructed by geologists on the basis of successive time intervals (e.g., Lower Cretaceous, Upper Cretaceous, Paleocene, etc.) depth is considered to be a linear function of time during each interval. Thus temperature also is approximated as a succession of linear functions of time during the successive time intervals. Under these conditions the system [Eq. (3)] cannot be easily integrated, as k_{1i} becomes a complex function of the time t. The easiest way of solving the system is by numerical integration, with successive Δt increments. This is done by using the computer, and the program includes the adjustment of the various parameters x_{i0}, E_{1i}, A_{1i} and y_0, which is in fact the calibration of the model to the particular type of organic matter.

For this calibration, only the first set of reactions (formation of oil) is considered. At the beginning, values measured on comparable recent sediments (e.g., Black Sea sediments for type-II kerogen) are used for y_0, and approximate values for A_{1i} (Tissot, 1969). In respect to the E_{1i}, it is necessary to consider all activation energies from a few kcal mol^{-1}, corresponding to the rupture of weak bonds, as in adsorption, to about 80 kcal mol^{-1}, corresponding to breaking carbon–carbon bonds. The objective of the adjustment should be to determine the frequency, P_i, of each i-type of bonds, or the amount x_{i0} of labile organic material involved in the ith reaction. The calibration is based on values of extractable organic matter (hydrocarbons + resins + asphaltenes) naturally present in cores taken at different depths in the sedimentary basin and/or on comparable figures resulting from laboratory experiments at higher temperature. The adjustment is based on the method of the least squares. The results concerning the three main types of kerogen I, II and III, are shown in Table V.4.1 and Figure V.4.2.

These figures are based on the data obtained from the source rocks of the Uinta, Paris and Douala basins which were originally used to define the three main types of kerogen. They offer a first approach to calculate the amount of oil

Table V.4.1. Distribution of activation energies and genetic potential of the principal kerogen types

	Activation energies		Kerogen types				
Class	Average value (kcal mol^{-1})	Type I		Type II		Type III	
		x_{i0}	A	x_{i0}	A	x_{i0}	A
E_{11}	10	0.024	4.75 10^5	0.022	1.27 10^5	0.023	5.20 10^3
E_{12}	30	0.064	3.04 10^{16}	0.034	7.47 10^{16}	0.053	4.20 10^{16}
E_{13}	50	0.136	2.28 10^{25}	0.251	1.48 10^{27}	0.072	4.33 10^{25}
E_{14}	60	0.152	3.98 10^{30}	0.152	5.52 10^{29}	0.091	1.97 10^{32}
E_{15}	70	0.347	4.47 10^{32}	0.116	2.04 10^{35}	0.049	1.20 10^{33}
E_{16}	80	0.172	1.10 10^{34}	0.120	3.80 10^{35}	0.027	7.56 10^{31}
Genetic potential of kerogen $x_0 = \sum_i x_{i0}$		0.895		0.695		0.313	

A is expressed as 10^6 yr^{-1}

value of y_0 Type I : 0.051
                Type II : 0.035
                Type III: 0.018

4.3 Genetic Potential of Source Rocks. Transformation Ratio

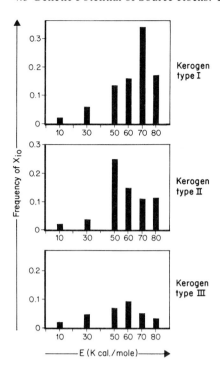

Fig. V.4.2. Distribution of the activation energies involved in the degradation of the three main types of kerogen (Tissot and Espitalié, 1975)

and gas generated from other source rocks of comparable type. However, due to the variability of kerogen composition, a specific adjustment based on laboratory assays, which provide a direct calibration of the model for each particular source rock, is generally preferable.

Calibration of the second set of reactions — formation of gas — has been made from laboratory experiments on cracking, including the results of McNab et al. (1952), and Johns and Shimoyama (1972). Consideration of a single reaction with an activation energy of 50 kcal mol^{-1} seems convenient to account for gas generation in the deep parts of the sedimentary basins.

4.3 Genetic Potential of Source Rocks. Transformation Ratio

The total amount of hydrocarbons which can be produced by a certain kerogen, provided it is heated to a sufficient temperature over a sufficient time, is the quantity

$$x_0 = \sum_i x_{i0}$$

that appears in Table V.4.1. This quantity is equivalent to the *genetic potential of the kerogen*, defined in Chapter II.7.2. The value depends on the type of kerogen, i.e., on its original chemical composition. A *source rock* could be defined as a rock whose genetic potential is above a threshold value, e. g., 0.25 or 0.30 related to the unit weight of kerogen, or (if we consider an average value of 1% organic carbon in rock) 0.25 to 0.30% related to unit weight of rock.

At any time, the stage of evolution is measured by using the *transformation ratio r*, defined in Chapter II.7.2, which is the ratio of the kerogen already transformed to the genetic potential:

$$r = \frac{\Sigma x_{i0} - \Sigma x_i}{\Sigma x_{i0}} = \frac{x_0 - \Sigma x_i}{x_0}.$$

The transformation ratio is zero at shallow depths and progressively increases to 1, which is reached when all labile organic material has been expelled leaving a carbonaceous residue.

4.4 Validity of the Model

In various sedimentary basins, comparison between the figures computed by using the model and the corresponding amounts of petroleum generated through burial shows excellent agreement, with a quadratic deviation lower than 10^{-2}, and a correlation coefficient better than 0.9. The model has been used to simulate experimental heating — either isothermal or with a regular rate of temperature increase — during various times from an hour to one year, again with satisfactory agreement. Furthermore, the same set of constants A_i and E_i, shown in Table V.4.1, is sufficient to account for all conditions of kerogen degradation (Tissot and Espitalié, 1975) including (a) natural evolution in sedimentary basins at relatively low temperature (50–150°C) over a period of 10 to 400 million years; (b) artificial evolution through laboratory experiments (180–250°C) (c) high-temperature (400–500°C) retorting of oil shales. The fact that a single model with the same set of constants is able to simulate such different situations is a confirmation of the validity of the hypothesis made on kinetics (statistics of bonds and activation energies, first-order reactions, etc.).

The timing of oil generation provided by the model may also in some cases be checked against geological data. An example from northern Sahara (Algeria) has been shown in Chapter II.7.6.

4.5 Significance of the Activation Energies in Relation to the Type of Organic Matter

The parameters E_{1i} used in the model are called activation energies for the sake of simplification. They have a role similar to that of activation energy, but they are not strictly activation energies, as the latter are normally defined in respect to a particular and single reaction. This is not the case in the model, where we consider the "formation of petroleum" from numerous parallel and/or successive reactions.

4.5 Significance of the Activation Energies

Many types of bonds are originally present in kerogen, with distinct rupture energies, and particularly:

- weak bonds corresponding to physical or chemical adsorption (hydrogen bonds, etc.),
- carbonyl and carboxyl bonds,
- ether and sulfur bonds,
- carbon–carbon bonds.

Furthermore, the rupture energy of most types of bonding may vary according to neighboring functional groups or substituents, length of chains, etc. Consequently, consideration of a distribution of the activation energies E_{1i} from 0 to 80 kcal mol^{-1}, as we did in Section 2.2, is probably closer to the effective mechanisms than a hypothetical measurement of each individual type of bond. Therefore, the best representation of kerogen composition may be the histogram of activation energies, derived from Table V.4.1, presented in Figure V.4.2 for type-I, -II and -III kerogens. With increasing burial and temperature (and decreasing $\frac{1}{T}$) the various bonds corresponding to the successive activation energies E_i are progressively broken, roughly in order of increasing E_i. This is suggested by the temperature dependence of the reaction parameters k_i in Figure V.4.3.

The genetic potential $x_0 = \Sigma x_{i0}$, and the distribution of the activation energies change according to the type of kerogen (Tissot and Espitalié, 1975):

a) Type-I kerogen contains a large proportion of labile organic material x_0, and thus its genetic potential is high. The distribution of the activation energies include few with low values corresponding to weak bonds. Most values are

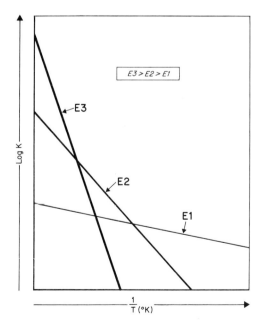

Fig. V.4.3. Temperature dependence of the reaction parameters k_i involved in kerogen degradation. The figure shows that the various bonds corresponding to the successive activation energies are progressively broken, in order of increasing activation energy (Tissot and Espitalié, 1975)

grouped around 70 kcal mol^{-1}, and may correspond to carbon–carbon bonds. Therefore, this particular kerogen requires higher temperatures for generating oil than the other types.

b) Type-II kerogen contains slightly less of the labile organic material resulting in a genetic potential slightly lower than in type I. However, the distribution of activation energies is wide and includes lower values than type I with a mode at 50 kcal mol^{-1}. Thus, the main formation of petroleum starts at somewhat lower temperature than in the previous type.

c) Type-III kerogen contains still less of the labile organic material than the others, and its genetic potential is low. As a result, the total amount of oil generated is comparatively small. The distribution of the activation energies is smooth with a maximum frequency at 60 kcal mol^{-1}.

A high concentration of organic matter related to type I or II results in a rich oil shale with a high oil yield. The Green River shales belong to type I and their values of x_0 is 0.8 to 0.9, i.e., 80 to 90% of the organic matter is able to be converted to oil. The Toarcian shales of Western Europe belong to type II, and the corresponding value of x_0 is 0.6, i.e., 60% of the organic matter can be converted to oil. Type-III organic matter, on the other hand, may be concentrated to form certain coals or carbonaceous shales, but with a low oil yield ($x_0 = 0.25$) and no commercial oil shales.

The relationship between the three main types of kerogen, as defined by their chemical properties, and the respective distribution of activation energies, may be used for a quick calibration of the model. If one knows the specific kerogen type (I, II or III) from optical examination or elemental analysis, the appropriate values of A_i and E_i from Table V.4.1 can be used in the model.

A review of the apparent activation energies proposed in the literature for petroleum generation, carbonization of coal, or oil-shale retorting shows that the values range from 8 to 65 kcal mol^{-1}. These apparent activation energies were computed on the basis of a single reaction, although it deals in fact with a set of successive and/or parallel reactions. As the true activation energies E_i, corresponding to the breaking of various bonds, are affected roughly in order of ascending values with increasing temperature, the apparent activation energy is close to the lowest E_i at low temperature and close to the highest E_i at high temperature. This behavior is confirmed by experimental data obtained from laboratory pyrolysis of oil shales by Weitkamp and Gutberlet (1968): they observed a progressive increase of the apparent activation energy from 20 to 60 kcal mol^{-1}, when the conversion rate of kerogen increased from 0 to 80%. This consideration is sufficient to explain the conclusion that apparent activation energies related to the beginning of oil formation are on the average 10–15 kcal mol^{-1} (Tissot, 1969; Connan, 1974) whereas apparent activation energies related to oil shale pyrolysis or carbonization of coal are about 50 to 65 kcal mol^{-1} (Abelson, 1963; Hanbaba and Jüntgen, 1969, etc.).

4.6 Application of the Mathematical Model to Petroleum Exploration

The mathematical model of kerogen degradation provides quantitative value of oil and gas generated as a function of time. Therefore, it is directly applicable for evaluation of the oil and gas potential of a sedimentary basin, and for determination of the timing of petroleum formation for comparison with the age of structural or stratigraphic traps.

The data requested for use of the model are presented in Figure V.4.4: (a) a direct calibration of the E_i distribution on laboratory assays, which in fact represents the chemical composition of the organic matter, or alternatively identification by optical or chemical methods of the kerogen type and subsequent use of the corresponding distribution shown in Table V.4.1; (b) the burial curve, i.e., the depth versus time relationship, from the time of source rock deposition to the present time, and (c) the geothermal gradient, for computation of the temperature versus time relationship. The gradient may vary with geological periods according to geotectonic conditions, and its determination will be discussed in the following paragraphs.

Provided such data are available, the amount of oil and gas per ton of rock generated in any place of the basin can be computed as a function of time (Fig. V.4.4). In order to cover the whole sedimentary basin, it is convenient to divide up the volume of the basin into a three-dimensional grid (as shown in the previous chapter, Fig. V.3.4) and to make the computation for each unit. The elementary units may be defined in different ways: in platform basins with gentle folding, a grid based on latitude and longitude is generally adequate; in mobile areas where folding is fairly strong, elementary units delineated by isobath curves are more nearly homogeneous in respect to depth and temperature history. The most

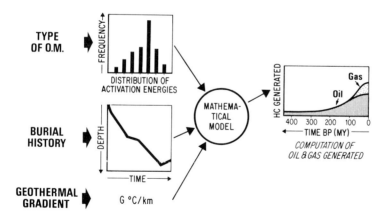

Fig. V.4.4. Principle of use of the kinetic model of oil and gas generation. The data provided are the type of organic matter (or the related distribution of activation energies), the burial history and the geothermal gradient. The model calculates the quantities of oil and gas generated, as a function of time (Tissot et al., 1980)

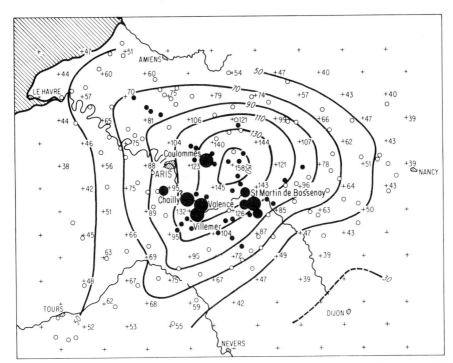

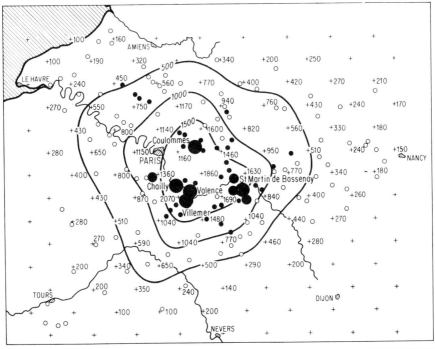

● Oil field ● Single producing well • Oil show ○ Dry hole

4.6 Application of the Mathematical Model to Petroleum Exploration

effective way to use this geochemical modeling is obviously to introduce this simulation into the basin geological modeling presented in the previous chapter.

The application of the mathematical model to oil generation in Jurassic source rocks of the Paris Basin is shown in Figure V.4.5. It can be expressed by the fraction of kerogen, which has been converted to petroleum, i.e., $r\,x_0$ in grams petroleum per kilogram of organic matter. Alternatively, if the distribution of organic carbon across the basin is also known, it can be related to the whole rock and expressed as grams petroleum per ton of rock. From the figure, it is clear that all known oil fields fall inside the area, where more than 110 g petroleum has been generated per kilogram of organic carbon, and inside the area, where more than 1500 g petroleum has been generated per ton of rock. Regarding the timing of oil generation, most petroleum has been formed in this basin during Cretaceous and Tertiary.

The example of oil and gas generation in Paleozoic source rocks of the Illizi basin, Algeria, is shown in Figure V.4.6 and deals with a more advanced stage of thermal evolution. Geological history of the basin includes two periods of sedimentation and subsidence separated by Hercynian folding and erosion. During Paleozoic subsidence, the Silurian and Devonian source rocks were buried to a great depth (about 3000 m) in the southwestern part of the basin and the kerogen was deeply degraded (transformation ratio $r \cong 1$) to generate gas. This gas subsequently migrated into anticlines formed by Hercynian folding. However, a deep post-Hercynian erosion reached the lower Paleozoic reservoirs and precluded conservation of gas. In the other parts of the basin, Paleozoic subsidence was moderate and only initiated oil generation. Folding and erosion were followed by an important Mesozoic subsidence, particularly in the northeastern part of the basin. The new burial exceeded the maximum pre-Hercynian depths in all places of the basin, except the southwestern part, where kerogen was definitely unable to evolve any more. In all other areas, kerogen degradation continued generating abundant oil and also gas in the northern and eastern parts of the basin where Cretaceous sedimentation was very thick. The amount of petroleum formed has been calculated, with consideration of the respective timing of generation. Thus, oil and gas potential may be described as: (a) low for both oil and gas in the southwest, due to erosion after generation; (b) good for oil in the central and eastern parts of the basin with some gas associated in the east; (c) good for gas in the northeast, with some oil associated in the less buried or colder areas.

The example of the Hassi Messaoud area in northern Sahara, Algeria, has been presented in Chapter II.7.6. There, purely geological considerations allow an independent confirmation of the timing of petroleum generation (Figs. II.7.5 and II.7.6). The mathematical model shows that the main phase of oil generation from Silurian source rocks is not reached until Cretaceous time, and extends

Fig. V.4.5. Application of the mathematical model to the generation of oil from Lower Jurassic source rocks of the Paris Basin. *Top:* the oil generated is expressed in g petroleum per kg organic carbon. *Bottom:* the oil generated is expressed in g petroleum per t rock (Deroo et al., 1969; Tissot and Pelet, 1971)

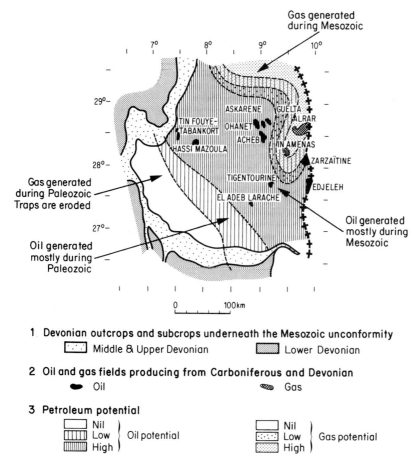

Fig. V.4.6. Hydrocarbon potential of the Lower Devonian beds in the Illizi Basin (Algeria), as determined by use of the mathematical model (Tissot and Espitalié, 1975)

mostly over Cretaceous and Tertiary. In this area all traps are equally prospective, whatever age they are, from Paleozoic to Cretaceous.

4.7 Reconstruction of the Ancient Geothermal Gradient

4.7.1 Heat Flow and Geothermal Gradient

The temperature increase with increasing depth results from a heat flow from the inside to the outside of the earth. The temperature T is a function of depth z and the geothermal gradient is $G = \dfrac{dT}{dz}$. At a given place of the earth's crust the heat flow Φ and the geothermal gradient G satisfy the relation:

$$G = \xi \cdot \Phi,$$

4.7 Reconstruction of the Ancient Geothermal Gradient

where $1/\xi$ is the thermal conductivity. G is expressed as the temperature increase per unit of depth, e.g., degrees Celsius km^{-1}. It may range from 10 to 80°C km^{-1}, with an average value of about 30°C km^{-1}. Φ is expressed in μcal cm^{-2} s^{-1} (10^{-6} calories per square centimeter per second). It may range from less than 1 to 8 μcal cm^{-2} s^{-1}, with an average value approximately 1–2 μcal cm^{-2} s^{-1}.

Thermal conductivity varies with the composition of rocks. For instance, the values are quite high in salt (Fig. III.2.3). The variations cause a variability of geothermal gradient, whereas values of heat flow are less scattered. Table V.4.2 shows the average values of heat flow in continental and oceanic basins according to geotectonic conditions.

Heat flow is low in Precambrian shields (0.9 μcal cm^{-2} s^{-1}) and in areas where orogeny and/or magmatism are of Paleozoic age or older (1.2 to 1.3 μcal cm^{-2} s^{-1}). It is somewhat higher in areas where orogeny and/or magmatism is of Mesozoic or Tertiary age (1.9 μcal cm^{-2} s^{-1}); also in Tertiary volcanic areas (2.2 μcal cm^{-2} s^{-1}); and moreover in thermal areas where the flow may exceed 3 μcal cm^{-2} s^{-1}. Thus high values of heat flow and geothermal gradient are reported from synorogenic or postorogenic basins of the Alpine belt (Pannonian basin extending over Hungary, Yugoslavia and Romania) and of the circum-Pacific belt: Okhotsk and Japan seas, Sakhalin, North Sumatra, Fiji, etc.

Oceanic areas also show a large range of heat-flow variation, from oceanic ridges to oceanic basins and troughs. Furthermore, the heat flow is quite high along the axial zone of the ridges, i.e., along the crest and rift valleys, where values as high as 8 μcal cm^{-2} s^{-1} are measured. It decreases progressively away from the crest. Within 150 km on both sides of the crest of the East Pacific ridge, half of the recorded values are still in the range of 3 to 8 μcal cm^{-2} s^{-1}; beyond that distance, the heat flow is always lower than 3 μcal cm^{-2} s^{-1}. A similar observation has been made along the mid-Atlantic ridge in regard to a 50-km zone on both sides of the crest. The present oceanic expansion being approximately 3 to 6 cm yr^{-1} in the East Pacific and 1 to 2 cm yr^{-1} in the Atlantic, the belt with a high heat flow corresponds approximately to an oceanic basement younger than 5 million years.

More precisely, Le Pichon and Langseth (1969) have computed the average values of heat flow as a function of distance to the axis of oceanic ridges (Table V.4.3). To measure the distance D they used, instead of kilometers, a unit corresponding to magnetic anomaly No. 5, which is definitely dated as slightly less than 10 million years. Thus, distances are in fact computed in terms of age of the oceanic crust, and oceans with different rates of expansion can be compared. From Table V.4.3, it is clear that high heat flow (2 or 3 times the average value in oceanic basins) are related to a belt with an age lower than 5 million years. Then the heat flow decreases with increasing age and is no longer distinct from the average oceanic basin where the age of the basement reaches 15 to 30 million years, depending on the area.

This view is confirmed by heat-flow values recorded in narrow elongated basins corresponding to Upper Tertiary and/or Recent expansion (in the Aden Gulf, the Red Sea, and the California Gulf; 3.9, 3.4 and 3.4 μcal cm^{-2} s^{-1} respectively) and also in continental grabens of similar origin (Baikal Lake and Rhine-Graben), where flows as high as 2.5 and 4 μcal cm^{-2} s^{-1}, respectively, have been measured.

Table V.4.2. Average values of geothermal flux. (After Lee and Uyeda, 1965)

Geotectonic Provinces	Average geothermal flux (10^{-6} cal cm^{-2} s^{-1})	Number of measurements
Precambrian shields	0.92	26
Areas stable since Precambrian[a]	1.32	16
Areas with Paleozoic orogenies	1.23	21
Areas with Mesozoic or Cenozoic orogenies	1.92	19
Areas of Cenozoic volcanism	2.16	11
Oceanic basins	1.28	273
Oceanic ridges	1.82	338
Oceanic troughs	0.99	21

[a] Excluding Australia (Cenozoic volcanism).

Table V.4.3. Geothermal flux on oceanic ridges as a function of normalized distance to the axes of the ridge (Le Pichon and Langseth, 1969)

A. East Pacific ridge

Normalized distance (D)	Flux	
	Average value (μcal cm^{-2} s^{-1})	Standard deviation
0 < D < 0.54	3.31	1.94
0.54 < D < 1.8	2.00	1.29
1.8 < D < 3.0	1.47	1.03

B. Atlantic Indian ridge

Normalized distance (D)	Flux	
	Average value (μcal cm^{-2} s^{-1})	Standard deviation
0 < D < 0.46	2.72	2.33
0.46 < D < 1.4	1.45	0.94
1.4 < D < 3.1	1.10	1.14

4.7.2 General Rules for Reconstitution of Ancient Geothermal Gradient

From the observed distribution of geothermal data, some general rules may be deduced for the reconstitution of paleogeothermy, according to the geotectonic setting of the sedimentary basins.

4.7.2.1 Stable Platforms

Present and ancient geothermal gradients can be about the same in sedimentary basins where no tectonic or magmatic activity has been known since deposition of the potential source rocks. This normally occurs in areas which, after possible folding and magmatic intrusions, are incorporated into a stable platform. Then, subsequent sedimentation of potential source and reservoir rocks results in a new sedimentary basin, now generally flat-lying or gently folded. In this case, the present geothermal gradient is generally applicable to source rock evolution.

This situation occurs in undisturbed Paleozoic basins resting on folded or even flat-lying pre-Cambrian basement, such as several Paleozoic basins of Australia, and in the northern and eastern Sahara in Algeria; in Mesozoic undisturbed basins resting on Paleozoic or older folded basement, like most of the West Canadian basin, the Paris Basin, etc. In such cases the values of geothermal gradient are often 25 to 35 °C km^{-1}.

4.7.2.2 Orogenic Basins

The similarity of past and present geothermal gradients is not acceptable in basins where the last orogeny and/or magmatic intrusion is contemporaneous with, or younger than, source rock deposition. This situation occurs in synorogenic or postorogenic intramontane basins such as the Pannonian basin in Central Europe, associated with the Carpathes alpine folding, or in Permo-Carboniferous basins associated with the Hercynian orogeny in Western Europe and Eastern Australia.

In these basins, geothermal gradient and flux may be high and quite variable with respect to location during the periods of orogenic activity or immediately following it. For instance, values as high as 50 °C km^{-1} are now observed in several parts of the Pannonian Tertiary basin, where Pliocene source rocks have been brought to sufficiently high temperature for generating oil, despite the short period of time.

4.7.2.3 Stable Continental Margins

The basins of deposition located along the stable continental margins resulting from oceanic opening and expansion are worth a special consideration. From the data reported in Section 4.7.1 and Table V.4.3, it can be assumed that during the first 5 million years after creation of a new oceanic basement, the first sediments deposited on the new basement have been subjected to a heat flow two or three times higher than the average. Then the flow progressively decreases, to reach the average value after 15 or 20 million years. Under such circumstances, moderately buried source rocks may reach the principal zone of oil formation within 5 or 10 million years after deposition.

This situation occurs mainly at the beginning of ocean spreading, when continental masses are still close to the hot axial zone, and provide abundant material for a thick sedimentation over this zone followed by quick burial. In later stages of oceanic evolution, when the oceanic spreading has already moved the continents far away from the axial zone, this hot axial zone generally receives a reduced and slow sedimentation. Examples of thick sedimentation associated with a high heat flow may occur in Cretaceous coastal basins of West Africa and South America as a result of the opening of the South Atlantic Ocean; in Jurassic and Cretaceous basins resulting from the opening of the North Atlantic Ocean; and basins of Western and Northwestern Australia; and in Mio-pliocene sediments of the Red Sea, Aden Gulf, etc.

These considerations, deduced from the present distribution of heat flows, are in agreement with observations made on the kerogen of ancient sediments of the Aquitaine basin, whose opening is linked to the expansion of the Atlantic Ocean. Robert (1971) observes the reflectance of the organic matter as a function of depth in bore holes, and finds an abrupt increase correlated with a lower Cretaceous stage. Such a shift is explained by the influence of a considerable heat flow over the period of time associated with the opening of the Aquitaine basin.

4.7.2.4 Evolution of Heat Flow

The heat flow through undisturbed sedimentary basins is usually either constant or decreasing as a function of geological time. This is also true with respect to orogenic and magmatic activity: heat flows are high when this activity is young and decrease when it becomes older. After a sufficient time, it remains constant, as present values in Precambrian or Paleozoic folded basins are comparable. The same general scheme — high values, then decrease to reach a rather constant state — is, for other reasons, valid for stable margins related to oceanic opening, as discussed in Section 4.7.2.3.

However, the reverse may happen, and present geothermal gradients may be, in certain cases, higher than the ancient ones. In particular, continental grabens are associated with rather complex mechanisms, including possible rifting, rising of thermal waters, etc. In the Rhine-Graben, heat flows and geothermal gradients are generally high and variable with the location. They may reach peak values like $80°C\,km^{-1}$ in the Landau bore hole. However, the stage of kerogen degradation, as well as vitrinite reflectance, is not consistent with such a high value (Doebl et al., 1974). Mathematical simulation shows that geothermal gradient was moderate until late Tertiary, and then increased substantially up to the present value, probably as a result of faulting and rising of thermal waters.

4.7.3 Evaluation of the Present Geothermal Gradient

In many basins of stable platforms, the present value of geothermal gradient is a satisfactory evaluation of the paleo-gradient (Sect. 4.7.2.1). Determination of the present gradient requires a sufficient number of temperature measurements in bore holes. Electric, acoustic or nuclear logging devices are usually equipped with a maximum reading thermometer. However, the time between drilling and logging is usually restrained to a minimum, so that thermal equilibrium with the formation is not established, and the formation temperatures are generally underestimated, resulting in too low gradient values.

Better results are obtained if extended testing is made on reservoir beds containing oil or gas, followed by long pressure build-up. Then the equilibrium is reached, resulting in accurate formation-temperature measurements and computation of gradient.

4.7.4 Evaluation of the Ancient Geothermal Gradient

McKenzie (1978, 1981) studied the time-temperature history in sedimentary basins formed by extension, followed by subsidence. The stretching model calculates the subsidence and the temperature gradient, within a subsiding basin with compaction. The principal results may also apply to some other basins produced by certain thermal disturbances of lithosphere.

Reconstitution of ancient geothermal data may also be attempted by measuring physical or chemical properties of kerogen and simulating its evolution by use of a mathematical model until satisfactory adjustment is made with the measured value. This technique requires the knowledge of the burial versus time curve, based on geological reconstitution. It is best achieved by integrating this calculation into the geological modeling presented in the previous chapter.

The parameters which may be used include vitrinite reflectance, electron spin resonance of kerogen (ESR), hydrocarbon compositions, etc. Some authors have tried the direct use of such measurements as "maximum reading thermometers". The variations of these properties all result from degradation of kerogen, which is governed by kinetic laws, and thus is a function of both time and temperature. Thus, it is not possible to relate any of these properties of kerogen to maximum temperature only, as they result from the whole thermal history. Furthermore, the different parameters may be linked to different aspects of the evolution of kerogen structure, and thus time and temperature may influence the various indices in different ways.

Among the physical properties of kerogen, which can be used for reconstitution of thermal history, one of them, vitrinite reflectance, is discussed here.

The qualitative aspects of the increase of vitrinite reflectance with burial depth are discussed in Chapter V.1. A model of vitrinite degradation has been developed on the basis of a sequence of Logbaba shales in the Douala basin (Cameroon). There, the organic matter consists mainly of vitrinite and of amorphous organic cement with the same physical properties (elemental composition, IR spectra, reflectance etc.). Thus it has been possible to calibrate a model

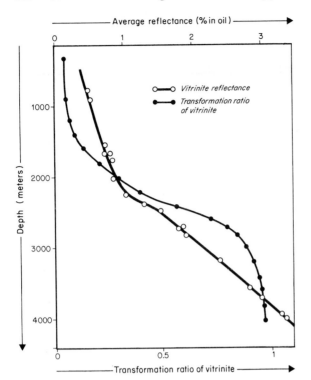

Fig. V.4.7. Changes of the transformation ratio of vitrinite and of its reflectance in the Logbaba wells, Douala Basin (Tissot and Espitalié, 1975)

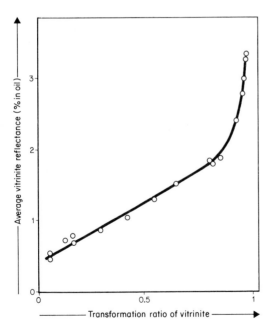

Fig. V.4.8. Relationship between the transformation ratio of vitrinite and its reflectance (Tissot and Espitalié, 1975)

4.7 Reconstruction of the Ancient Geothermal Gradient

(adjusted along the methods outlined in Sect. 4.2) to simulate the degradation of the organic matter of vitrinite type. Furthermore, it is possible to establish a relationship between reflectance and transformation ratio of vitrinite, by measuring both parameters on the same samples. Figure V.4.7 shows both parameters as a function of depth in the sedimentary sequence of the Douala basin, and Figure V.4.8 shows the relationship between reflectance and degradation of vitrinite.

Thus, it becomes possible to make a quantitative interpretation, in terms of thermal history, of a well located in any basin, provided the rocks contain some vitrinite. The interpretation is based on measurements made at different depths in a bore hole, and reconstitution of burial for each sample, made on purely geological information. Each reflectance measurement is converted into the corresponding transformation ratio, according to Figure V.4.8. Independently, the model of vitrinite evolution is used to calculate a transformation ratio, based on the burial history, as a function of geothermal gradient G. Then, an adjustment is made on variable G, until minimizing the quadratic deviation between the computed values of the transformation ratio and the values deduced from observations through Figure V.4.8. This procedure can be used either to adjust a simple law expressing G as a function of time, or to calculate an average value of G which is the best evaluation of the past geothermal gradient over the period of maximum burial.

This method has been used with success in several types of basins, with ancient gradients equal, higher or lower than the present one (Table V.4.4). In the Rhine Graben, the calculated values are in agreement with the observations made by Doebl et al. (1974) that the present geothermal gradient could not have lasted for

Table V.4.4. Present and ancient geothermal gradients

Basin	Well	Geothermal gradients ($°C\ km^{-1}$)		Period of time
		presently	computed	
Paris Basin (France)	Essises	33	31.0	From middle Jurassic
Aquitaine Basin (France)	LA 104	31	33.0	From middle Cretaceous
	PTS I	33	35.4	
	RSE I	31	32.0	
	Nat I	26	25.4	
Illizi Basin (Algeria)	Irarraren	33	32.9	Paleozoic
	Oudoume	25	26.4	Paleozoic to Cretaceous
Rhine-Graben (W. Germany)	Landau 2	77	58	Tertiary
	Sandhausen 1	41	30	Tertiary
Douala Basin (Cameroon)	Logbaba	32	47	Cretaceous Paleocene

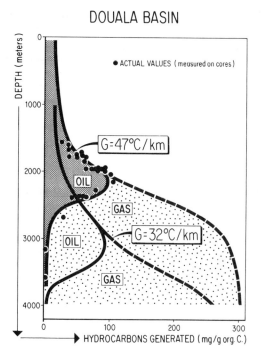

Fig. V.4.9. Comparison of the calculated amount of oil and gas generated in the Douala Basin, by using two different values of the geothermal gradient G. Calculation with the present gradient $G = 32\,°C\,km^{-1}$ differ markedly from the observed values, whereas calculation with the computed past gradient $G = 47\,°C\,km^{-1}$ shows a good agreement with actual values measured on cores

a very long time. The high values presently observed are probably a result of rising of thermal waters through a fault system. This interpretation would also explain the variability of the observed geothermal gradient over relatively short horizontal distances.

In the Douala Basin, the ancient geothermal gradient was rather high, probably due to the vicinity of the midoceanic ridge in the early stage of the basin. During the first 5 million years heat flow and geothermal gradient were probably well over average, progressively decreasing to reach a standard value after ca. 20 million years. Figure V.4.9 shows the amount of oil and gas generated at Logbaba, computed by the model of hydrocarbon generation using the burial versus depth curve and two different values of the geothermal gradient G: the present $G = 32\,°C\,km^{-1}$ and the computed past value $G = 47\,°C\,km^{-1}$. It is obvious that the observed quantities of oil generated (measured on core material) are in agreement with the simulation using the calculated value of average past geothermal gradient ($G = 47\,°C\,km^{-1}$) and differ markedly from the quantities based on the present gradient ($G = 32\,°C\,km^{-1}$).

4.8 Migration Modeling

The other major aspect of geochemical modeling is migration from source rocks to reservoirs and traps, and possibly to the surface. Although the mechanisms of migration are less completely understood than those of generation of hydrocarbons, a first attempt to simulate shale compaction and water expulsion by using a

4.8 Migration Modeling

numerical model was made by Smith (1971). More recently, Ungerer et al. (1983) and Durand et al. (1983) presented an integrated model of oil generation and migration, where the generation part is the kinetic model described above. The migration part is based on the conclusion, expressed in Part III, that the principal mode of fluid migration in sedimentary basins is a polyphasic mode with separate hydrocarbon and water phases.

The model presented by Ungerer et al. (1983) and Durand et al. (1983) is diphasic: fluid movements are governed by the Darcy's law applied to polyphasic flows with the help of the relative permeability concept. The main driving force is compaction due to the sedimentary load, which is supported partly by the solid mineral phase and partly by the pore fluids, water and hydrocarbons, thus increasing internal pressure. Another cause of pressure rise in source rocks is the conversion of solid kerogen into fluid hydrocarbons with a related increase of the average specific volume of the organic phases. Capillary pressures, and the change of oil viscosity with temperature are also accounted for. As presented by these authors, the model is two-dimensional, i.e., it works along a geological section. This section is represented by a grid of finite elements, each element having a specific lithology and organic content. The model calculates for each element and each interval of time: porosity, hydrocarbon saturation, pore fluid pressure, and velocity of fluid flows.

Durand et al. (1983) applied this scheme to part of the Viking Graben in the North Sea, and to the Mahakam delta in eastern Kalimantan, Indonesia. The results from the Viking Graben are presented in Figures V.4.10 and V.4.11. The section represented is bordered by two major faults, on both eastern and western sides, and comprises a tilted block of Jurassic beds underneath the major unconformity. The source rocks belong to Kimmeridge Clay, Heather and Dunlin formations, and also to Brent coals, all of Jurassic age. Sand reservoirs belong to Brent and Statfjord formations. The two major faults bordering the section are considered as impermeable barriers to fluid flow. The oil saturations are presented in Figure V.4.10 at 65 m.y.b.p. (Maestrichtian), when an accumulation is beginning to form in the Upper Brent sands, and also at the present time, when accumulations are located in Jurassic sandstones on the structural high. In source rocks, high saturations are observed in the syncline. It is noted that the onset of oil expulsion from source rocks occurs beyond 20% oil saturation. The model also calculates the excess pressure versus hydrostatic pressure, shown in Figure V.4.11 for the same intervals of time. Important excess pressure is predicted beyond 3000 m depth, and this has been effectively observed in wells. At a given depth, this excess pressure is more important on the structural high, but the maximum values are computed where burial of the Heather formation is the deepest.

The same model used in a very different situation, i.e., the Mahakam delta in Indonesia, has been able to calculate the location of the oil pools, the hydrocarbon reserves in place and the distribution of abnormal pressures, all values being in agreement with observed data. The fact that the model is able to account for very different geological histories, such as the North Sea and the Mahakam delta, suggests that the hypotheses and approximations used in the model are reasonably satisfactory.

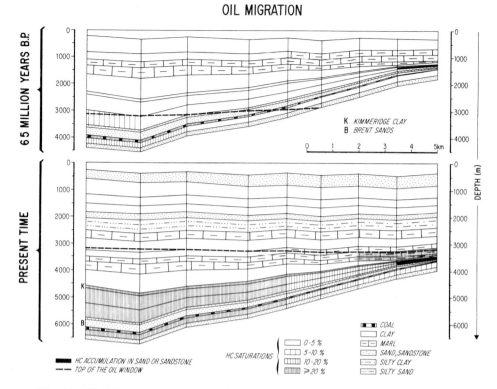

Fig. V.4.10. Hydrocarbons generation and migration modeling in part of the Viking Graben, North Sea. The section is bordered by two major faults on the eastern *(right)* and western *(left)* sides, and comprises a tilted block of Jurassic beds underneath the major unconformity. The model calculates oil saturation in all beds for each interval of time: *(top)* situation at 65 m.y.b.p. (Maestrichtian) which is the beginning of oil expulsion; *(bottom)* situation at the present time, showing accumulation in Jurassic sands (Durand et al., 1983)

Polyphasic models of oil migration certainly represent a significant advance, as most fundamental studies have now concluded that this is the principal mode of hydrocarbon migration. There are, however, specific difficulties inherent to this type of model. Darcy's law is widely used in reservoir engineering, but the conditions in source rocks are very different with respect to heterogeneity, and also extension of the relative permeability concept. Thus, the results presented here should be considered as a first approach. They have shown that, in certain geological situations, the model is able to generate data in agreement with observed facts: location of petroleum accumulations, distribution of pressure including overpressured zones. A present restriction is the two-dimensional character which is acceptable in grabens, such as the Viking Graben, or elongated basins. This should soon be overcome by the increasing availability of high speed computers, which will make the use of three-dimensional models possible.

4.9 Conclusion

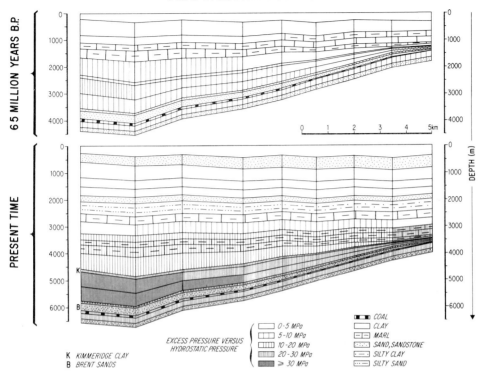

Fig. V.4.11. Excess pressure in part of the Viking Graben. The section is the same as shown in Figure V.4.10. The model calculates the excess pressure versus hydrostatic pressure for the same intervals of time (65 m.y.b.p. and 0 m.y.) (Durand et al., 1983)

4.9 Conclusion

It is now possible to simulate thermal degradation of the main types of kerogen by using a mathematical model. The model is calibrated against each type of organic matter by using a certain distribution of activation energies. A single model is able to simulate the whole thermal history from low temperatures in sedimentary basins to high temperatures in the oil shale industry.

Although migration mechanisms are less completely understood than kinetics of kerogen degradation, it is already possible to simulate migration and calculate hydrocarbon saturation and fluid pressure in each individual bed, as a function of time and location in the basin.

It is realized that modeling of geological processes is still in an initial phase and will rapidly develop, especially with respect to truly three-dimensional simulation procedures. However, some present and future applications of modeling can be listed as follows.

a) determination of prospective areas for oil and gas in sedimentary basins;
b) timing of oil and gas generation and migration, for comparison with the age of formation of traps and impermeable seals;
c) evaluation of the ultimate reserves of a sedimentary basin;
d) distribution of abnormal pressures; history of water flow in the basin;
e) reconstitution of ancient geothermals gradients and thermal history of sediments; this application is not limited to petroleum exploration but concerns geology in general;
f) simulation of retorting conditions of oil shales.

The number of geological/geochemical models used in petroleum exploration is certainly growing fast. Chenet et al. (1983) and Ungerer et al. (1983) discussed the influence of the various subsidiary models (subsidence and compaction, thermal history, hydrocarbon generation, fluid movements, thermodynamics of phase separation and exchange) in a comprehensive model of basin evolution. Bishop et al. (1983) estimated successively the quantities of oil and gas provided by the source rock within the drainage area, the quantities of gas lost by diffusion and dissolution and the trap volume. Then, by comparing these values, they made a probabilistic estimation of trapped hydrocarbon volume and nature (only oil, oil and gas, or only gas). A different type of model, mainly on a statistical, probabilistic basis was presented by Nederlof (1980) and Sluijk and Nederlof (1983). It is designed to calculate the chances that a given trap contains more than a certain volume of oil and/or gas. It draws from a world-wide collection of geological and geochemical data which form the statistical basis and calibration set for a prospect appraisal.

Summary and Conclusion

Geochemical models of petroleum generation offer a quantitative approach to exploration problems. They provide the amount of oil and gas generated (as kg per ton of rock) from the different source rocks in the various parts of the basin. Furthermore, these results are expressed as a function of time, and the timing of petroleum generation can be compared with the age of traps.

The data required as input to the model are the type of kerogen, the burial history (depth versus time curve) and the geothermal gradient. On stable platforms, the ancient gradient can frequently be approximated by the present gradient. In other basins, the ancient geothermal gradient can be calculated by using a model based on vitrinite reflectance.

Migration models provide information on hydrocarbon and water movement. They define the areas and sedimentary beds of hydrocarbon accumulation, and the occurrence of abnormal pressures. They also offer a clue to the problem of ultimate reserves of a sedimentary basin.

Chapter 5
Habitat of Petroleum

Four selected case histories of geologically different petroleum regions will be presented here: the Arabian Carbonate Platform, young delta areas, i.e., the Niger and the Mahakam Delta, the Linyi Basin of China and the Deep Basin of Western Canada. These case histories show the application of the new knowledge about the principles of petroleum generation, migration and accumulation. They demonstrate the understanding of source rock maturation and hydrocarbon generation as part of the geological evolution of a sedimentary basin. Following the considerations about the origin of hydrocarbons in each of the basins described, the most likely pathways and modes of migration into the reservoir rocks and traps are discussed. Finally the quality of the hydrocarbons found in traps is briefly outlined, with the implications with respect to the overall geological setting.

The four case histories selected encompass the most important prototypes of petroleum habitats:

- the Arabian Carbonate Platform, one of the most important petroleum provinces in the world, with Mesozoic marine carbonate source rocks and reservoirs, with a simple tectonic history and consequently simple structural features with an enormous horizontal scale;
- young delta areas, i.e., the Niger and the Mahakam Delta, Tertiary petroleum provinces composed of detrital rock sequences and heavily influenced by terrestrially derived sediments with a rather fast subsidence history;
- the Linyi Basin of China, an assymetric graben system of Tertiary age with a sedimentary filling of detrital rocks, representing limnic, fluviatile to lacustrine environments;
- the Deep Basin of Western Canada, a gas-prone area consisting of a thick Mesozoic section of interlayered sands and shales with numerous coal seams representing a marine to brackish near-shore environment. The gas occurrence in the Deep Basin is of special importance because the situation is the reverse of a conventional gas field, i.e., the gas-bearing zones correspond to the less porous and less permeable rocks situated in a downdip position, whereas more porous and permeable rocks are updip and above.

5.1 Habitat of Petroleum in the Arabian Carbonate Platform

The Arabian Carbonate Platform with the Arabian Basin in a central position is one of the most important petroleum provinces in the world. It can be considered largely to be a foreland-type basin like the Western Canada Basin and Eastern Venezuela (Demaison, 1977). These three basins have in common that they are large basins with a relatively simple asymmetric geometry showing little tectonic complexity and one long, gently dipping flank. They contain by far the largest petroleum accumulations on our planet. The truly gigantic heavy oil and tar sand accumulations at the eastern rim of the Western Canada basin are found at the updip eastern edge of the homocline where the sediments wedge out on the Canadian shield. In Eastern Venezuela the heavy oil belt North of the Orinoco River occurs in a comparable geologic setting where Cretaceaous and Tertiary sediments wedge out toward the south of the Brasilian shield.

A cross-section going from Iran in the East with the folded belt and overthrusting of the Zagros Mountains toward the West through the Persian-Arabian Gulf area and into Saudi Arabia toward the Arabian shield in the West again shows an overall similarity. However, one is tempted to say that the geologic setting on the Arabian Carbonate Platform was more advantageous in so far as there was no major erosion and/or ingression of meteoric waters after the oil had accumulated. Therefore, the petroleum accumulations of the Arabian Carbonate Platform were not largely converted into heavy oils and tars, although there are reports on tar mats in existing fields. This area now holds some of the largest oil fields of the world: Ghawa (Saudi Arabia), estimated ultimate recovery of 14×10^9 m^3 (83×10^9 barrels); Burgan (Kuwait), estimated ultimate recovery of 11.5×10^9 m^3 (72×10^9 barrels); Safaniya-Khafji (Saudi Arabia), estimated ultimate recovery of 5×10^9 m^3 (30×10^9 barrels). These fields and several others in this general area belong to the category of so-called "giant oil fields". Any field with an estimated ultimate recovery of 0.5×10^9 barrels (80×10^6 m^3) is often considered to be such a giant oil field (see Chap. V.6.5). In this connection, however, it should be remembered that the amount of heavy oil in place in the Athabasca deposits in Western Canada alone is far higher than in even the Ghawar field in Saudi Arabia.

The area under consideration here, the Persian-Arabian Gulf region and Saudi Arabia is schematically shown in Figure V.5.1. For many years little information was available about source rocks, hydrocarbon generation and migration in this area. Only recently information was made available for the public. The following discussion on the habitat of oil in the Arabian Carbonate Platform is largely based on two papers by Murris (1980) and Ayres et al. (1982).

The physical boundary of the Arabian Carbonate Platform to the southwest is the outcrop of Precambriam crystalline rocks in the southwestern half of the Arabian Peninsula. The limits to the northeast are less well-defined, but are usually considered to be the folded belt of the Zagros Mountains. This boundary would also roughly separate the Cretaceous and Jurassic shelf sediments in the southwest from open marine sediments and Alpine flysch to the northeast. The southeastern end of the basin is formed by crystalline rocks that were uplifted parallel to transform faults. The area under consideration was not a basin in the

5.1 Habitat of Petroleum in the Arabian Carbonate Platform

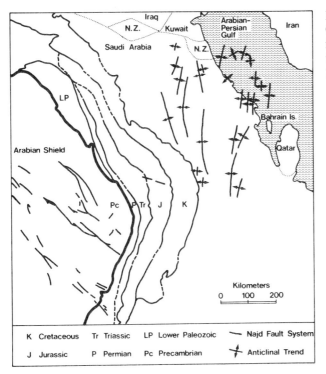

Fig. V.5.1. The Arabian Carbonate Platform and its principal tectonic features. (After Ayres et al., 1982)

geometric sense during the Jurassic and Cretaceous periods while the thickest and most important sediments were being deposited. Instead it was a broad, flat carbonate shelf developing out into the Tethys from the Afro-Arabian continent. Two principal types of depositional systems alternated in time (Mūrris, 1980):

1. ramp-type mixed carbonate-clastic units; during regressive periods clastics were brought into the basin;
2. differentiated carbonate shelves; transgressive periods were dominated by carbonate cycles. Starved euxinic basins were separated from carbonate-evaporite platforms by high-energy margins.

The structural tectonic development of the Arabian Platform can be subdivided into three stages (Murris, 1980). The first stage ended with the Turonian and was characterized by stable platform conditions. During the second stage from Turonian to Maestrichtian there was orogenic activity and the formation of a foredeep along the margin of the Tethys ocean. During the third stage from late Cretaceous to Miocene the platform regained its stability.

Broad and elongated anticlinal structure developed with a north–south direction as drape folds over rejuvenated basement uplifts that resulted originally from an east–west directed early (Precambrian?) compressional stress field (Ayres et al., 1982). These structures and also nearly circular structures in the eastern part of the area, stemming from the movement of Infracambrian Hofmuz salt, contain the large oil field of the Arabian Platform.

There is now ample evidence (Murris, 1980; Ayres et al., 1982; Welte, 1982) that the main source rocks in the Arabian Carbonate Platform are Middle Jurassic to Upper Jurassic organic-rich carbonates. Regionally, where sufficient Upper Cretaceous and Tertiary overburden has been accumulated, Lower to Middle Cretaceous organic rich carbonates and marls also reached sufficient maturity to generate appreciable amounts of petroleum. The accepted source rocks of Jurassic and Cretaceous age contain mainly type-II kerogen. The source rocks belong to the transgressive periods where on the differentiated carbonate shelves euxinic basins could have been developed. Names for the stratigraphic equivalents of source rock units vary regionally. Well-known representatives are the Upper Jurassic Hanifa source rock and the Aptian Shuaiba source rock (Murris, 1980).

A rather detailed description of a Jurassic source rock section (Callovian to Oxfordian) in Northern Saudi Arabia has been presented by Ayres et al. (1982). The unit was initially identified from geochemical analyses of well cuttings and core samples. The total organic carbon content averages around 3% to 5% by weight. Visual analyses showed the kerogen to be dominantly amorphous. Elemental analysis of kerogen (Ayres et al., 1982) and pyrolysis indices show that the organic matter belongs to type II.

An interesting technique was applied to map the regional occurrence of these source rocks throughout Northern Saudi Arabia (Ayres et al., 1982). Source rocks, being very rich in kerogen and extractable bitumen, have distinct geophysical wire line characteristics, showing high electrical resistivities com-

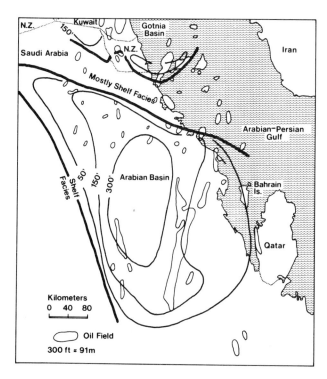

Fig. V.5.2. Distribution and isopachs of a Jurassic source rock section (Callovian to Oxfordian) in Northern Saudi Arabia with more than 1% total organic carbon. (After Ayres et al., 1982)

bined with low sonic velocities (Chap. III.2.4.5). Neutron logs respond similarly to sonic logs with respect to the organic facies. Gamma-ray response was generally high but variable. This variation in gamma-ray response of source rocks is more pronounced in carbonate source rocks than in shaly source rocks which are more steady as, for instance, documented in the case of the Kimmeridgian of the North Sea. Log responses in these uniform source lithologies of Saudi Arabia were calibrated to geochemical analytic data and used to estimate and map source rock richness. Figure V.5.2 shows the distribution and isopachs of the source rock with a total organic carbon content which is 1% or greater by weight in the Arabian Basin and the adjacent Gotnia Basin in the north.

The timing of maturation and hydrocarbon generation is related to burial and temperature exposure of the source rocks. In carbonate-type source rocks the vitrinite reflectance technique to determine source rock maturation frequently cannot be used because vitrinite due to the lack of terrestrial detrital input is scarce or absent in most carbonate sequences. The same problem has been encountered with Arabian carbonate source rocks. Consequently the degree of maturation of the kerogen has been estimated by Ayres et al. (1982) from atomic H/C ratios and from calculated maturities based on Lopatin's time–temperature index method (Lopatin, 1971; Waples, 1980). The use of H/C ratios for a determination of maturities is only possible when the type of kerogen is well known and uniform over the area under consideration. A map of the H/C ratios of kerogen within Callovian and Oxfordian source rock facies is shown in Figure V.5.3. The H/C ratios between 1.3 and 0.8 of this kerogen have been interpreted

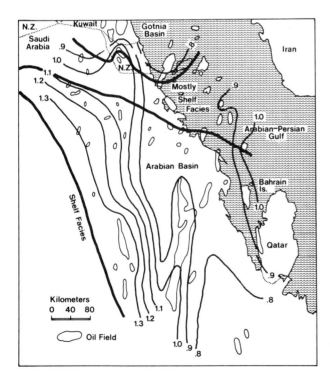

Fig. V.5.3. Map of H/C ratios of a Jurassic source rock section (Callovian to Oxfordian) in Northern Saudi Arabia. (After Ayres et al., 1982)

to correspond more or less to the oil window equivalent to vitrinite reflectance values of 0.6% to about 1.3%.

In general there seems to be agreement (Murris, 1980; Ayres et al., 1982; Welte, 1982) that the Middle to Upper Jurassic source rocks are mature over most of the area of the Arabian Carbonate Platform. The maturation history, i.e., the evolution of the onset and main phase of hydrocarbon generation in different parts of that area can be inferred from measurements and computations based on geothermal data and burial histories. In terms of absolute age values depending on local burial history and temperatures, Middle to Upper Jurassic source rocks reached the stage of maximum oil generation 45 to 70 million years ago, while Lower to Middle Cretaceous source rocks are even at present immature in certain regions.

Petroleum generated from these source rocks is found today in reservoir rocks which were formed by regressive sands and by the high-energy ooidal grainstones terminating the carbonate cycles or in rudist "reefs" and algal boundstones along the shelf margins (Murris, 1980). Leaching and cementation processes played an important role in determining the final reservoir quality of the carbonate rocks. The principal regional sealing formations are the Upper Jurassic Hith Anhydrite and the Cretaceous Nahr Umr Shale. They control largely the vertical migration of petroleum. Regionally vertical fractures and faults must have provided migration avenues from source rocks through several hundreds of meters of relatively tight carbonate rocks before petroleum was emplaced in a reservoir and trapped by an effective seal. The Ghawar Field can be cited as an example for such extensive vertical migration (Steineke et al., 1958). However, it must be emphasized at this point that in spite of this obvious vertical migration, the outstanding feature of migration on the Arabian Carbonate Platform is a substantial lateral migration (ARAMCO-staff, 1959; Murris, 1980).

According to Ayres et al. (1982), the oils in Mesozoic reservoirs within Saudi Arabia are similar in composition. The well-known difference between Arabian light and heavy (Safaniya) oils may be due to maturation differences. This overall uniformity of oils from numerous fields and production zones suggests a common or similar source. This is supported by correlation analyses that link oils and source rocks.

The quality of most oils is such that they fall into the aromatic-intermediate class (see also Fig. IV.2.1). Some oils have to be classified as paraffinic-naphthenic (Abu Dhabi, Qatar). A few oils apparently have suffered biodegradation and lost an important part of the saturated fraction. The sulfur content of the oils is generally high (1–4%) and may reach values up to 8% by weight. Stable carbon isotope ratios vary by less than 2.0% and the content in vanadium and nickel porphyrins is low.

Murris (1980) points out what makes the Middle East such an extraordinary rich petroleum habitat. It is not exceptionally rich or thick source rocks; it is not extraordinary reservoirs or seals. It is, however, the horizontal scale of the basin that provides the three essential ingredients of a petroleum province, source rocks, reservoir rocks and seals with an uncommonly wide extent. The Mesozoic depositional platform was 2000 to 3000 km wide and at least twice as long. The degree of differentiation, facies changes, and variations in the sedimentation-related morphology of the platform were minor. Therefore, source rocks, reservoirs and seals all have very wide lateral extent and rather uniform consistency. Likewise the structures are also very wide and have gentle slopes. This has a twofold effect in that no fracturing in general damages the seal, and large trap volumes are attached to even relatively small vertical closure. The horizontal extent of a closure of 1000 km^2 or even more is no exception.

Finally, because of these wide horizontal features migration efficiency is very high: large areas of mature source rocks could be drained over long geological periods and not much oil was lost along the paths of migration and to the surface.

The major difference to Western Canada and Eastern Venezuela is the fact that the flanks of the Arabian Basin were not cut open by erosion. Leakage to the surface was therefore much less important and likewise biodegradation and conversion to tar is relatively rare.

5.2 Habitat of Petroleum in Young Delta Areas

Large young delta areas represent a special setting for the origin, migration and accumulation of petroleum. The essential features and processes for the occurrence of petroleum are among the characteristics of a delta, provided that there is an organic rich source formation. The geologic setting is such that the marine shales of the delta front and the prodelta area, receiving mainly land-derived organic matter, can act as source rocks. The interlayered mixed sands and shales of the paralic sequence of the delta plan provide reservoir opportunities and sealing cap rocks in the direct neighborhood of the delta front. Furthermore, young deltas during their build-up and sediment accumulation undergo comparatively strong subsidence without uplifting. This is in turn essential for source rock maturation, generation of hydrocarbons, and migration. Thus the main geologic features and processes required for petroleum occurrences are suitably combined in space and time in delta areas; i.e., they offer a compact scenario of petroleum generation, migration and accumulation.

For the special importance of young deltas three reasons can be cited: the general fall of the sea level throughout the world oceans from about +350 m in Upper Cretaceous time to +60 m in Middle Miocene; the development and rapid spreading of an angiosperm flora that conquered the continents starting with the Upper Cretaceous; finally the Alpine folding, mainly occurring during late Cretaceous and Tertiary. The fall of the sea level created large drainage patterns and funneled the water run-off from the continents to the oceans. The new angiosperm flora provided great amounts of terrestrial organic matter of higher plant origin to be transported by huge river systems. The Alpine folding finally gave rise to a strong relief in certain areas, providing abundant clastic rock debris.

In delta areas the common geological rule that younger sediments are on top of older sediments is modified somewhat, encompassing also a lateral component with respect to the age of sediments. Generally speaking, due to the prograding delta, older sediments are on the landward side, and going in the opposite direction toward the sea, sediments become progressively younger. The progradation of the delta in time causes a shift of facies patterns as the delta moves out into the sea. This results in a well-known diachronism: the facies of continental, fluviatile sediments, the mixed sediments of the deltaic plain and those of the marine prodelta shales cut across time stratigraphic units. Due to their nature, deltas are situated at the edge of a land mass and very often at the junction between the continental and oceanic crust (Audley-Charles et al., 1977). Therefore, the depositional and structural evolution of the prograding delta is more strongly influenced by basement tectonics than is normally the case with other basins.

The Tertiary Niger Delta on the western edge of the African continent and the Tertiary Mahakam Delta on the eastern coast of Kalimantan (Borneo), Indonesia are next described. Both deltas are equatorial and provide abundant organic material from the tropical forests and swamps of the backland and also autochthonous organic material from the shallow waters of the delta front and thus ensure the deposition of potential source rocks mainly in front of the delta.

5.2.1 The Tertiary Niger Delta

The Tertiary Niger Delta can be considered as one of the best-known delta areas. The geology and the hydrocarbon habitat of the Niger Delta have been described in a series of papers (Weber and Daukoru, 1976; Evamy et al., 1978; Ekweozor and Okoye, 1980; Ejedawe, 1981) in considerable detail. The Niger Delta represents a regressive clastic sequence which reaches a maximum thickness of 9,000 to 12,000 m of sediment. It covers an area of about 75,000 km^2 and measures across the Tertiary delta well over 300 km. The principal geological features are shown in a SSW to NNE cross section (Fig. V.5.4). The formation of an initial basin at the site of the present delta is very probably related to Cretaceous rifting when South America started to be separated from Africa. Basement tectonics at the junction of oceanic and continental crust determined the original sites of the main river and thus controlled the depocenter and the development of the delta. This development of the delta has been dependent on the balance between the rate of sedimentation and the rate of subsidence. This balance and the resulting sedimentary patterns seem to be influenced by basement tectonics of individual fault blocks (Evamy et al., 1978). The thick wedge of sediments in the Niger Delta can be subdivided into three units: mainly marine shales at the bottom (Akata Formation), an overlying paralic sequence consisting of interbedded sands and shales (Agbada Formation), and a topmost continental unit composed of fluviatile gravels and sands (Benin Formation). These three formations are also shown in Figure V.5.4. They are strongly diachronous and cut across time stratigraphic units. As mentioned before, this latter effect is caused by the advancement of the delta toward the sea, mainly during Middle and Upper Tertiary, and the accompanying facies shift.

The marine shales of the Akata Formation are generally over-pressured (under-compacted) and are considered to be the main source of the hydrocarbons

5.2 Habitat of Petroleum in Young Delta Areas

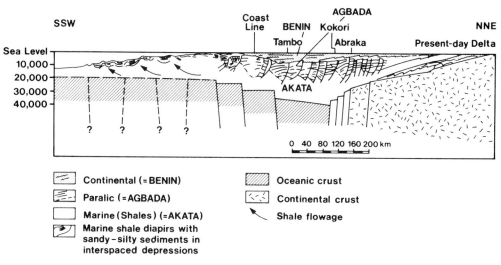

Fig. V.5.4. Cross-section through the Niger Delta and its principal geological features. (After Evamy et al., 1978)

found throughout the delta. The paralic sequence of the Agbada Formation with the alternating sand and shale layers contains nearly all petroleum accumulations. It plays, however, a minor role as a source formation. The great majority of these petroleum accumulations are found in traps associated with so-called "growth faults". These growth faults (see also Fig. III.4.9) are synsedimentary gravitational faults that remain active over extended periods of time. Therefore more and faster sedimentation is possible in the downthrown block relative to the upthrown block (Weber and Daukoru, 1976). Growth faults are also well known from the Gulf Coast area in the US.

Ejedawe (1981) has studied the distribution of oil reserves in the Niger Delta and its relation to the geological evolution of the prograding Tertiary delta, especially with respect to the tectonic and sedimentologic control. He found that a prolific belt (Fig. V.5.5), representing an area of preferential concentration of giant and medium-sized oil fields is close to the continental edge. It was concluded that the prolific belt either coincides with the transition zone, where the oceanic and continental crusts are thought to meet, or is floored by oceanic crust. In any case the prolific belt is the zone of maximum sedimentary thickness and of a high rate of subsidence. It contains a proper blend of sands and shales encompassing source rocks, reservoir rocks and seals. The enormous sedimentary thickness allows source rocks at greater depth to be in the oil- or gas-generating window. Landward there are fewer shales to provide potential source rocks and seals, seaward there are too few reservoir sands. There is agreement that the marine Akata shales constitute the major source rocks for the hydrocarbons of the Niger Delta. The deeply buried lower parts of the paralic sequence (Agbada Formation) cannot be discounted as sources for oil and gas (Weber and Daukoru, 1975; Ekweozor and Okoye, 1980), however they seem to be of lesser importance as compared to the Akata Formation. A wide variety of shale samples from the

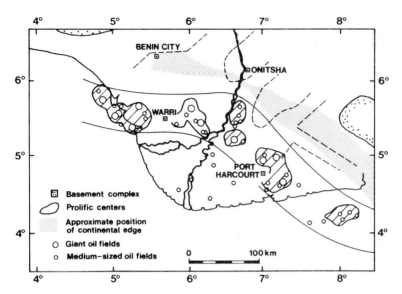

Fig. V.5.5. Map showing relations of prolific belt, position of continental edge, and major oil fields of Niger Delta. (After Ejedawe, 1981)

Akata and Agbada Formation showed organic carbon contents of about 1% with very few values in excess of 2% C_{org}. The quality of kerogen is mainly type III with varying admixtures of type II. Occasionally liptinite enrichments can occur. Following the terminology used by Shell (Shell Standard Legend, 1976) it would correspond to humic and mixed types of organic matter.

The threshold of intense hydrocarbon generation as deduced from a few wells based on analyses of extractable organic matter was found in the offshore area at approximately 2900 m and in the onshore area at considerably greater depth around 3300 m (Ekweozor and Okoye, 1980). A correction for migrated hydrocarbons obviously was not made, so that the main zone of hydrocarbon generation in reality could be somewhat deeper offshore and onshore. In addition, it has to be kept in mind that the temperature distribution in the Tertiary delta is variable, resulting in different depth levels of the top of the "oil kitchen" (Evamy et al., 1978) throughout the delta. A detailed temperature survey throughout the delta encompassing all three sedimentary units has shown that the present-day temperature gradient varies consistently with the sandstone/shale ratio in a given well (Fig. V.5.6). The temperature gradient increases with diminishing sand percentage from less than 1.0°F/100 ft (1.84°C/100 m) in continental sands, to about 1.5°F/100 ft (2.73°C/100 m) in the paralic section, to a maximum of about 3.0°F/100 ft (5.47°C/100 m) in the continuous marine shales. This temperature pattern is related to freely moving ground waters especially in the shallower sandstones of the continental sands and the existence of a high pressure zone at the bottom. Freely moving ground waters act as cooling agents and high pressure zones show a rather rapid temperature increase.

The thickest continental sandstones (Benin Formation) and the thickest paralic sequence (Agbada Formation) with alternating sands and shales is found in the

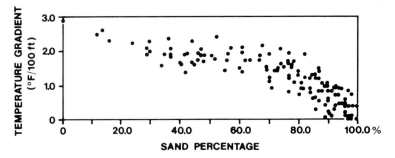

Fig. V.5.6. Relation between geotemperature gradient and sand percentage in the Niger Delta. (After Evamy et al., 1978)

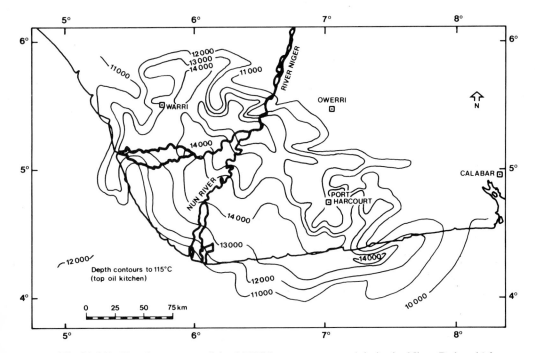

Fig. V.5.7. Depth contours of the 115°C isotemperature plain in the Niger Delta. (After Evamy et al., 1978)

middle part of the delta (see Figs. V.5.4 and V.5.5). Following the above concept of temperature distributions, it is not surprising to find on isotemperature plots the deepest subsurface contours for an isotemperature plain (e.g., 115°C) in the middle part of the delta (Fig. V.5.7). The temperature of 115°C has been refered to as the top of the oil kitchen in the Tertiary delta by Evamy et al. (1978). It is possible to use a certain temperature value here rather than a maturity indicator for defining the maturity of source rocks as maturation and oil generation are very young in the delta and related to a rapid, equally young subsidence in that area:

There was no additional uplift and erosion. It is interesting to note that Evamy et al. (1978) and Ekweozor and Okoye (1980) report the same depth difference of approximately 400 m with respect to the top of the oil kitchen or onset of intense hydrocarbon generation between given offshore and onshore areas, although they do not agree on absolute depth of the principal zone of oil generation. In the onshore area maturation is achieved at greater depth.

The mature source rocks have apparently generated light and paraffinic oils. The initial paraffin wax content may differ with the facies of the source rock, i. e., more input of terrestrial higher plants or, more precisely, a higher liptinite content in kerogen would cause the wax content to rise. Among Nigerian crude oils in principle three variations are found: paraffinic light oils, intermediate paraffinic oils with a certain wax content (both belong to the paraffinic class — Chap. IV.2) and biodegraded heavier oils. The degree of biodegradation varies. The general coincidence of the boundary between bacterially degraded and unaltered paraffinic oils, within a rather narrow temperature range (65° to 80°C), according to Evamy et al. (1978), implies that little or no subsidence with simultaneous increase in geotemperature of the oil-bearing reservoirs has occurred after emplacement and degradation of the oils. This distinct oil distribution pattern may be explained as follows: differences among the original paraffinic oils are either due to source rock heterogeneity and/or possibly migrational changes and there is no long-distance lateral migration of oil, whereas there may be extensive vertical migration. Thus the original oil distribution pattern in the delta has been largely preserved except for the alteration of light primary oils by bacterial degradation in zones that were invaded by meteoric waters and had formation temperatures not higher than 65° to 80°C. There the primary light paraffinic oils are transformed into medium-gravity low pour-point crudes. Philippi (1977) has reported a similar finding (see also p. 464).

Weber and Daukoru (1976) and Evamy et al. (1978) have emphasized the role of growth faults as migration avenues for hydrocarbons from the undercompacted Akata shales into Agbada reservoirs. The faults, at greater depth where they become more horizontal in the plastic marine shales, are not considered to provide effective migration paths. However, that part of the faults which is more vertical may very well provide an effective migration path (see also Fig. III.4.9). The growth faults thus help migration in a double sense, they can bring the overpressured mature source rock into juxtaposition with the downthrown reservoir sequence (Agbada), depending on the throw, and through the permeable vertical part of the fault they can provide migration avenues. Where the hydrocarbon kitchen lies well below the top of the continuous shales, the oil apparently has only a remote chance of migrating through the shales into the overlying reservoirs of the Agbada Formation.

5.2.2 The Tertiary Mahakam Delta

The Tertiary Mahakam Delta and its hydrocarbon occurrences have only been investigated over the last few years. The hydrocarbon habitat, mainly with respect to source rock and migrational phenomena, has been described by

Magnier et al. (1976), Combaz and De Matharel (1978), Durand and Oudin (1980), Vandenbroucke and Durand (1983) and by Schoell et al. (1983). These publications served as the main source of information for the following considerations on the Mahakam Delta.

The Tertiary Mahakam Delta is located in Eastern Kalimantan, at the eastern edge of the Kutei Basin. The deltaic system going landward from the delta front from the offshore area into the hinterland measures approximately 50 km and the "length" is nearly 100 km following the coast line (Fig. V.5.8). Deltaic sedimentation is known to have occurred at least since the Middle Miocene, i.e., for 15 million years. The eastward prograding sedimentary sequence is about 6000 to 8000 m thick, about 4000 m of which have been drilled. The exact position of the basement and the age of the first deltaic deposits are not known. The Mahakam Delta is considerably smaller in size and has accumulated less sediments than the Niger Delta. Since the Middle Miocene three major deltaic complexes have accumulated, separated by two marine transgressions. Underneath the interbedded clays and sands of the deltaic plain there are strongly marine-influenced or marine argillaceous sediments of the prodelta. These latter sediments, especially massive silty prodelta shales, are frequently overpressured. The alternating clays and sands of the deltaic plain in the Mahakam Delta are comparable to the paralic sequence (Agbada Formation) in the Niger Delta and the massive prodelta shales

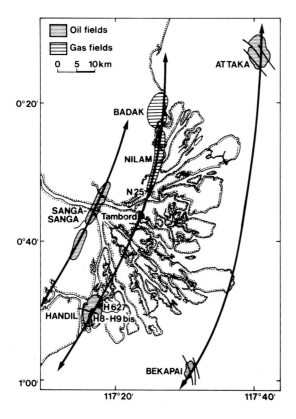

Fig. V.5.8. The Tertiary Mahakam Delta in Eastern Kalimantan (Indonesia) with major anticlinal axis and its oil and gas fields. (After Vandenbroucke et al., 1983)

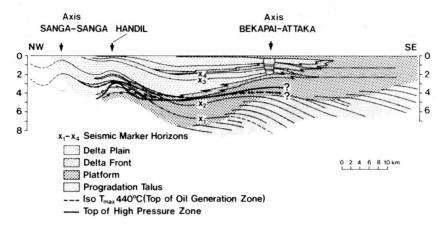

Fig. V.5.9. Cross-section through the Mahakam Delta (Indonesia) and its principal geological features. (After Durand et al., 1983)

to the marine shales (Akata Formation) in Nigeria respectively. As in the Niger Delta, the source rocks are found in the frequently overpressured deeper prodelta shales and the reservoir rocks among the alternating sand and shales of the plain of the Mahakam Delta.

In the Niger Delta the dominating structural features are growth faults. This is not so in the Mahakam Delta. There, the most important structures are major anticlinal axes which roughly parallel the coast line (Fig. V.5.8) and have a decreasing amplitude from west to east (Fig. V.5.9). The oil and gas fields are aligned along these anticlinal axes (Fig. V.5.8). The source rocks for oil and gas are reported to be mainly shaly prodelta sediments and also coals (Combaz and Matharel, 1978; Durand and Oudin, 1980; Vandenbroucke and Durand, 1983). The average organic carbon content is usually above 2% C_{org}. The argillaceous sediments in front of the delta contain ca. 2–3% C_{org}. Coaly layers may reach up to 80% in C_{org}. Practically all the kerogen is type III, and there is no difference in chemical composition between the organic matter of shales and that of coaly layers (Vandenbroucke and Durand, 1983).

In the Handil Oil Field (Figs. V.5.8 and V.5.9) the main zone of oil formation was determined by Durand and Oudin (1980) as occurring at T_{max} values between 440° and 455°C as based on Rock-Eval analyses. This corresponds to a vitrinite reflectance value between 0.7% and 1.1% of type-III kerogen. The zone of hydrocarbon formation in the Handil field was found to be far below the hydrocarbon-bearing reservoirs, between 500 and 1000 m deeper. The zone of hydrocarbon formation followed the contours of the structure, i.e., it was deeper at the flanks and highest at the crest of the structure (Fig. V.5.9). The top of the oil window at Handil is around 2500 m. Many excellent potential source rocks in the Mahakam Delta are still immature.

Similarly as in the Niger Delta, the source rocks reach sufficient maturity for hydrocarbon generation in the overpressured shales or at least in the so-called transition zone with intermediate pressures. The thickness of this transition zone is about 300 m in Handil field and probably about 700 m in Nilam and Badak field. It is argued convincingly that gaseous and very light hydrocarbons can migrate out of the overpressured zone more easily than heavier hydrocarbons and nonhydrocarbons. Then the light hydrocarbons on their way upward extract heavier compounds and carry them along. In the "hot" overpressured shales there seems to be a single supercritical hydrocarbon phase that, when moving upward to lower pressures and temperatures, progressively loses the heavier end of its hydrocarbons by retrograde condensation. In this manner the inverse specific gravity relationship of hydrocarbon occurrences in Handil can be explained, i.e., light products at the top and heavier at the bottom.

In the Mahakam Delta fracture planes seem to be of only minor importance for secondary migration as compared to the Niger Delta. Where fractures or faults are absent in the Mahakam Delta the sealing efficiency and the distance between the top of the mature zone in the source sequence (top oil kitchen) and the reservoir zone control the amount of hydrocarbons which can be expelled. If this barrier is thick, few hydrocarbons, except the more volatile compounds, can pass through.

Therefore it was concluded that in the Mahakam Delta the nature of the pooled hydrocarbons, i.e., oil or gas, and the amount that accumulated in different fields, was dependent on the relative position of the hydrocarbon kitchen with respect to the reservoir. A thicker shale interval between the hydrocarbon-generating zone and the reservoir sand would allow fewer hydrocarbons to accumulate than only a thin barrier between the mature source and the reservoir. The relative position of different reservoir beds vs. the underlying overpressured source zone would, due to retrograde condensation increasingly, allow lighter hydrocarbon mixtures to accumulate in reservoirs, the farther they are above the source. Thus, unlike in the Niger Delta, the differences among the pooled hydrocarbons are not seen as consequences of source rock quality but as related to migration. Oil–source rock correlation studies as based on steroid and triterpenoid hydrocarbons and stable isotopes (Schoell et al., 1983) have shown that the oils are sourced from humic kerogens (type III). The quality of the pooled hydrocarbons varies. Due to retrograde condensation there are gas pools and condensates at shallower depth and also light paraffinic oils. A slight biodegradation of primary light paraffinic oils may occur but does not seem to be frequent.

Summarizing the hydrocarbon habitat in the Niger and Mahakam Delta many similarities between the two Tertiary delta areas can be observed. The sedimentological succession with source rocks, reservoir rocks and seals is nearly identical in both cases. The source rocks are the more marine shales in front of the delta and at the bottom of the sedimentary sequence. The reservoir rocks are the alternating sands and shales of the paralic sequence in the Niger Delta and the comparable alternating sequence of the deltaic plain in the Mahakam Delta. In both deltas the mature source rocks are generally deep and the top of the oil window is in or close to the overpressured zone. The thermal gradients in both areas are non-linear and vary regionally. The type of kerogen is dominated by the terrestrial influx of organic matter, i.e., source rocks contain mainly type-III kerogen. Migration from source rocks to reservoir rocks is mainly vertical in both examples. The primary oils in both delta areas are light paraffinic oils. Generation, migration and accumulation of hydrocarbons in both instances is very young and apparently still in progress. Although the two deltas differ in size, their approximate reserves per area are comparable with each of about 10^5 m^3 oil equivalent per km^2. The Niger Delta contains about 6×10^9 m^3 oil equivalent and the Mahakam Delta about 0.5×10^9 m^3 oil equivalent.

Finally, it should be mentioned that are also some dissimilarities between the petroleum habitat of the two delta areas. Whereas in the Mahakam Delta coals play a major role as a source for hydrocarbons, there is no indication for this in Nigeria. The typical traps in Nigeria are growth faults which are, however, not found in the Mahakam Delta. Consequently faults are of importance for migration in Nigeria but not in the Mahakam Delta area. Conversely a gas-phase migration and retrograde condensation are mechanisms which prevail in the Mahakam Delta, but seem to be of no or minor importance in Nigeria.

5.3 The Linyi Basin in the People's Republic of China

The following information on the Linyi Basin is based on the first results provided by the joint research project (being continued at present) between the Geological Research Institute of the Shengli Oil Field (Shandong Province, People's Republic of China) and the Institute of Petroleum and Organic Geochemistry at the Nuclear Research Center in Jülich (KFA-Jülich), West-Germany[13]. The information is based on unpublished geological data of the Shengli Oil Field and geochemical and geological results obtained during the cooperation of both sides at the KFA Jülich. The case history of the Linyi Basin makes use of data, especially during the discussion of source rock maturation, obtained through geological-geochemical basin modeling by computer simulation.

The Linyi Basin is located on the North China plain, south of the Gulf of Bohai near the mouth of the Yellow River (Fig. V.5.10) and extends over an area of 3969 km^2. It represents a asymmetric graben system controlled by normal faults that came into existence at the beginning of Oligocene. The strike of the basin, especially of the central graben and the accompanying fault zones is SW–NE. The basin is mainly filled with detrital rocks, shales, silt- and sandstones but also contains minor amounts of carbonates. The basin filling in general represents a limnic, fluviatile to lacustrine environment. The graben system of the Linyi Basin is an extensional structure. The fault-controlled subsidence during the basin evolution in Oligocene was interrupted during mot of Miocene by a period of uplift or nondeposition and then followed again by subsidence. The total thickness of these sediments in the deepest part of the graben amounts to about 5000 m or more.

---

13 This cooperation was started in 1979 with the authorization of the Ministry of Petroleum Industry of the People's Republic of China and the Ministry of Research and Technology of the Federal Republic of Germany

5.3 The Linyi Basin in the People's Republic of China

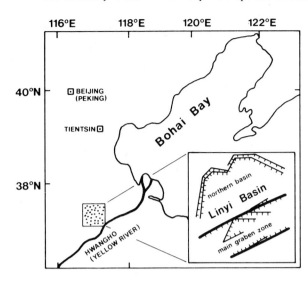

Fig. V.5.10. Location map of the Linyi Basin (North China) and main structural features

area investigated

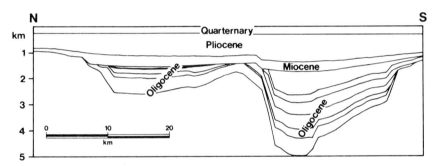

Fig. V.5.11. Typical cross-section through the Linyi Basin. This is a computer-plotted cross-section as taken from basin modeling. Fault zones bordering the main graben features are not depicted

Volcanism is restricted to the period of main tectonic activity during Oligocene time. A typical N–S cross-section through the basin is shown in Figure V.5.11. This cross-section represents a computer plotted profile and is based on an input data file employing a three-dimensional grid system covering the basin. Therefore, the fault zones bordering the individual parts of the basin, especially on the flanks of the deep central part of the graben, are not shown. This cross-section is a typical result of geological modeling (Chap. V.3).

Among the lower Oligocene shaly sediments there are units exhibiting excellent source rock properties with organic carbon values in excess of 2%. They represent a fluvio-lacustrine depositional environment with regional occurrences of typical lake facies. At a maturity level of about 0.5% vitrinite reflectance, source rock samples from these sedimentary units have hydrogen index values of 200 to 500 mg hydrocarbons/g C_{org} as based on Rock-Eval analyses. The kerogens are classified as type II or type II–III, which is in good correspondance with the type of kerogen found in the Dongying depression farther east (Zhou Guangjia, 1980). A maceral analysis under the microscope revealed very high liptinite contents: frequently more than 70% of total macerals identified were liptinites. This microscopic observation is in good agreement with geochemical analyses, for instance with high hydrogen index values. Especially source rock samples of lower Oligocene age are characterized by mixed algae (mainly Botryococcus), and degraded spores. Their kerogen is mainly type II and is considered to be a good source of oil.

The fact that type II and type III are the predominant types of kerogen, and not type I as in the Uinta Basin (another type of intracontinental lacustrine basin), seems to be related to the water circulation pattern and the influx of detrital organic material from the surrounding land. The predominance of type-I kerogen is probably tied to quiet undisturbed lakes such as the Eocene Green River in the western United States.

The terrigenous influence on the organic matter in the Linyi Basin sediments is obvious from gas chromatographic investigation of the extractable saturated hydrocarbon fractions. All chromatograms are dominated by long-chain n-alkanes derived from higher plant waxes. Partial microbial reworking is indicated in some samples by a slight shift of the n-alkane maximum to lower carbon numbers. The algal contribution recognized by kerogen microscopy cannot be seen in most of the saturated hydrocarbon distributions.

Basin evolution and subsidence largely determine the temperature history and therefore maturation of source rocks. The maturation of source rocks and the hydrocarbon generation in the Linyi Basin has been reconstructed with the help of geological-geochemical modeling based on a computer simulation following the concept of Welte and Yükler (1980). For the reconstruction of the following source rock maturation history a changing heat flow during basin evolution with 1.85 HFU (heat flow unit = 10^{-6} cal cm^{-2} s^{-1}) in the beginning and 1.40 HFU toward the end (Quaternary) was assumed. The changes in heat flow during

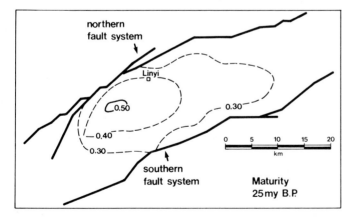

Fig. V.5.12. Computer-plotted maturity map of an Oligocene source rock in the central part of the graben of the Linyi Basin 25 million years before present. Maturity is given in isoreflectance lines of vitrinite (calculated by computer)

5.3 The Linyi Basin in the People's Republic of China

basin evolution follow the concept of McKenzie (1978). Three successive stages of the maturity development of an Oligocene source rock in the Linyi Basin are shown in Figure V.5.12 through V.5.14. The isoreflectance lines of vitrinite shown in these maps are each constructed from well over 100 control points from the data matrix provided by the computer model. The time slices with 25 and 16 million years before present are chosen arbitrarily. Depending on the time steps used in the model simulation any other time slices can be depicted as a maturity map for a given source rock. The computer-derived maturity data are calibrated by a comparison between measured and calculated vitrinite reflectance values for the present time in areas of the basin where there is well control. From Figure V.5.12 through Figure V.5.14 it can be seen that the onset of oil generation with

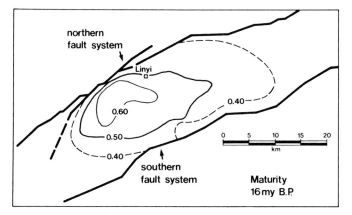

Fig. V.5.13. Computer-plotted maturity map of an Oligocene source rock in the central part of the graben of the Linyi Basin 16 million years before present. Maturity is given in isoreflectance lines of vitrinites (calculated by computer)

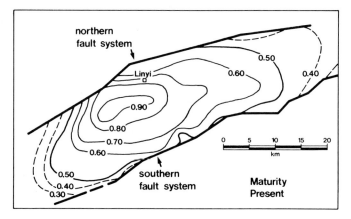

Fig. V.5.14. Computer-plotted maturity map of an Oligocene source rock in the central part of the graben of the Linyi Basin at present. Maturity is given in isoreflectance linés of vitrinite (calculated by computer)

respect to the source rock shown was only about 25 million years ago in the very center of the graben. Then the mature area with steadily increasing oil generation expanded slowly over the next 10 million years only reaching peak oil generation in the center of the basin over the most recent million years (Fig. V.5.14). Knowledge derived from such source rock maturity maps can be combined with information about the spatial distribution of carrier and reservoir rocks and structural features throughout the basin. In this way main migration avenues, changes in drainage patterns through time, and the availability of potential traps can be determined. The alternating sand-shale sequence in the Linyi Basin and the structural setting of the graben system offer numerous possibilities for oil accumulations. The fault system at the edge of the graben provides good migration avenues for oil derived from deeply buried source rocks in the center part of the graben.

The composition of the crude oils found in Linyi Basin reservoirs is in good agreement with an origin from limnic organic matter. Biodegradation of varying degrees is not uncommon among the crude oils of this basin. According to liquid chromatography the non-biodegraded oils are of the paraffinic type with saturated hydrocarbons in the stripped crude oils (C_{15+}-cut) from 37 to 69% (average about 55%) whereas aromatics and N, S, O-compounds are less abundant (11–31%). The asphaltene contents are relatively high and range from 2 to 8%. Many oils show a slight but distinct off-even predominance of higher n-alkanes above 20 C-atoms. The sterane distribution pattern strongly favoring C_{29} and C_{27} over C_{28} compounds is consistent with a significant contribution from terrestrial organic matter.

The Linyi Basin is a good example for the formation and occurrence of petroleum in a limnic, fluviatile to lacustrine environment with terrestrial influences. The computer simulation of the basin evolution following the concept of Welte and Yükler (1980) permits, contrary to concepts based on geothermal gradients, a regional reconstruction of the paleotemperature history for each sedimentary unit throughout the basin and thus detailed timing of source rock maturation.

5.4 Habitat of Gas in the Deep Basin of Western Canada

The Deep Basin in Western Canada is located east of the tectonically disturbed belt of the Rocky Mountains in the provinces of British Columbia and Alberta (Fig. V.5.15). It represents the deepest part of the huge asymmetric Western Canada Basin. The Deep Basin is approximately 650 km long and reaches a width of about 130 km. A schematic SSW–NNE cross-section shows the principal geological features of the basin (Fig. V.5.16). Paleozoic rocks, mainly carbonates, rest unconformably on the Precambrian basement which dips gently to the SW. They are overlain by a thick Mesozoic sequence largely consisting of Cretaceous dark shales with interlayered sandstones and conglomerates. Coal seams are frequent, particularly in the deeper part of the Lower Cretaceous. The thickness of the Mesozoic increases from approx. 300 m in Eastern Alberta to

5.4 Habitat of Gas in the Deep Basin of Western Canada

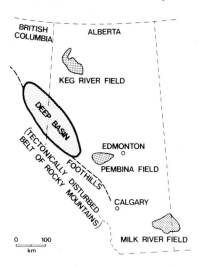

Fig. V.5.15. Location of Western Canada Deep Basin

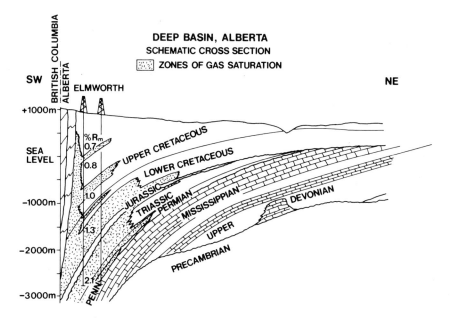

Fig. V.5.16. Schematic cross-section through Deep Basin showing ones of gas saturation

more than 4000 m near the Foothills and the overthrust belt which forms the border of the basin to the west. The strata in general dip southwestward into the Deep Basin, where both porosity and permeability of the sandstones decrease significantly due to greater compaction, higher clay content, and more intense diagenesis. Gas occurs only in the deepest part of the basin, over an area of approx. 67,000 km², where almost the entire Mesozoic section at a depth level exceeding 1000 m below the surface is gas-saturated as derived from resistivity logs (Masters, 1979) and geochemical analyses of several wells (Welte et al., 1982). These gas-bearing zones correspond to the less porous and less permeable rocks situated in a downdip position. The same strata in an updip position east of a transition zone, exhibiting higher porosities and permeabilities, are saturated with water. Thus, the situation is the reverse from a conventional gas field, where above the gas an impermeable seal would be expected, instead there are water-saturated reservoir-type strata.

Throughout the Mesozoic, in the Triassic, the Jurassic and primarily in Cretaceous rocks some 12 pay zones have been encountered with average porosities of 10% and permeabilities of about 0.5 md (Masters, 1979). The better parts of these pay zones are often conglomeratic and exhibit permeabilities which range from 50 md to several darcys. To produce the gas the wells generally have to be stimulated by hydraulic fracturing techniques. Recoverable gas in the Deep Basin may very well be around 50 Tcf (50×10^{12} cubic feet equivalent to 1.416×10^9 m^3, STP) or even more (Masters, 1983 personal communication).

In the following some results of combined geological and geochemical research are presented that show that the gas occurrences in the Deep Basin can best be explained as a dynamic situation between generation of gas from coals and carbonaceous shales on one hand, and losses to the shallower upper layers of the rock section and going updip on the other hand. The Mesozoic rock section in the area of the Elmworth Gas Field, due to its richness in organic carbon and type of

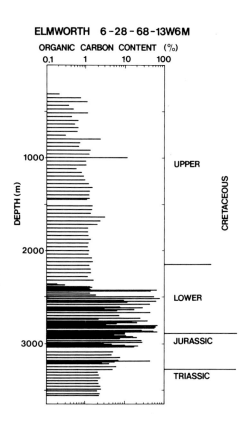

Fig. V.5.17. Organic carbon content (note logarithmic scale) plotted vs. depth for cuttings samples of an Elmworth well

5.4 Habitat of Gas in the Deep Basin of Western Canada

organic matter, represents probably one of the finest source sections for gas. A profile monitoring the organic carbon content as derived from cuttings analyses of a typical Elmworth well is shown in Figure V.5.17. Down to 2400 m depth the organic carbon content averages around 1%. Then a very rich zone follows with values up to 80% which reflects, in essence, the presence of coal seams in this part of the section (2400–3200 m). From 3200 m to 3555 m (T.D.) the C_{org} level is around 2%.

The type of the organic matter was determined by Rock-Eval pyrolysis. The diagram showing the hydrogen index plotted versus oxygen index (Fig. V.5.18) corresponds to the basic diagram represented in Figure V.1.12. The data of this well plot fairly close to the type-III curve indicates a hydrogen-poor kerogen which is mainly derived from higher terrestrial plants. Microscopic investigation revealed low amounts of liptinite macerals, but high contents of vitrinite (up to 70%) which is known to be a good gas generator, and inertinite. Thus, the microscopic data agree with the results from Rock-Eval pyrolysis. Therefore, it is concluded that this kerogen is not able to generate large amounts of oil, but is a good source of gas if it is in the right stage of maturation.

Figure V.5.19 shows vitrinite reflectance values (% R_m) for the same Elmworth well. The maturity of the organic matter is approx. 0.7% mean vitrinite reflectance (% R_m) in the uppermost part of the well and shows only a slight increase down to 1500 m where it reaches 0.8%. At this point there is a

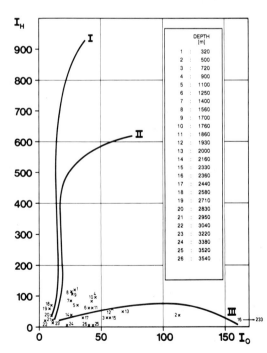

Fig. V.5.18. Van Krevelen-type diagram derived from Rock-Eval pyrolysis data (hydrogen index I_H and oxygen index I_O) for selected rock samples of an Elmworth well

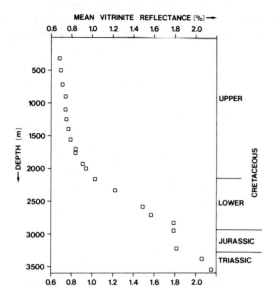

Fig. V.5.19. Mean vitrinite reflectance plotted against depth for rock samples of an Elmworth well

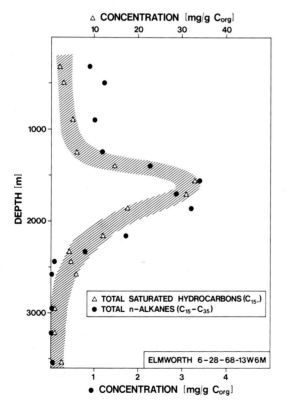

Fig. V.5.20. Total saturated hydrocarbon (C_{15+}) and total n-alkane (C_{15}–C_{35}) concentrations plotted against depth for rock samples of Elmworth well

5.4 Habitat of Gas in the Deep Basin of Western Canada 633

progressive change in the reflectance gradient and maturity increases quickly to 2.1% at total depth (T.D.). It is known that type-III kerogen normally starts to produce liquid hydrocarbons at a maturity level around 0.7% R_m. Oil generation reaches a maximum around 0.9% R_m and then decreases with further maturation until the bottom of the "oil window" (at approx. 1.3 to 1.4% R_m). Significant amounts of gas are thought to be generated at maturities exceeding 0.9 to 1.1% R_m. The bottom of the "dry gas window" is not yet reached in this well at T.D.

Concentrations of C_{15+}-saturated hydrocarbons and C_{15}–C_{35} n-alkanes as determined from cutting samples, both normalized to organic carbon are depicted in Figure V.5.20. It exhibits the typical shape of a generation curve as it would be predicted based on theoretical considerations. This indicates that, in general, the C_{15+} hydrocarbons remained at the place where they have been generated. There was no major redistribution, i.e., migration, in vertical direction. In this connection, it is important to remember that the (liquid) C_{15+} hydrocarbons which have been generated during maturation of the source rock at shallower depth are being cracked with further advancing maturation and converted to smaller molecules (e.g., gas) at greater depth. Therefore, hydrocarbon concentrations of the C_{15+} fraction pass through a maximum, as shown in this figure. Similar curves exist for both the total extractable aromatics fraction and individual aromatic compounds. As even the low boiling aromatics which are much more soluble in water than the corresponding saturates, also show this

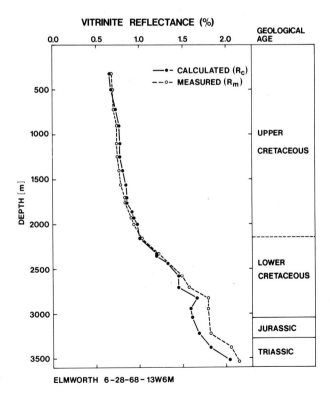

Fig. V.5.21. Calculated and measured vitrinite reflectance plotted against depth for selected rock samples of an Elmworth well. The calculated R_c-value is derived from the methylphenanthrene index (MPI) derived from the bitumen of these rock samples

typical distribution curve, it is suggested that no long-range vertical water flow has occurred since the time of intense hydrocarbon generation. Such water flow would also be unlikely because of the very low porosity and permeability of the rocks.

A further independent argument against a large vertical migration of the heavier hydrocarbons is provided by the Methylphenanthrene Index (MPI) (Fig. V.5.21). This chemical parameter, which is presented in Chapter V.1.3 permits the definition of the maturity of an oil or rock extract in terms of calculated vitrinite reflectance (R_c). In Figure V.5.21 the hydrocarbon internal maturity (R_c) is compared with the mean vitrinite reflectance of the kerogen (% R_m). There is obviously an excellent agreement between the two curves. Any invasion of a more mature oil or condensate from greater depths would have caused a positive deviation of the R_c-values from the vitrinite reflectance curve.

With decreasing carbon number the concentration profiles of hydrocarbons in Elmworth wells become broader and broader and eventually lose any similarity to a generation curve. Contrary to a previous figure (Fig. V.5.20), depicting a typical generation curve in the depth profile of C_{15+} hydrocarbons in the shale, the propane concentration profile is rather broad and remarkably constant between 1000 m and 2400 m in depth, below which it decreases stepwise (Fig. V.5.22). The propane distribution pattern as presented in this figure should show a definite trend towards lower and lower concentrations going from 2000 m depth upward if

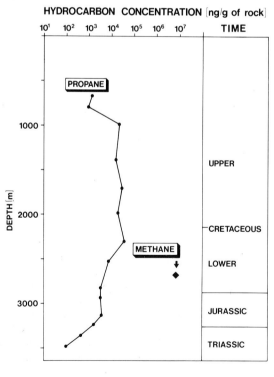

Fig. V.5.22. Propane concentration (absolute values obtained by combined thermovaporization-hydrogen stripping) plotted against depth for selected rock samples of an Elmworth well. Also shown in this figure is the calculated methane concentrations as derived from a pressure core barrel gas analysis

it would be influenced by generation rates. Maturities and temperatures above 2000 m are too low to warrant gas generation.

The absolute concentration of total light alkanes (C_2–C_5) reaches a maximum of 1.2×10^5 ng g^{-1}, or 120 g per metric ton of rock (Fig. V.5.23), which is considered to be a very high value. It becomes even more spectacular when considering the fact that methane is not included here. Hence, these geochemical analyses verify the conclusions based on resistivity log interpretations by Masters (1979) concerning the gas saturation. As the bulk of the light hydrocarbons is thought to be generated at maturity levels exceeding 0.9% R_m, corresponding to depth ranges below 2000 m, it has to be assumed that a considerable part of these hydrocarbons (e.g., C_1–C_3) has migrated over an appreciable distance into the overlying strata, contrary to the heavy C_{15+} hydrocarbons which apparently remain more or less at the site of their origin.

The concentrations displayed in Figs. V.5.22 and V.5.23 were obtained using a combined thermovaporization hydrogen stripping method developed by Schaefer et al. (1978a). These data are in good agreement with gas analyses including methane (see Fig. V.5.22) of a pressurized core-barrel sample taken from another Elmworth well between 2500 m and 3000 m depth. The measured amount of methane and higher hydrocarbons from the pressurized core-barrel samples makes it possible to extrapolate the absolute methane yield (which cannot be measured from cuttings quantitatively) from the propane concentration at a given maturity. This calculated value is 6.7×10^6 ng g^{-1} or about 11 m^3 methane per metric ton of coal, which is only the residual amount of gas present

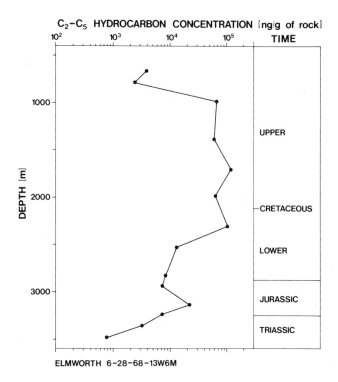

Fig. V.5.23. C_2–C_5 hydrocarbon concentration (absolute values obtained by combined thermovaporization-hydrogen stripping) plotted against depth for selected rock samples of an Elmworth well

today. Yet, a coal is thought to have produced 8 to 10 times as much methane upon reaching this maturation stage (approx. 1.3% R_m). This total volume generated cannot be stored by the coal itself and therefore migrates into adjacent strata.

Thus, the absolute concentration of methane and other gaseous hydrocarbons like ethane, propane and butane per rock volume and available pore space of the different sedimentary units becomes an important parameter to understand the problem of gas generation and migration. Furthermore, knowing that this gas is generated from coals and type-III kerogen, the methane concentrations normalized to content of C_{org} of these rocks inform us about the relative importance of the various gas generators, i.e., the sources of the gas. Following this line of thought ethane concentrations were determined in detail in coal-containing rock sections. Figure V.5.24 may serve as an example where organic carbon content and ethane concentrations have been analyzed for a coal-containing 100 m interval.

The figure shows a section of the Lower Cretaceous in the deeper part of the well where a Notikewin Coal Seam is overlain by Harmon Shales grading upward into shaly sands which are followed by the Cadotte Sandstone. The C_{org}-normalized ethane content remains fairly constant over the whole depth interval. Therefore, a certain volume of coal contains much more ethane than the same volume of shale than was to be expected. A similar curve has been found for propane. It can be assumed that methane concentrations follow the same trend, i.e., that the absolute methane concentration is highest in the coal. The fact that the C_{org}-normalized values for ethane are in the same order of magnitude for different rock sections with varying lithology and also for coals is interpreted as an

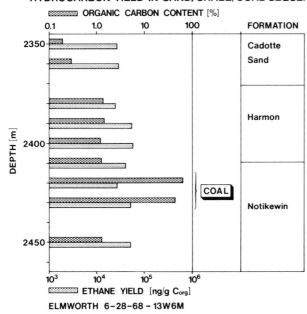

Fig. V.5.24. Organic carbon and ethane yield (obtained by hydrogen stripping) plotted against depth for a sand/shale/coal sequence of an Elmworth well

5.4 Habitat of Gas in the Deep Basin of Western Canada

indication that the gas generation process is still active and that concentration gradients for gas from the coals toward neighboring rocks are maintained. The concentration gradients would have been eliminated by diffusion if the coals had ceased to generate an appreciable amount of methane. These concentration gradients must be the driving force for an active gas diffusion going on today.

Hence, the conclusion is that in the Elmworth Gas Field there is even at present a zone of active gas generation mainly in the deeper part of the rock column exceeding 2000 m. In this zone between about 2000 m and 3500 m depth temperatures range from about 80°C to 120°C and maturities from 0.9% R_m to 2.0% R_m.

Based on detailed pressure studies of the Lower Cretaceous Cadomin Sandstone reservoir in the Elmworth Gas Field, Gies (1982) arrived independently at the same conclusion that the gas accumulation is not in static equilibrium but is in a dynamic state with ongoing updip gas migration. The pressure gradients for regional water and the actual gas as determined by Gies (1982) are shown in Figure V.5.25. It can be seen that the actual downdip gas pressures are greater than those predicted by a constructed hypothetical static gas pressure gradient based on a continuous gas saturation over a depth range of 800 m. The conclusion derived from these pressure data in the Cadomin Sandstone reservoir is that gas must be migrating upward in response to the pressure drop in an updip direction.

In summary the following observations have been made in the Elmworth Gas Field. In the tight, low porosity central part of the Deep Basin nearly the entire Mesozoic rock section is gas-saturated. The gas saturation decreases rapidly in the shallower part of the rock column and updip toward the east where more

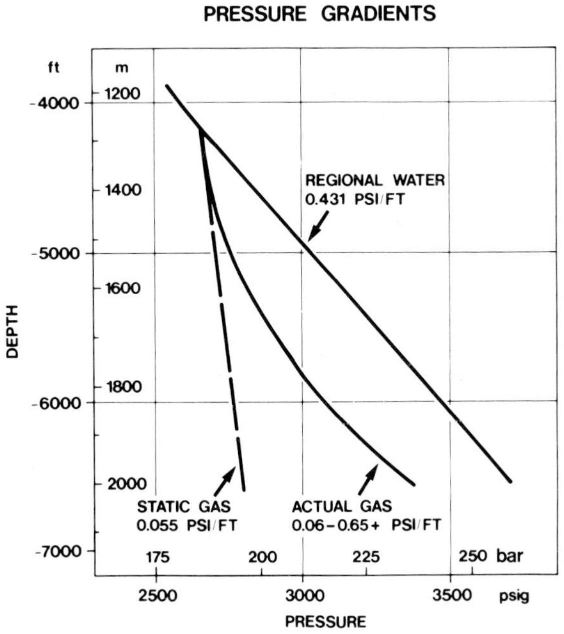

Fig. V.5.25. Pressure gradients for regional water, "static gas" and actual gas for the Elmworth area. (After Gies, 1982)

porous and permeable rocks are found. Detailed geochemical analyses show that there is no major redistribution, i.e., migration, of heavier hydrocarbons and no flow of water in the tight part of the rock section. However, there is evidence for massive redistribution, i.e., migration of gaseous hydrocarbons, and it can be shown that the main transportation mechanism must be diffusion inside the tight rock section. First balance calculations have shown that present-day gas saturation profiles in the center part of the Elmworth Gas Field can be simulated when assuming diffusion coefficients for methane in the range of 1.0×10^{-6} to 2.0×10^{-5} cm^2 s^{-1} whereby increasingly high diffusion coefficients have been adopted with increasing depth and temperature. Obviously the diffusional losses and other losses by conventional buoyancy-driven gas transport mainly toward more porous updip-situated strata are compensated by a continuing gas generation from coals and organic-rich shaly source rocks in the rock section. Temperatures between 80°C and 120°C seem to be high enough to guarantee an ongoing coalification process and hence gas generation. Balance calculations show the coal measures to be important sources for the gas. Cumulative coal thicknesses of Jurassic to Lower Cretaceous coals in the Elmworth region may range up to 70 m and more (Masters, 1983, pers. comm.).

All in all, the Elmworth gas field is a dynamic situation where gas is continuously being generated in the center part of the Deep Basin and lost toward the surface and the more porous edge. In the inner core of the gas-generating rock column diffusion processes seem to be the predominating modes of transportation.

Summary and Conclusion

Four selected case histories of geologically different petroleum regions are discussed with respect to the principles of petroleum generation, migration and accumulation: the Arabian Carbonate Platform, young delta areas, i.e., the Niger and the Mahakam Delta, the Linyi Basin of China and the Deep Basin of Western Canada.

The Arabian Carbonate Platform contains numerous giant oil fields. The main source rocks are marine Middle Jurassic to Upper Jurassic organic rich carbonates containing type-II kerogens. These source rocks became mature 45 to 70 million years ago. Petroleum generated from these source rocks is found today mainly in regressive sands and ooidal grainstones terminating the carbonate cycles or in rudist reefs and algal boundstones along the shelf margins. The principal regional seals are the Upper Jurassic Hith Anhydrite and the Cretaceous Nahr Umr Shale. The overall uniformity of most oils suggests a similar source environment. The oils belong mainly to the aromatic-intermediate class. Some oils have to be classified as paraffinic-naphthenic. The reason for the extraordinarily rich petroleum habitat in the Arabian Carbonate Platform is the enormous horizontal scale of the basin, providing source rocks, reservoir rocks and seals with an uncommonly wide extent. Because of these wide horizontal features migration efficiency was very high: large areas of mature source rocks could be drained over long geological periods. Regionally vertical fractures and faults provided migration avenues from source rocks through several hundreds of meters of relatively tight carbonate rocks before petroleum was emplaced in a reservoir or reached a carrier rock.

Young delta areas, like the Tertiary Niger and Mahakam Delta offer a compact scenario of petroleum generation, migration and accumulation. Source rocks are provided by the marine shales of the delta front and the prodelta area, receiving mainly land derived organic matter. In both deltas the mature source rocks are generally deep and the top of the oil window is in or close to the overpressured zone. The geothermal gradients are non-linear and vary regionally. Source rocks contain mainly type-III kerogen. In the Mahakam Delta coals seem to be a major source rock. The reservoir rocks and seals are found in both deltas adjacent to the source rocks in the alternating sands and shales of the deltaic plain. The primary oils found in these reservoirs of both deltas are light paraffinic oils. Certain oils in the Niger Delta are biodegraded. In the Mahakam Delta biodegradation is only of minor importance. An inverse specific gravity relationship is observed among oils of the Mahakam Delta. It can be explained by a migration of a supercritical hydrocarbon phase out of the overpressured shales that, when moving upward to lower pressures and temperatures loses progressively the heavier end of its hydrocarbons by retrograde condensation. In general, migration from source rocks to reservoir rocks in both delta areas is mainly vertical. The typical traps in Nigeria are growth faults. In the Mahakam Delta oil and gas are found in anticlinal axes which roughly parallel the coast line.

The Linyi Basin of China is an assymetric graben system of Tertiary age with a sedimentary filling of mainly detrital rocks, representing limnic, fluviatile to lacustrine environments. Among the lower Oligocene shaly sediments there are units with excellent source rock properties. The kerogens of these source rocks are classified as type II or type II–III. The terrigenous influence on the organic matter in the Linyi Basin sediments is obvious. The temperature history and maturation are largely determined by basin evolution and subsidence. Based on a computer simulation of the basin evolution, a regional reconstruction of the paleotemperature history of each individual source rock and consequently of its maturation history was performed. Computer-derived source rock maturity maps can be combined with information about the spatial distribution of carrier and reservoir rocks and structural features throughout the basin. The alternating sandshale sequence in the Linyi Basin and the structural setting of the graben system offer numerous possibilities for oil accumulations. The fault system at the edge of the graben provides good migration avenues for oil derived from deeply buried source rocks in the central part of the graben. The crude oils found are paraffinic with a relatively high asphaltene content. Aromatics and N, S, O-compounds are less abundant.

The Deep Basin of Western Canada is a gas-prone area consisting of a thick Mesozoic section of interlayered sands and shales representing a marine to brackish near-shore environment. The clastic Mesozoic rock section contains numerous shaly zones which are rich in organic matter and also a suite of coal strata. This section, containing mainly type-III kerogen, is the ideal gas generator. Maturity ranges from about 0.5% vitrinite reflectance to about 2.0% in the deeper part of the section. Apparently the mature section is still in an active phase of hydrocarbon generation. Due to the tightness of the rocks, hydrocarbon transport mechanism seems to be dominated by diffusion processes. The light hydrocarbon distribution patterns observed throughout the wells suggest a dynamic trapping mechanism. Light hydrocarbons are lost at the top of the mature hydrocarbon generating zone and are replenished in the middle part of the section where rich source rocks are found.

The gas occurrence in the Deep Basin is of special importance because the situation is the reverse from a conventional gas field, i. e., the gas-bearing zones correspond to the less porous and less permeable rocks situated in a downdip position, whereas more porous and permeable rocks are updip and above. The Elmworth Gas Field in the Deep Basin is a dynamic situation where gas is continuously being generated in the center part of the basin and lost toward the surface and the more porous edge.

Chapter 6
The Distribution of World Oil and Gas Reserves and Geological–Geochemical Implications

6.1 Introduction

The observed distribution of known reserves of oil and gas has been studied by exploration geologists and also by petroleum economists. The former have considered the distribution in relation to geological history of the earth, and particularly basin types and global tectonics: Brod (1965), Uspenskaya (1966, 1972), Halbouty (1970), Olenine (1977), Meyerhoff (1979), Bois et al. (1980, 1982), Klemme (1980) and Sokolow (1980). Petroleum economists were more interested in predicting ultimate reserves world or national, based on potential future discoveries and/or increase of engineering and economic feasibility of recovery: Desprairies (1977), Nehring (1978), Halbouty and Moody (1979), Desprairies and Tissot (1980) and Boy de la Tour et al. (1981).

In this respect, there is an obvious need for a systematic classification and a fixed nomenclature to designate the various solid, liquid or gaseous petroleum products and their reserves. In this chapter we shall refer to *natural gas* for the fraction of petroleum which is in gaseous phase under reservoir conditions, or is in solution in crude oil and becomes a separate gaseous phase under surface conditions. Natural gas may include non-hydrocarbon constituents.

Conventional oil refers to all petroleum products naturally occurring in a liquid phase, which can be explored and produced by conventional methods (primary or secondary recovery), including the additional quantities which may result from enhanced recovery procedures.

Non-conventional oil refers to petroleum accumulations which cannot be produced by conventional methods; they include:

- heavy oils and tar sands, which require enhanced oil recovery or mining techniques to be produced;
- petroleum accumulations located in offshore areas with a water depth greater than 400 m, and in polar areas;
- synthetic oil which can be obtained by retorting oil shales or coal: hydrocarbons and associated compounds are not naturally present in rock, but must be produced by chemical reactions.

Only two categories of non-conventional oil will be discussed in this chapter, i.e., the occurrences of heavy oil and a geological perspective of the petroleum potential of deep oceanic basins. Very little is known on crude oil reserves in polar areas. Oil shales reserves are discussed in Chapter II.9. Finally, coal liquefaction is not within the scope of this book.

A special remark concerns the age of the oil and gas reserves. The age of the reservoir rock is generally well known and is used throughout Section 6.2. The actual age of the source rock is not always available. In most cases, crude oil reservoir and source rock belong to the same geological period, e. g., Jurassic, Cretaceous, etc. There are, however, situations where oil has migrated from a source rock into a reservoir rock from a different period. In this case, the age assignment of the source rock used in Sections 6.3 and 6.4 has been made according to Bois et al. (1982). A comparable treatment is not possible for the age of gas source rock, due to the greater mobility of gas and the lack of an adequate compilation.

6.2 Geological Setting of Oil and Gas Reserves

In this description we shall follow the worldwide compilation recently prepared by Bois et al. (1980, 1982). The oil and gas accumulations are grouped by *petroleum zones,* a concept defined by Bois (1975). The petroleum zone is composed of accumulations which are located within the same productive sequence, with the same types of traps, and contain hydrocarbons of similar chemical composition. They are likely to derive from the same group of source beds.

The most relevant aspects of petroleum zones are the following:
- the environment of deposition, which is mainly responsible for the nature and abundance of organic matter;
- the nature of sedimentation, particularly the clastic terrigenous input, or the carbonate and/or evaporite autochthonous sedimentation responsible for deposition of reservoirs and seals;
- the rate of subsidence and the geothermal gradient which are responsible for conversion of kerogen into oil and gas; in this respect the major break is between the platform sedimentation (in cratonic or pericratonic areas) with a rate of subsidence in the order of 10 or 20 m m. y.$^{-1}$, and the sedimentation of rapidly subsiding basins in mobile belts (foredeep, intradeep, backdeep, marginal basins and associated troughs) with a rate of subsidence in the order or 50 or 100 m m. y.$^{-1}$, or even more;
- tectonic events which provide most of the traps but may also, in association with subsequent erosion, result in destruction of hydrocarbon accumulations by opening the traps and breaking the seals.

6.2.1 Paleozoic

Petroleum found in Paloezoic reservoirs amounts to 14% of the conventional oil and 29% of the gas reserves presently discovered. Most petroleum zones are located in North America (56% of the oil, 35% of the gas) and Europe (32% of the oil and 33% of the gas), whereas North Africa accounts for 11% of the oil and the Middle East for 28% of the gas.

6.2 Geological Setting of Oil and Gas Reserves

The geological setting during Paleozoic time was quite different from the present situation; furthermore many changes have occurred over a period of ca. 350 million years. Most oil source rocks and reservoirs, however, were deposited in platform environments, formed by shallow seas on the edge of Precambrian cratons. Carbonate reservoirs predominate in North America, with reef developments, over sandstones. Carbonate and sandstone reservoirs are found in Eastern Europe (USSR), whereas successive sandstone blankets are the reservoirs of North Africa (Algeria). Well known source rocks are Simpson (Middle Ordovician), Woodford (Devonian to Mississippian) and Phosphoria (Permian) formations in the United States, the Domanik formation (Late Devonian) in the USSR, and the Silurian shales of Sahara, Algeria.

Some petroleum zones are associated with the late Paleozoic mobile belts of North America and Western Europe; in the latter area, Carboniferous coal measures are the source rock for the gas found in late Carboniferous and Permian reservoirs. Finally, when source rocks are associated with older mobile belts, no significant reserves of oil or gas have been found.

The timing of petroleum generation and migration from several Paleozoic source rocks has been discussed in Chapter II.7.6. It has been shown that in many places, such as Hassi Messaoud (northern Sahara, Algeria) and Leduc (Alberta, Canada), subsidence remained moderate over Paleozoic platforms and the principal stage of oil formation was only reached in Mesozoic, or even Tertiary

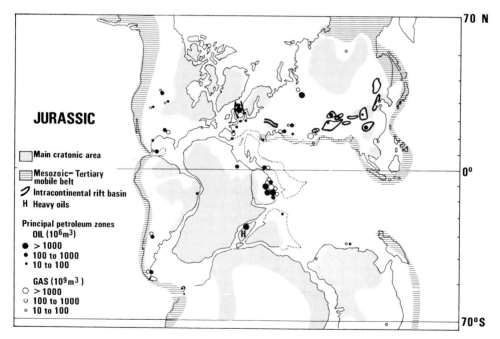

Fig. V.6.1. World distribution of petroleum zones (or groups of petroleum zones) in Jurassic reservoirs. *Size of the circles* is related to the importance of total recoverable reserves. Adapted with changes from Bois et al. (1982); situation of the continents at 170 m.y. modified from Smith (1973). *H* heavy oils and tar sands

time. This situation prevailed in Alberta and other basins of Western North America, North Africa and the Middle East. In Western Europe the coal measures, although buried at depth at the end of Paleozoic, generated most of the present gas during late Mesozoic or Tertiary. In Eastern Europe, however, and in other places of North America and North Africa (Ahnet-Mouydir basin) hydrocarbons were already generated by the end of Paleozoic.

A special remark concerns the widespread occurrence of Permian or Permo-Triassic evaporites, which seal off the Permian reservoirs and allowed huge quantities of gas to be preserved in the southern North Sea, North Germany and Netherlands, in Texas and in Southwestern Iran and other places in the Middle East. Indeed, gas is so much subject to escape toward the surface, including by diffusion (Leythaeuser et al., 1982), that preservation of such large accumulations requires an outstanding seal.

6.2.2 Mesozoic (Figs. V.6.1 and V.6.2)

Petroleum found in Mesozoic reservoirs amounts to 54% of the conventional oil and 44% of the gas reserves presently discovered (the figure for oil would be still higher, should heavy oil and tar sands be considered). The major petroleum

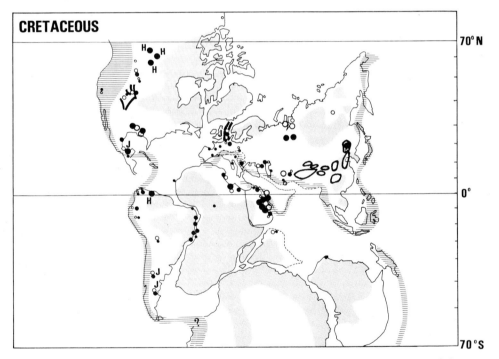

Fig. V.6.2. World distribution of petroleum zones (or groups of petroleum zones) in Cretaceous reservoirs. *Size of the circles* is related to the importance of total recoverable reserves. Adapted with changes from Bois et al. (1982); situation of the continents at 100 m.y. modified from Smith (1973). *H:* heavy oils and tar sands; *J:* oil or gas partly or wholly derived from Jurassic source rocks

zones are located in the Jurassic and/or Cretaceous of the Middle East, Western Siberia and the Gulf of Mexico for conventional oil, and Western Canada for heavy oils. The Triassic contribution is minor, as the main petroleum accumulations concerned with Triassic reservoirs are in fact derived from source rocks of different age (Algeria, Alaska).

The geological setting during Jurassic and Cretaceous has changed markedly since the Paleozoic and is shown in Figures V.6.1 and V.6.2. The major feature is the occurrence of an east–west ocean, comprising the Mesogean Ocean and the early Central Atlantic Ocean, from South East Asia to the Caribbean. It separates a northern hemisphere (North America, Eurasia) from the massive Gondwana land, which was progessively breaking up.

In the meantime, Jurassic and moreover Cretaceous periods are marked by several major transgressions over continental margins and intracontinental depressions. These areas include the margins of the Mesogean Ocean, the depressions located above former Paleozoic mobile belts, respectively. These transgressions created many shallow, frequently landlocked epicontinental seas. In such situation, abundance of nutrients favors algal blooms, and insufficient renewal of water enhances preservation of the organic matter.

Examples of marine source rocks associated with these major transgressions are numerous. Along the margins of the Mesogean were deposited the Oxfordian-Kimmeridgian carbonate source rocks of the Arabian platform, the middle Cretaceous carbonate source rocks of the Arabian Gulf, and the Cretaceous source beds of the Syrte basin, Libya. Altogether the Mesogean realm contains ca. 80% of the oil and half the gas from the Mesozoic.

The transgression over some former Paleozoic mobile belts has caused the deposition of the Jurassic (including Kimmeridgian) source beds and oil shales of Western Europe, the Jurassic-Cretaceous of Mexico, and the Cretaceous of Venezuela. From the Arctic Sea transgressions covered large areas in Western Siberia during the Upper Jurassic, resulting in thick shaly source beds. These shales generated large amounts of oil and thermal gas. The well-known Bazhenov formation with its huge amounts (between 300 and 600 billion tons) of disseminated bitumen (Bois and Monicard, 1981) is contained in this area, which also represents a former Paleozoic mobile belt.

Transgressions over the other passive margins were responsible for Cretaceous source rocks in West Africa, along the eastern coast of South America, and for Jurassic and Cretaceous source beds in Western Australia.

Transgressions over the foredeep of Mesozoic mobile belts caused the deposition of the Lower Cretaceous source rocks of the North Slope, Alaska, and the Lower to Middle Cretaceous shales, which are among the source beds of the Alberta province, W. Canada. A comparable situation probably prevailed in Columbia and Ecuador with the Cretaceous Napo formation.

Furthermore a worldwide rise of the sea level during Jurassic and Cretaceous reached an elevation ca. 350 m above present in the Late Cretaceous (85 m.y.): Pitman (1978) and Vail (1978). This situation resulted in an important decrease of the clastic input from the continent, and a large extension of carbonate sedimentation. Thus carbonate source rocks (usually marls or argillaceous limestone) and carbonate reservoirs are more frequent in Jurassic and Cretaceous than in other periods. Under favorable conditions, such source rocks may be very rich (less dilution by mineral particles) and contain a particularly favorable type of organic matter (type-II kerogen, derived from marine plankton and deposited in a reducing environment), able to generate large amounts of oil.

In many mobile belts or margins, there was rapid subsidence and sedimentation during Cretaceous and hence fast burial and rather early generation of oil in late Cretaceous to early Tertiary time. A particular aspect of the Mesozoic petroleum reserves is the frequent occurence of rich source beds over some limited intervals of time during the Upper Jurassic period (roughly Kimmeridgian) and during the Early to Middle Cretaceous period (approximately Albo-Aptian and Cenomanian to Coniacian). A possible explanation for these worldwide features is presented in Section 6.4.

6.2.3 Tertiary (Fig. V.6.3)

Petroleum found in Tertiary reservoirs amounts to 32% of the conventional oil and 27% of the gas reserves in the world. The major reserves are associated with Tertiary mobile belts. However, large deltas, such as Gulf Coast, Nigeria, Mackenzie, Mahakam, etc. also contain a large proportion of Tertiary gas (ca. 35%) and oil (ca. 20%).

The geological setting changes progressively toward the present configuration. The east–west-trending ocean is no longer the major feature, as the opening of the Atlantic Ocean has created passages for a global oceanic circulation, with the cold oxygenated polar waters reaching at depth the low latitudes and preventing the existence of large confined seas.

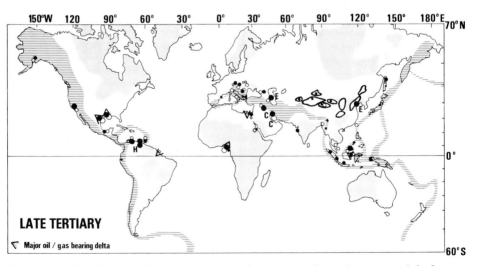

Fig. V.6.3. Distribution of petroleum zones (or groups of petroleum zones) in Late Tertiary reservoirs. Adapted with changes from Bois et al. (1982); present situation of the continents. *H:* heavy oils and tar sands; *C:* oil or gas partly or wholly derived from Cretaceous source rocks; *E:* from Early Tertiary source rocks

The Mesogean realm (sensu lato from Caribbean to South East Asia) again contains a large proportion of the Tertiary reserves (83% of oil and 86% of gas). However, these figures cannot be readily compared with the figures for the Mesozoic, as they cover very different situations. Among the three major petroleum provinces (Midle East, Caribbean, and Gulf Coast) two result from geological conditions not strictly dependent on the evolution of the Mesogean realm during Tertiary. In the Middle East (20% of the oil, 13% of the gas) the oil originates from Cretaceous source rocks and was subsequently transferred by a second migration in large elongated anticlinal traps formed during late Tertiary and sealed by evaporite deposits. In the Gulf Coast, a large deltaic system is responsible for ca. 10% of the oil and 30% of the gas in the Tertiary.

In the other areas of the Mesogean realm, the occurrence of the petroleum zones is largely controlled by the development of the Alpine orogenic belt, from the Caribbean to Indonesia and New Zealand: a thick clastic sedimentation and a high rate of subsidence — sometimes over 100 m m. y.$^{-1}$ — are common features sometimes associated with overpressured shales and mud volcanoes. Foredeep basins offered a favorable situation in the areas of continental collision such as the Caribbean, where the Orinoco basin includes the largest reserves of heavy oils and tar sands in the world. Other foredeep basins are in Europe or Western Asia, such as the Molasse basin, Romania and the Northern Caucasus. Furthermore, intramontane basins were formed in Europe and Western Asia: e.g., Vienna, Pannonian and Caspian basins, and represent another favorable situation. The Tertiary petroleum zones of Burma and Indonesia are located in back deeps or marginal basins.

The Western American mobile belts were very active during Tertiary and thus they provided relatively deep and restricted basins of deposition where marine organic matter was preserved: California, Alaska, Peru. Comparable situations may still exist there today.

There are two main reasons for the development of large deltaic systems during Tertiary time. The mountain-building resulting from the various tectonic phases of the Alpine cycle provided abundant opportunity for erosion and thus a source for clastic material. The general fall of the sea level from a maximum of +350 m above present in Late Cretaceous (85 m.y.) to +60 m in Middle Miocene (15 m.y.; since that time glacial fluctuations may have dominated sea-level changes). Thus, huge deltaic systems, such as Gulf Coast, Niger and Mackenzie deltas, are a new feature of the petroleum geology in the Tertiary era. Although the associated organic matter is from terrestrial origin (kerogen type III), i.e., it has a comparatively lower potential for oil than for gas, the enormous quantity which accumulates in this situation is responsible for ca. 35% of the gas, and ca. 20% of the oil known in the Tertiary.

Along the passive margins, other types of petroleum zones are found in the Gulf of Suez, in the Cambay basin (India) and the Gippsland basin (SE Australia). In China, oil is in internal rift basins which were filled by nonmarine, or paralic (Bohai basin) sedimentation.

6.2.4 Conclusions on Geological Setting of Petroleum Reserves

From the previous discussion, it is clear that the distribution of Paleozoic oil and gas reserves is mostly controlled by platform sediments deposited with a low rate of subsidence in cratonic or pericratonic areas. Furthermore, when such sediments became later incorporated in mobile belts, only those associated with late Paleozoic or younger belts offer some petroleum reserves.

During Mesozoic and Tertiary, a general trend is observed marked by a decrease of the role of platform deposits and an increase of the influence of filling-up basins with a high rate of clastic sedimentation in convergence zone (foredeep, backdeep and marginal basins, and associated troughs).

More generally, the basins associated with mobile belts include ca. 40% of the oil and gas reserves in the world, but their importance increases from the Paleozoic, where they control only 20% of the oil and 40% of the gas, to the Tertiary, where they control 70% of the oil and 55% of the gas.

The significance of these observations will be discussed in Section 6.4, particularly the increasing importance of mobile belts and the decreasing importance of platform sediments for petroleum occurrence, with increasing age of the Earth.

6.3 Age Distribution of Petroleum Reserves

This question has been discussed by Tissot (1979) and Bois et al. (1982). The age distribution of the initial in-place crude oil is shown in Figure V.6.4, i.e., the total amount of the oil already produced, plus the recoverable reserves, plus the non-recoverable fraction of oil in all discovered fields. The surface of the blocks is proportional to the total initial in-place oil: the age considered is the age of the *source rock*, not that of the *reservoir* rock. To convert recoverable oil to in-place oil and vice versa, a world recovery factor of 0.25 is assumed, i.e., 25% of the initial oil in place is recovered. Obviously this figure may vary from less than 5% with very viscous oils and poor reservoirs to more than 50% under very favorable conditions.

Consideration of the in-place, not the recoverable, crude oil is necessary to include heavy and extra-heavy oils (tar sands): for some of them the present recovery factor may be approximately zero, unless mining techniques are considered. Some of the heavy and extra-heavy oils accumulations belong to the largest in the world.

The most striking remarks concern the concentration of oil reserves over certain periods of time and the existence of two major cycles of source beds. Prolific petroleum source rocks deposited during Jurassic and Cretaceous periods are responsible for ca. 70% of the world's conventional crude oil, although this time interval amounts to only 22% of the time elapsed since the Precambrian. Furthermore, a detailed examination of the age of the source beds shows that the time span is mainly condensed to between 180 and 85 m.y., i.e., only 17% of the geological time since the Precambrian (Tissot, 1979).

6.3 Age Distribution of Petroleum Reserves

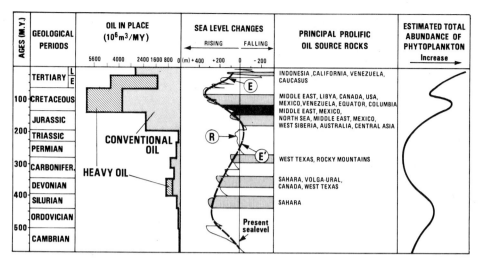

Fig. V.6.4. Age distribution of the initial conventional oil in-place and heavy oil derived from the various geological periods. The age considered is that of the source rock: it is compared with the major sea-level changes (*E*: absolute; *R*: relative; *E'*: general trend in Paleozoic and Mesozoic time; see Sect. 6.4.1) and the estimated abundance of phytoplankton. Some of the major prolific source rocks are also indicated. Adapted from Tissot (1979), with data from Bois et al. (1982) on oil reserves, Vail (1978) and Pitman (1978) on sea-level changes, Tappan and Loeblich (1970) on abundance of phytoplankton

Two cycles are clearly apparent in the age distribution of the sources for oil reserves: a Paleozoic cycle and a Jurassic to Tertiary cycle separated by a clear minimum in Triassic. However, the Paleozoic cycle is only responsible for ca. 13% of the reserves of conventional oil, with a maximum frequency in the range of 600 million m^3 m.y.$^{-1}$. The Jurassic to Tertiary cycle is responsible for more than 85% of the world reserves, with a maximum frequency in the range of 4000 million m^3 m.y.$^{-1}$ during Cretaceous. The figure is even more spectacular, if heavy oils are included as their major reserves are concentrated over the second cycle (Orinoco, W. Canada): then, the maximum frequency reaches 6000 million m^3 m.y.$^{-1}$ over the Cretaceous period.

The age distribution of gas is shown in Figure V.6.5. In this case, the actual figure of reserves is considered, as the world average recovery factor is ca. 0.80 of the in-place gas. The age considered is that of the *reservoir rock* due to the greater mobility of gas and to the difficulty of identifying the proper source rock, especially of some dry gas. World reserves of coal are also plotted in the same figure, for comparison.

The most obvious difference between the age distribution of crude oil and coal is that the rate of coal accumulation (in billion tons per million year) during several Paleozoic periods, such as late Carboniferous and Permian, was comparable (or even higher) to the rate of accumulation during the Mesozoic-Tertiary eras.

The significance of the age distribution of gas is difficult to interpret, due to insufficient knowledge of related source beds, and also due to the fact that

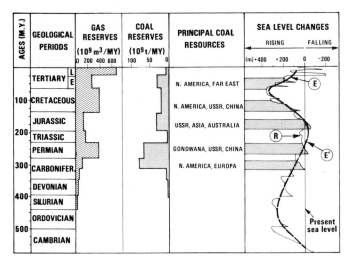

Fig. V.6.5. Age distribution of the world reserves of gas and coal, compared with the major sea-level changes. The age considered for gas is that of the reservoir rock. Some of the major coal provinces are indicated. Adapted from Tissot (1979), with data from Bois et al. (1982) on gas and coal reserves, Vail (1978) and Pitman (1978) on sea-level changes (curves E, R and E' are the same as in Figure V.6.4)

biogenic gas and thermal gas are generated under very different geological situation but they are not distinguished in the evaluation of reserves. The Cretaceous peak, however, may be interpreted along the same lines as the peak observed in the distribution of oil (Sect. 6.4.4, below), whereas the Permian peak may be a result of a widespread salt cover ensuring an exceptionally effective seal of giant gas accumulations.

6.4 Significance of the Age and Geotectonic Distribution of Petroleum and Coal

The data on geological setting and age distribution of world petroleum reserves (reported in Sects. 6.2 and 6.3) can be interpreted in terms of generation, maturation and destruction of crude oil accumulations. Five aspects will be discussed:

1. Existence of two major cycles of source beds which can be correlated with the major cycles of sea-level changes, and thus with the main events of global tectonics (Sect. 6.4.1).
2. The conditions for generation and preservation, or destruction, of oil accumulations explain how the younger cycle (Mesozoic-Tertiary) is about 10 times more productive (per million year) than the older cycle (Paleozoic) if the calculation is based on the presently existing reserves (Sect. 6.4.2).

3. The same factors also explain the progressive change in the geological setting of crude oil, from platform sediments in Paleozoic to mobile belts in Tertiary (Sect. 6.4.2).
4. A more detailed study of major source rock occurrences shows a frequent relationship with major transgression phases (Sect. 6.4.3).
5. Prolific petroleum source beds deposited over short periods of Jurassic and Cretaceous correspond to an association of several favorable conditions over large areas of sedimentation (oceanic anoxic events, Sect. 6.4.4).

6.4.1 Significance of the Two Major Cycles

The two major cycles (Paleozoic and Mesozoic-Tertiary) of the crude oil distribution according to the age of the source rocks have been presented in Section 6.3 and Figure V.6.4.

Recent developments in marine geology and geophysics made possible a comparison of these events with global sea-level changes (Tissot, 1979). Pitman (1978) has shown that absolute rise or fall of sea level are caused by the volume changes of the mid-oceanic ridge system. In turn, these variations are mainly due to change in spreading rate and to creation or destruction of a ridge system. The model used by Pitman takes into account the position of the ridges and the related spreading rates for calculating the absolute sea-level changes shown by curve (E) in Figures V.6.4 and V.6.5: the sea level was ca. +350 m above present in the Late Cretaceous (85 m.y.) and fell to +60 m in the Middle Miocene (15 m.y.). The rate of fall was in the order of 0.4 to 0.7 cm per 1000 years. Since that time, the successive glaciations have been the major cause for sea-level changes, with variations reaching 1 m per 1000 years.

Vail (1978) studied the seismic records on many continental shelves and margins, especially the onlap of coastal deposits. From these data, he evaluated the relative changes of sea level, i.e., its rise or fall with respect to the land surface. They are shown by curve (R) in Figures V.6.4 and V.6.5. Vail defined two major cycles of sea-level rise and fall, the first one from Cambrian to Triassic (580–200 m.y.), the second one from Jurassic to the present time (200–0 m.y.). Both curves (E) and (R) show a good agreement with respect to general trend and average values. Thus it is possible to extend the curve (E) of the absolute sea-level changes back into early Mesozoic and Paleozoic time by curve (E') averaging the general trend of (R).

Furthermore, the estimated abundance of phytoplankton (redrawn in Fig. V.6.4 from Fig. I.2.1) also shows two distinct maxima corresponding to the highest sea level, i.e., the largest extension of epicontinental seas, which are more productive than deep oceans.

A comparison of the different curves in Figure V.6.4 shows that the two major cycles featured by the age distribution of oil can be correlated with the two major cycles of sea-level rise and fall (curve E–E'), and also with two maxima of the estimated abundance of phytoplankton, which is the main primary producer of marine organic matter.

6.4.2 Relative Productivity of the Paleozoic and the Mesozoic-Tertiary Cycles; Causes for Different Geotectonic Setting of Their Reserves

There is a difference of one order of magnitude with respect to oil reserves between the older and the younger cycle: the average amount of crude oil per million years is lower by about 1:10 in the older cycle. We do not consider this difference as resulting from different geological or geochemical mechanisms being active during Paleozoic time. The distribution of coal (Fig. V.6.5) suggests that the conditions for production, preservation and accumulation of the organic matter were equivalent, at least during Upper Paleozoic. On the contrary, we consider the present distribution of crude oil to result from a dynamic situation where Paleozoic oil fields may have been subsequently altered or destroyed by thermal, tectonic and other influences.

First a greater thermal maturation, due to burial at great depth for long periods of time, has changed part of the Paleozoic oil into gas. Moreover, the preservation of oil and gas accumulations or their destruction might be the controlling factor. Tectonic events, such as faulting, folding and subsequent erosion, are responsible for breaking the seals and opening the traps (Fig. V.6.6; case A). Finally, physical leakage, including diffusion, is another way of destruction of old accumulations. All these factors in one or the other way are responsible for the difference of one order of magnitude observed between the reserves of the older and younger cycles.

Furthermore, the chances for conversion to gas, or for the destruction of Paleozoic oil fields were particularly important in *mobile belts:* cracking to gas

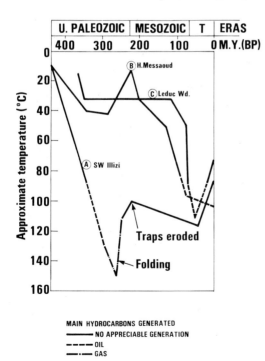

Fig. V.6.6. Influence of the rate of subsidence on hydrocarbon generation in three Paleozoic source rocks. *A:* Silurian, Southwest Illizi Basin, Algeria: the fast temperature increase during Paleozoic led to cracking of the kerogen and the oil previously generated; later Hercynian folding and faulting was followed by a deep erosion of the traps located updip from the area presented here. No commercial accumulation is reported. *B:* Silurian, Hassi Messaoud area, Algeria: due to the low rate of subsidence, no significant quantities of hydrocarbons were generated during Paleozoic. The bulk of oil was generated in response to the faster subsidence which occurred during Jurassic and Cretaceous. *C:* Devonian, Leduc–Woodbend area, Western Canada: the situation is comparable to B, with a still faster rate of subsidence during Cretaceous

6.4 Significance of the Age and Geotectonic Distribution of Petroleum and Coal

because subsidence was important: destruction because folding and subsequent erosion were particularly active in that geotectonic setting. Thus, even if mobile belts originally contained a large part of Paleozoic oil fields, as they do in the younger Mesozoic or Tertiary series, this fraction was extensively destroyed. In particular, no oil accumulation has actually survived where source rocks are associated with mobile belts older than late Paleozoic or Mesozoic.

On the contrary, source beds deposited on platforms (cratonic or pericratonic basins) were buried at a low rate and either reached the stage of oil formation by the end of Paleozoic (Devonian of the Volga-Ural) or even remained immature until they were buried through a younger cycle of sedimentation (Fig. V.6.6 B and C): The Silurian source rocks of Hassi Messaoud (Algeria) generated oil when the Jurassic and Cretaceous beds were deposited over the Northern Sahara platform (see also Chap. II.7.6); the Devonian source rocks of Leduc (Alberta) generated oil when the area received a thick filling-up sedimentation of Late Cretaceous age, as part of the foredeep basin of the Rocky Mountains mobile belt. In both cases the preservation, over a long period of time (250 to 300 m.y.), of the genetic potential of the source rocks has been the essential factor for the present occurrence of these oils. These considerations explain why platform sedimentation with a low rate of subsidence controls the distribution of oils from Paleozoic source rocks (ca. 80% of the oil).

In the young Tertiary beds, the situation in almost reversed: platform sediments deposited at a low rate of sedimentation, e.g., 10 or 20 m m.y.$^{-1}$, are not sufficiently buried to have reached the principal zone of oil formation.

Table V.6.1. Approximate timing of the beginning (taken at 20% oil generated) and the end (taken at 50% oil cracked into gas) of oil generation. The table refers to type-II kerogen, with a constant rate of subsidence S and a geothermal gradient $G = 35°C\ km^{-1}$. (For further details, see Tissot et al., 1980)

Rate of subsidence S (m m.y.$^{-1}$)	Timing of oil generation (m.y. after deposition)		Duration of oil-window (m.y.)	Geological examples	
	Begins at	Ends at		Source-rock Basin	Average subsidence S (m m.y.$^{-1}$)
10	190	300	110	Paleozoic E. Illizi	6
20	100	170	70	Jurassic Paris Basin	15
50	45	70	25	Cretaceous	30 to
100	25	37	12	W. Canada	80
200	12	18	6	Late Tertiary Los Angeles B.	200 to
400	7	10	3	Pannonian B.	300

Greater burial depth and faster subsidence is needed for oil formation. This is found in basins with rapid filling-up sedimentation, along the Alpine or Western American mobile belts, or in internal troughs, such as several Tertiary basins in China. These considerations explain why the basins with a fast rate of subsidence associated with mobile belts contain 70% of the oil from Tertiary source beds (Table V.6.1 and Fig. V.6.3).

6.4.3 Major Source Beds and Transgression Phases

Worldwide transgressions over continental platforms have occurred several times during geological history. Transgressions and regressions depend on: (1) the rate of absolute sea-level rise or fall and (2) the rate of subsidence minus sedimentation rate, the resultant of which controls the movement of the continental platform. Thus worldwide transgressions may be caused by changes of the rate of sea level rise or fall (Pitman, 1978). They appear as second-order cycles of the curve R (Fig. V.6.4) drawn by Vail (1978).

Shallow epicontinental seas transgressing over platforms and continental depressions often represent favorable conditions for source rock deposition. There, primary productivity of phytoplankton is high. Furthermore, transgressions over continental depressions may form landlocked seas, surrounded by emergent lands, providing mineral nutrients and favoring algal blooms. In turn, blooms result in an abundant bacterial consumption of oxygen, which is not sufficiently renewed, thus providing anoxic conditions. Under these circumstances, organic-rich source beds containing the oil-prone type-II kerogen are deposited.

In fact, the comparison of many well-known prolific source beds with the second-order cycles of curve R confirms that they were mainly deposited during periods of global marine transgression. In the Sahara (Algeria) the rich Silurian source rocks were deposited by a large transgression over the glacial topography inherited from the Ordovician continent. In Western Europe, the successive Jurassic transgressions over the Triassic continent deposited the source rocks of the North Sea and the Toarcian and Kimmeridgian oil shales (France, Germany, United Kingdom). Along the margins of the Mesogean Ocean, the transgression over the Arabian platform deposited the prolific carbonate source rocks of Oxfordian and Kimmeridgian age. In Western Siberia, a large transgression of the Arctic Sea over a former Paleozoic mobile belt deposited the Late Jurassic source beds including the very rich Bazhenov formation. In Western Canada, a marine transgression from the Arctic Sea over the foredeep of the Rocky mountains mobile belt deposited the lower to Middle Cretaceous source beds of the Western Canada Basin. Many other examples could be cited (Fig. V.6.4), including some of the major petroleum reserves, to demonstrate a direct relationship between the major source rock occurrences and the major transgression phases (Tissot, 1979).

6.4.4 Oceanic Anoxic Events

Prolific petroleum source rocks, responsible for a large fraction of the petroleum reserves, were deposited over relatively short periods of time, e.g., parts of Jurassic and Cretaceous. Schlanger and Jenkyns (1976) and Arthur and Schlanger (1979) reported a double worldwide occurrence of such sediments in lower and middle Cretaceous and named them Oceanic Anoxic Events. Such events are characterized by the occurrence of rich to very rich organic beds (up to 40% organic carbon) containing marine planktonic organic matter preserved in a reducing environment. Their lithologic composition varies from shales to carbonates.

During the first major Phanerozoic cycle (570 to 200 m.y. b.p.) some of these oceanic anoxic events are suspected, e.g., during late Devonian, but insufficient data and uncertain paleogeographic reconstruction make their assessment difficult. During the second major cycle (200 to 0 m.y. b.p.) three organic-rich events seem to have a wide, perhaps worldwide, significance:

- A late Jurassic event is present around the Northern seas: Alaska, Norwegian Sea, North Sea, northern end of Atlantic Ocean, Western Siberia; in the Gulf of Mexico, on the Arabian platform, etc.
- A Lower Cretaceous event, covering mainly Aptian and lower Albian (ca. 115 to 105 m.y. b.p.), is present in the South and Northeast Atlantic deep basins (Tissot et al., 1979, 1980), and also on the Manihiki Plateau in the South Central Pacific, on the Shatsky Rise in the Western North Pacific, and on the continental margin west of Australia (Schlanger and Jenkyns, 1976). Along the Tethys margins, comparable organic-rich intervals are known in Italy, Switzerland and the Middle East.
- A middle Cretaceous event, often of Cenomanian-Turonian age, sometimes extending from Late Albian to Coniacian (ca. 100–85 m.y. b.p.), is possibly the most widespread event. It is known in most of the Atlantic deep basins, except the Cape Basin (Tissot et al., 1979, 1980; de Graciansky et al., 1982) on the Shatsky and Hess Rises in the northern Pacific and on the continental margin west of Australia (Schlanger and Jenkyns, 1976). Organic rich beds are also present on some Tethys margins (Spain, Italy, North Africa, Middle East), in the Caribbean and South America (Venezuela, Trinidad, Colombia, Equator), in the Gulf of Mexico, etc.

The principal causes for the occurrence of these organic-rich beds seem to be a high fertility of the sea, highly favorable conditions for organic matter preservation, and a low rate of dilution by mineral constituents (Bois et al., 1982).

All three events correspond to major phases of transgression, and the last one approximately corresponds to the highest stand of the absolute sea level, due to the intense activity and fast spreading of the oceanic ridges ca. 115 to 85 m.y. b.p. and the related increase in the volume of the midoceanic ridges (Pitman, 1978). In turn, this situation determined a worldwide transgression over the continental platform: the global extension of shallow seas approximately doubled over that period of time. The favorable influence of such a situation was discussed in Section 6.4.3 above.

Furthermore, the high stand of the sea level, covering plains and lowlands, decreased the importance of detrital sedimentation, which is the major factor for dilution of organic matter by mineral constituents. This situation resulted in many carbonate source rocks (marl, argillaceous limestone) being deposited in shallow seas (Venezuela, Middle East), whereas highly organic rich shales were laid down in deep oceanic basins (Central Atlantic; Angola basin).

The lack of global oceanic circulation, which usually provides a renewal of dissolved oxygen in sea waters, was an important factor for organic matter preservation. The main oceanic belt was oriented east–west, from Southeast Asia to the Caribbean through Tethys and Central Atlantic. Thus there was no opportunity (except in the Pacific ocean) for a general north–south circulation which presently brings at low latitudes the deep, oxygenated polar waters. Locally, the existence of physiographic barriers such as the Falkland Plateau and the Rio Grande-Walvis Ridge prevented oceanic waters from circulating into the young basins resulting from the beginning of South Atlantic opening, such as the Cape and Angola Basins.

Furthermore, a warm equable climate, with high sea-water temperatures up to high latitudes, favored a high organic fertility and a decrease of the content of dissolved oxygen. All these factors contributed to a widespread oxygen minimum layer covering a much wider depth range than usual; this situation may explain the occurrence of organic-rich beds on plateau and rise of the Pacific ocean (Schlanger and Jenkyns, 1976). Furthermore, the major tectonic events determined many large-sized restricted or barred basins, along the Tethys (Middle East) and in the Central and South Atlantic.

6.5 Richness of Sedimentary Basins. Role of Giant Fields and Giant Provinces

The role of giant fields and giant provinces in the world distribution of hydrocarbons has been discussed by Halbouty et al. (1970), Perrodon (1978, 1980) and Nehring (1978). The relative richness of the various sedimentary basins is compared by Perrodon (1978). Giant and supergiant fields are usually defined as follows:

- giant field: ultimate recovery > 80 million m^3 or 0.5 billion bbl of oil;
- supergiant field: ultimate recovery > 800 million m^3 or 5 billion bbl of oil.

It should be pointed out that the definition is dependent on the physical characteristics of the reservoir (porosity, permeability, surface properties) and the fluids (composition, viscosity, etc.) which together control the ratio of recovery, and also on the state of art in production.

6.5 Richness of Sedimentary Basins

6.5.1 Giant Provinces

The concept of a petroleum province, composed of one or several sedimentary basins having common geological features and a comparable history, is used by Perrodon (1980). It generally includes several petroleum zones. For instance the Western Canada province covers the Alberta basin and its extensions in British Columbia and Saskatchewan; it includes the Paleozoic crude oils, the Cretaceous oils and also the heavy oils and tar sands of Athabasca and other locations.

Nehring (1978) and Perrodon (1978) pointed out that about 20 giant provinces (Table V.6.2: each containing more than 1.6 billion m^3 or 10 billion bbl) contain over 85% of the total production plus recoverable reserves of conventional oil, although they cover only 20% of the total surface of sedimentary basins. Among them, the Arabian-Iranian province (hereafter called Middle East), contains ca. 80 billion m^3 (ca. 500 billion barrels) and thus accounts for about half of the past production plus known reserves of conventional oil. If heavy oils and tar sands are also considered, Eastern Venezuela and Western Canada provinces fall in the same order of magnitude and the three provinces together account for three quarters of the world reserves. The giant provinces contain all 33 supergiant fields and 227 of the 272 known giant oil fields (Nehring, 1978: Table V.6.2).

6.5.2 Oil and Gas Richness

Perrodon (1978) defined the average richness of sedimentary basins as the amount of oil or gas (past production plus recoverable reserves) discovered per square kilometer. There may be some discussion whether the figure should be related to the surface or volume of sediments. However, it should be remembered that the total thickness of the sedimentary column does not influence directly the interval of that column which belongs to the oil generation zone. The situation is the same for wet gas, but more questionable for dry gas from the metagenesis zone, which may extend to greater depths if the sedimentary column is very thick. However, evaluation of the sedimentary volume is difficult and we shall use the richness in $m^3 \, km^{-2}$ for both oil and gas. Following Perrodon (1980, p. 340) we shall consider three classes of provinces: very rich with more than $20 \times 10^3 \, m^3 \, km^{-2}$; rich with 2 to $20 \times 10^3 \, m^3 \, km^{-2}$; poor below $2 \times 10^3 \, m^3 \, km^{-2}$.

Table V.6.2 shows that all 18 major oil provinces are rich or very rich. Their average richness is $13 \times 10^3 \, m^3 \, km^{-2}$ for oil and $6.10^6 \, m^3 \, km^{-2}$ for gas. These figures can be readily compared with the average of disseminated oil in good marine source rocks, which may range from $10^5 \, m^3 \, km^{-2}$ for a mature 10-m-thick source rock containing 2% organic carbon to $10^6 \, m^3 \, km^{-2}$ for a 20-m-thick source rock containing 10% organic carbon. The ratio of accumulated to disseminated oil is in the order of 1 to 10 and 1 to 100 respectively, which are acceptable figures for the efficiency of migration, at basin scale.

There are, however, large differences in oil richness, even between giant provinces: Maracaïbo, Reforma–Campeche, California, Middle East, Sirte and North Caucasus provinces are very rich and have more than $20 \times 10^3 \, m^3 \, km^{-2}$. If heavy oils are considered, Eastern Venezuela and Western Canada fall into the

Table V.6.2. Giant oil provinces of the world. (Adapted from Nehring, 1978; Perrodon, 1978; and Bois et al., 1980, with changes)

Province	Surface (10^3 km^2)	Production + proved reserves		Oil richness 10^3 m^3 km^{-2}	Gas richness 10^6 m^3 km^{-2}	Total richness 10^3 m^3 oil eq. km^{-2}	Number of giant oil fields[a]
		Conventional oil (10^9 m^3)	Gas (10^{12} m^3)				
Arabian–Iranian (Middle East)	2300	77	20	33	9	42	79
Tampico–Reforma–Campeche	250	11	2	44	8	52	4[b]
West Siberia	3200	8.4	17	3	5	8	14
Midcontinent + West Texas	1000	7.8	6.5	8	6.5	14.5	16
Gulf Coast + Mississippi + E. Texas	600	7.3	10	12	17	29	17
Maracaïbo	60	6.4	1	106	17	123	7
Volga–Ural	500	5.7	2.8	11	6	17	8
Sirte, Libya	200	5.4	0.9	27	5	32	15
North Caucasus + Kopet Dag	150	3.3	1	22	7	29	9
North Sea	500	3.0	2.5	6	5	11	14
Niger Delta	150	3.0	1.5	20	10	30	6
California	60	3	0.9	50	15	65	12
N. China (Bohai, Sung Liao)	—	2.5	0.2	—			1
West Canada	1300	2.3	2.6	2	2	4	8
Orinoco–Trinidad	120	2	0.3	17	2	19	8
Sumatra	210	2	0.5	10	2	12	4
Sahara, Algeria	600	1.6	3.5	3	6	9	4
North Slope, Alaska	180	1.6	0.3	9	4	13	1
Total 18 giant Provinces	11380+	153.2	73.5	13	6	19	227
Total world		ca. 175 (ca. 154×10^9 metric t)	105	—	—	—	272
18 Giant oil provinces / Total world		87.5%	70%	—	—	—	83%

[a] Excluding heavy oil accumulations. (After Nehring, 1978)
[b] Present figure probably higher

same group. All other giant provinces show a moderate richness, from 2 to 20×10^3 m³ km⁻² including Western Siberia, Gulf Coast, Ural–Volga, Niger delta, and the North Sea. Non-giant provinces usually show a moderate (2×10^3 to 20×10^3 m³ km⁻²) or low ($< 2.10^3$ m³ km⁻²) richness.

If the total hydrocarbon (oil plus gas) richness is now considered, large deltaic provinces such as Gulf Coast and Niger Delta join the group of the very rich provinces (Table V.6.2). This fact is possibly due to an important contribution of terrigenous organic matter (type-III kerogen) which has a limited potential for oil generation, but is able to produce large amounts of gas.

6.5.3 Concentrated Versus Scattered Distribution of Oil Reserves. Role of Giant Fields

Nehring (1978) pointed out that known recoverable crude oil reserves are heavily concentrated in giant fields. This situation is clearly reflected in Figure V.6.7, where the number of fields decreases abruptly as a function of their size, whereas the fraction of world known reserves increases dramatically with the size of the fields. In particular, there are between 20,000 and 30,000 fields smaller than 16 million m³ (100 million bbl) and they only account for ca. 10% of known oil reserves; on the contrary 908 fields larger than 16 million m³ amount to ca. 90% of oil reserves. Furthermore, 272 giant fields contain three quarters of the known oil reserves, and only 33 supergiant fields concentrate half of the reserves.

There are, however, some differences in the degree of areal concentration of oil reserves and also in the role of giant fields from one province to another, as observed by Perrodon (1980). A *concentrated* distribution of oil reserves is

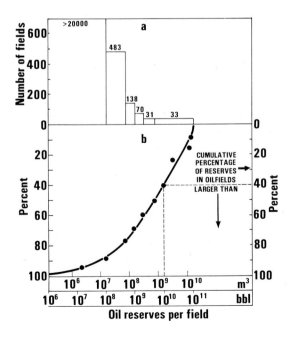

Fig. V.6.7a and b. Size distribution of oil fields. (a) Number of oil fields per class of size (the definition of the classes is based on the oil reserves per field). (b) Distribution of oil reserves as a function of the size of the fields. For instance 40% of the world reserves are concentrated in fields larger than 10 billion bbl.

defined by the occurrence of the major fraction of reserves in a small number of fields: there are large differences in size between fields. The extreme case is presently the North Slope of Alaska, where Prudhoe Bay encompasses the bulk of the known reserves in that area. Obviously, this situation may change, as a result of further exploration. Some more intensely drilled petroleum provinces, however, also show a concentrated distribution of oil reserves, e.g., the Ural–Volga province where the Romashkino field contains more than one third of the reserves; the West Siberian province where Samotlor field contains 40% of the oil reserves and Urengoy field one third of gas reserves; and the Algerian Sahara province, where Hassi Messaoud field contains two thirds of oil reserves and Hassi R'Mel field more than one third of gas reserves. A concentrated distribution of reserves is usually met in geological situations where a structural or stratigraphic feature has favored large drainage areas such as a regional uplift, a regional unconformity with weathered beds or with a basal sandstone acting as an avenue for migration etc. An extreme case of concentrated distribution is offered by the huge heavy oil accumulations of Western Canada and Eastern Venezuela with the possible conjunction of a regional unconformity for migration and an in-situ degradation responsible for sealing off and trapping.

A *scattered* distribution is defined by a wide distribution of reserves among a large number of fields; the largest field may contain only 5 or 10% of the total reserves. This situation is met in deltaic basins or stable intracratonic platforms:Niger delta and Gulf Coast provinces, where the 10 major fields amount to only one third of the oil reserves; Paris basin where 11 million tons are spread over 18 fields; Michigan basin where more than 600 fields total slightly over 100 million tons (Perrodon, 1980).

6.6 Ultimate World Oil and Gas Resources

As of 1.1.1982, the total past production of oil was ca. 62 billion tons, whereas proved reserves of conventional crude oil were evaluated to be ca. 92 billion tons, making a total of 154 billion tons (ca. 175 billion m^3). Beyond this value, additional reserves of conventional crude oil may be expected for two reasons (Fig. V.6.8): either there can be additional production from known fields through enhanced oil recovery (EOR) or new fields can be discovered. On Figure V.6.8, the ratio of oil recovery is supposed to change progressively from 25% (present worldwide average) to 40%. This figure, however, is more a target for improving existing technologies or introducing new ones than an actual prediction of the level which will be reached within a certain period of time. Along the other axis of Figure V.6.8 there are new discoveries. It should be pointed out, however, that any evaluation of future oil discoveries is entirely speculative.

Different procedures have been applied for evaluation of not-yet-discovered reserves of *conventional oil*, i.e., excluding heavy oils and tar sands, offshore areas with great water depth (beyond 400 m) and polar areas:

a) Consultation of numerous experts, followed by a statistical treatment was used by Desprairies (1977), updated by Desprairies and Tissot (1980). It was

6.6 Ultimate World Oil and Gas Resources

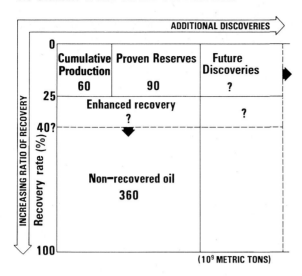

Fig. V.6.8. Discovery of additional new fields and enhanced oil recovery (EOR) are two possible ways for increasing the world reserves of crude oil. The two possibilities may result in additional reserves of comparable order of magnitude. (Adapted, with changes, from Boy de la Tour and Le Leuch, 1981)

concluded that ultimate resources (including past production plus known reserves) were in the 250–300 billion tons bracket.

b) Nehring (1978) emphasized the predominant role of giant oil fields and made a worldwide evaluation mainly based on that observation. He suggests that about 130–185 giant fields might be discovered, amounting to 17–37 billion tons, plus an equivalent amount from non-giant fields. The additional gain from EOR is estimated to be 56–97 billion tons, making a total of ultimate recoverable crude oil resources of 230–310 billion tons. Half of these resources are supposed to be located in the Middle East.

c) Halbouty and Moody (1979) by comparison of unknown with known provinces, reviewed the different basins of the world and quoted an expected range of 40 to 345 billion tons for undiscovered resources, with a most probable value of 141 billion tons, making a total ultimate resource of ca. 300 billion tons. In this evaluation, the share of the Middle East is smaller, as the most probable value for ultimate resources in this area is 100 billion tons, i.e., one third of the world resources.

d) Bois (in Desprairies and Tissot, 1980) and Bois et al. (1980) used the results of exploration in United States as a model for an advanced stage of exploration drilling. Giant fields amount to 36% of reserves in United States, whereas they represent 75% of them in the rest of the world. Thus, if the rate of exploration becomes comparable in the rest of the world to that in the United States, a doubling of the world reserves outside the USA could be anticipated, without discovering any new giant field. The figure would be 270 to 300 billion tons, to which should be added any further discoveries of giant fields (17 to 37 billion tons; according to Nehring, see above).

e) Bois (in Desprairies and Tissot, 1980) used again the "American model" by comparing the volume of Phanerozoic (post-Precambrian) sediments in the world with that in United States. He found the ratio 610:44 and estimated the world resources at 320 billion tons.

f) The Federal Geological Survey in Hannover, W. Germany, prepared a survey of energy resources for the 11th World Energy Congress (BGR, 1980) and evaluated the ultimate oil resources at ca. 354 billion tons.

Based on these different approaches, there seems to be some consensus to place the world ultimate resources of conventional oil in the range of 250 to 400 billion tons. However, it should be kept in mind that such a consensus is by no means an argument for the validity of these resource estimates.

The total amount of natural gas production plus proved reserves is estimated to be ca. 105×10^{12} m^3. Western Siberia, North America and Middle East contain about 2:3 of proved reserves. The ultimate resources were estimated by Meyerhoff (1979) to ca. 200×10^{12} m^3 and by BGR (1980) to ca. 290×10^{12} m^3. Furthermore, it can be anticipated that new hydrocarbon discoveries will include a larger proportion of gas accumulations. First, drilling at greater depths, especially in onshore basins where targets at moderate depths have been extensively prospected, will mostly investigate the late catagenesis or metagenesis zones. Then, offshore exploration will be focused on continental margins with its important terrestrial runoff, including organic matter (type III) which has a moderate potential for oil, but can generate large amounts of gas.

6.7 Paleogeography as a Clue to Future Oil and Gas Provinces

Among the resources of conventional oil, the fraction related to offshore basins is gradually increasing due to improvements in offshore technology and intensified exploration efforts: about 12% of the cumulative production of oil has been obtained from the sea, whereas more than 20% of proved reserves and possibly 60% of the future discoveries are located in the offshore basins. Furthermore, one of the major contributions to unconventional crude oil is supposed to come from deep marine basins (more than 400 m of water depth).

The cost of exploration drilling at sea, and that of production increase at a fast rate with increasing water depth. The only acceptable targets in deep waters will be provided by rich provinces containing giant or supergiant fields. Such large accumulations require prolific and widespread source rocks associated with good reservoirs and seals. Paleogeography of the organic and mineral sedimentation provides a way of understanding the distribution of source rocks and reservoirs, and thus is a clue to the exploration of continental margins and deep oceanic basins.

For instance, interpretation of the geochemical data obtained from Deep Sea Drilling (DSDP) core material allowed Tissot et al. (1979, 1980) and Rullkötter et al. (1982) to identify the principal organic facies present in the Cretaceous black shales of the Atlantic basins, to reconstruct their paleogeographic setting and to evaluate their respective petroleum potential. The three major organic facies observed in the Atlantic basins are the following:

a) Planktonic organic matter (type-II kerogen), generally occurs in areas of high productivity (see above, Sect. 6.4.4) where preservation was ensured by

6.7 Paleogeography as a Clue to Future Oil and Gas Provinces

anoxic or near-anoxic conditions of deposition. The related potential for oil and gas is high.

b) Terrestrial organic matter mainly derived from higher plants (type-III kerogen) and less degradable is preserved even in places where an oxic water-column prevents the deposition of planktonic organic matter. It results from a rapid influx and sedimentation of terrigenous material including slumps and mass flows. The potential for oil is only moderate, whereas the potential for gas is good.

c) Residual organic matter, either oxidized in subaerial environments and/or recycled from older sediments, provides an inert fraction of kerogen. It is very widespread and may predominate in some environments, such as hemipelagic sediments. It has no potential for oil or gas.

Geochemical logs were prepared for each of the DSDP wells, using in particular the Rock-Eval pyrolysis to identify the different types of organic matter and evaluate their hydrocarbon source potential. Figure V.6.9 shows, as an example, the data related to site 361 (Cape Basin) and site 364 (Angola Basin). A first occurrence of organic-rich black shales, of Aptian to Lower Albian age, is marked in both Cape and Angola basins by a large input of marine planktonic

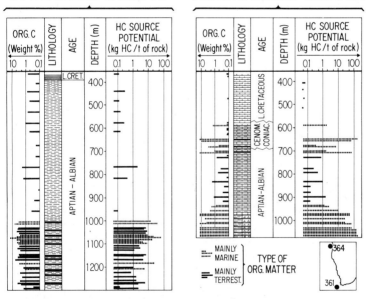

Fig. V.6.9. Examples of geochemical well logs prepared from cores of the Deep Sea Drilling Program (DSDP) in the Cape and Angola Basins. The hydrocarbon source potential is obtained from Rock Eval pyrolysis; t: metric ton. A first occurrence of black shales, with marine organic matter, occurs in Aptian to lower Albian beds of both site 361 (beyond 1000 m depth) and site 364 (beyond 930 m). A second occurrence of black shales at site 364 comprises a succession of discrete organic rich layers of Albian to Coniacian age with marine organic matter (ca. 600–700 m). (Tissot et al., 1980)

organic matter (type II). However, an important contribution of terrestrial organic matter (type III) in the Cape Basin dilutes the planktonic input and reduces the hydrocarbon source potential, as compared to the Angola Basin. A second occurrence of black shales is recorded in the Angola Basin during late Albian to Coniacian time. There, a succession of discrete layers contains a rich planktonic organic matter (type II), while no comparable deposition is known in the Cape Basin. These facts are interpreted as a witness of anoxic conditions in the deep oceanic Cape and Angola basins during Aptian and Lower Albian time, when they were barred and separated from the open ocean. Later the Cape Basin became open to oceanic circulation: the conditions were no longer appropriate for preservation of planktonic material. On the contrary, the Angola basin remained protected, at least temporarily, by the Walvis–Rio Grande Ridge and intermittent anoxic conditions were established during late Albian to Coniacian time.

Comparable observations were made in the North and Central Atlantic Ocean, and the distribution of organic matter types is reported for two different periods of time in Figure V.6.10. From these data, Tissot et al. (1980) observed that statistical regularities of distribution occur on a regional basis, although individual beds may not be correlated. Such regularities of distribution suggest that they reflect paleogeographic conditions. Based on the respective conditions for preservation of marine planktonic (type II), terrestrial (type III), and residual organic matter, they sketched the distribution of depositional environments through Cretaceous time in the Atlantic basins as follows (Fig. V.6.11); these concepts were subsequently precised and enlarged upon by De Graciansky et al. (1982).

a) During *Aptian* and part of *Albian,* the Cape and Angola basin appear as anoxic environments, protected by ridges from the open ocean. Another anoxic subbasin is observed, at least periodically, in the north-central Atlantic along the African margin. This situation was possibly related to upwelling currents as is presently the case in places along the Western coast of Africa and South America, or to restrictions to circulation of oxic waters. In other areas, such as the northern and northwestern Atlantic, oxic conditions prevailed except for short periods of time, and the continent was the only source of organic material, with an important proportion of residual organic matter.
b) During *Cenomanian,* anoxic conditions continued, at least periodically, along the northwest African margin; a temporary extension over large parts of the north and northwest Atlantic is observed around Late Cenomanian. In the South Atlantic, the Angola Basin also remained anoxic, but the Cape Basin became open to circulation of oxic waters.
c) During *Turonian* and moreover *Coniacian,* the anoxic environment was restricted to the Angola Basin, protected by the Rio Grande–Walvis ridge and to the Cape Verde area, the latter possibly in connection with upwelling conditions (Arthur and Natland, 1979).
d) During the later part of Cretaceous, all Atlantic basins were open to circulation of oxic waters and no further favorable environment for planktonic material preservation is found in the Atlantic.

6.7 Paleogeography as a Clue to Future Oil and Gas Provinces

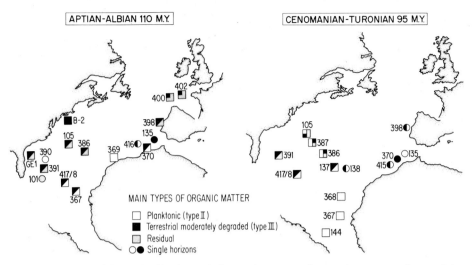

Fig. V.6.10a and b. Distribution of the main types of organic matter observed in Cretaceous sediments of the Central and North Atlantic. (a) Late Aptian–Albian time (position of the continents is only approximate at 110 m.y.) (b) Late Cenomanian–Turonian time (position of the continents is only approximate at 95 m.y.) (Tissot et al., 1979, with changes)

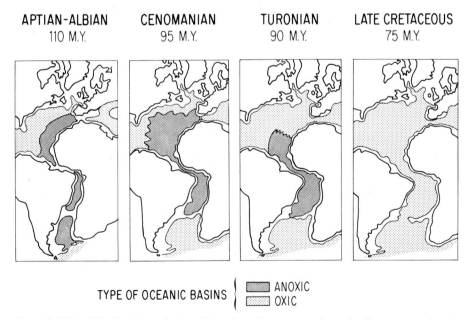

Fig. V.6.11. Distribution of depositional environments through Cretaceous time in Atlantic basins. In some places, or at certain times, anoxic conditions were probably intermittent, particularly in Coniacian time (Tissot et al., 1980)

Paleogeography, interpreted in the framework of plate tectonics, will afford a clue to petroleum exploration of continental margins. It provides us with the rationale for the distribution and characteristics of the various organic facies, and also of the reservoirs. Combined with the use of basin modeling, it offers a comprehensive approach to oil and gas exploration.

Summary and Conclusion

The age and geotectonic distribution of world petroleum reserves are interpreted in terms of sedimentary cycles, sea-level changes, anoxic events and environments of deposition, and subsequent tectonic history. The role of giant fields and giant provinces, and the richness of sedimentary basins are also discussed.

Paleogeography and its consequences on sedimentology of the source rocks and reservoirs appear to be an important contribution to the exploration of future offshore petroleum provinces. Combined with the use of basin modeling, it offers a comprehensive approach to oil and gas exploration.

References to Part V

Abelson, P. P.: Organic geochemistry and the formation of petroleum. 6th World Pet. Cong. Proc. **1,** 397–407 (1963)

Allred, V. D.: Kinetics of oil shale pyrolysis. Chem. Eng. Progr. **62,** 55–60 (1966)

Alpern, B.: Le pouvoir réflecteur des charbons français. Applications et répercussions sur la théorie de A. Duparque. Ann. Soc. Géol. Nord **89,** 143–166 (1969)

Alpern, B.: Classification pétrographique des constituants organiques fossiles des roches sédimentaires. Rev. Inst. Fr. Pét. **26,** 1233–1267 (1971)

Alpern, B.: Pétrographie de la matière organique des sédiments, relations avec la paléotempérature et le potentiel pétrolier. Coll. Int. Sept. 1973. Paris: Editions du C.N.R.S., 1975

Alpern, B.: Pétrographie du kérogène. In: Kerogen. Durand, B. (ed.). Technip, Paris, 1980, pp. 339–371

Alpern, B., Durand, B., Espitalié, J., Tissot, B.: Localisation, caractérisation et classification pétrographique des substances organiques sédimentaires fossiles. In: Advances in Organic Geochemistry 1971. von Gaertner, H. R., Wehner, H. (eds.). Oxford-Braunschweig: Pergamon Press, 1972, pp. 1–28

Alpern, B., Durand B., Durand-Souron C.: Propriétés optiques de résidus de la pyrolyse des kérogènes. Rev. Inst. Fr. Pét. **33,** 867–890 (1978)

Anonymous: International Handbook of Coal Petrography. Int. Committee for Coal Pet. Suppl., 2nd ed., 1972, 1976

Arabian American Oil Company Staff (Aramco): Ghawar oil field, Saudi Arabia. Am. Assoc. Pet. Geol. Bull. **43,** 434–454 (1959)

Arefjer, O. A., Makushina, V. M., Petror, A. A.: Asphaltenes as indicators of geochemical history of crude oils (in Russian). Izv. Akad. Nauk SSR, Ser. geol., **4,** 124–130 (1980)

Arthur, M. A., Natland, J. H.: Carbonaceous sediments in the North and South Atlantic. The role of salinity in stable stratification of early Cretaceous basins. In: Deep drilling results in the Atlantic Ocean: continental margins and paleoenvironment: Am. Geoph. Union, Maurice Ewing Ser., Vol. 3, pp. 375–401 (1979)

Arthur M. A., Schlanger, S. O.: Cretaceous "oceanic anoxic events" as causal factors in development of reef – reservoired giant oil fields: Am. Assoc. Pet. Geol. Bull. **63,** 870–885 (1979)

Audley-Charles, M. G., Curray, J. R., Evans, G.: Location of major deltas. Geology **5,** 341–344 (1977)

Ayres, M. G., Bilal, M., Jones, R. W., Slentz, L. W., Tartir, M., Wilson, O.: Hydrocarbon habitat in main producing areas, Saudi Arabia. Am. Assoc. Pet. Geol. Bull. **66,** 1–9 (1982)

Bailey, N. J. L., Evans, C. R., Milner, C. W. D.: Applying petroleum geochemistry to search for oil: examples from Western Canada Basin. Am. Assoc. Pet. Geol. Bull. **58,** 2284–2294 (1974)

Barker, C.: Pyrolysis techniques for source-rock evaluation. Am. Assoc. Pet. Geol. Bull. **58,** 2394–2361 (1974)

Bishop, R. S., Gehman, H. M., Young, A.: Concepts for estimating hydrocarbon accumulation and dispersion. Am. Assoc. Pet. Geol. Bull. **67,** 337–348 (1983)

Bois, C.: Petroleum-zone concept and the similarity analysis; contribution to resource appraisal. In: Methods of estimating the volume of undiscovered oil and gas resources. Am. Assoc. Pet. Geol. Stud. Geol. **1,** 87–89 (1975)

Bois, C., Bouché, P., Pelet, R.: Histoire geologique et répartition des réserves d'hydrocarbures dans le monde. In: Colloque C2, Energy Resources. Burolett, P. F., Ziegler, V., (eds.). 26th Int. Geol. Congress, Paris, Rev. Inst. Fr. Pét. **35,** 273–298 (1980)

Bois, C., Bouché, P., Pelet, R.: Global geological history and the distribution of hydrocarbon reserves. Am. Assoc. Pet. Geol. Bull. **66,** 1248–1270 (1982)

Bois, C., Monicard, R.: Pétrole: peut-on encore découvrir des gisements géants? La Recherche **124,** 854–862 (1981)

Bordenave, M., Combaz, A., Giraud, A.: Influence de l'origine des matières organiques et de leur degré d'evolution sur les produits de pyrolyse du kerogène. In: Advances in Organic Geochemistry 1966. Hobson, G. D., Speers, G. C. (eds.). Oxford: Pergamon Press, 1970, pp. 388–405

Bostick, N. H.: Thermal alteration of clastic organic particles (phytoclasts) as an indicator of contact and burial metamorphism in sedimentary rocks, Ph. D. thesis, Stanford Univ. Calif., 1970

Boy de la Tour, X., Le Leuch, H.: Nouvelles techniques de mise en valeur des ressources d'hydrocarbures. Séminaire sur le perfectionnement des techniques d'extraction et de transformation d'énergies primaires. Rev. Inst. Fr. Pét. **36,** 251–280 (1981)

Braun, R. L., Rothman, A. J.: Oil shale pyrolysis: kinetics and mechanism of oil production. Fuel **54,** 129–131 (1975)

Bray, E. E., Evans, E. D.: Distribution of n-paraffins as a clue to recognition of source beds. Geochim. Cosmochim. Acta **22,** 2–15 (1961)

Brod, I. O.: The petroleum basins in the world (in Russian). Moscow, Nedra (1965)

Brooks, J. D.: The use of coals as indicators of oil and gas. Aust. Pet. Exp. Ass. J. 1970, 35–40 (1970)

Chenet, P. Y., Bessis, F. G., Ungerer, P. M., Nogaret, E.: How mathematical geological models can reduce exploration risks. 11th World Petroleum Congress, London, Special Paper 7 (1983)

Chiarelli, A., du Rouchet, J.: Importance des phénomènes de migration verticale des hydrocarbures. Rev. Inst. Fr. Pét. **32,** 2 (1977)

Claret, J., Tchikaya, J. B., Tissot, B., Deroo, G., van Dorsselaer, A.: Un exemple d'huile biodegradée à basse teneur en soufre: le gisement d'Emeraude (Congo). In: Advances in Organic Geochemistry 1975. Campos, R., Goni, J. (eds.). Madrid, 1977, pp. 509–522

Claypool, G. E., Reed, P. R.: Thermal analysis technique for source-rock evaluation: quantitative estimate of organic richness and effects of lithologic variation. Am. Assoc. Pet. Geol. Bull. **60,** 608–612 (1976)

Combaz, A.: Les palynofacies. Rev. Micropaleontol. 3 (1964)

Combaz, A.: Les kerogenes vus au microscope. In: Kerogen. Durand B. (ed.). Technip, Paris, 1980, pp. 55–111

Combaz, A., De Matharel, M.: Organic Sedimentation and Genesis of Petroleum in Mahakam Detta, Borneo. Am. Assoc. Pet. Geol. Bull. **62,** No. 9, 1684–1695 (1978)

Connan, J.: Time–temperature relation in oil genesis. Am. Assoc. Pet. Geol. Bull. **58,** 2516–2521 (1974)

Connan, J.: Time–temperature relation in oil genesis: reply. Am. Assoc. Pet. Geol. Bull. **60,** 885–887 (1976)

Cornford, C.: Personal communication (1977)
Correia, M.: Relations possibles entre l'état de conservation des éléments figurés de la matière organique (microfossiles palynoplanctologiques) et l'existence de gisements d'hydrocarbures. Rev. Inst. Fr. Pét. **22,** 1285–1306 (1967)
Correia, M., Peniguel, G.: Étude microscopique de la matière organique. Ses applications à l'exploration pétrolière. Bull. Centre Rech. Pau **9,** 99–127 (1975)
De Graciansky, P. C., Brosse, E., Deroo, G., Herbin, J. P., Montadert, L., Müller, C., Schaaf, A., Sigal, J.: Les formations d'âge Crétacé de l'Atlantique Nord et leur matière organique: paléogéographie et milieux de dépôt. Rev. Inst. Fr. Pét. **37,** 275–335 (1982)
Demaison, G. J.: Relationship of coal rank to paleotemperatures in sedimentary rocks. In: Petrographie de la matière organique des sediments, relation avec la paléotemperature. Alpern, B. (ed.). Paris: Editions du Centre National de Recherche Scientifique, 1975, pp. 217–224
Demaison, G. J.: Tar sands and supergiant oil fields. Am. Assoc. Pet. Geol. Bull. **61,** 1950–1961 (1977)
Dembicki, H., Jr., Meinschein, W. G., Hattin, D. E.: Possible ecological and environmental significance of the predominance of even-carbon number C_{20}–C_{30} n-alkanes. Geochim. Cosmochim. Acta **40,** 203–208 (1976)
Deroo, G.: Corrélations huiles brutes — roches méres à l'échelle des bassins sédimentaires. Bull. Centre Rech. Pau **10,** 317–335 (1976)
Deroo, G., Durand, B., Espitalié, J., Pelet, R., Tissot, B.: Possibilité d'application des modèles mathématiques de formation du pétrole à la prospection dans les bassins sédimentaires. In: Advances in Organic Geochemistry 1968. Schenck, P. A., Havenaar, I. (eds.). Oxford: Pergamon Press, 1969, pp. 345–354
Desprairies, P.: Les limites de l'approvisionnement pétrolier mondial. World Energy Conference. Technip, Paris, 1977
Desprairies, P., Tissot, B.: Les limites de l'approvisionnement pétrolier mondial. In: 11th World Energy Conference. Munich, RT 3, pp. 549–573 (1980)
Doebl, F., Heling, D., Homann, W., Karweil, J., Teichmüller, M., Welte, D.: Diagenesis of Tertiary clayey sediments and included dispersed organic matter in relationship to geothermics in the Upper Rhine Graben. In: Approaches to Taphrogenesis. Stuttgart: Schweizerbartsche Verlagsbuchhandlung, 1974, pp. 192–207
Dow, W. G.: Kerogen studies and geological interpretations. J. Geochem. Explor. **7,** 79–99 (1977)
Durand, B., Espitalié, J.: Analyse géochimique de la matière organique extraite des roches sédimentaires: II. Extraction et analyse quantitative des hydrocarbures de la fraction C_1–C_{15} et des gaz permanents. Rev. Inst. Fr. Pét. **25,** 741–751 (1970)
Durand, B., Espitalié, J.: Formation and evolution of C_1 and C_{15} hydrocarbons and permanent gases in the Toarcian clays of the Paris Basin. In: Advances in Organic Geochemistry 1971. v. Gaertner, H. R., Wehner, H. (eds.). Pergamon Press, Oxford, 1972, pp. 455–468
Durand, B., Espitalié, J.: Evolution de la matière organique au cours de l'enfouissement des sédiments. C. R. Acad. Sci. Ser. D **276,** 2253–2255 (1973)
Durand, B., Oudin, J. L.: Exemple de migration des hydrocarbures dans une série deltaique: le delta de la Mahakam, Kalimantan, Indonesie. 10th World Pet. Cong. Proc. **2,** 3–11 (1980)
Durand, B., Marchand, A., Amiell, J., Combaz, A.: Etude de kérogènes par résonance paramagnétique électronique. In: Advances in Organic Geochemistry 1975. Campos. R., Goni, J. (eds.) 1977, pp. 753–779
Durand, B., Ungerer, Ph., Chiarelli, A., Oudin, J. L.: Modelisation de la migration de l'huile. Application à deux exemples de bassins sédimentaires. 11th World Petroleum Congress, London, PD 1, paper 3 (1983)

Ejedawe, J. E.: Patterns of incidence of oil reserves in Niger Delta Basin. Am. Assoc. Pet. Geol. Bull. **65**, 1574–1585 (1981)

Ekweozor, C. M., Okoye, N. V.: Petroleum source-bed evaluation of Tertiary Niger Delta. Am. Assoc. Pet. Geol. Bull. **64**, 1251–1259 (1980)

Ensminger, A., van Dorsselaer, A., Spyckerelle, C., Albrecht, P., Ourisson, G.: Pentacyclic triterpanes of the hopane type as ubiquitous geochemical markers: origin and significance. In: Advances in Organic Geochemistry 1973. Tissot, B., Bienner, F. (eds.). Technip, Paris, 1974, pp. 245–260

Erdman, G. J.: Relations Controlling Oil and Gas Generation in Sedimentary Basins. 9th World Pet. Cong. Proc. London: Applied Science Publishers, 1975, pp. 139–148

Espitalié, J., Laporte, J. L., Madec, M., Marquis, F., Leplat, P., Paulet, J., Boutefeu, A.: Méthode rapide de caractérisation des roches mères, de leur potentiel pétrolier et de leur degré d'évolution. Rev. Inst. Fr. Pét. **32**, 23–42 (1977)

Evamy, B. D., Haremboure, J., Kamerling, P., Knapp, W. A., Molloy, F. A., Rowlands, P. H.: Hydrocarbon habitat of Tertiary Niger Delta. Am. Assoc. Pet. Geol. Bull. **62**, 1–39 (1978)

Fausett, D. W.: Mathematical study of an oil shale retort. Ph. D. thesis, Univ. Wyoming, 1974

Forsman, J. P., Hunt, J. M.: Insoluble organic matter (kerogen) in sedimentary rocks of marine origin. In: Habitat of Oil. Weeks, L. G. (ed.). Tulsa: Am. Assoc. Pet. Geol., 1958, pp. 747–778

Galimov, E. M.: Carbon isotopes in oil and gas geology (in Russian). Moscow: Nedra, 1973. Engl. transl. Washington: NASA TT F-682

Gehman, H. M., Jr.: Organic matter in limestones. Geochim. Cosmochim. Acta **26**, 885–897 (1962)

GeoChem Laboratories Inc.: Vitrinite reflectance (Commercial brochure). Houston, Texas (1975)

Gies, R. M.: Origin, migration, and entrapment of natural gas in the Alberta Deep Basin, Part II. 67th Annual Meeting of the AAPG, Valgary/Alberta, June 27–29 (1982)

Gijzel, P. van: Review of the UV-fluorescence of fresh and fossil exines and exosporia. In: Sporopollenin. Brooks, J., Grant, P. R., Muir, M. D., van Gijzel, P., Shaw, G. (eds.). London: Academic Press, 1971, pp. 659–685

Gransch, J. A., Eisma, E.: Characterization of the insoluble organic matter of sediments by pyrolysis. In: Advances in Organic Geochemistry 1966. Hobson, G. D., Speers, G. C. (eds.). Oxford: Pergamon Press, 1970, pp. 407–426

Greiner, A. C., Spyckerelle, C., Albrecht, P.: Aromatic hydrocarbons from geologic sources – I. New naturally occurring phenanthrene and chrysene derivatives. Tetrahedron **32**, 257–260 (1976)

Gruenfeld, M.: Extraction of dispersed oils from water for quantitative analysis by infrared spectrophotometry. Envir. Sci. Technol. **7**, 636–639 (1973)

Gutjahr, C. C. M.: Carbonization of pollen grains and spores and their application. Leidse Geolog. Mededel. **38**, 1–30 (1966)

Hacquebard, P. A., Donaldson, J. R.: Coal metamorphism and hydrocarbon potential in the Upper Paleozoic of the Atlantic Provinces. Canada. Can. J. Earth Sci. **7**, 1139–1163 (1970)

Hagemann, H. W.: Petrographische und palynologische Untersuchung der organischen Substanz (Kerogen) in den liassischen Sedimenten Luxemburgs. In: Advances in Organic Geochemistry 1973. Tissot, B., Bienner, F. (eds.). Paris: Technip, 1974, pp. 29–37

Hagemann, H. W.: Personal communication: On oxidation rims of reworked organic particles (1977)

Halbouty, M. T., Moody, J. D.: World ultimate reserves of crude oil. 10th World Pet. Cong. Proc. **2**, 291–301 (1979)

Halbouty, M. T., Meyerhoff, A. A., King, R. E., Dott, R. H., Sr., Klemme, H. D., Shabad, T.: World's giant oil and gas fields, geologic factors affecting their formation and basin classification. Am. Assoc. Pet. Geol. Mem. **14**, 502–555 (1970)

Hanbaba, P., Jüntgen, H.: Zur Übertragbarkeit von Laboratoriums-Untersuchungen auf geochemische Prozesse der Gasbildung aus Steinkohle und über den Einfluß von Sauerstoff auf die Gasbildung. In: Advances in Organic Geochemistry 1968. Schenck, P. A., Havenaar, I. (eds.). Oxford: Pergamon Press, 1969, pp. 459–471

Heacock, R. L., Hood, A.: Process for measuring the live carbon content of organic samples. U.S. Patent 3508877, April 28, 1970

Heroux, Y., Chagnon, A., Bertrand, R.: Compilation and correlation of major thermal maturation parameters. Am. Assoc. Pet. Geol. Bull. **63**, 2128–2144 (1979)

Hood, A., Castaño, J. R.: Organic metamorphism. Its relationship to petroleum generation and application to studies of authigenic minerals. Coordinating Comm. Offshore Prospecting Techn. Bull. **8**, 85–118 (1974)

Hood, A., Gutjahr, C. C. M., Heacock, R. L.: Organic metamorphism and the generation of petroleum. Am. Assoc. Pet. Geol. Bull. **59**, 986–996 (1975)

Huang, W.-Y., Meinschein, W. G.: Sterols as ecological indicators. Geochim. Cosmochim. Acta **43**, 739–745 (1979)

Hubbard, A. B., Robinson, W. E.: A thermal decomposition study of Colorado oil shale. Rep. Invest. U.S. Bureau of Mines, 4744, 1950

Huc, A. Y.: Mise en évidence de provinces géochimiques dans les schistes bitumineux du Toarcien de l'Est du Bassin de Paris. Etude de la fraction organique soluble. Rev. Inst. Fr. Pét. **31**, 333–353 (1976)

Huck, G., Karweil, J.: Physikalische Probleme der Inkohlung. Brennst.-Chem. **36**, 1–11 (1955)

Hunt, J.: Origin of gasoline range alkanes in the deep sea. Nature (London) **254**, 411–413 (1975)

Johns, W. D., Shimoyama, A.: Clay minerals and petroleum forming reactions during burial and diagenesis. Am. Assoc. Pet. Geol. Bull. **56**, 2160–2167 (1972)

Jonathan, D., L'Hote, G., du Rouchet, J.: Analyse géochimique des hydrocarbures légers par thermovaporisation. Rev. Inst. Fr. Pét. **30**, 65–88 (1975)

Jones, P. S., Smith, H. M.: Relationships of oil composition and stratigraphy in the Permian basin of West Texas and New Mexico. In: Fluids in Subsurface Environment. Am. Assoc. Pet. Geol. Mem. **4**, 101–224 (1965)

Kalbassi, H., Welte, D. H.: Die Zusammensetzung der extrahierbaren Kohlenwasserstoffe aus Spül- und Kernproben der Bohrung Scheibenhardt 2 im Oberrheintal, unpublished information from Diplom-Arbeit. Kalbassi, H. (ed.). Aachen: Rhein.-Westfäl. Techn. Hochschule, 1975

Klemme, D. H.: The geology of future petroleum resources. In: Colloque C2, Energy Resources. Burollet, P. F., Ziegler, V. (eds.). 26th Int. Geology Congress, Paris. Rev. Inst. Fr. Pét. **35**, 337–349 (1980)

Koons, C. B., Bond, J. G., Peirce, F. L.: Effects of depositional environment and postdepositional history on chemical composition of lower Tuscaloosa oils. Am. Assoc. Pet. Geol. Bull. **58**, 1272–1280 (1974)

Krevelen, D. W. van: Coal. Amsterdam: Elsevier, 1961

Laplante, R. E.: Hydrocarbon generation in Gulf Coast Tertiary sediments. Am Assoc. Pet. Geol. Bull. **58**, 1281–1289 (1974)

Lee, W. H. K., Uyeda, S.: Review of heat flow data. In: Terrestrial Heat. A.G.U. Geophys. Mon. **8**, 87–190 (1965)

Le Pichon, X., Langseth, M. G.: Heat flow from the mid-ocean ridges and sea-floor spreading. Tectonophysics **8,** 4–6, 319–344 (1969)

Leplat, P., Noel, R.: Contribution à l'étude de l'état de diagénèse des roches à kérogène par le dosage automatique rapide du CO_2 et du C_2H_4 libérés par pyrolyse de 400 à 900°C. In: Advances in Organic Geochemistry 1973. Tissot, B., Bienner, F. (eds.). Paris: Technip, 1974, pp. 107–115

Le Tran, K.: Etude géochimique de l'hydrogène sulfuré adsorbé dans les sédiments. Bull. Cent. Rech. Pau-SNPA **5,** 321–332 (1971)

Le Tran, K.: Analyse et étude des hydrocarbures gazeux occlus dans les sédiments. Exemples d'application à l'exploration pétrolière. Bull. Cent. Rech. Pau-SNPA **9,** 223–243 (1975)

Leythaeuser, D., Hollerbach, A., Hagemann, H. W.: Source rock/crude oil correlation based on distribution of C_{27+}-cyclic hydrocarbons. In: Advances in Organic Geochemistry 1975. Campos, R., Goni, J. (eds.). Madrid, 1977, pp. 3–20

Leythaeuser, D., Altebäumer, F. J., Schaefer, R. G.: Effect of an igneous intrusion on maturation of organic matter in Lower Jurassic Shales from NW-Germany. In: Advances in Organic Geochemistry 1979. Douglas, A. G., Maxwell, J. R. (eds.). Pergamon Press, Oxford, pp. 133–139

Leythaeuser, D., Schaefer, R. G., Weiner, B.: Generation of low molecular weight hydrocarbons from organic matter in source beds as a function of temperature and facies. Chem. Geol. **25,** 95–108 (1979a)

Leythaeuser, D., Schaefer, R. G., Cornford, C., Weiner, B.: Generation and migration of light hydrocarbons (C_2–C_7) in sedimentary basins. Org. Geochem. **1,** 191–204 (1979b)

Leythaeuser, D., Schaefer, R. G., Yükler, A.: Role of diffusion in primary migration of hydrocarbons. Am. Assoc. Pet. Geol. Bull. **66,** 408–429 (1982)

Lopatin, N. V.: The main stage of petroleum formation (in Russian). Izv. Akad. Nauk SSSR, Ser. Geol. 69–76 (1969)

Lopatin, N. V.: Temperature and geologic time factors in coalification. Izv. Akad. Nauk. Uzb. SSR, Ser. Geol. **3,** 95–106 (1971)

Lopatin, N. V., Bostick, N. H.: Geological factors in coal catagenesis. In: Nature of Organic Matter in Recent and Ancient Sediments (in Russian). Symp. Nauka: Moscow, 1973, pp. 79–90

Louis, M.: Les corps optiquement actifs et l'origine du pétrole. Rev. Inst. Fr. Pét. **23,** 299–314 (1968)

Mackenzie, A. S.: Application of biological markers in petroleum geochemistry. In: Advances in Petroleum Geochemistry. Brooks, J., Welte, D. H. (eds.). Academic Press, London, in press

Mackenzie, A. S., Maxwell, J. R.: Assessment of thermal maturation in sedimentary rocks by molecular measurements. In: Organic Maturation Studies and Fossil Fuel Exploration. Brooks, J. (ed.). Academic Press, London, 1981, pp. 239–254

Mackenzie, A. S., Patience, R. L., Maxwell, J. R., Vandenbroucke, M., Durand, B.: Molecular parameters of maturation in the Toarcian Shales, Paris Basin, France – I. Changes in the configurations of acyclic isoprenoid alkanes, steranes and triterpanes. Geochim. Cosmochim. Acta **44,** 1709–1721 (1980)

Mackenzie, A. S., Hoffmann, C. F., Maxwell, J. R.: Molecular parameters of maturation in the Toarcian shales, Paris Basin, France – III. Changes in aromatic steroid hydrocarbons. Geochim. Cosmochim. Acta **45,** 1345–1355 (1981)

Mackenzie, A. S., Lamb, N. A., Maxwell, J. R.: Steroid hydrocarbons and the thermal history of sediments. Nature (London) **295,** 223–226 (1982a)

Mackenzie, A. S., Brassell, S. C., Eglinton, G., Maxwell, J. R.: Chemical fossils: the geological fate of steroids. Science **217,** 491–504 (1982b)

Mackenzie, D.: Some remarks on the development of sedimentary basins. Earth Planet. Sci. Lett. **40,** 25 (1978)

Mackenzie, D.: The variation of temperature with time and hydrocarbon maturation in sedimentary basins formed by extension. Earth Planet. Sci. Lett. **55,** 87–98 (1981)

Magnier, Ph., Oki, T., Witoelar Kartaadiputra, L.: The Mahakam Delta, Kalimantan, Indonesia. 9th World Pet. Cong. Proc. **2,** 239–250 (1976)

Maier, C. G., Zimmerley, S. R.: The chemical dynamics of the transformation of the organic matter to bitumen in oil shale. Univ. Utah Bull. **14,** 7, 62–81 (1924)

Mair, B.: Terpenoids, fatty acids and alcohols as source materials for petroleum hydrocarbons. Geochim. Cosmochim. Acta **28,** 1303–1321 (1964)

Marchand, A., Libert, P., Combaz, A.: Essai de caractérisation physico-chimique de la diagénèse de quelques roches organiques biologiquement homogènes. Rev. Inst. Fr. Pét. **24,** 3–20 (1969)

Masran, T. C., Pocock, S. A.: The classification of plant-derived particulate organic matter. Proc. 5th Internat. Palynol. Conference, Cambridge, Brooks, J. (ed.). Academic Press, London, pp. 145–175

Masters, J. A.: Deep basin Gas trap, Western Canada. Am. Assoc. Pet. Geol. Bull. **63,** 152–181 (1979)

Maxwell, J. R., Mackenzie, A. S., Volkman, J. K.: Configuration at C-24 in steranes and sterols. Nature (London) **286,** 694–697 (1980)

McCartney, J. T., Teichmüller, M.: Classification of coals according to degree of coalification by reflectance of the vitrinite component. Fuel **51,** 64–68 (1972)

McIver, R. D.: Composition of kerogen–clue to its role in the origin of petroleum. 7th World Pet. Cong. Proc. **2,** 25–36 (1967)

McNab, J. G., Smith, P. V., Betts, R. L.: The evolution of petroleum. Ind. Eng. Chem. **44,** 2556–2563 (1952)

Meyerhoff, A.: Proved and ultimate reserves of natural gas and natural gas liquids in the world. 10th World Pet. Cong. Proc. **2,** 303–311 (1979)

Moldowan, J. M., Seifert, W. K.: First discovery of botryococcane in petroleum. J. Chem. Soc. Chem. Commun. 912–914 (1980)

Murris, R. J.: Middel East: Stratigraphic evolution and oil habitat. Am. Assoc Pet. Geol. Bull. **64,** 597–618 (1980)

Nederlof, M. H.: The use of habitat of oil models in exploration prospect appraisal: Proc. 10th World Pet. Cong. Proc. **2,** 13–22 (1980)

Nehring, R.: Giant oil fields and world oil resources. Rand Corp. Report R-2284 CIA (1978)

Oberlin, A., Boulmier, J. L., Durand, B.: Electron microscope investigation of the structure of naturally and artificially metamorphosed kerogen. Geochim. Cosmochim. Acta **38,** 647–649 (1974)

Olenine, V. B.: Petroleum areas subdivision according to genetic principles (in Russian) Moscow, Nedra (1977)

Ottenjann, K., Teichmüller, M., Wolf, M.: Spektrale Fluoreszenz-Messungen an Sporiniten mit Auflicht-Anregung, eine mikroskopische Methode zur Bestimmung des Inkohlungsgrades gering inkohlter Kohlen. Fortschr. Geol. Rheinl. Westf. **24,** 1–36 (1974)

Perrodon, A.: Coup d'œil sur les provinces géantes d'hydrocarbures. Rev. Inst. Fr. Pét. **33,** 493–513 (1978)

Perrodon, A.: Geodynamique pétrolière, Paris, Masson and Elf Aquitaine, 381 p. (1980)

Philippi, G. T.: On the depth, time and mechanism of petroleum generation. Geochim. Cosmochim. Acta **29,** 1021–1049 (1965)

Philippi, G. T.: The deep sub-surface temperature controlled origin of the gaseous and gasoline-range hydrocarbons of petroleum. Geochim. Cosmochim. Acta **39**, 1353–1373 (1975)

Philippi, G. T.: On the depth, time and mechanism of origin of the heavy to medium-gravity naphthenic crude oils. Geochim. Cosmochim. Acta **41**, 33–52 (1977)

Philippi, G. T.: Correlation of crude oils with their oil source formation, using high resolution GLC C_6–C_7 component analysis. Geochim. Cosmochim. Acta **45**, 1495–1513 (1981)

Pitman, W. C.: Relation between eustacy and stratigraphic sequences of passive margins. Geol. Soc. Am. Bull. **89**, 1389–1403 (1978)

Pusey, W. C.: How to evaluate potential gas and oil source rock. World Oil **176**, 71–75 (1973)

Radke, M., Welte, D. H.: The Methylphenanthrene Index (MPI): a maturity parameter based on aromatic hydrocarbons. In: Advances in Organic Geochemistry 1981. Bjorøy M. et al. (eds.). John Wiley, Chichester, 1983

Radke, M., Welte, D. H., Willsch, H.: Geochemical study on a well in the Western Canada Basin: relation of the aromatic distribution pattern to maturity of organic matter. Geochim. Cosmochim. Acta **46**, 1–10 (1982a)

Radke, M., Willsch, H., Leythaeuser, D., Teichmüller, M.: Aromatic components of coal: relation of distribution pattern to rank. Geochim. Cosmochim. Acta **46**, 1831–1841 (1982b).

Reed, W. E.: Molecular compositions of weathered petroleum and comparison with its possible source. Geochim. Cosmochim. Acta **41**, 237–247 (1977)

Robert, P.: Étude pétrographique des matières organiques insolubles par la mesure de leur pouvoir reflecteur — Contribution à l'exploration pétrolière et à la connaissance des bassins sédimentaires. Rev. Inst. Fr. Pét. **26**, 105–135 (1971)

Robert, P.: The optical evolution of kerogen and geothermal histories applied to oil and gas exploration. In: Kerogen. Durand, B. (ed.). Technip, Paris, 1980, pp. 385–414.

Rogers, M. A.: Application of organic facies concepts to hydrocarbon source evaluation. 10th World Pet. Cong. Proc. **2**, 23–30 (1979)

Ronov, A. B.: Organic carbon in sedimentary rocks (in relation to presence of petroleum). Translation in Geochemistry, No. 5, 510–536 (1958)

Rullkötter, J., Cornford, C., Welte, D. H.: Geochemistry and petrography of organic matter in Northwest African Continental margin sediments: quantity, provenance, depositional environment and temperature history. In: Geology of the Northwest African Continental Margin. Von Rad, V., Hinz, K., Sarnthein, M., Seibold, E. (eds.). Springer, Heidelberg, 1982, pp. 686–703

Rullkötter, J., Welte, D. H.: Oil-oil and oil-condensate correlation by low eV GC-MS measurements of aromatic hydrocarbons. Advances in Organic Geochemistry 1979. Douglas, A. G., Maxwell, J. R. (eds.). Pergamon Press, Oxford, 1980, pp. 93–102

Rullkötter, J., Wendisch, D.: Microbial alteration of 17α(H)-hopanes in Madagascar asphalts: removal of C-10 methyl group and ring opening. Geochim. Cosmochim. Acta **46**, 1545–1553 (1982)

Scalan, R. S., Smith, J. E.: An improved measure of the odd-even predominance in the normal alkanes of sediment extracts and petroleum. Geochim. Cosmochim. Acta **34**, 611–620 (1970)

Schaefer, R. G., Leythaeuser, D.: Analysis of trace amounts of hydrocarbons (C_2–C_8) from rock and crude oil samples and its application in petroleum geochemistry. In: Advances in Organic Geochemistry 1979. Douglas, A. G., Maxwell, J. R. (eds.). Pergamon Press, Oxford, 1980, pp. 149–156

Schaefer, R. G., Leythaeuser, D.: Generation and migration of low-molecular weight hydrocarbons in sediments from Site 511 of DSDP/IPOD Leg 71, Falkland Plateau, South Atlantic. In: Advances in Organic Geochemistry 1981. Bjorøy, M. et al. (eds.). J. Wiley, Chichester 1983, pp. 164–174

Schaefer, R. G., Leythaeuser, D., Weiner, B.: Single-step capillary column gas chromatographic method for extraction and analysis of sub-parts per billion (10^9) amounts of hydrocarbons (C_2-C_8) from rock and crude oil samples and its application in organic geochemistry. J. Chromatogr. **167,** 355–363 (1978b)

Schaefer, R. G., Weiner, B., Leythaeuser, D.: Determination of sub-nanogram per gram quantities of light hydrocarbons (C_2-C_9) in rock samples by hydrogen stripping in the flow system of a capillary gas chromatograph. Anal. Chem. **50,** 1848–1854 (1978a)

Schlanger, S. O., Jenkins, H. C.: Cretaceous anoxic events: causes and consequences. Geologie en Mijnbouw **55,** 179–184 (1976)

Schoell, M.: The hydrogen and carbon isotopic composition of methane from natural gases of various origins. Geochim. Cosmochim. Acta **44,** 649–661 (1980)

Schoell, M., Teschner, M., Wehner, H., Durand, B., Oudin, J. L.: Maturity related biomarker and stable isotope variations and their application to oil/source rock correlation in the Mahakam Delta, Kalimantan. In: Advances in Organic Geochemistry 1981. Bjorøy, M., et al. (eds.). J. Wiley, Chichester, 1983, pp. 156–163

Seifert, W. K.: Steranes and terpanes in kerogen pyrolysis for correlation of oils and source rocks. Geochim. Cosmochim. Acta **42,** 473–484 (1978)

Seifert, W. K., Moldowan, J. M.: Application of steranes, terpanes and monoaromatics to the maturation, migration and source of crude oils. Geochim. Cosmochim. Acta **42,** 77–95 (1978)

Seifert, W. K., Moldowan, J. M.: The effect of biodegradation on steranes and terpanes in crude oils. Geochim. Cosmochim. Acta **43,** 111–126 (1979)

Seifert, W. K., Moldowan, J. M.: The effect of thermal stress on source-rock quality as measured by hopane stereochemistry. In: Advances in Organic Geochemistry 1979. Douglas, A. G., Maxwell, J. R. (eds.). Pergamon Press, Oxford, 1980, pp. 229–237

Seifert, W. K., Moldowan, J. M.: Paleoreconstruction by biological markers. Geochim. Cosmochim. Acta **45,** 783–794 (1981)

Seifert, W. K., Moldowan, J. M., Jones, R. W.: Application of biological marker chemistry to petroleum exploration. 10th World Pet. Cong. Proc. **2,** 425–438 (1980)

Seifert, W. R.: On the indigenous nature of bitumen in source rocks: kerogen pyrolysis with biological marker GC-MS, paper presented at the 8th Int. Cong. Org. Geochem. Moscow, May 1977

Shell Standard Legend: Exploration and production departments. Royal Dutch/Shell Group of Companies. Shell Intern. Pet. Maatschappij B. V., The Hague (1976)

Shi, J.-Y., Mackenzie, A. S., Alexander, R., Eglinton, G., Gowar, A. P., Wolff, G. A., Maxwell, J. R.: A biological marker investigation of petroleums and shales from the Shengli oilfield, the People's Republic of China. Chem. Geol. **35,** 1–31 (1982)

Silverman, S. R.: Carbon isotope evidence for the role of lipids in petroleum formation. J. Am. Oil. Chem. Soc. **44,** 691–695 (1967)

Sluijk, D., Nederlof, M. H.: Worldwide geological and geochemical experience as a systematic basis for prospect appraisal. In: Basin Geochemistry. Demaison, G., Murris, R. J. (eds.). Am. Assoc. Pet. Geol. Mem. **35,** (1983)

Smith, J. E.: Migration accumulation and retention of petroleum in the earth. 8th World Pet. Cong. Proc. **2,** 13–26 (1971)

Smith, A. G., Briden, J. C., Drewry, G. E.: Phanerozoic world maps. In: Hughes, N. F. (ed.), Organisms and continents through time. London Paleont. Assoc. Spec. Papers in Paleontology **12,** 1–42 (1973)

Snowdon, L. R., Roy, K. J.: Regional organic Metamorphism in the Mesozoic strata of the Sverdrup Basin. Can. Pet. Geol. Bull. **23,** 131–148 (1975)

Sokolov, B. A.: Evolution and hydrocarbon potential of sedimentary basins (in Russian). Moscow, Nauka (1980)

Stach, E.: Textbook of Coal Petrology. 2nd completely revised ed. by Stach, E., Mackowsky, M. T., Teichmüller, M., Taylor, G. H., Chandra, D., Teichmüller, R. Berlin-Stuttgart: Gebrüder Borntraeger, 1975

Stahl, W.: Kohlenstoff-Isotopenverhältnisse von Erdgasen. Erdöl u. Kohle, Erdgas, Petrochem. **28,** 188–191 (1975)

Stahl, W., Carey, Jr., B. D.: Source-rock identification by isotope analyses of natural gases from fields in the Val Verde and Delaware Basins, West Texas. Chem. Geol. **16,** 257–267 (1975)

Stahl, W., Wollanke, G., Boigk, H.: Carbon and nitrogen isotope data of upper Carboniferous and Rotliegend natural gases from north Germany and their relationship to the maturity of the organic source material. In: Advances in Organic Geochemistry 1975. Campos, R., Goni, J. (eds.). Madrid, 1977, pp. 539–560

Staplin, F. L.: Sedimentary organic matter, organic metamorphism, and oil and gas occurrence. Can. Pet. Geol. Bull. **17,** 47–66 (1969)

Steineke, M., Bramkamp, R. A., Sander, N. J.: Stratigraphic relations of Arabian Jurassic Oil. In: Habitat of oil. Weeks, L. G. (ed.). Am. Assoc. Pet. Geol., 1294–1329 (1958)

Tappan, H., Loeblich, A. R. Jr.: Symp. Palynology of the Late Cretaceous and Early Tertiary. Geol. Soc. Am., Michigan State Univ., 247–340 (1970)

Teichmüller, M.: The importance of coal petrology in prospecting for oil and natural gas. In: Stach, E., Mackowsky, M. T., Teichmüller, M., Taylor, G. H., Chandra, D., Teichmüller, R. (eds.) Textbook of Coal Petrology (3rd edn.). Gebr. Borntraeger, Berlin-Stuttgart, 1982, pp. 399–412

Teichmüller, M.: Anwendung kohlenpetrographischer Methoden bei der Erdöl- und Erdgasprospektion. Erdöl u. Kohle **24,** 69–76 (1971)

Teichmüller, M.: Entstehung und Veränderung bituminöser Substanzen in Kohlen in Beziehung zur Entstehung und Umwandlung des Erdöls. Fortschr. Geol. Rheinl. Westf. **24,** 54–112 (1974)

Teichmüller, M.: Durand, B.: Fluorescence microscopical rank studies on liptinites and vitrinites in peat and coals, and comparison with the results of the Rock-Eval pyrolysis. Int. J. Coal Geol. **2,** 197–230 (1983)

Thode, H. G.: Sulfur isotope ratios in petroleum research and exploration: Williston Basin. Am. Assoc. Pet. Geol. Bull. **65,** 1527–1535 (1981)

Thompson, K. F. M.: Light hydrocarbons in subsurface sediments. Geochim. Cosmochim. Acta **43,** 657–672 (1979)

Thompson, Th.: Plate tectonics in oil and gas exploration of continental margins. Am. Assoc. Pet. Geol. Bull. **60,** 1463–1501 (1976)

Tissot, B.: Premières données sur les mécanismes et la cinétique de la formation du pétrole dans les sédiments. Simulation d'un schéma réactionnel sur ordinateur. Rev. Inst. Fr. Pét. **24,** 470–501 (1969)

Tissot, B.: Effect on prolific petroleum source rocks and major coal deposits caused by sealevel changes. Nature (London) **277,** 462–465 (1979)

Tissot, B., Espitalié, J.: L'évolution thermique de la matière organique des sédiments: Applications d'une simulation mathématique. Rev. Inst. Fr. Pét. **30,** 743–777 (1975)

Tissot, B., Pelet, R.: Nouvelles données sur les mécanismes de genèse et de migration du pétrole. Simulation mathématique et application à la prospection. 8th World Pet. Cong. Proc. **2,** 35–46 (1971)

Tissot, B., Durand, B., Espitalié, J., Combaz, A.: Influence of nature and diagenesis of organic matter in formation of petroleum. Am. Assoc. Petr. Geol. Bull. **58**, 3, 499–506 (1974)

Tissot, B., Espitalié, J., Deroo, G., Tempere, C., Jonathan, D.: Origine et migration des hydrocarbures dans le Sahara oriental (Algérie). Advances in Organic Geochemistry 1973. Tissot, B., Bienner, F. (eds.). Paris: Technip, 1974, pp. 315–334

Tissot, B., Pelet, R., Roucaché, J., Combaz, A.: Utilisation des alcanes commes fossiles géochimiques indicateurs des environnements géologiques. In: Advances in Organic Geochemistry, 1975. Campos, R., Goni, J. (eds.). Madrid, 1977, pp. 117–154

Tissot, B., Deroo, G., Hood, A.: Geochemical study of the Uinta Basin: formation of petroleum from the Green River formation. Geochim. Cosmochim. Acta **42**, 1469–1485 (1978)

Tissot, B., Demaison, G., Masson, P., Delteil, J. R., Combaz, A.: Paleoenvironment and petroleum potential of Middle Cretaceous black shales in Atlantic Basins. Am. Assoc. Pet. Geol. Bull. **64**, 2051–2063 (1980)

Tissot, B. P., Deroo, G., Herbin, J. P.: Organic matter in Cretaceous sediments of the North Atlantic: contribution to sedimentology and paleogeography. In: Deep drilling results in the Atlantic Ocean: continental margins and paleoenvironment. Maurice Ewing Series 3, American Geophysical Union, Washington, pp. 362–401 (1979)

Tissot, B. P., Bard, J. F., Espitalié, J.: Principal factors controlling the timing of petroleum generation. In: Facts and principles of world petroleum occurrence. Miall, A. D. (ed.). Can. Soc. Pet. Geol. Mem. **6**, 143–152 (1980)

Ungerer, P., Bessis, F., Chenet, P. Y., Ngokwey, J. M., Nogaret, E., Perrin J. F.: Geological deterministic models and oil exploration: principles and practical examples. Am. Assoc. Pet. Geol. Bull. **67**, 185 (abstract only) (1983)

Uspenskaya, N. Y., Tabasaranski, Z. A.: Petroleum provinces and regions of the USSR (in Russian). Moscow, Nedra (1966)

Uspenskaya, N. Y., Tauson, N. M.: Petroleum provinces and regions of the foreign countries (in Russian). Moscow, Nedra (1972)

Vail, P. R., Mitchum, R. M. Jr., Thompson S.: Seismic stratigraphy and global changes of sea level, part 4: Global cycles of relative changes of sea level: Am. Assoc. Pet. Geol. Mem. **26**, 83–97 (1978)

Vandenbroucke, M., Durand, B.: Detecting migration phenomena in a geological series by means of C_1–C_{35} hydrocarbon amounts and distributions. In: Advances in Organic Geochemistry 1981. Bjorøy, M. et al. (eds.). J. Wiley, Chichester, 1983, pp. 147–155

Vassoevich, N. B., Korchagina, Yu. I., Lopatin, N. V., Chernyshev, V. V.: Principal phase of oil formation. Moscow. Univ. Vestnik **6**, 3–27, 1969 (in Russian); Engl. transl. Int. Geol. Rev. **12**, 1276–1296 (1970)

Waples, D. W.: Time and temperature in petroleum formation: application of Lopatin's method to petroleum exploration. Am. Assoc. Pet. Geol. Bull. **64**, 916–926 (1980)

Waples, D. W., Haug, P., Welte, D. H.: Occurrence of a regular C_{25} isoprenoid hydrocarbon in Tertiary sediments representing a lagoonal-type, saline environment. Geochim. Cosmochim. Acta **38**, 381–387 (1974)

Weber, K. J., Daukoru, E.: Petroleum Geology of the Niger Delta. 9th World Pet. Cong. Proc. **2**, 209–222 (1976)

Weitkamp, A. W., Gutberlet, L. C.: Application of a micro-retort to problems in shale pyrolysis. Am. Chem. Soc. Div. Pet. Chem. Preprints 13, 2, F71–F85 (1968)

Welte, D. H.: Petroleum exploration and organic geochemistry. J. Geochem. Explor. **1**, 117–136 (1972)

Welte, D. H.: Information on source rocks of the Arabian-Persian Gulf Region. Unpublished information, KFA-Jülich (1982)

Welte, D. H.: Neue Wege in der Kohlenwasserstoffexploration. Erdöl u. Kohle, Erdgas, Petrochem. **35,** 503–508 (1982)

Welte, D. H., Yükler, A.: Evolution of sedimentary basins from the standpoint of petroleum origin and accumulation — an approach for a quantitative basin study. Organic Geochemistry **2,** 1–8 (1980)

Welte, D. H., Yükler, A.: Petroleum origin and accumulation in basin evolution — a quantitative model. Am. Assoc. Pet. Geol. Bull. **65,** 1387–1396 (1981)

Welte, D. H., Waples, D. W.: Über die Bevorzugung geradzahliger n-Alkane in Sedimentgesteinen. Naturwissenschaften **60,** 516–517 (1973)

Welte, D. H., Hagemann, H. W., Hollerbach, A., Leythaeuser, D., Stahl, W.: Correlation between petroleum and source rock. 9th World Pet. Cong. Proc. **2,** 179–191 (1975)

Welte, D. H., Yükler, M. A., Radke, M., Leythaeuser, D.: Application of organic geochemistry and quantitative basin analysis to petroleum exploration. In: Origin and Chemistry of Petroleum. Atkinson, G., Zuckerman, J. J. (eds.). Pergamon Press, 1981, pp. 67–88

Welte, D. H., Kratochvil, H., Rullkötter, J., Ladwein, H., Schaefer, R. G.: Organic geochemistry of crude oils from the Vienna Basin and an assessment of their origin. Chem. Geol. **35,** 33–68 (1982)

Williams, J. A.: Characterization of Oil Types in Williston Basin. Am. Assoc. Pet. Geol. Bull. **58,** 1243–1252 (1974)

Zhou Guangjia: Property, maturation and evolution of organic matter in the source rock of continental origin. Int. Meeting on Petroleum Geology, Peking (1980)

Subject Index

abietic acid 37, 116, 117
abnormal pressure 301, 302
Abu Dhabi crude oil 614
accumulation of petroleum 293, 294, 341, 345, 349, 351–356, 358–365
acids
  in crude oil 121, 401, 403, 431, 434, 437
  fatty, see fatty acids
  in organisms 34–36, 45–49
  in sediments 107–110, 121, 125, 431, 433
  unsaturated 35, 47, 49
acritarch 14–16, 23, 136
activation energy 221, 540, 584, 585, 587–592
adamantane 390, 392
Aden Gulf 597, 600
adrenosterone 124
adsorption 307, 351
  capacity of a source rock 497
  of organic matter 58–60
aerobic, oxic 76–78, 92, 201, 464, 663–665
Africa, North 152, 153, 161, 162, 166, 168, 170, 180, 221–227
Agbada formation, Nigeria 617, 618, 620
age 223–227, 454–457
Ahnet-Mouydir basin, Algeria 221, 224, 227, 644
Akata shales, Nigeria 616, 618, 620
alanine 32
Alaska 563–565, 645, 647, 655, 658, 660
Albert shales, Canada 258
Alberta, see Canada
Aleksinac shales, Jugoslavia 142, 263
algae 16, 17, 32, 33, 36, 47–53, 105, 122, 151, 230, 237, 256, 426, 428, 430, 442, 502, 504, 505
  blue green 4, 6, 14, 15, 18, 21, 50
  brown 33, 47
  green 6, 14, 15, 18, 47

  red 6, 47
  as a source of hydrocarbons 47–49, 105, 122, 426–431
  as a source of kerogen 151–153, 157, 256, 441, 442
  unicellular 11
algal mat 142
Algeria, see Africa, North, and also Sahara
alginic acid 33
alginite 153, 157, 159, 504
aliphatic chains in kerogen 147, 148, 151, 153, 154, 250, 252, 258
alkanes
  branched 110–116, 182, 183, 385–389, 464, 466
  cyclic, see cycloalkanes
  normal, see n-alkanes
  (paraffins) in crude oil 315, 316, 382–384, 411, 416–419
alkenes (olefins) 48, 261, 385, 389, 390
alkylbenzene, alkylnaphthalene, alkylphenanthrene 193, 393–395, 412, 534–536
allochthonous organic matter 55, 230
alteration of petroleum 422, 423, 459–469, 474, 478
Amadeus basin, Australia 227
Amazon River, delta, fan 13, 57, 84, 441
amino acids
  in organisms 31, 32
  in sediments 58, 59, 76, 80, 83, 91
ammonia 27, 86, 87
amorphous organic matter, kerogen 158, 498, 503–505
Amposta crude oil, Spain 106, 107, 433
amyrin 120, 121, 123
anaerobic, anoxic 76–80, 92, 123, 201, 442, 443, 465, 478, 654–656, 663–665
analytical methods
  for chemical analysis of extractable bitumen 378, 379, 526–528

for chemical analysis of kerogen 140–147, 508, 509, 523–525
for crude oil characterization 375–379
for humic material 82–84
for light hydrocarbons 527–530
for optical examination of kerogen 133–139, 498–507, 515–520
for organic carbon 495–497
for source rock characterization by pyrolysis 509–513, 520–523
for vitrinite reflectance 505–507, 516–519
anchimetamorphism 72
androsterone 124
angiosperms 18–20, 43, 104
anhydrite, see evaporite
animals 41, 121
anisotropy of vitrinite 166, 242, 245
anoxic, see anaerobic
anteisoalkanes (3-methylalkanes) 47, 110, 386, 388, 389, 428, 430, 431
anthracene 393
anthracite 71, 72, 141, 145, 229, 235, 237, 238, 241
anticlinal structures, anticlines 352, 360–362, 611, 621, 647
API gravity 462, 471, 475–479
API research projects 6 and 48 376, 377
Appalachian basin 227, 467
aquathermal pressure, see water expansion
Aquitaine Basin, France 103, 104, 106, 117, 180–182, 184, 205, 206, 210–212, 348, 396, 420–422, 441, 447, 531, 533, 552, 566, 567, 603
arborinol, arborinone, isoarborinol 95, 99, 117, 121, 431
archaebacteria 114–116, 129, 130, 408, 431
Arctic Islands, of Canada 158, 423
argillaceous sediments 301
argon in natural gas 199, 207
aromatic
  hydrocarbons
    in crude oil 379–381, 383, 384, 393–397, 406–408, 411, 412, 416–418, 441–443, 464, 465, 472
    generation of 183, 186, 187, 193, 215, 441–443
    in sediments 126, 127
  sheets
    in asphaltenes 404, 405, 408

in kerogen 142–145, 148, 166, 172
steroids 119, 120, 123, 125, 186, 187, 396, 397, 435, 436, 537–540, 556, 557
triterpenoids 118–120, 123–125, 186, 187, 396, 397
aromatic-asphaltic crude oils 418, 419, 421–423, 446, 474, 475
aromatic-intermediate crude oils 418, 419, 421, 422, 441, 446, 614
aromatic-naphthenic crude oils 418, 419, 421, 423, 474, 475
aromatic/saturated hydrocarbon ratio 187
aromaticity 146, 237–239
aromatics
  as correlation parameter 555–557, 560, 566, 567
  as maturation parameter 435, 436, 534–536, 538–539
aromatization 237, 537–539
Arrhenius relation, formula, plot 584, 587
aspartic acid 32
asphalt 471, see also tar sands
  mat, see tar mat
  plug or seal 482
Asphalt Ridge, Utah 475
asphaltenes
  in alteration of crude oil 407–408, 423, 461–463, 465, 478, 479
  in crude oil 380, 381, 403–408, 410, 411, 413, 416, 418–423
  elemental composition 150, 176, 404, 408
  generation 175, 176, 189, 196, 217, 408
  in heavy oils 404, 406, 407, 421–423, 472, 473, 475, 476, 478, 479
  in migration 150, 306–309, 332, 333, 408, 413
  in source rock bitumen 175, 176, 189, 404
  metals in 410
  molecular weight 405, 406
  pyrolysis 407, 551
  separation of 378, 379, 404
  as a source of fossil molecules 130, 407, 408, 551
  structure of 150, 404–407
associations, natural 50, 51
Athabasca 347, 354, 422, 423, 465, 471–478, 480–482

Subject Index

Atlantic Ocean 25, 26, 58, 102, 118, 123, 146, 155–157, 443, 445, 600, 646, 655, 656, 662–665
atmosphere 4–7
autochthonous organic matter 55, 230
Australia, western 600, 645, 655
Autun boghead, France 135, 151, 256, 257, 261, 262
azaarenes 402
Azov-Kuban basin, USSR 178, 219

bacteria 4, 14, 16–18, 23, 31, 40, 45, 46, 50–53, 74–80, 82, 90–92, 108, 110, 114–116, 118, 124, 129, 130, 201, 203, 204, 229, 233, 234, 464–467, 478
   aerobic 21, 76, 77, 90, 464
   anaerobic 76, 77–80, 465, 478
   autotrophic 11
   degradation of oil by 463–467, 478, 620
   generation of gas by 77, 79, 80, 89, 90, 201–204
   green 6
   heterotrophic 44
   membranes of 114–116, 118, 130, 403
   methanogenic 77, 79, 80, 201
   red 6
   in reservoirs 463–466, 478
   as a source of fossil molecules 108, 110, 114–116, 118, 129, 130, 425, 428–430, 442
   sulfate-reducing 77–79, 201, 204, 446, 464, 465, 478
bacteriohopane tetrol, tetrahydroxybacteriohopane 125, 403
Baikal lake, USSR 597
Bakken shale, in Williston basin 321, 339
Baltic Sea 102
bark of plants 45, 49
barrier, effect of, during migration 352, 361
base, of a crude oil 416
basin
   of deposition 56, 57, 69
   modeling 572, 576–581, 625–627
   study 571–573, 575
basins, hydrodynamics of 348
Baxterville crude oil, gulf Coast 410
Bazhenov formation, W. Siberia 645, 654
Beaufort-Mackenzie basin, delta 158, 646, 647

Bell Creek oil field, Montana 363, 364
Bellshill, Lake oil field, Alberta 478, 479
Bemolanga heavy oil, Malagasy 421
Benin formation, Nigeria 618
benthos 53, 75, 92
benzanthracene 393
benzene 309, 312, 314, 315, 317, 393–395, 465
benzofluorene 393
benzopyrene 126, 127, 315
benzoquinoline 401, 402
benzothiophene, dibenzothiophene 395–397, 399, 400, 421, 446, 449, 472–474, 478, 555, 556, 566, 567
Bering Sea 80, 81, 85, 87
Bibi-Eibat crude oil, USSR 403
bile acids 40, 99, 120, 121, 125
bimacerite 499
biodegradation of crude oil 459, 463–467, 473, 474, 478–480, 614, 620
biodegraded oils, correlation of 407, 551, 560, 561
biogenic gas, biogenic methane 79, 80, 89, 90, 199–204, 209–211
biological markers, see geochemical fossils
biomass, composition of 31–53
biopolymers 75, 90, 91
biosphere, evolution of 14–19
bi-phytane 116
bitumen, extractable organic matter 14, 96, 97, 131, 132, 140, 176–180, 193–196, 248–251, 308, 317, 318, 330–332, 471, 495, 513, 514, 525, 526, 549–551
   from coal 248–251
   in experimental evolution 193–196
   generation of 176–180
   indigenous 551, 561
   ratio 177–180, 219, 223, 526
bituminous coal 71, 229, 234, 235, 237, 238
Black sea 10–12, 31, 51, 53, 56, 57, 78, 87, 88, 92, 102, 118, 447, 507
black shales 258, 264, 265, 663, 664
boghead coals 134, 135, 141, 151, 230, 256–258
Bohai Gulf basin, P. R. of China 624, 647
bonds
   breaking in kerogen degradation 169, 174, 175, 189–191, 584–586, 591–592
   type, distribution in kerogen 151, 153, 154, 591

Boscan Crude oil, Venezuela 129, 410, 474, 475
Botryococcus 48, 134, 135, 151, 157, 256, 506
Bouxwiller shales, France 108, 118, 121, 122
Bowen-Surat, basin, Australia 389
branched alkanes, see alkanes, branched
Brazil 441, 445
British Columbia 27, 80, 81, 84, 85
brown coal, lignite 70, 71, 141, 229, 231, 233–235, 241, 243, 250
bubble point 354
bubbles of hydrocarbons 310
buoyancy 341–347, 351
Bureau of Mines, analysis of crude oil 377, 378
Burgan oil field, Kuwait 464, 610
burial, burial depth 162, 163, 165–168, 174–176, 178–187, 191, 212, 215–217, 576, 585, 587, 593, 595, 613, 652–654
Burma 445, 647
butane 79, 199, 314, 319, 529

calibration of numerical models 588, 593
California 103, 179, 180, 184, 222, 223, 335, 409, 421, 452, 453, 464, 477, 647, 649, 657, 658
  Gulf of 61, 597
  Southern, sea, shelf, sediments 61, 76, 77, 90, 95, 96, 98, 101, 108, 224
calorific value 234
campesterol 40, 41
Canada, Western Canada Basin, Alberta 152, 155, 160, 180, 181, 203, 204, 213, 222, 224, 225, 320, 347, 354, 355, 364, 365, 391, 396, 409, 420–422, 440, 441, 449, 452, 457, 460–462, 465, 466, 471, 473–478, 481, 534, 556, 628, 629, 643, 645, 649, 652–654, 657, 658
cannel coals 230
capillary pressure 310, 337, 341–347, 351
cap-rock, see seal
carbazole, carbazole derivatives 127, 401, 402, 550
carbohydrates 30–33, 45, 46, 70, 74, 76, 83, 94, 233
carbon: if not stated otherwise refers to organic carbon
carbon
  annual production 3, 9, 25, 29

  budget and cycle 7, 9–12
  content in ancient sediments 97
  content in coal 234–236
  content in Recent sediments 95, 96
  content (Corg) in source rocks 97, 495–497, 612, 618, 622, 625, 631, 655, 657, 663
  dioxide
    generation of 163, 173–175, 199, 202–204, 207, 209, 214, 216, 246, 247
    from kerogen pyrolysis 219, 220, 510–512
    oxygen index 510–512
    in primary migration 316
  isotopes, see isotopes
  preference index, CPI, odd-even predominance 101, 103, 182, 184, 185, 249–251, 433, 434, 531–533
  ratio C_R/C_T 513
carbonaceous shales 630
carbonate
  oil shale 255
  sequence 433, 455–449
Carbonate Triangle, Alberta 471, 480
carbonates 79, 106, 193, 301, 305, 307, 312
  as reservoirs 353, 358, 359, 365, 614
  as source rocks 305, 338, 339, 445, 446, 497, 609, 612, 613, 654, 656
carbonization
  of coal, see coalification
  of palynomorphs 515
carbonyl, carboxyl groups in kerogen 143, 148–151, 153–154, 167, 170, 174
carboxylic acids 311, 401, 403
β-carotane, β-carotene, perhydro-β-carotene 41, 126, 193, 428, 430
carotenoids 41, 47, 126, 426
carrier bed, carrier rock 293, 294, 341–344, 346, 350
carveol, carvone 36
case histories 609–640
Caspian Sea, basin 51, 53, 647
catagenesis 11, 70, 71, 141, 161, 163, 166–170, 175, 181, 189, 196, 200, 201, 204, 205, 209–212, 215–217, 450, 543, 544, 546
  end of 72
catalysts, catalytic effect 107, 192, 193, 197, 329
Caucasus, USSR 207, 647, 657

cellulose 33, 44–46, 232, 233
cementation 214, 360
Cerro Negro heavy oil, Venezuela 475
Challenger Knoll 326
characterization of potential source rocks 540–546
Chattanooga Shales 157
chemosynthesis 11, 12
China, People's Republik of 259, 262, 445, 624–628, 647, 654, 658
chiral centers 538
chitin 32, 33
chlorophyll 4, 37, 42, 112–114, 127, 128, 192, 410
chloroplasts 4
cholanic acid 117, 401
cholesterol, cholestane, cholestene 40–42, 47, 99, 117, 390, 426, 562
cholic acid 40
chromatographic adsorption during migration 339
chromatography
  gas 378, 379, 527, 528
  liquid 378, 379
  thin layer 378, 379
chrysene 126, 127, 393
citronellal, citronellol 36
classification
  of coals 234, 235
  of crude oils 415–423
  of kerogens 151–158
clastic sequence 445, 446, 448
clay
  dehydration 328, 329
  minerals 72, 107, 192, 193, 196, 197, 307, 308, 320, 328 329
climate 231, 443, 656
Clostridia 79
closure of trap 357
cloud point 261, 384
clusters of aromatic sheets in kerogen 142, 144, 166, 172
coal
  association with crude oil 252, 445, 622
  classification of, see classification
  elemental composition 141, 235–237
  formation 229–241
  humic 229, 230, 241
  liquids 404, 407, 471, 472
  macerals 229, 230, 234, 237, 238, 241–244

  petrography 241–245
  ranks 234–239, 241, 243–245
  sapropelic 229, 230
  as a source of gas 206, 211–212, 221, 245–248, 630, 631
  as a source of liquid hydrocarbons 248–252, 622
  structure, chemical 237, 238
coal-bearing strata 231
coalification, carbonization of coal 234–241, 245, 246, 592
coaly kerogen 498, 499
coastal environment 231
coccolithophorids 15, 16, 51
Cold Lake heavy oil, Alberta 472, 478–480, 482
collinite 244, 504
colloidal solution 311, 312, 325
color, of spores and pollen 167, 515, 516, 543
combustion, in situ 482, 483
compaction 299, 301–305, 327, 329, 335, 336
  delayed 302
  waters 325, 335, 336
computer simulation in basin studies 572, 577–581, 583, 585–588, 605–607
condensate 199, 217, 338, 463, 550
  early-mature 338
condensation
  polycondensation 81–84, 175, 234, 237, 245
  retrograde 468, 623
conifers 18, 19, 234
coniferyl alcohol 37, 49; see also lignin
connate water 353
conodonts 16
Conroe crude oil, Texas 392
continental
  environment, of deposition 445
  margin, rise, shelf, slope 24, 25, 31, 59, 61, 154, 441, 442, 645, 646, 654, 662
    active 447
    passive, stable 443, 599, 645, 647
  organic matter, see terrestrial organic matter
  plants, see terrestrial plants
conversion factor, organic carbon to organic matter 496
Cooper, Basin, Australia 389
coorongite 88, 141, 151, 256, 257

copepods 22, 23, 45, 48, 49
core samples 574
correlation
   index, CI 377, 378
   oil-oil, oil-source rock 330, 333, 339, 407, 424, 459, 548–570, 614, 623
coumaryl alcohol 44
CPI, see carbon preference index
C_R/C_T 513, 522, 523
cracking 175, 180, 181, 190, 193, 200, 201, 205, 206, 217, 218, 245, 388, 389, 460, 586
cratonic, pericratonic 642, 653, 660
critical height, of oil or gas column 343, 344, 349, 350
crude oil, see also oil, see also petroleum
   classification 415–423
   composition 330, 332, 335, 375–414, 472
   correlation, see correlation
cuticles 43, 50, 229, 237
cuticular waxes 102
cutin, cutinite 42, 43, 49, 139, 504, 520
cuttings 573, 574
   gas 527, 528
Cyanophyceae, blue-green algae 21, 40
cyclic sedimentation 231
cycloalkanes, naphthenes
   C_{27+} cyclic molecules 561–563
   in bitumen 187, 533, 534
   in crude oil 382–384, 390–393, 416–422, 464–466
   formation of 183
cycloalkylaromatics, see naphthenoaromatics
cyclobutane 392
cycloheptane 392
cyclohexane, methylcyclohexane 95, 98, 390, 392
cycloparaffins, see cycloalkanes
cyclopentane, methylcyclopentane 95, 98, 390, 392, 530

Darcy's law 301, 605, 606
Darius crude oil, Iran
deasphalting 459, 461–463, 467, see also asphaltene separation
decalin 390
decarboxylation 106, 107, 114, 192, 193, 204, 245, 426, 433
Deep Basin of W. Canada 628–639

Deep Sea Drilling Project, DSDP 95, 96, 116, 123, 155, 156, 197, 202, 530, 662, 663
deep sea sediments, hydrocarbons in 96, 326, 333
degradation of curde oil 391, 422, 423, 459, 463–467, 470, 478–480, 614, 620
degraded oils 391, 418–422, 470, 474
degraded oils, correlation of, see biodegraded oils
dehydration, of clay minerals 328, 329
dehydroabietic acid 117
delta areas 327, 355, 362, 605, 615–624, 646, 647, 660
density
   of crude oil 342–344, 377, 411, 413, see also API gravity and specific gravity
   change with depth and age 451–457
   of heavy oil and tar 471, 475, 476, 478, 479
   of shale oil 261–263, 404, 471, 472
   of water 303, 304
depth 85, 178–180, 215, 222, 223, 327, 450–454, 460, see also burial
   of oil and gas fields 450–452
Desulfovibrio 77, 78, 446
diagenesis 11, 69, 70, 71, 75, 90, 94, 141, 174, 181, 200, 201, 203, 215, 216, 305, 360, 450, 543, 544, 546
   early 70
   end of 71
   of kerogen 162, 170, 174
diameters
   of molecules 206, 306, 309
   of pores, see pore diameter
diasterane, diasterene 121
diatoms 15, 16, 21–23, 27, 32, 34, 45–47, 50, 51
dibenzofuran 401, 403
dibenzothiophene, see benzothiophene
diffusion
   losses by 339, 340, 467, 468
   in migration 309, 310, 318–320, 351, 637, 638
dihydrophytol 113, 114
Dinoflagellates 15, 16, 21–23, 27, 45, 47, 51
dinosterol 122, 125
diploptene 39
dismigration 293
disproportionation 460, 461

Subject Index

distances of migration 347, 354, 355, 479, 480, 614, 620
distillate 376
distillation 191, 376, 377, 415
diterpenes, diterpanes, diterpenoids 37, 116, 117, 250, 426
dolomitization 305
Domanik formation, USSR 643
Douala Basin, Cameroon 88, 141, 148, 149, 152, 154, 161, 168, 178–180, 182, 183, 219, 220, 348, 451, 511, 512, 517, 525, 601–604
drainage area 353, 354, 481, 615
dry gas formation 167, 175, 200, 205, 217, 221

East Shetland basin, North Sea 539, 540
East Texas 362, 464
effusion 351
E_h, oxidation-reduction, redox potential 59, 70, 77, 232, 233
electron
　diffraction 139, 142–145, 509
　spin resonance, ESR, 508, 523–525
elemental composition
　of biomass 51
　of coal 235–237
　of kerogen 140, 141, 151, 165, 170, 171, 508, 511, 523, 524
　of resins and asphaltenes 404
ellagic acid 45
Elmworth gas field, W. Canada 629–638
Emba, South, USSR 447
Emeraude oil field, Congo 421, 474, 560
environment of deposition 11, 12, 21–30, 106, 114, 153, 424, 426–433, 440–449
environmental studies 437
enzymes 31, 32, 76
epicontinental sea 443, 651, 654
epimetamorphism 72
Equisetum 104
ergostane, ergosterol 40, 41, 46, 47, 95, 562
Ermelo oil shales, South Africa 257, 263, 264
essential oils 36, 49, 110, 111
ester, ester bonds 84, 148, 154, 172
estradiol, estrone 124
ethane 79, 199, 202, 314, 319, 529, 636
ether bonds 148, 154, 404
eubacteria 114

eukaryotes 15, 114, 116
euphotic zone 10, 25, 26, 59
European, North-west European basin 206, 212, 251
euxinic 611, 612
evaporation of hydrocarbons 465
evaporite 106, 305, 433, 447, 611, 614, 644, 647
even predominance, of n-alkanes 106–108, 432, 433, 533, 570
evolution
　of the biosphere 14–20
　path
　　of asphaltenes 408
　　of kerogen 151–158, 218–221, 511, 512
evolutionary state of plants 230
exinite 153 157, 237, see also liptinite
experimental
　evolution
　　of kerogen 169–174, 193–196, 590
　　of pure compounds 192, 193
　　of sediments 193
　generation of hydrocarbons 192–196
exploration
　philosophy 572
　targets 571
　well 573
expulsion of oil 252, 336
exsudatinite 503
extract, see bitumen
extra-heavy oils, see heavy oils

facies
　characteristics, of crude oil and bitumen 568–570
　organic, in source-rocks 572, 662, see also kerogen type and palynofacies
Falkland Plateau 656
farnesol 37, 114
Fasken crude oil, Texas 384
Fastnet basin, Ireland 529
fats 34, 35
fatty acids 34, 35, 46–49, 79–81, 100, 106–108, 110, 192, 193, 403, 433
　iso- and anteiso- 46, 47
　unsaturated 49
faults 355, 360–363, 617, 620, 623, 624
fermentation 79
ferns 124, 524
fichtelite 116, 117, 390

fingerprint pattern 549
Fischer assay 260
fish 53
Florida Bay 108
fluid
  flow 301, 304, 309, 310
    models of 348, 578, 605
  pressure, see pressure
fluids, monophasic, polyphasic 310, 605
fluoranthene 126, 127, 393
fluorene 393, 394
fluorenone 401, 403
fluorescence 134, 136, 137, 139, 243, 498,
  499, 502–504, 506, 507, 519, 520, 543
foraminifera 16, 17, 50
formation water, see water
fossilization of organic matter 11–13, 21
fractions of crude oil, key fractions 377,
  378, 415, 416
framboids 136, 138
free radicals 162, 163, 193, 523–525
friedelin 120, 121, 123
Frigg field, North Sea 210, 212
fringelite 129
fucosterol 41, 47
fulvic acids 82–88, 91, 94
functional
  analysis of kerogen 142
  groups, in kerogen and coal 148, 170,
    172, 173, 238, 239
fungi 16, 17, 44, 233
fungisterol 40
Fushun oil shale, P. R. of China 259, 262
fusinite 139, 245, 504

Gabon 108, 427, 430, 445, 514
gallic acid 45
Gamba crude oil, Gabon 445
gammacerane 39, 390
gas
  biogenic 79, 80, 89, 90, 200–203, 215,
    216
  cap 353, 354
  carbon isotope composition of, see
    isotope
  chromatography, see chromatography
    mass spectrometry, GC-MS 436, 437
  composition 199
  in cuttings, see cutting gas
  depth limit, -destruction, -floor 213,
    214, 218
  dissolved in formation waters 333–335
  distribution in sedimentary basins 213,
    214
  formation 180, 199–221, 245–248,
    328, 628–638
  inorganic origin 207, 208
  non-hydrocarbon 203, 204, 206–208
  /oil ratio, GOR 353, 354, 462, 463
  phase migration 334, 337
  reserves 649, 650, 658, 662
  source rocks 219–221, 251, 252
Gato Ridge crude oil, California 402
genetic potential 159, 218, 219, 511–513,
  589–591
geochemical fossils, biological markers
  93–130, 424–437, 527, 548–550, 554,
  555, 558, 562–566
  acyclic isoprenoids 111–116
  branched alkanes 110, 111
  diterpenoids, 116, 117
  from asphaltenes 130, 407, 551
  from kerogen 129, 130, 551
  indicators of despositional environ-
    ment 426–432
    of early diagenesis 432, 433
    of thermal maturation 433–436
  in oil-oil correlation 548–550, 554,
    555, 558
  in oil-source rock correlation 562–566
  n-alkanes 100–110
  porphyrins 128, 129
  significance 424, 425
  steroids and triterpenoids 117–126
geopolymers 70, 75, 91
geostatic pressure, lithostatic-, petrostatic-
  299, 300
geothermal
  flux, see heat flow
  history 599, 600
  gas 207
  gradient 215, 216, 223, 224, 296–298,
    303, 304, 450, 451, 460, 572, 593, 596–
    604
    ancient 601–604
geraniol, geranic acid 36, 111
Ghawar oil field, Saudi Arabia 610, 614
giant field, giant province 656–661
gibberellins 37, 38
Gippsland basin, Australia 389, 647
globules of hydrocarbons 310
Glossopteris flora 230

glucose 3, 33
glutamic acid 32
glycerides 34
glycerol 34, 35, 114, 115
  ethers 114, 116
glycine 32
Golden Lane, Mexiko 360
Gondwana 230, 232, 264, 645
graben 289, 297, 597, 600, 603
graphite 72, 201, 235, 241
graptolites 16, 136
gravity, see API and specific gravity
  segregation 462, 463
Great Plains of Canada and United States 202, 204
Greece 106, 421, 433, 447
Green River formation, shales 118, 122, 140–142, 152, 153, 161, 179, 255–259, 261, 263, 265, 420, 428, 430, 512, 584
Groningen gas field 206, 211, 212, 328, 339, 360
Grosnyi crude oil, USSR 403
growth faults 327, 355, 362, 617, 620, 624
Gulf Coast 84, 298, 301, 302, 327, 335, 355, 362, 363, 421, 441, 451–453, 646, 647, 658–660
Gulf of Mexico 57, 59, 60, 96, 98, 101, 118, 203, 326, 327, 348, 645, 655
gymnosperms 19, 20
gypsum, see evaporite

Handil oil field, Indonesia 620–623
Harmattan East crude oil, Alberta 396
Hassi Messaoud, Algeria 225, 226, 316, 327–329, 339, 385, 595, 643, 652, 653, 660
Hassi R'Mel gas field, Algeria 660
H/C, O/C diagram, see Van Krevelen diagram, see also elemental composition of kerogen
heat flow 296, 297, 596–598
heavy oils 419–423, 465, 470–483, 641, 643, 644, 646, 648, 649, 657, 660
  API gravity of 471, 475–479
hcight of oil or gas column 343–345, 637
helium in natural gas 199, 207, 208
hemin 127, 128
Hempel distillation 415
heptane 384, 386–389
  isomers 386, 388
  value 530

herbaceous kerogen 498, 503, 505
heterocompounds (N. S. O.) 399–408
  generation 189, 217
  in migration 312, 313, 333, 408
heterocycles
  in asphaltenes 404, 405
  in humic acids 82
  in kerogen 147, 148
hexane, hexane isomers 385–389, 530
hexene 385
high pressure of fluid, overpressure, abnormal pressure, excess pressure 300, 302, 311, 321, 328, 605–607, 616, 621–623, 647
high-sulfur crude oil, see sulfur
high-wax crude oil, see wax
higher plants, see terrestrial higher plants
histogram of vitrinite reflectance 244, 245, 516, 519
Hodonin crude oil, Czechoslovakia 392
hopane 40, 46, 50, 117, 118, 121, 122, 124, 125, 426, 429, 430, 434–436, see also bacteriohopane
hopene 117, 124, 390
hormons 125
humic
  acids 82–88, 91, 94, 158, 240, 442, 503
  coal, see coal
  kerogen, organic matter 134–136, 158, 233, 240, 243, 498, 499, 503–505
humification 229, 234, 237
humin 74, 82, 85, 86, 88, 91, 94, 131
huminite 230, 234, 237, 240, 243
hydrobarbon
  composition, change with depth 180–188, 200, 205, 215, 250, 251, 451
  formation 93, 94, 176–188, 201–206, 215–218, 248–252
  gas, see gas, see also light hydrocarbons
  identification in crude oil 376–379
  /organic carbon ratio 96, 97, see also hydrocarbon ratio
  phase migration 320–323, 334, 336–339
  ratio 177–180, 223, 526
  ring analysis of saturated 186, 187, 391, 392
  type distribution 183, 188, 382–384
hydrocarbons
  in ancient sediments 91, 94, 95, 97, 98, see also geochemical fossils
  in coal 248–251

in crude oil 382–397
in cuttings 527–530
in deep oceanic sediments 95, 96, 202, 326, 333
in heavy oils 472, 475
isotopic ratios 190–192, 200–203, 205, 209–211
in natural gas 199, 201–205, 209–211
in organisms 36–39, 41, 46–49, 110–112, 114–115, 118, 425
in Recent sediments 80, 81, 94–96, 98; see also geochemical fossils
in source rocks 97, 176–188, 527–540
hydrodynamics, hydrodynamic fluid flow, hydrodynamic gradient 341, 344–349
hydrogen bonds in asphaltene-resin interaction 406
hydrogen
  content
    in asphaltenes and resins 404
    in coal 234, 236, 240
    in kerogen 140, 141, 151–153, 157, 158, 162, 163, 166
  index, of kerogen 510–512
  origin in natural gas 206, 207
  sulfide 78, 79, 199, 204–207, 448, 453, 455
  transfer 461
hydrolysis 32, 33
hydrolyzable organic compounds in sediments 80, 85, 88, 233
hydrophilic, hydrophobic 311
hydrostatic 299, 300, 341, 342, 345, 346
hydrotroilite 79
hydroxy acids 42, 43, 45, 49
hydrous pyrolysis 196

igneous intrusion 197
illite 196, 328, 329
Illizi Basin, Algeria 221, 224, 348, 349, 556, 557, 595, 596, 603, 652, 653,
immature
  crude oil 441, 450, 477
  stage of evolution 163, 215, 216, 495, 517, 518, 521, 533, 545
indane 396
indole 402
Indonesia 108, 109, 224, 389, 409, 419–421, 440, 441, 444, 445, 647
inertinite 139, 155, 237, 240, 242–244, 498–500, 503, 504

infrared spectroscopy 143, 146, 156, 167, 171, 173, 220, 508
injection pressure 342
in situ combustion, see combustion
insolubilization 85–89, 91
interfacial tension 310, 342–344
internal surface 306, 307
intrusion 146, 197
Irati shale, Brazil 257, 258, 262, 264
iron
  iron sulfides 6–8, 70, 79, 127, 134, 136, 138, 447
  oxides 72
isoalkanes (2-methylalkanes) 47, 109, 110, 385, 388, 389, 428, 430
isoarborinol, see arborinol
isomerization 434–436, 537–540
isomers in crude oil and sediments 388, 434, 537–540
isooctane 385
isoprene 34, 36, 111
isoprenoid acids 403
isoprenoids 4, 107, 111–116, 129, 130, 182, 185, 186, 250, 332, 389, 425, 426, 466, 533
  in biodegradation 466
isotope
  fractionation during catagenesis and metagenesis 189–192
  ratio
    of bitumen and crude oil 189, 350, 351, 548, 550, 551, 553, 554
    of gas, of methane 89, 90, 190–192, 199–203, 205, 209–212
    of kerogen 89, 90, 189, 201, 508, 548, 551
    of organic matter in young sediments 89, 90
isotopes
  of carbon 51, 61, 89, 90, 189–192, 199–203, 205, 209–212, 350, 351, 463, 508, 548, 550, 551
  of hydrogen 89, 90, 190, 191, 200, 548, 550
  of nitrogen 206
  of sulfur 548, 550, 553, 554
Italy, North, see Po Valley

Jobo heavy oil, Venezuela 475
John Creek crude oil, Kansas 386

kaolinite 329
kerogen
  aliphatic chains in 147, 148, 151, 153, 154, 156
  amorphous 136, 158, 498, 499, 503–505
  aromatic sheets in 142–144, 147, 148
  catagenesis 163, 165, 170, 175, 216, 217
  classification 151–156, 498
  color 167
  concentrates 132, 135–137
  correlation to crude oil 548, 551
  definition 131, 132
  diagenesis 162, 165, 170, 174, 216
  electron diffraction of 142–145
  elemental composition of 140, 141, 151, 165, 166, 170, 171
  evolution paths 151–155, 216
  experimental evolution of 169–174
  formation of 59, 88, 90, 91
  functional analysis of 142, 154
    groups in 142, 147–149, 154, 172
  heterocycles in 147, 148
  hydrocarbon yield of different types 181–183, 219, 220, 440–442
  infrared spectroscopy of 143, 146, 155, 156, 167, 171–173
  internal surface area 307
  isolation 132, 133
  matrix as a trap for biological molecules 116, 129, 130, 186, 195, 551
  maturation of, see kerogen diagenesis, catagenesis, metagenesis
  metagenesis 165, 166, 172, 176, 216–218
  microscopic constituents of 133–139, 498–505
  nitrogen in 141, 149
  nuclei 147, 148, 151, 153, 154
  in oil shales 255–257, 262, 263
  oxidation of 140, 142
  oxygen in 141, 148, 149
  pyrolysis of 142, 170–173, 193–196, 551, see also pyrolysis and Rock Eval
  in relation to coal 157, 158, 241, 252
  in relation to humin 74, 88, 89, 131
  source of fossil molecules 129, 186, 195, 551
  structure of 147–150, 174–176, 238–239
  sulfur in 141, 150

  thermal analysis 147, 168, 169
  type 151–159, 181–183, 219, 220, 257, 426, 441–443, 445, 497–514, 517–519, 589, 591, 592
  van Krevelen diagram of 151, 152, 155, 158, 161, 257, 429, 512
  weight loss of 170, 171, 173
Kerosene shales, Australia 257, 258, 260, 262, 264
key fraction 416
kinetics 539, 542, 546, 584–587
Kirkuk oil field, Irak 360
Kivu, lake 79
Kukersite, USSR 141, 257, 258, 261, 263
Kutei basin, Indonesia 621
Kuwait crude oil 409, 421, 464

Lacq field, France 210, 212
lacustrine basin, lake 151, 231, 258, 259, 443, 445, 624–626, 628
lagoon 61, 78, 447
La Luna formation and crude oil, Venezuela 339, 410
lamination, in organic rich sediments 92, 259
lanosterol 39, 40
leaves 45, 49, 234
Leduc oil field, Canada 224, 227, 461, 643, 652, 653
level of organic metamorphism, LOM 71, 515, 542–544
Libya 365, 416, 645, 657, 658
light hydrocarbons 450, 527–530, 574
lignin 31, 32, 37, 42, 44–46, 50, 220, 233, 442, 443
lignite, see brown coal
limestone as source rock 496, see also carbonates
limonene 111
linoleic acid 35
Linyi basin, P. R. of China 624–628
lipids 31, 33, 35, 45–48, 60, 76, 91, 94, 98
liptinite 139, 237, 242, 243, 246, 252, 498, 499, 502–504, 516, 518, 519
lithification 304, 305, 334
Lloydminster heavy oil, Alberta 476, 478, 479
Logbaba wells 329, see also Douala basin
LOM, see level of organic metamorphism
Lopatin method 584, 585
Louisiana 96, 101, 160, 451, 464

Los Angeles basin, California 103, 178–180, 184, 222, 223, 335, 361, 653
lupane 38, 39, 119–121, 123
lycopane 112, 116, 126, 389
lycopene 41, 112
Lycopodium 203
lysine 32

maceral 133, 229, 230, 234, 237, 239–242
 groups 242, 499
Mackenzie delta, see Beaufort-Mackenzie basin
Magellan Basin, South America 104, 105, 419, 429, 431, 440, 441, 444, 445, 514
magmatic activity, intrusion, see intrusion
Mahakam delta, Indonesia 173, 252, 327, 338, 605, 620–624, 646, 647
Maillard reaction 83
Malossa field, Italy 209
Mannville shales, Western Canada 152, 155, 421, 422
Maracaibo petroleum province, Venezuela 657, 658
marine organic matter 7, 51, 153, 157, 159, 219, 385, 426, 429, 431, 440–443
marsh gas 79
mass
 balance 587
 fragmentogram 538, 539, 554, 555, 558, 559, 564
 spectrometry (MS) 376, 378, 379, 551, 560
mathematical model of hydrocarbon generation 585–589
maturation (thermal alteration or evolution) of crude oils, see thermal alteration
maturation (thermal evolution) of kerogen, of organic matter, of source rock 215–218, 495, 515, 541–546, 572, 575
 of source rocks, chemical indicators 523–540
  pyrolysis methods 520–523
  optical indicators 515–520
McMurray, Alberta 422
Medicine Hat gas field, W. Canada 80, 335
Mediterranean sea 57, 106, 433, 447
Melekess heavy oil, USSR 480, 481
membranes 42, 46, 114–116, 118, 125, 130, 431, 432

mercaptans 399
Mesogea 647
Messel shale, Germany 108, 118, 121, 122, 125, 141, 194, 195, 257, 258, 262, 431
meta-anthracite 71, 72, 241, 243
metagenesis 71, 72, 141, 166, 172, 175, 181, 200, 201, 205, 210, 211, 215–218, 543, 544, 546
metals
 in asphaltenes 405
 in crude oils 408–410
metamorphism 72
meteoric water 348, 463–465
methane
 biogenic, see biogenic gas
 in coal 245 248
 floor 213, 214, 218
 generation of 201–205, 207, 209–212, 246–248
 isotopic composition of, see isotope ratio of methane, see also gas
 in young sediments 79, 80, 90, 201–203, 216
Methanobacteria 79, 128
methanogenic bacteria 79, 112, 115
methanol-acetone-benzene (MAB) extract 175, 189
methylcyclohexane 95, 387, 390
methylcyclopentane 95, 392, 530
methylphenanthrene 393
 index 534–536
Mexico 360, 363, 409, 441, 447, 645
micellar solution 311, 312, 334, 335
micelle 306, 311, 312, 317
micrinite 243, 246, 504
microbial
 activity in young sediments 70, 75–80, 89, 90, 201
 alteration, see biodegradation
 generation of gas, see bacteria, generation of gas by
 organic matter 108, 109, 114–116, 153, 428–430, 443
 populations in peat 233
microfractures, microfissures 321–323, 337, 338
microreservoirs 522
microscopic methods 133–139, 498–507
Middle East 339, 354, 365, 398, 409, 420, 421, 423, 439–441, 443, 447, 451, 452,

Subject Index

455, 481, 610–615, 642, 644, 645, 647, 649, 657, 658, 661, 662
Midland coal measures, Great Britain 252
Midway Sunset crude oil, California 431
migration, see also primary migration and secondary migration
   changes in composition during 330–333
   distance of, see distance
   of petroleum, definitions 293–295
   vertical 355
Mihai field, Pannonian basin 207
mineral matrix, retention effect of 196
Mississippi delta 98, 108
mobile belts 647, 648
modeling
   geochemical basin modeling 583–608
   geological basin modeling 572, 576–581
moisture content of coal 234
Molasse Basin, Germany 209, 211, 647
molecular
   diameters of petroleum compounds, see diameter
   sieves 376
   solution 312–318, 325, 334, 335
   weight of resins and asphaltenes 403, 405, 406
monosaccharides 32
monoterpenes 36
montmorillonite, see smectite
Morichal heavy oil, Venezuela 475
Morondava basin, Malagasy 558, 559
Moscow lignites 160, 223
mosses 124, 425
Mowry shale, Wyoming 177
mud volcanoes 647
myrcene 36, 111

Nagoaka plain, Japan 326, 333
n-alkanes
   in ancient sediments and crude oils 100–110, 182, 184, 306, 309, 313–319, 384–387, 425–432, 471, 531–533
   biodegradation of 464–466
   even numbered 101, 106, 107, 385, 432, 433
   generation of 182–184
   as geochemical fossils 100–110
   in living organisms 47–50

long chain, from waxes 108, 109
   odd numbered C_{15}–C_{17} 105, 106, 426
   C_{25}–C_{33} 100–105, 194, 385, 426, 428
   in oil shales or coal 104, 249–251
   in Recent sediments 100–103, 108
nannoplankton 15, 21
naphthalene 393, 394
naphthene index 533
naphthenes, see cycloalkanes
naphthenic
   acids 311, 403
   crude oils 416–419, 421
naphthenoaromatic hydrocarbons, in crude oil 381–397
Napo formation, Colombia and Ecuador 645
natural gas, see gas
nickel, in crude oils 408–410, 474
Niger delta, Nigeria 60, 84, 103, 327, 348, 355, 362, 363, 441, 616–620, 622–624, 646, 647, 658–660
nitrate 23, 27, 28
nitrogen compounds in crude oil 402
nitrogen
   in coal 237, 247, 248
   in crude oil 402, 403, 404
   in kerogen 141, 149, 176, 206
   in natural gas 199, 204, 206, 207, 209, 211, 213
   in young sediments 86, 87
non-hydrocarbons, see N.S.O. compounds, asphaltenes and resins
   gas, see gas, non-hydrocarbons
non-marine organic matter, see terrestrial organic matter
North Sea 210–213, 297, 348, 363, 420, 421, 440, 441, 644, 654, 655, 658, 659
North Smyer crude oil, Texas 386
Norton Sound, Alaska 317
Norwegian Sea 102, 108, 655
N.S.O.-compounds 189, 403–408, 415, 416, 419, 420, 465, 467, 475
nuclear magnetic resonance (NMR) spectroscopy 146, 163
nuclei in kerogen 148, 151, 153, 154
nutrients 25, 27, 28

O/C, oxygen to carbon ratio, see elemental composition of kerogen and Van Krevelen diagram

ocean spreading, oceanic ridges 597, 600, 651, 655
  water, chemical composition of 27
oceanic basins
  geothermal flux in 597, 598, 600
  petroleum potential of 662–666
odd-even predominance (OEP), see carbon preference index
Officina basin, Venezuela 252, 445
oil, see also crude oil and petroleum
  column, height of, length of 343–346, 349
  family 548, 552, 553
  fields, distribution of depth 451, 452
  formation or generation, main stage, zone or principal phase of 163, 175, 180, 181, 217, 227, 327, 336, 517
  gravity, see API and specific gravity
  phase migration, see hydrocarbon phase migration
  reserves
    distribution of 641–660
    resources 657–662
  richness 657–659
  saturation 353, 354
oil shale 254–265, 592, 645, 654
  composition of organic matter 256–258, 262, 263
  composition of shale oil 261–263
  density 259
  environments of deposition 258, 259
  oil yield 260–263
  organic richness 255
  pyrolysis, retorting 259, 260, 590, 608
Oklahoma, compaction of Paleozoic sediments 302
oleanane 38, 39, 124
olefins, see alkenes
oleic acid 35
onocerin 39
Ontario Lake, Canada 87
optical activity, optical rotation 191, 393, 464, 553
organic carbon, see carbon
organic matter
  accumulation and preservation of 3, 9, 10, 12, 13, 55, 57, 59, 60
  composition of 131, 132
  dissolved 13, 55, 57–61
  of ocean sediments 96, 530, 662–665
  particulate 13, 55, 57–61

  production of 3, 25, see also carbon production
  quantity in sediments 96, 97
  in relation to sediment grain size 56
  in sedimentation and early diagenesis 74–92
  source of 12, 45–53
  type of, see kerogen type
organisms
  autotrophic 3, 6, 16
  composition of 45–50
  heterotrophic 5, 6, 16
Oriente Basin, Colombia, Ecuador and Peru 409, 655
Orinoco delta, Venezuela 98, 325
Orinoco heavy oil belt, Eastern Venezuela 423, 472–481, 647
Orinoco-Trinidad petroleum province 658
ostracods 16
overpressure, see high pressure
oxic, see aerobic
oxidation of methane in near surface conditions 90
oxidation
  of organic matter, or particles, or kerogen 139, 155–157, 500, 663
  of pooled oil 465, 467, 478
oxidizing conditions 232–234
oxygen
  in atmosphere and hydrosphere 7
  compounds in crude oil 401, 403
  in crude oil 403, 404
  diffusion of 55
  index of kerogen 510–513
  in kerogen 141–143, 146, 148–150, 154, 166, 172
  minimum, – vertical distribution in ocean 26, 28, 57, 59
  in reservoir waters 463
  in young sediments 76, 77, 86

Pacific Ocean 58, 76, 80, 81, 326
paleogeography of source rocks 572, 575, 655, 662–666
paleohydrodynamics 347
paleotemperatures 524, 525, 601–604
Paleozoic source rocks 225–227, 643, 644, 652, 653
palmitic acid 35, 47, 49
palynofacies 136

palynomorphs, carbonization of 515
Pannonian basin, Central Europe 184, 185, 207, 224, 225, 419, 440, 441, 451, 452, 539, 540, 597, 599, 647
Panuco-Ebano, Mexico 207
paraffin, see alkane
paraffinic crude oils 416–419, 423, 441, 445, 620, 628
  -naphthenic crude oil 416–420, 423, 441, 614
paraffinicity, 351
paralic 231, 445
Paris basin, France 109, 123, 125, 127, 133, 134, 136, 137, 140, 141, 148–150, 152–154, 162–171, 176–183, 186–188, 193–196, 219, 220, 222, 223, 256–258, 261, 262, 265, 330, 348, 420, 440, 441, 447, 511, 512, 517–519, 532, 539, 540, 594, 595, 603, 653, 660
Peace River heavy oils, Alberta 422, 472, 479, 480, 482
peat 70, 71, 229–235, 241
pectin 33
Peedee Belemnite, PDB 89
Pembina crude oil, Alberta 396
pentacyclic triterpanes, see triterpanes
pentane 388, 529
peptide bond 32, 83, 86, 149
Peridineans, see Dinoflagellates
permeability 294, 297, 301, 305, 358, 359
Permian Basin, West Texas 205, 207, 347, 398, 409, 421, 455, 549
perylene 126, 127
petroleum, see also crude oil and also oil
  accumulation 351–354, 360–365
    favorable zone of 571, 572, 575
  alteration 459–468
  composition, change with age and depth 451–457
  distribution, age and geotectonic 642–656
  formation
    depth and time 93, 180, 215, 222–227
    general scheme of 215–218
  habitat 610–640
  potential 218, 511–513, 583, 588, 589, 593–596
  prospects, locating of 571–573, 575, 593–596, 608
  reserves, resources 657–662

petrostatic pressure, see geostatic pressure
pH 70, 232
phenanthrene 393, 394, 534
phenols 49
phenylalanine 193
phosphate 23, 26–28
phospholipids 34, 42, 46 48
Phosphoria formation, Wyoming 643
photosynthesis 3, 12
physiography 443
phytadiene 114
phytane 111–114, 250, 389, 466
  predominance over pristane 106, 107, 432, 433, 568
phytanic acid 114
phytobenthos 22
phytol 4, 37, 98, 112–114, 128
phytoplankton, plankton 14, 15, 20, 23, 31, 45, 47, 51, 98, 122, 153, 426, 443, 503, 505, 649, 651, 654, 655, 662–665
Piceance Basin, Colorado 259
pigments 34, 41, 42
plankton, see phytoplankton and zooplankton
plants, higher, terrestrial, see terrestrial higher plants
platform, stable platform 599, 609–615, 643, 645, 648, 653–655
Po Valley, basin, Italy 203, 209, 211, 335
polar compounds in sediments and petroleum 129, 130, 311, 312, 330, 333, 350, 437
pollen 17, 49, 50, 57, 136, 137, 167, 234, 503–505, 515
pollution 434, 437
polyaromatic hydrocarbons, polycyclic aromatic hydrocarbons (PAH) 126, 127, 394, 560, 567
polycondensation 81–85, 234, 237, 245, 246
polyhydroxyquinone 129
polymerization 81, 90, 93, 191, 234, 237
polysaccharides 3, 32, 80, 91
polyterpenes 126
Ponca City, crude oil, Oklahoma 111, 376, 377, 386, 389, 392
Porcupine Basin, Ireland 566, 568, 569
pore
  diameter, radius 306–308, 310, 342–346, 358

water
   free and bound 307, 308, 311
   salinity 300
porosity 294, 297, 301, 302, 304, 306, 307, 358–360
   primary and secondary 305, 359, 360
porphin 4, 42, 127
porphyrins 42, 128, 129, 350, 351, 402, 410, 435, 436, 539
pour point 413
Powder River basin, Wyoming 363, 364
pressure 196, 223, 299–303, 306, 308–311, 337 see also abnormal pressure, capillary pressure, geostatic pressure, hydrostatic
Probistip oil shale, Jugoslavia 502, 506
primary migration 293, 294, 296–340
   change of composition during 330–333
   in colloidal or micellar solution 311, 312
   diffusion controlled 318–320
   gas phase 337, 338
   hydrocarbon phase 320–323
   in molecular solution 312–318
   time and depth 325–329, 335–337
principal phase of oil formation, see oil formation
pristane 49, 98, 111–114, 250, 389, 466
   isomers 434, 538, 539
   to phytane ratio 429, 434
productivity, primary 21, 23–25, 28, 29, 56, 57
prokaryotes 15, 41, 118, 122, 125, 130, 425, 430
proline 32
propane 79, 199, 202, 314, 319, 529, 634
prospects, see petroleum prospects
proteins 31, 45, 46, 79, 233
Prudhoe Bay, Alaska 563–565, 660
Psilopsida 18, 43
Pteridophyta 230
pycnocline 60
pyrene 126, 393
pyridine 127, 401, 402
pyrite 79, 134, 136, 138, 508
pyrobitumen 460, 463
pyrolysis 130, 142, 190, 255, 407, 509–513, 520–523, 551, 574
pyrolysis-fluorescence 513
pyrrole 42, 127, 402

Qatar crude oil 385, 421
Quass method 132
quaternary carbon atom 388
quinoline 127, 401, 402, 404
Quiriquire oil field, Venezuela 328, 339, 350, 351

racemic mixtures of hydrocarbons 464
radioactivity 207
Radiolarii 16, 27
Rainbow area, W. Canada 365
rank of coalification 234–241, 243, 245
rate of chemical reactions 584, 585, 592
ratios, see bitumen ratio, hydrocarbon ratio, transformation ratio
recrystallization 305, 360
Red Sea 224, 597, 600
Red Wash crude oil, Utah 396
reducing conditions, environments 106, 114, 153, 157, 159, 232, 234, 433, 655, see also anaerobic
reduction on sulfate 11, 77–79, 446–449
reef 364, 365, 461, 462
reflectance
   of huminite and vitrinite 70–72, 141, 161–163, 166–168, 209, 212, 234–237, 241–245, 249–252, 499, 505, 506, 516–520, 522, 530, 539, 543, 544
   of exinite/liptinite and inertinite 162, 163, 171, 242, 243, 499, 516, 517, 519
Reforma-Campeche petroleum province, Mexico 657–660, 662
refractive index 377
regression 56, 61
reserves, resources 583, 608, 657–662
reservoir
   bitumen 463
   migration in, see secondary migration
   rock 357–360
residue, solid, in reservoirs 461
resinite, resin-bearing tissues 234, 252, 504
resins
   constituent of crude oil 189, 378–381, 403–408, 411–413, 416, 418, 420–423, 472–476, 478, 479, 504, 505
   organic particles from plants 37, 49, 237
resistivity of formations 321, 612
retene 116, 117

retention effect of mineral matrix in pyrolysis 196
retorting process of oil shales 260
retrograde condensation 468, 623
reworked material 156, 237, 243, 499, 516
Rhine Graben, Rhine Valley 106, 107, 118, 121, 122, 296, 297, 329, 354, 362, 431, 433, 435, 441, 562, 597, 600, 603
Rhone delta 84
rhythmic sedimentation 231
rift 445, 597
ring analysis, of saturated hydrocarbons 391, 392
Rio Grande Ridge 656, 664
Rock Eval, method of pyrolysis 509, 522
rock fracturing 311, 321–323, 337
Rocky Mountains 207, 213, 455, 653
Romania 363, 647
Romashkino oil field, USSR 660
Rotliegend formation, North Sea, Germany and Netherlands 211, 212, 360
Rousse condensate field, France 566, 567
Ruhr area, Germany 535, 569

Saar area, Germany 249, 252, 298
Safaniya oil field, Saudi Arabia 610, 614
Sahara, North Africa 117, 120, 125, 152, 153, 161, 162, 165–170, 186, 225–227, 330, 335, 347, 378, 511, 512, 557, 643, 653, 654, 658, 660
salinity
  of pore water 300
  of sea water 25, 27
salt 212, 226, 227, 297, 299, 362–364, 650
  domes 362, 363, 465
Samotlor oil field, USSR 660
sampling program, for source rock identification 573
sand
  sandstone reservoirs 358–360
  /shale sequence, see clastic sequence
San Pedro oil field, Argentina 328
sapropelic
  coal 229, 230
  kerogen, organic matter 135, 136, 138, 158, 503, 505
Sargasso Sea 29
saturated hydrocarbons
  in crude oils 379–382, 417
  enrichment in migration 332, 333

generation of 181–187, 193–195, 215, 441, 442
saturates/aromatics ratio, change with depth 187
saturation pressure 354
Saudi Arabia 421, 447, 610–615
sclerotinite 504
Scotian shelf 84
screening techniques for source rocks 574
sea level changes 651, 654
sea water 12, 25–28
seal, cap-rock 294, 351, 352, 354, 357, 362, 363, 365
secondary migration 293, 294, 299, 341–355, 572
  distance of 354, 355, 479, 480, 614, 620
sedimentary
  basins, distribution of gas in 213, 214
  processes 55
sedimentation
  rate 12
  cycle 225, 226, 653
sediments
  hydrocarbons in, organic matter in, see hydrocarbons and organic matter
  methane in young- 90
seeds 49
seeps
  in coal mines 252
  of gas 317, 326
  of oil 326, 465
semifusinite 245, 504
separation-migration 355, 468
sesquiterpenes 37, 112, 114
sesterterpenes 112, 114
shale oil 260–263, 404, 471, 472
shales
  compaction and porosity 301–304
  diffusion in 319, 320
  pore diameters in 306, 307
  as source rocks 338, 339, 446, 447, 616–620, 622, 623, 625, 626
shark liver oil 38
Shatsky Rise 326, 655
shelf 24, 25, 31, 611, 612
Shengli oil field, P. R. of China 624
Siberia, W. Siberia, USSR 80, 202–204, 645, 654, 655, 658, 659
Sicily, Italy 259, 421, 447
side wall core samples 574
silico flagellates 15, 22

Simpson formation 643
sinapyl alcohol 44
Sirte basin, Libya 645, 657, 658
sitosterol, sitostane 40, 41, 50, 125, 426
smectite, montmorillonite 193, 196, 328, 329
soils, gas analysis in 90
solubility of petroleum
  in gas 337, 338
  in water 312–315, 317
solubilization of organic matter 11, 12
solution migration 312–318, 335, 338
source rock
  carbonates as, see carbonates
  case histories of gas source rocks 209–212, 631–637
  case histories of oil source rock 478, 479, 612–614, 616–620, 622–623, 625–628
  characterization 495, 540–546
    by C_{15+}-hydrocarbons 513, 514, 525–527, 531–540
    by kerogen analysis 508, 509, 523–525
    by light hydrocarbons 527–530
    by optical methods 498–507, 515–520
    by pyrolysis 509–513, 520–523
  coal as 251, 252
  conditions of deposition 11, 13, 14, 61
  /crude oil correlation, see correlation
  formation of petroleum in 176–189, 215–217, 222–227
  gas versus oil 219–221
  genetic potential of 159, 218, 589, 590
  identification in basins 572–575
  in migration 293, 294, 308, 330–333, 336
  organic content of 55, 97, 495–497
specific
  gravity
    of crude oils 294, 413, 446, 449, 453, 462, 463, 467
    of heavy oils 471, 472, 475–477
  volume of water 304
spilling plane, spill point of accumulation 355, 360, 362
Spitsbergen Island, Norway 331
spores 17, 18, 34, 49, 50, 136, 229, 230, 234, 237, 243, 503–505, 507
  and pollen carbonization, color 167, 515, 516, 543

sporinite 504
sporopollenin 34, 36
Springhill formation, Magellan Basin 431
squalane 112, 116, 126, 389
squalene 38, 39, 112
Stanleyville basin, Zaire 258
stanols 118, 121
State Line crude oil, Wyoming 386
steam injection 482
stearic acid 37, 47, 49
steranes 118–125, 186, 195, 250, 393, 426–428, 430, 433–436, 441, 537–540, 558, 560, 562, 564, 565 628
  aromatic, see steroids, aromatic
  /hopane ratio 122, 426, 429
  isomerization 434–436, 537–540
sterenes 119, 122, 186, 389
stereochemistry, stereoisomers 113, 124, 425, 434–437, 537, 538
steroid acid 117, 120–122, 125, 127, 431
steroids 34, 38, 40, 41, 46, 50, 117–126, 186, 187, 195, 393, 396, 397, 425–428, 430, 433–436
  aromatic 117, 119, 120, 123, 125, 186, 187, 396, 397, 425, 433, 435, 546, 538, 539
  in sediments 117–126
sterols 40, 46, 47, 50, 118, 122, 123, 433, 442
stigmasterol, stigmastane 40, 41, 46, 50, 124, 426
stratigraphic traps 357, 360, 364, 365
structural traps 294, 357, 360–363
structured water, adsorbed on clay surfaces, see water, adsorbed
suberin 42, 43, 49
subsidence 160, 223, 224, 247, 572, 616, 624, 626, 647, 652–654
  rate of 616, 642, 647, 652, 653
Suez, Golf of 224, 647
sugars 32, 33, 81
sulfate 7–9, 214, 218, 465
  reduction by bacteria 77–79, 201, 204, 446, 447, 465, 478
sulfides 8, 70, 79, 148, 150, 399, 400, 447, 465
sulfur
  aromatic derivatives in crude oil 395–397, 399, 400, 411, 472, 473
  in asphaltenes 405, 408
  in biodegradation 465–467

in coal 238, 239
compounds in crude oil 398–400
content
in crude oil 351, 398, 411, 413, 417, 418, 420–422, 446, 448, 449, 454, 456
in heavy oils 473, 474
correlation with density 413, 449
high-sulfur crude oil 445–450
incorporation into sediment 77–79, 442, 447
isotopes, see isotopes
in kerogen 148, 150, 447
origin in crude oil 411, 413, 442, 447
Sumatra, Indonesia 444, 554, 558, 658
Sunnyside, Utah 475
surface area, internal, specific 307, see also internal surface
suspended matter 56, 57, 59
Sverdrup basin, N. Canada 528, 529
Swan Hills formation, W. Canada 339

TAI, see thermal alteration index
tannin 42, 44, 45
tar 421, 422, 465, 471
mats 464, 610
sand 422, 423, 465, 470–483
tasmanite 134, 141, 151, 256, 257, 262–265
Tasmanites 137, 162, 256
tectonic
events and destruction of accumulations 595, 652, 653
orogenic activity and heat flux 597–599
telinite 244, 504
Temblador oil field, Venezuela 479
temperature 162, 174, 222, 223, 296–299, 302–304, 306, 450, 460, 596, 601, see also thermal
measurements 601
profile 297, 298
versus time, see time-temperature relation
terpanes
generation from kerogen 195
in oil correlation 554, 555, 564, 565
terpenes, monoterpenes 34, 36, 110, 111
terpenoids 34, 36–40, 46, 50, 251
terpinene 36
terrestrial
or continental organic matter 7, 17, 18, 20, 31, 36, 38, 40, 41, 45, 46, 49–52, 102–104, 108, 122, 128, 154, 157, 159, 219–221, 229, 230, 233, 234, 385, 426, 431, 440–443
higher plants, see terrestrial organic matter
tertiary
carbon atom 388
migration 341
Tertiary deltas 615–624, 646, 647, 658, 659
Tethys 447, 611, 655, 656
tetraethers of glycerol 115, 116
tetrahydronaphthalene, tetralin 394, 396
tetrahymanol 39
tetraterpenes, C_{40}-terpenes 41, 112, 126
thermal
alteration, evolution 422, 423, 459–463, 467, 556
index TAI 516, 543
analysis of kerogen 147
cracking 190, 193, 205, see also cracking
conductivity of rock 296, 297, 299
history 596–600
maturation of kerogen, see maturation
waters 600
thermocline 60
thermodynamic equilibrium, isomerization equilibrium 193, 388, 460
thiabutane 399
thiacyclohexane, thiacyclopentane 399, 400
thin layer chromatography, see chromatography
thiols 399, 400
thiophene, thiophene derivatives in crude oil 395–397, 399, 400, 411, 412, 421, 446, 449, 472–474, 478, 550, 555, 556, 560, 566, 567
threshold (depth, temperature) to oil or gas generation 179, 181, 223
time
in basin modeling 586–588, 593
of hydrocarbon diffusion (to surface) 339, 340
of migration 320, 325–328, 461, 605, 608
of oil and gas formation 222–227, 584, 590, 595–596, 608
-temperature relationship in petroleum generation 222, 223, 460

timing of oil and gas formation, see time
Toarcian shales, see Paris basin
  of Western Europe 258, 261, 654
toluene 95, 314, 315, 393, 465
torbanite 134, 151, 256–258, 262
trace elements, in crude oil 84, 408–410
transformation ratio 177, 218, 219, 521, 526, 590
transgression 56, 61, 612, 621, 645, 651, 654, 655
trap 227, 294, 351, 357, 360–365, 459, 572, 596
  formation of 227, 575, 583, 608
tricyclodecane, see adamantane
triglyceride 35
trilobites 16
trimacerals 139
trimacerite 499, 500
triphenylene 126, 127
triterpanes 118–122, 124, 186, 195, 250, 393, 428, 434, 558, 562
triterpenes 38, 39, 42, 46, 120–125, 250, 393
triterpenoid alcohol, acid 121, 124, 125, 431
triterpenoids 38–40, 117–127, 130, 186, 426
  aromatic 117–120, 123, 125, 186, 396
  diagenesis of 121–126
troilite 79
Tunisia 106, 107, 409, 433
Tuscaloosa reservoir, Gulf Coast 552, 553
type of kerogen, of organic matter, see kerogen type

Uinta basin, Utah 104, 105, 108, 109, 152, 178–180, 182, 183, 219, 265, 386, 389, 391, 396, 420, 428, 430, 431, 441, 444, 445, 474, 480, 526, 562, 626, see also Green River formation
unconformity 225–227, 347, 355, 478, 481, 482
unsaturated hydrocarbons, see alkenes
upwelling 25, 26, 29, 59
urea adduction 379
Urengoy gas field, USSR 660
ursane 38, 39, 124
U. S. Bureau of Mines, see Bureau of Mines

vanadium, vanadyl 128, 129, 402, 408–410, 474
van Krevelen diagram 151, 152, 157, 159, 161, 216, 219, 220, 239–241, 257, 429, 511, 512, 523
Venezuela 302, 325, 328, 339, 350, 351, 398, 408–410, 421–423, 451–453, 472–481, 610, 615, 645, 655–658
Ventura basin, California 179, 184, 223
vertical migration, see migration, vertical
Victoria, Australia 231
Vienna Basin, Austria 451, 452, 560, 561, 647
Viking Graben, North Sea 297, 605–607
Viking shales, W. Canada 56, 157
viscosity 301, 407, 411, 413, 421, 449, 450
  of heavy oils 470, 471, 475–478, 482, 483
vitrinite 139, 153, 157, 230, 234, 237–244, 246, 498–501, 504
  reflectance, see reflectance
volatile matter 234–236, 245–248
volcanic gas 207
volcanoes, mud, see mud volcanoes
Volga-Ural basin 409, 455, 480, 481, 653, 658–660
Voltzia 106

Wabasca heavy oils, W. Canada 422, 472, 479, 480, 482
Walvis Bay 118
Walvis Ridge 656, 664
Wasson crude oil, Texas 376, 400
water
  adsorbed on clay surfaces, structured water 307, 308
  bound or fixed, in clay dehydration 328, 329
  connate 353
  expansion of 303, 304, 341
  flow of 297, 338, 344–346, 352
  generation of, from kerogen 163, 174, 216, 219, 220
  irreducible or residual amount of 353, 360
  meteoric, see meteoric water
  role in pyrolysis 196
  saturation 353
  washing 465, 467
wax 34, 237, 384, 385, 425, 442–445, 620, 626

esters 48–50
high-wax crude oil 252, 384, 385, 419, 440, 441, 459, 471
weathering, of organic matter 156
weight loss of kerogen 168–173
well cuttings and cores 573, 574
well-head samples of crude oil 376
West Africa, west coast of Africa 86, 104, 105, 117, 177, 363, 391, 409, 419, 421, 422, 440, 441, 451–453, 464, 474, 600, 645
Western Canada basin, see Canada
Western Siberia, see Siberia
wet gas generation 163, 175, 200, 217
wettability, oil-wet or water-wet 310, 320, 336
Whiterocks, Canyon asphalt, Utah 402, 474
Williston basin, USA 321, 329, 335, 339, 553

Wilmington oil field, California 361, 402
wood 32
Woodbine sandstone, Texas 362
Woodford shale 643
woddy kerogen 498, 505
Wüstrow field, Germany 212, 213

xanthophylls 41
xylene 95, 314, 393, 465

Yallourn lignite, Australia 194, 195

Zagros mountains 610
Zechstein formation, Germany and North Sea 106, 212
zooplankton 16, 20, 31, 45, 51–53, 98, 153, 157, 426

E. C. Dahlberg

Applied Hydrodynamics in Petroleum Exploration

1982. 117 figures. X, 161 pages
ISBN 3-540-90677-0

Contents: Fluids in the Subsurface Environment. – Hydrogeological Conditions. – Hydrodynamics Exploration Analysis. – The Potentiometric Surface. – Pressure-Depth Gradients. – Hydrocarbon Entrapment Potential Constructions *(U, V, Z)*. – Entrapment Potential Cross-Sections. – Hydrodynamic Mapping. – Appendix A: List of Symbols and Abbreviations Appearing in the Text. – Appendix B: Hubbert's Proof and Exercise Answers. – References. – Suggested Readings. – Index.

Springer-Verlag
Berlin
Heidelberg
New York
Tokyo

Written with the oil explorationist himself in mind, this book is the first unified presentation of hydrodynamic principles and methodology for predicting the location of subsurface oil and gas deposits. In it, the author has compiled theory, procedures and case histories from sources widely scattered in both space and time. Added to this are refinements developed by the author in the course of his international consulting work.
With the search for fossil fuels growing both more difficult and more critical every day, this book will prove to be the guide oil workers have sought in their efforts to pinpoint new fields.

A. D. Miall

Principles of Sedimentary Basin Analysis

1984. Approx. 387 figures. Approx. 550 pages
ISBN 3-540-90941-9

Contents: Introduction. – Collecting the data. – Stratigraphic correlation. – Facies analysis. – Basin mapping methods. – Depositional systems. – Burial history. – Regional and global stratigraphic cycles. – Sedimentation and plate tectonics. – Conclusions.

Principles of Sedimentary Basin Analysis provides geologists with a practical guide for the study of the geologic evolution of ancient sedimentary basins, using modern methods of facies and depositional systems analysis, seismic stratigraphy and a broad range of basin mapping techniques. A new approach to stratigraphy is demonstrated that explains the genesis of lithostratigraphic units. This book also contains a detailed treatment of the plate tectonic control of basin architecture and has a unique discussion of the widespread effects of eustatic sea level changes.

The emphasis throughout is on what geologists can actually see in outcrops, well records and cores, and what can be obtained using geophysical techniques. Profusely illustrated and containing numerous examples from around the world, this up-to-date textbook will be welcomed by the student and professional geologist alike.

Springer-Verlag
Berlin
Heidelberg
New York
Tokyo